Human
BIOLOGY

Fifth Edition

Human

B I O L O G Y

Sylvia S. Mader

WCB
McGraw-Hill

Boston, Massachusetts   Burr Ridge, Illinois   Dubuque, Iowa
Madison, Wisconsin   New York, New York   San Francisco, California   St. Louis, Missouri

WCB/McGraw-Hill

A Division of The **McGraw·Hill** *Companies*

TITLE: HUMAN BIOLOGY, 5/E

 This book is printed on recycled, acid free paper containing 10% postconsumer waste.

2 3 4 5 7 8 9 0 QPD/QPD 9 0 9 8

ISBN 0-697-27821-2

Publisher: Michael D. Lange
Sponsoring editor: Patrick Reidy
Developmental editor: Connie Balius-Haakinson
Marketing manager: Julie Joyce Keck
Project manager: Margaret B. Horn
Production supervisor: Mary E. Haas
Designer: K. Wayne Harms
Photo research coordinator: Lori Hancock
Art editor: Brenda A. Ernzen
Compositor: Christopher Reese
Typeface: 10/12 Palatino
Printer: Quebecor, Inc.

Library of Congress Cataloging-in-Publication Data

Mader, Sylvia S.
        Human Biology / Sylvia S. Mader — 5th ed.
                    p.   cm.
        Includes index.
        ISBN 0-697-27821-2
        1. Human biology.   I. Title.
QP36.M2   1998
612—dc21   97-13872
                                                                CIP

www.mhhe.com

for my family

Brief Contents

Contents

*These helpful aids appear in every chapter.

Readings

Preface

*H*uman Biology is suitable for use in one-semester biology courses that emphasize human physiology and the role that humans play in the biosphere. All students should leave college with a firm grasp of how their bodies normally function and how the human population can become more fully integrated into the biosphere. This knowledge can be applied daily and helps assure our continued survival as individuals and as a species. The application of biological principles to practical human concerns is now widely accepted as a suitable approach to the study of biology because it fulfills a great need. Human beings are frequently called upon to make decisions about their bodies and their environment. Wise decisions require adequate knowledge.

In this edition, as in previous editions, each chapter presents the topic clearly, simply, and distinctly so that students will feel capable of achieving an adult level of understanding. Detailed, high-level scientific data and terminology are not included because I believe that true knowledge consists of working concepts rather than technical facility. Students and instructors alike will find the text stimulating and a pleasure to read and study, especially since the illustration program is outstanding and there are many spectacular four-color illustrations that integrate art with micrographs.

Concepts Are Stressed

In this edition, the major topics are numbered, and the concepts listed on the chapter's opening page are grouped according to these topics. This numbering system, which is used in the text material and in the summaries, allows instructors to assign just certain portions of the chapter. It also allows students to study the chapter in terms of the concepts presented.

Applications Are Made

Educational theory tells us that students are most interested in knowledge of immediate practical application. This text is consistent with and remains true to this approach.

Each chapter now begins with a short story that applies chapter material to real-life situations. The readings stress applications and so does the running text material.

At the end of the chapter, certain Applying Your Knowledge questions ask students to relate the concepts of the chapter to events that occur in their own lives. Others are concerned with a bioethical issue. The issue is discussed at some length, and then questions are asked that can be used as a basis for class discussion.

Illustration Program

Many illustrations use both a micrograph and drawing along with boxed statements that emphasize key concepts, thereby increasing student comprehension. Common leaders and labels allow students to accurately relate the micrograph and the drawing.

In this edition, illustrations are even more closely correlated with the text material. Every illustration is on the same or the facing page to its reference to aid learning and studying.

Working Together Boxes

A theme for this edition is how systems work together to achieve homeostasis. Working Together boxes appear at the end of the systems chapters. These illustrations describe how each organ system works with the other systems to achieve its many functions, which are discussed throughout the chapter.

Readings

Two main concerns of this edition (health and ecology) are carried through the book by Health Focus and Ecology Focus readings. The Health Focus readings are designed to help students cope with a common health problem. The Ecology Focus readings draw attention to a particular environmental problem. The readings emphasize the concepts of the chapter and tie biological principles to practical human concerns.

New to this edition, a bioethical issue is described in detail at the end of each chapter, and questions are provided for class discussion.

New and Modified Chapters

All chapters have been modified or rewritten to improve the presentation. The following changes are of special interest.

Part Three, Movement and Support, is new to this edition and contains chapters 10 (Skeletal System) and 11 (Muscular System), which are both completely new chapters.

The chapters dealing with genetics and biotechnology have been extensively revised to include the latest research, techniques, and information. The AIDS supplement was completely rewritten.

Chapter 22 (Evolution) was rewritten to better present basic principles, the origin of life, and to update the human evolution section.

Technology Aids Pedagogy

Many technology aids are available for use with *Human Biology.*

For the Student

The Mader Home Page on the World Wide Web provides additional exercises to aid learning and resources that expand on the text's content and applications. The Applying Technology section in each chapter reminds students to visit the site and explore the many activities.

Interactive CD-ROMs bring biology to life. New to this edition, *The Dynamic Human* CD-ROM offers three-dimensional visuals that facilitate an understanding of human anatomy and physiology. *Explorations in Human Biology* has 16 modules, and each one allows students to study a particular high-interest biological topic. *Explorations in Cell Biology and Genetics* provides 17 exciting interactive activities. *The Life Science Animations* include 53 additional topics that can be studied in a visually appealing way. Numerous other aids are available, and all of these are listed on the Technology page (see page xviii of the Preface).

For the Instructor

The *Extended Lecture Outline* software makes the contents of the book available in a way that facilitates lecture preparation, and the *Visual Resource Library* on CD-ROM makes the text illustrations available for classroom use.

To help with the mechanics of teaching, there is testing software, which is the computerized version of the *Test Item File;* the computerized test bank is available in Windows and Macintosh formats.

NEW TO THIS EDITION

▶ Major topics are numbered on the chapter opening page, throughout the chapter, and in the summary of the chapter. Each illustration is on the same or facing page as its reference.

▶ Chapter opening stories now introduce the chapter in an appealing way that emphasize the relevancy of the topics to be studied.

▶ A bioethical issue is explained in detail, with questions for class discussion, at the end of each chapter. Health Focus and Ecology Focus readings occur in nearly every chapter.

▶ Working Together boxes appear at the end of the systems chapters. These illustrations describe how each organ system works with the other systems to achieve its many functions, which are discussed throughout the chapter. Each system can also be studied by using *The Dynamic Human* CD-ROM. For further explanation regarding this CD-ROM, see the Technology section on p. xviii.

▶ Part Three, Movement and Support, is new and contains chapters 10 (Skeletal System) and 11 (Muscular System), which are both completely new chapters. The AIDS supplement was updated, as were the genetics chapters. Chapter 22 (Evolution) was rewritten to better explain the principles and to update the section on human evolution.

▶ Technology aids are described and assigned a level of difficulty at the end of each chapter. New to this edition, *The Dynamic Human* CD-ROM is an interactive three-dimensional visual guide to human anatomy and physiology. *Explorations in Human Biology* and *Explorations in Cell Biology and Genetics* CD-ROMs offer exciting new ways to understand biological concepts. The *Life Science Animations* videotapes allow students to see processes in motion.

▶ Explore the Mader Home Page for even more information:

http://www.mhhe.com/sciencemath/biology/mader/

▶ Appendix G, Internet Guide, provides instructions for reaching the Mader Home Page.

Aids to the Reader

Human Biology includes a number of aids that have helped students study biology successfully and enjoyably.

Part Introduction

An introduction for each part highlights the central ideas of that part and specifically tells the student how the topics within each part contribute to biological knowledge.

Chapter Concepts

Each chapter begins with a list of concepts that organizes the content of the chapter into a few meaningful sentences. The concepts provide a framework for the content of the chapter.

Note that the concepts are grouped under the major headings and are page-referenced for student study.

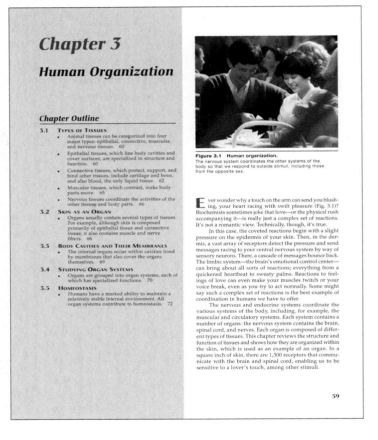

Figure 3.1 Human organization.
The nervous system coordinates the other systems of the body so that we respond to outside stimuli, including those from the opposite sex.

Internal Summary Statements

Internal summaries are highlighted to illustrate the chapter's key concepts. These appear at the ends of major sections and help students focus their study efforts on the basics.

Humans belong to the world of living things and are vertebrates. They have modified existing ecosystems to the point that they must now be seriously concerned about the continued existence of the biosphere.

Readings

Two types of readings are included in the text. The Health Focus readings give practical information concerning some particular topic of interest, such as how to do a breast exam for cancer, when not to give blood, and proper nutrition for healthy living. The Ecology Focus readings draw attention to some particular environmental problem, such as the need to preserve tropical rain forests, the relationship between ozone holes and skin cancer, and the possible ecological effects of biotechnology. The Instructor's Manual explains how to use the readings in the classroom.

New to this edition, a bioethical issue is described in detail at the end of each chapter, and questions are provided for class discussion.

Working Together Boxes

These boxes, which appear at the end of the systems chapters, describe how each system works with other organ systems to achieve its many functions, which are discussed in the chapter.

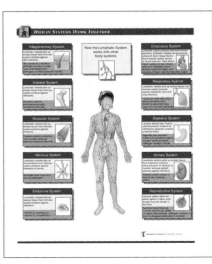

Illustrations and Tables

The illustrations and tables in *Human Biology* are consistent with multicultural educational goals. There are many integrative illustrations that include both a micrograph and a drawing combined with information formerly found in tables. Often it is easier to understand a given process by studying a drawing, especially when it is carefully coordinated with the text as in *Human Biology*.

Chapter Summaries

The summary is organized according to the major sections of the chapter. The sections are numbered as on the chapter opening page and in the chapter. The summary helps students review the important concepts and topics discussed in each chapter.

Selected Key Terms

Key terms are boldfaced in the chapter, defined in context, and also appear in the end-of-text glossary. Especially significant key terms appear in the selected key term list at the close of the chapter. Here each term is accompanied by a definition and a phonetic spelling, if needed.

Chapter Questions

The chapter questions allow students to test their ability to fulfill the study objectives.

Studying the Concepts

These questions, which are page-referenced and organized according to the major sections of the chapter, review the chapter.

Applying Your Knowledge to the concepts and a bioethical issue

In this section, three or four questions ask students to relate the concepts they have learned to matters of practical concern. A bioethical issue is described in a nonbiased manner, and then questions appropriate for class discussion are presented.

Testing Your Knowledge

These objective questions allow students to test their ability to answer recall-based questions. At least one question requires that students label a diagram or fill in a table. Answers to Applying Your Knowledge questions and Testing Your Knowledge questions appear in Appendix A.

Further Readings

The list of readings at the end of each part suggests references that can be used for further study of the topics covered in the chapters of that part. The references listed in the section were carefully chosen for readability and accessibility. New to this edition, references are followed by a short description and an indication of their level of rigor.

Appendices and Glossary

The appendices contain optional information. Appendix A is the Answer Key for the Applying Your Knowledge to the concepts and Testing Your Knowledge questions; Appendix B is a new presentation of the metric system; Appendix C is an expanded table of chemical elements; Appendix D reviews most of the drugs of abuse; and Appendix E gives pertinent information about the most frequent types of cancer. Appendix F is a new reference that defines acronyms given throughout the text; and Appendix G is a new appendix that gives information on accessing the Internet in general, and the Mader Home Page specifically.

The glossary defines all the boldfaced terms in the text. These terms are the ones most necessary for making the study of biology successful. Terms that are difficult to pronounce have a phonetic breakdown.

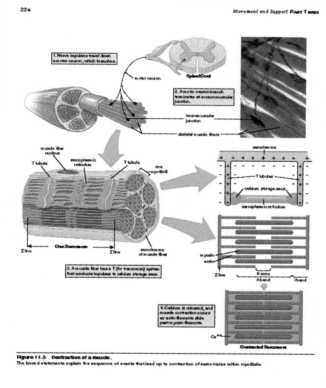

TECHNOLOGY

Several new state-of-the-art technology products are available that are correlated to this textbook. These useful and enticing supplements can assist you in teaching and can improve student learning.

EXPLORING THE INTERNET

http://www.mhhe.com/sciencemath/biology/mader/

The Mader Home Page allows students and teachers from all over the world to communicate. *Human Biology* has a complete text-specific site developed exclusively for users of the fifth edition. By visiting this site, students can access additional study aids, explore links to other relevant biology sites, catch up on current information, and pursue other activities. Appendix G in the text provides directions for accessing the Mader Home Page.

THE INTERNET PRIMER
by Fritz J. Erickson & John A. Vonk

This short, concise primer shows students and instructors how to access and use the Internet. The guide provides enough information to get started by describing the most critical elements of using the Internet.

THE DYNAMIC HUMAN CD-ROM

This guide to anatomy and physiology interactively illustrates the complex relationships between anatomical structures and their functions in the human body. Realistic, three-dimensional visuals are the premier feature of this exciting learning tool. The program covers each body system, demonstrating to the viewer the anatomy, physiology, histology, and clinical applications of each system as described in the Applying Technology section at the end of each chapter.

EXPLORATIONS IN HUMAN BIOLOGY CD-ROM; EXPLORATIONS IN CELL BIOLOGY AND GENETICS CD-ROM

These interactive CDs, by Dr. George B. Johnson, feature fascinating topics in biology. *Explorations in Human Biology* and *Explorations in Cell Biology and Genetics* have 33 different modules that allow students to study a high-interest biological topic in an interactive way. In this edition of *Human Biology*, the Explorations that correlate to the chapter are briefly described in the Applying Technology section, which appears at the end of each chapter.

LIFE SCIENCE ANIMATIONS VIDEOTAPES

Fifty-three animations of key physiological processes are available on videotapes. The animations bring visual movement to biological processes that are difficult to understand on the text page. In this edition of *Human Biology*, the videotapes are correlated to the chapters and are briefly described in the Applying Technology section, which appears at the end of each chapter.

BIOSOURCE VIDEODISC

BioSource Videodisc, by Wm. C. Brown Publishers and Sandpiper Multimedia, Inc., features 20 minutes of animations and nearly 10,000 full-color illustrations and photos, many from leading WCB/McGraw-Hill biology textbooks.

BIOETHICS FORUMS VIDEODISC

Bioethics Forums is an interactive program that explores societal dilemmas arising from recent breakthroughs in biology, genetics, and biomedical technology. The scenarios are fictional, but the underlying science and social issues are real. *Bioethics Forums* encourages students to explore the science behind decisions as well as the processes of ethical reasoning and decision-making.

VISUAL RESOURCE LIBRARY

Our electronic art image bank is a CD-ROM that contains hundreds of biological images from *Human Biology,* fifth edition. The CD-ROM contains an easy-to-use program that enables you to quickly view images and import the images into PowerPoint to create your own multimedia presentations. Also included are several video clips featuring key animated biological processes.

HEALTHQUEST CD-ROM

HealthQuest CD-ROM is an interactive CD-ROM designed to help students address the behavioral aspects of personal health and wellness. *HealthQuest* allows users to assess their current health and wellness status, determine their health risks and relative life expectancy, explore options and make decisions to improve the behaviors that impact their health.

VIRTUAL PHYSIOLOGY LABORATORY CD-ROM

Virtual Physiology Lab CD-ROM features ten simulations of the most common and important animal-based experiments ordinarily performed in the physiology component of your laboratory. This revolutionary program allows users to repeat laboratory experiments until they adeptly master the principles involved. The program contains video, audio, and text to clarify complex physiological functions.

THE SECRET OF LIFE VIDEO MODULES
WGBH, Boston and BBC-TV

WGBH has produced eight 15-minute video modules that illuminate the biological universe with unique stories and animation. Each module concludes with a series of stimulating questions for class discussion.

THE SECRET OF LIFE VIDEODISC
WGBH, Boston

A two-sided videodisc will be available as a companion to *Human Biology,* fifth edition. Topic coverage includes biotechnology, human reproduction, portraits of modern science and research, and human genetics.

Our CD-ROM products may be packaged with the text at a cost-savings. Contact your WCB/McGraw-Hill sales representative for details.

TECHNOLOGY
CORRELATIONS

The fifth edition of *Human Biology* has three technology learning tools that are correlated to the chapters. The Applying Technology section at the end of the chapter describes those that are appropriate to that chapter.

 The Dynamic Human is an interactive CD-ROM with three-dimensional visuals demonstrating the anatomy, physiology, histology, along with clinical applications, of each body system.

Explorations in Human Biology and *Explorations in Cell Biology and Genetics* are interactive CD-ROMs consisting of 33 different modules that cover key topics in biology.

Life Science Animations is a set of five videotapes containing 53 animations of processes integral to the study of biology.

Chapter 1 Chemistry of Life
Life Science Animations 1, 11   (Tape 1)

Chapter 2 Cell Structure and Function
Explorations in Cell Biology & Genetics 2, 3, 6, 8
Life Science Animations 2, 3, 4, 5, 6, 7, 11   (Tape 1)

Chapter 3 Human Organization
The Dynamic Human, Anatomical Orientation

Chapter 4 Digestive System and Nutrition
The Dynamic Human, Digestive System
Explorations in Human Biology 3, 7
Life Science Animations 33, 34, 35, 36   (Tapes 3 and 4)

Chapter 5 Composition and Function of Blood
The Dynamic Human, Lymphatic System
Life Science Animations 37   (Tape 4)
Explorations in Human Biology 12

Chapter 6 Cardiovascular System
The Dynamic Human, Cardiovascular System
Life Science Animations 32, 37, 38   (Tapes 3 and 4)

Chapter 7 Lymphatic System and Immunity
The Dynamic Human, Lymphatic System
Explorations in Human Biology 12
Life Science Animations 41, 42, 43, 44   (Tape 4)

Chapter 8 Respiratory System
The Dynamic Human, Respiratory System
Explorations in Human Biology 3, 6

Chapter 9 Urinary System and Excretion
The Dynamic Human, Urinary System

Chapter 10 Skeletal System
The Dynamic Human, Skeletal System

Chapter 11 Muscular System
The Dynamic Human, Muscular System
Life Science Animations 29, 30, 31   (Tape 3)
Explorations in Human Biology 4, 9

Chapter 12 Nervous System
The Dynamic Human, Nervous System
Explorations in Human Biology 8, 9
Life Science Animations 22, 23, 24, 25   (Tape 3)

Chapter 13 Senses
Explorations in Cell Biology & Genetics 4
Explorations in Human Biology 11
Life Science Animations 28   (Tape 3)

Chapter 14 Endocrine System
The Dynamic Human, Endocrine System
Explorations in Human Biology 11
Life Science Animations 28   (Tape 3)

Chapter 15 Reproductive System
The Dynamic Human, Reproductive System

Chapter 16 Sexually Transmitted Diseases
Explorations in Human Biology 13

Chapter 17 Development and Aging
Explorations in Human Biology 3, 15
Life Science Animations 21, 39   (Tapes 2 and 4)

Chapter 18 Chromosomal Inheritance
Explorations in Cell Biology & Genetics 10

Chapter 19 Genes and Medical Genetics
Explorations in Cell Biology & Genetics 12, 13
Explorations in Human Biology 1

Chapter 20 DNA and Biotechnology
Explorations in Cell Biology & Genetics 14, 16, 17
Life Science Animations 15, 16, 17   (Tape 2)

Chapter 21 Cancer
Explorations in Cell Biology & Genetics 5
Explorations in Human Biology 3, 6

Chapter 23 Ecosystems
Life Science Animations 51, 52   (Tape 5)

Chapter 24 Population Concerns
Explorations in Human Biology 16

More Teaching and Learning Aids

Transparencies

A set of 200 full color transparency acetates accompanies the text. These acetates contain key illustrations from the text.

Instructor's Manual/Test Item File

The *Instructor's Manual/Test Item File*, prepared by Jennifer Carr Burtwistle, is designed to assist instructors as they plan and prepare for classes using *Human Biology*. The first part of the *Instructor's Manual* pertains to the text chapters and the second part is the Test Item File.

Each chapter in the manual begins with a list of objectives, and the questions in the Test Item File are sequenced according to these objectives. Instructors can decide which objectives are best suited to their particular course and then easily locate the appropriate test questions.

The *Instructor's Manual* contains both an extended lecture outline and a general discussion, which together review in detail the contents of the text chapter. The technology section lists videos and computer software items that are available from outside sources and also those that are available from WCB/McGraw-Hill. The student activities section includes some activities that are based on technology accompanying this text. Additional Applying Your Knowledge Questions, with answers, are given in the *Instructor's Manual*.

The Test Item File for each chapter contains approximately 60 objective test questions and several essay questions. Some questions require more thought than others, and the questions are coded as to level of difficulty. These same questions are found in the computerized version of the Test Item File.

Visuals Testbank

This testbank contains black-and-white versions of 200 illustrations. The labels are deleted and copies can be run off for student quizzing or practice.

Study Guide

Jennifer Carr Burtwistle prepared the *Study Guide* that accompanies the text. Each text chapter has a corresponding *Study Guide* chapter that includes a listing of objectives, study questions, and a chapter test. Answers to the study questions and the chapter tests are provided to give students immediate feedback.

The objectives in the *Study Guide* are the same as those in the *Instructor's Manual*, and the practice test questions in the *Study Guide* are sequenced according to these objectives. Instructors who make their choice of objectives known to the students can thereby direct student learning in an efficient manner. Instructors and students who make use of the *Study Guide* should find that student performance increases dramatically.

Laboratory Manual

The author has also written the *Laboratory Manual* to accompany *Human Biology*. With few exceptions, each chapter in the text has an accompanying laboratory exercise in the manual (some chapters have more than one accompanying exercise). In this way, instructors are better able to emphasize particular portions of the curriculum if they wish. The 19 laboratory sessions in the manual are designed to further help students appreciate the scientific method and to learn the fundamental concepts of biology and the specific content of each chapter. All exercises have been tested for student interest, preparation time, and feasibility.

Laboratory Resource Guide

More extensive information regarding preparation is found in the *Laboratory Resource Guide*. The guide includes suggested sources for materials and supplies, directions for making up solutions and otherwise setting up the laboratory, expected results for the exercises, and suggested answers to all questions in the laboratory manual. It is free to all adopters of the laboratory manual.

Micrograph Slides

This ancillary provides 35mm slides of many photomicrographs and all electron micrographs in the text.

From WCB/McGraw-Hill

How to Study Science, 2nd Edition

by Fred Drewes, Suffolk County Community College

This excellent new workbook offers students helpful suggestions for meeting the considerable challenges of a college science course. It offers tips on how to take notes, how to get the most out of laboratories, and how to overcome science anxiety. The book's unique design helps students develop critical thinking skills while facilitating careful note taking. (ISBN 0–697–15905-1)

A Life Science Living Lexicon

by William N. Marchuk, Red Deer College

This portable, inexpensive reference helps introductory-level students quickly master the vocabulary of the life sciences. Not a dictionary, it carefully explains the rules of word construction and derivation, in addition to giving complete definitions of all important terms. (ISBN 0–697–12133–X)

Biology Study Cards

by Kent Van De Graaff, R. Ward Rhees,
and Christopher H. Creek, Brigham Young University

This boxed set of 300 two-sided study cards provides a quick yet thorough visual synopsis of all key biological terms and concepts in the general biology curriculum. Each card features a masterful illustration, pronunciation guide, definition, and description in context. (ISBN 0–697–03069–5)

Virtual Biology Laboratory CD-ROM

by John Beneski and Jack Waber, West Chester University

This CD-ROM is designed primarily for nonscience major students. The exercises (10 modules) are designed to expose students to the types of tools used by biologists, allow students to perform experiments without the use of wet lab setups, and support and illustrate topics and concepts from a traditional biology course. (ISBN 0–697–33427–9)

Life Science Living Lexicon CD-ROM

by William N. Marchuk, Red Deer College

A Life Science Living Lexicon CD-ROM contains a comprehensive collection of life science terms, including definitions of their roots, prefixes, and suffixes as well as audio pronunciations and illustrations. The Lexicon is student-interactive, providing quizzing and notetaking capabilities. It contains 4,500 terms, which can be broken down for study into the following categories: anatomy and physiology, botany, cell and molecular biology, genetics, ecology and evolution, and zoology.

Critical Thinking Case Study Workbook

by Robert Allen

This ancillary includes 34 critical thinking case studies that are designed to immerse students in the "process of science" and challenge them to solve problems in the same way biologists do. The case studies are divided into three levels of difficulty (introductory, intermediate, and advanced) to afford instructors greater choice and flexibility. An answer key accompanies this workbook. (ISBN 0-697-34250-6)

The AIDS Booklet

by Frank D. Cox

This booklet describes how AIDS and related diseases are commonly spread so that readers can protect themselves and their friends against this debilitating and deadly disease. This booklet is updated quarterly to give readers the most current information. Also visit the Mader Home Page for additional AIDS material. (ISBN 0–697–26261–8)

Chemistry for Biology

by Carolyn Chapman

This workbook is a self-paced introduction or review of the basic principles of chemistry that are most useful in other areas of science. (ISBN 0-697-24121-1)

Biology Startup

by Myles Robinson and Kathleen Pace, Grays Harbor College

Biology Startup is a five-disk Macintosh tutorial that helps nonmajors master challenging biological concepts such as basic chemistry, photosynthesis, and cellular respiration. This program can be a valuable addition to a resource center and is especially helpful as a refresher or for students who need additional assistance to succeed in an introductory biology course.

Acknowledgments

The personnel at WCB/McGraw-Hill have always lent their talents to the success of *Human Biology*. My editor, Michael Lange, directed the efforts of all. Connie Haakinson, my developmental editor, served as a liaison between the editor, me, and many other people. She met each new challenge in a prompt and most professional way.

The production team worked diligently toward the success of this edition. Margaret Horn was the project manager; Lori Hancock, the photo coordinator; and Wayne Harms, the designer. My thanks to each of them for a job well done!

The Reviewers

Many instructors have contributed not only to this edition of *Human Biology,* but also to previous editions. I am extremely thankful to each one, for they have all worked diligently to remain true to our calling and provide a product that will be the most useful to our students.

In particular, it is appropriate to acknowledge the help of the following individuals for the fifth edition:

Donald Jasper
Illinois Institute of Technology

Allan R. Stevens
Snow College

C. L. Swendsen
Warren Wilson College

Don Naber
University College, University of Maine

Tom Denton
Auburn University at Montgomery

Mary King Kananen
Penn State–Altoona

Gina Erickson
Highline Community College

Diane Merlos
Grossmont College

Arlene Marian
Niagara University

Lawton Owen
Kansas Wesleyan University

Donald A. Wheeler
Edinboro University of Pennsylvania

Craig Berezowsky
The University of British Columbia

Elaine Rubenstein
Skidmore College

Ronald F. Cooper

Dr. Charles Hummel
New York Institute of Technology

Valerie Vander Vliet
Lewis University

David Wolfrom
Paducah Community College

Al Avenoso
University of Houston–Downtown

James J. Greene
The Catholic University of America

Grant M. Barkley
Kent State University

Michael Emsley
George Mason University

Felix Baerlocher
Mount Allison University

Charlene L. Forest
Brooklyn College of CUNY

Isaac Elegbe
College of New Rochelle

Pat Selelyo
College of Southern Idaho

Joseph V. Martin
Rutgers University

Loretta M. Parsons
Skidmore College

James R. Philips
Babson College

Garry Davies
University of Alaska–Anchorage

Fritz Taylor
University of New Mexico

Lynette Rushton
South Puget Sound Community College

Alexander Varkey
Liberty University

Donald S. Emmeluth
Fulton-Montgomery Community College

Penelope ReVelle
Essex Community College

Patricia Matthews
Grand Valley State University

Joe Connell
Leeward Community College

Vaughn Rundquist
Montana State University–Northern

Arnold E. S. Gussin
St. Francis College

Robert H. Chesney
William Paterson College of New Jersey

Robert J. Ratterman
Jamestown Community College

Stephen R. Karr
Carson-Newman College

Hessell Bouma III
Calvin College

Walt Sinnamon
Southern Wesleyan University

Marirose T. Ethington
Genesee Community College

Robert S. Greene
Niagara University

Michelle A. Green
SUNY @ Alfred

Stanton F. Hoegerman
College of William and Mary

Dana Demmans
Finger Lakes Community College

Douglas J. Burks
Wilmington College of Ohio

Helen Cadwallader
Black Hawk College

J. D. Brammer
North Dakota State University

Rodney Mowbray
University of Wisconsin–LaCrosse

Edward W. Carroll
Marquette University

Char A. Beeanson
St. Olaf College

Barbara Wineinger
Vincennes University–Jasper

Charles Ellison
Willmington College–Cincinnati Branch

Lee H. Lee
Montclair State University

Robert Olson
Briar Cliff College

Benjamin C. Stark
Illinois Institute of Technology

Debra J. Martin
St. Mary's University of Minnesota

Ann L. Henninger
Wartburg College

Ted Johnson
St. Olaf College

Florence M. Dusek
Des Moines Area Community College

Madeline M. Hall
Cleveland State University

Albert C. Jensen
Central Florida Community College

William E. Dunscombe
Union County College

Kathleen Lauckner
UNLV-Harry Reid Center for Environmental Studies

Penny Bernstein
Kent State University–Stark Campus

Darryl L. Daley
Snow College

Doris M. Shoemaker
Dalton College

Kim R. Finer
Kent State University–Stark Campus

Lisa Danko
Mercyhurst College

James A. Gessaman
Utah State University

Debra Zehner
Wilkes University

Robert H. Tamarin
University of Massachusetts–Lowell

George A. Hudock
Indiana University

Allan Hunt
Elizabeth Community College

Dale Lambert
Tarrant County Junior College

Caren K. Shapiro
D'Youville College

Theresa Hoffman-Till
Northern Virginia Community College

David E. Dallas
Northeastern Oklahoma Agri. & Mechanical College

Orrie O. Stenroos
Lynchburg College in Virginia

Marcus Young Owl
California State University–Long Beach

Kathleen Lauber
Catonsville Community College

Dalia Giedrimiene
Saint Joseph College

M. L. Tiell
Mercy College

Gregory J. Stewart
State University of West Georgia

William P. Ventura
Pace University

Kathryn Sergeant Brown

Stephen Smith
Johnson Bible College

Carl M. Christenson
Indiana University Southeast

The *Dynamic* Human

Experience human biology in an entirely new dimension. **The Dynamic Human** CD-ROM interactively illustrates the complex relationships between anatomical structures and their functions in the human body. Realistic, three-dimensional visuals are the premier features of this exciting learning tool. After a brief introduction, **The Dynamic Human** covers each body system—demonstrating to the viewer the anatomy, physiology, histology, and clinical applications of each system.

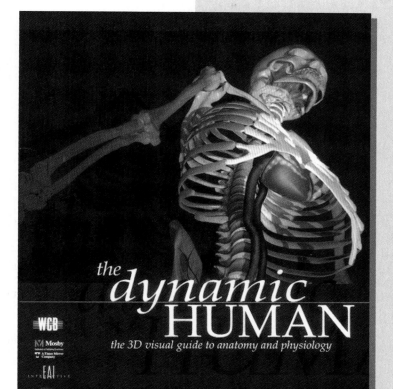

Contents:

Anatomical Orientation

Skeletal System

Muscular System

Nervous System

Endocrine System

Cardiovascular System

Lymphatic System

Digestive System

Respiratory System

Urinary System

Reproductive System

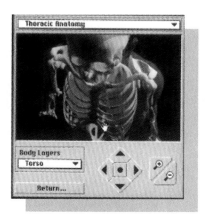

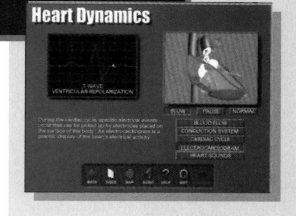

To order any of these products, contact your bookstore manager, or call-toll free: 1-800-338-3987.

Human Biology, fifth edition, has been correlated to **The Dynamic Human**. A "dancing man" icon informs the reader that information relating to a system can be found on **The Dynamic Human** CD-ROM.

See page xix for a chapter correlation list.

Windows Version 0-697-37910-8
Macintosh Version 0-697-37909-4

Look...

...at the great anatomy and physiology study tools WCB/McGraw-Hill has to offer!

WCB Life Science Animations Videotapes

Series of five videotapes containing animations of complex physiological processes. These animations make challenging concepts easier to understand.

Tape 1 **Chemistry, The Cell, and Energetics**
ISBN: 0-697-25068-7

Tape 2 **Cell Division, Heredity, Genetics, Reproduction, and Development**
ISBN: 0-697-25069-5

Tape 3 **Animal Biology #1**
ISBN: 0-697-25070-9

Tape 4 **Animal Biology #2**
ISBN: 0-697-25071-7

Tape 5 **Plant Biology, Evolution, and Ecology**
ISBN: 0-697-26600-1

Life Science Living Lexicon CD-ROM
by William Marchuk

ISBN: 0-697-37993-0

This interactive CD-ROM contains a complete lexicon of life science terminology. Conveniently assembled on an easy-to-use CD-ROM are components such as a glossary of common biological roots, prefixes, and suffixes; a categorized glossary of common biological terms; and a section describing the classification system.

To order any of these products, contact your bookstore manager, or call-toll free: 1–800–338–3987.

Introduction

A Human Perspective

Figure I.1   Researchers in the laboratory.
Experiments in the laboratory often lead to medical cures for human ills.

It was curiosity, more than anything, that led an English country doctor named Edward Jenner to discover the world's most successful vaccine. The year was 1796. Jenner had long pondered a rumor that for some reason, milkmaids managed to avoid the dreaded smallpox, a deadly disease caused by a virus.

After long hours in the lab, Jenner and others concluded that smallpox missed these women because they had already contracted a similar, though milder, form of the disease called cowpox. They reasoned that cowpox must provide immunity against smallpox.

Jenner decided to test his theory on a young boy. First, he injected the boy with cowpox, and then later, he injected him with smallpox. His volunteer survived, and Jenner became famous. His primitive vaccine paved the way for more sophisticated versions of vaccines also developed in a laboratory setting (Fig. I.1). Even so, smallpox is the single disease that medicine has truly eradicated from the face of the earth.

Like Jenner, all biologists strive to make sense of the biological world, and many of their findings are discussed in this text, which has two primary functions. The first is to explore human anatomy and physiology so you will know how the body functions. The second is to take a look at human evolution and ecology so you will understand the place of humans in nature. Both the human body and the environment are self-regulating systems that can be thrown out of kilter by misuse and mismanagement. An appreciation of the delicate balance present in both systems provides the perspective from which future decisions can be made. It is hoped that adequate information will better enable you to keep your body and the environment healthy.

When biologists do their work, they are answering specific questions. How do cells work? How do genes control who we are? Ecologically, what impact do humans have on the environment? These and millions of other questions challenge biologists who use the scientific method to come to conclusions they share with others. This chapter not only gives an overview of the book, it also explains the scientific method.

I.1   Biologically Speaking

You are about to launch on a study of human biology. Before you begin, it is appropriate to define who humans are and how they fit into the world of living things.

Who Are We?

Certain characteristics tell us who human beings are biologically speaking.

Human beings are highly organized. A **cell** is the basic unit of life, and human beings are multicellular since they are composed of many types of cells. Like cells form tissues, and tissues make up organs (Fig. I.2). Each type of organ is a part of an organ system. The different systems perform the specific functions listed in Table I.1. Together, the organ systems maintain **homeostasis,** an internal environment for cells that varies only within certain limits. For example, cells require a constant supply of nutrients and they give off waste products. The digestive system takes in nutrients, and the circulatory system distributes these to the cells. The waste

TABLE I.1	
Human Organ Systems	
System	**Function**
Digestive	Converts food particles to nutrient molecules
Circulatory	Transports nutrients to and wastes from cells
Immune	Defends against disease
Respiratory	Exchanges gases with the environment
Excretory	Eliminates metabolic wastes
Nervous	Regulates systems and internal environment
Musculoskeletal	Supports and moves organism
Endocrine	Regulates systems and internal environment
Reproductive	Produces offspring

products are excreted by the excretory system. The work of the nervous and endocrine systems is critical because they coordinate the functions of the other systems.

Human beings reproduce and grow. Reproduction and growth are fundamental characteristics of all living things. Just as cells come only from pre-existing cells, so living things have parents. When living things **reproduce,** they create a copy of themselves and assure the continuance of the species. (A species is a type of living thing.) Human reproduction requires that a sperm contributed by the male fertilize an egg contributed by the female. Growth occurs as the resulting cell develops into the newborn. Development includes all the changes that occur from the fertilized egg to death and, therefore, all the changes that occur during childhood, adolescence, and adulthood.

Humans have a cultural heritage. We are born without knowledge of civilized ways of behavior, and we gradually acquire these by adult instruction and imitation of role models. It is our cultural inheritance that makes us think we are separate from nature. But actually we are a product of **evolution,** a process of change that has resulted in the diversity of life, and we are a part of the **biosphere,** a network of life that spans the surface of the earth.

> Like other living things, humans are composed of cells, and when they reproduce, growth and development occur. Unlike other living things, humans have a cultural heritage.

How Do We Fit In?

Certain characteristics tell us how human beings fit into the world of living things.

Human beings are a product of an evolutionary process. Life has a history that began with the evolution of the first cell(s) about 3.5 billion years ago. It is possible to trace human ancestry from the first cell through a series of prehistoric ancestors until the evolution of modern-day humans. The presence of the same types of chemicals tells us that *human beings are related to all other living things.* DNA is the genetic material, and ATP is the energy currency in all cells, including human cells. It is even possible to do research with bacteria and have the results apply to humans.

The classification of living things mirrors their evolutionary relationships. It is common practice to classify living things into five major groups called **kingdoms** (Fig. I.2). *Humans are vertebrates in the animal kingdom.* **Vertebrates** have a nerve cord that is protected by a vertebral column whose repeating units (the vertebrae) indicate that we and other vertebrates are segmented animals. Among the vertebrates, we are most closely related to the apes, specifically the chimpanzee, from whom we are distinguished by our highly developed brains, completely upright stance, and the power of creative language.

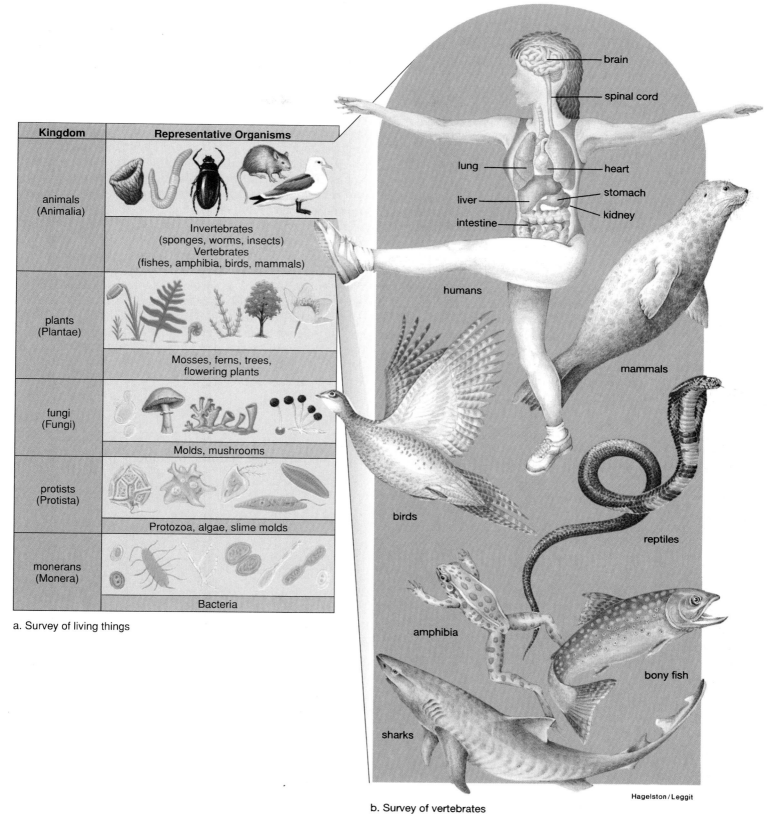

Kingdom	Representative Organisms
animals (Animalia)	Invertebrates (sponges, worms, insects) Vertebrates (fishes, amphibia, birds, mammals)
plants (Plantae)	Mosses, ferns, trees, flowering plants
fungi (Fungi)	Molds, mushrooms
protists (Protista)	Protozoa, algae, slime molds
monerans (Monera)	Bacteria

a. Survey of living things

brain

spinal cord

lung

heart

liver

stomach

intestine

kidney

humans

mammals

birds

reptiles

amphibia

bony fish

sharks

Hagelston / Leggit

b. Survey of vertebrates

Figure I.2   Classification and evolution of humans.
a. Living things are classified into five kingdoms, and humans are in the animal kingdom. **b.** Human beings are most closely related to the other vertebrates shown. The evolutionary tree of life has many branches; the vertebrate line of descent is just one of many.

Human beings are a part of the biosphere. All living things are a part of the biosphere, where living things live in the air, in the sea, and on land. In any portion of the biosphere, such as a particular forest or pond, the various populations interact with one another, and with the physical environment to form an **ecosystem** in which chemicals cycle and energy flows. All organisms of one type in a particular ecosystem belong to a *population*. A major part of the interactions between populations pertains to who eats whom. Plants produce organic food, and animals that eat plants may be food for other animals. Both plants and animals interact with the physical environment, as when they exchange gases with the atmosphere. As Figure I.3 shows, all living things are dependent upon solar energy and upon plants, which use this energy to convert inorganic nutrients into a form that is usable by all living things, including humans.

The interactions between populations in an ecosystem tend to keep the system relatively stable. Although a forest or pond changes—trees fall, ducks come and go, seeds sprout—each ecosystem remains recognizable year after year. We say it is in dynamic balance. In many cases, even the extinction of species (and their replacement by new species through evolution) still allows the dynamic balance of the system to be maintained.

If the ecosystem is big enough, it needs no raw materials from the outside. A big ecosystem just keeps cycling its raw materials, like water and nitrogen. The only input it needs is energy.

Humans Threaten the Biosphere

Human populations tend to modify existing ecosystems for their own purposes. For example, humans clear forests or grasslands in order to grow crops; later, they build houses on what was once farmland; and, finally, they convert small towns into cities. Human populations ever increase in size and require greater amounts of material goods and energy input each year (Fig. I.4). With each step, fewer and fewer original organisms remain, until at last ecosystems are completely altered. If this continues, only humans and their domesticated plants and animals will largely exist where once there were many diverse populations.

More and more ecosystems are threatened as the human population increases in size. As discussed in the Ecology Focus reading on page 7, presently, there is great concern among scientists and laypersons about the destruction of the world's rain forests due to logging and the large numbers of persons who are starting to live and to farm there.

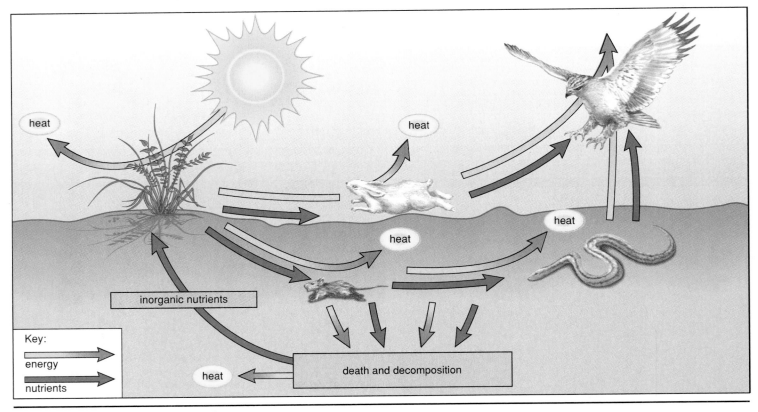

Figure I.3 Ecosystem organization.
Within an ecosystem, nutrients cycle (see blue arrows); plants make and use their own organic food, and this becomes food for several levels of animal consumers, including humans. When these organisms die and decompose, the inorganic remains are used by plants as they produce organic food. Energy flows (see yellow arrows); solar energy used by plants to produce organic food is eventually converted to heat by all members of an ecosystem (therefore, a constant supply of solar energy is required for life to exist).

But we are beginning to realize how dependent we are on intact ecosystems and the services they perform for us. For example, the tropical rain forests act like a giant sponge, which absorbs carbon dioxide, a pollutant that pours into the atmosphere from the burning of fossil fuels, like oil and coal. An increased amount of carbon dioxide in the atmosphere is expected to have many adverse effects, such as an increase in the average daily temperature.

An ever-increasing human population size is a threat to the continued existence of *Homo sapiens* when it means that the dynamic balance of the biosphere is upset. The recognition that the workings of the biosphere need to be preserved is one of the most important developments of our new ecological awareness.

Biodiversity: Going, Going, Gone When humans modify existing ecosystems, they reduce biodiversity. **Biodiversity** is the total number of species, the variability of their genes, and the ecosystems in which they live. The present biodiversity of our planet has been estimated to be as high as 80 million species, and so far, under 2 million have been identified and named. Extinction, the death of a species, occurs when a species is unable to adapt to a change in environmental conditions. It's estimated that presently we are losing from 24 to even 100 species a day due to human activities. For example, the existence of the species featured in the reading on page 7 is threatened because tropical rain forests are being reduced in size. As another example,

because of seaside development, pollution, and overfishing, 14 of the most valuable finfishes are becoming commercially extinct, meaning that too few remain to justify the cost of catching them.

Most biologists are alarmed over the present rate of extinction and believe the rate may eventually rival that of the five mass extinctions that have occurred during our planet's history. The dinosaurs became extinct during the last mass extinction, 65 million years ago. Everyone needs to realize that humans are totally dependent on other species for food, clothing, medicines, and various raw materials. Therefore, it is very shortsighted of us to allow other species to become extinct. Ecosystems and the species living in them should be preserved because only then can the human species continue to exist. And it takes from 2,000 to 10,000 generations for new species to evolve and to replace the ones that have died out. Because we are dependent upon the normal function and the present biodiversity of the biosphere, the existing species should be preserved.

> Humans belong to the world of living things and are vertebrates. They have modified existing ecosystems to the point that they must now be seriously concerned about the continued existence of the biosphere.

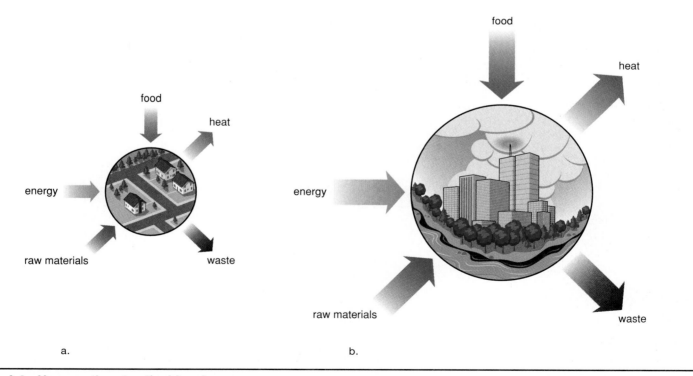

food

heat

energy

raw materials

waste

a.

food

heat

energy

raw materials

waste

b.

Figure I.4   Humans threaten the biosphere.
As cities grow in size, more materials, fuel, and food are taken from the environment, and more heat and waste are returned to the environment. **a.** Small town. **b.** Large city.

Ecology Focus

Tropical Rain Forests: Can We Live Without Them?

Figure IA   Tropical rain forest inhabitants.
These animals and plants make their homes in the Amazon basin.

The tropics are home for two-thirds of the plant species, 90% of the nonhuman primates, 40% of the birds of prey, and 90% of the insects that have been identified thus far. Many more species of organisms are estimated to exist but have not yet been discovered (perhaps as many as 30 million), and these are believed to live in the tropical rain forests that occur in a green belt spanning the planet on both sides of the equator.

If tropical rain forests are preserved, the rich diversity of plants and animals would continue to exist for scientific and pharmacological study (Fig. IA). One-fourth of the medicines we currently use come from tropical rain forests. For example, the rosy periwinkle from Madagascar has produced two potent drugs for use against Hodgkin disease, leukemia, and other blood cancers. It is hoped that many of the still-unknown plants will provide medicines for other human ills.

Tropical forests cover 6–7% of the total land surface of the earth —an area roughly equivalent to our contiguous 48 states. Every year humans destroy an area of forest equivalent to the size of Oklahoma (Fig. IB). At this rate, these forests and the species they contain will disappear completely in just a few more decades. Even if the forest areas now legally protected survive, 58–72% of all tropical forest species would still be lost.

The loss of tropical rain forests results from an interplay of social, economic, and political pressures. Many people already live in the forest, and as their numbers increase, more of the land is cleared for farming. People move to the forests because internationally financed projects build roads and open up the forests for exploitation. Small-scale farming accounts for about 60% of tropical deforestation, and decreasing percentages are due to commercial logging, cattle ranching, and mining. International demand for timber promotes destructive logging of rain forests in Southeast Asia and South America. The market for low-grade beef encourages their conversion to pastures for cattle. The lure of gold draws miners to rain forests in Costa Rica and Brazil.

The destruction of tropical rain forests gives only short-term benefits but is expected to cause long-term problems. The forests act like a giant sponge, soaking up rainfall during the wet season and releasing it during the dry season. Without them, a regional yearly regime of flooding followed by drought is expected to destroy property and reduce agricultural harvests. Worldwide, there could be changes in climate that would affect the entire human race.

However, studies show that if the forests were used as a sustainable source of nonwood products, such as nuts, fruits, and latex rubber, they would generate as much or more revenue while continuing to perform their various ecological functions and biodiversity could still be preserved. Brazil is exploring the concept of "extractive reserves," in which plant and animal products are harvested, but the forest itself is not cleared. Ecologists have also proposed "forest farming" systems, which mimic the natural forest as much as possible while providing abundant yields. But for such plans to work maximally, the human population size and the resource consumption per person must be stabilized.

Preserving tropical rain forests is a wise investment. Such action promotes the survival of most of the world's species—indeed, the human species, too.

Figure IB Burning of trees in a tropical rain forest.
It is estimated that an area the size of Oklahoma is being lost each year. While trees ordinarily take up carbon dioxide, burning releases carbon dioxide to the atmosphere.

I.2   The Process of Science

Science helps human beings understand the natural world. Science aims to be objective rather than subjective even though it is very difficult to make objective observations and to come to objective conclusions—we are often influenced by our own particular prejudices. Still, we should strive for objective observations and conclusions. Finally, scientific conclusions are subject to change whenever new findings so dictate. Quite often in science, new studies, which might utilize new techniques and equipment, tell us that previous conclusions need to be modified or changed entirely.

The ultimate goal of science is to understand the natural world in terms of **theories,** concepts based on the conclusions of observations and experiments (Fig. I.5). In a movie, a detective might claim to have a theory about the crime, or you might say that you have a theory about the win-loss record of your favorite baseball team, but in science, the word *theory* is reserved for a conceptual scheme supported by a large number of observations and not yet found lacking. Some of the basic theories of biology are as follows:

Name of Theory	Explanation
Cell	All organisms are composed of cells.
Biogenesis	Life comes only from life.
Evolution	All living things have a common ancestor, but each is adapted to a particular way of life.
Gene	Organisms contain coded information that dictates their form, function, and behavior.

Evolution is the unifying concept of biology because it pertains to various aspects of living things. For example, the theory of evolution enables scientists to understand the history of life, the variety of living things, and the anatomy, physiology, and development of organisms—even their behavior. Because the theory of evolution has been supported by so many observations and experiments for over a hundred years, some biologists refer to the *principle* of evolution. They believe this is the appropriate terminology for theories that are generally accepted as valid by an overwhelming number of scientists.

Scientists ask questions and carry on investigations that pertain to the natural world. The conclusions of these investigations are tentative and subject to change. Eventually, it may be possible to arrive at a theory that is generally accepted by all.

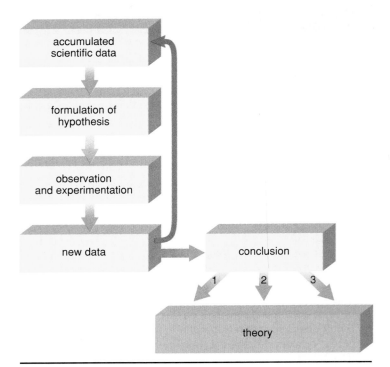

Figure I.5   Flow diagram for the scientific method.
Science is ongoing. The data gathered from previous observations and experiments plus new observations become the accumulated data used to formulate a hypothesis that leads to further observations and experimentation. Similar conclusions allow scientists to arrive at a theory.

The Scientific Method Has Steps

Scientists, including biologists, employ an approach to gathering information that is known as the **scientific method.** The approach of individual scientists to their work is as varied as they themselves are; still, for the sake of discussion, it is possible to speak of the scientific method as consisting of certain steps. Figure I.5 outlines the essential steps in the scientific method. **Data** are any observations and experimental results pertinent to the matter at hand. Previously made observations and/or experimental results (accumulated data) are used to formulate a **hypothesis** that becomes the basis for more observation and/or experimentation. The new data help a scientist come to a conclusion that either supports or does not support the hypothesis. Because hypotheses are always subject to modification, they can never be proven true; however, they can be proven false—that is, hypotheses are falsifiable. When the hypothesis is not supported by the data, it must be rejected; therefore, some think of the body of science as what is left after alternative hypotheses have been rejected.

Scientists working in the same area, such as cell biology, evolution, genetics, or any other area, may eventually suggest a theory, a biological concept that helps biologists understand the natural world.

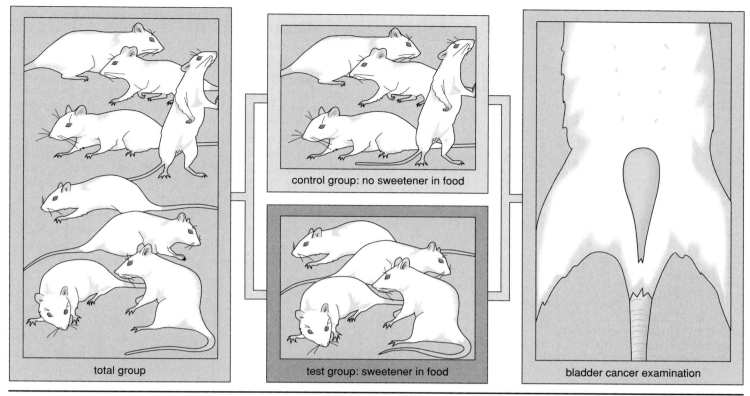

Figure I.6   Design of a controlled experiment.
From *left* to *right:* Genetically identical mice are randomly divided into a control group and the test groups. All groups are exposed to the same conditions, such as housing, temperature, and water supply. The control group is not subjected to the test (presence of sweetener S in the food). At the end of the experiment, all mice are examined for bladder cancer.

Scientists Use Controlled Experiments

When scientists are studying a phenomenon, they often perform controlled experiments in a laboratory. A controlled experiment contains a **control** group, which goes through all the steps of the experiment except the one being tested.

There are two major variables in a controlled experiment: the experimental variable and the dependent variable. The *experimental variable* is whatever is being tested, and the *dependent variable* is the result or change that is observed. The control sample is not subjected to the experimental variable.

Designing the Experiment

Suppose, for example, physiologists want to determine if sweetener S is a safe food additive (Fig. I.6). On the basis of available information, they might formulate a hypothesis that sweetener S is a safe food additive. Next, they might design the experiment described in Figure I.6 to test the hypothesis.

Test group: 50% of diet is sweetener S

Control group: diet contains no sweetener S

The researchers first place a certain number of randomly chosen inbred (genetically identical) mice into the various groups—say, 100 mice per group. If any of the mice are different from the others, it is hoped random selection has distributed them evenly among the groups. The researchers also make sure that all conditions, such as availability of water, cage setup, and temperature of the surroundings, are the same for both groups. The food for each group is exactly the same except for the amount of sweetener S.

At the end of the experiment, both groups of mice are to be examined for bladder cancer. Let's suppose that 50% of the mice in the test group are found to have bladder cancer, while none in the control group have bladder cancer. The results of this experiment do not support the hypothesis that sweetener S is a safe food additive.

Continuing the Experiment

Science is ongoing, and one experiment frequently leads to another. Physiologists might now wish to hypothesize that sweetener S is safe if the diet contains less than 50% sweetener S. They feed sweetener S to groups of mice at ever-greater concentrations:

Group 1: diet contains no sweetener S (the control)

Group 2: 5% of diet is sweetener S

Group 3: 10% of diet is sweetener S
↓
Group 11: 50% of diet is sweetener S

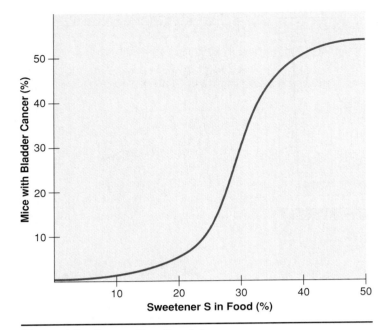

Figure I.7 **Presenting the data.**
Scientists often report mathematical data in the form of a table or a graph. Mathematical data are more decisive and objective than visual observations. The data in this instance suggest there is a correlation between the amount of sweetener S in food and the incidence of bladder cancer. Similar experiments will be repeated many times to test these results, and the results will be statistically analyzed to determine if they are significant or due to chance alone.

Usually, data obtained from experiments such as this are presented in the form of a table or a graph (Fig. I.7). Researchers might run a statistical test to determine if the difference in the number of cases of bladder cancer between the various groups is significant. After all, if a significant number of mice in the control group develop cancer, the results are invalid. Scientists prefer *mathematical data* because they are highly objective and not subject to individual interpretation.

On the basis of the results, the experimenters try to develop a recommendation concerning the safety of sweetener S in the food of humans. They might caution, for example, that the intake of sweetener S beyond 10% of the diet is associated with too great a risk of bladder cancer.

Many scientists work in laboratories, where they carry out controlled experiments.

I.3   Science and Social Responsibility

Science is objective and not subjective. It assumes that each person is capable of collecting data and seeing natural events in the same way and that the same theories and principles are applicable to past, present, and future events. Therefore, science seeks a natural cause for the origin and history of life. Doctrines of creation that have a mythical, philosophical, or theological basis are not a part of science because they are not subject to objective observations and experimentation by all. Many cultures have their own particular set of supernatural beliefs, and various religions within a culture differ as to the application of these beliefs. Such approaches to understanding the world are not within the province of science. Similarly, scientific creationism, which states that God created all species as they are today, cannot be considered science because explanations based on supernatural rather than natural causes involve faith rather than data.

There are many ways in which science has improved our lives. The discovery of antibiotics, such as penicillin, and of the polio, measles, and mumps vaccines, has increased our life span by decades. Cell biology research is helping us understand the causes of cancer. Genetic research has produced new strains of agricultural plants that have eased the burden of feeding our burgeoning world population.

Science also has effects we may find disturbing. For example, it sometimes fosters technologies that can be ecologically disastrous if not controlled properly. Too often we blame science for these developments and think that scientists are duty bound to pursue only those avenues of research that are consistent with our present system of values. But making value judgments is not a part of science. Ethical and moral decisions must be made by all people. The responsibility for how we use the fruits of science, including a given technology, must rest with people from all walks of life, not upon scientists alone. Scientists should provide the public with as much information as possible when such issues as the use of atomic energy, fetal research, and genetic engineering are being debated. Then they, along with other citizens, can help make decisions about the future role of these technologies in our society. All men and women have a responsibility to decide how to use scientific knowledge so that it benefits the human species and all living things.

The information presented in this text has been gathered using the scientific method. The text, while covering all aspects of biology, focuses on human biology. It is hoped that your study of biology will enable you to make wise decisions regarding your own individual well-being and also the well-being of all species, including our own.

SUMMARY

I.1 BIOLOGICALLY SPEAKING

Human beings, just like other organisms, are a product of the evolutionary process. They are members of the animal kingdom and are the vertebrates most closely related to other primates, including the apes. Like other living things, human beings reproduce, are highly organized, and maintain a fairly constant internal environment. They are members of the biosphere but have a cultural heritage that sometimes hinders the realization of their place in nature.

I.2 THE PROCESS OF SCIENCE

When studying the world of living things, biologists and other scientists use the scientific method, which consists of the following steps:

data, hypothesis, observation and

experimentation, new data, conclusions

I.3 SCIENCE AND SOCIAL RESPONSIBILITY

It is the responsibility of all to make ethical and moral decisions about how best to make use of the results of scientific investigations.

STUDYING THE CONCEPTS

1. Name five characteristics of human beings, and discuss each one. 2

2. Describe the five kingdom system of classification, and name types of organisms in each kingdom. 3

3. Give evidence that human beings are related to all other living things. 2

4. Human beings are dependent upon what services performed by plants? 4

5. What is homeostasis, and how is it maintained? Choose one organ system and tell how it helps maintain homeostasis. 2

6. Name the steps of the scientific method, and discuss each one. 8

7. How do you recognize a control group, and what is its purpose in an experiment? 9, 10

8. What is our social responsibility in regard to scientific findings? 10

APPLYING YOUR KNOWLEDGE

Concepts

1. Many industries today test various products on lower organisms, even as simple as bacteria, before they are put on the market for human consumption. What justification is there for assuming that bacteria can be used for these tests?

2. Homeostasis is the maintenance of a dynamic equilibrium by the body. How does a physician determine that your body is maintaining homeostasis?

3. What is (are) the reason(s) that humans are considered to be a severe threat to the biosphere, whereas other animals are not considered to be a threat?

Bioethical Issue

Scientists strive to be objective by carefully designing experiments and recording data. They are obligated to report the answers they find in experiments—even when they don't like those answers. After all, experiments that produce results other than the predicted ones can be just as informative as those that produce predicted results.

Still, scientists are only human. Rarely, the scientific community learns that a researcher has falsified data to get desired results. These incidents of dishonesty disturb people, laymen and scientists alike, and are used by critics in their arguments that scientists are naturally biased toward their theories.

Do you think science labs produce objective, trustworthy results? Should results be confirmed by a number of different researchers before the public believes in them? What does it take for you to have confidence in the results of an experiment?

TESTING YOUR KNOWLEDGE

1. Human beings are classified as _____ because they have a backbone.

2. Human beings are dependent upon plants because plants use _____ energy to make food.

3. To reproduce is to make a _____ of one's self.

4. Aside from our biological heritage, we also receive a _____ heritage, in large part from our parents but also from society as a whole.

5. Human beings are made up of cells; therefore they are said to be _____.

6. The _____ group in an experiment does not undergo the treatment being tested.

7. After considering previous findings, a scientist formulates a _____, which will be tested by observations and experimentation.

8. _____ has a responsibility to decide how scientific knowledge should be used.

SELECTED KEY TERMS

biodiversity Total number of species, the variability of their genes, and the ecosystems in which they live. 5

biosphere Portion of the surface of the earth where living organisms exist, including air, water, and land. 2

cell Structural and functional unit of an organism; the smallest structure capable of performing all the functions necessary for life. 2

control In experimentation, a sample that undergoes all the steps in the experiment except the one being tested. 9

data Facts that are derived from observations and experiments pertinent to the matter under study. 8

ecosystem Region in which populations interact with each other and with the physical environment. 4

evolution Changes that occur in the members of a species with the passage of time, often resulting in increased adaptation of organisms to the environment. 2

homeostasis Maintenance of a dynamic equilibrium, such that temperature, blood pressure, and other body conditions, remain within narrow limits. 2

hypothesis Statement that is capable of explaining present data and is to be tested by future observation and experimentation. 8

kingdom Classification category into which organisms are placed: Monera, Protists, Fungi, Plants, and Animals. 2

reproduce To make a copy similar to oneself, as when unicellular organisms divide or humans have children. 2

scientific method A step-by-step process for discovery and generation of knowledge, ranging from observation and hypothesis to theory and principle. 8

theory Concept consistent with conclusions based on a large number of experiments and observations, using the scientific method. 8

vertebrate Animal possessing a backbone composed of vertebrae. 2

FURTHER READINGS FOR INTRODUCTION

Barnard, C., et al. 1993. *Asking questions in biology.* Essex: Longman Scientific & Technical. First-year life science students are introduced to the skills of scientific observation and inquiry.

Carey, S. 1994. *A beginner's guide to scientific method.* Belmont, CA: Wadsworth Publishing. The basics of the scientific method are explained.

Cranbrook, E., and Edwards, D. S. 1994. *Belalong: A tropical rain forest.* London: The Royal Geographic Society, and Singapore: Sun Tree Publishing. Provides a very readable, well-illustrated account of biodiversity in a brunei rain forest.

deDuve, C. April 1996. The birth of complex cells. *Scientific American* 274(4):50. Article discusses the role of natural selection in the evolution of the complex cell.

Drewes, F. 1997. *How to study science.* 2d ed. Dubuque, Iowa: Wm. C. Brown Publishers. Supplements any introductory science text; shows students how to study and take notes and how to interpret text figures.

Frenay, A. C. F., and Mahoney, R. M. 1993. *Understanding medical terminology.* Dubuque, Iowa: Wm. C. Brown Publishers. A structural approach to the study of medical terminology.

Johnson, G. B. 1996. *How scientists think.* Dubuque, Iowa: Wm. C. Brown Publishers. Presents the rationale behind 21 important experiments in genetics and molecular biology that became the foundation for today's research.

Marchuk, W. N. 1992. *A life science lexicon.* Dubuque, Iowa: Wm. C. Brown Publishers. Helps students master life sciences terminology.

Margulis, L., et al. 1994. *The illustrated five kingdoms: A guide to the diversity of life on Earth.* New York: HarperCollins College Publishers. Introduces the kingdoms of organisms.

May, R. M. October 1992. How many species inhabit the earth? *Scientific American* 267(4):42. After more than 250 years, fewer than 2 million species have been catalogued.

McCain, G., and Segal, E. M. 1988. *The game of science.* 5th ed. Pacific Grove, Calif.: Brooks/Cole Publishing Co. Provides an awareness and understanding of the relationship of science and our culture.

Serafini, A. 1993. *The epic history of biology.* New York: Plenum Press. This is a history of biology from ancient Egyptian medicine to present-day biotechnology.

Young, J. Z. 1951. *Doubt and certainty in science.* Oxford: Clarendon Press. Discusses how a scientific theory develops.

Part

1

Human Organization

The human body is composed of cells, the smallest units of life. An understanding of cell structure, physiology, and biochemistry serves as a foundation for understanding how the human body functions.

Principles of inorganic and organic chemistry are discussed before a study of human cell structure is undertaken. The human cell is bounded by a membrane and contains organelles, which are also membranous. Membranes regulate entrance and exit of molecules and help cellular organelles carry out their functions.

The many cells of the body are specialized into tissues that are found within the organs of the various systems of the body. All body systems help maintain a dynamic constancy of the internal environment so that proper physical conditions exist for each cell.

Chapter 1

Chemistry of Life

Chapter Outline

1.1   ELEMENTS AND ATOMS
- All matter is composed of elements, each having one type of atom.   14

1.2   MOLECULES AND COMPOUNDS
- Atoms react with one another forming ions, molecules, and compounds.   16

1.3   SOME IMPORTANT INORGANIC MOLECULES
- Some important types of inorganic molecules in living organisms are water, acids, and bases.   19

1.4   MOLECULES OF LIFE
- The macromolecules found in cells are carbohydrates, lipids, proteins, and nucleic acids.   24
- Macromolecules arise when their specific monomers (unit molecules) join together.   24

1.5   CARBOHYDRATES
- Carbohydrates function as a ready source of energy in most organisms.   25
- Glucose is a simple sugar; starch, glycogen, and cellulose are chains of glucose.   25
- Cellulose lends structural support to plant cell walls.   26

1.6   LIPIDS
- Lipids are varied molecules.   27
- Fats and oils, which function in long-term energy storage, are composed of glycerol and three fatty acids.   27
- Sex hormones are derived from cholesterol, a complex ring compound.   28

1.7   PROTEINS
- Proteins help form structures (e.g., muscles and membranes) and function as enzymes.   29
- Proteins are chains of amino acids.   30

1.8   NUCLEIC ACIDS
- Genes are composed of DNA (deoxyribonucleic acid). RNA (ribonucleic acid) helps DNA specify protein synthesis and helps construct proteins.   34
- Nucleic acids are chains of nucleotides.   34

Figure 1.1 Chemicals and the body.
The human body is affected by the chemicals that we breathe and consume as nutrients. Even the sweeteners used in soft drinks can influence body metabolism as those with a metabolic disorder, such as phenylketonuria, know all too well.

Glance at the back of any diet Coke can, and you'll find an important warning: "Contains Phenylalanine." For most of us, this chemical—one of 20 naturally occurring amino acids—is harmless. That's because our bodies contain a liver enzyme that converts phenylalanine into a useful chemical for the body.

But people with PKU, or phenylketonuria, lack the necessary enzyme. In these people, phenylalanine builds up, and dangerous by-products flood the bloodstream. Brain damage or other problems follow.

Thus, doctors keep PKU sufferers on a strict diet, limiting their intake of phenylalanine. Coca-Cola and other companies help by labeling products containing the chemical (Fig. 1.1).

PKU is just one of hundreds of disorders caused by malfunctioning or absent enzymes. From simple inorganic molecules to complex organic macromolecules, such as enzymes, chemicals are essential to our being. As people with PKU know, lacking just one of these precious chemicals can be disastrous.

This chapter reviews the structure of atoms and how they join to form both inorganic and organic chemicals. Inorganic chemicals like salts have significant functions in the human body just as organic chemicals like proteins do. All chemicals whether inorganic or organic are composed of elements.

1.1 Elements and Atoms

Elements are basic substances that cannot be broken down further into simpler substances. Considering the variety of living and nonliving things in the world, it's quite remarkable that there are only 92 naturally occurring elements. It is even more surprising that over 90% of the human body is composed of just three elements: carbon, oxygen, and hydrogen.

Every element has a name and a symbol; for example, calcium has been assigned the atomic symbol Ca (Fig. 1.2*a*). Calcium is not one of the most abundant elements in cells, yet it has many important functions. It is necessary for strong bones and plays a role in nervous conduction and muscular contraction.

Common Elements in Living Things				
Element	Atomic Symbol	Atomic Number	Atomic Weight	Comment
hydrogen	H	1	1	These
carbon	C	6	12	elements
nitrogen	N	7	14	make up
oxygen	O	8	16	most
phosphorus	P	15	31	biological
sulfur	S	16	32	molecules.
sodium	Na	11	23	These
magnesium	Mg	12	24	elements
chlorine	Cl	17	35	occur mainly
potassium	K	19	39	as dissolved
calcium	Ca	20	40	salts.

a.

Figure 1.2 Elements and atoms.
a. The atomic symbol, atomic number, and atomic weight are given for the common elements in living things. **The atomic symbol for calcium is Ca, the atomic number is 20, and the atomic weight is 40. b.** An atom contains the subatomic particles called protons (p) and neutrons (n) in the nucleus (colored pink) and electrons (colored gray) in shells about the nucleus.

p = Protons
n = Neutrons
= Electrons

6p
6n

Carbon
$^{12}_{6}C$

b.

Atoms Have Structure

An **atom** is the smallest unit of an element that still retains the chemical and physical properties of an element. While it is possible to split an atom by physical means, an atom is the smallest unit to enter into chemical reactions. For our purposes, it is satisfactory to think of each atom as having a central *nucleus,* where subatomic particles called **protons** and **neutrons** are located, and *shells,* where **electrons** orbit about the nucleus (Fig. 1.2b). Most of an atom is empty space. If we could draw an atom the size of a football field, the nucleus would be like a gumball in the center of the field, and the electrons would be tiny specks whirling about in the upper stands.

Two important features of protons, neutrons, and electrons are their weight and charge:

Name	Charge	Weight
Electron	One negative unit	Almost no weight
Proton	One positive unit	One atomic unit
Neutron	No charge	One atomic unit

The atomic number of an atom tells you how many protons (+) and therefore how many electrons (–) an atom has when it is electrically neutral. For example, the atomic number of calcium is 20; therefore, when calcium is neutral, it has 20 protons and 20 electrons. How many electrons are there in each shell of an atom? The inner shell has the lowest energy level and can hold only two electrons; after that, each shell for the atoms noted in Fig. 1.2a can hold up to eight electrons. Using this information, calculation determines that calcium has four shells and the outermost shell has two electrons. As we shall see, an atom is most stable when the outermost shell has eight electrons. (Hydrogen with only one shell is an exception to this statement. Atoms with only one shell are stable when this shell contains two electrons.)

The subatomic particles are so light that their weight is indicated by special designations called atomic mass units. Notice in the chart above that protons and neutrons each have about one atomic unit of weight and electrons have almost no weight. Therefore, the atomic weight generally tells you the number of protons plus the number of neutrons. How could you calculate that carbon (C) has 6 neutrons? Carbon's atomic weight is 12, and you know from its atomic number that it has 6 protons. Therefore, carbon has 6 neutrons (Fig. 1.2b).

As shown in Figure 1.2b, the atomic number of an atom is often written as a subscript to the lower left of the atomic symbol. The atomic weight is often written as a superscript to the upper left of the atomic symbol. Therefore, carbon can be designated in this way:

$$^{12}_{6}C$$

All matter is composed of elements, each containing just one type of atom. Each atom has an atomic symbol, atomic number (number of protons), and atomic weight (number of protons and neutrons).

Isotopes Differ in Atomic Weight

The atomic weights given in the periodic table are the average weight for each kind of atom. This is because atoms of the same type may differ in the number of neutrons; therefore, their weight varies. Atoms that have the same atomic number and differ only in the number of neutrons are called **isotopes.** Isotopes of carbon can be written in the following manner, where the subscript stands for the atomic number and the superscript stands for the atomic weight:

$$^{12}_{6}C \qquad ^{13}_{6}C \qquad ^{14}_{6}C$$

Carbon-12 has six neutrons, carbon-13 has seven neutrons, and carbon-14 which has eight neutrons is radioactive.

Isotopes have many uses. Each type of food has its own proportion of isotopes, and this information allows biologists to study mummified or fossilized human tissue to know what ancient peoples ate. Most isotopes are stable, but radioactive isotopes break down and emit radiation in the form of radioactive particles or radiant energy. Because carbon-14 breaks down at a known rate, the amount of this atom remaining is often used to determine the age of fossils. Radioactive isotopes are widely used in biological and medical research; for example, because the thyroid gland uses iodine (I), it is possible to administer a dose of radioactive iodine and then observe later that the thyroid has taken it up (Fig. 1.3).

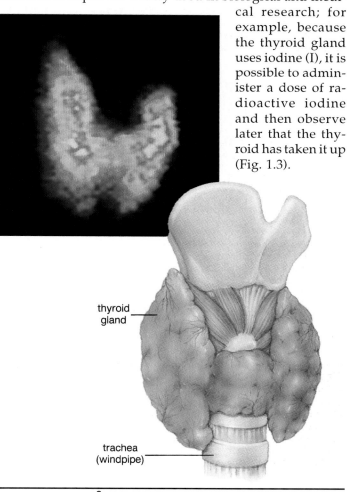

Figure 1.3 Use of radioactive iodine.
A scan of the thyroid gland 24 hours after the patient was administered radioactive iodine. The thyroid gland is located at the base of the neck.

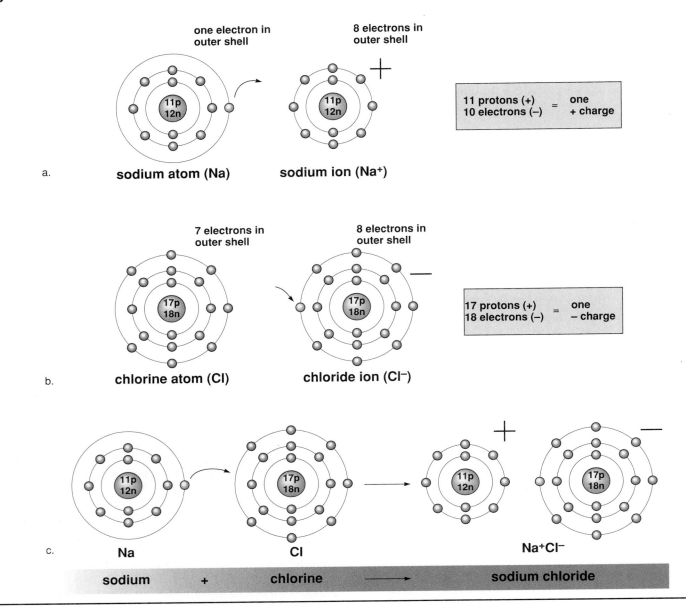

one electron in outer shell

8 electrons in outer shell

| 11 protons (+)
10 electrons (–) | = | one
+ charge |

a. **sodium atom (Na)**   **sodium ion (Na+)**

7 electrons in outer shell

8 electrons in outer shell

| 17 protons (+)
18 electrons (–) | = | one
– charge |

b. **chlorine atom (Cl)**   **chloride ion (Cl⁻)**

c. **Na**   **Cl**   **Na+Cl⁻**

sodium   +   chlorine   ⟶   sodium chloride

Figure 1.4   Ionic reaction.
a. When a sodium atom gives up an electron, it becomes a positive ion. **b.** When a chlorine atom gains an electron, it becomes a negative ion. **c.** When sodium reacts with chlorine, the compound sodium chloride (Na+Cl⁻) results. In sodium chloride, an ionic bond exists between the ions.

1.2   Molecules and Compounds

Atoms often bond with each other to form a chemical unit called a **molecule.** A molecule can contain atoms of the same kind, as when an oxygen atom joins with another oxygen atom to form oxygen gas. Or the atoms can be different, as when an oxygen atom joins with two hydrogen atoms to form water. When the atoms are different, a compound results.

Two types of bonds join atoms: the ionic bond and the covalent bond.

Opposite Charges Attract in Ionic Bonds

Recall that atoms are most stable when the outermost shell contains eight electrons. During an ionic reaction, atoms give up or take on an electron(s) in order to achieve a stable outermost shell.

Figure 1.4 depicts a reaction between a sodium (Na) and chlorine (Cl) atom in which chlorine takes an electron from sodium. **Ions** are particles that carry either a positive (+) or negative (–) charge. The sodium ion carries a positive

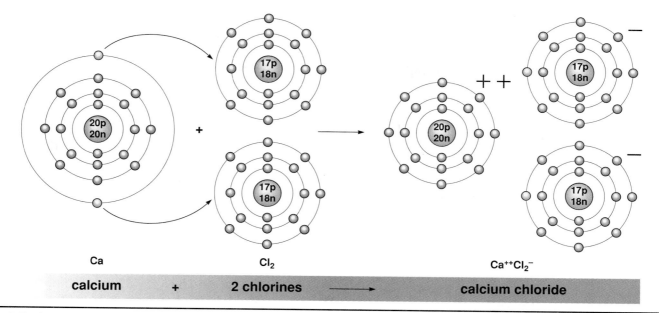

Figure 1.5 Ionic reaction.
The calcium atom reacts with two chlorine atoms to produce the compound calcium chloride ($Ca^{++}Cl_2^-$).

charge because it now has one more proton than electrons, and the chloride ion carries a negative charge because it now has one less proton than electrons. The attraction between oppositely charged sodium ions and chloride ions forms an **ionic bond.** Ionic bonds are typically found in inorganic compounds. We are quite familiar with the inorganic compound sodium chloride because it is table salt, which we use to enliven the taste of foods.

Figure 1.5 shows an ionic reaction between a calcium atom and two chlorine atoms. Notice that calcium with two electrons in the outermost shell reacts with two chlorine atoms. Why? Because with seven electrons already, each chlorine requires only one more electron to have a stable outermost shell. The resulting salt is called calcium chloride.

Significant ions in the human body are listed in Table 1.1. The balance of these ions in the body is important to our health. Too much sodium in the blood can cause high blood pressure; not enough calcium leads to rickets (a bowing of the legs) in children; too much or too little potassium results in heartbeat irregularities. Bicarbonate, hydrogen, and hydroxide ions are all involved in maintaining the acid-base balance of the body. If the blood is too acidic or too basic, the body's cells cannot function properly.

TABLE 1.1

Significant Ions in the Body

Name	Symbol	Special Significance
Sodium	Na^+	Found in body fluids; important in muscle contraction and nerve conduction.
Chloride	Cl^-	Found in body fluids.
Potassium	K^+	Found primarily inside cells; important in muscle contraction and nerve conduction.
Phosphate	PO_4^{-3}	Found in bones, teeth, and the high energy molecule ATP.
Calcium	Ca^{++}	Found in bones and teeth; important in muscle contraction.
Bicarbonate	HCO_3^-	Important in acid-base balance.
Hydrogen	H^+	Important in acid-base balance.
Hydroxide	OH^-	Important in acid-base balance.

An ionic bond is the attraction between oppositely charged ions. Ionic bonds are often seen in inorganic compounds.

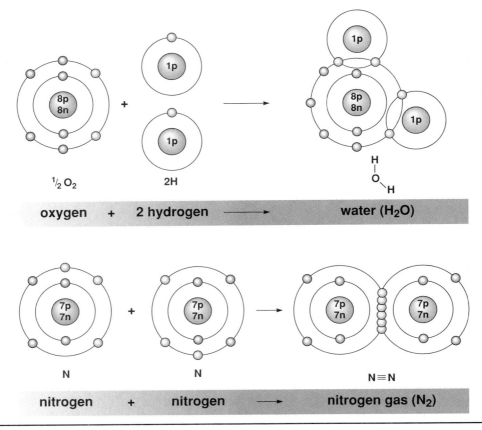

Figure 1.6   Covalent reactions.
After a covalent reaction, atoms share electrons, and each atom has eight electrons in the outermost shell. To show this, it is necessary to count the shared electrons as belonging to both bonded atoms.

Electrons Are Shared in Covalent Bonds

After covalent reactions, the atoms share electrons in **covalent bonds** instead of losing or gaining them. Covalent bonds, which are typically found in organic molecules, can be represented in a number of ways. The overlapping outermost shells in Figure 1.6 indicate that the atoms are sharing electrons. Just as two hands participate in a handshake, each atom contributes one electron to the pair that is shared. These electrons spend part of their time in the outermost shell of each atom; therefore, they are counted as belonging to both bonded atoms. When this is done, each atom has eight electrons in the outermost shell.

Structural formulas use straight lines to show the covalent bonds between the atoms. Each line represents a pair of shared electrons. Molecular formulas indicate only the number of each type of atom making up a molecule.

Structural formula: Cl — Cl

Molecular formula: Cl_2

More Than One Pair Is Shared in Double and Triple Bonds

Besides a single bond, in which atoms share only a pair of electrons, a double or a triple bond can form. In a double bond, atoms share two pairs of electrons, and in a triple bond, atoms share three pairs of electrons between them. For example, in Figure 1.6, each nitrogen atom (N) requires three electrons to achieve a total of eight electrons in the outermost shell. Notice that six electrons are placed in the outer overlapping shells in the diagram and that three straight lines are in the structural formula for nitrogen gas (N_2).

A covalent bond arises when atoms share electrons. In double covalent bonds, atoms share two pairs of electrons, and in triple covalent bonds, atoms share three pairs of electrons.

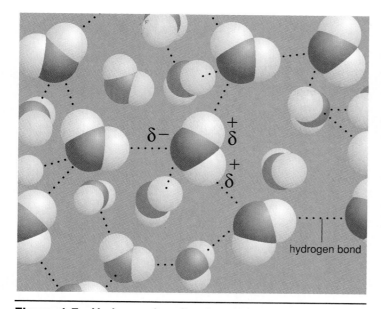

Figure 1.7   Hydrogen bonding between water molecules.
The polarity of the water molecules allows hydrogen bonds (dotted lines) to form between the molecules.

1.3   Some Important Inorganic Molecules

Inorganic molecules are characterized by the presence of a small number of atoms ionically bonded together (see Figs. 1.4 and 1.5). Water is an exception to this statement because its atoms are covalently bonded (see Fig. 1.6).

Water Is a Polar Molecule

Water is the most abundant molecule in living organisms, and it makes up about 60 to 70% of the total body weight of most organisms. The physical and chemical properties of water make life possible as we know it.

Sometimes covalently bonded atoms share electrons unevenly; that is, the electrons spend more time circling the nucleus of one atom than circling the other. In water, the electrons spend more time circling the larger oxygen (O) than the smaller hydrogen (H) atoms. As a result, the hydrogen atoms have a partial positive charge, and the oxygen atom has a partial negative charge.

Because the water molecule has charged atoms, it is called a *polar molecule*. Hydrogen bonding occurs between water molecules because they are polar (Fig. 1.7). A **hydrogen bond** occurs whenever a covalently bonded hydrogen is attracted

to a negatively charged atom some distance away. The hydrogen bond is represented by a dotted line in Figure 1.7 because it is relatively weak and can be broken rather easily.

In some covalent bonds, the electrons are shared unequally, and the result is a polar molecule. Hydrogen bonding can occur between polar molecules.

Water Has Unique Properties

Because of its polarity and hydrogen bonding, water has many characteristics beneficial to life. Hydrogen bonding causes water molecules to be cohesive and cling together. Without hydrogen bonding between molecules, water would boil much below 100°C and freeze much lower than 0°C. Because water remains a liquid between 100°C and 0°C, it is suitable for life.

1.   Water is the universal solvent and facilitates chemical reactions both outside of and within living systems.

When a salt such as sodium chloride (Na^+Cl^-) is put into water, the negative ends of the water molecules are attracted to the sodium ions, and the positive ends of the water molecules are attracted to the chloride ions. This causes the sodium ions and the chloride ions to separate and to dissolve in water:

The salt $Na^+ Cl^-$ dissolves in water

When ions and molecules disperse in water, they move about and collide, allowing reactions to occur. Therefore, water is a solvent that facilitates chemical reactions.

Those molecules that interact with water are said to be *hydrophilic*. Nonionized and nonpolar molecules that do not interact with water are said to be *hydrophobic*.

2.   Water molecules are cohesive and fill vessels.

Water molecules cling together because of hydrogen bonding, and yet, water flows freely. This property allows dissolved and suspended molecules to be evenly distributed throughout a system. Therefore, water is an excellent transport system both outside of and within living organisms. Human beings have internal vessels in which water serves to transport nutrients and wastes.

a.   b.   c.

Figure 1.8   Characteristics of water.
a. Water boils at 100°C. If it boiled at a lower temperature, life could not exist. **b.** It takes much body heat to vaporize sweat, which is mostly liquid water, and this helps keep bodies cool when the temperature rises. **c.** Ice is less dense than water, and it forms on top of water, making skate sailing possible.

3. The temperature of liquid water rises and falls slowly, preventing sudden or drastic changes.

A calorie of heat energy is needed to raise the temperature of one gram of water 1°C. This is about twice the amount of heat required for other covalently bonded liquids. The many hydrogen bonds that link water molecules cause water to absorb a great deal of heat before it boils (Fig. 1.8*a*). On the other hand, water holds heat, and its temperature falls slowly. Therefore, water protects organisms from rapid temperature changes and helps them maintain their normal internal temperature. This property also allows great bodies of water, such as oceans, to maintain a relatively constant temperature. Water is a good temperature buffer.

4. Water has a high heat of vaporization, keeping the body from overheating.

Converting one gram of the hottest water to steam requires an input of 540 calories of heat energy. Hydrogen bonds are broken when water is changed to steam; this accounts for the very large amount of heat needed for evaporation. This property also helps moderate the earth's tem-

perature so that life can continue to exist. It gives animals in a hot environment an efficient way to release excess body heat. When an animal sweats, body heat is used to vaporize the sweat, which is mostly liquid water, and therefore, the body cools (Fig. 1.8*b*).

5. Frozen water is less dense than liquid water so that ice floats on water.

As water cools, the molecules come closer together. They are densest at 4°C, but they are still moving about. At temperatures below 4°C, there is only vibrational movement, and hydrogen bonding becomes more rigid but also more open. This makes ice less dense. Bodies of water always freeze from the top down, making skate sailing possible (Fig. 1.8*c*). When a body of water freezes on the surface, the ice acts as an insulator to prevent the water below it from freezing. Aquatic organisms are protected, and they have a better chance of surviving the winter.

Water is a universal solvent because it is a polar molecule. Because of its polarity and hydrogen bonding, water has many characteristics that benefit life.

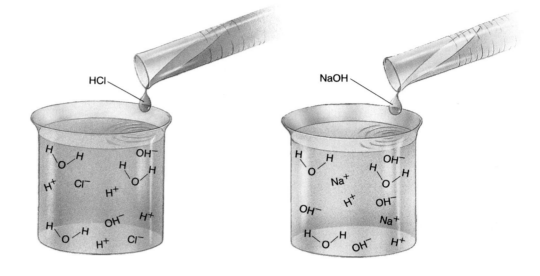

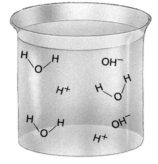

Figure 1.9   Dissociation of water molecules.
Dissociation produces an equal number of hydrogen ions (H⁺) and hydroxide ions (OH⁻). (These figures are for illustration and are not mathematically accurate.)

Figure 1.10   Addition of hydrochloric acid (HCl).
HCl releases hydrogen ions (H⁺) as it dissociates. The addition of HCl to water results in a solution with more H⁺ than OH⁻.

Figure 1.11   Addition of sodium hydroxide (NaOH), a base. NaOH releases OH⁻ as it dissociates.
The addition of NaOH to water results in a solution with more OH⁻ than H⁺.

How Acids Differ from Bases

When water dissociates, it releases an equal number of hydrogen ions (H⁺) and hydroxide ions (OH⁻) (Fig. 1.9). Only a few water molecules at a time are dissociated, and the actual number of ions is very small (10^{-7} moles/liter). A mole is a unit of scientific measurement for atoms, ions, and molecules.

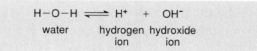

Acids (H⁺ Up)

Lemon juice, vinegar, tomato juice , and coffee are all familiar acids. What do they have in common? **Acids** are molecules that dissociate in water, releasing hydrogen ions (H⁺)[1]. For example, an important inorganic acid is hydrochloric acid (HCl), which dissociates in this manner:

$$HCl \rightarrow H^+ + Cl^-$$

Dissociation is almost complete; therefore, this is called a strong acid. If hydrochloric acid is added to a beaker of water (Fig. 1.10), the number of hydrogen ions increases.

Bases (H⁺ Down)

Milk of magnesia and ammonia are common bases that most people have heard of. **Bases** are molecules that either take up hydrogen ions (H⁺) or release hydroxide ions (OH⁻). For example, an important inorganic base is sodium hydroxide (NaOH), which dissociates in this manner:

$$NaOH \rightarrow Na^+ + OH^-$$

Dissociation is almost complete; therefore sodium hydroxide is called a strong base. If sodium hydroxide is added to a beaker of water (Fig. 1.11), the number of hydroxide ions increases.

---

1. A hydrogen atom contains one electron and one proton. A hydrogen ion has only one proton, so is often called a proton.

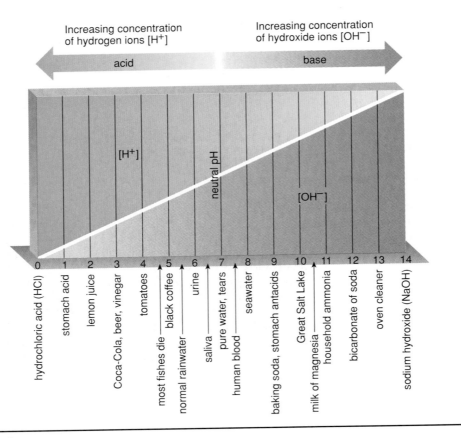

Figure 1.12   The pH scale.
The diagonal line indicates the proportionate concentration of hydrogen ions (H⁺) to hydroxide ions (OH⁻) at each pH value. Any pH value above 7 is basic, while any pH value below 7 is acidic.

pH Scale

The **pH scale**[2] is used to indicate the acidity and basicity (alkalinity) of a solution. A pH of exactly 7 is neutral pH. Pure water has an equal number of hydrogen ions (H⁺) and hydroxide ions (OH⁻), and therefore, one of each is released when water dissociates. One mole of pure water contains only 10^{-7} moles/liter of hydrogen ions, which is the source of the pH value for neutral solutions.

The pH scale was devised as a means of expressing the hydrogen ion concentration [H⁺] and consequently, the hydroxide ion concentration [OH⁻]; it eliminates the use of cumbersome numbers. For example,

$$1 \times 10^{-6} \ [H^+] = pH \ 6$$
$$1 \times 10^{-7} \ [H^+] = pH \ 7$$
$$1 \times 10^{-8} \ [H^+] = pH \ 8$$

The pH scale (Fig. 1.12) ranges from 0 to 14. As we move down the pH scale, each unit has 10 times the acidity of the previous unit, and as we move up the scale, each unit has 10 times the basicity of the previous unit. A pH of 7 has an equal concentration of hydrogen ions and hydroxide

ions. Above pH 7 there are more hydroxide ions than hydrogen ions, and below pH 7 there are more hydrogen ions than hydroxide ions.

In living things, pH needs to be maintained within a narrow range or there are health consequences. As discussed in the Ecology reading on page 23, there are environmental consequences of rain and snow becoming more acidic.

Buffers Keep pH Steady

The pH of our blood when we are healthy is always about 7.4. Normally, pH stability is possible because the body has built-in mechanisms to prevent pH changes. **Buffers** are the most important of these mechanisms. Buffers help keep the pH within normal limits because they are chemicals or combinations of chemicals that take up excess hydrogen ions (H⁺) or hydroxide ions (OH⁻).

Acids have a pH that is less than 7, and bases have a pH that is greater than 7. Buffers, which can combine with both hydrogen ions and hydroxide ions, help to keep the pH of internal body fluids near pH 7 (neutral).

---

2. pH is defined as the negative logarithm of the hydrogen ion concentration [H⁺].

Ecology Focus

The Harm Done by Acid Deposition

Normally, rainwater has a pH of about 5.6 because the carbon dioxide in the air combines with water to give a weak solution of carbonic acid. Rain falling in northeastern United States and southeastern Canada now has a pH between 5.0 and 4.0. We have to remember that a pH of 4 is ten times more acidic than a pH of 5 to comprehend the increase in acidity this represents.

There is very strong evidence that this observed increase in rainwater acidity is a result of the burning of fossil fuels, like coal and oil, as well as gasoline derived from oil. When fossil fuels are burned, sulfur dioxide and nitrogen oxides are produced, and they combine with water vapor in the atmosphere to form acids. These acids return to earth contained in rain or snow, a process properly called wet deposition, but more often called acid rain. Dry particles of sulfate and nitrate salts descend from the atmosphere during dry deposition.

Unfortunately, regulations that require the use of tall smokestacks to reduce local air pollution only cause pollutants to be carried far from their place of origin. Acid deposition in southeastern Canada is due to the burning of fossil fuels in factories and power plants in the Midwest. Tensions between Canada and the United States

have been eased by the 1990 Clean Air Act; it called for a ten-million-ton reduction in sulfur dioxide emissions and a two-million-ton reduction in nitrogen oxide emissions in the United States by the year 2000. Thereafter, sulfur dioxide emissions will be capped and no increase in total emissions will be allowed. Canada agreed to make comparable reductions in its emission of air pollutants, some of which find their way to the northeastern United States. In 1991, the two countries signed a formal air pollution agreement.

Acid deposition adversely affects lakes, particularly in areas where the soil is thin and lacks limestone (calcium carbonate, $CaCO_3$), a buffer to acid deposition. It leaches aluminum from the soil, carries aluminum into the lakes, and converts mercury deposits in lake bottom sediments to soluble and toxic methyl mercury. Lakes not only become more acidic, but they also show accumulation of toxic substances. In Norway and Sweden, at least 16,000 lakes contain no fish, and an additional 52,000 lakes are threatened. In Canada, some 14,000 lakes are almost fishless, and an additional 150,000 are in peril because of excess acidity. In the United States, about 9,000 lakes (mostly in the Northeast and upper Midwest) are threatened, one-third of them seriously.

In forests, acid deposition weakens trees because it leaches away nutrients and releases aluminum. By 1988, most spruce, fir, and other conifers atop North Carolina's Mt. Mitchell were dead from being bathed in ozone and acid fog for years. The soil was so acidic, new seedlings could not survive. Nineteen countries in Europe have reported woodland damage ranging from 5 to 15%

of the forested area in Yugoslavia and Sweden to 50% or more in the Netherlands, Switzerland, and the former West Germany. More than one-fifth of Europe's forests are now damaged.

These aren't the only effects of acid deposition. Reduction of agricultural yields, damage to marble and limestone monuments and buildings, and even illnesses in humans have been reported. Acid deposition has been implicated in the increased incidence of lung cancer and possibly colon cancer in residents of the East Coast. Tom McMillan, Canadian Minister of the Environment, says that acid rain is "destroying our lakes, killing our fish, undermining our tourism, retarding our forests, harming our agriculture, devastating our heritage, and threatening our health."

There are, of course, things that can be done. We could

a. whenever possible use alternative energy sources, such as solar, wind, hydropower, and geothermal energy.

b. use low-sulfur coal or remove the sulfur impurities from coal before it is burned.

c. require factories and power plants to use scrubbers, which remove sulfur emissions.

d. require people to use mass transit rather than driving their own automobiles.

e. reduce our energy needs through other means of energy conservation.

Sources:

G. Tyler Miller, Living in the Environment, *Wadsworth Publishing Company, Belmont, CA, 1992; and Lester R. Brown, et al.,* State of the World. *W.W. Norton and Company, Inc., New York, NY, 1993.*

1.4   Molecules of Life

The molecules associated with living things are called organic molecules. **Organic molecules** always contain carbon (C) and hydrogen (H). The chemistry of carbon accounts for the formation of the very large variety of organic molecules. A carbon atom has four electrons in the outermost shell. In order to achieve eight electrons in the outermost shell, a carbon atom shares electrons covalently with as many as four other atoms. Methane is a molecule in which a carbon atom shares electrons with four hydrogen atoms.

A carbon atom can also share with another carbon atom, and in so doing, a long chain can result:

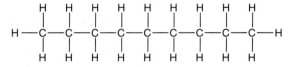

A carbon chain can also turn back on itself to form a ring compound:

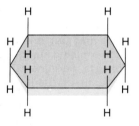

So-called functional groups can be attached to carbon chains. A functional group is a particular cluster of atoms that always behaves in a certain way. One functional group of interest is the acidic (carboxyl) group —COOH because it can give up a hydrogen (H^+) and ionize to —COO^-.

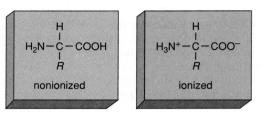

Whereas a hydrocarbon chain is *hydrophobic* (not attracted to water) because it is nonpolar, a hydrocarbon chain with an attached ionized group is *hydrophilic* (is attracted to water) because it is polar.

The molecules of life are divided into four classes: carbohydrates, lipids, proteins, and nucleic acids. Carbohydrates, lipids, and proteins are very familiar to you because certain foods are known to be rich in these molecules, as illustrated in Figures 1.13–1.15.

The molecules of life are macromolecules. Just as atoms can join to form a molecule, so molecules can join to form a macromolecule. The smaller molecules are called monomers, and the macromolecule is called a polymer. A polymer is a chain of monomers.

Polymer	Monomer
polysaccharide	monosaccharide
lipid (e.g., fat)	glycerol and fatty acid
protein	amino acid
nucleic acid	nucleotide

Figure 1.13   Foods rich in carbohydrates.
Breads, pasta, rice, corn, and oats all contain complex carbohydrates.

Figure 1.14   Foods rich in lipids.
Butter and oils contain fat, the most familiar of the lipids.

Figure 1.15   Foods rich in proteins.
Meat, eggs, and cheese have a high content of protein.

1.5 Carbohydrates

Carbohydrates first and foremost function for quick and short-term energy storage in all organisms, including humans. Carbohydrate molecules are characterized by the presence of the atomic grouping H—C—OH, in which the ratio of hydrogen atoms (H) to oxygen atoms (O) is approximately 2:1. Since this ratio is the same as the ratio in water, the name—hydrates of carbon—seems appropriate.

Carbohydrates Can Be Simple

If the number of carbon atoms in a molecule is low (from three to seven), then the carbohydrate is a simple sugar, or **monosaccharide.** These molecules are often designated by the number of carbon atoms they contain; for example, **glucose,** with six carbon atoms is called a **hexose** (Fig. 1.16). Other common hexoses are fructose, found in fruits, and galactose, a constituent of milk. These three hexoses (glucose, fructose, and galactose) all occur as ring structures with the molecular formula $C_6H_{12}O_6$, but the exact shape of the ring differs, as does the arrangement of the hydrogen (—H) and the hydroxyl groups (—OH) attached to the ring.

A **disaccharide** (di means two and saccharide means sugar) contains two monosaccharides. **Synthesis** (making) of a disaccharide is a condensation reaction because water is removed as the two monosaccharides join to form the disaccharide (Fig. 1.17). Degradation is a **hydrolysis** reaction because water is used to split a bond. When two glu-cose molecules join, maltose forms. When glucose and fructose join, the disaccharide sucrose forms. Sucrose, which is ordinarily derived from sugarcane and sugar beets, is commonly known as table sugar. Eating a candy bar provides quick energy because sucrose breaks down to glucose and fructose, which is converted to glucose in the liver. Cells use glucose as their primary energy source.

Starch and Glycogen are Complex Carbohydrates

Starch and **glycogen** are ready storage forms of glucose in plants and animals, respectively. Starch and glycogen are polysaccharides; that is, they are polymers of glucose formed just as a necklace might be formed from only one type of bead. The following equation shows how starch is synthesized:

$$\text{glucose molecules} \underset{\text{hydrolysis}}{\overset{\text{synthesis}}{\rightleftharpoons}} \text{starch} + H_2O \text{ molecules}$$
$$\text{(monomers)} \qquad\qquad\qquad \text{(polymer)}$$

Notice that this is a **condensation** reaction: as the polymer forms, water is removed.

Figures 1.18 and 1.19 show how the structure of starch differs from the structure of glycogen. Starch has fewer side branches, or chains of glucose that branch off from the main chain, than does glycogen. Why does the micrograph in Figure 1.18 feature starch granules inside plant cells? The micrograph illustrates that starch is the storage form of glucose inside plant cells. Flour, which we usually acquire by grinding wheat and use to bake bread and rolls, is high in starch. Figure 1.19, on the other hand, features glycogen granules inside the liver. After we eat a candy bar, glucose enters the bloodstream, and the liver stores glucose as glycogen. In between eating, the liver releases glucose so that the blood glucose concentration is always about 0.1%.

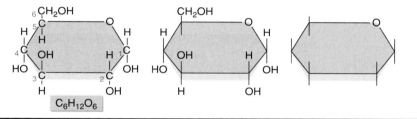

Figure 1.16 Three ways to represent the structure of glucose.
The *far left* structure shows the carbon atoms; $C_6H_{12}O_6$ is the molecular formula for glucose. The *far right* structure is the simplest way to represent glucose.

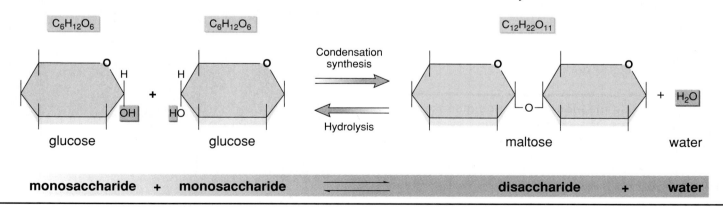

Figure 1.17 Condensation synthesis and hydrolysis of maltose, a disaccharide.
During condensation synthesis of maltose, a bond forms between the two glucose molecules and the components of water are removed. During hydrolysis, the components of water are added, and the bond is broken.

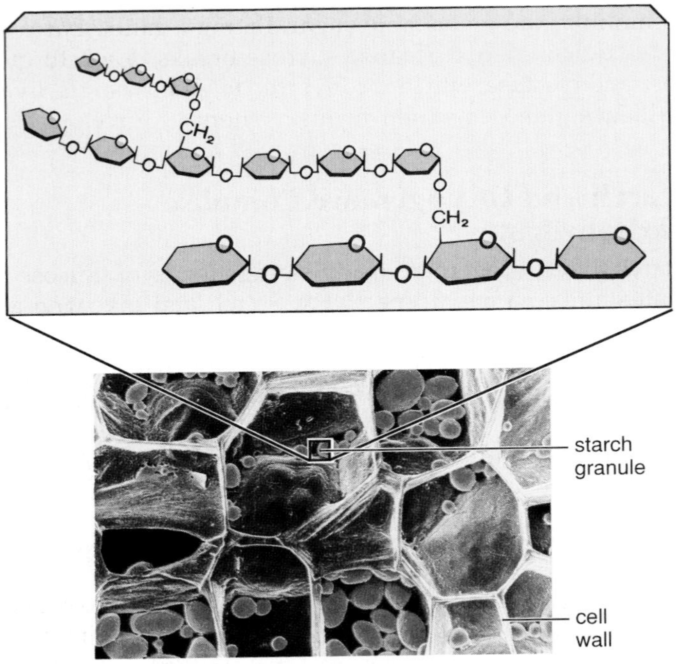

Figure 1.18   **Starch structure and function.**
Starch is a chain of glucose molecules that branches as indicated. The electron micrograph shows starch granules in plant cells. Starch is the storage form of glucose in plants.

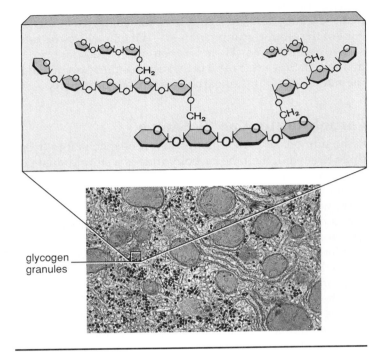

Figure 1.19   **Glycogen structure and function.**
Glycogen is a highly branched polymer of glucose molecules. The electron micrograph shows glycogen granules in liver cells. Glycogen is the storage form of glucose in animals.

Cellulose Provides Structural Support

The polysaccharide **cellulose** is found in plant cell walls and accounts in part for the strong nature of these walls. In cellulose (Fig. 1.20), the glucose units are joined by a slightly different type of linkage than that in starch or glycogen. (Observe the alternating position of the oxygen atoms in the linked glucose units.) While this might seem to be a technicality, actually it is important because we are unable to digest foods containing this type of linkage; therefore, cellulose largely passes through our digestive tract as fiber, or roughage. Recently, it has been suggested that fiber in the diet is necessary to good health and may even help to prevent colon cancer.

> Cells usually use the monosaccharide glucose as an energy source. The polysaccharides starch and glycogen are storage compounds in plant and animal cells, respectively, and the polysaccharide cellulose is found in plant cell walls.

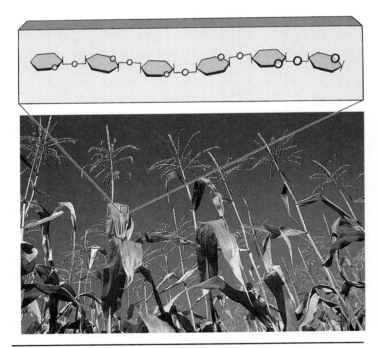

Figure 1.20   **Cellulose structure and function.**
Cellulose contains a slightly different type of linkage between glucose molecules than that in starch or glycogen. Plant cell walls contain cellulose, and the rigidity of the cell walls permits nonwoody plants to stand upright as long as they receive an adequate supply of water.

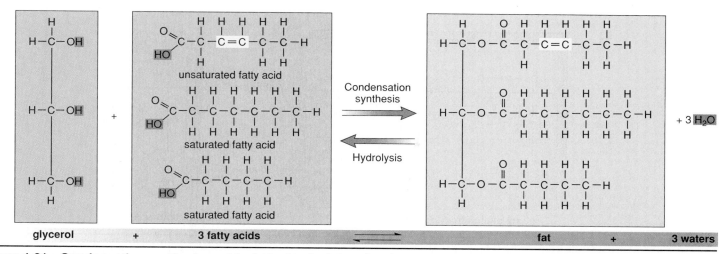

Figure 1.21 Condensation synthesis and hydrolysis of a fat molecule.
Fatty acids can be saturated (no double bonds between carbon atoms) or unsaturated (have double bonds between carbon atoms). When a fat molecule forms, three fatty acids combine with glycerol, and three water molecules are produced.

1.6 Lipids

Lipids are diverse in structure and function, but they have a common characteristic: they do not dissolve in water.

Fats and Oils Are Similar

The familiar lipids are the fats (e.g., lard and butter) and oils (e.g., corn oil and soybean oil). Fat has several functions in the body: it is used for long-term energy storage, it insulates against heat loss, and it forms a protective cushion around major organs.

A fat or an oil is formed when one glycerol molecule reacts with three fatty acid molecules. Again, this is a condensation reaction, and three water molecules result. A fat is sometimes called a **triglyceride** because of its three-part structure, and the term neutral fat is sometimes used because the molecule is nonpolar (Fig. 1.21).

Fatty Acids Are Saturated or Unsaturated

A fatty acid is a hydrocarbon chain that ends with the acidic group —COOH (Fig. 1.21). Most of the fatty acids in cells contain 16 or 18 carbon atoms per molecule, although smaller ones with fewer carbons are also known.

Fatty acids are either saturated or unsaturated. Saturated fatty acids have no double bonds between carbon atoms. The carbon chain is saturated, so to speak, with all the hydrogens it can hold. Saturated fatty acids account for the solid nature at room temperature of butter and lard, which are derived from animal sources. Unsaturated fatty acids have double bonds between carbon atoms wherever the number of hydrogens is less than two per carbon atom. Unsaturated fatty acids account for the liquid nature of vegetable oils at room temperature. Vegetable oils are hydrogenated to make margarine and products such as Crisco.

Soaps Are Emulsifiers

Strictly speaking, soaps are not lipids, but they are considered here as a matter of convenience. A soap is a salt formed from a fatty acid and an inorganic base. For example,

Unlike fats, soaps have a polar end (the charged end) in addition to the nonpolar end (the hydrocarbon chain represented by R). Therefore, a soap does mix with water. When soaps are added to oils, the oils, too, mix with water because a soap positions itself about an oil droplet so that its nonpolar ends project into the fat droplet, while its polar ends project outward.

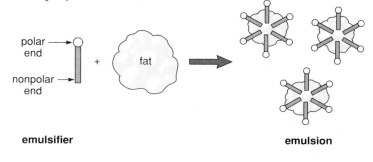

Now the droplet disperses in water, and it is said that emulsification has occurred. Emulsification occurs when dirty clothes are washed with soaps or detergents. Also, prior to the digestion of fatty foods, fats are emulsified by bile. A person who has had the gallbladder removed may have trouble digesting fatty foods because this organ stores bile for emulsifying fats prior to the digestive process.

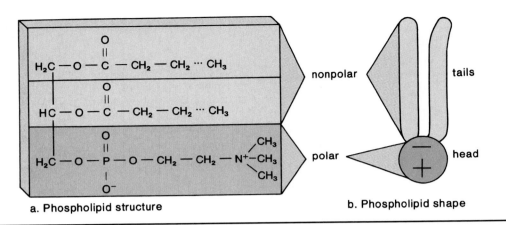

a. Phospholipid structure

b. Phospholipid shape

Figure 1.22   Phospholipid structure and shape.
a. Phospholipids are constructed like fats, except that they contain a phosphate group. This phospholipid includes a group that contains both phosphate and nitrogen. **b.** The polar portion of the phospholipid molecule (head) is soluble in water, whereas the two hydrocarbon chains (tails) are not. This causes the molecule to arrange itself as shown when exposed to water.

Phospholipids Have Polar Heads and Nonpolar Tails

Phospholipids, as their name implies, contain a phosphate group:

Essentially, phospholipids are constructed like fats, except that in place of the third fatty acid, there is a phosphate group or a grouping that contains both phosphate and nitrogen (Fig. 1.22). These molecules are not electrically neutral as are fats because the phosphate group can ionize. It forms the so-called "head" of the molecule while the rest of the molecule becomes the nonpolar "tails." The plasma membrane is a phospholipid bilayer in which the "heads" face outward into a watery medium and the tails face each other because they are water repelling.

Steroids Have Four Rings

Steroids are lipids having a structure that differs entirely from that of fats. Steroid molecules have a backbone of four fused carbon rings, but each one differs primarily by the arrangement of the atoms in the rings and the type of functional groups attached to them. Cholesterol (Fig. 1.23) is the precursor of several other steroids, such as the sex hormones estrogen and testosterone, which help to maintain sex characteristics in males and females.

Evidence has been accumulating for years that a diet high in saturated fats and cholesterol can lead to circulatory disorders. This type of diet leads to deposits of fatty material inside the lining of blood vessels and

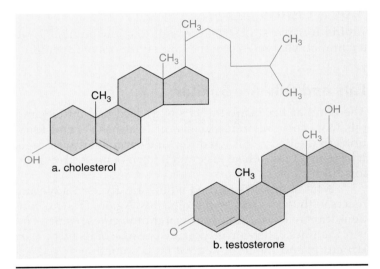

a. cholesterol

b. testosterone

Figure 1.23   Steroid diversity.
a. Cholesterol, like all steroid molecules, has four adjacent rings, but the effects of steroids on the body largely depend on the attached groups indicated in red. **b.** Testosterone is the male sex hormone.

reduced blood flow. As discussed in the Health reading on page 33, nutrition labels are now required to list the calories from fat per serving and the percent daily value from saturated fat and cholesterol.

Lipids include fats and oils for long-term energy storage and steroids (e.g., hormones), which are insoluble in water. Phospholipids (in plasma membrane) and soaps (act as emulsifiers) are related compounds that are soluble in water because they have a polar group.

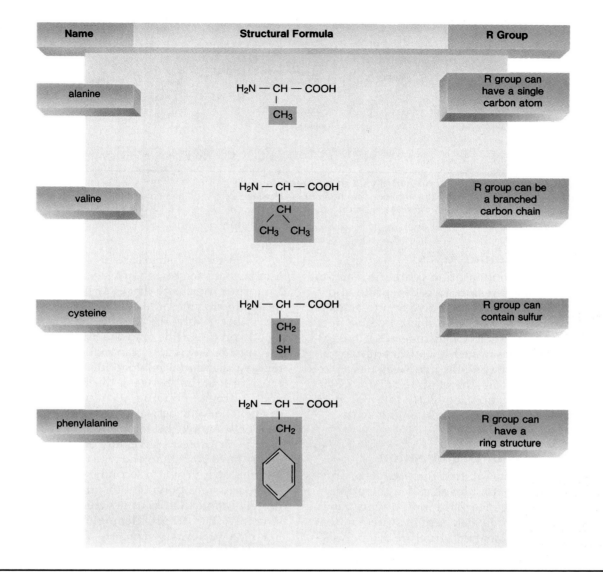

Name	Structural Formula	R Group
alanine	$H_2N — CH — COOH$ CH_3	R group can have a single carbon atom
valine	$H_2N — CH — COOH$ CH $CH_3 \quad CH_3$	R group can be a branched carbon chain
cysteine	$H_2N — CH — COOH$ CH_2 SH	R group can contain sulfur
phenylalanine	$H_2N — CH — COOH$ CH_2	R group can have a ring structure

Figure 1.24   Representative amino acids.
Amino acids differ from one another by their *R* group; the simplest *R* group is a single hydrogen atom (H). The *R* groups that contain carbon vary as shown.

1.7   Proteins

Protein macromolecules sometimes have a structural function. For example, in humans, the protein keratin makes up hair and nails, whereas collagen is found in all types of connective tissue, including ligaments, cartilage, bones, and tendons. The muscles contain proteins, which account for their ability to contract.

Some **proteins** are enzymes, necessary contributors to the chemical workings of the cell and therefore of the body. **Enzymes** speed chemical reactions; they work so quickly that a reaction that normally takes several hours or days without an enzyme takes only a fraction of a second with an enzyme.

Amino acids are the monomers in a protein. An amino acid has a central carbon atom bonded to a hydrogen atom and three groups. The name of the molecule is appropriate because one of these groups is an amino group ($—NH_2$) and another is an acidic group ($—COOH$). The other group is called an *R* group because it is the *Remainder* of the molecule. Amino acids differ from one another by their *R* group; the *R* group varies from a single hydrogen (H) to a complicated ring (Fig. 1.24).

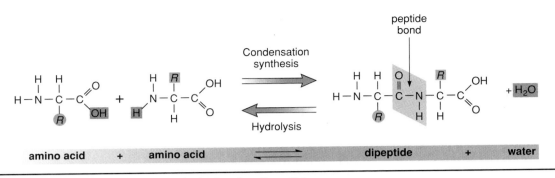

Figure 1.25   Condensation synthesis and hydrolysis of a dipeptide.
The two amino acids on the left-hand side of the equation are different, as signified by the difference in the *R*-group notations. As these amino acids join, a peptide bond forms, and a water molecule is produced. During hydrolysis, water is added, and the peptide bond is broken.

Peptide Bonds Join Amino Acids

Figure 1.25 shows that a condensation synthesis reaction between two amino acids results in a dipeptide and a molecule of water. A bond that joins two amino acids is called a **peptide bond.** The atoms associated with a peptide bond—oxygen (O), carbon (C), nitrogen (N), and hydrogen (H)—share electrons in such a way that the oxygen has a partial negative charge and the hydrogen has a partial positive charge. Therefore, the peptide bond is polar, and hydrogen bonding occurs frequently in a polypeptide. A **polypeptide** is a single chain of amino acids.

Proteins Have Levels of Organization

The structure of a protein has at least three levels of organization (Fig. 1.26*a–c*). The first level, called the primary structure, is the linear sequence of the amino acids joined by peptide bonds. Polypeptides can be quite different from one another. You will recall that the structure of a polysaccharide can be likened to a necklace that contains a single type "bead," namely, glucose. Polypeptides can make use of 20 different possible types of amino acids or "beads." Each particular polypeptide has its own sequence of amino acids. It can be said that each polypeptide differs by the sequence of its *R* groups and the number of amino acids in the sequence.

The secondary structure of a protein comes about when the polypeptide takes on a particular orientation in space. One common arrangement of the chain is the alpha (α) helix, or a right-handed coil, with 3.6 amino acids per turn. Hydrogen bonding between amino acids, in particular, stabilizes the helix.

The tertiary structure of a protein is its final three-dimensional shape. In muscles, the helical chains of myosin form a rod shape that ends in globular (globe-shaped) heads. In enzymes, the helix bends and twists in different ways. Invariably, the hydrophobic portions are packed mostly on the inside, and the hydrophilic portions are on the outside where they can make contact with water. The tertiary shape of a polypeptide is maintained by various types of bonding between the *R* groups; covalent, ionic, and hydrogen bonding all occur. One common form of covalent bonding between *R* groups is disulfide (S—S) linkages between two cysteine amino acids.

Some proteins have only one polypeptide, and some others have more than one type of polypeptide chain, each with its own primary, secondary, and tertiary structures. These separate polypeptides are arranged to give some proteins a fourth level of structure, termed the quaternary structure (Fig. 1.26*d*). Hemoglobin is a complex protein having a quaternary structure; most enzymes also have a quaternary structure.

The final shape of a protein is very important to its function. When proteins are exposed to extremes in heat and pH, they undergo an irreversible change in shape called *denaturation.* For example, we are all aware that the addition of acid to milk causes curdling and that heating causes egg white, a protein called albumin, to coagulate. Denaturation occurs because the normal bonding between the *R* groups has been disturbed. Once a protein loses its normal shape, it is no longer able to perform its usual function.

Proteins, which contain covalently linked amino acids, are important in the structure and the function of cells. Some proteins are enzymes, which speed chemical reactions.

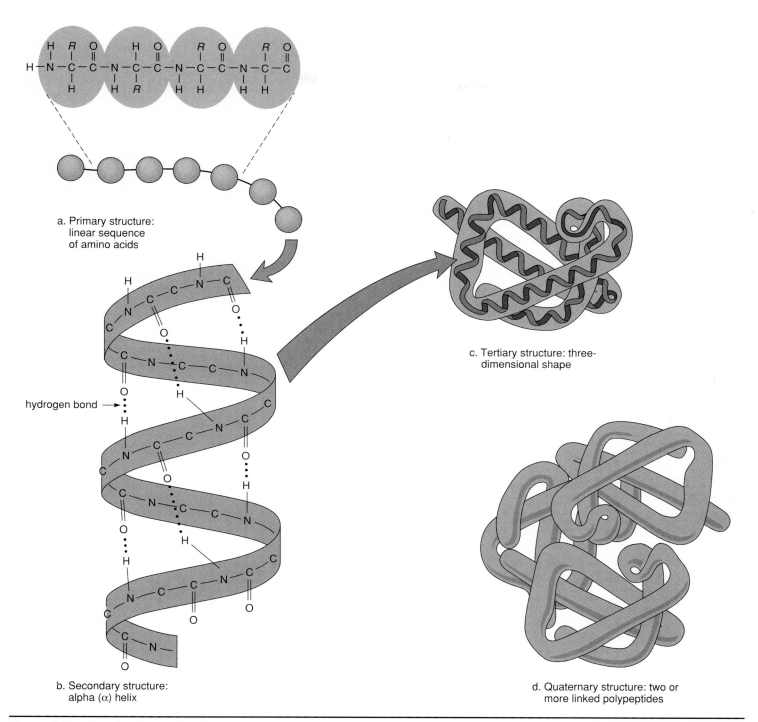

a. Primary structure:
   linear sequence
   of amino acids

hydrogen bond →

b. Secondary structure:
   alpha (α) helix

c. Tertiary structure: three-
   dimensional shape

d. Quaternary structure: two or
   more linked polypeptides

Figure 1.26   Levels of organization in protein structure.
a. Primary structure of a polypeptide is the order of the amino acids. **b.** Secondary structure is often an alpha (α) helix, in which hydrogen bonding occurs along the length of a polypeptide, as indicated by the dotted lines. **c.** In globular (globe-shaped) proteins, the tertiary structure is the twisting and turning of the polypeptide helix, which takes place because of covalent and other types of bonding between the *R* groups of the amino acids. **d.** The quaternary structure occurs when a protein contains two or more polypeptides.

Types of Proteins

In general there are two types of proteins: globular and fibrous. Our discussion thus far has centered on the structure of globular proteins (see Fig. 1.26) which, true to their name, have a spherical shape. Enzymes are globular proteins, which are essential to cellular metabolism, and we will explore this topic in the next chapter. No reaction occurs in a cell unless its particular enzyme is present. There are nonenzyme globular proteins also. The globular proteins in blood have numerous functions like transporting molecules, clotting blood, and fighting infections.

In terms of weight, the human body contains more muscle protein than any other kind (Fig. 1.27). There are two muscle proteins called myosin and actin. Each actin molecule is globular, but the molecules assemble to produce filaments. Myosin, on the other hand, is a fibrous rod attached to globular heads.

Fibrous proteins characteristically have a rod-like shape and a structural function. Usually, in a fibrous protein, three long helical polypeptides are wound around each other, much as hair can be braided. Covalent bonding like disulfide linkages occur between the polypeptides of the braid. Bundles of the braids are aligned to form rod-like units that are then arranged to produce the final structure like a hair shaft, a tendon, or even the outer ear.

Keratin

Keratin is a fibrous protein found in skin, nails, and hair. The keratinized cells of the skin form a waterproof coating that protects us from the drying effect of the air. The secondary and tertiary structure of keratin is well studied by discussing certain properties of hair. As expected, hydrogen bonding holds the helical shape of each polypeptide strand. But hair goes limp when wet because the amino acids form hydrogen bonds with water instead of each other. When wet hair is wound on curlers, the hydrogen bonds between amino acids will reform in a way that produces the curls. The curls, of course, disappear should they be exposed to moisture.

Perming the hair is more "permanent" because it goes one step further and affects the disulfide linkages between the polypeptides. In straight hair, the disulfide linkages are straight across between adjacent amino acids of the polypeptides. In curly hair, the disulfide linkages are at an angle between nonadjacent amino acids. The steps of a permanent are these: (1) the hair is wound on curlers; (2) the first solution breaks the present disulfide linkages; and (3) the second solution causes disulfide linkages to form between nonadjacent amino acids so that curls result.

Collagen and Elastin

Collagen and elastin are the proteins found in so-called connective tissues including tendons, ligaments, bone, and cartilage. As expected, elastin is more elastic than collagen. So you could surmise, for example, that bone probably contains collagen fibers, and cartilage like that of the nose and ears contain plenty of elastin fibers. Collagen fibers make

Figure 1.27   Muscles contain protein.
The total amount of muscle proteins in humans exceeds that of any other protein. About 40% of the body weight of a healthy adult human is muscle, and of that 5 to 6 kilograms are muscle proteins.

up the tendons and ligaments that hold muscle to bone and bone to bone, respectively. Aside from these structures, collagen and elastin are associated with skin and lend support to most internal organs. It is unfortunate that, as people age, collagen fibers (not the polypeptides) become increasingly crosslinked by, of all things, glucose molecules. Undoubtedly, this cross-linking contributes to the stiffening and loss of elasticity characteristic of aging tendons and ligaments. It also may account for the inability of organs, such as the blood vessels, the heart, and the lungs to function as they once did.

Marfan's syndrome is a relatively common human genetic disease that affects connective tissues rich in elastic fibers. This syndrome is named for a pediatrician Bernard-Jean Antonin Marfan, who in 1896 described a 5-year-old girl with unusually long fingers and toes, limited joint motion, and curvature of the spine. More life threatening than the skeletal deformities are the circulatory difficulties. A valve in the heart is leaky and allows a backward flow of blood, and the aorta (the major artery attached to the heart) is either enlarged at birth or becomes enlarged during childhood. Bursting of the aorta can cause instant death. Abe Lincoln, sixteenth president of the United States, was assassinated in 1865, years before Marfan syndrome was recognized. Still, there are those who have deduced that he did indeed have this condition. If Lincoln was in such bad health, then it is reasoned that had he not been shot, he would still have died soon thereafter. Today it is possible to diagnose Marfan syndrome and replace weakened sections of the aorta with synthetic grafts.

Health Focus

Nutrition Labels

As of May 1994, packaged foods have a nutrition label like the one depicted in Figure 1A. The nutrition information given on this label is based on the serving size (that is, 1¼ cup, 57 grams) of a cereal. A Calorie* is a measurement of energy. One serving of the cereal provides 220 Calories, of which 20 are from fat. At the bottom of the label, the recommended amounts of nutrients are based on a typical diet of 2,000 Calories for women and 2,500 Calories for men.

Fats are the nutrient with the highest energy content: 9 Cal/g compared to 4 Cal/g for carbohydrates and proteins. The body stores fat under the skin and around the organs for later use. A 2,000-Calorie diet should contain no more than 65 g (585 Calories) of fat. Dietary fat has been implicated in cancer of the colon, pancreas, ovary, prostate, and breast. Although saturated fat and cholesterol are essential nutrients, dietary consumption of saturated fats and cholesterol in particular should be controlled. Cholesterol and saturated fat contribute to the formation of deposits called plaques, which clog arteries and lead to cardiovascular disease, including high blood pressure.

For these reasons, it is important to know how a serving of the cereal will contribute to the maximum daily recommended amount of fat, saturated fat, and cholesterol. You can find this out by looking at the listing under % Daily Value: the total fat in one serving of the cereal provides 3% of the daily recommended amount of fat. How much will a serving of the cereal contribute to the maximum recommended daily amount of saturated fat? cholesterol?

Carbohydrates (sugars and polysaccharides) are the quickest,

most readily available source of energy for the body. Because carbohydrates aren't usually associated with health problems, they should comprise the largest proportion of the diet. Breads and

cereals containing complex carbohydrates are preferable to candy and ice cream containing simple carbohydrates because they are likely to contain dietary fiber (nondigestible plant material). Insoluble fiber has a laxative effect and seems to reduce the risk of colon cancer; soluble fiber combines with the cholesterol in food and prevents the cholesterol from entering the body proper.

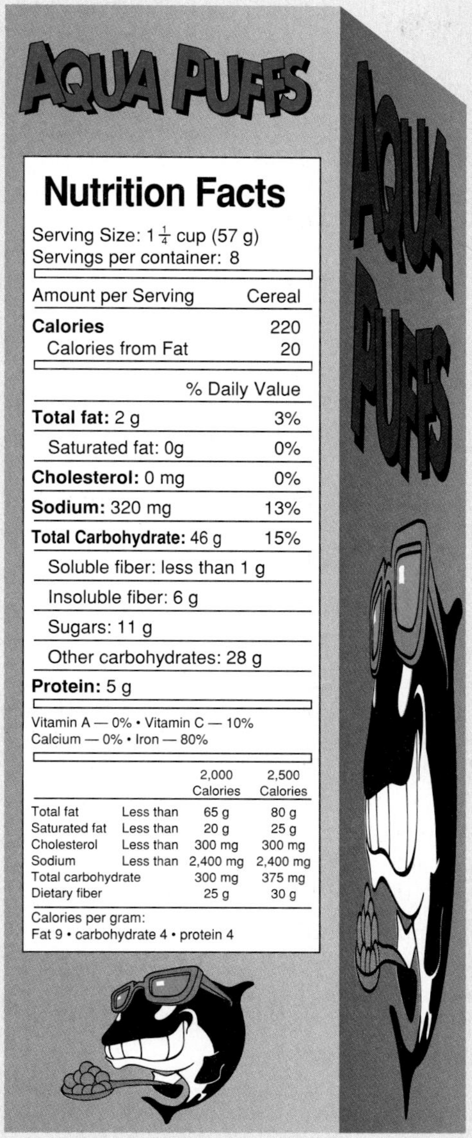

The body does not store amino acids for the production of proteins, which are found particularly in muscles but also in all cells of the body. Although it is not stated, a woman should have about 44 g of protein per day, and a man should have about 56 g of protein a day. Red meat is rich in protein, but it is usually also high in saturated fat. Therefore, it is considered good health sense to rely on protein from plant origins (e.g., whole-grain cereals, dark breads, legumes) more than is customary in the United States.

The amount of dietary sodium (as in table salt) is of concern because excessive sodium intake has been linked to high blood pressure in some people. It is recommended that the intake of sodium be no more than 2,400 mg per day. A serving of this cereal provides what percent of this maximum amount?

Vitamins are essential requirements needed in small amounts in the diet. Each vitamin has a recommended daily intake, and the food label tells what percent of the recommended amount is provided by a serving of this cereal.

Nutrition Facts

Serving Size: 1¼ cup (57 g)
Servings per container: 8

Amount per Serving	Cereal
Calories	220
Calories from Fat	20

	% Daily Value
Total fat: 2 g	3%
Saturated fat: 0g	0%
Cholesterol: 0 mg	0%
Sodium: 320 mg	13%
Total Carbohydrate: 46 g	15%
Soluble fiber: less than 1 g	
Insoluble fiber: 6 g	
Sugars: 11 g	
Other carbohydrates: 28 g	
Protein: 5 g	

Vitamin A — 0% • Vitamin C — 10%
Calcium — 0% • Iron — 80%

		2,000 Calories	2,500 Calories
Total fat	Less than	65 g	80 g
Saturated fat	Less than	20 g	25 g
Cholesterol	Less than	300 mg	300 mg
Sodium	Less than	2,400 mg	2,400 mg
Total carbohydrate		300 mg	375 mg
Dietary fiber		25 g	30 g

Calories per gram:
Fat 9 • carbohydrate 4 • protein 4

Figure 1A   Nutrition label on side panel of cereal box.

---

*A calorie is the amount of heat required to raise the temperature of one gram of water one degree centigrade. A Calorie (capital C) is 1,000 calories.

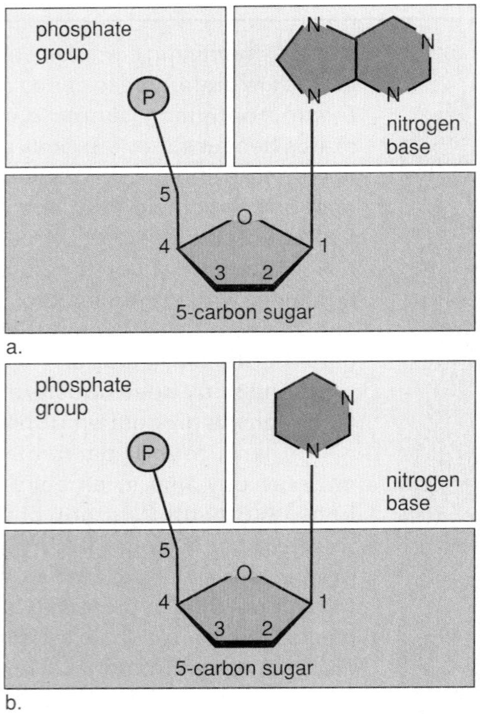

Figure 1.28 Generalized nucleotides.
All nucleotides contain a phosphate group (P), a pentose (five-carbon) sugar, and a nitrogen base. The base can have **(a)** two rings (purine base) or **(b)** one ring (pyrimidine base).

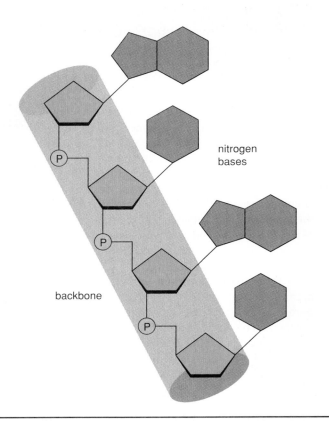

Figure 1.29 Generalized nucleic acid strand.
Each nucleic acid strand is a chain of nucleotide molecules. Each strand has a backbone made of sugar molecules and phosphate groups. The nitrogen bases project from the side of the backbone.

1.8   Nucleic Acids

Nucleic acids are important for the growth and reproduction of cells and organisms. Human genes are composed of a nucleic acid called **DNA (deoxyribonucleic acid).** The nucleic acid **RNA (ribonucleic acid)** works in conjunction with DNA to specify the order of amino acids when proteins are syntesized.

Nucleotides Join, Forming Nucleic Acids

DNA and RNA contain nucleotides. A **nucleotide** has three parts: a **pentose** (five-carbon) sugar, a phosphate group, and a nitrogen base. (The phosphate group symbolized by (P) is an acid, and this accounts for the name: nucleic acid.) For DNA, the sugar is deoxyribose, and for RNA, the sugar is ribose. The base is a single- or double-ring structure containing nitrogen; therefore, it is called a nitrogen base. (The designation base refers to the ability of nitrogen to take up hydrogen ions.) There are four different types of nucleotides in DNA because there are four different types of nitrogen bases—two with a single ring and two with a double ring. Figure 1.28 shows only generalized nucleotides.

Both DNA and RNA are polymers of nucleotides. Just like the other synthetic reactions we have studied in this section, nucleic acids are formed by condensation.

$$\text{nucleotide molecules} \rightleftharpoons \text{nucleic acid} + H_2O$$
$$\text{(monomers)} \qquad\qquad \text{(polymer)}$$

The nucleic acid polymer, which is often called a strand, has a backbone with this sequence: phosphate–sugar–phosphate–sugar–phosphate. The nitrogen bases project to one side of the backbone (Fig. 1.29). RNA is single stranded, and DNA is double stranded. The strands in DNA are held to each other by hydrogen bonding between the bases. This is only a brief description of the structure and the function of DNA and RNA, and their structures will be examined in more depth when genetics is considered.

Nucleic acids function in the growth and reproduction of cells and organisms. DNA is composed of nucleotides having the sugar deoxyribose, while RNA nucleotides contain the sugar ribose. Genes are composed of DNA, and with the help of RNA, DNA specifies protein synthesis.

ATP Is a Carrier of Energy

ATP (adenosine triphosphate) is a nucleotide that functions as an energy carrier in cells. In ATP, the base adenine is joined to the sugar ribose (together called adenosine), and there are three phosphate groups (triphosphate) instead of one. It is customary to draw the molecule as shown in Figure 1.30, with the three phosphate groups shown on the right. The wavy lines between the phosphate groups indicate that ATP is a high-energy molecule. When a phosphate group is removed, a large amount of energy is released.

ATP is the energy currency of cells because when it breaks down, a large amount of energy is released.

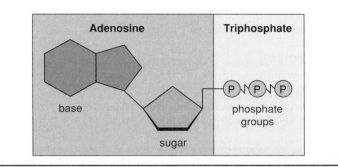

Figure 1.30   Structure of ATP.
ATP (adenosine triphosphate) is a nucleotide with three phosphate groups ⓟ. ATP breakdown involves the removal of a phosphate group and the release of much energy.

SUMMARY

1.1   Elements and Atoms

All matter is composed of some 92 elements. Each element is made up of just one type atom. An atom has a weight, which is dependent on the number of protons and neutrons in the nucleus, and its chemical properties are dependent on the number of electrons in the outermost shell.

1.2   Molecules and Compounds

Atoms react with one another by forming ionic bonds or covalent bonds. Ionic bonds are an attraction between charged ions. Atoms share electrons in covalent bonds, which can be single, double, or triple bonds.

Oxidation is the loss of electrons (hydrogen atoms), and reduction is the gain of electrons (hydrogen atoms).

1.3   Some Important Inorganic Molecules

Water, acids, and bases are important inorganic molecules. The polarity of water accounts for it being the universal solvent; hydrogen bonding accounts for it boiling at 100°C and freezing at 0°C. Because it is slow to heat up and slow to freeze, it is liquid at the temperature of living things.

Pure water has a neutral pH; acids increase the hydrogen ion concentration [H^+] but decrease the pH, and bases decrease the hydrogen ion concentration but increase the pH of water.

1.4   Molecules of Life

The chemistry of carbon accounts for the chemistry of organic compounds. Carbohydrates, lipids, proteins, and nucleic acids are macromolecules with specific functions in cells (Table 1.2). Macromolecules are polymers, each containing specific monomers.

TABLE 1.2

Organic Compounds Associated with Living Things

Macromolecule	Monomer	Function
Proteins	Amino acids	Enzymes speed up chemical reactions; structural components (e.g., muscle and membrane proteins)
Carbohydrates		
Starch	Glucose	Energy storage in plants
Glycogen	Glucose	Energy storage in animals
Cellulose	Glucose	Plant cell walls
Lipids		
Fats and Oils	Glycerol, 3 fatty acids	Long-term energy storage
Phospholipids	Glycerol, 2 fatty acids, phosphate group	Plasma membrane structure
Nucleic Acids		
DNA	Nucleotides with deoxyribose sugar	Genetic material
RNA	Nucleotides with ribose sugar	Protein synthesis

1.5   Carbohydrates

Glucose is the six-carbon sugar most utilized by cells for "quick" energy. Like the rest of the macromolecules to be studied, condensation synthesis joins two or more sugars, and a hydrolysis reaction splits the bond. Plants store glucose as starch, and animals store glucose as glycogen. Humans cannot digest cellulose, which forms plant cell walls.

1.6   Lipids

Lipids are varied in structure and function. Fats and oils, which function in long-term energy storage, contain glycerol and three fatty acids. Fatty acids can be saturated or unsaturated. Plasma membranes contain phospholipids that have a polarized end. Certain hormones are derived from cholesterol, a complex ring compound.

1.7 Proteins

The primary structure of a polypeptide is its own particular sequence of the possible twenty types of amino acids. The secondary structure is often an alpha (α) helix. The tertiary structure occurs when a polypeptide bends and twists into a three-dimensional shape. A protein can contain several polypeptides, and this accounts for a possible quaternary structure.

1.8 Nucleic Acids

Nucleic acids (DNA and RNA) are polymers of nucleotides, which have three parts: pentose sugar, nitrogen base, and phosphate. ATP, the high-energy molecule of cells, is a nucleotide with three phosphates.

STUDYING THE CONCEPTS

1. Name the subatomic particles of an atom; describe their charge, weight, and location in the atom.   14–15

2. Give an example of an ionic reaction, and explain it.   16

3. Diagram the atomic structure of calcium, and explain how it can react with two chlorine atoms.   17

4. Give an example of a covalent reaction, and explain it.   18

5. Relate the characteristics of water to its polarity and hydrogen bonding between water molecules.   19–20

6. On the pH scale, which numbers indicate a basic solution? An acidic solution? Why?   22

7. What are buffers, and why are they important to life?   22

8. Relate the variety of organic compounds to the bonding capabilities of carbon.   24

9. Name the four classes of organic molecules in cells, and relate them to macromolecules and also polymers.   24

10. Name some monosaccharides, disaccharides, and polysaccharides, and state some general functions for each. What is the most common monomer for polysaccharides?   25–26

11. How is a neutral fat synthesized? What is a saturated fatty acid? An unsaturated fatty acid? What is the function of fats?   27

12. Relate the structure of a phospholipid to that of a neutral fat. What is the function of a phospholipid?   28

13. What is the general structure and significance of cholesterol?   28

14. What are some functions of proteins? What is a peptide bond, a dipeptide, and a polypeptide?   29–32

15. Discuss the primary, secondary, and tertiary structures of globular proteins.   30

16. Discuss the structure and function of the nucleic acids, DNA and RNA.   34

APPLYING YOUR KNOWLEDGE

Concepts

1. Explain why you would expect the covalent bond to be stronger than an ionic bond or a hydrogen bond.

2. Many soft drinks sold today are carbonated, that is, the gas CO_2 has been added to them. Considering the equation $CO_2 + H_2O \rightarrow H_2CO_3$ (carbonic acid), explain why the pH of carbonated drinks moves toward neutral once the bottle is opened.

3. You often go many hours (from an evening meal at 6:00 p.m. to a morning meal at 8:00 a.m.) without eating, yet your blood sugar level remains relatively the same. How is this managed by your body?

4. Heating an enzyme to a certain temperature usually destroys its enzymatic activity even though the same sequence of amino acids remains. What levels of structure have been affected by heating?

Bioethical Issue

Like a well-tuned system, your body constantly balances acidic and basic chemicals to keep your pH level at a steady state. Normally, this system works fine.

But when a person gets sick or endures physical stress—as in, for example, childbirth—pH levels may dip or rise too far, endangering that person's life. So in most American hospitals, doctors routinely administer IVs, or intravenous infusions, of certain fluids to maintain a patient's pH level.

However, some people oppose IVs for philosophical reasons. For example, women preferring a natural childbirth may refuse an IV. That's relatively safe, as long as the woman is healthy.

Problems arise when hospital policy dictates an IV, even though a patient does not want one. Should a patient be allowed to refuse an IV? Or does a hospital have the right to insist, for health reasons, that patients accept IV fluids? And what role should doctors play—patient advocates or hospital representatives?

TESTING YOUR KNOWLEDGE

1. The atomic number is equal to the number of _____ in an atom.

2. An ion is negatively charged when it has more _____ than _____.

3. In a covalent bond, the atoms _____ electrons.

4. Water boils at a much higher temperature and freezes at a much lower temperature than other liquids because of _____ between the molecules.

5. _____ take up either hydrogen ions (H^+) or hydroxide ions (OH^-) and therefore act to stabilize the pH.

6. When an acid is added to water, the number of hydrogen ions _____ and the pH _____.

7. A hydrocarbon chain is nonpolar, but a hydrocarbon chain with an acidic group that _____ is polar.

8. The simple sugar _____ is the primary energy source of the body.

9. The polymer _____ is found in plant cell walls.

10. Fatty acids having no double bonds between carbon atoms are said to be _____.

11. A fat results from the union of a _____ molecule and three fatty acid molecules.

12. _____ are organic catalysts that speed chemical reactions.

13. A _____ bond joins two amino acids in a dipeptide.

14. The sequence, or order, of amino acids in a polypeptide is termed its _____ structure.

15. Nucleic acids are composed of _____.

16. DNA is _____ stranded, while RNA is single stranded.

17. Label this diagram of synthesis and hydrolysis.

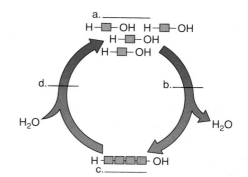

APPLYING TECHNOLOGY

Your study of chemistry is supported by these available technologies:

Exploring the Internet

The Mader Home Page provides further resources for studying this chapter.

`http://www.mhhe.com/sciencemath/biology/mader/`

(Click on *Human Biology*.)

Life Science Animations Video

Video #1   Chemistry, The Cell, and Energetics

Formation of an Ionic Bond (#1)   The atomic structure of sodium and chlorine is reviewed before these atoms are seen reacting to form sodium chloride. (1)*

ATP as an Energy Carrier (#11)   ATP structure and the ATP cycle are reviewed pictorially.   (1)

*Level of difficulty

SELECTED KEY TERMS

acid Solution in which pH is less than 7; a substance that contributes or liberates hydrogen ions in a solution.   21

amino acid Monomer of a protein; takes its name from the fact that it contains an amino group ($-NH_2$) and an acid group ($-COOH$).   29

atom Smallest unit of matter that cannot be divided by chemical means.   14

ATP (adenosine triphosphate) (ah-DEN-ah-zeen try-FOS-fayt) Compound having a nitrogen base, ribose, and three phosphate groups; a carrier of energy in cells.   35

base Solution in which pH is greater than 7; a substance that tends to lower the hydrogen ion concentration of a solution; alkaline; opposite of acid. Also, a term commonly applied to one of the components of a nucleotide.   21

buffer Substance or compound that prevents large changes in the pH of a solution.   22

carbohydrate Organic compound characterized by the presence of CH_2O groups; includes monosaccharides, disaccharides, and polysaccharides.   25

cellulose Polysaccharide composed of glucose molecules; the chief constituent of a plant's cell wall.   26

covalent bond (coh-VAY-lent) Chemical bond between atoms that results from the sharing of a pair of electrons.   18

disaccharide Sugar that contains two units of a monosaccharide; e.g., maltose.   25

DNA (deoxyribonucleic acid) Nucleic acid found in cells; the genetic material that specifies protein synthesis in cells.   34

electron Subatomic particle that has almost no weight and carries a negative charge; orbits in a shell about the nucleus of an atom.   15

enzyme Organic catalyst, usually a protein, that speeds up a reaction in cells due to its particular shape.   29

glucose Six-carbon sugar that organisms degrade as a source of energy during cellular respiration.   25

glycogen Storage polysaccharide found in animals that is composed of glucose molecules joined in a linear fashion but having numerous branches.   25

hexose Six-carbon sugar.   25

hydrogen bond Weak attraction between a hydrogen atom with a partial positive charge and an atom of another molecule with a partial negative charge.   19

hydrolysis (hy-DRAH-lih-sis) Splitting of a covalent bond by the addition of water.   25

ion Atom or group of atoms carrying a positive or negative charge.   16

ionic bond Bond created by an attraction between oppositely charged ions.   17

isotope One of two or more atoms with the same atomic number that differ in the number of neutrons and therefore in weight.   15

lipid (LIP-id) Organic compound that is insoluble in water; notably fats, oils, and steroids.   27

molecule Like or different atoms joined by a bond; the unit of a compound.   16

monosaccharide Simple sugar; a carbohydrate that cannot be decomposed by hydrolysis.   25

neutron Subatomic particle that has a weight of one atomic mass unit, carries no charge, and is found in the nucleus of an atom.   15

nucleotide Monomer of a nucleic acid that forms when a nitrogen base, a pentose sugar, and a phosphate join.   34

pentose Five-carbon sugar; deoxyribose is the pentose sugar found in DNA; ribose is the pentose sugar found in RNA.   34

peptide bond Covalent bond that joins two amino acids.   30

pH scale Measure of the hydrogen ion concentration $[H^+]$; any pH below 7 is acidic and any pH above 7 is basic.   22

phospholipid Molecule making up the bilayer of cellular membranes; has a hydrophilic head and hydrophobic tails.   28

polypeptide Polymer of many amino acids linked by peptide bonds.   30

protein Organic compound that is composed of either one or several polypeptides.   29

RNA (ribonucleic acid) Nucleic acid found in cells that assists DNA in controlling protein synthesis.   34

starch Storage polysaccharide found in plants that is composed of glucose molecules joined in a linear fashion.   25

synthesis To build up, such as the combining of two smaller molecules to form a larger molecule.   25

triglyceride Neutral fat composed of glycerol and three fatty acids.   27

Chapter 2

Cell Structure and Function

Chapter Outline

Figure 2.1   Racing cyclists.
Cycling or any human activity is dependent on the functioning of skeletal muscle cells (colored red in the insert). Oxygen reaches muscle cells by way of the capillaries (colored blue in this micrograph).

Sacrificing his usual Saturday morning snooze, Michael decided to join friends in a nearby 30-mile cycling event. He wasn't an experienced cyclist, but things were going well—until about halfway through the ride. Then, grinding his way up a hill, Michael simply ran out of steam. His legs felt rubbery. He gasped for air. He got off the bike and sat down.

What happened? Michael had forgotten to pace himself. His leg muscles worked so hard they used up oxygen faster than it could be resupplied. Without oxygen, cells switch to a kind of energy production called fermentation. One unwanted by-product of fermentation is lactic acid, which begins to collect in skeletal muscles and makes muscle contraction more difficult.

Eventually, lactic acid builds to the point that it causes muscle fatigue. In cycling lingo, this is called the dreaded "bonk." To recover, an athlete must slow down and breathe heavily, giving badly needed oxygen back to the body.

Our bodies are composed of many cells, and although each type is specialized for a particular function, all cells have a certain basic structure and metabolism. This chapter discusses how cells are constructed and carry on their activities. The cell is the fundamental unit of our bodies, and it is at the cellular level that we must understand how the body functions (Fig. 2.1). Michael's fatigue began with each individual muscle cell and not with the muscle as a whole. Because cells are microscopic, it is sometimes hard to imagine that individual muscle cells account for the functioning of an organ like a skeletal muscle.

The powerful electron microscopes developed during this century revealed the organelles, little bodies that carry on the metabolism of the cell. This chapter discusses the structure and function of the main organelles in the cell. It also briefly considers how a cell acquires the energy it needs for everyday and special activities like cycling.

2.1   Cell Size

All living things are made up of tiny fundamental units called **cells.** Because cells are so small, the study of cells did not begin until the invention of the first microscope in the seventeenth century. Then the *cell theory,* which states that *all living things are composed of cells, and new cells arise only from preexisting cells,* was formulated.

Regardless of a cell's size and shape, it must carry on the functions associated with life—interacting with the environment, obtaining chemicals and energy, growing, and reproducing. A few cells, like a hen's egg or a frog's egg, are large enough to be seen by the naked eye, but most are not. This is the reason a microscope is needed to see cells. Why are cells so small (most are less than one cubic millimeter)? An explanation for why cells are so small and why we are multicellular is explained by considering that nutrients enter a cell and wastes exit a cell at its surface. A large cell requires more nutrients and produces more wastes than a small cell. Therefore, a large cell that is actively

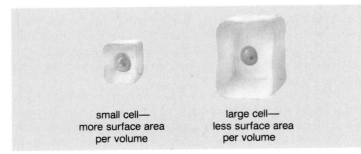

small cell—
more surface area
per volume

large cell—
less surface area
per volume

metabolizing cannot make do with proportionately less surface area than a small cell. Yet, as cells get larger in volume, the proportionate amount of surface area actually decreases.

We would expect, then, that there would be a limit to how large an actively metabolizing cell can become. Once a hen's egg is fertilized and starts actively metabolizing, it divides repeatedly without growth. Cell division restores the amount of surface area needed for adequate exchange of materials.

A cell needs a surface area that can adequately exchange materials with the environment. This explains why cells stay small.

Microscopy Reveals Cell Structure

Three types of microscopes are most commonly used: the *compound light microscope, transmission electron microscope,* and *scanning electron microscope.* Figure 2.2 depicts these microscopes, along with a micrograph of red blood cells viewed with each one.

In a compound light microscope, light rays passing through a specimen are brought to a focus by a set of glass lenses, and the resulting image is then viewed by the human eye. In the transmission electron microscope, electrons passing through a specimen are brought to a focus by a set of magnetic lenses, and the resulting image is projected onto a fluorescent screen or photographic film.

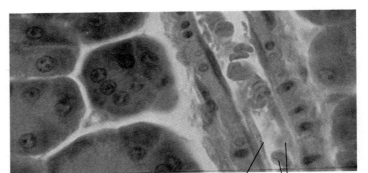

blood
vessel

red blood
cells

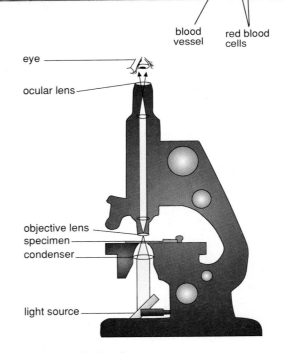

eye

ocular lens

objective lens
specimen
condenser

light source

Compound light microscope

Figure 2.2   Blood vessels and red blood cells viewed with three different types of microscopes.

The magnification produced by an electron microscope is much higher than that of a light microscope. Also, the ability of the electron microscope to make out detail in enlarged images is much greater. In other words, the electron microscope has a higher resolving power, that is, the ability to distinguish between two adjacent points. The resolving power of the

eye:	0.2 mm	=	200 μm	=	200,000 nm[1]	
light microscope:	.0002 mm	=	.200 μm	=	200 nm	
(1,000 ×)						
electron microscope:	.00001 mm	=	.0001 μm	=	10 nm	
(50,000 ×)						

[1] See Metric System in Appendix B

A scanning electron microscope provides a three-dimensional view of the surface of an object. A narrow beam of electrons is scanned over the surface of the specimen, which has been coated with a thin layer of metal. The metal gives off secondary electrons, which are collected to produce a television-type picture of the specimen's surface on a screen.

A picture obtained using a light microscope sometimes is called a photomicrograph, and a picture resulting from the use of an electron microscope is called a transmission electron micrograph (TEM) or a scanning electron micrograph (SEM), depending on the type of microscope used.

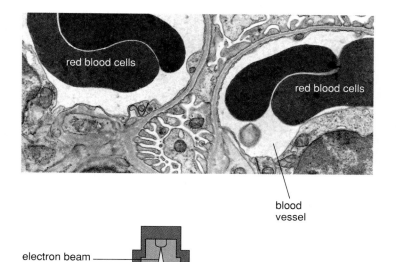

red blood cells

red blood cells

blood vessel

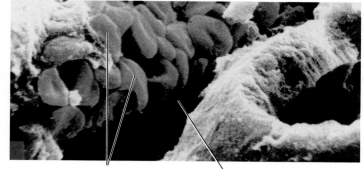

red blood cells   blood vessel

electron beam

condenser

specimen

objective lens

projector lens

observation or photograph

Transmission electron microscope

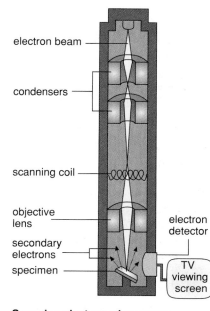

electron beam

condensers

scanning coil

objective lens

secondary electrons

specimen

electron detector

TV viewing screen

Scanning electron microscope

2.2   Cellular Organelles

In an animal cell (e.g., human cell) the **organelles** are small, membranous cellular inclusions that compartmentalize the cell, keeping various cellular activities separated from one another (Table 2.1 and Fig. 2.3).

The nucleus is a membrane-bound organelle that contains DNA within the threadlike chromatin. Chromatin condenses and coils to become the chromosomes during cell division.

The nucleolus is a special region of chromatin where a form of RNA called ribosomal RNA is produced and where the subunits of ribosomes are assembled. The subunits of ribosomes pass through the nuclear envelope at the nuclear pores. Some remain free within the **cytoplasm,** the portion of the cell between the nucleus and the plasma membrane. The **plasma membrane** (also called the cell membrane) is the outer boundary of the cell. Other ribosomes attach to the exterior surface of the endoplasmic reticulum, a membranous system of saccules and tubules just outside the nucleus.

DNA in the nucleus specifies the synthesis of proteins at the ribosomes. Strings of ribosomes in the cytoplasm that are all manufacturing the same protein are called polyribosomes. Endoplasmic reticulum that has attached ribosomes is called rough endoplasmic reticulum (RER), and that portion that does not have attached ribosomes is called smooth endoplasmic reticulum (SER). Proteins that are formed at RER pass into the interior of an endoplasmic reticulum tubule before being packaged in a vesicle that moves to the Golgi apparatus. The Golgi apparatus, which is a series of saccules, processes proteins and then repackages them for secretion at the plasma membrane or for inclusion in lysosomes. Lysosomes are vesicles that contain powerful hydrolytic enzymes that can fuse with and digest materials in other vesicles or even parts of the cell.

Mitochondria are the powerhouses of the cell because they convert the energy of organic molecules, usually glucose, into the energy within ATP molecules. ATP is the energy currency of cells. When cells synthetically produce something like a protein they "spend" the energy within ATP molecules.

The centrioles are small barrel-shaped structures that lie at right angles to each other outside the nucleus in the major microtubule organizing center of the cell. Microtubules are thin cylinders found free within the cytoplasm and also in cilia and flagella.

It is now known that microtubules and filaments, such as actin filaments, form a cytoskeleton that consists not only of microtubules, but also filaments. The cytoskeleton helps maintain the shape of the cell and is involved in the movement of the cell and its organelles.

The human cell has a central nucleus and an outer plasma membrane. Various organelles are found within the cytoplasm, the portion of the cell between the nucleus and the plasma membrane.

TABLE 2.1

Structures in Animal Cells

Name	Composition	Function
Plasma membrane	Phospholipid bilayer with embedded proteins	Selective passage of molecules into and out of cell
Nucleus	Nuclear envelope surrounding nucleoplasm, chromatin, and nucleolus	Storage of genetic information
Nucleolus	Concentrated area of chromatin, RNA, and proteins	Ribosomal formation
Ribosome	Protein and RNA in two subunits	Protein synthesis
Endoplasmic reticulum (ER)	Membranous saccules and canals	Synthesis and/or modification of proteins and other substances, and transport by vesicle formation
Rough ER	Studded with ribosomes	Protein synthesis
Smooth ER	Having no ribosomes	Various; lipid synthesis in some cells
Golgi apparatus	Stack of membranous saccules	Processing, packaging, and distributing molecules
Vacuole and vesicle	Membranous sacs	Storage of substances
Lysosome	Membranous vesicle containing digestive enzymes	Intracellular digestion
Mitochondrion	Inner membrane (cristae) within outer membrane	Cellular respiration
Cytoskeleton	Microtubules, actin filaments	Shape of cell and movement of its parts
Cilia and flagella	9 + 2 pattern of microtubules	Movement of cell
Centriole	9 + 0 pattern of microtubules	Formation of basal bodies

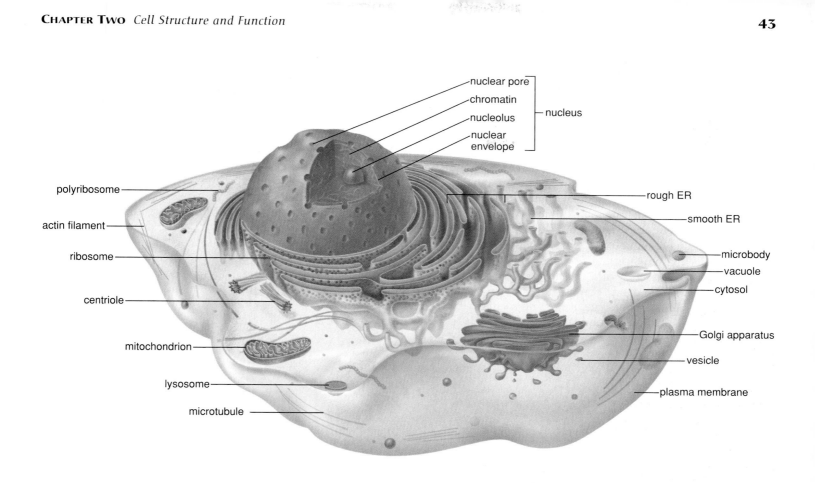

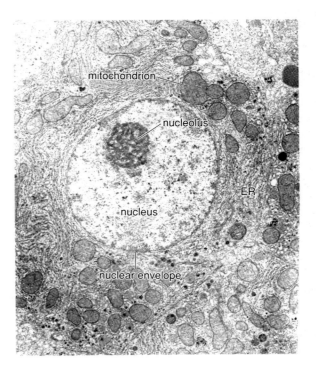

Figure 2.3 Animal cell.
This generalized representation is based on electron micrographs.

Plasma Membrane Is a Bilayer

An animal cell is surrounded by an outer *plasma membrane.* The plasma membrane marks the boundary between the outside of the cell and the inside of the cell, termed the *cytoplasm.* Plasma membrane integrity and how it functions are necessary to the life of the cell.

The plasma membrane is a phospholipid bilayer with attached or embedded proteins. The structure of a phospholipid is such that the molecule has a polar head and nonpolar tails (Fig. 2.4*a*). The polar heads, being charged, are hydrophilic (water loving) and face outward, toward the cytoplasm on one side and the tissue fluid on the other side, where they will encounter a watery environment. The nonpolar tails are hydrophobic (not attracted to water) and face inward toward each other, where there is no water. When phospholipids are placed in water, they naturally form a circular bilayer because of the chemical properties of the heads and the tails. At body temperature, the phospholipid bilayer is a liquid; it has the consistency of olive oil, and the proteins are able to change their position by moving laterally. The fluid-mosaic model, a working description of membrane structure, says that the protein molecules have a changing pattern (form a mosaic) within the fluid phospholipid bilayer (Fig. 2.4*b*).

Short chains of sugars are attached to the outer surface of some protein and lipid molecules (called glycoproteins and glycolipids, respectively). It is believed that these carbohydrate chains, specific to each cell, mark it as belonging to a particular individual and account for such characteristics as blood type or why a patient's system sometimes rejects an organ transplant. Other glycoproteins have a special configuration that allows them to act as a receptor for a chemical messenger like a hormone. Some plasma membrane proteins form channels through which certain substances can enter cells or are carriers involved in the passage of molecules through the membrane.

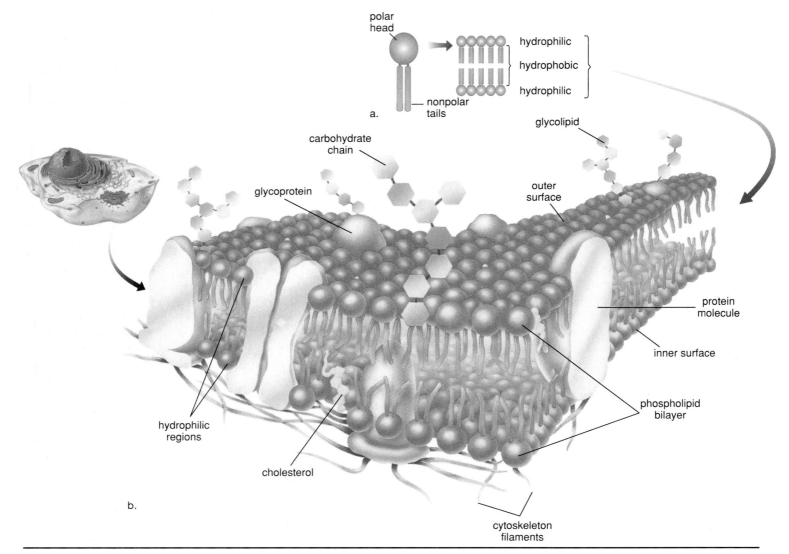

Figure 2.4 Fluid-mosaic model of the plasma membrane.
a. The membrane is composed of a phospholipid bilayer. The polar heads of the phospholipids are at the surfaces of the membrane; the nonpolar tails make up the interior of the membrane. **b.** Proteins are embedded in the membrane. Some of these function as receptors for chemical messengers, as conductors of molecules through the membrane, and as enzymes in metabolic reactions. Carbohydrate chains of glycolipids and glycoproteins are involved in cell-to-cell recognition.

How the Plasma Membrane Functions

The plasma membrane keeps a cell intact. It allows only certain molecules and ions to enter and exit the cytoplasm freely; therefore, the plasma membrane is said to be **selectively permeable.** Small molecules that are lipid soluble, such as oxygen and carbon dioxide, can pass through the membrane easily. Certain other small molecules, like water, are not lipid soluble but still cross the membrane passively by moving through a protein channel. Still other molecules and ions require the use of a carrier to enter a cell.

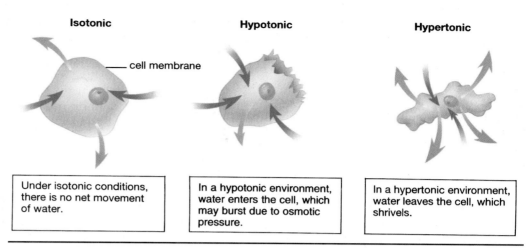

Figure 2.5   Tonicity.
The arrows indicate the movement of water.

Under isotonic conditions, there is no net movement of water.

In a hypotonic environment, water enters the cell, which may burst due to osmotic pressure.

In a hypertonic environment, water leaves the cell, which shrivels.

The plasma membrane, composed of phospholipid and protein molecules, is selectively permeable and regulates the entrance and exit of molecules and ions into and from the cell.

Diffusion   Diffusion is the random movement of molecules from the area of higher concentration to the area of lower concentration until they are equally distributed. To illustrate diffusion, imagine opening a perfume bottle in the corner of a room. The smell of the perfume soon permeates the room because the molecules that make up the perfume move to all parts of the room. Another example is putting a tablet of dye into water. The water eventually takes on the color of the dye as the tablet diffuses.

The chemical and physical properties of the plasma membrane allow only a few types of molecules to enter and exit a cell simply by diffusion. Lipid-soluble molecules such as alcohols can diffuse through the membrane because lipids are the membrane's main structural components. Gases can also diffuse through the lipid bilayer; this is the mechanism by which oxygen enters cells and carbon dioxide exits cells. As an example, consider the movement of oxygen from the alveoli (air sacs) of the lungs to blood in the lung capillaries. After inhalation (breathing in), the concentration of oxygen in the alveoli is higher than that in the blood; therefore, oxygen diffuses into the blood.

When molecules simply diffuse down their concentration gradients without assistance, across plasma membranes, no cellular energy is involved.

Molecules diffuse down their concentration gradients. A few types of small molecules can simply diffuse through the plasma membrane, and no carrier protein or cellular energy is involved.

Osmosis   Osmosis is the diffusion of water across a plasma membrane. It occurs whenever there is an unequal concentration of water on either side of a selectively permeable membrane. Normally, body fluids are **isotonic** to cells (Fig. 2.5)—there is an equal concentration of substances (solutes) and water (solvent) on both sides of the plasma membrane, and cells maintain their usual size and shape. Intravenous solutions medically administered usually have this tonicity. **Tonicity** is the degree to which a solution's concentration of solute versus water causes water to move into or out of cells.

Solutions that cause cells to swell or even to burst due to an intake of water are said to be hypotonic solutions. If red blood cells are placed in a **hypotonic** solution, which has a higher concentration of water (lower concentration of solute) than do the cells, water enters the cells and they swell to bursting. The term lysis is used to refer to disrupted cells; hemolysis, then, is disrupted red blood cells.

Solutions that cause cells to shrink or to shrivel due to a loss of water are said to be hypertonic solutions. If red blood cells are placed in a **hypertonic** solution, which has a lower concentration of water (higher concentration of solute) than do the cells, water leaves the cells and they shrink. The term crenation refers to red blood cells in this condition.

These changes have occurred due to osmotic pressure. *Osmotic pressure* is the force exerted on a selectively permeable membrane because water has moved from the area of higher to lower concentration of water (higher concentration of solute).

In an isotonic solution, a cell neither gains nor loses water. In a hypotonic solution, a cell gains water. In a hypertonic solution, a cell loses water and the cytoplasm shrinks.

Transport by Carriers   Most solutes do not simply diffuse across a plasma membrane; rather, they are transported by means of protein carriers within the membrane. During **facilitated transport,** a molecule (e.g., an amino acid or glucose) is transported across the plasma membrane from the side of higher concentration to the side of lower concentration. The cell does not need to expend energy for this type of transport because the molecules are moving down their concentration gradient.

During **active transport,** a molecule is moving contrary to the normal direction; that is, from lower to higher concentration (Fig. 2.6). For example, iodine collects in the cells of the thyroid gland; sugar is completely absorbed from the gut by cells that line the digestive tract; and sodium (Na^+) is sometimes almost completely withdrawn from urine by cells lining kidney tubules. Active transport requires a protein carrier and the use of cellular energy obtained from the breakdown of ATP. When ATP is broken down, energy is released, and in this case the energy is used by a carrier to carry out active transport. Therefore, it is not surprising that cells involved in active transport, such as kidney cells, have a large number of mitochondria near the membrane at which active transport is occurring.

Proteins involved in active transport often are called pumps because just as a water pump uses energy to move water against the force of gravity, proteins use energy to move substances against their concentration gradients. One type of pump that is active in all cells but is especially associated with nerve and muscle cells moves sodium ions (Na^+) to the outside of the cell and potassium ions (K^+) to the inside of the cell.

The passage of salt (Na^+Cl^-) across a plasma membrane is of primary importance in cells. The chloride ion (Cl^-) usually crosses the plasma membrane because it is attracted by positive charged sodium ions (Na^+). First, sodium ions are pumped across a membrane; then, chloride ions simply diffuse through channels that allow their passage. Chloride ion channels malfunction in persons with **cystic fibrosis,** and this leads to the symptoms of this inherited (genetic) disorder.

Endocytosis and Exocytosis   During *endocytosis,* cells take in substances by vesicle formation. A portion of the plasma membrane invaginates to envelop the substance, and then the membrane pinches off to form an intracellular vesicle. During *exocytosis*, the vesicle often formed by the Golgi apparatus fuses with the plasma membrane as secretion occurs. This is the way that insulin leaves insulin-secreting cells, for instance.

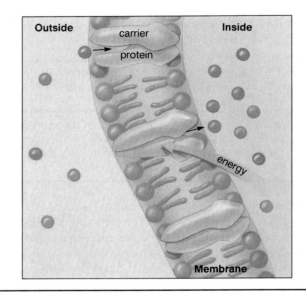

Figure 2.6   Active transport through a plasma membrane.
Active transport is apparent when a molecule crosses the plasma membrane toward the area of higher concentration. An expenditure of ATP energy is required, presumably for a carrier protein to transport a molecule across the plasma membrane against its concentration gradient.

TABLE 2.2

Passage of Molecules into and out of Cells

	Name	Direction	Requirement	Examples
Passive Transport Means	DIFFUSION	Toward lower concentration	Concentration gradient	Lipid-soluble molecules, water, and gases
	FACILITATED TRANSPORT	Toward lower concentration	Carrier and concentration gradient	Sugars and amino acids
Active Transport Means	ACTIVE TRANSPORT	Toward greater concentration	Carrier plus energy	Sugars, amino acids, and ions
	ENDOCYTOSIS	Toward inside	Vesicle formation	Macromolecules
	EXOCYTOSIS	Toward outside	Vesicle fuses with plasma membrane	Macromolecules

Nucleus Controls the Cell

The **nucleus,** which has a diameter of about 5 μm, is a prominent organelle in the eukaryotic cell. The nucleus is of primary importance because it stores genetic information. Although each cell gets a copy of every gene, only certain ones are turned on in a particular cell. Genes contain DNA, which works with RNA to direct synthesis of proteins. The proteins of a cell determine its structure and how it functions.

When you look at the nucleus in an electron micrograph, you cannot see genes, but you can see chromatin. **Chromatin** looks grainy, but actually it is a threadlike material that undergoes coiling into rodlike structures called **chromosomes,** just before the cell divides. Chromatin is immersed in a semifluid medium called the *nucleoplasm.*

A nucleus also has one or two darker bodies called nucleoli (sing., **nucleolus**). A type of RNA, called ribosomal RNA (rRNA), is produced in nucleoli, and there, rRNA joins with proteins from the cytoplasm to form the subunits of ribosomes. (Ribosomes are granules found in the cytoplasm.)

The nucleus is separated from the cytoplasm by a double membrane known as the **nuclear envelope** (Fig. 2.7). Protein fibers associated with the inner membrane of the nuclear envelope (called the nuclear lamina) help maintain the shape of the nucleus, provide chromatin attachment sites, and may funnel substances toward or away from the nuclear pores. **Nuclear pores** are open-

ings in the nuclear envelope of sufficient size (100 nm) to permit the passage of proteins into the nucleus and ribosomal subunits out of the nucleus. High-power electron micrographs show that a complex of eight proteins is associated with each pore.

In the nucleus, chromatin contains DNA, which directs protein synthesis in the cytoplasm. The nucleus contains one or more nucleoli, where rRNA is made.

Granule-like Particles

Ribosomes look like small, dense granules in low-power electron micrographs, but they are actually composed of two subunits. Each of these subunits has its own particular mix of rRNA and proteins. We have already mentioned that rRNA joins with proteins within the nucleolus, but the two subunits are not assembled into one ribosome until they reach the cytoplasm.

Ribosomes are the site of protein synthesis in the cytoplasm. They can be attached to the endoplasmic reticulum (discussed next) or lie free within the cytoplasm. When several ribosomes are making the same protein, they are arranged in a functional group called a **polyribosome.**

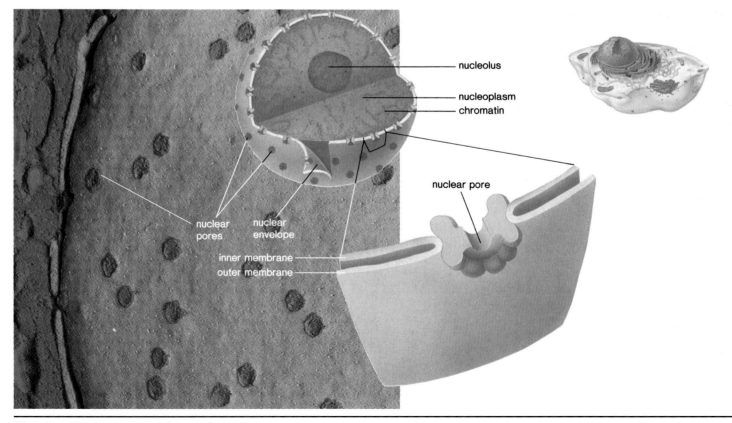

Figure 2.7 Anatomy of the nucleus.
The nucleoplasm contains chromatin, composed of DNA, protein, and at least one nucleolus, where ribosomal RNA (rRNA) is produced. Nucleoplasm is a semifluid medium contained by the nuclear envelope.

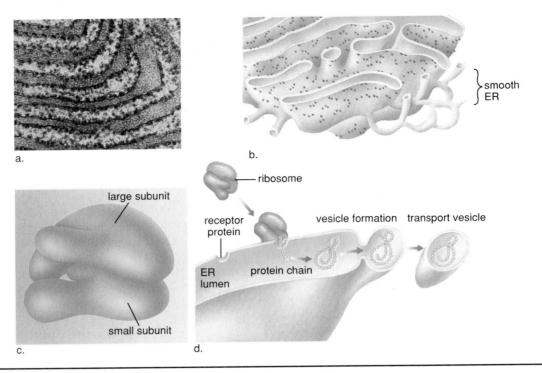

Figure 2.8   Rough endoplasmic reticulum (ER).
a. Electron micrograph of a mouse hepatocyte (liver cell) shows a cross section of many flattened vesicles with ribosomes attached to the side that abuts the cytosol. **b.** Shows the three dimensions of the organelle. **c.** Model of a single ribosome illustrates that each one is actually composed of two subunits. **d.** Method by which the ER acts as a transport system.

Membranous Canals and Vacuoles

The endoplasmic reticulum, the Golgi apparatus, vacuoles, and lysosomes are structurally and functionally related membranous structures. They work together to produce, transport, store, or secrete cellular products.

Endoplasmic Reticulum Produces, Modifies, and Transports

The **endoplasmic reticulum (ER)** forms a membranous system of tubular canals, which is continuous with the nuclear envelope and branches throughout the cytoplasm (see Fig. 2.3). If ribosomes are attached to ER, it is called **rough ER;** if ribosomes are not present, it is called **smooth ER.** Both types of ER are involved in synthesis and modification of macromolecules.

Smooth ER produces different molecules in different cells. It is abundant in the testes and adrenal cortex, both of which produce steroid hormones. In the liver, smooth ER is involved in the detoxification of drugs, including alcohol. A **vacuole** is a large membrane-enclosed sac; a vesicle is a small vacuole. Special vacuoles called *peroxisomes* are often attached to smooth ER, and these contain enzymes capable of detoxifying drugs.

Rough ER specializes in protein synthesis (Fig. 2.8). The ribosomes attached to rough ER make proteins for export from the cell. The proteins enter the lumen (interior space) of rough ER, where they may be modified. After arriving at the lumen of smooth ER, a vesicle pinches off and then carries the protein to the Golgi apparatus, where it is further processed. This is how the ER serves as a transport system.

Golgi Apparatus Modifies, Processes, and Packages

The **Golgi apparatus** (Fig. 2.9) is named for the person who first discovered its presence in cells. It is composed of a stack of about a half-dozen or more saccules (flattened vacuoles), which look like hollow pancakes. One side of the stack, called the inner face, is directed toward the nucleus and the ER. The other side of the stack, called the outer face, is directed toward the plasma membrane. Vesicles occur at the edges of the saccules.

The Golgi apparatus functions in the packaging, storage, and distribution of proteins produced by the ER. After the proteins are received at its inner face, they are often modified by the addition of a carbohydrate chain or a phosphate group. Finally, they are often packaged in secretory vesicles that move to the plasma membrane where their contents are discharged. For example, this is the way hormones are secreted into blood by the glands that produce them.

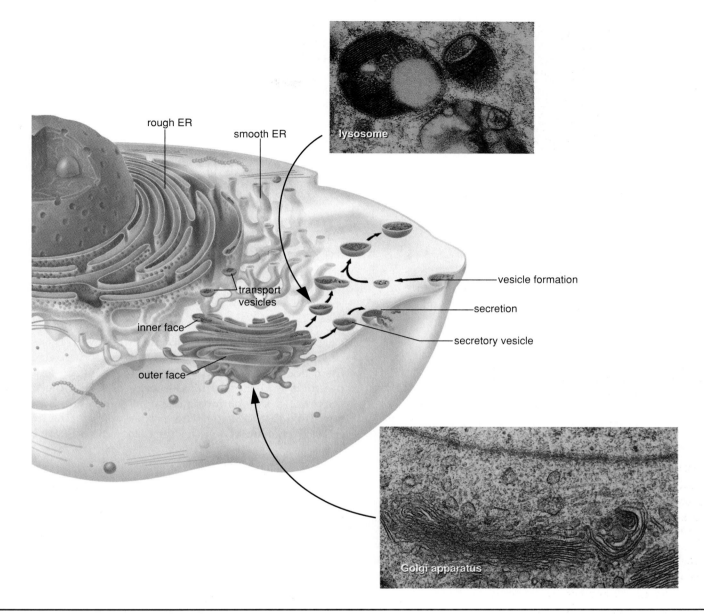

Figure 2.9 Structure and function of the Golgi apparatus.
The Golgi apparatus, a stack of several saccules, receives protein-containing vesicles from smooth ER. Protein molecules are modified as they move through the Golgi apparatus and are then repackaged into secretory vesicles. These vesicles take the protein to the plasma membrane, where they are discharged. The Golgi apparatus also produces lysosomes. Macromolecules can enter a cell by endocytosis (vesicle formation), and when this vesicle fuses with lysosomes, the macromolecules are digested.

Lysosomes Digest

Lysosomes, vesicles formed by the Golgi apparatus, contain *hydrolytic enzymes.* Sometimes macromolecules are brought into a cell by endocytosis at the plasma membrane. Then a lysosome fuses with the vesicle and digests its contents into simpler molecules, which enter the cytoplasm. Also, parts of a cell are often digested by the cell's own lysosomes. Normal cell rejuvenation most likely takes place in this way, but autodigestion is also important during development. For example, the fingers of a human embryo are at first webbed, but they are freed from one another by lysosomal action.

Occasionally, a child is born with a metabolic disorder involving a missing or inactive lysosomal enzyme. In these cases, the lysosomes fill to capacity with macromolecules that cannot be broken down. In *Tay-Sachs disease,* the cells become so full of these lysosomes the child dies.

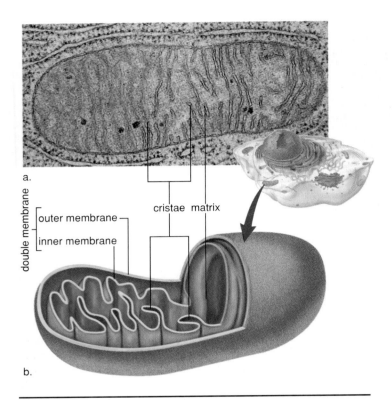

a.

double membrane

outer membrane
inner membrane

cristae   matrix

b.

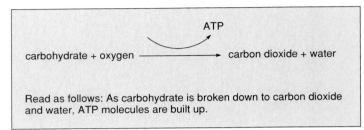

Figure 2.10   Mitochondrion structure.
A mitochondrion has a double membrane, and the cristae are infoldings of the inner membrane. The cristae project into the matrix, a space filled with a gel-like fluid.

Mitochondria Convert Energy

Most mitochondria (sing., **mitochondrion**) are between 0.5 μm and 1.0 μm in diameter and 7 μm in length, although the size and the shape can vary. Mitochondria are bounded by a double membrane. The inner membrane is folded to form little shelves called *cristae*, which project into the *matrix*, an inner space filled with a gel-like fluid (Fig. 2.10).

Mitochondria are the site of ATP (adenosine triphosphate) production involving complex metabolic pathways. ATP molecules are the common carrier of energy in cells. A shorthand way to indicate the chemical transformation that occurs in mitochondria is as follows:

ATP

carbohydrate + oxygen ⟶ carbon dioxide + water

Read as follows: As carbohydrate is broken down to carbon dioxide and water, ATP molecules are built up.

Mitochondria are often called the powerhouses of the cell: just as a powerhouse burns fuel to produce electricity, the mitochondria convert the chemical energy of glucose products into the chemical energy of ATP molecules. In the process, mitochondria use up oxygen and give off carbon dioxide and water. The oxygen you breathe in enters cells and then mitochondria; the carbon dioxide you breathe

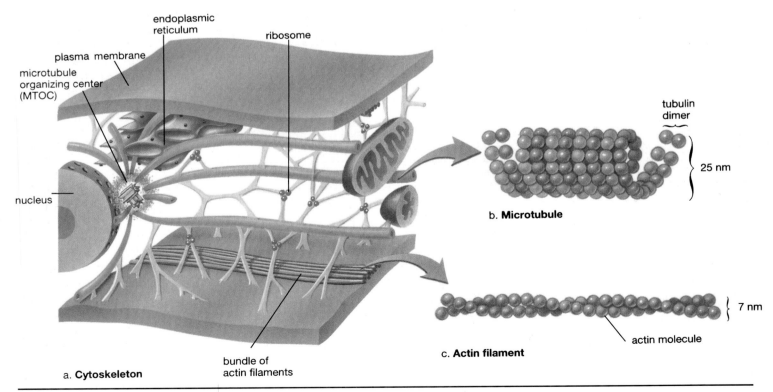

endoplasmic reticulum

ribosome

plasma membrane

microtubule organizing center (MTOC)

nucleus

tubulin dimer

25 nm

b. Microtubule

7 nm

actin molecule

c. Actin filament

bundle of actin filaments

a. **Cytoskeleton**

Figure 2.11   The cytoskeleton.
a. The cytoskeleton both anchors the organelles and allows them to move. **b.** A microtubule is a cylinder composed of tubulin protein molecules. Microtubules radiate from the microtubule organizing center (MTOC) and assist intracellular movement of organelles. **c.** An actin filament contains molecules of the protein actin. Bundles of actin filaments lie close to the plasma membrane. Actin filaments also form networks within the cytoplasm.

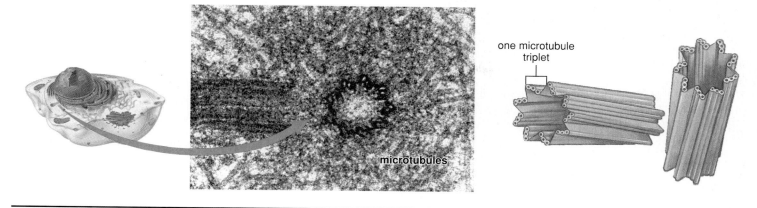

one microtubule triplet

microtubules

Figure 2.12 Centrioles.

Centrioles are composed of nine microtubule triplets. They lie at right angles to one another within the microtubule organizing center, which is believed to assemble microtubules at the time of cell division.

out is released by mitochondria. Because gas exchange is involved, it is said that mitochondria carry on **aerobic cellular respiration.**

The matrix of a mitochondrion contains enzymes for breaking down glucose products. ATP production then occurs at the cristae. The protein complexes that aid in the conversion of energy are located in an assembly-line fashion on these membranous shelves.

Every cell uses a certain amount of ATP energy to synthesize molecules, but many cells use ATP to carry out their specialized function. For example, muscle cells use ATP for muscle contraction, which produces movement, and nerve cells use it for the conduction of nerve impulses, which make us aware of our environment.

Mitochondria are the sites of aerobic cellular respiration, a process that provides ATP molecules to the cell.

Cytoskeleton Maintains Cell Shape

Several types of filamentous protein structures form a **cytoskeleton** that helps maintain the cell's shape and either anchors the organelles or assists their movement as appropriate (Fig. 2.11). The cytoskeleton includes microtubules and actin filaments.

Microtubules are shaped like thin cylinders and are several times larger than actin filaments. Each cylinder contains 13 rows of tubulin, a globular protein, arranged in a helical fashion. Remarkably, microtubules can assemble and disassemble. In many cells, the regulation of microtubule assembly is under the control of a microtubule organizing center (MTOC), which lies near the nucleus. Microtubules radiate from the MTOC, helping to maintain the shape of the cell and acting as tracks along which organelles move. It is well known that during cell division, microtubules form spindle fibers, which assist the movement of chromosomes.

Actin filaments are long, extremely thin fibers that usually occur in bundles or other groupings. Actin filaments have been isolated from various types of cells, especially those in which movement occurs. Microvilli, which project from certain cells and can shorten and extend, contain actin filaments. Actin filaments, like microtubules, can assemble and disassemble.

The cytoskeleton contains microtubules and actin filaments. Microtubules (13 rows of tubulin protein molecules arranged to form a hollow cylinder), and actin filaments (thin actin strands) maintain the shape of the cell and also direct the movement of cell parts.

Centrioles and Microtubules

In animal cells, **centrioles** are short cylinders with a 9 + 0 pattern of microtubules. There are nine outer microtubule triplets and no center microtubules (Fig. 2.12). There is always one pair of centrioles lying at right angles to one another near the nucleus (see Fig. 2.3). Before a cell divides, the centrioles duplicate, and the members of the new pair are also at right angles to one another. During cell division, the pairs of centrioles separate so that each daughter cell gets one pair of centrioles.

Centrioles are part of a microtubule organizing center that also includes other proteins and substances. Microtubules begin to assemble in the center, and then they grow outward, extending through the entire cytoplasm. In addition, centrioles may be involved in other cellular processes that use microtubules, such as movement of material throughout the cell or the formation of the spindle, a structure that distributes the chromosomes to daughter cells during cell division. Their exact role in these processes is uncertain, however. Centrioles also give rise to basal bodies that direct the formation of cilia and flagella.

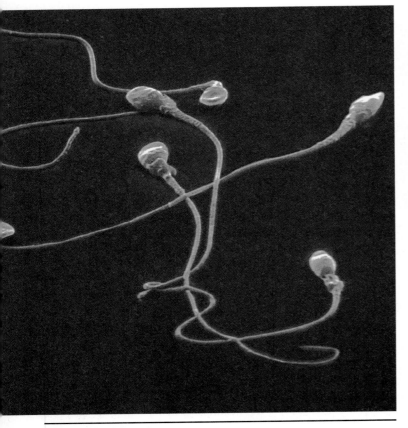

Figure 2.13 Sperm cells.
Sperm cells use long, whiplike flagella to move about.

Cilia and Flagella Make Cells Move

Cilia and **flagella** are projections of cells that can move either in an undulating fashion, like a whip, or stiffly, like an oar. Cilia are short (2–10 μm) while flagella are longer (usually no longer than 200 μm). Cells that have these organelles are capable of self-movement or moving material along the surface of the cell. For example, sperm cells, carrying genetic material to the egg, move by means of flagella (Fig. 2.13). The cells that line our respiratory tract are ciliated. These cilia sweep debris trapped within mucus back up the throat, and this action helps keep the lungs clean.

Each cilium and flagellum has a basal body at its base, which lies in the cytoplasm. **Basal bodies,** like centrioles, have a 9 + 0 pattern of microtubule triplets. They are believed to organize the structure of cilia and flagella even though cilia and flagella have a 9 + 2 pattern of microtubules. In cilia and flagella, there are nine microtubule doublets surrounding two central microtubules. This arrangement is believed to be necessary to their ability to move.

Centrioles give rise to basal bodies that organize the pattern of microtubules in cilia and flagella.

2.3 Cellular Metabolism

Cellular **metabolism** includes all the chemical reactions that occur in a cell. Quite often these reactions are organized into metabolic pathways.

$$A \xrightarrow{1} B \xrightarrow{2} C \xrightarrow{3} D \xrightarrow{4} E \xrightarrow{5} F \xrightarrow{6} G$$

The letters, except *A* and *G*, are *products* of the previous reaction and the *reactants* for the next reaction. *A* represents the beginning reactant(s), and *G* represents the end product(s). The numbers in the pathway refer to different enzymes. *Every reaction in a cell requires a specific enzyme.* In effect, no reaction occurs in a cell unless its enzyme is present. For example, if enzyme number 2 in the diagram is missing, the pathway cannot function; it will stop at *B*. Since enzymes are so necessary in cells, their mechanism of action has been studied extensively.

Metabolic pathways contain many enzymes that perform their reactions in a sequential order.

Enzymes and Coenzymes

When an enzyme speeds up a reaction, the reactant(s) that participate(s) in the reaction is called the enzyme's *substrate(s).* Enzymes are often named for their substrate(s) (Table 2.3). Enzymes have a specific region, called an **active site,** where the substrates are brought together so that they can react. An enzyme's specificity is caused by the shape of the active site, where the enzyme and its substrate(s) fit together in a specific way, much as the pieces of a jigsaw puzzle fit together (Fig. 2.14). After one reaction is complete, the product or products are released, and the enzyme is ready to catalyze another reaction. This can be summarized in the following manner:

$$E + S \rightarrow ES \rightarrow E + P$$

(where E = enzyme, S = substrate, ES = enzyme-substrate complex, and P = product).

Environmental conditions such as an incorrect pH or high temperature can cause an enzyme to become denatured. A denatured enzyme no longer has its usual shape and is therefore unable to speed up its reaction.

TABLE 2.3	
Enzymes Named for their Substrates	
Substrate	**Enzyme**
Lipid	Lipase
Urea	Urease
Maltose	Maltase
Ribonucleic acid	Ribonuclease
Lactose	Lactase

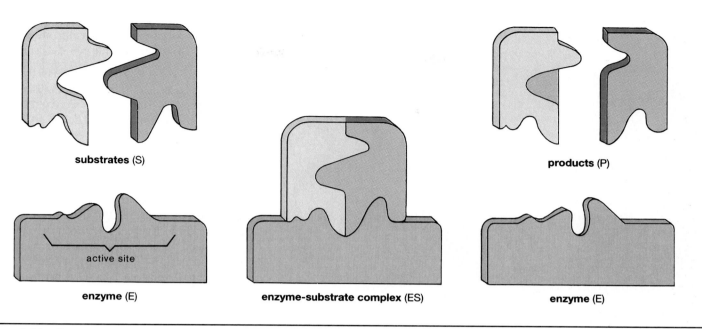

substrates (S)

products (P)

active site

enzyme (E)   enzyme-substrate complex (ES)   enzyme (E)

Figure 2.14 Enzymatic action.
An enzyme has an active site where the substrates and enzyme fit together in such a way that the substrates are oriented to react. Following the reaction, the products are released.

Many enzymes require cofactors. Some cofactors are inorganic, such as copper, zinc, or iron. Some cofactors are organic, nonprotein molecules and are called **coenzymes.** These cofactors assist the enzyme and may even accept or contribute atoms to the reaction. It is interesting that vitamins are often components of coenzymes. The vitamin niacin is a part of the coenzyme NAD, which removes hydrogen (H) atoms from substrates and therefore is called a *dehydrogenase.* Hydrogen atoms are sometimes removed by NAD as molecules are broken down. NAD that is carrying hydrogen atoms is written as $NADH_2$ because NAD removes two hydrogen atoms at a time. As we shall see, the removal of hydrogen atoms releases energy that can be used for ATP buildup.

Enzymes are specific because they have an active site that accommodates their substrates. Enzymes often have organic, nonprotein helpers called coenzymes. NAD is a dehydrogenase, a coenzyme that removes hydrogen from substrates.

Cellular Respiration Uses Metabolic Pathways

Cellular respiration is an important part of cellular metabolism because it accounts for ATP buildup in cells. Cellular respiration includes *aerobic* (requires oxygen) cellular respiration and fermentation, an *anaerobic* (does not require oxygen) process. During both processes, the energy released as glucose breakdown occurs is used to build up ATP molecules, the common energy carrier in cells.

Aerobic Cellular Respiration Uses Oxygen

During *aerobic cellular respiration*, glucose is broken down to carbon dioxide and water. Even though it is possible to write an overall equation for the process, aerobic cellular respiration does not occur in one step. Glucose breakdown requires three subpathways: *glycolysis*, the *Krebs cycle*, and the *electron transport system*. The location of these subpathways is as follows:

Glycolysis—occurs in the cytoplasm, outside a mitochondrion

Krebs cycle—occurs in the matrix of a mitochondrion

Electron transport system—occurs on the cristae of mitochondrion

Glycolysis and the Krebs cycle are a series of reactions in which the product of the previous reaction becomes the substrate for the next reaction. Every reaction that occurs during glycolysis and the Krebs cycle requires a specific enzyme. Each pathway resembles a conveyor belt in which a beginning substrate continuously enters at the start and, after a series of reactions, end products leave at the termination of the belt. It is important to realize, too, that these two pathways and the electron transport system occur at the same time. They can be compared to the inner workings of a watch, in which all parts are synchronized.

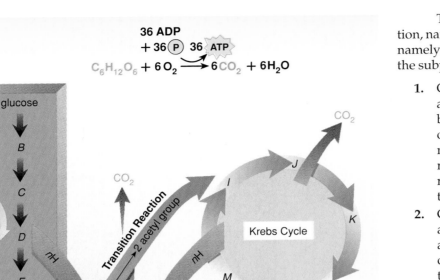

$$36 \text{ ADP} + 36 \text{ P } 36 \text{ ATP}$$
$$C_6H_{12}O_6 + 6 O_2 \longrightarrow 6 CO_2 + 6 H_2O$$

Figure 2.15   Aerobic cellular respiration.
The overall reaction shown at the *top* actually requires three
subpathways: glycolysis, the Krebs cycle, and the electron transport
system. As the reactions occur, a number of hydrogen (H) atoms
and carbon dioxide (CO_2) molecules are removed from the various
substrates. Oxygen (O_2) acts as the final acceptor for hydrogen
atoms ($2e^- + 2H^+$) and becomes water (H_2O).

TABLE 2.4	
Overview of Aerobic Cellular Respiration	
Name of Pathway	**Result**
Glycolysis	Removal of H from substrates produces **2 ATP**
Transition reaction	Removal of H from substrates releases **2 CO₂**
Krebs cycle	Removal of H from substrates releases **4 CO₂**
	Produces **2 ATP** after 2 turns
Electron transport system	Accepts H from other pathways and passes electrons on to O_2, producing **H_2O**
	Produces **32 ATP**

The reactants of aerobic cellular respiration, namely glucose and oxygen, and products,
namely carbon dioxide and water, are related to
the subpathways in the manner described next.

1. Glucose, a C_6 molecule, is to be
    associated with **glycolysis,** the
    breakdown of glucose to two molecules
    of pyruvate (pyruvic acid), a C_3
    molecule. During glycolysis, energy is
    released as hydrogen (H) atoms are
    removed. This energy is used to form
    two ATP molecules (Fig. 2.15).

2. Carbon dioxide, CO_2, is to be
    associated with the transition reaction
    and the Krebs cycle, both of which
    occur in mitochondria. During the
    transition reaction, pyruvate is
    converted to a C_2 acetyl group after
    CO_2 comes off. Because the transition
    reaction occurs twice per glucose
    molecule, two molecules of CO_2 are
    released. Hydrogen (H) atoms are also
    removed at this time.

    The acetyl group enters the **Krebs
    cycle,** a cyclical series of reactions that
    give off two CO_2 molecules and
    produce one ATP molecule. Since the
    Krebs cycle occurs twice per glucose
    molecule, altogether four CO_2 and
    two ATP are produced per glucose
    molecule. Hydrogen (H) atoms are
    removed from the substrates and
    added to NAD, forming $NADH_2$ as
    the Krebs cycle occurs.

3. Oxygen, O_2, and water, H_2O, are to be associated
    with the **electron transport system.** The electron
    transport system begins with $NADH_2$, the coenzyme
    that carries most of the hydrogen (H) atoms to the
    system, but after that it consists of molecules that
    carry electrons. High energy electrons are removed
    from the hydrogen atoms, leaving behind hydrogen
    ions (H^+), and then the electrons are passed from one
    molecule to another until the electrons are received
    by oxygen which becomes $O^=$. At this point, $2H^+$
    combine with ionic oxygen to give water. As the
    electrons are passed down the system, their energy is
    released to allow the buildup of ATP.

4. ATP is to be associated with glycolysis, the Krebs
    cycle, and the electron transport system. Altogether,
    36 ATP result from the breakdown of one glucose
    molecule (Table 2.4).

Aerobic cellular respiration requires glycolysis,
which takes place in the cytoplasm; the Krebs cycle,
which is located in the matrix of the mitochondria;
and the electron transport system, which is located
on the cristae of the mitochondria.

Fermentation is Anaerobic

Fermentation is an anaerobic process. When oxygen is not available to cells, the electron transport system soon becomes inoperative because oxygen is not present to accept electrons. In this case, most cells have a safety valve so that some ATP can still be produced. Glycolysis operates as long as it is supplied with "free" NAD; that is, NAD that can pick up hydrogen atoms. Normally, $NADH_2$ takes hydrogens to the electron transport system and thereby becomes "free" of hydrogen atoms. However, if the system is not working due to lack of oxygen, $NADH_2$ passes its hydrogen atoms to pyruvate as shown in the following reaction:

$$NADH_2 \searrow NAD$$
$$pyruvate \longrightarrow lactate$$

The Krebs cycle and electron transport system do not function as part of fermentation. When oxygen is available again, lactate (lactic acid) can be converted back to pyruvate, and metabolism can proceed as usual.

Fermentation takes less time than aerobic cellular respiration, but since glycolysis alone is occurring, it produces only 2 ATP per glucose molecule. Also, fermentation results in the buildup of lactate. Lactate is toxic to cells and causes muscles to cramp and fatigue. If fermentation continues for any length of time, death follows.

It is of interest to know that fermentation takes its name from yeast fermentation. Yeast fermentation produces alcohol and carbon dioxide (instead of lactate). When yeast is used to leaven bread, it is the carbon dioxide that produces the desired effect. When yeast is used to produce alcoholic beverages, it is the alcohol that humans make use of.

Fermentation is an anaerobic process, a process that does not require oxygen but produces very little ATP per glucose molecule and results in lactate or alcohol and carbon dioxide buildup.

SUMMARY

2.1 Cell Size

Cells are quite small, and it usually takes a microscope to see them. Small cubes, like cells, have a more favorable surface/volume ratio than do large cubes. Only inactive eggs are large enough to be seen by the naked eye; once development begins, cell division results in small-size cells.

2.2 Cellular Organelles

A cell is surrounded by a plasma membrane, which regulates the entrance and exit of molecules and ions. Some molecules, such as water and gases, diffuse through the membrane. The direction in which water diffuses is dependent on its concentration within the cell compared to outside the cell.

Table 2.1 lists the cell organelles we have studied in the chapter. The nucleus is a large organelle of primary importance because it controls the rest of the cell. Within the nucleus lies the chromatin, which condenses to become chromosomes during cell division.

Proteins are made at the rough ER before being modified and packaged by the Golgi apparatus into vesicles for secretion. During secretion, a vesicle discharges its contents at the plasma membrane. Golgi-derived lysosomes fuse with incoming vesicles to digest any material enclosed within, and lysosomes also carry out autodigestion of old parts of cells.

Mitochondria are the powerhouses of the cell. During the process of aerobic cellular respiration, mitochondria convert carbohydrate energy to ATP energy.

Microtubules and actin filaments make up the cytoskeleton, which maintains the cell's shape and permits movement of cell parts. Centrioles are a part of the microtubule organizing center, which is associated with the formation of microtubules in general and the spindle that appears during cell division. Centrioles also produce basal bodies that give rise to cilia and flagella.

2.3 Cellular Metabolism

Cellular metabolism is the sum of all biochemical pathways of the cell. In a pathway, a series of reactions proceed in an orderly step-by-step manner. Each of these reactions requires a specific enzyme. Sometimes enzymes require coenzymes, nonprotein portions that participate in the reaction. NAD is a coenzyme.

Aerobic cellular respiration (the breakdown of glucose to carbon dioxide and water) includes three pathways: glycolysis, the Krebs cycle, and the electron transport system. If oxygen is not available in cells, the electron transport system is inoperative, and fermentation (an anaerobic process) occurs. Fermentation makes use of glycolysis only, plus one more reaction in which pyruvate is reduced to lactate.

STUDYING THE CONCEPTS

1. Describe the structure and biochemical makeup of a plasma membrane. 44

2. What are three mechanisms by which substances enter and exit cells? Define isotonic, hypertonic, and hypotonic solutions. 45–46

3. Describe the nucleus and its contents, including the terms DNA and RNA in your description. 47

4. Describe the structure and function of endoplasmic reticulum. Include the terms rough and smooth ER and ribosomes in your description. 48

5. Describe the structure and function of the Golgi apparatus and its relationship to vesicles and lysosomes. 48

6. Describe the structure of mitochondria, and relate this structure to the pathways of aerobic cellular respiration. 50

7. Describe the composition of the cytoskeleton. 50

8. Describe the structure and function of centrioles, cilia, and flagella. 51–52

9. Discuss and draw a diagram for a metabolic pathway. Discuss and give a reaction to describe the specificity theory of enzymatic action. Define coenzyme. 52

10. Name and describe the events within the three subpathways that make up aerobic cellular respiration. Why is fermentation necessary but potentially harmful to the human body? 53–55

APPLYING YOUR KNOWLEDGE

Concepts

1. In examining the constituents of the plasma membrane of two individual people, which of the following constituents would probably show the greatest difference between the two: phospholipid, protein, glycoprotein, cholesterol? Give the reasoning for your answer.

2. Under certain pathological conditions, the cell digests itself; this is referred to as autolysis. Which specific cellular organelle is probably involved in this process? Explain.

3. A microtubule is a hollow cylinder composed of 13 rows of protein molecules. What evidence do you have each row is not an actin filament?

4. The process of cellular respiration results in 36 ATP, and the process of fermentation results in 2 ATP. Where is the rest of the chemical energy following fermentation?

Bioethical Issue

Today, about 50,000 people in the United States are waiting to receive organ transplants. Years may pass—and some patients will die—before the right donors appear. It is a sad situation.

But now, researchers say there may be a solution to the nation's organ shortage: animal transplants. In 1995, a man suffering from AIDS was the first American to receive bone marrow cells taken from a baboon. As in any transplant, the idea was to replace the patient's damaged cells with those of a healthy donor.

Many scientists and others heralded the surgery as revolutionary. Some, however, expressed concern. Could a dangerous, unknown baboon virus infect humans via transplants? And how can we ensure the rights of animals used to sustain human life?

What do you think? Are animal transplants a wise alternative to waiting for a human donor? Are they ethical?

APPLYING TECHNOLOGY

Your study of cell structure and function is supported by these available technologies:

Exploring the Internet
The Mader Home Page provides further resources for studying this chapter.

`http://www.mhhe.com/sciencemath/biology/mader/`

(Click on *Human Biology*.)

Life Science Animations Video

Video #1: Chemistry, The Cell, and Energetics.

Journey Into a Cell (#2)   The structure and function of all major organelles are reviewed with reference to a cross section of the animal cell.   (1)*

Endocytosis (#3)   Vesicle formation occurs as a particle enters the cell.   (1)

Cellular Secretion (#4)   A protein made at the rough endoplasmic reticulum is repackaged into a vesicle at the Golgi apparatus. When this vesicle fuses with the plasma membrane, secretion occurs. Lysosomes are also seen budding from the Golgi apparatus.   (1)

Glycolysis (#5)   The glycolytic pathway is described in depth, and the molecular structure of each participant is shown.   (4)

Oxidative Respiration (#6)   The Krebs cycle is described in depth, and the molecular structure of each participant is shown.   (4)

The Electron Transport Chain and the Production of ATP (#7)   Chemiosmotic ATP synthesis in a mitochondrion is described in a straightforward and simplified manner.   (2)

ATP as an Energy Carrier (#11)   ATP structure and the ATP cycle are reviewed pictorially.   (1)

Explorations in Cell Biology & Genetics CD-ROM

Cell Size (#2)   The rate of diffusion varies as students change the cell size, cell shape, the number of dimples, and number of villi (projections that increase cell surface area).   (3)

Active Transport (#3)   Amino acid transport rate varies as students change the expenditure of ATP and amino acid concentration.   (4)

Thermodynamics (cell chemistry) (#6)   An enzyme reaction rate graph varies as students change the enzyme concentration, temperature, and pH for three different enzymes.   (3)

Oxidative Respiration (#8)   The rate of ATP production in a mitochondrion varies as students change the availability of food and oxygen molecules to the amount of ATP already present in the cell.   (2)

TESTING YOUR KNOWLEDGE

For questions 1–5, match the organelles in the key to their functions.

Key: a. mitochondria

b. nucleus

c. Golgi apparatus

d. rough ER

e. centrioles

1. packaging and secretion _____
2. cell division _____
3. powerhouses of the cell _____
4. protein synthesis _____
5. control center for cell _____
6. Label only the parts of the cell that are involved in protein synthesis and modification. Explain your choices.

7. Microtubules and actin filaments are a part of the _____ , the framework of the cell that provides its shape and regulates movement of organelles.

8. Water enters a cell when it is placed in a _____ solution.

9. Substrates react at the _____ , located on the surface of their enzyme.

10. During aerobic cellular respiration, most of the ATP molecules are produced at the _____ , a series of carriers located on the _____ of mitochondria.

11. Fermentation of a glucose molecule produces only _____ ATP compared to the _____ ATP produced by aerobic cellular respiration.

12. Complete the following diagram by labeling the pathways and by adding ATP, CO_2, O_2, and H_2O where needed.

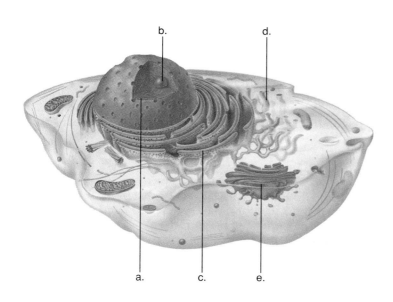

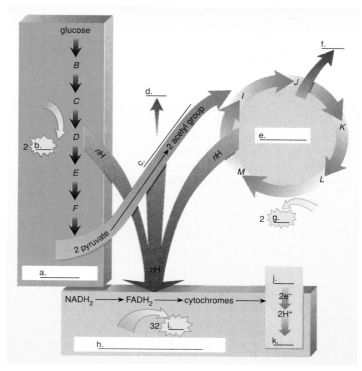

SELECTED KEY TERMS

actin filament Extremely thin fiber found within the cytoplasm that is composed of the protein actin; involved in the maintenance of cell shape and the movement of cell contents.   50

active site Region on the surface of an enzyme where the substrate binds and where the reaction occurs.   52

active transport Transfer of a substance into or out of a cell from a region of lower concentration to a region of higher concentration by a process that requires a carrier and an expenditure of energy.   46

aerobic cellular respiration Complete breakdown of glucose to carbon dioxide and water in the presence of oxygen.   50

centriole Short cylinder in animal cells that contains microtubules in a 9 + 0 pattern; associated with the formation of basal bodies.   51

chromosome Rodlike structure in the nucleus seen during cell division; contains the hereditary units, or genes.   47

cilium Short, hairlike extension from a cell, occurring in large numbers and used for cell mobility.   52

coenzyme Nonprotein organic molecule that aids the action of the enzyme to which it is loosely bound.   53

cytoplasm Semifluid medium between the nucleus and the plasma membrane that contains the organelles.   42

cytoskeleton Internal framework of a cell that helps maintain the shape of the cell, anchor the organelles, and allow the cell and its organelles to move.   50

diffusion Movement of molecules from a region of higher concentration to a region of lower concentration.   45

electron transport system Chain of electron carriers in the cristae of mitochondria. The electrons release energy as they pass down the chain, and this is used to produce ATP.   54

endoplasmic reticulum (ER) (en-doh-PLAZ-mik reh-TIK-yoo-lum) Membranous system of tubules, vesicles, and sacs in cells, sometimes having attached ribosomes. Rough ER has ribosomes; smooth ER does not.   48

facilitated transport Passive transfer of a substance into or out of a cell along a concentration gradient by a process that requires a carrier.   46

fermentation Anaerobic breakdown of carbohydrates that results in two ATP products such as alcohol and lactic acid.   55

flagellum Slender, long extension that propels a cell through a fluid medium, occuring as one or two on a cell.   52

glycolysis (gly-KOL-uh-sis) Metabolic pathway found in the cytoplasm that participates in aerobic cellular respiration and fermentation; it converts glucose to two molecules of pyruvate.   54

Golgi apparatus Organelle consisting of concentrically folded saccules that functions in the packaging, storage, and distribution of cellular products.   48

Krebs cycle Cyclical metabolic pathway found in the matrix of mitochondria that participates in aerobic cellular respiration; breaks down acetyl groups to carbon dioxide and hydrogen.   54

lysosome Membrane-bounded organelle containing digestive enzymes.   49

metabolism All of the chemical reactions within a cell (or an organism) including break-down reactions and synthetic reactions.   52

microtubule Cytoskeletal element composed of 13 rows of globular proteins called tubulin; also found in multiple units within other structures, such as the centriole, cilia, flagella, as well as spindle fibers.   50

mitochondrion Organelle where ATP is produced during aerobic cellular respiration; the powerhouse of the cell.   50

nuclear envelope Double membrane that surrounds the nucleus and is continuous with the endoplasmic reticulum.   47

nucleolus A special region found inside the nucleus where rRNA is produced for ribosome formation.   47

nucleus (NOO-klee-us) Region of a eukaryotic cell, containing chromosomes, that controls the structure and function of the cell.   47

organelle Specialized membranous structure within cells (e.g., nucleus, mitochondria, and endoplasmic reticulum) that performs specific functions.   42

osmosis Movement of water from an area of higher concentration of water to an area of lower concentration of water across a selectively permeable membrane.   45

plasma membrane Membrane that surrounds the cytoplasm of cells and regulates the passage of molecules and ions into and out of the cell.   42

ribosome Minute particle that is attached to endoplasmic reticulum or occurs loose in the cytoplasm and is the site of protein synthesis.   47

tonicity Degree to which a solution's concentration of solute versus water causes water to move into or out of cells. In isotonic solutions, cells neither gain nor lose water; in hypotonic solutions, cells gain water; in hypertonic solutions, cells lose water.   45

Chapter 3

Human Organization

Chapter Outline

Figure 3.1   Human organization.
The nervous system coordinates the other systems of the body so that we respond to outside stimuli, including those from the opposite sex.

Ever wonder why a touch on the arm can send you blushing, your heart racing with swift pleasure (Fig. 3.1)? Biochemists sometimes joke that love—or the physical rush accompanying it—is really just a complex set of reactions. It's not a romantic view. Technically, though, it's true.

In this case, the coveted reactions begin with a slight pressure on the epidermis of your skin. Then, in the dermis, a vast array of receptors detect the pressure and send messages racing to your central nervous system by way of sensory neurons. There, a cascade of messages bounce back. The limbic system—the brain's emotional control center—can bring about all sorts of reactions; everything from a quickened heartbeat to sweaty palms. Reactions to feelings of love can even make your muscles twitch or your voice break, even as you try to act normally. Some might say such a complex set of reactions is the best example of coordination in humans we have to offer.

The nervous and endocrine systems coordinate the various systems of the body, including, for example, the muscular and circulatory systems. Each system contains a number of organs: the nervous system contains the brain, spinal cord, and nerves. Each organ is composed of different types of tissues. This chapter reviews the structure and function of tissues and shows how they are organized within the skin, which is used as an example of an organ. In a square inch of skin, there are 1,300 receptors that communicate with the brain and spinal cord, enabling us to be sensitive to a lover's touch, among other stimuli.

Since the nervous and endocrine systems coordinate the other systems of the body, they play a pivotal role in determining our responses to not only outside stimuli but also internal stimuli. Homeostasis, the dynamic equilibrium of the internal environment, is an absolute necessity for our continued existence. All systems of the body contribute to homeostasis, from the digestive system, which provides nutrient molecules, to the muscular system, which enables us to bring food to the mouth. This chapter introduces a significant type of box called *working together boxes* which appear throughout the text. The working together boxes tell how the systems of the body help each other maintain homeostasis.

3.1 Types of Tissues

A **tissue** is composed of similarly specialized cells that perform a common function in the body. The tissues of the human body can be categorized into four major types: *epithelial tissue*, which covers body surfaces and lines body cavities; *connective tissue*, which binds and supports body parts; *muscular tissue*, which moves body parts; and *nervous tissue*, which receives stimuli and conducts impulses from one body part to another.

Cancers are classified according to the type of tissue from which they arise. *Carcinomas,* the most common type, are cancers of epithelial tissues; *sarcomas* are cancers arising in muscle or connective tissue (especially bone or cartilage); *leukemias* are cancers of the blood; and *lymphomas* are cancers of reticular connective tissue. The chance of cancer in a particular tissue shows a positive correlation to the rate of cell division; new blood cells arise at a rate of 2.5×10^6 cells per second, and epithelial cells also reproduce at a high rate.

Epithelial Tissue Covers

Epithelial tissue, also called epithelium, consists of tightly packed cells that form a continuous layer or sheet lining the entire body surface and most of the body's inner cavities. On the external surface, it protects the body from injury, drying out, and possible pathogenic invasion. A **pathogen** is a disease-causing agent like a virus or an infectious bacterium. On internal surfaces, epithelial tissue may be specialized for other functions in addition to protection. For example, epithelial tissue secretes mucus along the digestive tract and sweeps up impurities from the lungs by means of **cilia.** It efficiently absorbs molecules from kidney tubules and from the intestine because of minute cellular extensions called **microvilli.**

There are three types of epithelial tissue (Fig. 3.2). **Squamous epithelium** is *composed of flattened cells* and is found lining the lungs and blood vessels. **Cuboidal** epithelium contains *cube-shaped cells* and is found lining the kidney tubules. **Columnar epithelium** has cells *resembling rectangular pillars* or *columns,* and nuclei are usually located near the bottom of each cell. This epithelium is found lining the digestive tract. Ciliated columnar epithelium is found lining the oviducts, where it propels the egg toward the uterus or womb.

An epithelium can be simple or stratified. *Simple* means the tissue has a single layer of cells, and *stratified* means that the tissue has layers of cells piled one on top of the other. The walls of the smallest blood vessels, called capillaries, are composed of a single layer of epithelial cells. The permeability of capillaries allows exchange of substances between the blood and tissue cells. The nose, mouth, esophagus, anal canal, and vagina are all lined by stratified squamous epithelium. As we shall see, the outer layer of skin is also stratified squamous epithelium, but the cells have been reinforced by keratin, a protein that provides strength.

Pseudostratified epithelium appears to be layered; however, true layers do not exist because each cell touches the base line. The lining of the windpipe, or trachea, is called *pseudostratified ciliated columnar epithelium.* A secreted covering of mucus traps foreign particles, and the upward motion of the cilia carries the mucus to the back of the throat, where it may either be swallowed or expectorated. Smoking can cause a change in mucous secretion and inhibit ciliary action, and the result is a chronic inflammatory condition called bronchitis.

A so-called *basement membrane* often joins an epithelium to underlying connective tissue. We now know that the basement membrane is glycoprotein, reinforced by fibers that are supplied by the connective tissue, the type of tissue discussed in the next section of this chapter.

An epithelium sometimes secretes a product, in which case it is described as glandular. A **gland** can be a single epithelial cell, as in the case of mucous-secreting goblet cells found within the columnar epithelium lining the digestive tract, or a gland can contain many cells. Glands that secrete their product into ducts are called *exocrine glands,* and those that secrete their product directly into the bloodstream are called *endocrine glands.* The pancreas is both an exocrine gland, because it secretes digestive juices into the small intestine via ducts, and an endocrine gland, because it secretes insulin into the bloodstream.

Epithelial tissue is named according to the shape of the cell. These tightly packed protective cells can occur in more than one layer, and the cells lining a cavity can be ciliated and/or glandular.

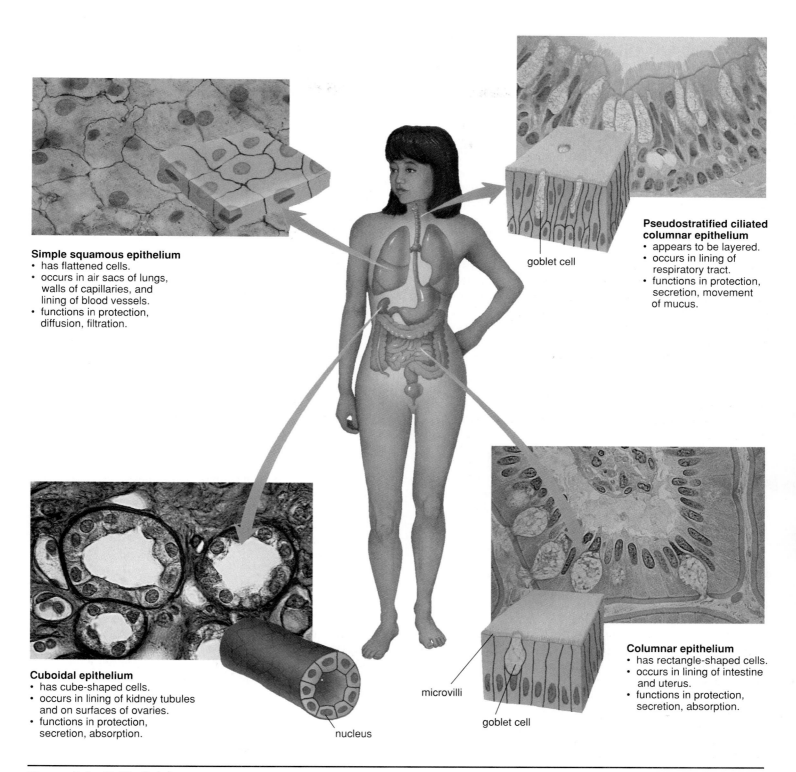

Simple squamous epithelium
- has flattened cells.
- occurs in air sacs of lungs, walls of capillaries, and lining of blood vessels.
- functions in protection, diffusion, filtration.

Pseudostratified ciliated columnar epithelium
- appears to be layered.
- occurs in lining of respiratory tract.
- functions in protection, secretion, movement of mucus.

goblet cell

Cuboidal epithelium
- has cube-shaped cells.
- occurs in lining of kidney tubules and on surfaces of ovaries.
- functions in protection, secretion, absorption.

nucleus

microvilli

goblet cell

Columnar epithelium
- has rectangle-shaped cells.
- occurs in lining of intestine and uterus.
- functions in protection, secretion, absorption.

Figure 3.2 Epithelial tissue.
The three types of epithelial tissue—squamous, cuboidal, and columnar—are named for the shape of their cells. They all have a protective function, as well as the other functions noted.

Junctions Help Communication

For cells of a tissue to act in a coordinated manner, it is beneficial for the plasma membrane of adjoining cells to interact. The junctions that occur between cells help cells function as a tissue (Fig. 3.3). A *tight junction* forms an impermeable barrier because adjacent plasma membrane proteins actually join, producing a zipperlike fastening. In the intestine, the gastric juices stay out of the body, and in the kidneys, the urine stays within kidney tubules because epithelial cells are joined by tight junctions.

A *gap junction* forms when two adjacent plasma membrane channels join. This lends strength, but it also allows ions, sugars, and small molecules to pass between the two cells. Gap junctions in heart and smooth muscle ensure synchronized contraction. In an *adhesion junction* (desmosome), the adjacent plasma membranes do not touch but are held together by intercellular filaments firmly attached to buttonlike thickenings. In some organs—like the heart, stomach, and bladder, where tissues get stretched—adhesion junctions hold the cells together.

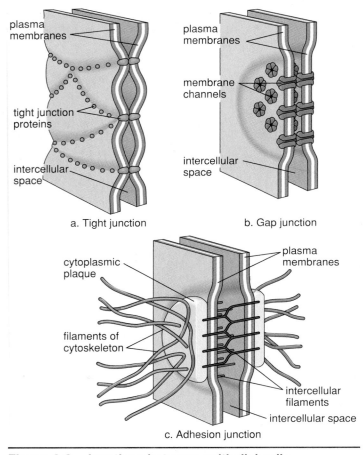

Figure 3.3   Junctions between epithelial cells.
Epithelial tissue cells are held tightly together by **(a)** tight junctions; **(b)** gap junctions that allow materials to pass from cell to cell; and **(c)** adhesion junctions allow tissues to stretch.

Connective Tissue Connects

Connective tissue binds organs together, provides support and protection, fills spaces, produces blood cells, and stores fat. The body uses this stored fat for energy, insulation, and organ protection. As a rule, connective tissue cells are widely separated by a **matrix,** which is found between cells and consists of a noncellular material that varies in consistency from solid to semifluid to fluid. The matrix may have fibers of which there are three types: *Collagen* (white) *fibers* contain collagen, a protein that gives them flexibility and strength. *Elastic* (yellow) *fibers* contain elastin, a protein that is not as strong as collagen but is more elastic. *Reticular fibers* are very thin fibers which are also composed of collagen. Reticular fibers are highly branched and form delicate supporting networks.

Loose Connective Tissue Supports

Loose connective tissue binds structures together (Fig. 3.4a). The cells of this tissue, which are mainly *fibroblasts,* are located some distance from one another and are separated by a semifluid, jellylike matrix, which contains many white collagen fibers and yellow elastic fibers. The collagen fibers occur in bundles and are strong and flexible. The elastic fibers form networks that when stretched return to their original length. Loose connective tissue commonly lies beneath epithelial layers.

Adipose tissue (Fig. 3.4b), is a type of loose connective tissue in which the enlarged fibroblasts store fat and the intercellular matrix is reduced. The fibroblasts of reticular connective tissue are called *reticular cells,* and the matrix contains only reticular fibers. This tissue, also called *lymphoid tissue,* is found in lymph nodes, the spleen, thymus, and red bone marrow. These organs are a part of the immune system because they store and/or produce white blood cells, particularly lymphocytes. All types of blood cells are produced in red bone marrow.

Fibrous Connective Tissue Binds

Fibrous connective tissue has a semifluid matrix produced by fibroblasts and contains many white collagen fibers packed closely together. This type of tissue has more specific functions than loose connective tissue. For example, fibrous connective tissue is found in **tendons,** which connect muscles to bones, and in **ligaments,** which connect bones to other bones at joints. Tendons and ligaments take a long time to heal following an injury because their blood supply is relatively poor.

Loose connective tissue and fibrous connective tissue, which bind and support body parts, differ according to the type and the abundance of fibers in the matrix.

Cartilage Is Flexible

In **cartilage,** the cells lie in small chambers called lacunae (sing., **lacuna**), separated by a matrix that is solid yet flexible. Unfortunately, because this tissue lacks a direct blood supply, it heals very slowly. There are three types of cartilage, distinguished by the type of fiber in the matrix.

Hyaline cartilage (Fig. 3.4*c*), the most common type of cartilage, contains only very fine collagen fibers. The matrix has a white, translucent appearance. Hyaline cartilage is found in the nose and at the ends of the long bones and the ribs, and it forms rings in the walls of respiratory passages. The fetal skeleton also is made of this type of cartilage. Later, the cartilaginous fetal skeleton is replaced by bone.

Elastic cartilage has more elastic fibers than hyaline cartilage. For this reason, it is more flexible and is found, for example, in the framework of the outer ear.

Fibrocartilage has a matrix containing strong collagen fibers. Fibrocartilage is found in structures that withstand tension and pressure, such as the pads between the vertebrae in the backbone and the wedges found in the knee joint.

Bone Is Rigid

Bone is the most rigid connective tissue. It consists of an extremely hard matrix of inorganic salts, chiefly calcium salts, deposited around protein fibers, especially collagen fibers. The inorganic salts give bone rigidity, and the protein fibers provide elasticity and strength, much as steel rods do in reinforced concrete.

Compact bone makes up the shaft of a long bone (Fig. 3.4*d*). It consists of cylindrical structural units called osteons (Haversian systems). The central canal of each osteon is surrounded by rings of hard matrix. Bone cells, called *osteocytes*, are located in spaces called lacunae between the rings of matrix. Blood vessels in the central canal carry nutrients that allow bone to renew itself. The nutrients can reach all of the cells because *canaliculi* (minute canals) containing thin processes of the osteocytes connect them with one another and with the central canals.

The ends of a long bone contain spongy bone, which has an entirely different structure. **Spongy bone** contains numerous bony bars and plates, separated by irregular spaces. Although lighter than compact bone, spongy bone still is designed for strength. Just as braces are used for support in buildings, the solid portions of spongy bone follow lines of stress.

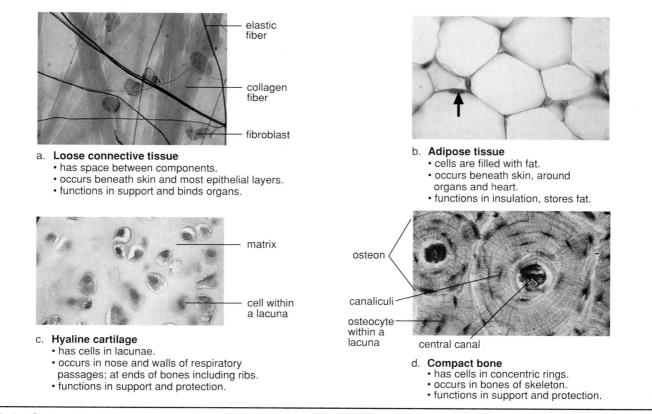

a. **Loose connective tissue**
- has space between components.
- occurs beneath skin and most epithelial layers.
- functions in support and binds organs.

b. **Adipose tissue**
- cells are filled with fat.
- occurs beneath skin, around organs and heart.
- functions in insulation, stores fat.

c. **Hyaline cartilage**
- has cells in lacunae.
- occurs in nose and walls of respiratory passages; at ends of bones including ribs.
- functions in support and protection.

d. **Compact bone**
- has cells in concentric rings.
- occurs in bones of skeleton.
- functions in support and protection.

Figure 3.4   Connective tissue examples.
a. In loose connective tissue, cells called fibroblasts are separated by a jellylike matrix, which contains both collagen and elastic fibers. **b.** Adipose tissue cells have nuclei (arrow) pushed to one side because the cells are filled with fat. **c.** In hyaline cartilage, the flexible matrix has a white, translucent appearance. **d.** In compact bone, the hard matrix contains calcium salts. Concentric rings of osteocytes in lacunae form an elongated cylinder called an osteon (Haversian system), which has a central canal that contains blood vessels and nerve fibers.

Blood Has a Liquid Matrix

The functions of blood include transporting molecules, regulating the tissues, and protecting the body. Blood transports nutrients and oxygen to cells and removes carbon dioxide and other wastes. It helps distribute heat and also plays a role in fluid, ion, and pH balance. Various components of blood, as discussed below, help protect us from disease, and its ability to clot prevents fluid loss.

If blood is transferred from a person's vein to a test tube and prevented from clotting, it separates into two layers (Fig. 3.5). The upper liquid layer, called plasma, represents about 55% of the volume of whole blood and contains a variety of inorganic and organic substances dissolved or suspended in water (Table 3.1). The lower layer consists of red blood cells (*erythrocytes*), white blood cells (*leukocytes*), and blood platelets (*thrombocytes*). Collectively, these are called the formed elements and represent about 45% of the volume of whole blood. Formed elements are manufactured in the red bone marrow of the skull, ribs, vertebrae, and ends of long bones.

The **red blood cells** are small, biconcave, disk-shaped cells without nuclei. The presence of the red pigment hemoglobin makes the cells red, and in turn, makes the blood red. Hemoglobin is composed of four units; each is composed of the protein globin and a complex iron-containing structure called heme. The iron forms a loose association with oxygen, and in this way red blood cells transport oxygen.

White blood cells may be distinguished from red blood cells by the fact that they are usually larger, have a nucleus, and without staining would appear to be translucent. White blood cells characteristically appear bluish because they have been stained that color. White blood cells, which fight infection, function primarily in two ways. Some white blood cells are phagocytic and engulf infectious pathogens, while other white blood cells produce antibodies, molecules that combine with foreign substances to inactivate them.

Platelets are not complete cells; rather, they are fragments of giant cells present only in bone marrow. When a blood vessel is damaged, platelets form a plug that seals the vessel and along with injured tissues release molecules that help the clotting process.

Blood is unlike other types of connective tissue in that the matrix (i.e., plasma) is not made by the cells. Some people do not classify blood as connective tissue; instead, they suggest a separate tissue category for blood called vascular tissue.

Blood is a connective tissue in which the matrix is plasma.

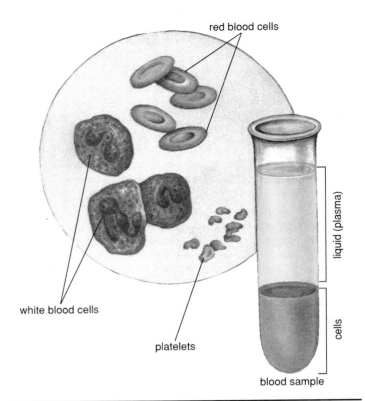

Figure 3.5   Blood, a liquid tissue.
Blood is classified as connective tissue because the cells and platelets are separated by a matrix called plasma. Blood services the body's tissues by transporting nutrients, gases, and wastes. The red blood cells transport oxygen, the platelets assist blood clotting, and the white blood cells help fight infections.

TABLE 3.1	
Blood Plasma	
Water (92% of Total)	
Solutes (8% of Total)	
Inorganic ions (salts)	Na^+, Ca^{++}, K^+, Mg^{++}; Cl^-, HCO_3^-, $HPO_4^=$, $SO_4^=$
Gases	O_2, CO_2
Plasma proteins	Albumin, globulins, fibrinogen
Organic nutrients	Glucose, fats, phospholipids, amino acids, etc.
Nitrogenous waste products	Urea, ammonia, uric acid
Regulatory substances	Hormones, enzymes

Muscular Tissue Contracts

Muscular (contractile) **tissue** is composed of cells that are called *muscle fibers*. Muscle fibers contain actin filaments and myosin filaments, whose interaction accounts for the movements we associate with animals. There are three types of vertebrate muscles: *skeletal, smooth,* and *cardiac.*

Skeletal muscle, also called voluntary muscle (Fig. 3.6*a*), is attached by tendons to the bones of the skeleton, and when it contracts, body parts move. Contraction of skeletal muscle is under voluntary control and occurs faster than the other muscle types. Skeletal muscle fibers are cylindrical and quite long—sometimes they run the length of the muscle. They arise during development when several cells fuse, and the result is one fiber with multiple nuclei. The nuclei are located at the periphery of the cell, just inside the plasma membrane. The fibers have alternating light and dark bands that give them a **striated** appearance. These bands are due to the placement of actin filaments and myosin filaments in the cell.

Smooth muscle is so named because the cells lack striations. The spindle-shaped cells form layers in which the thick middle portion of one cell is opposite the thin ends of adjacent cells. Consequently, the nuclei form an irregular pattern in the tissue (Fig. 3.6*b*). Smooth muscle is not under voluntary control and therefore is said to be involuntary. Smooth muscle, found in the walls of viscera (intestine, stomach, and other internal organs) and blood vessels, contracts more slowly than skeletal muscle but can remain contracted for a longer time. When the smooth muscle of the intestine contracts, food moves along its lumen (central cavity). When the smooth muscle of the blood vessels contracts, blood vessels constrict, helping to raise blood pressure.

Cardiac muscle (Fig. 3.6*c*) is found only in the walls of the heart. Its contraction pumps blood and accounts for the heartbeat. Cardiac muscle combines features of both smooth muscle and skeletal muscle. It has striations like skeletal muscle, but the contraction of the heart is involuntary for the most part. Cardiac muscle cells also differ from skeletal muscle cells in that they have a single, centrally placed nucleus. The cells are branched and seemingly fused one with the other, and the heart appears to be composed of one large interconnecting mass of muscle cells. Actually, cardiac muscle cells are separate and individual, but they are bound end to end at *intercalated disks,* areas where folded plasma membranes between two cells contain desmosomes and gap junctions.

All muscular tissue contains actin filaments and myosin filaments; these form a striated pattern in skeletal and cardiac muscle, but not in smooth muscle.

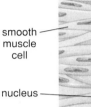
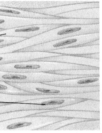

striation
nucleus

a.

Skeletal muscle
• has striated cells with multiple nuclei.
• usually attached to skeleton.
• functions in voluntary movement.

smooth muscle cell

nucleus

b.

Smooth muscle
• has spindle-shaped cells, each with a single nucleus.
• occurs in walls of internal organs.
• functions in movement of substances in lumens of body.

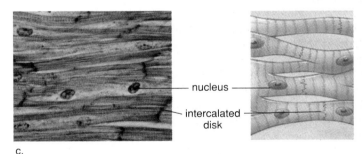

nucleus

intercalated disk

c.

Cardiac muscle
• has branching striated cells, each with a single nucleus.
• occurs in the wall of the heart.
• functions in the pumping of blood.

Figure 3.6 Muscular tissue.
a. Skeletal muscle is voluntary and striated. **b.** Smooth muscle is involuntary and nonstriated. **c.** Cardiac muscle is involuntary and striated. Cardiac muscle cells branch and fit together at intercalated disks.

Nervous Tissue Conducts Impulses

The brain and the spinal cord contain conducting cells termed neurons. A **neuron** is a specialized cell that has three parts: dendrites, cell body, and an axon (Fig. 3.7). A dendrite is a process that conducts signals toward the cell body. The cell body contains the nucleus. An axon is a process that typically conducts nerve impulses away from the cell body.

Axons and dendrites are also called *neuron fibers.* Fibers can be quite long, and outside the brain and the spinal cord, long fibers, bound by connective tissue, form **nerves.** Nerves conduct impulses from receptors to the spinal cord and the brain, where the phenomenon called sensation occurs. They also conduct nerve impulses away from the spinal cord and the brain to the muscles and glands, causing them to contract and secrete, respectively.

In addition to neurons, nervous tissue contains **neuroglial cells.** These cells maintain the tissue by supporting and protecting neurons. They also provide nutrients to neurons and help to keep the tissue free of debris.

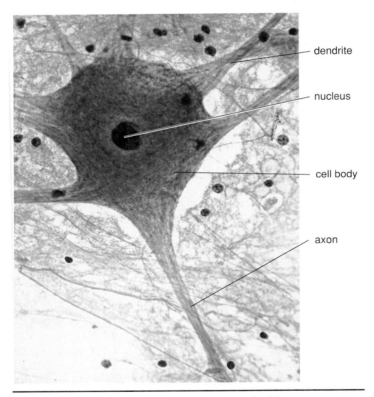

— dendrite

— nucleus

— cell body

— axon

Figure 3.7　Photo of a neuron surrounded by neuroglial cells.
Conduction of the nerve impulse is dependent on neurons, each of which has the three parts indicated: a dendrite conducts signals toward the cell body; an axon typically conducts nerve impulses away from the cell body; and the cell body contains a nucleus.

3.2　Skin as an Organ

The skin can be used as an example of an organ, a structure that is composed of two or more tissues (Fig. 3.8). Skin covers the body, protecting underlying parts from physical trauma, microbial invasion, and water loss. Skin helps to regulate body temperature, and because it contains receptors, skin also helps us to be aware of our surroundings and to communicate with others by touch.

Skin Has Layers

The **epidermis** of skin is made up of stratified squamous epithelium. New cells derived from basal cells become flattened and hardened as they push to the surface. Hardening occurs because the cells produce keratin, a waterproof protein. *Dandruff* occurs when the rate of keratinization is two or three times the normal rate. A thick layer of dead keratinized cells, arranged in spiral and concentric patterns, form fingerprints and footprints. Specialized cells in the epidermis called *melanocytes* produce melanin, the pigment responsible for skin color.

The **dermis** is a layer of fibrous connective tissue beneath the epidermis. The number of collagen and elastic fibers decreases with exposure to the sun, and the skin becomes less supple and is prone to wrinkling. *Hair follicles* begin in the dermis and continue through the epidermis where the hair shaft extends beyond the skin. Epidermal cells form the root of hair, and their division causes a hair to grow. The cells become keratinized and dead as they are pushed farther from the root. Each hair follicle has one or more *oil (sebaceous) glands,* which secrete sebum, an oily substance that lubricates the hair within the follicle and the skin itself. If the sebaceous glands fail to discharge, the secretions collect and form "whiteheads" or "blackheads." The color of blackheads is due to oxidized sebum.

Contraction of the *arrector pili muscles* attached to hair follicles cause the hairs to "stand on end" and cause goose bumps to develop. *Sweat (sudoriferous) glands* are quite numerous and are present in all regions of skin. A sweat gland begins as a coiled tubule within the dermis, but then it straightens out near its opening. Some sweat glands open into hair follicles, but most open onto the surface of the skin.

Receptors are specialized nerve endings in the dermis that respond to external stimuli. There are receptors for touch, pressure, pain, and temperature. The fingertips contain the most touch receptors, and these add to our ability to use our fingers for delicate tasks. The dermis also contains nerve fibers and blood vessels. When blood rushes into these vessels, a person blushes, and when blood is minimal in them, a person turns "blue."

Skin has two layers: the epidermis is the protective layer, and the dermis contains many structures that account for the other functions of the skin.

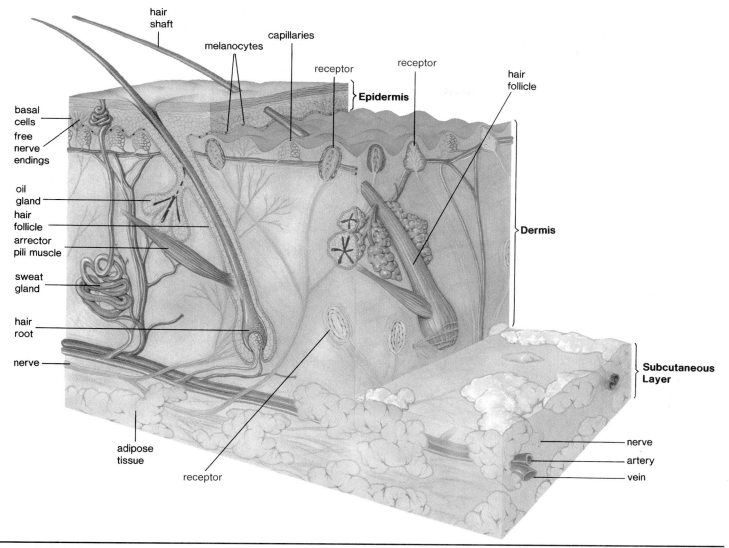

Figure 3.8 Human skin anatomy.
Skin consists of two tissue layers, the epidermis and dermis. A subcutaneous layer lies below the dermis.

The **subcutaneous layer,** which lies below the dermis, is composed of loose connective tissue, including adipose tissue, which stores fat. Fat represents stored energy that can be called upon when necessary. Adipose tissue helps to thermally insulate the body from either gaining heat from the outside or losing heat from the inside. A well-developed subcutaneous layer gives the body a rounded appearance and provides protective padding against external assaults. Excessive development of the subcutaneous layer accompanies obesity.

Skin Cancer on the Increase

Malignant melanoma arises from melanocytes in the skin. A melanoma is a darkly pigmented spot that resembles a nonmalignant mole. Individuals with light skin who burn easily seem to be especially at risk for this type of cancer. Activation of the immune system is the latest therapy for malignant melanoma, a cancer that can lead to death.

Two common types of skin cancer, *basal cell carcinoma* and *squamous cell carcinoma,* are likely to occur in all persons exposed to sunlight. Ultraviolet radiation causes epidermal cells to become cancerous. Precancerous dark patches of skin become rough and scaly, with a reddish base. In basal cell carcinoma, epidermal cells invade the dermis and form ulcers. Both types of cancer can usually be surgically removed.

In recent years, there has been a great increase in the number of persons with skin cancer, and physicians believe this is due to sunbathing or even to the use of tanning machines. These professionals strongly recommend that everyone stay out of the sun and refrain from using tanning beds. If persons must be in the sun, sunscreens should be used. A lotion with a sun protection factor (SPF) of 15 means that 15 minutes in the sun with protection is equivalent to one minute without protection. Even higher SPF strengths are available.

Ecology Focus

Ozone Depletion Threatens the Biosphere

The earth's atmosphere is divided into layers. The troposphere envelops us as we go about our day-to-day lives. When ozone (O_3) is present in the troposphere (called ground-level ozone), it is considered a pollutant because it adversely affects a plant's ability to grow and our ability to breathe oxygen (O_2). In the stratosphere, some 50 kilometers above the earth, ozone forms a shield that absorbs much of the ultraviolet (UV) rays of the sun so that less rays strike the earth.

UV radiation causes mutations that can lead to skin cancer and can make the lens of the eyes develop cataracts. It also is believed to adversely affect the immune system and our ability to resist infectious diseases. Crop and tree growth is impaired, and UV radiation also kills off small plants (phytoplankton) and tiny shrimplike animals (krill) that sustain oceanic life. Without an adequate ozone shield, our health and food sources are threatened.

Depletion of the ozone layer within the stratosphere in recent years is, therefore, of serious concern. It became apparent in the 1980s that some worldwide depletion of ozone had occurred and that there was a severe depletion of some 40–50% above the Antarctic every spring. Severe depletions of the ozone layer are commonly called "ozone holes." Detection devices now tell us that there is an ozone hole above the Arctic as well, and ozone holes could also occur within northern and southern latitudes, where many people live. Whether or not these holes develop depends on prevailing winds, weather conditions, and the type of particles in the atmosphere. A United Nations Environment Program report predicts a 26% rise in cataracts

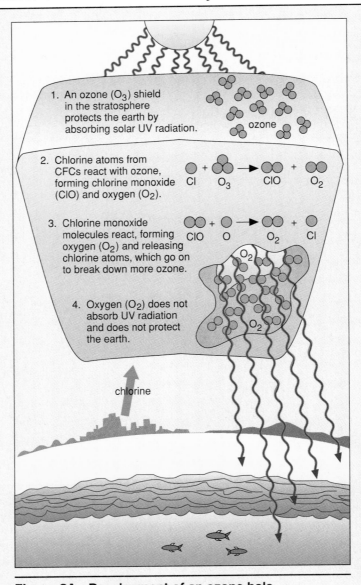

Figure 3A   Development of an ozone hole.
The development of an ozone hole due to the release of chlorine atoms from CFCs.

1. An ozone (O_3) shield in the stratosphere protects the earth by absorbing solar UV radiation.

ozone

2. Chlorine atoms from CFCs react with ozone, forming chlorine monoxide (ClO) and oxygen (O_2).

$Cl + O_3 \rightarrow ClO + O_2$

3. Chlorine monoxide molecules react, forming oxygen (O_2) and releasing chlorine atoms, which go on to break down more ozone.

$ClO + O \rightarrow O_2 + Cl$

4. Oxygen (O_2) does not absorb UV radiation and does not protect the earth.

chlorine

and nonmelanoma skin cancers for every 10% drop in the ozone level. A 26% increase translates into 1.75 million additional cases of cataracts and 300,000 more nonmelanoma skin cancers every year, worldwide.

The cause of ozone depletion can be traced to the release of chlorine atoms (Cl) into the stratosphere (Fig. 3A). Chlorine atoms combine with ozone and strip away the oxygen atoms, one by one. One atom of chlorine can destroy up to 100,000 molecules of ozone before settling to the earth's surface as chloride years later. These chlorine atoms come from the breakdown of chlorofluorocarbons (CFCs), chemicals much in use by humans. The best known CFC is Freon, a heat transfer agent found in refrigerators and air conditioners. CFCs are also used as cleaning agents and foaming agents during the production of styrofoam found in coffee cups, egg cartons, insulation, and paddings. Formerly, CFCs were used as propellants in spray cans, but this application is now banned in the United States and several European countries.

Most countries of the world have agreed to stop using CFCs by the year 2000, but the United States halted production in 1995. Scientists are now searching for CFC substitutes that will not release chlorine atoms (nor bromine atoms) to harm the ozone shield.

3.3 Body Cavities and Their Membranes

The internal organs are located within specific body cavities (Fig. 3.9). During human development, there is a large ventral cavity called a **coelom,** which becomes divided into the thoracic (chest) and abdominal cavities. Membranes divide the thoracic cavity into the pleural cavities, containing the right and left lungs, and the pericardial cavity, containing the heart. The thoracic cavity is separated from the abdominal cavity by a horizontal muscle called the diaphragm. The stomach, liver, spleen, gallbladder, and most of the small and large intestines are in the upper portion of the abdominal cavity. The lower portion contains the rectum, the urinary bladder, the internal reproductive organs, and the rest of the large intestine. Males have an external extension of the abdominal wall, called the scrotum, containing the testes.

The dorsal cavity also has two parts: the cranial cavity within the skull contains the brain; and the vertebral column, formed by the vertebrae, contains the spinal cord.

Membranes Line and Cover

In this context, we are using the term *membrane* to refer to a thin lining or covering composed of an epithelium overlying a loose connective tissue layer. Membranes line the cavities of the body and internal spaces of organs and tubes that open to the outside.

Mucous membranes line the tubes of the digestive, respiratory, urinary, and reproductive systems. The epithelium of this membrane contains goblet cells that secrete mucus. This mucus ordinarily protects the body from invasion by bacteria and viruses; hence, more mucus is secreted and expelled when a person has a cold and has to blow her/his nose. In addition, mucus usually protects the walls of the stomach and small intestine from digestive juices, but this protection breaks down when a person develops an ulcer.

Serous membranes line the thoracic and abdominal cavities and the organs that they contain. They secrete a watery fluid that keeps the membranes lubricated. Serous membranes support the internal organs and compartmentalize the large thoracic and abdominal cavities. This helps to hinder the spread of any infection.

The pleural membranes are serous membranes that line the pleural cavity and lungs. *Pleurisy* is a well-known infection of these membranes. The peritoneum lines the abdominal cavity and its organs. In between the organs, there is a double layer of peritoneum called mesentery. *Peritonitis* is a life-threatening infection of the peritoneum. Peritonitis is likely if an inflamed appendix bursts before it is removed.

Synovial membranes line freely movable joint cavities. They secrete synovial fluid into the joint cavity; this fluid lubricates the ends of the bones so that they can move freely. In *rheumatoid arthritis,* the synovial membrane becomes inflamed and grows thicker, restricting movement.

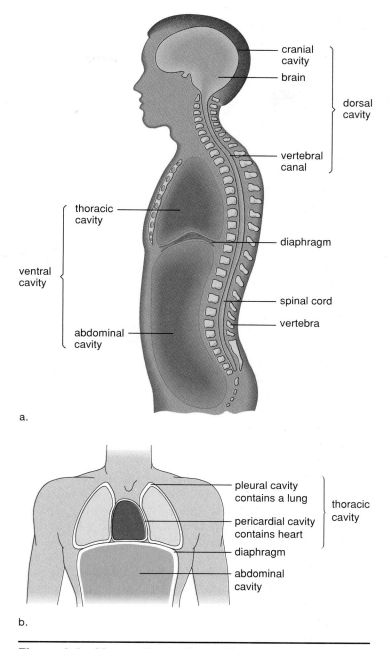

a.

b.

Figure 3.9   Mammalian body cavities.
a. Side view. There is a dorsal cavity, which contains the cranial cavity and the vertebral canal. The brain is in the cranial cavity, and the spinal cord is in the vertebral canal. There is a well-developed ventral cavity, which is divided by the diaphragm into the thoracic cavity and the abdominal cavity. The heart and lungs are in the thoracic cavity, and most other internal organs are in the abdominal cavity. **b.** Frontal view of the thoracic cavity.

The *meninges* are membranes found within the dorsal cavity. They are composed only of connective tissue and serve as a protective covering for the brain and spinal cord. *Meningitis* is a life-threatening infection of the meninges.

3.4   Studying Organ Systems

The body contains a number of systems that work together to maintain homeostasis. The next page introduces a box that will be used at the end of each organ system chapter. In this chapter, the box reviews the general functions of the body's organ systems. The corresponding boxes in other chapters will show how a particular organ system interacts with all the other systems.

Maintenance of the Body

The internal environment of the body consists of the blood within the blood vessels and the *tissue fluid* that surrounds the cells. Five systems add substances to and remove substances from the blood: the circulatory, lymphatic, respiratory, digestive, and urinary systems.

The *circulatory system* consists of the heart and blood vessels that carry blood through the body. *Blood* transports nutrients and oxygen to the cells, and removes their waste molecules that are to be excreted from the body. Blood also contains cells produced by the lymphatic system.

The *lymphatic system* consists of lymphatic vessels, lymph fluid, lymph nodes, and other lymphoid organs. This system protects the body from disease by purifying lymph and supporting lymphocytes, the white blood cells that produce antibodies. Lymphatic vessels absorb fat from the digestive system and collect excess tissue fluid, which is returned to the blood circulatory system.

The *respiratory system* consists of the lungs and the tubes that take air to and from the lungs. The respiratory system brings oxygen into the lungs and takes carbon dioxide out of the lungs.

The *digestive system* consists of the mouth, esophagus, stomach, small intestine, and large intestine (colon) along with the associated organs: teeth, tongue, salivary glands, liver, gallbladder, and pancreas. This system receives food and digests it into nutrient molecules, which can enter the cells of the body.

The *urinary system* contains the kidneys and the urinary bladder. This system rids the body of nitrogenous wastes and helps regulate the fluid level and chemical content of the blood.

> The circulatory system, lymphatic system, respiratory system, digestive system, and the urinary system all perform specific processing and transporting functions to maintain the normal conditions of the body.

Integumentary System

The skin is sometimes called the *integumentary system* because it contains accessory organs such as hair, nails, sweat glands, and sebaceous glands. The skin provides external support and protects underlying tissues, helps regulate body temperature, contains receptors, and even synthesizes certain chemicals that affect the rest of the body.

Support and Movement

The skeletal system and the muscular system give the body support and are involved in the ability of the body and its parts to move.

The *skeletal system*, consisting of the bones of the skeleton, protects body parts. For example, the skull forms a protective encasement for the brain, as does the rib cage for the heart and lungs. The skeleton, as a whole, serves as a place of attachment for the skeletal muscles. Contraction of muscles in the *muscular system* accounts for movement of the body and also body parts.

> The skeletal system and the muscular system support the body and permit movement.

Integration and Coordination

The *nervous system* consists of the brain, spinal cord, and associated nerves. The nerves conduct nerve impulses from receptors to the brain and spinal cord. They also conduct nerve impulses from the brain and spinal cord to the muscles and glands, allowing us to respond to both external and internal stimuli.

The *endocrine system* consists of the hormonal glands that secrete chemicals that serve as messengers between body parts. Both the nervous and endocrine systems help maintain a relatively constant internal environment by coordinating and regulating the functions of the body's other systems. The endocrine system also helps maintain the proper functioning of male and female reproductive organs.

> The nervous and endocrine systems coordinate and regulate the activities of the body's other systems.

Reproduction and Development

The *reproductive system* involves different organs in the male and female. The male reproductive system consists of the testes, and other glands, and various ducts that conduct semen to and through the penis. The female reproductive system consists of the ovaries, oviducts, uterus, vagina, and external genitalia.

> The reproductive system in males and in females carries out those functions that give humans the ability to reproduce.

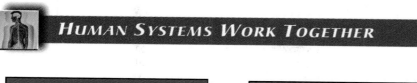

HUMAN SYSTEMS WORK TOGETHER

Integumentary System

External support and protection of body.

Respiratory System

Gaseous exchange between external environment and blood.

Circulatory System

Transport of nutrients to body cells and transport of wastes away from cells.

Skeletal System

Internal support and protection; body movement; production of blood cells.

Lymphatic System/Immunity

Immunity; absorption of fats; drainage of tissue fluid.

Muscular System

Body movement; production of body heat.

Digestive System

Breakdown and absorption of food materials.

Nervous System

Regulation of all body activities; learning and memory.

Urinary System

Maintenance of volume and chemical composition of blood.

Endocrine System

Secretion of hormones for chemical regulation of all tissues.

Reproductive System

Production of sperm and egg; transfer of sperm to female system where development occurs.

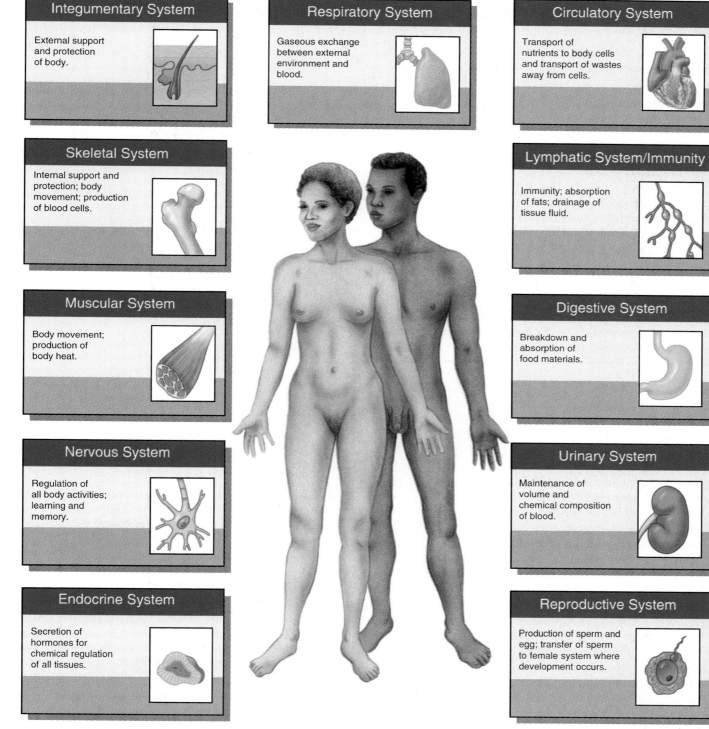

Dynamic Human

3.5   Homeostasis

Homeostasis means that the internal environment remains within normal limits or values, regardless of the conditions in the external environment. In humans, for example:

1. The blood glucose concentration remains at about 100 mg/100 ml.
2. The pH of blood is always near 7.4.
3. Blood pressure in the brachial artery averages near 120/80.
4. Body temperature averages around 37°C (98.6°F).

The ability of the body to keep the internal environment within a certain range allows humans to live in a variety of habitats, such as the Arctic regions, the deserts, or the tropics.

This internal environment consists of tissue fluid, which bathes all the cells of the body. Tissue fluid is refreshed when molecules such as oxygen and nutrients exit blood and wastes enter blood (Fig. 3.10). Tissue fluid remains constant only as long as blood composition remains constant. Although we are accustomed to using the word *environment* to mean the external environment of the body, it is important to realize that it is the internal environment of tissues that is ultimately responsible for our health and well-being.

The internal environment of the body consists of tissue fluid, which bathes the cells.

Most systems of the body contribute to maintenance of a constant internal environment. The circulatory system conducts blood to and away from capillaries, the smallest of the blood vessels, whose thin walls permit exchanges to occur. At the capillaries, blood pressure aids the movement of water out of capillaries, and osmotic pressure aids the movement of water into capillaries. Blood pressure is created by the pumping of the heart, while osmotic pressure is maintained by the protein content of plasma. The formed elements also contribute to homeostasis. Red blood cells transport oxygen and participate in the transport of carbon dioxide. White blood cells fight infection, and platelets participate in the clotting process. The lymphatic system is accessory to the circulatory system. Lymphatic capillaries collect excess tissue fluid, and this is returned via lymphatic veins to the circulatory veins.

The digestive system takes in and digests food, providing nutrient molecules that enter blood and replace the nutrients that are constantly being used by the body cells. The respiratory system adds oxygen to and removes carbon dioxide from the blood. The chief regulators of blood composition are the liver and the kidneys. They monitor the chemical composition of plasma (see Table 3.1) and alter it as required. Immediately after glucose enters the blood, it can be removed by the liver for storage as glycogen. Later,

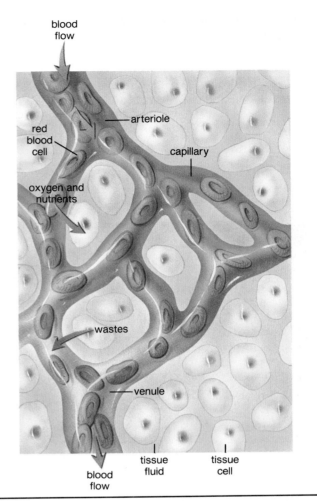

Figure 3.10   Tissue fluid composition.
Cells are surrounded by tissue fluid, which is continually refreshed because oxygen and nutrient molecules constantly exit, and waste molecules continually enter, the bloodstream as shown.

the glycogen can be broken down to replace the glucose used by the body cells; in this way, the glucose composition of blood remains constant. The hormone insulin, secreted by the pancreas, regulates glycogen storage. The liver also removes toxic chemicals, such as ingested alcohol and other drugs. The liver makes urea, a nitrogenous end product of protein metabolism. Urea and other metabolic waste molecules are excreted by the kidneys. Urine formation by the kidneys is extremely critical to the body, not only because it rids the body of unwanted substances, but also because it offers an opportunity to carefully regulate blood volume, salt balance, and the pH of the blood.

Most systems of the body contribute to homeostasis, that is, maintaining the relative constancy of the internal environment.

How the Nervous and Endocrine Systems Coordinate

The nervous system and endocrine system are ultimately in control of homeostasis. The endocrine system is slower acting than the nervous system, which rapidly brings about a particular response.

Previously, we mentioned that the liver is involved in homeostasis because it stores glucose as glycogen. But actually there is a hormone produced by an endocrine gland that regulates storage of glucose by the liver. When the glucose content of the blood rises after eating, the pancreas secretes insulin, a hormone that causes the liver to store glucose as glycogen. Now the glucose level falls, and the pancreas no longer secretes insulin. This is called control by **negative feedback** because the response (low blood glucose) negates the original stimulus (high blood glucose). In some instances, an endocrine gland is sensitive to the blood level of a hormone whose concentration it regulates. For example, the pituitary gland produces a hormone that stimulates the thyroid gland to secrete its hormone. When the blood level of this hormone rises to a certain level, the pituitary gland no longer stimulates the thyroid gland.

A negative feedback system can regulate itself because it has a sensing device, which is a mechanism by which the system detects a particular condition. For example, consider the feedback mechanism that functions to maintain the room temperature of a house. In this feedback system, the thermostat is a device that is sensitive to room temperature. The furnace produces heat, and when the temperature of a room reaches a certain point, the thermostat signals a switching device that turns the furnace off. On the other hand, when the temperature falls below that indicated on the thermostat, it signals the switching device, which turns the furnace on again.

Figure 3.11*a* shows that in the body there are receptors that fulfill the role of sensing devices. When a receptor is stimulated, it signals a regulator center that then turns on an effector. The effector brings about a response that negates the original conditions that stimulated the receptor. In the absence of suitable stimulation, the receptor no longer signals the regulator center.

Figure 3.11*b* gives an actual example involving the nervous system. When blood pressure rises, receptors signal a regulator center, which then sends out nerve impulses to the arterial walls, causing them to relax, and the blood pressure now falls. Therefore, the receptors are no longer stimulated, and the system shuts down. Notice that negative feedback control results in a fluctuation above and below a mean. Thus, there is a dynamic equilibrium of the internal environment.

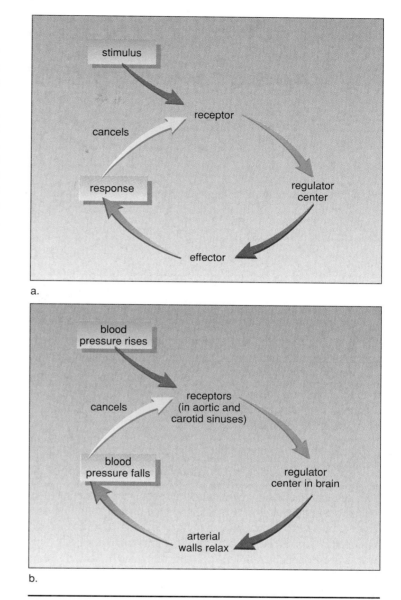

a.

b.

Figure 3.11 Negative feedback control.
a. A stimulus causes a receptor to signal a regulator center in the brain. The regulator center signals effectors to respond, and response cancels the stimulus. **b.** For example, when blood pressure rises, special receptors in blood vessels signal a particular center in the brain. The brain signals the arteries to relax, and blood pressure falls.

Homeostasis of internal conditions is a self-regulatory mechanism that usually results in slight fluctuations above and below a mean.

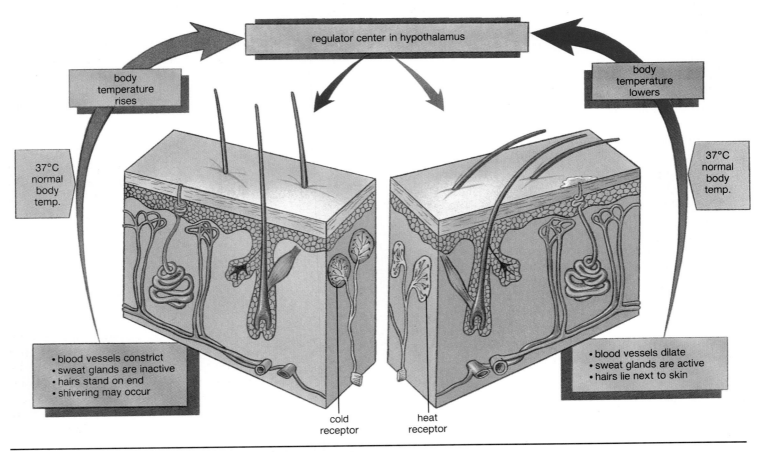

Figure 3.12   **Homeostasis and control of body temperature.**
When the body temperature rises above normal, the regulator center directs the blood vessels to dilate and the sweat glands to be active so that the body temperature returns to 37°C. When the body temperature lowers below normal, the regulator center directs the blood vessels to constrict, causing the hairs to stand on end, and shivering to occur. The body temperature now returns to 37°C.

How Body Temperature Is Regulated

Although the skin senses a change in external temperature, the regulator center for body temperature, located in the hypothalamus, is sensitive only to the temperature of the blood. When the body temperature falls below normal, the regulator center directs (via nerve impulses) the blood vessels of the skin to constrict (Fig. 3.12). This conserves heat. Also, the arrector pili muscles pull hairs erect, and a layer of insulating air is trapped next to the skin. If body temperature falls even lower, the regulator center sends nerve impulses to the skeletal muscles, and shivering occurs. Shivering generates heat, and gradually body temperature rises to 37°C. If the body temperature is normal, the regulator center is not active.

When the body temperature is higher than normal, the regulator center is activated and then directs the blood vessels of the skin to dilate. This allows more blood to flow near the surface of the body, where heat can be lost to the environment. In addition, the regulator center activates the sweat glands, and the evaporation of sweat also helps to lower body temperature. Gradually, body temperature decreases to 37°C.

The temperature of the human body is maintained at about 37°C due to the action of a regulator center in the hypothalamus.

SUMMARY

3.1 Types of Tissues

Human tissues are categorized into four groups. Epithelial tissue covers the body and lines its cavities. The different types of epithelial tissue (squamous, cuboidal, and columnar) can be stratified and have cilia or microvilli. Also, columnar cells can be pseudostratified. Epithelial cells sometimes form glands that secrete either into ducts or into blood.

Connective tissues, in which cells are separated by a matrix, often bind body parts together. Loose connective tissue has both white and yellow fibers and may also have fat (adipose) cells. Fibrous connective tissue, such as tendons and ligaments, contains closely packed collagen fibers. Both cartilage and bone have cells within lacunae, but the matrix for cartilage is more flexible than that for bone, which contains calcium salts. In bone, the lacunae lie in concentric circles within an osteon (or Haversian system) about a central canal. Blood is a connective tissue in which the matrix is a liquid called plasma.

Muscular tissue is of three types. Both skeletal and cardiac muscle are striated; both cardiac and smooth muscle are involuntary. Skeletal muscle is found in muscles attached to bones, and smooth muscle is found in internal organs. Cardiac muscle makes up the heart.

Nervous tissue has one main type of conducting cell, the neuron, and several types of neuroglial cells. Each neuron has dendrites, a cell body, and an axon. The brain and spinal cord contain complete neurons, while the nerves contain only neuron fibers. Neurons and their fibers are specialized to conduct nerve impulses.

3.2 Skin as an Organ

Tissues are joined together to form organs, each one having a specific function. Skin is a two-layered organ that waterproofs and protects the body. The epidermis contains a germinal layer that produces new epithelial cells that become keratinized as they move toward the surface. The dermis, a largely fibrous connective tissue, contains epidermally derived glands and hair follicles, nerve endings, and blood vessels. Receptors for touch, pressure, temperature, and pain are present. Sweat glands and blood vessels help control body temperature. A subcutaneous layer, which is made up of loose connective tissue containing adipose cells, lies beneath the skin.

3.3 Body Cavities and Their Membranes

The internal organs occur within cavities; the thoracic cavity contains the heart and lungs; the abdominal cavity contains organs of the digestive, urinary, and reproductive systems, among others. Membranes line body cavities and internal spaces of organs. As an example, mucous membrane lines the tubes of the digestive system; serous membrane lines the thoracic and abdominal cavities and covers the organs they contain.

3.4 Studying Organ Systems

The skin is sometimes considered to be the integumentary system. The circulatory, lymphatic, respiratory, digestive, and urinary systems perform processing and transporting functions that maintain the normal conditions of the body. The skeletal and muscular systems support the body and permit movement. The nervous and endocrine systems coordinate the other systems, and the reproductive system allows humans to make more of their own kind.

3.5 Homeostasis

Homeostasis is the dynamic equilibrium of the internal environment. All organ systems contribute to the constancy of tissue fluid and blood. Special contributions are made by the liver, which keeps blood glucose constant, and the kidneys, which regulate the pH. The nervous and hormonal systems regulate the other body systems. Both of these are controlled by a feedback mechanism, which results in fluctuation above and below the desired levels. Body temperature is regulated by a center in the hypothalamus.

STUDYING THE CONCEPTS

1. Name the four major types of tissues. 60
2. Name the different kinds of epithelial tissue, and give a location and function for each. 60
3. What are the functions of connective tissue? Name the different kinds, and give a location for each. 62–64
4. What are the functions of muscular tissue? Name the different kinds, and give a location for each. 65
5. Nervous tissue contains what type of cell? Which organs in the body are made up of nervous tissue? 66
6. Describe the structure of skin, and state at least two functions of this organ. 66–67
7. Distinguish between plasma membrane and body membrane. 69
8. In what cavities are the major organs located? 69
9. What is homeostasis, and how is it achieved in the human body? 72–73
10. Specifically describe how body temperature is maintained at about 37°C. 74

APPLYING YOUR KNOWLEDGE

Concepts

1. The human fetal skeleton is composed of cartilage, much of which is later converted to bone. What advantage(s) is there to the fetal skeleton being composed of cartilage?

2. What is the advantage of having some structures composed of skeletal muscle and others composed of smooth muscle?

3. Some people wish to remove hair from certain areas of their bodies. They shave, they pull out the hair, and they use electrolysis; but only electrolysis is a permanent method of hair removal. Please explain.

4. "Goose bumps" often occur when you are cold. Explain what is happening and what purpose, if any, is served by this phenomenon.

Bioethical Issue

One January day in 1983, Nancy Cruzan lost control of her car on a quiet Missouri road. By the time a medical team arrived, Nancy had probably been without oxygen for 20 minutes. Already, she had suffered permanent brain damage.

Nancy spent the next seven years in a so-called "persistent vegetative state," which means she was essentially unconscious. A feeding tube kept her alive. Doctors had no hope that she would ever recover.

In the first U.S. Supreme Court case involving a right to die, Nancy's parents asked the Court to allow doctors to disconnect her life-support system. The Cruzans argued that Nancy would not want to live life this way. In 1990, the family won the right to let Nancy die.

The Cruzan case sparked a firestorm of right-to-die discussions. Should a family be allowed to disconnect the life-support system of a patient in a coma or vegetative state? Should doctors be legally required to pursue life support? And, if a person writes a "living will"—a document that specifically requests the right to die under certain conditions—should everyone abide by it?

TESTING YOUR KNOWLEDGE

1. Most organs contain several different types of _____.

2. Kidney tubules are lined by cube-shaped cells called _____ epithelium.

3. Pseudostratified ciliated columnar epithelium contains cells that appear to be _____, have projections called _____, and are _____ in shape.

4. Both cartilage and blood are classified as _____ tissue.

5. Cardiac muscle is _____ but involuntary.

6. Nerve cells are called _____.

7. In skin, the _____ lies beneath the epidermis.

8. Mucous membrane contains _____ tissue overlying _____ tissue.

9. Outer skin cells are filled with _____, a waterproof protein that strengthens them.

10. The heart and lungs are located in the _____ cavity.

11. Homeostasis is maintenance of the relative _____ of the internal environment, that is, blood and _____ fluid.

12. Give the name, the location, and the function for each of these tissues.
    a. Type of epithelial tissue
    b. Type of muscular tissue
    c. Type of connective tissue

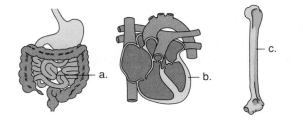

APPLYING TECHNOLOGY

Your study of human organization is supported by these available technologies:

Exploring the Internet

The Mader Home Page provides further resources for studying this chapter.

`http://www.mhhe.com/sciencemath/biology/mader/`

(Click on *Human Biology*.)

Dynamic Human CD-ROM

Anatomical Orientation. The planes of the body and directional terminology are reviewed; cross sections of various regions show the placement of organs.

SELECTED KEY TERMS

adipose tissue A connective tissue in which fat is stored. 62

blood Connective tissue, composed of cells separated by plasma, which transports substances in the cardiovascular system. 64

bone Connective tissue having a hard matrix of mineral salts deposited around protein fibers. 63

cardiac muscle Specialized type of muscle tissue found only in the heart. 65

cartilage Connective tissue in which the cells lie within lacunae separated by a flexible matrix. 63

cilium Hairlike projection used for locomotion by many unicellular organisms and that has various purposes in higher organisms. 60

coelom (SEE-lum) Embryonic body cavity lying between the digestive tract and body wall that is completely lined by mesoderm; in humans, the embryonic coelom becomes the thoracic and abdominal cavities. 69

columnar epithelium Pillar-shaped cells usually with the nuclei near the bottom of each cell and found lining the digestive tract, for example. 60

compact bone Hard bone consisting of osteons (Haversian systems) cemented together. 63

connective tissue A type of tissue characterized by cells separated by a matrix that often contains fibers. 62

cuboidal epithelium Cube-shaped cells found lining the kidney tubules, for example. 60

dermis Inner layer of skin that lies beneath the epidermis. 66

elastic cartilage Cartilage composed of elastic fibers allowing greater flexibility. 63

epidermis Outer layer of skin, composed of stratified squamous epithelium. 66

epithelial tissue Type of tissue that covers the external surface of the body and lines its cavities. 60

fibrocartilage Cartilage with a matrix of strong collagenous fibers. 63

gland Epithelial cell or group of epithelial cells that are specialized to secrete a substance. 60

homeostasis Maintenance of a dynamic equilibrium, such that temperature, blood pressure, and other body conditions remain within narrow limits. 72

hyaline cartilage Cartilage whose cells lie in lacunae separated by a white translucent matrix containing very fine collagen fibers. 63

ligament Fibrous connective tissue that joins bone to bone at a joint. 62

matrix Substance that fills the space between cells in connective tissues or inside organelles. 62

muscular (contractile) tissue Type of tissue that contains cells capable of contracting; skeletal muscles are attached to the skeleton, smooth muscle is found within walls of internal organs, and cardiac muscle makes up the heart. 65

negative feedback Self-regulatory mechanism that is activated by an imbalance and results in a fluctuation above and below a mean. 73

neuroglial cell One of several types of cells found in nervous tissue that support, protect, and nourish neurons. 66

neuron Nerve cell that characteristically has three parts: dendrites, cell body, and axon. 66

pathogen Disease-causing agent. 60

platelet Cell fragment that is necessary to blood clotting; thrombocyte. 64

red blood cell (erythrocyte) Cell that contains hemoglobin and carries oxygen from the lungs to the tissues in vertebrates. 64

skeletal muscle Contractile tissue that comprises the muscles attached to the skeleton; also called striated muscle. 65

smooth (visceral) muscle Contractile tissue that comprises the muscles found in the walls of internal organs. 65

spongy bone ;Porous bone found at the ends of long bones where blood cells are formed. 63

squamous epithelium Flat cells found lining the lungs and blood vessels, for example. 60

striated Having bands; cardiac and skeletal muscle are striated with light and dark bands. 65

subcutaneous layer Tissue layer that lies just beneath the skin and contains adipose tissue. 67

tendon Fibrous connective tissue that joins muscle to bone. 62

white blood cell (leukocyte) Cell of which there are several types, each having a specific function in protecting the body from invasion by foreign substances and organisms. 64

FURTHER READINGS FOR PART ONE

Applegate, E. J. 1995. *The anatomy and physiology learning system.* Philadelphia: W. B. Saunders Publishing. Designed for a one-semester introductory course, this text provides fundamental information for students with minimal science background.

Becker, W. M., and Deamer, D. W. 1996. *The world of the cell.* 3d ed. Redwood City, Calif.: Benjamin/Cummings Publishing. Presents an overview of cell biology.

Chapman, C. 1994. *Basic chemistry for biology.* Dubuque, Iowa: Wm. C. Brown Publishers. This introductory chemistry text is useful to beginning students.

Clemente, C. D. 1996. *Anatomy: A regional atlas of the human body.* 4th ed. Philadelphia: Lea and Febiger. This atlas contains both drawings and photographs of anatomical regions of the human body.

Enger, E. D., et al. 1994. *Foundation of allied health sciences.* 3d ed. Dubuque, Iowa: Wm. C. Brown Publishers. This text provides a basic scientific background for careers in the allied health sciences.

Fox, S. I. 1996. *Human physiology.* 5th ed. Dubuque, Iowa: Wm. C. Brown Publishers. This is an introductory physiology text.

Gunstream, S. E. 1995. *Anatomy and physiology.* 2d ed. Dubuque, Iowa: Wm. C. Brown Publishers. This is an introductory anatomy and physiology text-workbook.

Hole, J. W. Jr. 1995. *Essentials of human anatomy and physiology.* 5th ed. Dubuque, IA: Wm. C. Brown Publishers. An introductory text that is written in an easily understood manner.

Lasic, D. D. May/June 1996. Liposomes. *Science & Medicine* 3(3):34. Liposomes can be used to deliver drugs or genes for gene therapy.

Mader, S. S. 1997. *Understanding anatomy and physiology.* 3d ed. Dubuque, Iowa: Wm. C. Brown Publishers. A text that emphasizes the basics for beginning allied health students.

Marieb, E. N. 1995. *Essentials of human anatomy and physiology.* 4th ed. Redwood City, Calif.: Benjamin/Cummings Publishing. This introductory text presents anatomy and physiology to students in the allied health career fields.

National Geographic Society. 1994. *The incredible machine.* Washington, D.C.: National Geographic Society. A colorfully illustrated reference of the structures and functions of the human body.

Packer, L. 1994. Vitamin E is nature's master antioxidant. *Science & Medicine* 1(1E):54. Vitamin E is proving useful as an antioxidant in reducing oxidative destruction of membrane lipids—a normal process of aging.

Rothman, J. E., and Orci, L. March 1996. Budding vesicles in living cells. *Scientific American* 274(3):70. Article discusses the mechanisms by which cellular vesicles transport molecules to the right destination.

Scanlon, V. C., and Sanders, T. 1995. *Essentials of anatomy and physiology.* 2d. ed. Philadelphia: F. A. Davis Company. This introductory text is designed for students with diverse educational backgrounds.

Schwartz, A. T., et al. 1994. *Chemistry in context: Applying chemistry to society.* Dubuque, Iowa: Wm. C. Brown Publishers. This introductory text is designed for students in the allied health fields.

Seeley, R. R., et al. 1996. *Essentials of anatomy & physiology.* 2d ed. St. Louis: Mosby-Year Book, Inc. Acquisition of basic anatomical and physiological facts is the focus of this introductory text.

Sheeler, P. 1996. *Essentials of human physiology.* Dubuque, Iowa: Wm. C. Brown Publishers. This text is suitable for majors in nursing, physical and health education, physical therapy, and nutrition science.

Sloane, E. 1994. *Anatomy and physiology: An easy learner.* Boston: Jones and Bartlett Publishers. Core facts of anatomy and physiology are presented in outline form.

Tate, P., et al. 1994. *Understanding the human body.* St. Louis: Mosby-Year Book, Inc. The relationship between structure and function is stressed in this introductory text.

Tortora, G. J., and Grabowski, S. R. 1996. *Principles of anatomy and physiology.* 8th ed. New York: HarperCollins College Publishers. An introductory anatomy and physiology text that presents basic information in an easy to understand manner.

Van De Graaff, K. M. 1995. *Human anatomy.* 4th ed. Dubuque, Iowa: Wm. C. Brown Publishers. This introductory text provides a balanced presentation of anatomy at various levels.

Vander, A. J., et al. 1994. *Human physiology: The mechanisms of body function.* 6th ed. New York: McGraw-Hill, Inc. Presents the principles of human physiology with an emphasis on physiological mechanisms.

Wardlaw, G., et al. 1994. *Contemporary nutrition.* 2d ed. St. Louis: Mosby-Year Book, Inc. This text gives a clear understanding of nutritional information found on product labels.

Part

Maintenance of the Body

All of the systems of the body help maintain homeostasis, resulting in a dynamic equilibrium of the internal environment. Our internal environment is the blood within blood vessels and the fluid that surrounds the cells of the tissues. The heart pumps the blood and sends it in vessels to the tissues, where exchange of materials occurs with tissue fluid. The composition of blood tends to remain relatively constant as a result of the actions of the digestive, respiratory, and excretory systems. Nutrients enter the blood at the small intestine, external gas exchange occurs in the lungs, and metabolic waste products are excreted at the kidneys. The immune system prevents pathogens from taking over the body and interfering with its proper functioning.

Chapter 4

Digestive System and Nutrition

Chapter Outline

4.1   THE DIGESTIVE SYSTEM
- The human digestive system is an extended tube with specialized parts between two openings, the mouth and the anus.   80
- Food is ingested and then digested to small molecules that are absorbed. Nondigestible materials are eliminated.   80

4.2   THREE ACCESSORY ORGANS
- The pancreas, the liver, and the gallbladder are accessory organs of digestion because their activities assist the digestive process.   88

4.3   DIGESTIVE ENZYMES
- The products of digestion are small molecules, such as amino acids and glucose, that can cross plasma membranes.   90
- The digestive enzymes are specific and have an optimum temperature and pH at which they function.   90

4.4   WORKING TOGETHER
- The digestive system works with the other systems of the body to maintain homeostasis.   93

4.5   NUTRITION
- Proper nutrition supplies the body with energy and nutrients, including the essential amino acids and fatty acids, and all vitamins and minerals.   93

In the early 1980s, an Australian medical resident named Barry Marshall firmly believed that bacteria play a role in ulcers. But physicians have always blamed the open sores on stress or prescription-drug side effects. Marshall set out to prove the bacterial link. One morning in 1984, he walked into his lab, stirred a beaker full of beef soup and *H. pylori,* and gulped the concoction. After five days, he began to vomit. His stomach grew inflamed. With further research, Marshall and others demonstrated that *H. pylori* is responsible for at least 70% of ulcers. Stress may still aggravate these cases, but it's not the direct cause.

We have only to consider the frequency of TV commercials concerned with treating gastrointestinal ills in order to conclude that the proper functioning of the digestive system is critical to our everyday lives. This chapter reviews both the anatomy and physiology of our internal tubular digestive tract and its accessory organs. The liver is an accessory organ with a myriad of functions besides its role in digestion, and we will examine many of these. Today, we recognize that in a sense "we are what we eat," and therefore, a knowledge of nutrition is essential. This chapter ends with a discussion of the basic principles of nutrition.

4.1   The Digestive System

Digestion takes place within a tube called the digestive tract, which begins with the mouth and ends with the anus (Fig. 4.1). The functions of the digestive system are to ingest the food, to digest it to small molecules that can cross plasma membranes, to absorb these nutrient molecules, and to eliminate nondigestible remains.

Mouth Receives

The oral cavity of the mouth is bounded by the lips and cheeks. The lips actually extend from the base of the nose to the start of the chin; the portion we are used to calling "the lips" is poorly keratinized so that the red color of blood shows through. When humans eat, food passes through the lips into the oral cavity. Most people enjoy eating food largely because they like the texture and taste of foods. Receptors called taste buds are found primarily on the tongue. These are activated by the presence of food in the mouth. Taste buds initiate nerve impulses, which travel by way of cranial nerves to the brain. Still, what we call taste is largely due to stimulation of olfactory (smell) receptors in the nose. If you have a cold and your nose is blocked, food has little taste.

The roof of the mouth separates the nasal cavities from the oral cavity. The roof has two parts: an anterior **hard palate** and a posterior **soft palate** (Fig. 4.2*a*). The hard palate contains several bones, but the soft palate is composed entirely of muscle. The soft palate ends in a cone-shaped projection called the *uvula*. The tonsils are at the sides of the oral cavity, at the base of the tongue, and in the nose (called adenoids). The tonsils play a minor role in protecting the body from disease-causing organisms.

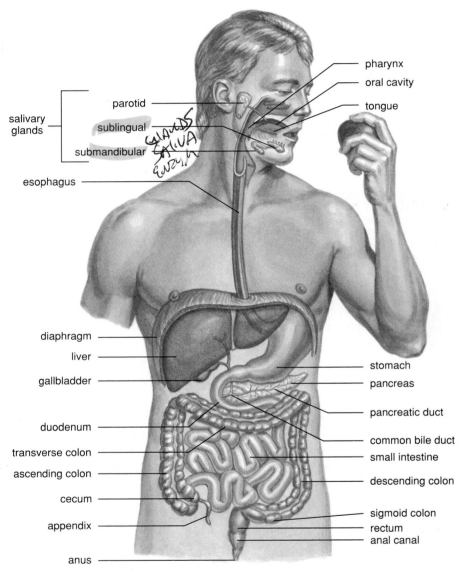

Figure 4.1   **Digestive system.**
Trace the path of food from the mouth to the anus. Notice that the large intestine consists of the cecum; ascending, transverse, descending, and sigmoid colons; plus the rectum and anal canal. Note also the location of the accessory organs of digestion: the pancreas, the liver, and the gallbladder.

Dynamic Human | Digestive System

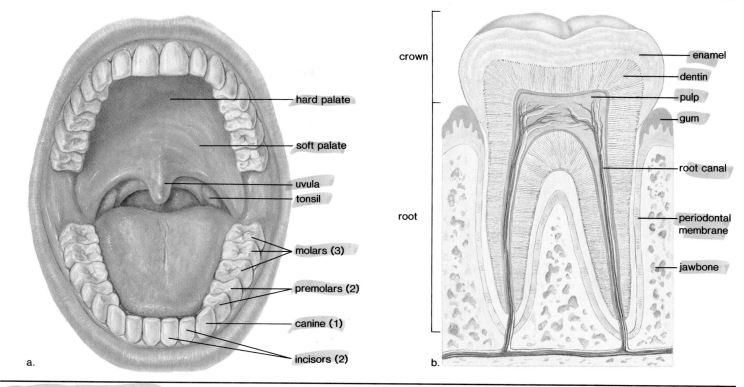

Figure 4.2 Mouth and teeth.

a. The chisel-shaped incisors bite; the pointed canines tear; the fairly flat premolars grind; and the flattened molars crush food. The last molar, called a wisdom tooth, may fail to erupt, or if it does, it is sometimes crooked and useless. Often dentists recommend the extraction of the wisdom teeth. **b.** Longitudinal section of a tooth. The crown is the portion that projects above the gum line and is sometimes replaced by a dentist. When a root canal is done, the nerves are removed. When the periodontal membrane is inflamed, the teeth can loosen.

The three pairs of **salivary glands** are exocrine glands that send their juices (saliva) by way of ducts to the mouth. The *parotid glands* lie at the sides of the face immediately below and in front of the ears. These glands swell when a person has the mumps, a viral infection most often seen in children. Each parotid gland has a duct that opens onto the inner surface of the cheek at the location of the second upper molar. The *sublingual glands* lie beneath the tongue, and the *submandibular glands* lie beneath the posterior floor of the oral cavity. The ducts from these glands open into the mouth under the tongue. You can locate all these openings if you use your tongue to feel for small flaps on the inside of your cheek and under your tongue. Saliva contains an enzyme called *salivary amylase* that begins the process of digesting starch.

Teeth Chew Food

With our teeth we chew food into pieces convenient for swallowing. During the first two years of life, the 20 deciduous, or baby, teeth appear. These are eventually replaced by 32 adult teeth (Fig. 4.2). Each tooth has two main divisions, a crown and a root. The crown has a layer of enamel, an extremely hard outer covering of calcium compounds; dentin, a thick layer of bonelike material; and inner pulp, which contains the nerves and the blood vessels. Dentin and pulp are also found in the root.

Tooth decay, or *dental caries*, commonly called a cavity, occurs when bacteria within the mouth metabolize sugar and give off acids, which erode teeth. Two measures can prevent tooth decay: eating a limited amount of sweets and daily brushing and flossing of teeth. It also has been found that fluoride treatments, particularly in children, can make the enamel stronger and more resistant to decay. Gum disease is more apt to occur with aging. Inflammation of the gums (*gingivitis*) can spread to the periodontal membrane, which lines the tooth socket. A person then has **periodontitis**, characterized by a loss of bone and loosening of the teeth so that extensive dental work may be required. Stimulation of the gums in a manner advised by your dentist is helpful in controlling this condition.

The tongue, which is composed of striated muscle and an outer layer of mucous membrane, mixes the chewed food with saliva. It then forms this mixture into a mass called a *bolus* in preparation for swallowing.

The salivary glands send saliva into the mouth, where the teeth chew the food and the tongue forms it into a bolus for swallowing.

Pharynx Swallows

The **pharynx** is a region that receives food from the mouth and air from the nose (Table 4.1). The food passage and air passage cross in the pharynx because the trachea (windpipe) is superior to the esophagus, a long muscular tube that takes food to the stomach.

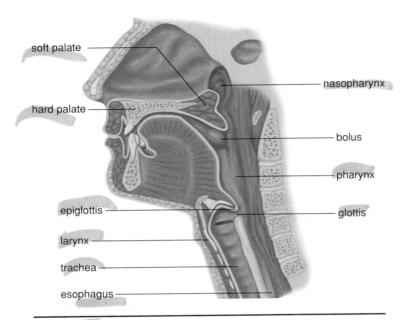

soft palate

nasopharynx

hard palate

bolus

pharynx

epiglottis

glottis

larynx

trachea

esophagus

Figure 4.3   Swallowing.
When food is swallowed, the soft palate closes off the nasopharynx and the epiglottis covers the glottis, forcing the bolus to pass down the esophagus. Therefore, you do not breathe when swallowing.

Swallowing is a process that occurs in the pharynx (Fig. 4.3). Swallowing is a *reflex action,* that is, it is usually performed automatically, without conscious thought. During swallowing, food normally enters the esophagus because the air passages are blocked. Unfortunately, we have all had the unpleasant experience of having food "go the wrong way." The wrong way may be either into the nasal cavities or into the trachea. If it is the latter, coughing will most likely force the food up out of the trachea and into the pharynx again. Usually during swallowing, the soft palate moves back to close off the *nasopharynx,* which leads to the nasal cavities. The trachea moves up under the **epiglottis,** which then covers the **glottis,** the opening to the larynx (voice box). Therefore, breathing temporarily stops. The up and down movement of the *Adam's apple,* the frontal part of the larynx, is easy to observe when a person swallows.

The air passage and the food passage cross in the pharynx. When you swallow, the air passage usually is blocked off, and food must enter the esophagus.

Esophagus Conducts

The **esophagus** is a muscular tube that passes from the pharynx through the thoracic cavity and diaphragm into the abdominal cavity where it joins the stomach. The esophagus is ordinarily collapsed, but it opens and receives the bolus when swallowing occurs. A rhythmic contraction of the digestive tract, called **peristalsis,** pushes the food along.

TABLE 4.1
Path of Food

Organ	Function of Organ	Special Feature(s)	Function of Special Feature(s)
Mouth	Receives food; digestion of starch	Teeth Tongue	Chewing of food Formation of bolus
Esophagus	Passageway		
Stomach	Storage of food; acidity kills bacteria; digestion of protein	Gastric glands	Release gastric juices
Small intestine	Digestion of all foods; absorption of nutrients	Intestinal glands Villi	Release fluids Absorb nutrients
Large intestine	Absorption of water; storage of nondigestible remains		

Occasionally, peristalsis begins even though there is no food in the esophagus. This produces the sensation of a lump in the throat.

The esophagus plays no role in the chemical digestion of food. Its sole purpose is to conduct the food bolus from the mouth to the stomach. The entrance of the esophagus to the stomach is marked by a constrictor called the cardiac or gastroesophageal sphincter, although the muscle in this sphincter is not as developed as in a true sphincter. **Sphincters** are muscles that encircle tubes and act as valves; tubes close when sphincters contract, and they open when sphincters relax. When the cardiac sphincter relaxes, the bolus is able to pass into the stomach. Normally, the cardiac sphincter prevents the acidic contents of the stomach from backing up into the esophagus. Heartburn, which feels like a burning pain rising up into the throat, occurs when some of the stomach contents escape into the esophagus. When vomiting occurs, a contraction of the abdominal muscles and diaphragm propels the contents of the stomach upward through the esophagus.

The esophagus conducts the bolus from the pharynx to the stomach. Peristalsis begins in the esophagus and occurs along the entire length of the digestive tract.

Wall of the Digestive Tract Has Layers

The wall of the esophagus in the abdominal cavity is comparable to that of the digestive tract, which has these layers (Fig. 4.4):

Mucosa (mucous membrane layer)   A layer of epithelium supported by connective tissue and smooth muscle lines the **lumen** (central cavity) and contains glandular epithelial cells that secrete mucus. Digestive glands, if present, are in this layer.

Submucosa (submucosal layer)   A broad band of loose connective tissue that contains blood vessels.

Muscularis (smooth muscle layer)   Two layers of smooth muscle make up this section. The inner, circular layer of cells encircles the gut; the outer, longitudinal layer lies in the same direction as the gut.

Serosa (serous membrane layer)   The rest of the digestive tract has a serosa, a very thin, outermost layer of squamous epithelium supported by connective tissue. The serosa secretes a serous fluid that keeps the outer surface of the intestines moist so that the organs of the abdominal cavity slide against one another. The esophagus has an outer layer composed only of loose connective tissue called the adventitia.

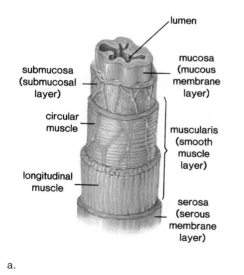

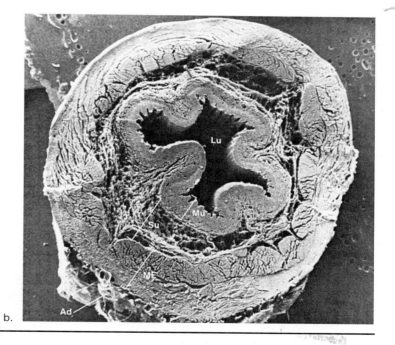

a.   b.

Figure 4.4   Wall of the digestive tract.
a. Several different types of tissues are found in the wall of the digestive tract. Note the placement of circular muscle inside longitudinal muscle. **b.** Scanning electron micrograph of the wall of the esophagus (Lu = central lumen; Mu = mucosa; Su = submucosa; ME = muscularis externa; and Ad = adventitia).

Micrograph from R. G. Kessel and R. H. Kardon, Tissues and Organs: A Text Atlas of Scanning Electron Microscopy, 1979 W. H. Freeman and Company.

Stomach Stores

The **stomach** (Fig. 4.5) is a thick-walled, J-shaped organ that lies on the left side of the body beneath the diaphragm. The stomach is continuous with the esophagus above and the duodenum of the small intestine below. The stomach stores food and aids in digestion. The wall of the stomach has deep folds, which disappear as the stomach fills to an approximate capacity of one liter. Its muscular wall churns, mixing the food with gastric juice. The term *gastric* always refers to the stomach.

The columnar epithelial lining of the stomach has millions of gastric pits, which lead into **gastric glands.** The gastric glands produce gastric juice. Gastric juice contains an enzyme called *pepsin*, which digests protein, plus hydrochloric acid (HCl) and mucus. HCl causes the stomach to have a high acidity of about pH 2, and this is beneficial because it kills most bacteria present in food. Although HCl does not digest food, it does break down the connective tissue of meat and activates pepsin. The cells that secrete gastric juice are acid resistant, and the rest of the wall of the stomach is protected by a thick layer of mucus secreted by goblet cells in its lining. If, by chance, HCl penetrates this mucus, the wall can begin to break down, and an ulcer results. An *ulcer* is an open sore in the wall caused by the gradual disintegration of tissue. It's been shown that an ulcer is due to a bacterial (*Helicobacter pylori*) infection that impairs the ability of epithelial cells to produce protective mucus.

Alcohol is absorbed in the stomach, but there is no absorption of food substances. Normally, the stomach empties in about 2–6 hours. When food leaves the stomach, it is a thick, soupy liquid called *chyme*. Chyme leaves the stomach and enters the small intestine by way of the *pyloric sphincter*. The pyloric sphincter repeatedly opens and closes, allowing chyme to enter the small intestine in small squirts only.

The stomach can expand to accommodate large amounts of food. When food is present, the stomach churns, mixing food with acidic gastric juice.

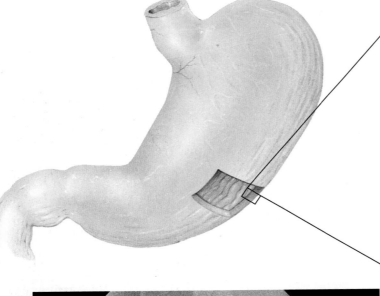

a.

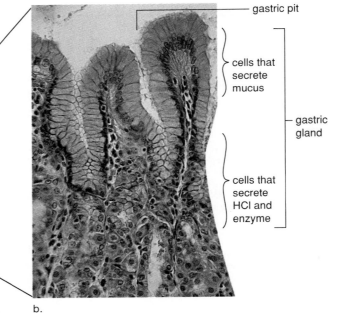

- gastric pit
- cells that secrete mucus
- gastric gland
- cells that secrete HCl and enzyme

b.

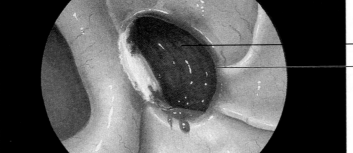

c.

- stomach lining
- ulcer

Figure 4.5   Anatomy and histology of the stomach.
a. The stomach has a folded, thick wall that expands as the stomach fills with food. **b.** The inner mucous membrane layer secretes mucus and contains gastric glands, which secrete a gastric juice active in the digestion of protein. **c.** View of a bleeding ulcer by using an endoscope (a tubular instrument bearing a tiny lens and a light source) that can be inserted into the abdominal cavity.

Small Intestine Absorbs

The **small intestine** is named for its small diameter (compared to that of the large intestine); but perhaps it should be called the long intestine. In life, the small intestine averages about 3.0 meters in length compared to the large intestine which is about 1.5 meters in length. (After death, the small intestine becomes as much as 6 meters long.)

The first 25 cm of the small intestine is called the **duodenum.** Ducts from the liver and pancreas join to form one duct that enters the duodenum (see Fig. 4.1). The small intestine receives bile from the liver and digestive enzymes from the pancreas via this duct. Bile from the liver emulsifies fat, and the pancreatic enzymes continue the process of digestion. The intestinal juices are normally basic because pancreatic juice contains sodium bicarbonate ($NaHCO_3$). The intestinal wall provides digestive enzymes that complete the process of digestion.

The wall of the small intestine contains fingerlike projections called **villi,** which give the intestinal wall a soft, velvety appearance (Fig. 4.6). Each villus has an outer layer of columnar epithelium and contains blood vessels and a small lymphatic vessel called a **lacteal.** The lymphatic system is an adjunct to the circulatory system—its vessels carry a fluid called lymph to the circulatory veins.

The columnar epithelial cells of the villi contain microvilli, thousands of microscopic projections. In electron micrographs, the microvilli give the cells a fuzzy border, called a brush border. The microvilli bear the intestinal digestive enzymes, which are therefore called brush-border enzymes. The microvilli greatly increase the surface area of the villus for the absorption of nutrients. Sugars and amino acids pass through the mucosa and enter a blood vessel. The components of fats (glycerol and fatty acids) rejoin in smooth endoplasmic reticulum and are combined with proteins in the Golgi apparatus before they enter a lacteal.

The small intestine is specialized to absorb the products of digestion. It is quite long (3.0 meters) and has fingerlike projections called villi, where nutrient molecules are absorbed into the circulatory (glucose and amino acids) and lymphatic (fats) systems.

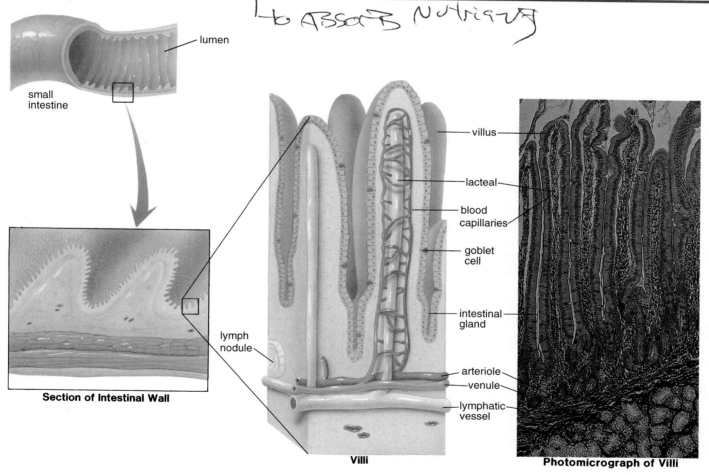

Figure 4.6   Anatomy of small intestine.
The wall of the small intestine has folds that bear fingerlike projections called villi. The products of digestion are absorbed by villi, which contain blood vessels and a lacteal.

Digestive Secretions Are Regulated

It's been known for some time that even the thought of food can cause the nervous system to promote the secretion of digestive juices. But in this century, investigators discovered that hormones are also involved in the secretion of digestive juices. A *hormone* is a substance produced by one set of cells that affects a different set of cells, the so-called target cells. Hormones are usually transported by the bloodstream. When a person has eaten a meal particularly rich in protein, the stomach produces the hormone gastrin. Gastrin enters the bloodstream, and soon the stomach is churning, and the secretory activity of gastric glands is increasing. A hormone produced by the duodenal wall, GIP (gastric inhibitory peptide), works opposite from gastrin: it inhibits gastric gland secretion.

Cells of the duodenal wall produce two other hormones that are of particular interest—secretin and CCK (cholecystokinin). Acid, especially hydrochloric acid (HCl) present in chyme, stimulates the release of secretin, while partially digested protein and fat stimulate the release of CCK. Soon after these hormones enter the bloodstream, the pancreas increases its output of pancreatic juice, and the liver increases its output of bile. The gallbladder contracts to release bile (Fig. 4.7).

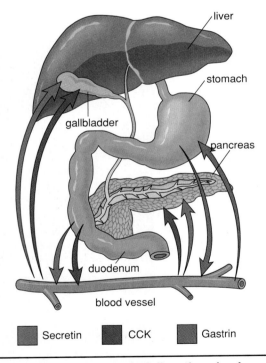

Secretin   **CCK**   **Gastrin**

Figure 4.7 Hormonal control of digestive gland secretions.
Gastrin, produced by the lower part of the stomach, enters the bloodstream and thereafter stimulates the upper part of the stomach to produce more digestive juices. Secretin and CCK, produced by the duodenal wall, stimulate the pancreas to secrete its digestive juices and the gallbladder to release bile.

Large Intestine Eliminates

The **large intestine,** which includes the cecum, the colon, the rectum, and the anal canal, is larger in diameter than the small intestine (6.5 cm compared to 2.5 cm), but it is shorter in length (1.5 meters compared to 3 meters) (see Fig. 4.1). The large intestine absorbs water, salts, and some vitamins. It also stores nondigestible material until it is eliminated at the anus.

The *cecum,* which lies below the entrance of the small intestine, is the blind end of the large intestine. The cecum has a small projection called the vermiform **appendix** (*vermiform* means wormlike) (Fig. 4.8). In humans, the appendix, like the tonsils, may play a role in immunity. This organ is subject to inflammation, a condition called appendicitis. If inflamed, it is wise to remove the appendix before the fluid content rises to the point that the appendix bursts, a situation that may cause *peritonitis,* a generalized infection of the lining of the abdominal cavity. Peritonitis can lead to death.

The **colon** includes the *ascending colon,* which goes up the right side of the body to the level of the liver; the *transverse colon,* which crosses the abdominal cavity just below the liver and the stomach; the *descending colon,* which passes down the left side of the body; and the *sigmoid colon,* which enters the *rectum,* the last 20 cm of the large intestine. The rectum opens at the **anus,** where **defecation,** the expulsion of *feces,* occurs (Fig. 4.9). Feces are three-quarters water and one-quarter solids. Bacteria, fiber (roughage), and other nondigestible materials are in the solid portion. The brown color of feces is due to bacteria breaking down bilirubin, and the odor is due to breakdown products as bacteria work on the nondigested remains. This action also produces gases.

For many years, it was believed that facultative bacteria (bacteria that can live with or without oxygen), such as *Escherichia coli,* were the major inhabitants of the colon, but new culture methods show that over 99% of the colon bacteria are obligate anaerobes (bacteria that die in the presence of oxygen). Not only do the bacteria break down nondigestible material, but they also produce some vitamins and other molecules that can be absorbed and used by us. In this way, they perform a service for us.

Water is considered unsafe for swimming when the coliform (nonpathogenic intestinal) bacterial count reaches a certain number. A high count is an indication that a significant amount of feces has entered the water. The more feces present, the greater the possibility that disease-causing bacteria are also present.

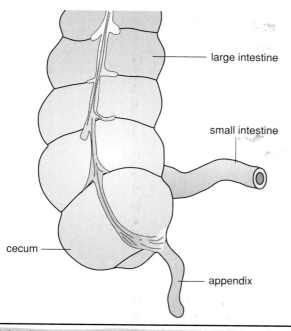

Figure 4.8   Junction of the small intestine and the large intestine.
The cecum is the blind end of the ascending colon. The appendix is attached to the cecum.

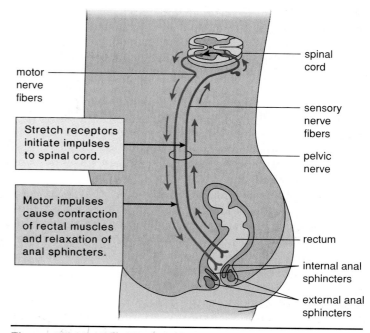

Figure 4.9   Defecation reflex.
The accumulation of feces in the rectum causes it to stretch, which initiates a reflex action resulting in rectal contraction and expulsion of the fecal material.

Polyps

The colon is subject to the development of *polyps,* small growths arising from the epithelial lining. Polyps, whether benign or cancerous, can be removed individually. If colon cancer is detected while still confined to a polyp, the expected outcome is a complete cure. Some investigators believe that dietary fat increases the likelihood of colon cancer because dietary fat causes an increase in bile secretion. It could be that intestinal bacteria convert bile salts into substances that promote the development of cancer. On the other hand, fiber in the diet seems to inhibit the development of colon cancer. Dietary fiber absorbs water and adds bulk, thereby diluting the concentration of bile salts and facilitating the movement of substances through the intestine. Regular elimination reduces the time that the colon wall is exposed to any cancer-promoting agents in feces.

Diarrhea and Constipation

Two common everyday complaints associated with the large intestine are *diarrhea* and *constipation.* The major causes of diarrhea are infection of the lower tract and nervous stimulation. In the case of infection, such as food poisoning caused by eating contaminated food, the intestinal wall becomes irritated, and peristalsis increases. Water is not absorbed, and the diarrhea that results rids the body of the infectious organisms. In nervous diarrhea, the nervous system stimulates the intestinal wall, and diarrhea results. Prolonged diarrhea can lead to dehydration because of water loss and to disturbances in the heart's contraction due to an imbalance of salts in the blood.

When a person is constipated, the feces are dry and hard. One reason for this condition is that socialized persons have learned to inhibit defecation to the point that the desire to defecate is ignored. Two components of the diet that can help prevent constipation are water and fiber. Water intake prevents drying out of the feces, and fiber provides the bulk needed for elimination. The frequent use of laxatives is discouraged. If, however, it is necessary to take a laxative, a bulk laxative is the most natural because, like fiber, it produces a soft mass of cellulose in the colon. Lubricants, like mineral oil, make the colon slippery, and saline laxatives, like milk of magnesia, act osmotically—they prevent water from being absorbed and, depending on the dosage, may even cause water to enter the colon. Some laxatives are irritants; they increase peristalsis to the degree that the contents of the colon are expelled.

Chronic constipation is associated with the development of hemorrhoids, enlarged and inflamed blood vessels at the anus.

The large intestine does not produce digestive enzymes; it does absorb water and salts. In diarrhea, too little water has been absorbed by the large intestine; in constipation, too much water has been absorbed.

4.2   Three Accessory Organs

The pancreas, liver, and gallbladder are accessory organs of digestion. Figure 4.1 shows how the pancreatic duct from the pancreas and the common bile duct from the liver and gallbladder join before entering the duodenum.

Pancreas Produces Enzymes

The **pancreas** lies deep in the abdominal cavity, resting on the posterior abdominal wall. It is an elongated and somewhat flattened organ that has both an endocrine and an exocrine function. As an endocrine gland it secretes insulin and glucagon, hormones that help keep the blood glucose level within normal limits. We are now interested in its exocrine function. Most pancreatic cells produce pancreatic juice, which contains sodium bicarbonate ($NaHCO_3$) and digestive enzymes for all types of food. *Pancreatic amylase* digests starch, *trypsin* digests protein, and *lipase* digests fat. In cystic fibrosis, a thick mucus blocks the pancreatic duct, and the patient must take supplemental pancreatic enzymes by mouth for proper digestion to occur.

Liver Produces Bile

The **liver,** which is the largest organ in the body, lies mainly in the upper right section of the abdominal cavity, under the diaphragm (see Fig. 4.1). The liver has two main lobes, the right lobe and the smaller left lobe, which crosses the midline and lies above the stomach. The liver contains approximately 100,000 lobules that serve as the structural functional units of the liver (Fig. 4.10). Triads consisting of these three structures are located between the lobules: (1) a branch of the hepatic artery that brings oxygenated blood to the liver; (2) a branch of the hepatic portal vein that transports nutrients from the intestines; and (3) a bile duct that takes bile away from the liver. The central veins within the lobules enter the hepatic vein. Note in Figure 4.11 that the liver lies between the *hepatic portal vein* and the *hepatic vein*, which enters the *vena cava*.

In some ways, the liver acts as the gatekeeper to the blood. As the blood from the intestines passes through the liver, it removes poisonous substances and works to keep the contents of the blood constant. It also removes and stores iron and the fat-soluble vitamins A, D, E, and K. The liver makes the plasma proteins from amino acids, and these have important functions within blood itself.

The liver maintains the blood glucose level at about 100 mg/100 ml = 0.1%, even though a person eats intermittently. Any excess glucose that is present in the hepatic portal vein is removed and stored by the liver as glycogen. Between eating, glycogen is broken down to glucose, which enters the hepatic vein, and in this way, the blood glucose level remains constant.

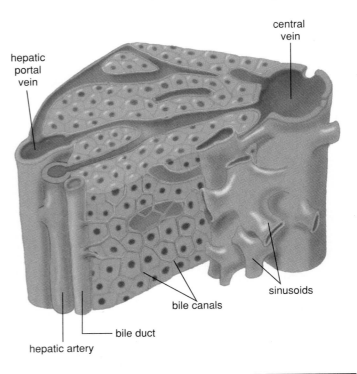

Figure 4.10   Hepatic lobules.
The liver contains over 100,000 lobules shown here in cross section. Each lobule contains many cells that perform the various functions of the liver. They remove from and/or add materials to blood and deposit bile in bile ducts.

If, by chance, the supply of glycogen is depleted, the liver will convert glycerol (from fats) and amino acids to glucose molecules. The conversion of amino acids to glucose necessitates deamination, the removal of amino acids. By a complex metabolic pathway, the liver then combines ammonia with carbon dioxide to form urea:

$$2\,NH_3 \;+\; CO_2 \;\rightarrow\; H_2N-\overset{\overset{\textstyle O}{\|}}{C}-NH_2$$

ammonia   carbon   urea
                    dioxide

Urea is the usual nitrogenous waste product from amino acid breakdown in humans. After its formation in the liver, urea is excreted by the kidneys.

The liver produces **bile** which is stored in the gallbladder. Bile has a yellowish green color because it contains the bile pigments bilirubin and biliverdin, which are derived from the breakdown of hemoglobin, the red pigment found in red blood cells. Bile also contains bile salts (derived from cholesterol), which emulsify fat in the duodenum of the small intestine. When fat is emulsified, it breaks up into droplets, providing a much larger surface area, which can be acted upon by a digestive enzyme from the pancreas.

Kroiwthis

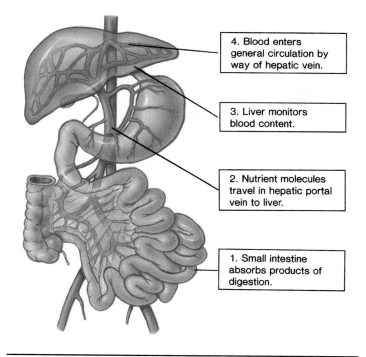

4. Blood enters general circulation by way of hepatic vein.

3. Liver monitors blood content.

2. Nutrient molecules travel in hepatic portal vein to liver.

1. Small intestine absorbs products of digestion.

Figure 4.11 Hepatic portal system.
The hepatic portal vein takes the products of digestion from the digestive system to the liver, where they are processed before entering the circulatory system proper.

Altogether, the following are significant functions of the liver:

1. Detoxifies blood by removing and metabolizing poisonous substances.

2. Stores iron and the fat-soluble vitamins A, D, E, and K.

3. Makes plasma proteins, such as albumins and fibrinogen, from amino acids.

4. Stores glucose as glycogen after eating and breaks down glycogen to glucose to maintain the glucose concentration of blood between eating periods.

5. Produces urea from the breakdown of amino acids.

6. Destroys old red blood cells and converts hemoglobin to breakdown products (bilirubin and biliverdin). These are excreted along with bile salts in bile.

7. Produces bile, which is stored in the gallbladder before entering the small intestine, where bile salts emulsify fats.

Serious Liver Disorders

Jaundice, hepatitis, and cirrhosis are three serious diseases that affect the entire liver and hinder its ability to repair itself. Therefore, they are life-threatening diseases. When a person has *jaundice,* there is a yellowish tint to the whites of the eyes and also to the skin of light-pigmented persons. Bilirubin is deposited in the skin due to an abnormally large amount in the blood. In *hemolytic jaundice,* red blood cells are broken down in abnormally large amounts; in *obstructive jaundice,* bile ducts are blocked or liver cells are damaged. Obstructive jaundice often occurs when crystals of cholesterol come out of solution and form gallstones. The stones may be so numerous that passage of bile along a bile duct is blocked, and the gallbladder must be removed.

Jaundice can also result from *hepatitis,* inflammation of the liver. Viral hepatitis occurs in several forms. Hepatitis A is usually acquired from sewage-contaminated drinking water. Hepatitis B, which is usually spread by sexual contact, can also be spread by blood transfusions or contaminated needles. The hepatitis B virus is more contagious than the AIDS virus, which is spread in the same way. Thankfully, however, there is now a vaccine available for hepatitis B. Hepatitis C, which is usually acquired by contact with infected blood and for which there is no vaccine, can lead to chronic hepatitis, liver cancer, and death.

Cirrhosis is another chronic disease of the liver. First the organ becomes fatty, and liver tissue is then replaced by inactive fibrous scar tissue. Cirrhosis of the liver is often seen in alcoholics due to malnutrition and to the excessive amounts of alcohol (a toxin) the liver is forced to break down.

The liver has amazing generative powers and can recover if the rate of regeneration exceeds the rate of damage. During liver failure, however, there may not be enough time to let the liver heal itself. Liver transplantation is usually the preferred treatment for liver failure, but artificial livers have been developed and tried in a few cases. One type is a cartridge that contains liver cells. The patient's blood passes through cellulose acetate tubing of the cartridge and is serviced in the same manner as with a normal liver. In the meantime, the patient's liver has a chance to recover.

Gallbladder Stores Bile

The **gallbladder** is a pear-shaped, muscular sac attached to the ventral surface of the liver (see Fig. 4.1). About 1,000 ml of bile are produced by the liver each day, and any excess is stored in the gallbladder. In the gallbladder, water is reabsorbed so that bile becomes a thick, mucus-like material. When needed, bile leaves the gallbladder and proceeds to the duodenum via the common bile duct.

4.3   Digestive Enzymes

The digestive enzymes are **hydrolytic enzymes,** which catalyze breakdown by the introduction of water at specific bonds. Digestive enzymes are like other enzymes we have studied. They are proteins with a particular shape that fits their substrate. They also have an optimum pH, which maintains their shape, thereby enabling them to speed up their specific reaction.

The various digestive enzymes present in the digestive juices mentioned previously help break down carbohydrates, proteins, and fats, the major components of food. Starch is a carbohydrate, and its digestion begins in the mouth. Saliva from the salivary glands has a neutral pH and contains salivary **amylase,** the first enzyme to act on starch:

$$\text{starch} + \text{H}_2\text{O} \xrightarrow{\substack{\text{salivary}\\\text{amylase}}} \text{maltose}$$

In this equation, salivary amylase is written above the arrow to indicate that it is neither a reactant nor a product in the reaction. It merely speeds the reaction in which its substrate, starch, is digested to many molecules of maltose. Maltose molecules cannot be absorbed by the intestine; additional digestive action in the small intestine converts maltose to glucose, which can be absorbed.

Protein digestion begins in the stomach. Gastric juice secreted by gastric glands has a very low pH—about 2—because it contains hydrochloric acid (HCl). Pepsinogen, a precursor that is converted to the enzyme **pepsin** when exposed to HCl, is also present in gastric juice. Pepsin acts on protein to produce peptides:

$$\text{protein} + \text{H}_2\text{O} \xrightarrow{\text{pepsin}} \text{peptides}$$

Peptides vary in length, but they always consist of a number of linked amino acids. Peptides are usually too large to be absorbed by the intestinal lining, but later they are broken down to amino acids in the small intestine.

Starch, proteins, and fats are all enzymatically broken down in the small intestine. Pancreatic juice, which enters the duodenum, is basic (alkaline) because it contains sodium bicarbonate (NaHCO$_3$). It also contains enzymes for the digestion of all types of food. One pancreatic enzyme, **pancreatic amylase,** digests starch:

$$\text{starch} + \text{H}_2\text{O} \xrightarrow{\substack{\text{pancreatic}\\\text{amylase}}} \text{maltose}$$

Another pancreatic enzyme, **trypsin,** digests protein:

$$\text{protein} + \text{H}_2\text{O} \xrightarrow{\text{trypsin}} \text{peptides}$$

Trypsin is secreted as trypsinogen, which is converted to trypsin in the duodenum.

Lipase, a third pancreatic enzyme, digests fat molecules in the fat droplets after they have been emulsified by bile salts:

$$\text{fat} \xrightarrow{\text{bile salts}} \text{fat droplets}$$

$$\text{fat droplets} + \text{H}_2\text{O} \xrightarrow{\text{lipase}} \text{glycerol} + 3 \text{ fatty acids}$$

The end products of lipase digestion, glycerol and fatty acid molecules, are small enough to cross the cells of the intestinal villi, where absorption takes place. As mentioned previously, glycerol and fatty acids enter the cells of the villi, and within these cells, they are rejoined and packaged as lipoprotein droplets before entering the lacteals (see Fig. 4.6).

Peptidases and **maltase,** two enzymes secreted by the small intestine, complete the digestion of protein to amino acids and starch to glucose, small molecules that cross into the cells of the villi. Peptides, which result from the first step in protein digestion, are digested to amino acids by peptidases:

$$\text{peptides} + \text{H}_2\text{O} \xrightarrow{\text{peptidases}} \text{amino acids}$$

Maltose, which results from the first step in starch digestion, is digested to glucose by maltase:

$$\text{maltose} + \text{H}_2\text{O} \xrightarrow{\text{maltase}} \text{glucose} + \text{glucose}$$

Other disaccharides, each of which has its own enzyme, are digested in the small intestine. The absence of any one of these enzymes can cause illness. For example, many people, including as many as 75% of African Americans, cannot digest lactose, the sugar found in milk, because they do not produce lactase, the enzyme that converts lactose to its components, glucose and galactose. Drinking untreated milk often gives these individuals the symptoms of *lactose intolerance* (diarrhea, gas, cramps), caused by a large quantity of nondigested lactose in the intestine. In most areas, it is possible to purchase milk made lactose-free by the addition of synthetic lactase or *Lactobacillus acidophilus* bacteria, which break down lactose.

Table 4.2 lists some of the major digestive enzymes produced by the digestive tract, salivary glands, or the pancreas. Each type of food is broken down by specific enzymes.

Digestive enzymes present in digestive juices help break down food to the nutrient molecules: glucose, amino acids, fatty acids, and glycerol. The first two are absorbed into the blood capillaries of the villi, and the last two re-form within epithelial cells before entering the lacteals as lipoprotein droplets.

TABLE 4.2

Major Digestive Enzymes

Food	Digestion	Enzyme	Optimum pH	Produced by	Site of Action
Starch	Starch + H_2O → maltose	Salivary amylase	Neutral	Salivary glands	Mouth
		Pancreatic amylase	Basic	Pancreas	Small intestine
	Maltose + H_2O → glucose*	Maltase	Basic	Small intestine	Small intestine
Protein	Protein + H_2O → peptides	Pepsin	Acidic	Gastric glands	Stomach
		Trypsin	Basic	Pancreas	Small intestine
	Peptides + H_2O → amino acids*	Peptidases	Basic	Small intestine	Small intestine
Fat	Fat droplets + H_2O → glycerol* + fatty acids*	Lipase	Basic	Pancreas	Small intestine

*Absorbed by villi.

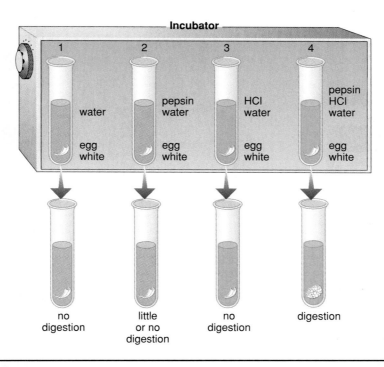

Figure 4.12   Digestion experiment.
Knowing that the correct enzyme, optimum pH, and the correct substrate must be present for digestion to occur, explain the results of this experiment.

Best Conditions for Digestion

Laboratory experiments can define the necessary conditions for digestion. For example, the four test tubes described in Figure 4.12 can be prepared and observed for the digestion of egg white, a protein digested in the stomach by the enzyme pepsin.

After all tubes are placed in an incubator at body temperature for at least one hour, the results depicted are observed. Tube 1 is a control tube; no digestion has occurred in this tube because the enzyme and HCl are missing. (If a control gives a positive result, then the experiment is in-

validated.) Tube 2 shows limited or no digestion because HCl is missing, and therefore the pH is too high for pepsin to be effective. Tube 3 shows no digestion because although HCl is present, the enzyme is missing. Tube 4 shows the best digestive action because the enzyme is present and the presence of HCl has resulted in an optimum pH. This experiment supports the belief that for digestion to occur, the substrate and enzyme must be present and the environmental conditions must be optimum. The optimal environmental conditions include a warm temperature and the correct pH.

HUMAN SYSTEMS WORK TOGETHER

Integumentary System

Digestive tract provides nutrients needed by skin.

Skin helps to protect digestive organs; helps to provide vitamin D for Ca^{++} absorption.

Skeletal System

Digestive tract provides Ca^{++} and other nutrients for bone growth and repair.

Bones provide support and protection; hyoid bone assists swallowing.

Muscular System

Digestive tract provides glucose for muscle activity; liver metabolizes lactic acid following anaerobic muscle activity.

Smooth muscle contraction accounts for peristalsis; skeletal muscles support and help protect abdominal organs.

Nervous System

Digestive tract provides nutrients for growth, maintenance, and repair of neurons and neuroglial cells.

Brain controls nerves, which innervate smooth muscle and permit tract movements.

Endocrine System

Stomach and small intestine produce hormones.

Hormones help control secretion of digestive glands and accessory organs; insulin and glucagon regulate glucose storage in liver.

How the Digestive System works with other body systems

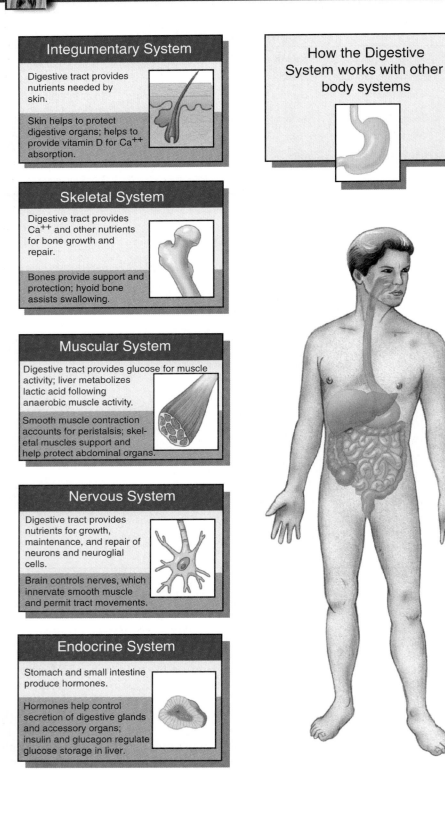

Circulatory System

Digestive tract provides nutrients for plasma protein formation and blood cell formation; liver detoxifies blood, makes plasma proteins, destroys old red blood cells.

Blood vessels transport nutrients from digestive tract to body; blood services digestive organs.

Lymphatic System/Immunity

Digestive tract provides nutrients for lymphoid organs; stomach acidity prevents pathogen invasion of body.

Lacteals absorb fats; Peyer's patches prevent invasion of pathogens; appendix contains lymphoid tissue.

Respiratory System

Breathing is possible through the mouth because digestive tract and respiratory tract share the pharynx.

Gas exchange in lungs provides oxygen to digestive tract and excretes carbon dioxide from digestive tract.

Urinary System

Liver synthesizes urea; digestive tract excretes bile pigments from liver and provides nutrients.

Kidneys convert vitamin D to active form needed for Ca^{++} absorption; compensate for any water loss by digestive tract.

Reproductive System

Digestive tract provides nutrients for growth and repair of organs and for development of fetus.

Pregnancy crowds digestive organs and promotes heartburn and constipation.

Dynamic Human | Digestive System

4.4   Working Together

The box on page 92 tells how the digestive system works with other systems in the body to maintain homeostasis.

4.5   Nutrition

The body requires three major classes of *macronutrients* in the diet: carbohydrate, protein, and fat. These supply the energy and the building blocks that are needed to synthesize cellular contents. *Micronutrients*—especially vitamins and minerals—are also required because they are necessary to cellular metabolism.

Several modern nutritional studies suggest that certain nutrients can protect against heart disease, cancer, and other serious illnesses. These studies include an analysis of the eating habits of healthy people in the United States and from around the world, especially those with lower rates of heart disease and cancer. The result has been the dietary recommendations illustrated by a food pyramid (Fig. 4.13).

The bulk of the diet should consist of bread, cereal, rice, and pasta as energy sources. Whole grains are preferred over those that have been milled because they contain fiber (roughage which is nondigestible carbohydrates) and vitamins and minerals. Next, it is best to rely on vegetables and fruits as nutrients since these too are a rich source of fiber, vitamins, and minerals. It is easy to see, then, that a largely vegetarian diet is recommended.

Animal products, especially meat, need only be minimally included in the diet; fats and sweets should be used sparingly. Dairy products and meats tend to be high in saturated fats, and an intake of saturated fats increases the risk of cardiovascular disease (see Lipids, p. 98). It is possible to purchase low-fat dairy products, but there is no way to take much of the fat out of beef. Beef meat, in particular, contains a relatively high fat content. It is ironic that the affluence of people in the United States contributes to a poor diet and, therefore, possible illness. Only comparatively rich people can afford fatty meats from grain-fed cattle and carbohydrates that have been highly processed to remove fiber and to add sugar and salt.

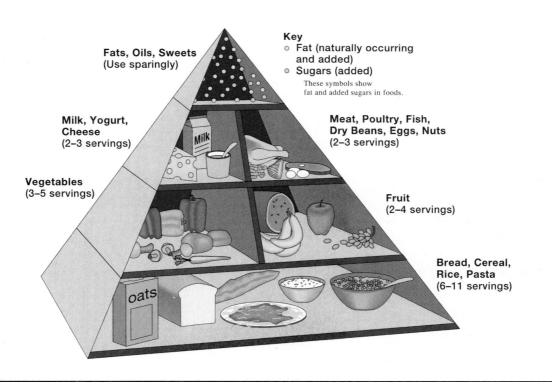

Figure 4.13   Ideal American diet.
The U.S. Department of Agriculture uses a pyramid to show the ideal diet because it emphasizes the importance of including grains and fruits and vegetables in the diet. Meats and dairy products are needed in limited amounts; fats, oils, and sweets should be used sparingly.

Source: Data from the U.S. Department of Agriculture.

Carbohydrates Are Quick Energy

The quickest, most readily available source of energy for the body is glucose. Carbohydrates are digested to simple sugars, which are or can be converted to glucose. Glucose is stored by the liver in the form of glycogen. Between eating periods, the blood glucose level is maintained at about 0.1% by the breakdown of glycogen or by the conversion of glycerol (from fats) or amino acids to glucose. If necessary, amino acids are taken from the muscles—even from the heart muscle. While body cells can utilize fatty acids as an energy source, brain cells require glucose. For this reason alone, it is necessary to include carbohydrate in the diet. As indicated in Fig. 4.13, it is recommended that carbohydrates make up the bulk (at least 58%) of the diet. Further, it is assumed that these carbohydrates are complex and not simple carbohydrates. Complex sources of carbohydrates include preferably whole-grain pasta, rice, bread, and cereal (Fig 4.14). Potatoes and corn, although considered vegetables, are also sources of carbohydrates.

Simple carbohydrates (e.g., sugars) are labeled "empty calories" by some dieticians because they contribute to energy needs and weight gain without supplying any other nutritional requirements. Table 4.3 gives suggestions on how to reduce the consumption of dietary sugar (simple carbohydrates). In contrast to simple sugars, complex carbohydrates are likely to be accompanied by a wide range of other nutrients and by **fiber**, which is nondigestible plant material.

The intake of fiber is recommended because it decreases the risk of colon cancer, a major type of cancer, and cardiovascular disease, the number one killer in the United States. Insoluble fiber, such as that found in wheat bran, has a laxative effect and may guard against colon cancer, because any cancer-causing substances are in contact with the intestinal wall for a limited amount of time. Soluble fiber, such as that found in oat bran, combines with bile acids and cholesterol in the intestine and prevents them from being absorbed. The liver now removes cholesterol from the blood and changes it to bile acids, replacing those that were lost. While the diet should have an adequate amount of fiber, a high-fiber diet can be detrimental. Some evidence suggests that the absorption of iron, zinc, and calcium is impaired by a diet too high in fiber.

Complex carbohydrates, which contain fiber, should form the bulk of the diet.

TABLE 4.3
Reducing Dietary Sugar

To reduce dietary sugar, the following suggestions are recommended.

1. Eat fewer sweets, such as candy, soft drinks, ice cream, and pastry.
2. Eat fresh fruits or fruits canned without heavy syrup.
3. Use less sugar—white, brown, or raw—and less honey and syrups.
4. Avoid sweetened breakfast cereals.
5. Eat less jelly.
6. Drink pure fruit juices, not imitations.
7. When cooking, use spices like cinnamon instead of sugar to flavor foods.
8. Do not put sugar in tea or coffee.

Figure 4.14 Complex carbohydrates. To meet our energy needs, dieticians recommend consuming foods rich in complex carbohydrates, like those shown here, rather than foods consisting of simple carbohydrates, like candy and ice cream. The latter provide little but calories, which cause weight gain and supply few other nutrients.

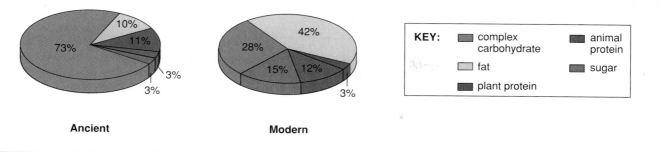

Figure 4.15   Ancient versus modern diet of native Hawaiians.
Among those native Hawaiians who have switched back to the native diet, the incidence of cardiovascular disease, cancer, and diabetes has dropped.

Proteins Supply Building Blocks

Foods rich in protein include red meat, fish, poultry, dairy products, legumes (i.e., peas and beans), nuts, and cereals. Following digestion of protein, amino acids enter the bloodstream and are transported to the tissues. Ordinarily, amino acids are not used as an energy source. Most are incorporated into structural proteins found in muscles, skin, hair, and nails. Others are used to synthesize such proteins as hemoglobin, plasma proteins, enzymes, and hormones.

Adequate protein formation requires 20 different types of amino acids. Of these, eight are required from the diet in adults (nine in children) because the body is unable to produce them. These are termed the **essential amino acids.** The body produces the other 11 amino acids by simply transforming one type into another type. Some protein sources, such as meat, are *complete;* they provide all 20 types of amino acids. Vegetables and grains supply us with amino acids, but each vegetable or grain alone is an *incomplete* protein source because at least one of the essential amino acids is absent. Absence of one essential amino acid prevents utilization of the other 19 amino acids. Soybeans and tofu, made from soybeans, are rich in amino acids, but it is wise to combine foods to acquire all the essential amino acids. For example, the combinations of cereal with milk, or beans, a legume, with rice, a grain, will provide all the essential amino acids.

Amino acids are not stored in the body, and a daily supply is needed. However, it does not take very much protein to meet the daily requirement. Two servings of meat a day (equal in total quantity to a deck of cards) is usually enough. Some meats (e.g., hamburger) are high in protein but also high in fat. Everything considered, it is probably a good idea to depend on protein from plant origins (e.g., whole-grain cereals, dark breads, legumes) to a greater extent than is often the custom in the United States. This can be illustrated by the health statistics of native Hawaiians who no longer eat as their ancestors did (Fig. 4.15).

<table>
<tr><td colspan="1">TABLE 4.4</td></tr>
<tr><td>Reducing Lipids</td></tr>
</table>

To reduce dietary fat, the following suggestions are recommended.

1. Choose poultry, fish, or dry beans and peas as a protein source.
2. Remove skin from poultry before cooking, and place on a rack so that fat drains off.
3. Broil, boil, or bake rather than fry.
4. Limit your intake of butter, cream, hydrogenated oils, shortenings, and tropical oils (coconut and palm oils)*
5. Use herbs and spices to season vegetables instead of butter, margarine, or sauces. Use lemon juice instead of salad dressing.
6. Drink skim milk instead of whole milk, and use skim milk in cooking and baking.
7. Eat nonfat or low-fat foods.

To reduce dietary cholesterol, the following suggestions are recommended.

1. Avoid cheese, egg yolks, liver, and certain shellfish (shrimp and lobster). Preferably, eat white fish and poultry.
2. Substitute egg whites for egg yolks in both cooking and eating.
3. Include soluble fiber in the diet. Oat bran, oatmeal, beans, corn, and fruits such as apples, citrus fruits, and cranberries are high in soluble fiber.

*Although coconut and palm oils are from plant sources, they are saturated fats.

The modern diet depends on animal rather than plant protein and is 42% fat. A statistical study showed that the island's native peoples now have a higher than average death rate from cardiovascular disease and cancer. Diabetes is also common in persons who follow the modern diet. But the health of those who have switched back to the ancient diet has improved immensely! Table 4.4 gives guidelines for reducing the amount of lipids in the diet.

Health Focus

Fat Is Everyone's Issue

Fat tastes good. Unfortunately, eating too much fat of any kind increases the risk for heart disease and cancer. It also increases the risk for an expanding waistline: protein and carbohydrates have four calories per gram; each gram of fat contains a whopping nine calories. Government guidelines recommend 30% of total daily calories be from fat, but people actually need far less than this. The body's requirements would be met if fat comprised just 10% of total calories.

FAT FACTS

All dietary fats fall into three categories: saturated, monounsaturated, and polyunsaturated. Saturation refers to how many hydrogen atoms a fat contains. Saturated fats contain the maximum amount of hydrogen possible; fats missing a single pair of hydrogen atoms are monounsaturated; and those missing more than one pair are polyunsaturated.

One way to distinguish saturated fats from unsaturated ones is to check their consistency at room temperature. Saturated fats, such as butter and coconut oil, remain relatively firm, while unsaturated fats, such as vegetable oils, are liquid.

However, a normally soft or liquid unsaturated fat can be transformed into a harder, more saturated fat by hydrogenation, a chemical process by which hydrogen atoms are added to an unsaturated oil. This is generally done by food manufacturers to keep fat from becoming rancid and to ensure that products remain firm at room temperature. Partial hydrogenation gives margarine its texture. Because the soft or liquid form is less hydrogenated than the stick variety, it has a lower transformed fatty acid content.

Saturated fatty acids have long been linked to cardiovascular disease because they, along with cholesterol, contribute to the formation of plaque, which is a fatty deposit on arterial walls. Cholesterol is carried from the liver to the cells by plasma proteins called low-density lipoproteins (LDLs) and is carried away from the cells to the liver by high-density lipoproteins (HDLs). Therefore, LDL is the type of lipoprotein associated with the development of plaque. For reasons that are not well understood, saturated fats raise low-density lipoprotein (LDL) cholesterol in the blood. On the other hand, substituting poly- and monounsaturated fats lowers LDL cholesterol. But transformed fats found in margarine can raise LDL cholesterol almost as much as saturated fatty acids, and they can decrease levels of beneficial HDL cholesterol. So the burning question— "which is better, butter or margarine?"—may be moot. *The important thing is to reduce all dietary fats. The best advice is to choose liquid forms of monounsaturated fats (olive, canola, peanut oils) for cooking and use butter or margarine sparingly, if at all. As substitutes for butter and margarine, try jam on bagels, nonfat sour cream on potatoes, and lemon juice and herbs on vegetables.*

FAKE FAT HAS FEW FRIENDS

An old adage says that there is no such thing as a free lunch. Nor is there such a thing as a fat-free food that's good for you, according to nutrition researchers at the Harvard School of Public Health, at least not if it contains *olestra.*

In January 1996, olestra became the first calorie-free fat substitute approved by the Food and Drug Administration (FDA). For now, the FDA has limited its use to potato chips and other salty snacks. Cincinnati-based Procter & Gamble, the product's manufacturer, wants the agency to approve it as an ingredient in oil, ice cream, salad dressings, and cheese.

Although olestra is made from real fatty acids and looks, tastes, and acts like real fat, it isn't one. Because the digestive system cannot break it down, olestra travels the length of the gut without being absorbed or contributing any calories to the day's total.

Unfortunately, as olestra exits, it takes with it the fat-soluble vitamins A, D, E, and K; it also reduces circulating levels of carotenoids, antioxidants found in fruits and vegetables.

Not only that, olestra-containing foods cause what has been politely called "anal leakage" (underwear staining). Some people who consume it also develop diarrhea, intestinal cramping, and flatulence (gas).

It doesn't take a giant bag of chips to wash out key nutrients—a handful will do. In one study, the amount of olestra found in six potato chips (about 3 grams) caused a 20% decline in beta-carotene levels and a 38% decrease in lycopene, another key carotenoid. Procter & Gamble fortifies olestra-containing foods with vitamins A, D, E, and K but doesn't take concerns about lost carotenoids as seriously.

Although optimistic scientists once saw beta carotene as a promising way to lower the risk for both cancer and heart disease, this has not proved true in large studies where it has been administered as a supplement. Nevertheless, "it is clear that a diet high in fruits and vegetables helps prevent cancer and heart disease. While carotenoids may not explain the entire relationship, they no doubt play some role," says researcher Meir J. Stampfer, professor of epidemiology and nutrition at the Harvard School of Public Health.

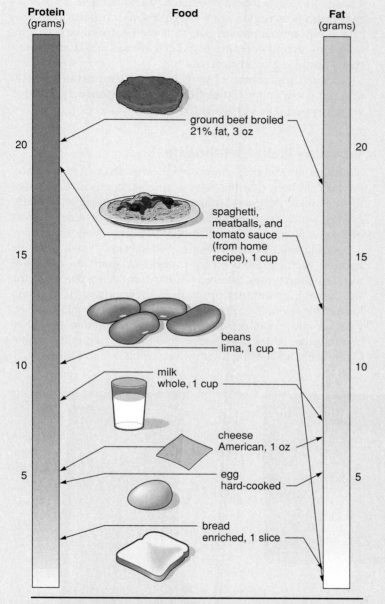

| Protein (grams) | Food | Fat (grams) |

Figure 4A   Protein-rich foods are frequently high in fat.

- ground beef broiled 21% fat, 3 oz
- spaghetti, meatballs, and tomato sauce (from home recipe), 1 cup
- beans lima, 1 cup
- milk whole, 1 cup
- cheese American, 1 oz
- egg hard-cooked
- bread enriched, 1 slice

not only lowered LDL cholesterol as much as polyunsaturated varieties, but that they did not lower good HDL cholesterol as polys supposedly did.

Armed with this new information, many heart disease experts told their patients it was time for an oil change, advising them to use monounsaturated varieties such as olive or canola instead of poly-rich corn or safflower oils. A decade later, however, an analysis of 14 studies comparing the effects of monos and polys on cholesterol suggest that both of these fats leave HDL levels unchanged.

"People should be less concerned about choosing monos over polys and more concerned about reducing saturated and total fat," said the study's author, Christopher Gardner, a research fellow at Stanford University.

Further, polyunsaturated fats are nutritionally essential because they are the only type of fat that contains linoleic acid, a fatty acid the body cannot make. Vegetable oils and fish are the best dietary sources of polyunsaturates.

THE WAY IT IS

Although government guidelines say that dietary fat should be 30% or less of total calories, many nutrition experts believe that 20–25% is as high as people should go. Some say that overweight people—especially those with high LDL cholesterol—should consume no more than 15–20% of a reduced calorie intake from fat to encourage weight loss.

The current American diet contains about 7% of calories as polyunsaturated fat, which the National Cholesterol Education Program considers a reasonable level. Intake of polyunsaturated fat should not exceed 10% of total calories, according to the government guidelines. Monounsaturated fat can comprise up to an additional 15% of total calories. The major portion of monos should come from olive, canola, or other vegetable oils. Americans currently consume most of their monos from animal products such as butter, whole milk, cheese, ice cream, beef, pork, lamb, and poultry, which are also rich in harmful saturated fat (Fig. 4A). Certain plant products such as palm and coconut oils are also high in saturated fats.

The best way to lower dietary fat is to eat fewer animal products and drink skim or low-fat milk. Also, try to cut out packaged baked goods, snack foods, french fries, and crackers, which are very high in trans fatty acids. Instead, eat plenty of fruits, vegetables, whole grains, and legumes—all of which are dense with nutrients, as well as fiber.

People need not deny themselves an occasional ice-cream cone, slice of cake, or juicy steak. The key is to view these foods as special treats, not as routine parts of a diet. If a low-fat diet is eaten day in and day out, the question of whether to use an occasional pat of butter or dollop of margarine should be of little concern.

WHERE WE WERE

Until the mid-1980s, heart disease experts believed that polyunsaturated fat was the best type a person could consume, because it lowered serum cholesterol. Monounsaturated fats were thought to have no effect on blood lipids. Then, in 1985, a highly publicized study revealed that monounsaturated fats

Excerpted from the June 1996 issue of the Harvard Health Letter, © *1996, President and Fellows of Harvard College.*

Lipids Are Stored Energy

Fat and cholesterol are both lipids. Fat is present not only in butter, margarine, and oils, but also in many foods high in protein (see Fig. 4A). The body can alter ingested fat to suit the body's needs, except it is unable to produce linoleic acid, which is a polyunsaturated fatty acid. Saturated fatty acids contain the maximum number of hydrogen atoms; polyunsaturated fatty acids have double bonds because they have fewer hydrogen atoms.

The current guidelines suggest that fat account for no more than 30% of our daily calories. The chief reason is that an intake of fat not only causes weight gain, it also increases the risk of cancer and cardiovascular disease as discussed in the reading on page 96. Dietary fat apparently increases the risk of colon, hepatic, and pancreatic cancers. Although recent studies suggest no link between dietary fat and breast cancer, others still believe that the matter deserves further investigation.

Cardiovascular disease is often due to arteries occluded (blocked) by fatty deposits, called plaque, that contain saturated fats and cholesterol. Cholesterol is carried in the blood by two types of lipoproteins: low density lipoprotein (LDL) and high density lipoprotein (HDL). LDL is thought of as being "bad" because it carries cholesterol from the liver to the cells, while HDL is thought of as being "good" because it carries cholesterol to the liver, which takes it up and converts it to bile salts. Saturated fats, whether in butter or margarine, can raise LDL cholesterol levels, while monounsaturated (one double bond) fats and polyunsaturated (many double bonds) fats lower LDL cholesterol levels. Olive oil and canola oil contain monounsaturated fats; corn oil and safflower oil contain polyunsaturated fats. These oils have a liquid consistency and come from plants. Saturated fats, which are solids at room temperature, usually have an animal origin; two well-known exceptions are palm oil and coconut oil, which are saturated and come from the plants mentioned.

Nutritionists stress that it is more important for the diet to be low in fat rather than be overly concerned about which type fat is in the diet.

Vitamins Help Metabolism

Vitamins are organic compounds (other than carbohydrate, fat, and protein) that the body is unable to produce but uses for metabolic purposes. Many vitamins are portions of coenzymes, which are enzyme helpers. For example, niacin is part of the coenzyme NAD, and riboflavin is part of another dehydrogenase, FAD. Coenzymes are needed in only small amounts because each can be used over and over again. Not all vitamins are coenzymes; vitamin A, for example, is a precursor for the visual pigment that prevents night blindness. It has been known for several years that if vitamins are lacking in the diet, various symptoms develop (Fig. 4.16). Altogether there are 13 vitamins, which are divided into those that are fat soluble (Table 4.5) and those that are water soluble (Table 4.6).

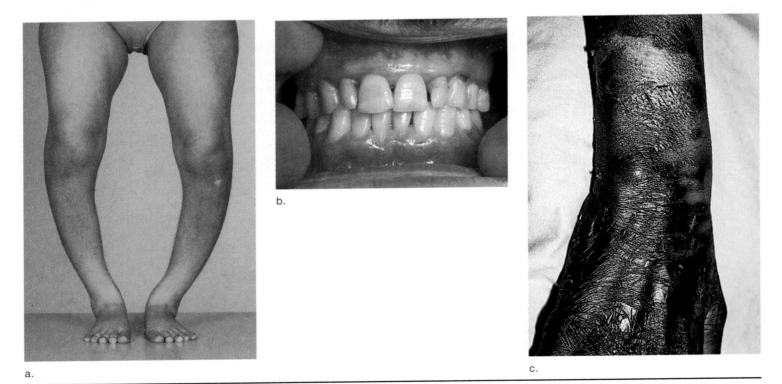

a.   c.

b.

Figure 4.16   Illnesses due to vitamin deficiency.
a. Bowing of bones (rickets) due to vitamin D deficiency. **b.** Bleeding of gums (scurvy) due to vitamin C deficiency. **c.** Dermatitis (pellagra) of areas exposed to light due to niacin (vitamin B_3) deficiency.

TABLE 4.5

Fat-Soluble Vitamins

Vitamin	Functions	Food Sources	Conditions with	
			Too Little	**Too Much**
Vitamin A	Antioxidant synthesized from beta-carotene; needed for synthesis of visual pigments and maintenance of healthy skin, hair, and mucous membranes, and for proper bone growth	Deep yellow/orange and leafy, dark green vegetables, fruits, cheese, whole milk, butter, and eggs	Night blindness, impaired growth of bones and teeth	Headache, dizziness, nausea, hair loss
Vitamin D	A group of steroids needed for development and maintenance of bones and teeth	Milk fortified with vitamin D, fish liver oil; also made in the skin when exposed to sunlight	Rickets, bone decalcification and weakening	Calcification of soft tissues, diarrhea, and possible renal damage
Vitamin E	Antioxidant that prevents oxidation of vitamin A and polyunsaturated fatty acids	Leafy green vegetables, fruits, vegetable oils and margarine, nuts, wheat germ and whole-grain breads and cereals	Unknown	Diarrhea, nausea, headaches, fatigue, muscle weakness
Vitamin K	Needed for synthesis of substances active in clotting of blood	Leafy green vegetables, cabbage and cauliflower; also made by bacteria in intestines of adults	Easy bruising and bleeding	Can interfere with anticoagulant medication

TABLE 4.6

Water-Soluble Vitamins

Vitamin	Functions	Food Sources	Conditions with	
			Too Little	**Too Much**
Vitamin C	Antioxidant; needed for forming collagen; helps maintain capillaries, bones and teeth; aids in absorption of iron and synthesis of hormones	Citrus fruits, leafy green vegetables, tomatoes, potatoes, cabbage	Scurvy, wounds heal slowly, lowered resistance to infection	Gout, kidney stones, diarrhea, decreased copper
Thiamine (Vitamin B_1)	Part of coenzyme needed for cellular respiration; also promotes activity of the nervous system	Whole-grain cereals, dried beans and peas, sunflower seeds, and nuts	Beriberi, muscular weakness, heart enlarges	Can interfere with absorption of other vitamins
Riboflavin (Vitamin B_2)	Part of coenzymes, such as FAD; aids cellular respiration, including oxidation of protein and fat	Nuts, dairy products, whole-grain cereals, poultry, and leafy green vegetables	Dermatitis, blurred vision, growth failure	Unknown
Niacin (Nicotinic acid)	Part of coenzymes NAD and NADP; needed for cellular respiration, including oxidation of protein and fat	Peanuts, poultry, whole-grain cereals, leafy green vegetables, and beans	Pellagra, diarrhea, and mental disorders	High blood sugar and uric acid, vasodilation, etc.
Folacin (Folic acid)	Coenzyme needed for production of hemoglobin and the formation of DNA	Dark leafy green vegetables, nuts, beans, whole-grain cereals	Megaloblastic anemia, spina bifida	May mask B_{12} deficiency
Vitamin B_6	Coenzyme needed for the synthesis of hormones and hemoglobin; CNS control	Whole-grain cereals, bananas, beans, poultry, nuts, leafy green vegetables, and avocados	Rarely, convulsions, vomiting, seborrhea, muscular weakness	Insomnia
Pantothenic acid	Part of coenzyme A needed for oxidation of carbohydrates and fats; aids in the formation of hormones and certain neurotransmitters	Nuts, beans, dark green vegetables, poultry, fruits, and milk	Rarely, loss of appetite, mental depression, muscle spasms, numbness	Unknown
Vitamin B_{12}	Complex, cobalt-containing compound; part of the coenzyme needed for synthesis of nucleic acids and myelin	Dairy products, fish, poultry, eggs, fortified cereals	Pernicious anemia	Unknown
Biotin	Coenzyme needed for metabolism of amino acids and fatty acids	Generally in foods, especially eggs; also made by bacteria in the intestines of humans	Skin rash, nausea, fatigue	Unknown

Some Vitamins Are Antioxidants

Over the past 20 years, numerous statistical studies have been done to determine whether a diet rich in fruits and vegetables is protective against cancer. Cellular metabolism generates free radicals, unstable molecules that carry an extra electron. The most common free radical in cells is oxygen in the unstable form, O_3^-. In order to stabilize themselves, free radicals donate an electron to another molecule like DNA, or proteins, including enzymes, and lipids, which are found in plasma membranes. Such attacks most likely damage these cellular molecules and thereby may lead to disorders, perhaps even cancer. Also, it appears that plaque formation in arteries may begin when arterial linings are injured by oxidized LDL cholesterol.

Vitamins C, E, and A are believed to defend the body against free radicals, and therefore, they are termed antioxidants. These vitamins are especially abundant in fruits and vegetables. The new dietary guidelines shown in Figure 4.13 suggest that we eat five servings of fruits and vegetables a day. It is not difficult to achieve the goal of having five servings of fruits and vegetables a day because a serving can be salad greens, raw or cooked vegetables, dried fruit, and fruit juice in addition to traditional apples and oranges and such.

Dietary supplements may provide a potential safeguard against cancer and cardiovascular disease, but it is important to realize that it is not appropriate to take supplements instead of improving intake of fruits and vegetables. There are many beneficial compounds in fruits that cannot be obtained from a vitamin pill. They enhance each other's absorption or action and also perform independent biological functions.

Vitamin D Makes Bones Strong

Skin cells contain a precursor cholesterol molecule that is converted to vitamin D after UV exposure. Since only a small amount of UV radiation is needed to change the precursor molecule to vitamin D, the skin should not be exposed unnecessarily. Vitamin D leaves the skin and is modified first in the kidneys and then in the liver until finally it is activated into a form called calcitriol. Calcitriol circulates throughout the body, regulating calcium uptake and metabolism. It promotes the absorption of calcium by the intestines, and the absence of vitamin D leads to rickets in children (see Fig. 4.16a). Rickets, characterized by a bowing of the legs, is caused by defective mineralization of the skeleton. Most milk today is fortified with vitamin D, which helps prevent the occurrence of rickets.

Vitamins are essential to cellular metabolism; many are protective against identifiable illnesses and conditions.

Minerals Are Needed Too

In addition to vitamins, various minerals are required by the body (Table 4.7). As discussed in the reading on page 101, plants are a good source of **minerals,** which are inorganic substances most often present in soil. Minerals are divided into macrominerals, which are recommended in amounts more than 100 mg per day, and microminerals (trace elements), which are recommended in amounts less than 20 mg per day. The macrominerals sodium, magnesium,

---

TABLE 4.7

Minerals: Their Role in the Body and Their Food Sources

Mineral	Major Role(s) in Body	Good Food Sources
Macrominerals (more than 100 mg/day needed)		
Calcium (Ca)	Strong bones and teeth; nerve conduction; muscle contraction	Dairy products, leafy green vegetables
Phosphorus (P)	Strong bones and teeth	Meat, dairy products, whole grains
Potassium (K)	Nerve conduction, muscle contraction	Many fruits and vegetables
Sodium (Na)	Nerve conduction, pH balance	Table salt
Chloride (Cl)	Water balance	Table salt
Magnesium (Mg)	Protein synthesis	Whole grains, leafy green vegetables
Microminerals (less than 20 mg/day needed)		
Zinc (Zn)	Wound healing, tissue growth	Whole grains, legumes, meats
Iron (Fe)	Hemoglobin synthesis	Whole grains, legumes, eggs
Fluorine (F)	Strong bones and teeth	Fluoridated drinking water, tea
Copper (Cu)	Hemoglobin synthesis	Seafood, whole grains, legumes
Iodine (I)	Thyroid hormone synthesis	Iodized table salt, seafood

Ecology Focus

Our Crops Deserve Better

When plants photosynthesize, they convert carbon dioxide from the air and water from the soil into a carbohydrate, namely glucose. Carbohydrates are organic food for plants and all other living things. Animals, including humans, either eat plants directly or eat animals that have eaten plants. When provided with a source of nitrogen and sulfur, plants have the metabolic capability to modify and change carbohydrates into all the other types of organic molecules they require. This makes it possible for vegetarians to not only acquire carbohydrates but also amino acids from plants. The usual slimness of vegetarians testifies that most plants are not an abundant source of fat, and this fact can only make the vegetarian diet even more appealing.

Plant cell metabolism utilizes most of the same vitamins and minerals that humans require. Tables 4.5, 4.6, and 4.7 show that plants are an important source of vitamins and minerals as well as organic food. Minerals are found in the soil in low concentrations, but plants are able to take them up and concentrate them. As the root system of a plant grows, it branches and branches again so that the roots are exposed to a tremendous amount of soil. It has been estimated that a rye plant has roots totaling about 900 kilometers in length. Water enters the roots by diffusion, but active transport is needed to acquire some minerals and to concentrate minerals within the organs of a plant. A plant uses a great deal of ATP for active transport.

We are lucky that plants can concentrate minerals, for animals, like ourselves, often are dependent on them for supplying such elements as potassium for cardiac contraction, phosphorus for strong bones, and manganese for metabolic reactions. Once plants have taken up minerals, they are often incorporated into other molecules, including amino acids, phospholipids, and nucleotides.

Unfortunately, the health of plants is affected by the very same pollutants that affect the health of humans. We have already mentioned that the expected increase in UV radiation due to the depletion of ozone will adversely affect the ability of plants to carry on photosynthesis. Ground-level ozone in smog destroys leaves and roots. Acid rain, which causes minerals to be leached from the soil, can even cause the death of plants. Global warming may cause the extinction of many plant species. Global warming is due to the buildup of carbon dioxide in the atmosphere from the combustion of gasoline, coal, and oil. Carbon dioxide acts like the glass of a greenhouse—it allows the sun's rays to pass through but traps the heat and doesn't allow it to escape. An increase in the average annual temperature is predicted.

The best climate for growing crops may shift to more rocky northern terrains with a resultant loss in crop yield. These problems don't begin to compare to the reduction in crop yield due to soil erosion, however. Soil erosion occurs when the wind blows and the rain washes away the topsoil because farmers have plowed the land in straight rows and/or left it without adequate cover. The U.S. Department of Agriculture estimates that erosion is causing a steady drop in the productivity of farmland equivalent to the loss of 1.25 million acres per year. Soil erosion can be halted by the implementation of proven techniques (e.g., contour farming, drip irrigation, no-till farming) that might also halt desertification, the further degradation of the land to desert conditions (Fig. 4B).

Considering how dependent we are on plants, we must remember that plants as well as animals, including humans, are affected when the quality of the environment is reduced.

Figure 4B   Desertification.

Proper management allows land to be productive for many years. Mismanagement can lead to desert conditions and the loss of land fertility. In this photograph, crops are being covered by windblown sand, and therefore, land productivity will be reduced.

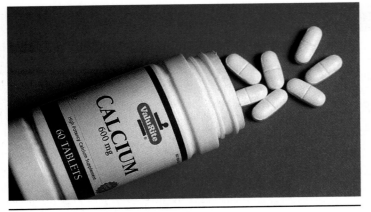

Figure 4.17   Calcium in the diet.
Many over-the-counter calcium supplements now are available to boost the amount in the diet. Some forms of calcium are more readily absorbed than other forms.

TABLE 4.8
Reducing Dietary Sodium
To reduce dietary sodium, the following suggestions are recommended.
1. Use spices instead of salt to flavor foods.
2. Add little or no salt to foods at the table, and add only small amounts of salt when you cook.
3. Eat unsalted crackers, pretzels, potato chips, nuts, and popcorn.
4. Avoid hot dogs, ham, bacon, luncheon meats, smoked salmon, sardines, and anchovies.
5. Avoid processed cheese and canned or dehydrated soups.
6. Avoid brine-soaked foods, such as pickles or olives.

phosphorus, chlorine, potassium, and calcium are constituents of cells and body fluids and are structural components of tissues. For example, calcium is needed for the construction of bones and teeth and for nerve conduction and muscle contraction.

The microminerals seem to have very specific functions. For example, iron is needed for the production of hemoglobin, and iodine is used in the production of thyroxin, a hormone produced by the thyroid gland. As research continues, more and more elements are added to the list of microminerals considered to be essential. During the past three decades, for example, very small amounts of molybdenum, selenium, chromium, nickel, vanadium, silicon, and even arsenic have been found to be essential to good health.

Occasionally, individuals do not receive enough iron (especially women), calcium, magnesium, or zinc in their diet. Adult females need more iron in the diet than males (18 mg compared to 10 mg) because they lose hemoglobin each month during menstruation. Stress can bring on a magnesium deficiency, and due to its high-fiber content, a vegetarian diet may make zinc less available to the body. However, a varied and complete diet usually supplies enough of each type of mineral.

Calcium and Bone Disease

There has been much public interest in the taking of calcium supplements (Fig. 4.17) to counteract *osteoporosis,* a degenerative bone disease that afflicts an estimated one-fourth of older men and one-half of older women in the United States. Osteoporosis develops because bone-eating cells called osteoclasts are more active than bone-forming cells called osteoblasts. Therefore, the bones are porous, and they break easily because they lack sufficient calcium. Due to recent studies that show consuming more calcium does slow bone loss in elderly people, the guidelines have been revised. A calcium intake of 1,000 mg a day is recom-

mended for men and women who are premenopausal or who use estrogen replacement therapy, and 1,500 mg a day is recommended for postmenopausal women who do not use estrogen replacement therapy. To achieve this amount, supplemental calcium is most likely necessary.

Estrogen replacement therapy and exercise in addition to calcium supplements are effective means to prevent osteoporosis. Medications are also available that slow bone loss while increasing skeletal mass. Etidronate disodium needs to be taken cyclically or else weak, abnormal bone instead of normal bone develops. Fosamax is a new drug that can be taken continuously but may accumulate in the skeleton. Therefore, long-term studies are needed to determine if the drug can safely be given for a long time.

Presently, calcium supplements, estrogen therapy for women, and exercise are the best ways to prevent osteoporosis.

Too Much Sodium

The recommended amount of sodium intake per day is 500 mg, although the average American takes in 4,000–4,700 mg every day. In recent years, this imbalance has caused concern because high sodium intake has been linked to hypertension (high blood pressure) in some people. About one-third of the sodium we consume occurs naturally in foods; another one-third is added during commercial processing; and we add the last one-third either during home cooking or at the table in the form of table salt.

Clearly, it is possible for us to cut down on the amount of sodium in the diet. Table 4.8 gives recommendations for doing so.

Excess sodium in the diet can lead to hypertension; therefore, excess sodium intake should be avoided.

Analyzing Eating Disorders

Authorities recognize three primary eating disorders: obesity, bulimia, and anorexia nervosa. Although they exist in a continuum as far as body weight is concerned, there is much overlap between them.

Obesity is defined as a body weight of more than 20% above the ideal weight for the sex and age of the person. It most likely is caused by a combination of factors, including genetic, hormonal, metabolic, and social factors. The social factors include the eating habits of other family members. There are all sorts of diets available, and it is important to choose one that still supplies all necessary nutrients. Dieting combined with exercise is usually recommended for weight loss because muscle burns more calories than any other type of tissue. There should be a lifelong commitment to a properly planned program —otherwise the person may get into a cycle of weight gain followed by weight loss. Consulting a physician is a good idea in order to bring body weight down to normal and keep it there permanently.

> **Obesity has many complex causes that possibly can be detected by a physician.**

Bulimia can coexist with either obesity or anorexia nervosa. People, usually young women, who are afflicted, have the habit of eating to excess and then purging themselves by some artificial means, such as vomiting or laxatives. Frequent vomiting can lead to a sore throat and an ulcerated stomach lining, but even so, the binge-and-purge cycle is very hard to break. Bulimic individuals are usually depressed, but whether the depression causes or is caused by bulimia has not been determined, and the possibility of a hormonal disorder is still being considered.

Anorexia nervosa is diagnosed when an individual, typically an adolescent girl, is extremely thin but still claims to "feel fat" and continues to diet. In addition to eating only low-calorie foods, further weight loss is assured by self-induced vomiting, taking laxatives, or by intense exercise. Anorexia nervosa is accompanied by low blood pressure, an irregular heartbeat, constipation, and constant chilliness. Menstruation ceases as the body weight plunges. It is possible that anorexic individuals have an incorrect body image that makes them think they are fat. It is also possible they have various psychological problems, including a desire to suppress their sexuality. Also, some clinicians suggest that not eating may be a way for individuals to exert some control in their lives. Others, such as parents or spouses, may control most of their activities, but they cannot control eating.

> **Both bulimia and anorexia nervosa are serious disorders that require the assistance of competent medical personnel.**

SUMMARY

4.1 The Digestive System

The salivary glands send saliva into the mouth, where the teeth chew the food and the tongue forms a bolus for swallowing.

The air passage and food passage cross in the pharynx. When one swallows, the air passage is usually blocked off and food must enter the esophagus where peristalsis begins.

The stomach expands and stores food. While food is in the stomach, it churns, mixing food with the acidic gastric juices.

The walls of the small intestine have fingerlike projections called villi where nutrient molecules are absorbed into the circulatory and lymphatic systems.

The large intestine consists of the cecum, colons (ascending, transverse, descending, and sigmoid), and the rectum, which ends at the anus.

The large intestine does not produce digestive enzymes; it does absorb water, salts, and some vitamins.

4.2 Three Accessory Organs

The three accessory organs of digestion—the pancreas, liver, and gallbladder—send secretions to the duodenum via ducts. The pancreas produces pancreatic juice, which contains digestive enzymes for carbohydrate, protein, and fat.

The liver produces bile, which is stored in the gallbladder. The liver receives blood from the small intestine by way of the hepatic portal vein. It has numerous important functions, and any malfunction of the liver is a matter of considerable concern.

4.3 Digestive Enzymes

Digestive enzymes are present in digestive juices and break down food into the nutrient molecules glucose, amino acids, fatty acids, and glycerol (see Table 4.2). Glucose and amino acids are absorbed into the blood capillaries of the villi. Fatty acids and glycerol rejoin to produce fat, which enters the lacteals.

Digestive enzymes have the usual enzymatic properties. They are specific to their substrate and speed up specific reactions at body temperature and optimum pH.

4.4 Working Together

The digestive system works with the other systems of the body in the ways described in the box on page 92.

4.5 Nutrition

The nutrients released by the digestive process should provide us with an adequate amount of energy, essential amino acids and fatty acids, and all necessary vitamins and minerals.

The bulk of the diet should be carbohydrates (like bread, pasta, and rice) and fruits and vegetables. These are low in saturated fatty acids and cholesterol molecules, whose intake is linked to cardiovascular disease. The vitamins A, E, and C are antioxidants that protect cell contents from damage due to free radicals.

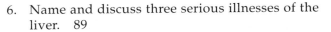

STUDYING THE CONCEPTS

1. List the organs of the digestive tract, and state the contribution of each to the digestive process.   80–86

2. Discuss the absorption of the products of digestion into the lymphatic and circulatory systems.   85

3. Name and state the functions of the hormones that assist the nervous system in regulating digestive secretions.   86

4. Name the accessory organs, and describe the part they play in the digestion of food.   88–89

5. Choose and discuss any three functions of the liver.   88–89

6. Name and discuss three serious illnesses of the liver.   89

7. Discuss the digestion of starch, protein, and fat, listing all the steps that occur to bring about digestion of each of these.   90–91

8. What is the chief contribution of each of these constituents of the diet: a. carbohydrates; b. proteins; c. fats; d. fruits and vegetables?   94–98

9. Why should the amount of saturated fat be curtailed in the diet?   96–98

10. Name and discuss three eating disorders.   103

TESTING YOUR KNOWLEDGE

1. When swallowing, the _____ covers the opening to the larynx.

2. The _____ takes food to the stomach, where _____ digestion is started.

3. The products of digestion are absorbed into the cells of the _____, fingerlike projections of the intestinal wall.

4. a. Label each organ indicated in the diagram.
   b. For the arrows, use either glucose, amino acids, lipids, or water.

5. The pancreas sends digestive juices to the _____, the first part of the small intestine.

6. After eating, the liver stores glucose as _____.

7. The gallbladder stores _____, a substance that _____ fat.

8. In the mouth, salivary _____ digests starch to _____.

9. Pancreatic juice contains _____ for digesting protein, _____ for digesting starch, and _____ for digesting fat.

10. Whereas a(n) _____ pH is optimum for pepsin, a(n) _____ pH is optimum for the enzymes found in pancreatic juice

11. The diet should include a complete protein source, one that includes all the _____.

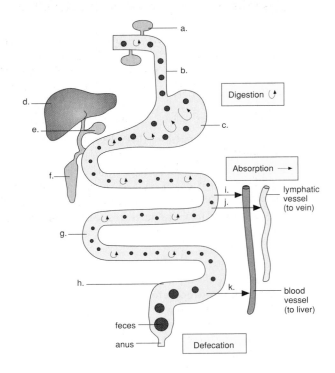

APPLYING YOUR KNOWLEDGE

Concepts

1. Perhaps you have experienced the undesirable result of laughing while swallowing a drink and part of the drink has exited by way of your nose. How does this occur?

2. Enzymes are named for the substrate they act upon and the action they catalyze. Succinic acid dehydrogenase is the name of a specific protein. Tell as much as you can about this molecule from its name.

3. The type of food one eats directly affects the texture and color of one's feces. A pale-colored feces often results from eating mainly vegetables and little red meat. Explain this.

4. A friend of yours is overweight and wants to lose about 50 pounds. Therefore, she has decided to eliminate fats from her food intake. What advice do you have for her?

Bioethical Issue

Glance down any health-food store shelf, and you'll find dozens of bottled vitamins and minerals touting physical benefits. In particular, a growing number of stores have begun to sell bottles of "phytochemicals": isolated substances, such as magnesium or selenium, found in fruits and vegetables.

Many of these supplements may be beneficial. But public health advocates worry that the products sometimes do more harm than good. For example, scientists recently cut short a study of beta-carotene's cancer-prevention qualities. Why? They learned that participants taking beta-carotene actually experienced "higher" rates of cancer than those without the supplement. This finding was a major surprise.

Is it ethical for a company to advertise natural supplements as cancer preventers, even if researchers have little evidence? Should the government check the accuracy of these claims, or should you as a consumer be responsible for weighing the health effects of these products?

APPLYING TECHNOLOGY

Your study of the digestive system and nutrition is supported by these available technologies:

Exploring the Internet

The Mader Home Page provides further resources for studying this chapter.

`http://www.mhhe.com/sciencemath/biology/mader/`

(Click on *Human Biology*.)

Dynamic Human: Digestive System CD-ROM

In *Gross Anatomy,* major organs are reviewed with some presented in 3-D; in *Explorations,* certain organs are studied in detail; in *Histology,* sections of cadavers are highlighted to reveal organs, and microscopic slides are highlighted to reveal tissues; in *Clinical Concepts,* barium radiography and endoscopy are shown, and gallstones and ulcers are explained.

Explorations in Human Biology CD-ROM

Life Span and Life Style (#3)   Life expectancy varies as students change the amount of smoking, drinking, fat in diet, and exercise.   (2)*

Diet and Weight Loss (#7)   Students develop daily diets and then see how weight is gained in relation to the diet and various activity patterns.   (4)

Life Science Animations Video

Video #4:   Animal Biology, Part 2

Peristalsis (#33)   Movement of a bolus occurs as muscular contraction occurs along the esophagus.   (1)

Digestion of Carbohydrates (#34)   A review of the digestive system is given before carbohydrate digestion occurs in the mouth and small intestine; sugars are seen entering the villi.   (2)

Digestion of Proteins (#35)   Digestion of proteins occurs in the stomach and small intestine; peptides and amino acids are seen entering the villi.   (4)

Digestion of Lipids (#36)   A review of lipids is given before fat digestion occurs in the small intestine, and micelles (coloidal particles) are seen entering the villi.   (2)

---

*Level of difficulty

SELECTED KEY TERMS

amylase (AM-i-lays) Starch-digesting enzyme secreted by the salivary glands (salivary amylase) and the pancreas (pancreatic amylase).   90

anorexia nervosa Eating disorder caused by the fear of becoming obese; includes loss of appetite and inability to maintain a normal minimum body weight.   103

anus Outlet of the digestive tract.   86

appendix Small, tubular projection that extends outward from the cecum of the large intestine.   86

bile Secretion of the liver that is temporarily stored in the gallbladder before being released into the small intestine, where it emulsifies fat.   88

bulimia Eating disorder of binge eating followed by purging.   103

colon Large intestine that extends from the cecum to the rectum.   86

defecation Discharge of feces from the rectum through the anus.   86

duodenum (doo-uh-DEE-num) First portion of the small intestine into which secretions from the liver and pancreas enter.   85

epiglottis (ep-uh-GLAHT-us) Structure that covers the glottis and closes off the air tract during the process of swallowing.   82

esophagus (i-SAHF-uh-gus) Tube that transports food from the pharynx to the stomach.   82

essential amino acids Amino acids required in the human diet because the body cannot make them.   95

fiber Plant material that is nondigestible; cellulose and lignin.   94

gallbladder Saclike organ associated with the liver that stores and concentrates bile.   89

gastric gland Gland within the stomach wall that secretes gastric juice.   84

glottis (GLAHT-us) Opening for airflow into the larynx.   82

hard palate Bony, anterior portion of the roof of the mouth.   80

hydrolytic enzyme Enzyme that catalyzes a reaction in which the substrate is broken down with the addition of water.   90

lacteal (LAK-tee-ul) Lymphatic vessel in a villus of the intestinal wall.   85

large intestine Last major portion of the digestive tract, extending from the small intestine to the anus and consisting of the cecum, the colon, the rectum, and the anal canal.   86

lipase Fat-digesting enzyme secreted by the pancreas.   90

liver Large organ in the abdominal cavity that has many functions, such as production of proteins and bile and detoxification of harmful substances.   88

lumen Cavity inside any tubular structure, such as the lumen of the digestive tract.   83

maltase Enzyme produced in small intestine that breaks down maltose to two glucose molecules.   90

mineral Homogeneous inorganic substance; certain minerals are required for normal metabolic functioning of cells and must be in the diet.   100

obesity Excess adipose tissue; exceeding desirable weight by more than 20%.   103

pancreas Elongated, flattened organ in the abdominal cavity that secretes digestive enzymes into the duodenum (exocrine function) and hormones into the blood (endocrine function).   88

pepsin Protein-digesting enzyme secreted by gastric glands.   90

peptidase Intestinal enzyme that breaks down short chains of amino acids to individual amino acids that are absorbed across the intestinal wall.   90

periodontitis Inflammation of the gums.   81

peristalsis (per-uh-STAWL-sus) Rhythmic contraction that serves to move the contents along in tubular organs, such as the digestive tract.   82

pharynx (FAR-ingks) Portion of the digestive tract between the mouth and the esophagus that serves as a passageway for food and also air on its way to the trachea.   82

salivary gland Gland associated with the oral cavity that secretes saliva.   81

small intestine Long, tubelike chamber of the digestive tract between the stomach and large intestine.   85

soft palate Entirely muscular posterior portion of the roof of the mouth.   80

sphincter Muscle that surrounds a tube and closes or opens the tube by contracting and relaxing.   83

stomach Muscular sac that mixes food with gastric juices to form chyme, which enters the small intestine.   84

trypsin Protein-digesting enzyme secreted by the pancreas.   90

villus Fingerlike projection that lines the small intestine and functions in absorption.   85

vitamin Essential requirement in the diet, needed in small amounts; often a part of a coenzyme.   98

Chapter 5

Composition and Function of Blood

Chapter Outline

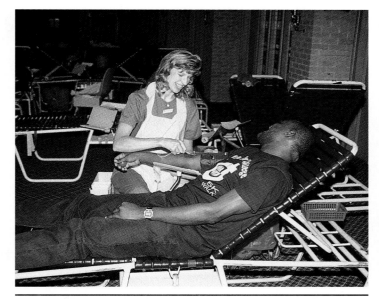

Figure 5.1   Giving blood.
Can your blood help save someone's life? It can if you are healthy and the components of your blood are just as they should be.

In 1961, a Boston researcher named Victor Herbert went on one of the blandest diets ever. Everything Herbert ate—from Thanksgiving dinner to simple applesauce—was boiled three times. Why? To remove all traces of folic acid, an essential vitamin, from Herbert's food.

Herbert had a hunch that diets low in folic acid could lead to anemia, a condition in which red cells or other elements in the blood malfunction, failing to carry sufficient oxygen. Anemic patients grow weak and, if untreated, can suffer permanent nervous system damage. The transport of oxygen is so critical that an insufficient amount of hemoglobin may make it necessary to give a patient blood donated by someone else (Fig. 5.1). Not only does blood transport oxygen, it also transports nutrients and hormones to the body's cells. In the tissues, blood also takes away carbon dioxide and other wastes given off by cell.

Herbert's folic acid theory proved correct. After some five months of boiled food, a test showed that Herbert's red blood cell activity was deteriorating. For the first time, researchers realized folic acid's link to anemia.

Today, doctors routinely prescribe folic acid to pregnant women. Some manufacturers add the vitamin to food. As for Herbert, his work led to academic success. Not to mention a new appreciation for spicy food!

Disorders of blood, from anemia to AIDS, can be devastating to the body. This chapter covers the functions of the blood and its components.

FORMED ELEMENTS	Function and Description	Source
Red Blood Cells (erythrocytes)	Transport O_2 and help transport CO_2	Red bone marrow
4 million–6 million per mm³ blood	7–8 μm in diameter Bright-red to dark-purple biconcave disks without nuclei	
White Blood Cells (leukocytes)	Fight infection	Red bone marrow
Granular leukocytes		
• Basophil 20–50 per mm³ blood	10–12 μm in diameter Spherical cells with lobed nuclei; large, irregularly shaped, deep-blue granules in cytoplasm	
• Eosinophil 100–400 per mm³ blood	10–14 μm in diameter Spherical cells with bilobed nuclei; coarse, deep-red, uniformly sized granules in cytoplasm	
• Neutrophil 3,000–7,000 per mm³ blood	10–14 μm in diameter Spherical cells with multilobed nuclei; fine, pink granules in cytoplasm	
Agranular leukocytes		
• Lymphocyte 1,500–3,000 per mm³ blood	5–17 μm in diameter (average 9–10 μm) Spherical cells with large round nuclei	
• Monocyte 100–700 per mm³ blood	10–24 μm in diameter Large spherical cells with kidney-shaped, round, or lobed nuclei	
• **Platelets** (thrombocytes) 150,000–300,000 per mm³ blood	Aid clotting 2–4 μm in diameter Disk-shaped cell fragments with no nuclei; purple granules in cytoplasm	Red bone marrow

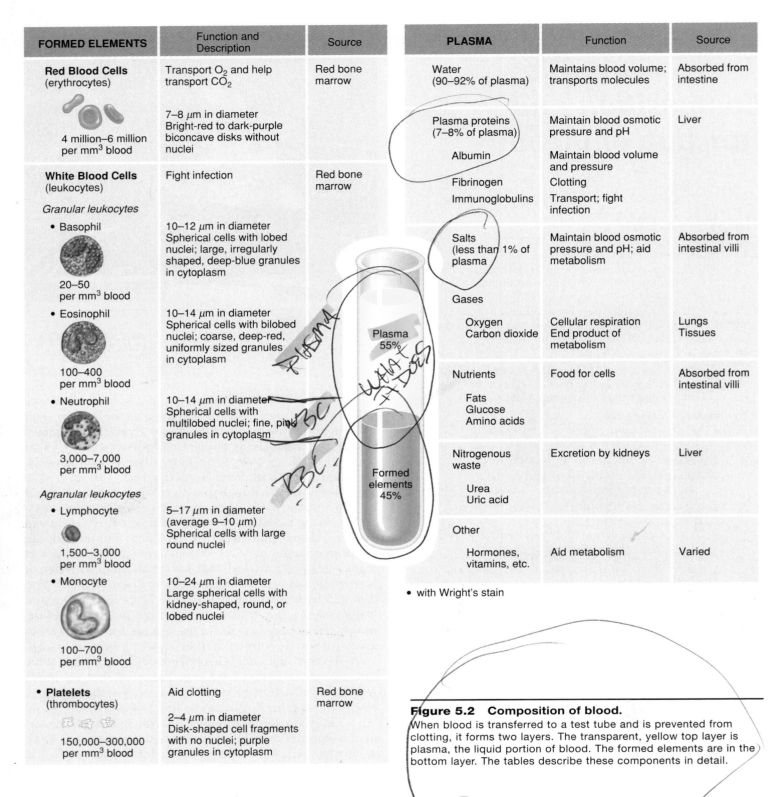

PLASMA	Function	Source
Water (90–92% of plasma)	Maintains blood volume; transports molecules	Absorbed from intestine
Plasma proteins (7–8% of plasma)	Maintain blood osmotic pressure and pH	Liver
Albumin	Maintain blood volume and pressure	
Fibrinogen	Clotting	
Immunoglobulins	Transport; fight infection	
Salts (less than 1% of plasma)	Maintain blood osmotic pressure and pH; aid metabolism	Absorbed from intestinal villi
Gases		
Oxygen	Cellular respiration	Lungs
Carbon dioxide	End product of metabolism	Tissues
Nutrients	Food for cells	Absorbed from intestinal villi
Fats Glucose Amino acids		
Nitrogenous waste	Excretion by kidneys	Liver
Urea Uric acid		
Other		
Hormones, vitamins, etc.	Aid metabolism	Varied

• with Wright's stain

Figure 5.2   Composition of blood.
When blood is transferred to a test tube and is prevented from clotting, it forms two layers. The transparent, yellow top layer is plasma, the liquid portion of blood. The formed elements are in the bottom layer. The tables describe these components in detail.

The transport of oxygen relies on the respiratory pigment, hemoglobin, which is contained within the red blood cells. If a person has anemia, there is not enough red blood cells or not enough hemoglobin in the red blood cells. Then, mitochondria don't get enough oxygen, resulting in a lack of ATP, which causes the person to have a tired, run-down feeling.

White blood cells fight infection. Each type of white blood cell has its own function. Lymphocytes produce antibodies that combine with antigens, which are often borne on the surface of *pathogens* (infectious bacteria and viruses). The AIDS virus attacks a particular type of lymphocyte, and that's why a person with AIDS eventually succumbs to infectious diseases.

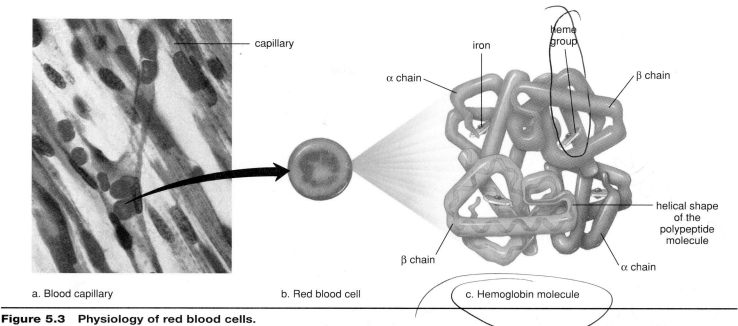

Figure 5.3   Physiology of red blood cells.
a. Red blood cells move single file through the capillaries. **b.** Each red blood cell is a biconcave disk containing many molecules of hemoglobin, the respiratory pigment. **c.** Hemoglobin contains four polypeptide chains, two of which are alpha (α) chains and two of which are beta (β) chains. There is an iron-containing heme group in the center of each chain. Oxygen combines loosely with iron when hemoglobin is oxygenated.

Aside from transport and defense, blood has certain regulatory functions. It helps regulate body temperature by removing heat from active muscles and taking it to the skin, where it can be given off. It also helps regulate the pH through the action of buffers in the blood. It helps maintain blood pressure by the presence of salts and proteins in the blood and by preventing fluid loss through blood clotting. A component of blood called platelets helps in the blood-clotting process.

If blood is transferred from a person's vein to a test tube and is prevented from clotting, it separates into two layers (Fig. 5.2). The lower layer consists of red blood cells (erythrocytes), white blood cells (leukocytes), and blood platelets (thrombocytes). Collectively, these are called the **formed elements.** Formed elements make up about 45% of the total volume of whole blood. The upper layer is plasma, which contains a variety of inorganic and organic molecules dissolved or suspended in water. Plasma accounts for about 55% of the total volume of whole blood.

The functions of blood contribute to homeostasis, a dynamic equilibrium of the internal environment. Cells are surrounded by tissue fluid whose composition must be kept within relatively narrow limits or the cells cease to function in an effective manner. Only if the composition of blood is within the range of normality can tissue fluid also have the correct composition. All of the functions of blood are necessary to keep tissue fluid relatively stable.

Blood functions to maintain homeostasis so that the environment of cells (tissue fluid) remains relatively stable.

5.1   Red Blood Cells

Red blood cells (erythrocytes) are small, biconcave disks that lack a nucleus when mature. They occur in great quantity; there are 4 to 6 million red blood cells per mm^3 of whole blood. Each red blood cell is packed with hemoglobin, a pigment that combines with oxygen. The absence of a nucleus provides more space for hemoglobin.

Hemoglobin

Hemoglobin is called a respiratory pigment because it carries oxygen, and it is red in color. When hemoglobin combines with oxygen, the combination is called oxyhemoglobin. Oxyhemoglobin gives arterial blood its bright red color. Each red blood cell contains about 200 million hemoglobin molecules. If this much hemoglobin were suspended within the plasma rather than enclosed within the cells, blood would be so thick the heart would have difficulty pumping it.

Each hemoglobin molecule contains four polypeptide chains that make up the protein globin, and each chain is associated with heme, a complex iron-containing group (Fig. 5.3). Although the iron portion of hemoglobin carries oxygen, the equation for oxygenation of hemoglobin is usually written as

$$Hb + O_2 \underset{\text{tissues}}{\overset{\text{lungs}}{\rightleftharpoons}} HbO_2$$

The hemoglobin on the right, which is combined with oxygen, is called oxyhemoglobin. Oxyhemoglobin has a bright red color. The hemoglobin on the left, which has given up oxygen to tissue fluid, is called deoxyhemoglobin.

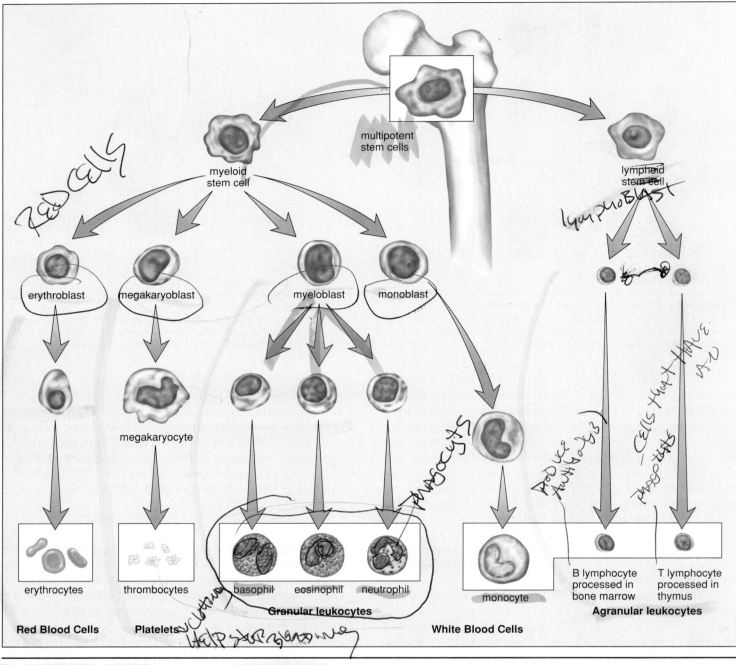

Figure 5.4   **Blood cell formation in red bone marrow.**
Multipotent stem cells give rise to specialized stem cells. The myeloid stem cell gives rise to still other cells, which become red blood cells, platelets, and all the white blood cells except lymphocytes. The lymphoid stem cell gives rise to lymphocytes.

Deoxyhemoglobin is a dark purplish color. Oxyhemoglobin forms in the lungs, where hemoglobin combines with oxygen. Deoxyhemoglobin forms in the tissues, where hemoglobin gives up oxygen.

Plasma carries only about 0.3 ml of oxygen per 100 ml, but whole blood carries 20 ml of oxygen per 100 ml. This shows that hemoglobin increases the oxygen-carrying capacity of blood more than 60 times. As we shall see, hemoglobin also assists in the transport of carbon dioxide.

Carbon monoxide, present in combustion gases from automobiles, furnaces, stoves, and cigarette smoke, combines with hemoglobin more readily than does oxygen, and it stays combined for several hours, making hemoglobin unavailable for oxygen transport. In New York City traffic, the blood concentration of carbon monoxide has been shown to reach 5.8%, a dangerous level when compared with the 1.5% that physicians consider safe. The reading on page 112 discusses this component of air pollution.

Life Cycle of Red Blood Cells

In infants, red blood cells are produced in the red bone marrow of all bones, but in adults, production primarily occurs in the red bone marrow of the skull bones, ribs, sternum, vertebrae, and pelvic bones.

The number of red blood cells produced increases whenever arterial blood carries a reduced amount of oxygen, as happens when an individual first takes up residence at a high altitude or loses red blood cells or full use of their lungs. Under these circumstances, the kidneys release an enzyme that converts a plasma protein to *erythropoietin*, a hormone that is carried in blood to red bone marrow. Once there, it stimulates further production of red blood cells. The liver and other tissues also produce erythropoietin. Erythropoietin, now mass-produced through biotechnology, is sometimes abused by athletes in order to raise their red blood cell counts and thereby increase the oxygen-carrying capacity of their blood.

All blood cells, including erythrocytes, are formed from special red bone marrow cells called *stem cells* (Fig. 5.4). A stem cell is ever capable of dividing and producing new cells that differentiate into specific type cells. Myeloid stem cells produce erythroblasts, which undergo various stages of maturation to become red blood cells. As they become mature, red blood cells lose their nucleus and acquire hemoglobin (Fig. 5.5). Possibly because they lack a nucleus, red blood cells live only about 120 days. As they age, they are destroyed in the liver and spleen, where they are engulfed by macrophages, which are large phagocytic cells. It is estimated that about 2 million red blood cells are destroyed per second, and therefore an equal number must be produced to keep the red blood cell count in balance.

When red blood cells are broken down, the hemoglobin is released. The globin portion of the hemoglobin is broken down into its component amino acids, which are recycled by the body. The iron is recovered and is returned to the bone marrow for reuse. The heme portion of the molecule undergoes chemical degradation and is excreted as bile pigments by the liver into the bile. These are the bile pigments bilirubin and biliverdin, which contribute to the color of feces. Chemical breakdown of heme is also what causes a bruise of the skin to change color from red/purple to blue to green to yellow.

Red blood cells are produced in red bone marrow after stem cells divide to form cells that undergo maturation in stages. Red blood cells live only 120 days and are destroyed by phagocytic cells in the liver and spleen.

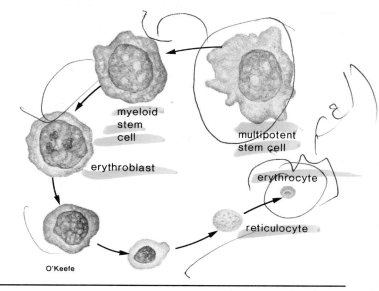

Figure 5.5 Maturation of red blood cells (erythrocytes). Red blood cells are made in the bone marrow, where stem cells continuously divide. During the maturation process, a red blood cell loses its nucleus, gains hemoglobin, and gets much smaller.

Anemia

When there is an insufficient number of red blood cells or the cells do not have enough hemoglobin, the individual suffers from **anemia** and has a tired, run-down feeling. In some types of anemia, the hemoglobin blood level is low. It may be that the diet does not contain enough iron or folic acids. Certain foods, such as whole-grain cereals, are rich in iron and folic acid, and the inclusion of these in the diet can help to prevent anemia.

In another type of anemia, called pernicious anemia, the digestive tract is unable to absorb enough vitamin B_{12}, found in dairy products, fish, eggs, and poultry. This vitamin is essential to the proper formation of red blood cells; without it, immature red blood cells tend to accumulate in the bone marrow in large quantities. A special diet and administration of vitamin B_{12} by injection is an effective treatment for pernicious anemia.

Hemolysis is the rupturing of red blood cells. Hemolytic anemia is seen in *sickle-cell disease* because the sickle-shaped red blood cells tend to rupture. Hemolytic disease of the newborn, which is discussed at the end of this chapter (p. 119), is also a type of hemolytic anemia.

Illness (anemia) results when the blood has too few red blood cells and/or not enough hemoglobin.

Ecology Focus

Carbon Monoxide, A Deadly Poison

Carbon monoxide (CO) is an air pollutant that primarily comes from the incomplete combustion of natural gas and gasoline. Figure 5A shows that transportation contributes most of the carbon monoxide to our cities' air. But power plants, factories, waste incineration, and home heating also contribute to the carbon monoxide level. Cigarette smoke contains carbon monoxide and is delivered directly to the smoker's blood and also to nonsmokers nearby.

Because carbon monoxide is a colorless, odorless gas, people can be unaware that it is affecting their systems. But it binds to iron 200 times more tightly than oxygen. Hemoglobin contains iron and so does cytochrome oxidase, the carrier in the electron transport system that passes electrons on to oxygen. When these molecules bind preferentially to carbon monoxide, they cannot perform their usual functions. The end result is that delivery of oxygen to the cells is impaired and so is the functioning of

mitochondria, the organelle that carries on cellular respiration.

Flushed red skin, especially on facial cheeks, is a first sign of carbon monoxide poisoning because hemoglobin bound to carbon monoxide is a brighter red than oxygenated hemoglobin. Marked euphoria, then sleepiness, coma, and death follow. Removing a person from the carbon monoxide source is not sufficient treatment because carbon monoxide, unlike oxygen, remains tightly bound to iron for many hours. A transfusion of red blood cells will help to increase the carrying capacity of the blood, and pure oxygen given under pressure will displace some carbon monoxide. Despite good medical care, some people still die each year from CO poisoning.

The level of carbon monoxide in polluted air may not be sufficient to kill people, but it does interfere with the ability of the body to function properly. The elderly and those with cardiovascular disease are especially at risk. One study found that even levels once thought to be safe can cause angina patients to experience chest pains

because oxygen delivery to the heart is reduced.

We can all help prevent air pollution and reduce the amount of carbon monoxide in the air by doing the following:

- Don't smoke (especially indoors).

- Walk or bicycle instead of driving.

- Use public transportation instead of driving.

- Heat your home with solar energy—not furnaces.

- Support the development of more efficient automobiles, factories, power plants, and home furnaces.

- Support the development of alternative fuels. When the gas hydrogen is burned, for example, the result is water, not carbon monoxide and carbon dioxide.

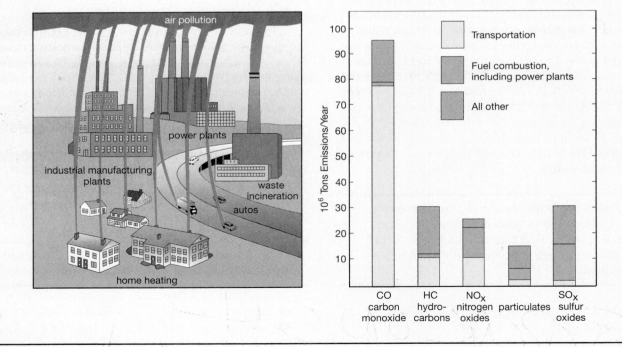

Figure 5A   Air pollution.
Air pollutants (carbon monoxide, hydrocarbons, nitrogen oxides, particulates, and sulfur oxides) enter the atmosphere from the sources noted.

5.2 White Blood Cells

White blood cells (leukocytes) differ from red blood cells in that they are usually larger, have a nucleus, lack hemoglobin, and without staining, are translucent. White blood cells are not as numerous as red blood cells. There are only 5,000–11,000 per mm³ of blood. White blood cells fight infection in ways that are discussed at greater length in chapter 7, which concerns immunity.

White blood cells are derived from stem cells in the red bone marrow, and they, too, undergo several maturation stages (see Fig. 5.4). There is an increase in the production of white blood cells whenever the body is invaded by pathogens. Each type of white blood cell seems capable of producing specific growth factors that circulate back to the bone marrow and stimulate their increased production.

Red blood cells are confined to the blood, but white blood cells are also found in lymph and tissue fluid. They are able to squeeze through pores in the capillary wall and therefore, they are found in tissue fluid and lymph (Fig. 5.6). When there is an infection, white blood cells greatly increase in number. Many white blood cells live only a few days—they probably die while engaging pathogens. Others live months or even years.

White blood cells fight infection. They defend us against pathogens that have invaded the body.

Types of White Blood Cells

It is possible to divide white blood cells into the **granular leukocytes** and the **agranular leukocytes.** Both types of cells have granules in the cytoplasm surrounding the nucleus, but they are more visible upon staining in granular leukocytes. The granules contain various enzymes and proteins, which help white blood cells defend the body. There are three types of granular leukocytes and two types of agranular leukocytes. They differ somewhat by the size of the cell and the shape of the nucleus (see Fig. 5.2), and they also differ in their functions.

Granular Leukocytes

Neutrophils are the most abundant of the white blood cells. They have a multilobed nucleus joined by nuclear threads; therefore, they are also called *polymorphonuclear.* They have granules that do not significantly take up the stain eosin, a pink to red stain, or a basic stain that is blue to purple. (This accounts for their name, neutrophil.) Neutrophils are the first type of white blood cells to respond to an infection, and they engulf pathogens during phagocytosis.

Eosinophils have a bilobed nucleus, and their large, abundant granules take up eosin and become a red color. (This accounts for their name, eosinophil.) Not much is known specifically about the function of eosinophils, but they are known to increase in number when there is a parasitic worm infection or in the case of allergic reactions.

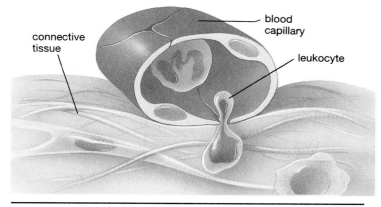

Figure 5.6 Mobility of white blood cells.
White blood cells can squeeze between the cells of a capillary wall and enter the tissues of the body.

Basophils have a U-shaped or lobed nucleus. Their granules take up the basic stain and become a dark blue color. (This accounts for their name, basophil.) Basophils enter the tissues and are believed to become mast cells, which release the histamine associated with allergic reactions. Histamine dilates blood vessels and causes contraction of smooth muscle.

Agranular Leukocytes

The agranular leukocytes include monocytes, which have a kidney-shaped nucleus, and lymphocytes, which have a spherical-shaped nucleus. These cells are responsible for *immunity* which is specific resistance to particular pathogens and their toxins (poisonous substances). Pathogens display antigens on their outer surface that mark them as being foreign to the body.

Monocytes are the largest of the white blood cells, and after taking up residence in the tissues, they differentiate into even larger macrophages. Macrophages phagocytize pathogens, old cells, and cellular debris. They also stimulate other white blood cells to defend the body.

The **lymphocytes** are of two types, B lymphocytes and T lymphocytes. B lymphocytes protect us by producing antibodies that combine with *antigens.* T lymphocytes, on the other hand, directly destroy any cell that bears antigens. B lymphocytes and T lymphocytes are discussed more fully in chapter 7.

White blood cells are divided into the granular leukocytes and the agranular leukocytes. Each type of white blood cell has a specific role to play in defending the body against disease.

Leukemia

Leukemia is characterized by an abnormally large number of immature white blood cells that fill the red bone marrow and prevent red blood cell development. Anemia results, and the immature white cells offer little protection from disease. The cause of leukemia, a type of cancer, is unknown, but proper chemotherapy has been most successful, particularly in acute childhood leukemia.

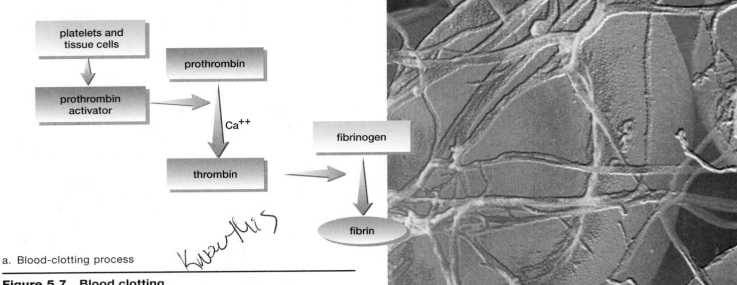

a. Blood-clotting process

Figure 5.7   Blood clotting.
a. Platelets and damaged tissue cells release prothrombin activator, which acts on prothrombin in the presence of calcium ions (Ca^{++}) to produce thrombin, which acts on fibrinogen to form fibrin.
b. A scanning electron micrograph shows a red blood cell caught in the fibrin threads of a blood clot.

b. Blood clot

5.3   Blood Clotting

Platelets (thrombocytes) result from fragmentation of certain large cells, called *megakaryocytes,* in the red bone marrow. Platelets are produced at a rate of 200 billion a day, and the blood contains 150,000–300,000 per mm^3. These formed elements are involved in the process of blood **clotting,** or coagulation.

There are at least 12 clotting factors in the blood that participate in the formation of a blood clot. We will discuss the roles played by platelets, prothrombin, and fibrinogen. **Fibrinogen** and **prothrombin** are proteins manufactured and deposited in blood by the liver. Vitamin K, found in green vegetables and also formed by intestinal bacteria, is necessary for the production of prothrombin, and if by chance this vitamin is missing from the diet, hemorrhagic disorders develop.

Blood Clotting Has Steps

When a blood vessel in the body is damaged, platelets clump at the site of the puncture and partially seal the leak. They and the injured tissues release a clotting factor called *prothrombin activator* that converts prothrombin to thrombin. This reaction requires calcium ions (Ca^{++}). **Thrombin,** in turn, acts as an enzyme that severs two short amino acid chains from each fibrinogen molecule. These activated fragments then join end to end, forming long threads of *fibrin.* Fibrin threads wind around the platelet plug in the damaged area of the blood vessel and provide the framework for the clot. Red blood cells also are trapped within the fibrin threads; these cells make a clot appear red (Fig. 5.7). A fibrin clot is present only temporarily. Clot

dissolution occurs when fibrin threads absorb plasminogen and this reaction occurs:

$$plasminogen \xrightarrow{\text{plasminogen activator}} plasmin$$

Plasminogen activator comes from the wall of the blood vessel, and *plasmin* is an enzyme that can digest fibrin threads. Today, there are commercially produced plasminogen activators that are administered to patients in order to dissolve a blood clot that is blocking an artery to the heart. These drugs include tPA (tissue plasminogen activator), a biotechnology product; streptokinase from bacteria; and urokinase from kidney cells.

You can prevent blood from clotting in a test tube by adding a chemical such as citrate that absorbs calcium ions (Ca^{++}). However, if the blood is allowed to clot, a yellowish fluid is expressed from the clot. This fluid is called **serum,** and it contains all the components of plasma except fibrinogen, which was used up during clot formation. Table 5.1 reviews the many different terms we have used to refer to various body fluids related to blood.

TABLE 5.1 BODY FLUIDS	
Name	*Composition*
Blood	Formed elements and plasma
Plasma	Liquid portion of blood
Serum	Plasma minus fibrinogen
Tissue fluid	Plasma minus most proteins
Lymph	Tissue fluid within lymphatic vessels

Health Focus

What to Know When Giving Blood

THE PROCEDURE

After you register to give blood, you are asked private and confidential questions about your health history and your lifestyle, and any questions you may have are answered.

Your temperature, blood pressure, and pulse will be checked, and a drop of your blood is tested to ensure that you are not anemic.

You will have several opportunities prior to giving blood and even afterwards to let Red Cross officials know whether your blood is safe to give to another person.

All of the supplies, including the needle, are sterile and are used only once—for YOU. You *cannot* get infected with HIV (the virus that causes AIDS) or any other disease from donating blood.

When the actual donation is started, you may feel a brief "sting." The procedure takes about 10 minutes, and you will have given about a pint of blood. Your body replaces the liquid part (plasma) in hours and the cells in a few weeks.

After you donate, you are given a card with a number to call if you decide after you leave that your blood may not be safe to give to another person.

An area is provided in which to relax after donating blood. Most people feel fine while they give blood and afterward. A few may have an upset stomach, a faint or dizzy feeling, or a bruise, redness, and pain where the needle was. Very rarely, a person may faint, have muscle spasms, and/or suffer nerve damage.

Your blood is tested for syphilis, AIDS antibodies, hepatitis, and other viruses. You are notified if tests give a positive result. Your blood won't be used if it could make someone ill.

THE CAUTIONS

\ \ \ \ D O N O T G I V E B L O O D / / / /

if you have
> ever had hepatitis;
>
> had malaria or have taken drugs to prevent malaria in the last 3 years;
>
> been treated for syphilis or gonorrhea in the last 12 months.

if you have AIDS or one of its symptoms:
> unexplained weight loss (4.5 kilograms or more in less than 2 months);
>
> night sweats;
>
> blue or purple spots on or under skin;
>
> long-lasting white spots or unusual sores in mouth;
>
> lumps in neck, armpits, or groin for over a month;
>
> diarrhea lasting over a month;
>
> persistent cough and shortness of breath;
>
> fever higher than 37°C for more than 10 days.

if you are at risk for AIDS; that is, if you have
> taken illegal drugs by needle, even once;
>
> taken clotting factor concentrates for a bleeding disorder such as hemophilia;
>
> tested positive for any AIDS virus or antibody;
>
> been given money or drugs for sex, since 1977;
>
> had a sexual partner within the last 12 months who did any of the above things;
>
> (for men): had sex *even once* with another man since 1977; within the last 12 months had sex with a female prostitute;
>
> (for women): had sex with a male or female prostitute within the last 12 months; *or* had a male sexual partner who had sex with another man *even once* since 1977.

DO NOT GIVE BLOOD to find out whether you test positive for antibodies to the viruses (HIV) that cause AIDS. Although the tests for HIV are very good, they aren't perfect. HIV antibodies may take weeks to develop after infection with the virus. If you were infected recently, you may have a negative test result yet be able to infect someone. **It is for this reason that you must not give blood if you are at risk of getting AIDS or other infectious diseases.**

Courtesy of the American Red Cross.

5.4   Plasma

Plasma is the liquid portion of blood; about 92% of plasma is water. There are also various salts (ions) and organic molecules in plasma. The salts, which are simply dissolved in plasma, help maintain the pH and osmotic pressure of the blood. Small organic molecules like glucose, amino acids, and urea can also dissolve in plasma. Glucose and amino acids are nutrients for cells; urea is a nitrogenous waste product on its way to the kidneys for excretion. The large organic molecules in plasma include hormones and the plasma proteins (Table 5.2).

Plasma Proteins

Plasma proteins make up about 7% of plasma, and **albumin** is the most plentiful plasma protein. Most plasma proteins are made by the liver. Exceptions are antibodies and the protein-type hormones. The plasma proteins are able to take up and release hydrogen ions; therefore, they help buffer the blood and maintain the pH of blood around 7.40. They also help maintain the osmotic pressure, which helps keep water in the blood.

Certain plasma proteins combine with and transport large organic molecules in blood. For example, albumin transports the molecule bilirubin, a breakdown product of hemoglobin. Lipoproteins, whose protein portion contain a type of plasma protein called globulin, transport cholesterol. There are three types of globulins designated alpha, beta, and gamma globulins. Antibodies, which help fight infections by combining with antigens, are gamma globulins. Other plasma proteins also have specific functions. Fibrinogen is necessary to blood clotting, for example.

TABLE 5.2	
Blood Plasma	
Water (92% of Total)	
Solutes (8% of Total)	
Plasma proteins	Albumin, globulins, fibrinogen
Inorganic ions (salts)	Na^+, Ca^{++}, K^+, Mg^{++}; Cl^-, HCO_3^-, $HPO_4^=$, $SO_4^=$
Gases	O_2, CO_2
Organic nutrients	Glucose, fats, phospholipids, amino acids, etc.
Nitrogenous waste products	Urea, ammonia, uric acid
Regulatory substances	Hormones, enzymes

5.5   Exchange of Materials with Tissue Fluid

The internal environment consists of blood and **tissue fluid.** The composition of tissue fluid stays relatively constant because of exchanges with blood in the region of capillaries (Fig. 5.8). Water makes up a large part of tissue fluid, and any excess is collected by lymphatic capillaries, which are always found near cardiovascular capillaries.

Arterial End of the Capillary

When arterial blood enters the tissue capillaries, it is bright red because red blood cells are carrying oxygen. It is also rich in nutrients, which are dissolved in the plasma. At the arterial end of the capillary, blood pressure (40 mm Hg) is higher than the osmotic pressure of the blood (25 mm Hg). Blood pressure, you recall, is created by the pumping of the heart; the osmotic pressure is caused by the presence of salts and, in particular, by the plasma proteins that are too large to pass through the wall of the capillary. Since the blood pressure is higher than the osmotic pressure, fluid together with nutrients (glucose and amino acids) exit the capillary. Red blood cells and most all plasma proteins generally remain in the capillaries, but small substances leave the capillaries. Therefore, tissue fluid, created by this process, consists of all the components of plasma except the proteins.

Midsection

Along the length of the capillary, molecules follow their concentration gradient as diffusion occurs. Diffusion, you recall, is the movement of molecules from an area of greater concentration to an area of lesser concentration. In the tissues, the area of greater concentration for nutrients and oxygen is always blood because after these molecules have passed into tissue fluid, they are taken up and metabolized by the tissue cells. The cells use glucose ($C_6H_{12}O_6$) and oxygen (O_2) in the process of aerobic cellular respiration, and they use amino acids for protein synthesis. Following aerobic cellular respiration, the cells give off carbon dioxide (CO_2) and water (H_2O). Carbon dioxide and other waste products of metabolism leave the cell by diffusion. Since tissue fluid is always the area of greater concentration for these waste materials, they diffuse into the capillary.

Oxygen and nutrient molecules (e.g., glucose and amino acids) exit a capillary near the arterial end; waste molecules (e.g., carbon dioxide) enter a capillary near the venous end.

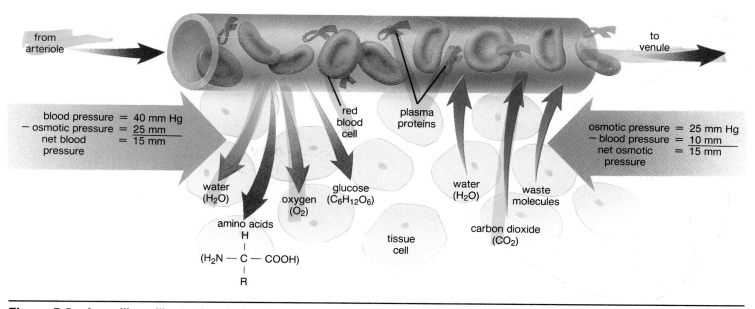

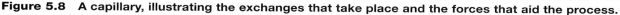

Figure 5.8   A capillary, illustrating the exchanges that take place and the forces that aid the process.
At the arterial end of a capillary, the blood pressure is higher than the osmotic pressure; therefore, water (H_2O) and nutrients tend to leave the bloodstream. In the midsection, molecules, including oxygen (O_2), follow their concentration gradients. At the venous end of a capillary, the osmotic pressure is higher than the blood pressure; therefore, water and wastes tend to enter the bloodstream. Notice that the red blood cells and usually plasma proteins are too large to exit a capillary.

Venous End of the Capillary

At the venous end of the capillary, blood pressure is much reduced (10 mm Hg), as can be verified by reviewing Figure 5.8. However, there is no reduction in osmotic pressure (25 mm Hg), which tends to pull fluid back into the capillary. As water enters a capillary, it brings with it additional waste molecules. Blood that leaves the capillaries is deep purple in color because red blood cells contain reduced hemoglobin.

Retrieving fluid by means of osmotic pressure is not completely effective. There is always some fluid that is not picked up at the venous end. This excess tissue fluid enters the lymphatic capillaries (Fig. 5.9). **Lymph** is tissue fluid contained within lymphatic vessels. The lymphatic system is a one-way system, and lymph is returned to the systemic venous blood when the major lymphatic vessels enter cardiovascular veins.

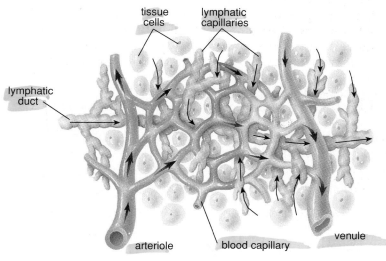

Figure 5.9   Lymphatic capillaries.
Arrows indicate that lymph is formed when lymphatic capillaries take up excess tissue fluid. Lymphatic capillaries lie near blood capillaries.

Exchange of nutrients for wastes occurs at the capillaries. Here lymphatic veins also collect excess tissue fluid.

5.6 Blood Typing

Blood typing involves the two types of molecules called antigens and antibodies. As mentioned previously, when a foreign substance acting as an *antigen* enters the body, an *antibody* reacts with it. The membrane of red blood cells contains possible antigens for a recipient of transfused blood, and the recipient's plasma may contain antibodies that will react with them. Such reactions are life threatening and should be avoided; hence the need to type blood.

ABO System

There are many different systems for typing the blood of humans, but the most common system is the ABO system. In the ABO system, it is determined whether type A or type B antigens are on the red blood cells. For example, if a person has type A blood, the A antigen is on his or her red blood cells. This molecule is not an antigen to this individual, although it can be an antigen to a recipient who does not have type A blood.

In the simplified ABO system, there are four types of blood: A, B, AB, and O (Table 5.3). Type O blood has neither the A antigen nor the B antigen on red blood cells; the other types of blood are designated by the antigen(s) present on red blood cells.

Within the plasma, there are antibodies to the antigens that are *not* present on the person's red blood cells. Therefore, for example, type A blood has an antibody called anti-B in the plasma. Type AB blood has neither anti-A nor anti-B antibodies in plasma—both antigens A and B are on the red blood cells. This is reasonable because if these antibodies were present in plasma, **agglutination,** or clumping of red blood cells, would occur. Agglutination of red blood cells can cause blood to stop circulating in small blood vessels, and this leads to organ damage. It also is followed by hemolysis, which may cause the death of the individual. For a recipient to receive blood from a donor, the recipient's plasma must not have an antibody that causes the donor's cells to agglutinate. For this reason, it is important to determine each person's blood type. Figure 5.10 demonstrates a way to use the antibodies derived from plasma to determine the blood type. If clumping occurs after a sample of blood is exposed to a particular antibody, the person has that type of blood.

In the ABO system, there are four types of blood: A, B, AB, and O. Type O blood has neither the A antigen nor the B antigen on red blood cells; the other types of blood are designated by the antigen(s) present on red blood cells.

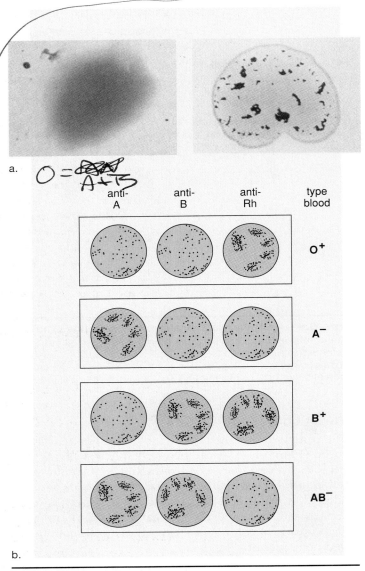

a.

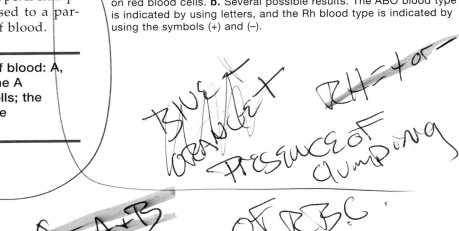

b.

Figure 5.10   Blood typing.

The standard test to determine ABO and Rh blood type consists of putting a drop of anti-A antibodies, anti-B antibodies, and anti-Rh antibodies separately on a slide. To each of these three antibody solutions, a drop of the person's blood is added. **a.** If agglutination occurs, as seen in the *top right* photo, the person has this antigen on red blood cells. **b.** Several possible results. The ABO blood type is indicated by using letters, and the Rh blood type is indicated by using the symbols (+) and (−).

The A,B, AB O Test (handwritten)

TABLE 5.3

The ABO System

PROTEIN (handwritten)

Blood Type	Antigen on Red Blood Cells	Antibody in Plasma	% U.S. African American	% U.S. Caucasian	% U.S. Asian	% North American Indian
A	A	Anti-B	27	41	28	8
B	B	Anti-A	20	9	27	1
AB	A, B	None	4	3	5	0
O	None	Anti-A and anti-B	49	47	40	92

Today, blood transfusions are a matter of concern not only because blood types should match, but also because each person wants to receive blood that is of good quality and free of infectious agents. Blood is tested for the more serious agents such as those that cause AIDS, hepatitis, and syphilis. Donors can help protect the nation's blood supply by knowing when not to give blood. This is the topic of the reading on page 115.

Rh System

Another important antigen in matching blood types is the **Rh factor.** Eighty-five percent of the U.S. population has this particular antigen on the red blood cells and are Rh$^+$ (Rh positive). Fifteen percent do not have this antigen and are Rh$^-$ (Rh negative). Rh$^-$ individuals normally do not have antibodies to the Rh factor, but they may make them when exposed to the Rh factor. It is possible to use anti-Rh antibodies for blood testing. When Rh$^+$ blood is mixed with anti-Rh antibodies, agglutination occurs.

The designation of blood type usually includes whether the person has the Rh factor (Rh$^+$) or does not have the Rh factor (Rh$^-$) on the red blood cells (Fig. 5.10).

During pregnancy, if the mother is Rh$^-$ and the father is Rh$^+$, the child may be Rh$^+$ (Fig. 5.11). The Rh$^+$ red blood cells may begin leaking across the placenta into the mother's circulatory system, as placental tissues normally break down before and at birth. The presence of these Rh antigens causes the mother to produce anti-Rh antibodies. In this or a subsequent pregnancy with another Rh$^+$ baby, anti-Rh antibodies produced by the mother (but usually not anti-A and anti-B antibodies discussed earlier) may cross the placenta and destroy the child's red blood cells. This is called hemolytic disease of the newborn (HDN) because hemolysis continues after the baby is born. Due to red blood cell destruction followed by heme breakdown, bilirubin rises in the blood. Excess bilirubin can lead to brain damage and mental retardation or even death.

The Rh problem is prevented by giving Rh$^-$ women an Rh immunoglobulin injection either midway through the first pregnancy or no later than 72 hours after giving birth to any Rh$^+$ child. This injection contains anti-Rh antibodies that attack any of the baby's red blood cells in the mother's blood before these cells can stimulate her immune system to produce her own antibodies. The injection is not beneficial if the woman has already begun to produce antibodies; therefore, the timing of the injection is most important.

The possibility of hemolytic disease of the newborn exists when the mother is Rh$^-$ and the father is Rh$^+$.

Child is Rh positive; mother is Rh negative

Red blood cells leak across placenta

Mother makes anti-Rh antibodies

Antibodies attack Rh-positive red blood cells in child

SUMMARY

Blood, which is composed of formed elements and plasma, has several functions. It transports hormones, oxygen, and nutrients to the cells and carbon dioxide and other wastes away from cells. It fights infections and has various regulatory functions. It maintains blood pressure, regulates body temperature, and keeps the pH within normal limits. All of these functions help maintain homeostasis.

5.1   Red Blood Cells

Red blood cells are small, biconcave disks that lack a nucleus. They are made in red bone marrow and are destroyed in the liver and spleen when they are old or abnormal. The production of red blood cells is controlled by oxygen concentration of the blood. When the oxygen concentration decreases, more red blood cells are produced. Red blood cells contain hemoglobin, the respiratory pigment, which combines with oxygen and transports it to the tissues.

5.2   White Blood Cells

White blood cells are larger than red blood cells, they have a nucleus, and they are translucent unless stained. Like red blood cells, they are produced in the red bone marrow. White blood cells are divided into the granular leukocytes and the agranular leukocytes. The granular leukocytes have conspicuous granules; in eosinophils, granules are red when stained with eosin, and in basophils, granules are blue when stained with a basic dye. The granules in neutrophils don't take up either dye significantly. Neutrophils are the most plentiful of the white blood cells, and they are able to phagocytize pathogens. Many neutrophils die within a few days when they are fighting an infection. The agranulocytes include the lymphocytes and the monocytes. The lymphocytes and monocytes function in specific immunity. On occasion, the monocytes become large phagocytic cells of great significance. They engulf worn-out red blood cells and pathogens at a ferocious rate.

5.3   Blood Clotting

When there is a break in a blood vessel, the platelets clump to form a plug. Blood clotting itself requires a series of enzymatic reactions involving blood platelets, prothrombin, and fibrinogen. In the final reaction, fibrinogen becomes fibrin threads, entrapping cells. The fluid that escapes from a clot is called serum and consists of plasma minus fibrinogen.

5.4   Plasma

Plasma is mostly water (92%) and the plasma proteins (7%). The plasma proteins, of which albumin is the most plentiful, maintain osmotic pressure, help regulate pH, and transport molecules. Some plasma proteins have specific functions: fibrinogen and prothrombin are necessary to blood clotting, and antibodies are gamma globulins.

Small organic molecules like glucose and amino acids are dissolved in plasma and serve as nutrients for cells.

5.5   Exchange of Materials with Tissue Fluid

At the arterial end of a capillary, blood pressure is greater than osmotic pressure; therefore, water leaves the capillary. In the midsection, oxygen and nutrients diffuse out of the capillary; carbon dioxide and other wastes diffuse into the capillary. At the venous end, osmotic pressure created by the presence of proteins exceeds blood pressure, causing water to enter the capillary.

5.6   Blood Typing

Red blood cells of an individual are not necessarily received without difficulty by another individual. For example, the membranes of red blood cells may contain type A, B, AB, or no antigens. In the plasma there are two possible antibodies: anti-A or anti-B. If the corresponding antigen and antibody are put together, clumping, or agglutination, occurs; in this way, the blood type of an individual may be determined in the laboratory. After determination of the blood type, it is theoretically possible to decide who can give blood to whom. For this, it is necessary to consider the donor's antigens and the recipient's antibodies.

Another important antigen is the Rh antigen. This particular antigen must also be considered in the transfusing of blood, and it is important during pregnancy because an Rh^- mother may form antibodies to the Rh antigen while carrying or after the birth of the child who is Rh^+. These antibodies can cross the placenta to destroy the red blood cells of any subsequent Rh^+ child.

STUDYING THE CONCEPTS

1. State the two main components of blood, and give the functions of blood.   109

2. What is hemoglobin, and how does it function?   109

3. Describe the life cycle of red blood cells, and tell how the production of red blood cells is regulated.   111

4. Name the five types of white blood cells; describe the structure and give a function for each type.   113

5. Name the steps that take place when blood clots. Which substances are present in blood at all times, and which appear during the clotting process?   114

6. Define blood, plasma, tissue fluid, lymph, and serum.   114

7. List and discuss the major components of plasma. Name several plasma proteins, and give a function for each.   116

8. What forces operate to facilitate exchange of molecules across the capillary wall?   116–17

9. What are the four ABO blood types? For each, state the antigen(s) on the red blood cells and the antibody(ies) in the plasma.   118

10. Explain why a person with type O blood cannot receive a transfusion of type A blood.   118

11. Problems can arise during childbearing if the mother is which Rh type and the father is which Rh type? Explain why this is so.   119

APPLYING YOUR KNOWLEDGE

Concepts

1. Some persons suffering from acute leukemia are given bone marrow transplants. Explain the reason for this type of treatment.

2. In a previous Olympics that was held at a relatively high altitude in Mexico City, certain South American long-distance runners won medals for the first time. Suggest a reason that this happened. Explain.

3. A couple gave birth to a second child who died of HDN (hemolytic disease of the newborn). They can't understand how this could happen because their first child was "perfectly normal." Explain the situation so they could understand how this could happen.

Bioethical Issue

During surgery or serious injuries, people often lose a lot of blood. Hospitals respond by pumping healthy blood from donors into a patient. In this way, blood transfusions save lives.

Unfortunately, transfusions occasionally harm patients as well. Donated blood is routinely tested for viruses that cause AIDS, hepatitis, or other diseases. But sometimes the offending pathogens slip by undetected. What's more, no practical screening tests exist for some types of deadly bacteria.

Does a transfusion patient have the right to sue a hospital that accidentally delivers tainted blood? What if the blood carries a disease that can't be tested for easily? If hospital personnel explain the risks of transfusion and obtain consent from a patient, should the hospital still be held liable for "bad" blood?

APPLYING TECHNOLOGY

Your study of the composition and function of blood is supported by these available technologies:

Exploring the Internet

The Mader Home Page provides further resources for studying this chapter.

http://www.mhhe.com/sciencemath/biology/mader/

(Click on *Human Biology*.)

Dynamic Human: Lymphatic System CD-ROM

In *Clinical Concepts,* students have an opportunity to mix different types of blood.

Life Science Animations Video

Video #4: Animal Biology, Part 2

A, B, O Blood Types (#37)   Clumping of red blood cells occurs when the recipient's plasma contains antibodies for antigens on the donor's red blood cells.   **(3)***

*Level of difficulty

TESTING YOUR KNOWLEDGE

1. The liquid part of blood is called _plasma_.
2. Red blood cells carry _O2_, and the function of white blood cells is to _Protect_
3. Hemoglobin that is carrying oxygen is called _Oxyhemoglobin_
4. Human red blood cells lack a _Nucleus_ and live only about _120_ days.
5. When a blood clot occurs, fibrinogen has been converted to _fibrin_ threads.
6. The most common granular leukocyte is the _____, a phagocytic white blood cell.
7. B lymphocytes produce _Antibodys_ that react with antigens.
8. At a capillary, _H0_, _O2_, and _____ leave the arterial end, and _CO2_ and _H0_ enter the venous end.

9. Type AB blood has the antigens _____ and _____ on red blood cells and _____ antibodies in plasma.
10. Hemolytic disease of the newborn can occur if the mother is _____ and the father is _____.
11. Add these labels to this diagram of a capillary: arterial end, plasma proteins, venous end, oxygen, nutrients, carbon dioxide, water.

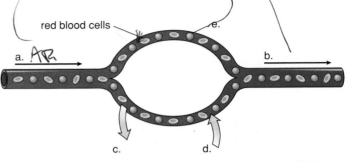

red blood cells

a. _Art_

b.

c.

d.

e.

SELECTED KEY TERMS

agglutination (uh-gloot-un-AY-shun) Clumping of cells, particularly in reference to red blood cells involved in an antigen-antibody reaction. 118

agranular leukocyte White blood cell that does not contain distinctive granules. 113

albumin Plasma protein of the blood having transport and osmotic functions. 116

anemia Inefficient oxygen-carrying ability of blood due to hemoglobin or mature red blood cell shortage. 111

basophil Granular leukocyte capable of being stained with a basic dye. 113

clotting Process of blood coagulation, usually when injury occurs. 114

eosinophil Granular leukocyte capable of being stained with the dye eosin. 113

fibrinogen Plasma protein that is converted into fibrin threads during blood clotting. 114

formed element Constituent of blood that is either cellular (red blood cells and white blood cells) or cellular in origin (platelets). 109

granular leukocyte White blood cell that contains distinctive granules. 113

hemoglobin Iron-containing protein in red blood cells that combines with and transports oxygen. 109

hemolysis Rupture of red blood cells. 111

lymph Fluid derived from tissue fluid that is carried in lymphatic vessels. 117

lymphocyte Specialized leukocyte that functions in specific defense; occurs in two forms—T lymphocyte and B lymphocyte. 113

monocyte Type of agranular leukocyte that functions as a phagocyte. 113

neutrophil Granular leukocyte that is the most abundant of the white blood cells; first to respond to infection. 113

phagocytosis (fag-oh-suh-TOH-sis) Taking in of bacteria and/or debris by engulfing; cell eating. 113

plasma Liquid portion of blood. 116

platelet Fragment of a megakaryocyte; formed element that is necessary to blood clotting. 114

prothrombin Plasma protein that is converted to thrombin during the steps of blood clotting. 114

red blood cell (erythrocyte) Formed element that contains hemoglobin and carries oxygen from the lungs to the tissues. 109

Rh factor One type of antigen on red blood cells. 119

serum Light yellow liquid left after clotting of blood. 114

thrombin Enzyme that converts fibrinogen to fibrin threads during blood clotting. 114

tissue fluid Solution that bathes and services every cell in the body. Also called interstitial fluid. 116

white blood cell (leukocyte) Formed element of which there are several types, each having a specific function in protecting the body from invasion by foreign substances and organisms. 113

Chapter 6

Cardiovascular System

Chapter Outline

Figure 6.1   Overexertion.
Athletes, like all persons, need to lead heart-healthy lives and undergo regular medical exams in order to prevent any possible cardiovascular mishap during times of extreme exertion.

Reggie Lewis and Sergei Grinkov couldn't have been more different. Lewis played basketball for the rowdy Boston Celtics. Grinkov, a native of Russia, quietly ice skated to Olympic medals with his wife. But both men stunned fans when they collapsed and died—Lewis on the basketball court, and Grinkov on the ice. In each case, a heart condition was to blame.

Celebrity deaths always win headlines. In particular, fans often find it hard to believe that strong athletes can die (Fig. 6.1). Still, both Grinkov and Lewis were probably candidates for heart attacks. Lewis had an enlarged heart, and Grinkov had a family history of both severe cardiovascular disease along with high blood cholesterol.

Heart attacks are caused by a variety of problems. Sometimes the heart begins to beat irregularly, reducing the output of blood. Other times, cholesterol plaque builds up in coronary arteries, preventing essential oxygen from reaching heart tissue. Mild heart attacks or those caught quickly may do little damage. Unfortunately, severe heart attacks—as the families of Lewis and Grinkov know—can be deadly.

Knowledge about the anatomy and physiology of the heart is one of the first steps toward keeping oneself fit. This chapter reviews the cardiovascular system and how it operates to circulate the blood about the body. In humans, the heart pumps blood to the lungs and to the tissues in two separate vascular circuits. Exchange of substances occurs only across the thin walls of capillaries; in the lungs, blood picks up oxygen and gets rid of carbon dioxide. This oxygen eventually goes to the tissues along with nutrients from the digestive tract. Circulation of the blood is so important that if the heart stops beating for only a few minutes, death results.

6.1   Blood Vessels

The circulatory system has three types of blood vessels: the **arteries** (and arterioles), which carry blood away from the heart; the **capillaries,** which permit exchange of material with the tissues; and the **veins** (and venules), which return blood to the heart (Fig. 6.2).

Arteries and Arterioles: Away from the Heart

Arteries have thick walls (Fig. 6.2*b*). The walls have an inner layer of specialized squamous epithelial cells called the endothelium; a thick middle layer of elastic tissue and smooth muscle; and an outer fibrous connective tissue layer. An artery can expand to accommodate the sudden increase in blood volume after each heartbeat, even though the walls are so thick that they are supplied with blood vessels.

Arterioles are small arteries just visible to the naked eye. The middle layer of arterioles has some elastic tissue but is composed mostly of smooth muscle, whose fibers encircle the arteriole. If these muscle fibers contract, the lumen of the arteriole gets smaller; if the fibers relax, the lumen of the arteriole enlarges. Whether arterioles are constricted or dilated affects blood pressure. The greater the number of vessels dilated, the lower the blood pressure.

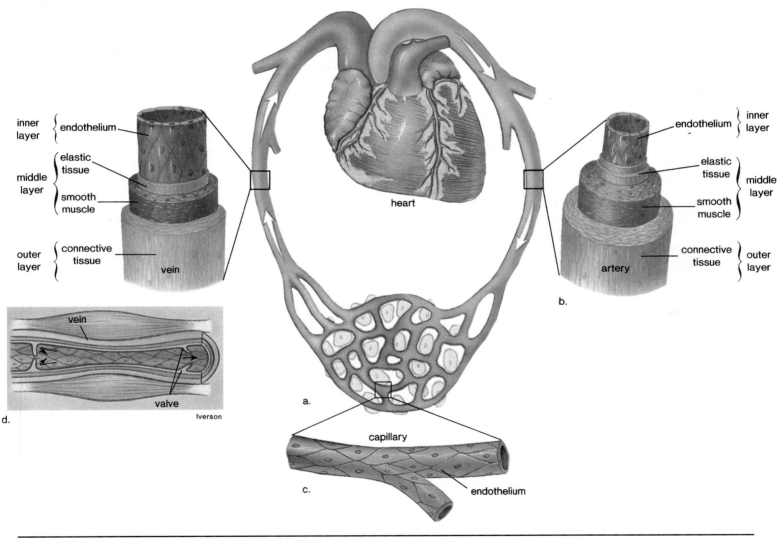

Figure 6.2   Blood vessels.
a. Blood leaving the heart moves from an artery to arterioles to capillaries to venules and then returns to the heart by way of a vein. **b.** Arteries have well-developed walls with a thick middle layer of elastic tissue and smooth muscle. **c.** Capillary walls are one cell thick. **d.** Veins have thinner walls, particularly because the middle layer is not as thick as in arteries. Veins have valves, which point toward the heart.

Capillaries: Exchange Takes Place

Arterioles branch into capillaries. Each capillary is an extremely narrow, microscopic tube with one-cell thick walls composed only of endothelium (Fig. 6.2c). *Capillary beds* (networks of many capillaries) are present in all regions of the body; consequently, a cut to any body tissue draws blood. The capillaries are a very important part of the human circulatory system because an exchange of substances takes place across their thin walls. Oxygen and nutrients diffuse out of a capillary into the tissue fluid that surrounds cells, and carbon dioxide and other wastes diffuse into the capillary. Some water also leaves a capillary; any excess is picked up by lymphatic vessels, which return it to the blood circulatory system.

Since the capillaries serve the cells, the heart and the other vessels of the circulatory system can be thought of as the means by which blood is conducted to and from the capillaries. Only certain capillary beds are open at any given time. Shunting of blood is possible because each capillary bed has a thoroughfare channel that allows blood to go directly from arteriole to venule (Fig. 6.3). Contracted sphincter muscles prevent the blood from entering the capillary vessels. After eating, for example, blood is shunted through the muscles of the body and diverted to the digestive system. This is why swimming after a heavy meal may cause cramping.

Veins and Venules: To the Heart

Veins and venules take blood from the capillary beds to the heart. First, the **venules** (small veins) drain blood from the capillaries and then join to form a vein. The walls of venules (and veins) have the same three layers as arteries, but the middle layer is poorly developed, and therefore, the walls are thinner (see Fig. 6.2d). Veins often have **valves,** which allow blood to flow only toward the heart when open and prevent the backward flow of blood when closed.

At any given time, more than half of the total blood volume is in the veins and the venules. If a loss of blood occurs, for example, due to hemorrhaging, nervous stimulation causes the veins to constrict, providing more blood to the rest of the body. In this way, the veins act as a blood reservoir.

Arteries and arterioles carry blood away from the heart; veins and venules carry blood to the heart; and capillaries join arterioles to venules.

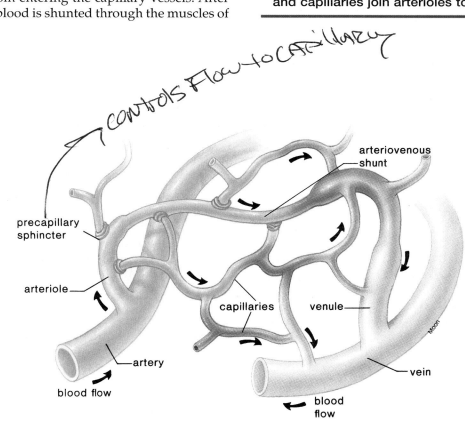

controls flow to capillary

Figure 6.3   Anatomy of a capillary bed.
A capillary bed forms a maze of capillary vessels that lie between an arteriole and a venule. Blood can move directly between the arteriole and the venule by way of a shunt. When sphincter muscles are closed, blood flows through the shunt. When sphincter muscles are open, the capillary bed is open, and blood flows through the capillaries. As blood passes through a capillary in the tissues, it gives up its oxygen (O_2). Therefore, blood goes from being oxygenated in the arteriole (red color) to being deoxygenated (blue color) in the vein.

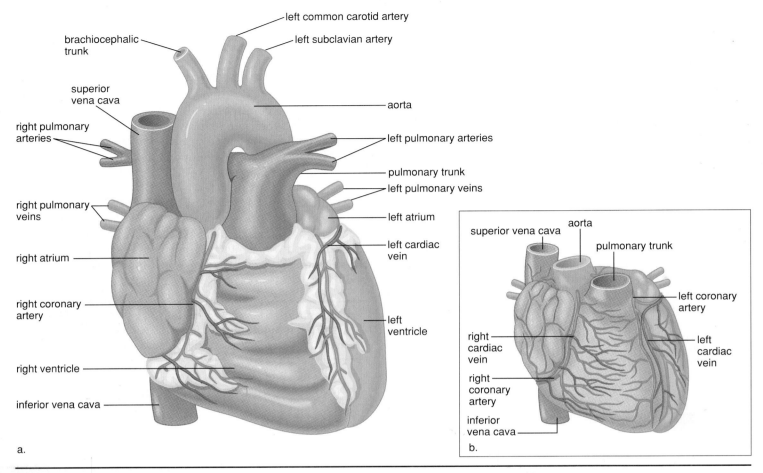

Figure 6.4   External heart anatomy.
a. The venae cavae bring deoxygenated blood to the right side of the heart from the body, and the pulmonary arteries take this blood to the lungs. The pulmonary veins bring oxygenated blood from the lungs to the left side of the heart, and the aorta takes this blood to the body.
b. The coronary arteries and cardiac veins pervade cardiac muscle. The coronary arteries are the first blood vessels to branch off the aorta. They bring oxygen and nutrients to cardiac cells.

6.2   The Heart

The **heart** is a cone-shaped, muscular organ about the size of a fist (Fig. 6.4). It is located between the lungs directly behind the sternum (breastbone) and is tilted so that the apex (the pointed end) is directed to the left. The major portion of the heart, called the **myocardium,** consists largely of cardiac muscle tissue. The muscle fibers of the myocardium are branched and tightly joined to one another. The heart lies within the *pericardium,* a thick, membranous sac that contains a small quantity of lubricating liquid. The inner surface of the heart is lined with endocardium, which consists of connective tissue and endothelial tissue.

Internally, a wall called the **septum** separates the heart into a right side and a left side (Fig. 6.5). The heart has four chambers: two upper, thin-walled atria (sing., **atrium**), sometimes called auricles, and two lower, thick-walled **ventricles.** The atria are much smaller and weaker than the muscular ventricles, but they hold the same volume of blood.

The heart also has valves, which direct the flow of blood and prevent its backward movement. The valves that lie between the atria and the ventricles are called the **atrioventricular valves.** These valves are supported by strong fibrous strings called *chordae tendineae.* The chordae, which are attached to muscular projections of the ventricular walls, support the valves and prevent them from inverting when the heart contracts. The atrioventricular valve on the right side is called the tricuspid valve because it has three flaps, or cusps. The valve on the left side is called the bicuspid (or the mitral) because it has two flaps. There are also **semilunar valves,** whose flaps resemble half moons, between the ventricles and their attached vessels. The pulmonary semilunar valve lies between the right ventricle and the pulmonary trunk. The aortic semilunar valve lies between the left ventricle and the aorta.

Humans have a four-chambered heart (two atria and two ventricles). A septum separates the right side from the left side.

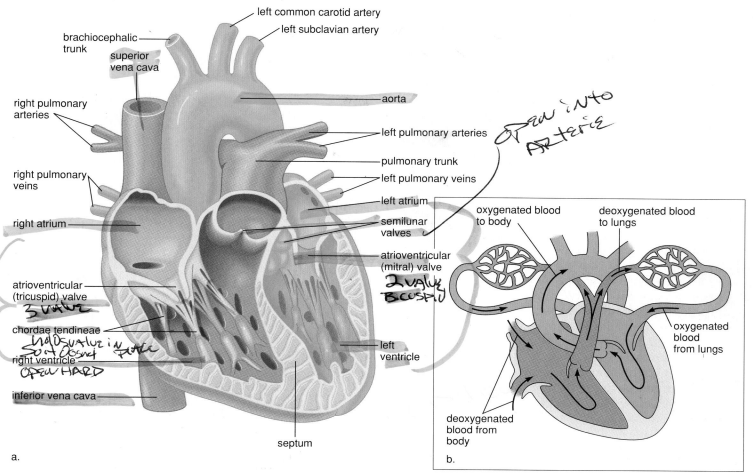

Handwritten annotations: *open into arterie*, *2 valve bicuspid*, *3 valve*, *chordsvalve in place soot doesnt*, *open hard*

Figure 6.5   Internal view of the heart.
a. The right side of the heart contains deoxygenated blood. The venae cavae empty into the right atrium, and the pulmonary trunk leaves the right ventricle. The left side of the heart contains oxygenated blood. The pulmonary veins enter the left atrium, and the aorta leaves from the left ventricle. **b.** This diagrammatic representation of the heart allows you to trace the path of the blood. On the right side of the heart, blood flows through the venae cavae, right atrium, right ventricle, and pulmonary arteries to the lungs. On the left side of the heart, blood flows through the pulmonary veins, left atrium, left ventricle, and aorta to the body. Restate this and put in the names of the valves where appropriate.

Taking Blood Through the Heart

We can trace the path of blood through the heart (Fig. 6.5) in the following manner:

The superior (anterior) **vena cava** and the inferior (posterior) vena cava, both carrying deoxygenated blood (low in oxygen and high in carbon dioxide), enter the right atrium.

The right atrium sends blood through an atrioventricular valve (the tricuspid valve) to the right ventricle.

The right ventricle sends blood through the pulmonary semilunar valve into the pulmonary trunk and the two **pulmonary arteries** to the lungs.

Four **pulmonary veins,** carrying oxygenated blood (high in oxygen and low in carbon dioxide) from the lungs, enter the left atrium.

The left atrium sends blood through an atrioventricular valve (the bicuspid, or mitral, valve) to the left ventricle.

The left ventricle sends blood through the aortic semilunar valve into the **aorta** to the body proper.

From this description, you can see that deoxygenated blood never mixes with oxygenated blood and that blood must go through the lungs in order to pass from the right side to the left side of the heart. In fact, the heart is a *double pump* because the right ventricle of the heart sends blood through the lungs, and the left ventricle sends blood throughout the body. The left ventricle has the harder job of pumping blood to the entire body, and its walls are thicker than those of the right ventricle, which pumps blood to the lungs.

The right side of the heart pumps blood to the lungs, and the left side of the heart pumps blood throughout the body.

When the Heart Beats

Each heartbeat is called a cardiac cycle (Fig. 6.6). First, the two atria contract at the same time; then the two ventricles contract at the same time. Then all chambers relax. The word **systole** refers to contraction of heart muscle, and the word **diastole** refers to relaxation of heart muscle. The heart contracts, or beats, about 70 times a minute, and each heartbeat lasts about 0.85 seconds. A normal adult rate can vary from 60 to 100 beats per minute.

When the heart beats, the familiar lub-dub sound occurs as the valves of the heart close. The lub is caused by vibrations occurring when the atrioventricular valves close and the ventricles contract. The dub is heard when the semilunar valves close. A heart murmur, or a slight slush sound after the lub, is often due to ineffective valves, which allow blood to pass back into the atria after the atrioventricular valves have closed. Rheumatic fever resulting from a bacterial infection is one cause of a faulty valve, particularly the bicuspid valve. If operative procedures are unable to open and/or restructure the valve, it can be replaced with an artificial valve.

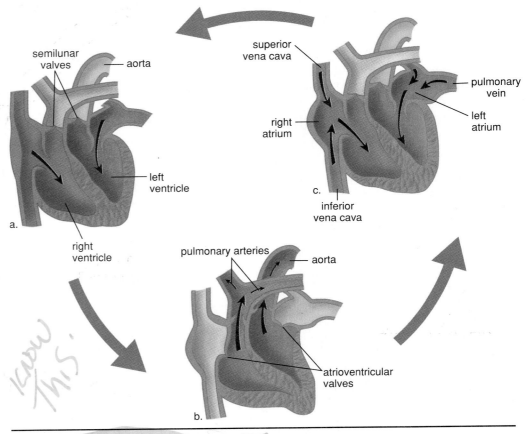

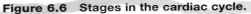

Figure 6.6 Stages in the cardiac cycle.
a. When the atria contract, the ventricles are relaxed and filling with blood. **b.** When the ventricles contract, the atrioventricular valves are closed, the semilunar valves are open, and blood is pumped into the pulmonary trunk and aorta. **c.** When the heart is relaxed, both atria and ventricles are filling with blood.

The surge of blood entering the arteries causes their elastic walls to stretch, but then they almost immediately recoil. This alternating expansion and recoil of an arterial wall can be felt as a **pulse** in any artery that runs close to the body's surface. It is customary to feel the pulse by placing several fingers on a radial artery, which lies near the outer border of the palm side of the wrist. A carotid artery, on either side of the trachea in the neck, is another accessible location to feel the pulse. Normally, the pulse rate indicates the rate of the heartbeat because the arterial walls pulse whenever the left ventricle contracts.

Conduction System Controls Heartbeat

Nodal tissue, which has both muscular and nervous characteristics, is located in two regions of the heart. The **SA (sinoatrial) node** is found in the upper dorsal wall of the right atrium; the other, the **AV (atrioventricular) node,** is found in the base of the right atrium very near the septum (Fig. 6.7a). The SA node initiates the heartbeat and automatically sends out an excitation impulse every 0.85 seconds; this causes the atria to contract. When the impulse reaches the AV node, the AV node signals the ventricles to contract by way of a bundle of fibers that branch and terminate in the more numerous and smaller Purkinje fibers. The SA node is called the **pacemaker** because it usually keeps the heartbeat regular. If the SA node fails to work properly, the heart still beats, but slower (40 to 60 beats per minute). To correct this condition, it is possible to implant an artificial pacemaker, which automatically gives an electric stimulus to the heart every 0.85 seconds so that it beats about 70 times per minute.

With the contraction of any muscle, including the myocardium, ionic changes occur; these can be detected with electrical recording devices. The pattern that results, called an **electrocardiogram** (ECG or EKG) (Fig. 6.7b), has an atrial phase and a ventricular phase. The first wave in the electrocardiogram, called the *P* wave, represents excitation and occurs just prior to contraction of the atria. The second wave, or the *QRS* complex, occurs just prior to ventricular contraction. The third, or *T*, wave occurs just before the ventricles relax. An examination of the electrocardiogram indicates whether the heartbeat has a normal or an irregular pattern.

Ventricular fibrillation characterized by uncoordinated contraction of the ventricles is most often due to occlusion of a coronary artery but can also be caused by an injury or drug overdose. It is the most common cause of sudden cardiac death in a seemingly healthy person. Once the ventricles are fibrillating, they have to be defibrillated by applying a strong electric current for a short period of time. Then the SA node may be able to reestablish a coordinated beat.

The conduction system of the heart includes the SA node, the AV node, and the Purkinje fibers. An ECG helps determine if the conduction system, and therefore the heartbeat, is regular.

Nervous System and Hormones Modify Heartbeat

A cardiac center in the brain can alter the beat of the heart by way of the autonomic system, which has two divisions: the parasympathetic system promotes those functions we associate with normal activities, and the sympathetic system, brings about those responses we associate with times of stress. For example, the parasympathetic system causes the heartbeat to slow down, and the sympathetic system, which releases norepinephrine, causes the heartbeat to speed up and produce a stronger beat. Various factors, such as the relative need for oxygen or blood pressure, determine which of these systems is activated.

Epinephrine, a hormone secreted by the adrenal glands, has such a profound effect on the heart that it is sometimes injected directly into a heart that has stopped beating in an attempt to stimulate its contraction. Thyroxin, a thyroid gland hormone, causes a sustained increase in heart rate; therefore, hyperthyroidism can lead to a weakened heart.

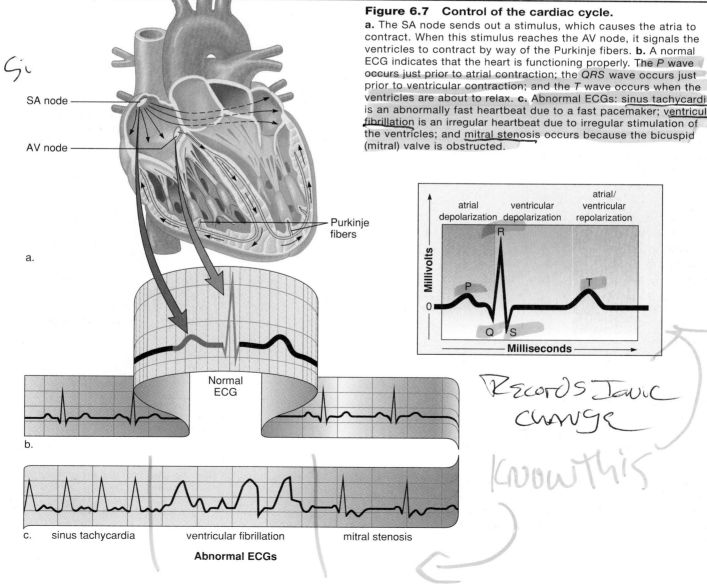

Figure 6.7 Control of the cardiac cycle.
a. The SA node sends out a stimulus, which causes the atria to contract. When this stimulus reaches the AV node, it signals the ventricles to contract by way of the Purkinje fibers. **b.** A normal ECG indicates that the heart is functioning properly. The *P* wave occurs just prior to atrial contraction; the *QRS* wave occurs just prior to ventricular contraction; and the *T* wave occurs when the ventricles are about to relax. **c.** Abnormal ECGs: sinus tachycardia is an abnormally fast heartbeat due to a fast pacemaker; ventricular fibrillation is an irregular heartbeat due to irregular stimulation of the ventricles; and mitral stenosis occurs because the bicuspid (mitral) valve is obstructed.

SA node

AV node

Purkinje fibers

a.

Normal ECG

b.

c.   sinus tachycardia   ventricular fibrillation   mitral stenosis

Abnormal ECGs

atrial depolarization   ventricular depolarization   atrial/ventricular repolarization

R

Millivolts

P

T

0

Q S

Milliseconds

[handwritten notes: CONTRACT 1st SOUND OPEN NO SOUND RELAX]

Blood Pressure Affects Swiftness of Flow

Blood pressure is the pressure of blood against the wall of a blood vessel. A sphygmomanometer is used to measure blood pressure, as described in Figure 6.8. The highest arterial pressure, called the *systolic pressure*, is reached during ejection of blood from the heart. The lowest arterial pressure is called the *diastolic pressure*. Diastolic pressure occurs while the heart ventricles are relaxing. Normal resting blood pressure for a young adult is said to be 120 mm of mercury (Hg) over 80 mm, or simply 120/80. The higher number is the systolic pressure, and the lower number is the diastolic pressure. Actually, 120/80 is the expected blood pressure in the brachial artery of the arm; blood pressure decreases with distance from the left ventricle (Fig. 6.9). Blood pressure is, therefore, higher in the arteries than in the arterioles. Further, there is a sharp drop in blood pressure when the arterioles reach the capillaries. The decrease can be correlated with the increase in the total cross-sectional area of the vessels as blood moves through arteries, arterioles, and then into capillaries. There are more arterioles than arteries and many more capillaries than arterioles.

The velocity of blood flow varies in different parts of the circulatory system. Blood pressure accounts for the velocity of the blood flow in the arterial system; therefore, as blood pressure decreases due to the increased cross-sectional area of the arterial system, so does velocity. Blood moves more slowly through the capillaries than it does through the aorta. This is important because the slow progress allows time for the exchange of substances between blood in the capillaries and the surrounding tissues.

Blood pressure cannot account for the movement of blood through the venules and the veins since they lie on the other side of the capillaries. Instead, movement of blood through the venous system is due to skeletal muscle contraction. When the skeletal muscles contract, they put pressure against the weak walls of the veins. This causes blood to move past a *valve* (Fig. 6.10). Once past the valve, blood cannot return. The importance of muscle contraction in moving blood in the venous system can be demonstrated by forcing a person to stand rigidly still for an hour or so. Frequently, fainting occurs because blood collects in the limbs, robbing the brain of oxygen. In this case, fainting is beneficial because the resulting horizontal position aids in getting blood to the head.

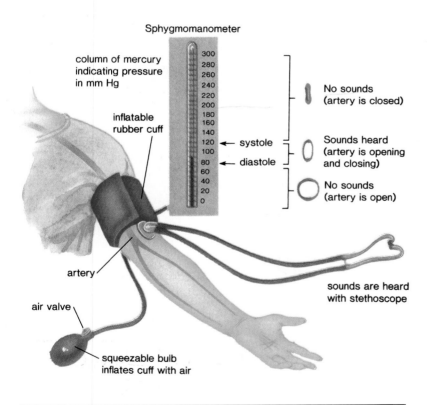

Figure 6.8   Determination of blood pressure using a sphygmomanometer.
The technician inflates the cuff with air, gradually reduces the pressure, and listens with a stethoscope for the sounds that indicate blood is moving past the cuff in an artery. This is systolic blood pressure. The pressure in the cuff is further reduced until no sound is heard, indicating that blood is flowing freely through the artery. This is diastolic pressure.

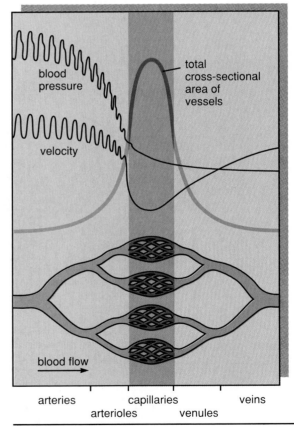

Figure 6.9   Velocity and blood pressure related to vascular cross-sectional area.
Capillaries have the greatest cross-sectional area, and blood is under the least pressure and has the least velocity. Skeletal muscle contraction, not blood pressure, accounts for the velocity of blood in the veins.

Blood flow gradually increases in the venous system (Fig. 6.9) due to a progressive reduction in the cross-sectional area as small venules join to form veins. The two venae cavae together have a cross-sectional area only about double that of the aorta. Respiratory movements also aid venous return. During inspiration, the pressure within the thoracic cavity is reduced because the rib cage moves upward and outward and the diaphragm moves downward. The downward movement of the diaphragm presses against the viscera in the abdominal cavity and the abdominal veins. Now venous blood flows from the abdominal cavity where pressure is higher to the thoracic cavity where pressure is lower.

Blood pressure accounts for the flow of blood in the arteries and the arterioles; skeletal muscle contraction accounts for the flow of blood in the venules and the veins.

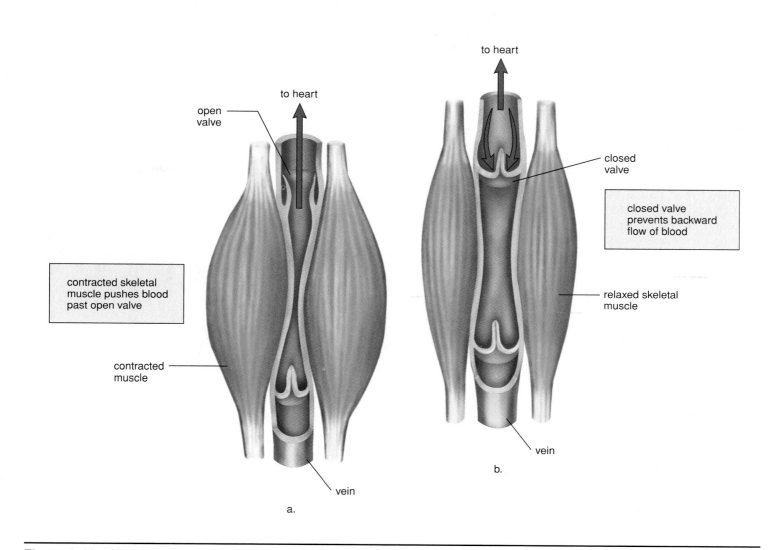

Figure 6.10   Skeletal muscle contraction moves blood in veins.
a. Muscle contraction exerts pressure against the vein, and blood moves past the valve. **b.** Blood cannot flow back once it has moved past the valve.

6.3   Vascular Pathways

The cardiovascular system, which is represented in Figure 6.11, includes two circuits: the **pulmonary circuit,** which circulates blood through the lungs, and the **systemic circuit,** which serves the needs of body tissues.

Pulmonary Circuit: Through the Lungs

The path of blood through the lungs can be traced as follows. Blood from all regions of the body first collects in the right atrium and then passes into the right ventricle, which pumps it into the pulmonary trunk. The pulmonary trunk divides into the right and left *pulmonary arteries,* which branch as they approach the lungs. The arterioles take blood to the pulmonary capillaries, where carbon dioxide is given off and oxygen is picked up. Blood then enters the pulmonary venules, which lead to the *pulmonary veins* that enter the left atrium. Since blood in the pulmonary arteries is deoxygenated and blood in the pulmonary veins is oxygenated, it is not correct to say that all arteries carry oxygenated blood and all veins carry deoxygenated blood. It is just the reverse in the pulmonary circuit.

The pulmonary arteries take deoxygenated blood to the lungs, and the pulmonary veins return oxygenated blood to the heart.

Systemic Circuit: Serving the Body

The systemic circuit includes all of the other arteries and veins shown in Figure 6.12. The largest artery in the systemic circuit is the *aorta,* and the largest veins are the *superior* and *inferior venae cavae.* The superior vena cava collects blood from the head, the chest, and the arms, and the inferior vena cava collects blood from the lower body regions. Both enter the right atrium. The aorta and the venae cavae serve as the major pathways for blood in the systemic circuit.

The path of systemic blood to any organ in the body begins in the left ventricle, which pumps blood into the aorta. Branches from the aorta go to the organs and major body regions. For example, this is the path of blood to and from the legs:

left ventricle—aorta—iliac artery—iliac arteriole, capillaries, venules—iliac vein—inferior vena cava—right atrium

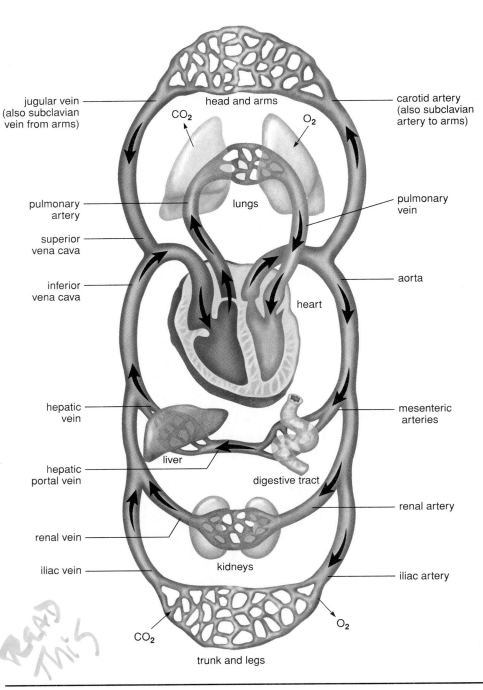

Figure 6.11   Cardiovascular system.
The blue-colored vessels carry deoxygenated blood, and the red-colored vessels carry oxygenated blood; the arrows indicate the direction of blood flow. Compare this diagram, useful for learning to trace the path of blood, to Figure 6.12 to realize that both arteries and veins go to all parts of the body. Also, there are capillaries in all parts of the body. No cell is located far from a capillary.

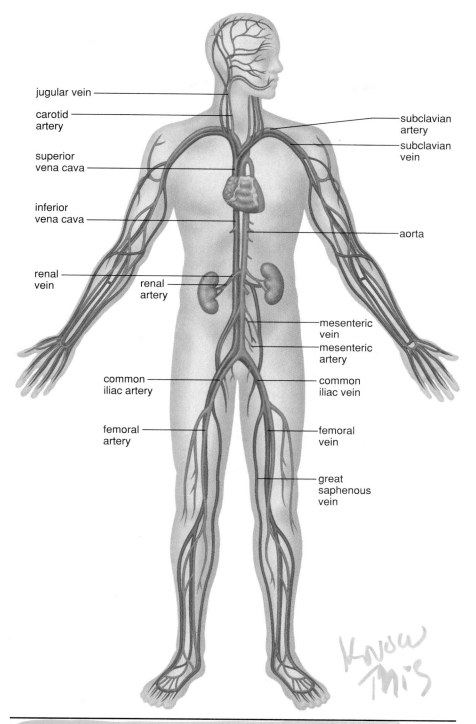

jugular vein

carotid artery

superior vena cava

inferior vena cava

renal vein

renal artery

common iliac artery

femoral artery

subclavian artery

subclavian vein

aorta

mesenteric vein

mesenteric artery

common iliac vein

femoral vein

great saphenous vein

Figure 6.12   Major arteries and veins of the systemic circuit.
A more realistic representation of the systemic circuit shows how the systemic arteries and veins are actually arranged in the body. The superior and inferior venae cavae take their names from their relationship to which organ?

Notice when tracing blood, you need mention only the aorta, the proper branch of the aorta, the region, and the vein returning blood to the vena cava. In most instances, the artery and the vein that serve the same organ are given the same name (Fig. 6.12). In the systemic circuit, arteries contain oxygenated blood and have a bright red color, but veins contain deoxygenated blood and appear a dark purplish color.

The **coronary arteries** (see Fig. 6.4b), serve the heart muscle itself. (The heart is not nourished by the blood in its chambers.) The coronary arteries are the first branch off the aorta. They originate just above the aortic semilunar valve and lie on the exterior surface of the heart, where they divide into diverse arterioles. Because they have a very small diameter, the coronary arteries may become blocked, as discussed on pages 134–35. The coronary capillary beds join to form venules. The venules converge to form the cardiac veins, which empty into the right atrium.

The body has a portal system called the *hepatic portal system*, which is associated with the liver. A portal system begins and ends in capillaries; in this instance, the first set of capillaries occurs at the villi of the small intestine and the second occurs in the liver. Blood passes from the capillaries of the intestinal villi into venules that join to form the *hepatic portal vein*, a vessel that connects the villi of the intestine with the liver. The *hepatic vein* leaves the liver and enters the inferior vena cava.

While Figure 6.11 is helpful in tracing the path of blood, remember that all parts of the body receive both arteries and veins, as illustrated in Figure 6.12.

The systemic circuit takes blood from the left ventricle of the heart to the right atrium of the heart. It serves the body proper.

Health Focus

Prevention of Cardiovascular Disease

All of us can take steps to prevent the occurrence of cardiovascular disease, the most frequent cause of death in the United States. There are genetic factors that predispose an individual to cardiovascular disease, such as family history of heart attack under age 55, male gender, and ethnicity (African Americans are at greater risk). Those with one or more of these risk factors need not despair, however. It means only that they need to pay particular attention to these guidelines for a heart-healthy life-style.

THE DON'TS

Smoking and Drug Abuse

Hypertension is well recognized as a major contributor to cardiovascular disease. When a person smokes, the drug nicotine, present in cigarette smoke, enters the bloodstream. Nicotine causes the arterioles to constrict and the blood pressure to rise. Restricted blood flow and cold hands are associated with smoking by most people. More serious is the need for the heart to pump harder to propel the blood through the lungs at a time when the oxygen-carrying capacity of the blood is reduced.

Stimulants, such as cocaine and amphetamines, can cause an irregular heartbeat and lead to heart attacks and strokes in people who are using drugs, even for the first time.

Obesity

Hypertension occurs more often in persons who are obese—those who are more than 20% above recommended weight. More tissues require servicing, and the heart sends the extra blood out under greater pressure in those who are overweight. Since it is very difficult for obese individuals to lose weight, it is recommended that weight control be a lifelong endeavor. Even a slight decrease in weight can bring with it a reduction in hypertension. A 4.5-kilogram weight loss doubles the chance that blood pressure can be normalized without drugs.

THE DO'S

Healthy Diet

It was once thought that a low-salt diet was protective against cardiovascular disease, and it still may be in certain persons. Theoretically, hypertension occurs because the more salty the blood, the greater the osmotic pressure and the higher the water content. In recent years, the emphasis has switched to a diet low in saturated fats and cholesterol as protective against cardiovascular disease. Cholesterol is ferried in the blood by two types of plasma proteins called LDL (low-density lipoprotein) and HDL (high-density lipoprotein). LDL (called "bad" lipoprotein) takes cholesterol from the liver to the tissues, and HDL (called "good" lipoprotein) transports cholesterol out of the tissues to the liver. When the LDL level in blood is abnormally high or the HDL level is abnormally low, cholesterol accumulates in the cells. When cholesterol-laden cells line the arteries, plaque develops, which interferes with circulation (Fig. 6A).

It is recommended that everyone know his or her blood cholesterol level. Individuals with a high blood cholesterol level (240 mg/100 ml) should be further tested to determine their LDL-cholesterol level. The LDL-cholesterol level together with other risk factors such as age, family history, general health, and whether the patient smokes will determine who needs dietary therapy to lower their LDL. Drugs are to be reserved for high-risk patients.

6.4   Circulatory Disorders

Cardiovascular disease (CVD) is the leading cause of untimely death in the Western countries. Modern research efforts have resulted in improved diagnosis, treatment, and prevention. This section discusses the range of advances that have been made in these areas. The reading on this emphasizes how to prevent CVD from developing in the first place.

Hypertension is Deadly

It is estimated that about 20% of all Americans suffer from *hypertension*, which is high blood pressure. Women of any age are considered to have hypertension if their blood pressure reading is 160/95 or above. For a man under age 45, a reading above 130/90 is hypertensive, and beyond age 45, a reading above 140/95 is considered hypertensive. While both systolic and diastolic pressures are considered important, it is the diastolic pressure that is emphasized when medical treatment is being considered.

Hypertension is sometimes called a silent killer because it may not be detected until a stroke or heart attack occurs. It has long been thought that a certain genetic makeup might account for the development of hypertension. Now researchers have discovered two genes that may be involved in some individuals. One gene codes for angiotensinogen, a plasma protein that is converted to a powerful vasoconstrictor in part by the product of the second gene. Persons with hypertension due to overactivity of these genes might one day be cured by gene therapy.

Atherosclerosis and Fatty Arteries

Hypertension also is seen in individuals who have *atherosclerosis* (formerly called arteriosclerosis), an accumulation of soft masses of fatty materials, particularly cholesterol, beneath the inner linings of arteries (Fig. 6A). Such deposits are called plaque, and as it develops, plaque tends to protrude into the lumen of the vessel and interfere

Evidence is mounting to suggest a role for antioxidant vitamins (A, E, and C) in the prevention of cardiovascular disease. Antioxidants protect the body from free radicals that may damage HDL cholesterol through oxidation or damage the lining of an artery, leading to a blood clot that can block the vessel. Nutritionists believe that the consumption of at least five servings of fruit and vegetables a day may be protective against cardiovascular disease.

Exercise

Those who exercise are less apt to have cardiovascular disease. One study found that moderately active men who spent an average of 48 minutes a day on a leisure-time activity such as gardening, bowling, or dancing had one-third fewer heart attacks than peers who spent an average of only 16 minutes each day. Exercise helps to keep weight under control, may help minimize stress, and reduces hypertension. The heart beats faster when exercising, but exercise slowly increases its capacity. This means that the heart can beat slower when we are at rest and still do the same amount of work. One physician recommends that his cardiovascular patients walk for one hour, three times a week, and in addition they are to practice meditation and yoga-like stretching and breathing exercises to reduce stress.

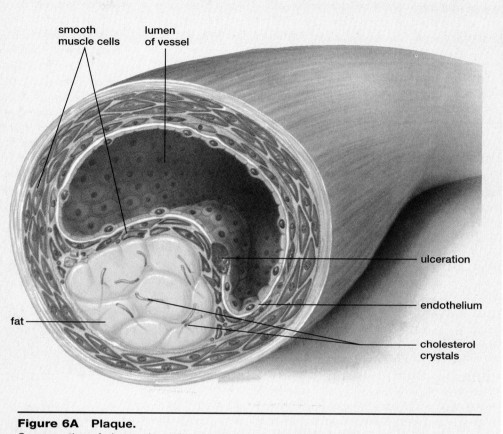

Figure 6A   Plaque.
Cross section of plaque shows its composition and indicates how it bulges out into the lumen of an artery, obstructing blood flow. Now a blood clot is more likely to occur and block the artery.

with the flow of blood. Atherosclerosis begins in early adulthood and develops progressively through middle age, but symptoms may not appear until an individual is 50 or older. To prevent its onset and development, the American Heart Association and other organizations recommend a diet low in saturated fat and cholesterol and rich in fruits and vegetables.

Plaque can cause a clot to form on the irregular arterial wall. As long as the clot remains stationary, it is called a *thrombus,* but when and if it dislodges and moves along with the blood, it is called an *embolus.* If *thromboembolism* is not treated, complications can arise, as mentioned in the following section.

In certain families, atherosclerosis is due to an inherited condition such as *familial hypercholesterolemia.* The presence of the disease-associated mutation can be detected, and this information is helpful if measures are taken to prevent the occurrence of the disease.

Stroke, Heart Attack, and Aneurysm

Stroke, heart attack, and aneurysm are associated with hypertension and atherosclerosis. A cardiovascular accident (CVA), also called a *stroke,* often results when a small cranial arteriole bursts or is blocked by an embolus. A lack of oxygen causes a portion of the brain to die, and paralysis or death can result. A person sometimes is forewarned of a stroke by a feeling of numbness in the hands or the face, difficulty in speaking, or temporary blindness in one eye.

A myocardial infarction (MI), also called a *heart attack,* occurs when a portion of the heart muscle dies due to a lack of oxygen. If a coronary artery becomes partially blocked, the individual may then suffer from *angina pectoris,* characterized by a radiating pain in the left arm. Nitroglycerin or related drugs dilate blood vessels and help relieve the pain. When a coronary artery is completely blocked, perhaps because of thromboembolism, a heart attack occurs.

An aneurysm is a ballooning of a blood vessel, most often the abdominal artery or the arteries leading to the brain. Atherosclerosis and hypertension can weaken the wall of an artery to the point that an aneurysm develops. If a major vessel like the aorta should burst, death is likely. Since capillaries are the region of exchange with tissues, it is possible to replace a portion of an artery with a plastic tube.

Stroke, heart attack, and an aneurysm are associated with both hypertension and atherosclerosis.

Dissolving Blood Clots

Medical treatment for a heart attack due to thromboembolism includes the use of tPA, a biotechnology drug. This drug converts plasminogen, a molecule found in blood, into plasmin, an enzyme that dissolves blood clots. In fact, tPA, which stands for *tissue plasminogen activator,* is the body's own way of converting plasminogen to plasmin. tPA is also being used for thrombolytic stroke patients but with limited success because some patients experience life-threatening bleeding in the brain. A better treatment might be new biotechnology drugs that act on the plasma membrane to prevent brain cells from releasing and/or receiving toxic chemicals caused by the stroke.

If a person has had symptoms of angina or has had a stroke, then aspirin may be prescribed. Aspirin reduces the stickiness of platelets and therefore lowers the probability that a clot will form. There is evidence that aspirin protects against first heart attacks, but there is no clear support for taking aspirin every day to prevent strokes in symptom-free people. Physicians warn that long-term use of aspirin might have harmful effects, including bleeding in the brain.

Treating Clogged Arteries

Surgical procedures are also available to treat clogged arteries. Each year thousands of persons have coronary bypass surgery. During this operation, surgeons take a segment of another blood vessel from the patient's body and stitch one end to the aorta and the other end to a coronary artery past the point of obstruction (Fig. 6.13). Another type of treatment has been tried. A laser is used to produce small pores in the ventricular musculature so that the blood from the ventricles can nourish the heart muscle.

In *angioplasty*, a cardiologist threads a plastic tube into an artery of an arm or a leg and guides it through a major blood vessel toward the heart. When the tube reaches the region of plaque in a coronary artery (see Fig. 6.4), a balloon attached to the end of the tube is inflated, forcing the vessel open. However, the artery may

not remain open because the trauma causes smooth muscle cells in the wall of the artery to proliferate and close it. Two lines of attack are being explored. Small metal devices—either coils or slotted tubes called stents—are expanded inside the artery to keep the artery open. When the stents are coated with heparin to prevent blood clotting and chemicals to prevent arterial closing, results have been promising.

Gene therapy for clogged arteries is another possibility. Someday, angioplasty might deliver a gene that prevents arteries from closing once they have been opened, or angioplasty could deliver a gene that codes for vegF, a growth factor that encourages new blood vessels to sprout out of an artery. If collateral blood vessels do form, they would transport blood past clogged arteries, making bypass surgery unnecessary.

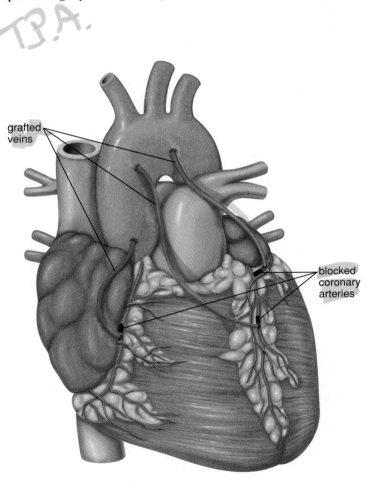

grafted veins

blocked coronary arteries

Figure 6.13 Coronary bypass operation.
During this operation, the surgeon grafts segments of another vessel, usually a small vein from the leg, between the aorta and the coronary vessels, bypassing areas of blockage. Patients who are ill enough to require surgery often receive two or three bypasses in a single operation.

HUMAN SYSTEMS WORK TOGETHER

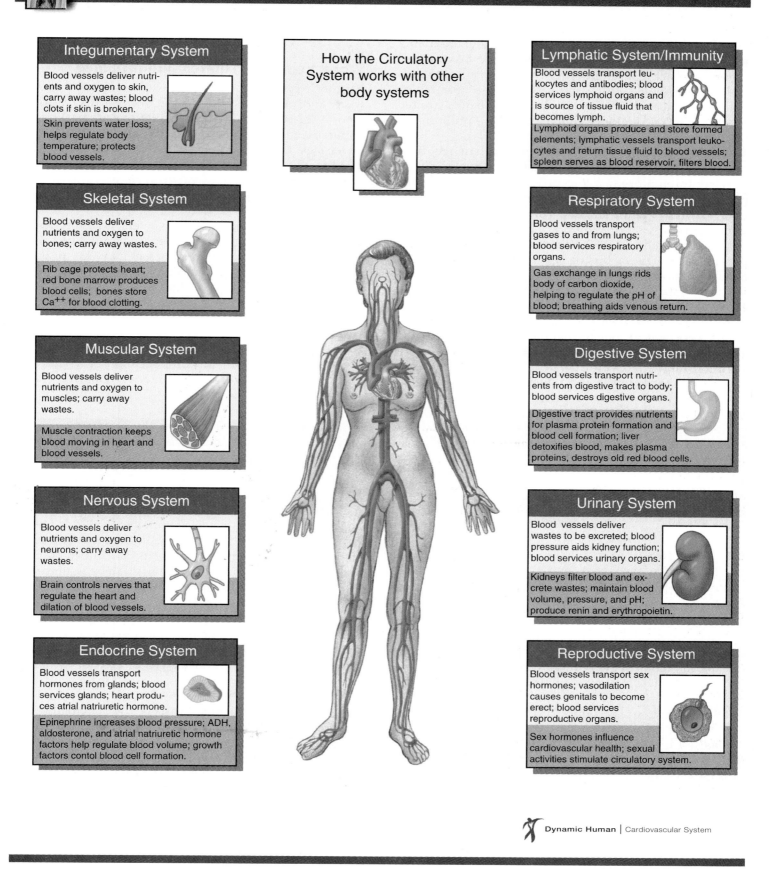

Integumentary System

Blood vessels deliver nutrients and oxygen to skin, carry away wastes; blood clots if skin is broken.

Skin prevents water loss; helps regulate body temperature; protects blood vessels.

Skeletal System

Blood vessels deliver nutrients and oxygen to bones; carry away wastes.

Rib cage protects heart; red bone marrow produces blood cells; bones store Ca^{++} for blood clotting.

Muscular System

Blood vessels deliver nutrients and oxygen to muscles; carry away wastes.

Muscle contraction keeps blood moving in heart and blood vessels.

Nervous System

Blood vessels deliver nutrients and oxygen to neurons; carry away wastes.

Brain controls nerves that regulate the heart and dilation of blood vessels.

Endocrine System

Blood vessels transport hormones from glands; blood services glands; heart produces atrial natriuretic hormone.

Epinephrine increases blood pressure; ADH, aldosterone, and atrial natriuretic hormone factors help regulate blood volume; growth factors contol blood cell formation.

How the Circulatory System works with other body systems

Lymphatic System/Immunity

Blood vessels transport leukocytes and antibodies; blood services lymphoid organs and is source of tissue fluid that becomes lymph.

Lymphoid organs produce and store formed elements; lymphatic vessels transport leukocytes and return tissue fluid to blood vessels; spleen serves as blood reservoir, filters blood.

Respiratory System

Blood vessels transport gases to and from lungs; blood services respiratory organs.

Gas exchange in lungs rids body of carbon dioxide, helping to regulate the pH of blood; breathing aids venous return.

Digestive System

Blood vessels transport nutrients from digestive tract to body; blood services digestive organs.

Digestive tract provides nutrients for plasma protein formation and blood cell formation; liver detoxifies blood, makes plasma proteins, destroys old red blood cells.

Urinary System

Blood vessels deliver wastes to be excreted; blood pressure aids kidney function; blood services urinary organs.

Kidneys filter blood and excrete wastes; maintain blood volume, pressure, and pH; produce renin and erythropoietin.

Reproductive System

Blood vessels transport sex hormones; vasodilation causes genitals to become erect; blood services reproductive organs.

Sex hormones influence cardiovascular health; sexual activities stimulate circulatory system.

Heart Transplants and Other Treatments

Persons with weakened hearts eventually may suffer from congestive heart failure, meaning the heart is no longer able to pump blood adequately, and blood backs up in the heart and lungs. Sometimes, it is possible to repair a weak heart. For example, a back muscle can be wrapped around a heart to strengthen it. One day it may be possible to use cardiac cell transplants, because researchers who have injected live cardiac muscle cells into animal hearts find that they will contribute to the pumping of the heart.

The difficulties with a heart transplant are, first, one of availability and, second, the tendency of the body to reject foreign organs. Recently, researchers, through biotechnology, altered the immune system of a strain of pigs with the hope that they will soon be a source of donor hearts.

On December 2, 1982, Barney Clark became the first person to receive a Jarvik-7, an artificial heart. The Jarvik-7, no longer approved for experimental use in humans, was driven by bursts of air received from a large external machine. Today, investigators have developed a heart pump that can be placed inside the body. The pump, powered by an external battery half the size of a videotape, is embedded in the abdominal wall. Blood moves from the left ventricle to the pump which sends it under pressure to the aorta. Newer and even smaller models are planned.

Defibrillators, implantable in the thorax, are also now available to treat an irregular heartbeat. These devices deliver a stimulus to the heart whenever it senses ventricular fibrillation.

When Veins Are Dilated and Inflamed

Varicose veins develop when the valves of veins become weak and ineffective due to the backward pressure of blood. Abnormal and irregular dilations are particularly apparent in the superficial (near the surface) veins of the lower legs. Crossing the legs or sitting in a chair so that its edge presses against the back of the knees can contribute to the development of varicose veins. Varicose veins also occur in the rectum, where they are called piles, or more properly, *hemorrhoids.*

Phlebitis, or inflammation of a vein, is a more serious condition, particularly when a deep vein is involved. Blood in the nonbroken but inflamed vessel may clot, in which case thromboembolism occurs. Pulmonary embolism, which occurs when a blood clot lodges in a pulmonary vessel, can cause death.

6.5   Working Together

The Working Together box on page 137 shows how the circulatory system works with other systems of the body to maintain homeostasis.

SUMMARY

6.1 Blood vessels

Blood vessels include arteries (and arterioles), which take blood away from the heart; capillaries, where exchange of substances with the tissues occurs; and veins (and venules), which take blood to the heart.

6.2 The Heart

The movement of blood in the circulatory system is dependent on the beat of the heart. During the cardiac cycle, the SA node (pacemaker) initiates the beat and causes the atria to contract.

The AV node conveys the stimulus and initiates contraction of the ventricles. The heart sounds, lub-dub, are due to contraction of the ventricles and closing of the atrioventricular valves, followed by the closing of the semilunar valves.

Blood pressure accounts for the flow of blood in the arteries, but because blood pressure drops off after the capillaries, it cannot cause blood flow in the veins. Skeletal muscle contraction pushes blood past a venous valve, which then shuts, preventing backward flow. The velocity of blood flow is slowest in the capillaries, where exchange of nutrients and wastes takes place.

6.3 Vascular Pathways

The circulatory system is divided into the pulmonary circuit and the systemic circuit. In the pulmonary circuit, two pulmonary arteries take blood from the right ventricle to the lungs, and the four pulmonary veins return it to the left atrium. To trace the path of blood in the systemic circuit, start with the aorta from the left ventricle. Follow its path until it branches to an artery going to a specific organ. It can be assumed that the artery divides into arterioles and capillaries, and that the capillaries lead to venules. The vein that takes blood to the vena cava most likely has the same name as the artery that delivered blood to the organ. In the adult systemic circuit, unlike the pulmonary circuit, the arteries carry oxygenated blood, and the veins carry deoxygenated blood.

6.4 Circulatory Disorders

Hypertension and atherosclerosis are two circulatory disorders that lead to heart attack and to stroke. Medical and surgical procedures are available to control cardiovascular disease, but the best policy is prevention by following a heart-healthy diet, getting regular exercise, maintaining a proper weight, and not smoking cigarettes.

6.5 Working Together

The circulatory system works with the other systems of the body in the ways described in the box on page 137.

STUDYING THE CONCEPTS

1. What types of blood vessels are there? Discuss their structure and function.   124

2. Trace the path of blood through the heart, mentioning the vessels attached to and the valves within the heart.   127

3. Describe the cardiac cycle (using the terms systole and diastole), and explain the heart sounds.   128

4. Describe the cardiac conduction system and an ECG. Tell how an ECG is related to the cardiac cycle.   128

5. What is blood pressure, and where in the body is the average normal arterial blood pressure about 120/80?   130

6. In which type of vessel is blood pressure highest? lowest? Velocity is lowest in which type of vessel, and why is it lowest? Why is this beneficial? What factors assist venous return of the blood?   130–31

7. Trace the path of blood in the pulmonary circuit as it travels from and returns to the heart.   132

8. Trace the path of blood to and from the kidneys in the systemic circuit.   132–33

9. What is atherosclerosis?   134   Name two illnesses associated with hypertension and thromboembolism.   135

10. Discuss the treatment for blocked cardiac blood vessels.   136

APPLYING YOUR KNOWLEDGE

Concepts

1. Obesity increases our chance of hypertension and/or a heart attack. What changes in the circulatory system occur due to extra body weight that can produce these results?

2. Patients with an organ transplant are usually given immunosuppressive drugs to help prevent rejection of the foreign cells in the transplanted organ. The level of drugs given is very critical. Explain the reason for this.

3. Research has shown that the atria of the heart can be rendered incapable of contracting and still the heart works quite well in pumping blood to all parts of the body. Explain the reason for this.

Bioethical Issue

We typically cheer on athletes, celebrities, and others who overcome physical conditions to excel. The injured climber who braves Mt. Everest in the dead of winter is a hero. The basketball star who shines despite a known illness is a winner.

Or is he? Some bioethicists question the philosophies of these enthusiasts. In particular, they point to the devastated spouses and children of risk-takers who in fact die.

Is it right for athletes or enthusiasts to chance death in order to fulfill an inner dream? Or is it selfish to take that chance, knowing your family could be left to suffer the consequences? On the flip side, do family members have the right to ask a mother or father to stay out of danger's way?

APPLYING TECHNOLOGY

Your study of the cardiovascular system is supported by these available technologies:

Exploring the Internet

The Mader Home Page provides further resources for studying this chapter.

`http://www.mhhe.com/sciencemath/biology/mader/`

(Click on *Human Biology*.)

Dynamic Human: Cardiovascular System CD-ROM

In *Anatomy,* the 3-dimensional viewer shows the heart in both anterior and posterior positions as well as in a cut-away; in *Explorations,* heart mechanics and blood vessel anatomy is explained; in *Histology,* microscopic slides show cardiac tissue and cross sections of an artery and vein; in

Clinical Concepts, heart murmur sounds are heard and myocardial infarction is explained.

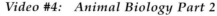

Life Science Animations Video

Video #4:   Animal Biology Part 2

Cardiac Cycle and Production of Heart Sounds (#32)   Heart sounds occur as blood flows and valves close within a heart that is actively pumping blood.   (2)*

Blood Circulation (#37)   The anatomy of the heart and the principles of blood circulation are reviewed as blood flow occurs.   (2)

Production of Electrocardiogram (#38)   An electrocardiogram is explained as the heart beats.   (3)

*Level of difficulty

TESTING YOUR KNOWLEDGE

1. Arteries are blood vessels that take blood _____ from the heart.

2. When the left ventricle contracts, blood enters the _____ .

3. The pulmonary veins carry blood _____ in oxygen.

4. The right side of the heart pumps blood to the _____ .

5. The _____ node is known as the pacemaker.

6. The arteries that serve the heart are the _____ arteries.

7. The pressure of blood against the walls of a vessel is termed _____ .

8. Blood moves in the arteries due to _____ and in the veins due to _____ .

9. Varicose veins develop when _____ become weak and ineffective.

10. Reducing the amount of _____ and _____ in the diet reduces the chances of plaque buildup in the arteries.

11. Label this diagram of the heart.

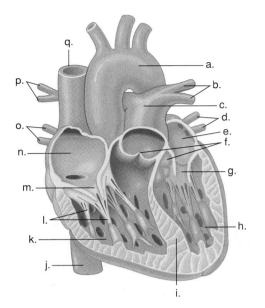

SELECTED KEY TERMS

aorta Major systemic artery that receives blood from the left ventricle. 127

arteriole Vessel that takes blood from an artery to capillaries. 124

artery Vessel that takes blood away from the heart to arterioles; characteristically possessing thick layers of connective and muscular tissue in the wall. 124

atrioventricular valve Valve located between the atrium and the ventricle. 126

atrium Chamber; particularly an upper chamber of the heart lying above a ventricle; either the left atrium or the right atrium. 126

AV (atrioventricular) node Small region of neuromuscular tissue that transmits impulses received from the SA node to the ventricular walls. 128

blood pressure Force of blood pushing against the inside wall of an artery. 130

capillary Microscopic vessel connecting an arteriole to a venule; thin walls permit substances to either exit or enter blood. 124

coronary artery Artery that supplies blood to the wall of the heart. 133

diastole Relaxation of a heart chamber. 128

electrocardiogram (ECG or EKG) Recording of the electrical activity associated with the heartbeat. 128

heart Muscular organ located in the thoracic cavity responsible for maintenance of blood circulation. 126

myocardium Cardiac muscle in the wall of the heart. 126

pacemaker See SA (sinoatrial) node. 128

pulmonary artery Blood vessel that takes blood away from the heart to the lungs. 127

pulmonary circuit That part of the circulatory system that takes deoxygenated blood to and oxygenated blood away from the gas-exchanging surfaces in the lungs. 132

pulmonary vein Blood vessel that takes blood to the heart from the lungs. 127

pulse Vibration felt in arterial walls due to expansion of the aorta following ventricular contraction. 128

SA (sinoatrial) node Small region of neuromuscular tissue that initiates the heartbeat; also called the pacemaker. 128

semilunar valve Valve with flap resembling a half moon located between the ventricles and their attached blood vessels. 126

septum Partition or wall that divides two areas; the septum in the heart separates the right half from the left half. 126

systemic circuit That part of the circulatory system that serves body parts and does not include the gas-exchanging surfaces in the lungs. 132

systole Contraction of a heart chamber. 128

valve Membranous extension of a vessel or the heart wall that opens and closes, ensuring one-way flow. 125

vein Vessel that takes blood to the heart from venules; characteristically having nonelastic walls. 124

vena cava Large systemic vein that returns blood to the right atrium of the heart; either the superior or inferior vena cava. 127

ventricle Cavity in an organ, such as a lower chamber of the heart; either the left ventricle or the right ventricle. 126

venule Vessel that takes blood from capillaries to a vein. 125

Chapter 7

Lymphatic System and Immunity

Chapter Outline

Figure 7.1 Allergies.
The immune system protects us from disease but has unwanted side effects such as reactions to environmental substances that will do no harm to the body.

Every spring, Paula C. knows what's coming. The flowers bloom. The wind blows. And she sniffles. Itching, sneezing, rubbing watery eyes with a tissue, Paula endures the onslaught of allergy season.

Ironically, Paula's agony stems from her immune system's admirable capability to fight off pathogens. Allergies are a hyperreactive response to dust and pollen. Sensing these foreign substances, antibodies kick into gear, replicating like crazy. The well-meaning molecules bind to certain proteins and release a flood of chemicals, such as histamines, that lead to allergy symptoms (Fig. 7.1).

For relief, allergy sufferers turn to antihistamines, or nonprescription pills that block histamine production. And so Paula swallows the medicine, grabs a tissue and waits—fitfully—for cold weather.

Usually, the immune system helps the body defend itself from disease. Sometimes the process of immunity has unwanted effects, as when Paula suffers from allergies or transplant patients have to take immunosuppressive drugs.

Immunity involves some defenses that are termed nonspecific because they occur for all types of pathogens (bacteria and viruses) that enter the body. Cut yourself and the skin becomes inflamed as white blood cells rush to the scene and begin their attack. Then, too, there are plasma proteins that latch onto bacterial surfaces and form pores so that water and salts enter until the bacteria burst. These proteins belong to the "complement system," so called because it complements certain immune responses.

Specific defenses might be slower to start, but they get the job done if you are invaded by a large number of the same type antigen. (An **antigen** is any foreign substance, usually protein or carbohydrate, that is capable of activating the immune system.) Get the flu and there is nothing to do but wait for your immune system to win the battle! Unseen in your body, lymphocytes are gearing up to produce just the right type of antibody that will make you well. The antibody combines with the antigen and in the process brings about its own destruction, since the complex is then phagocytized by a macrophage.

So despite the fact that Paula's immune system causes her to have allergies, it otherwise does a magnificent job of keeping her well. Unfortunately, pesticides have been found to surpass the immune system, as discussed in the Ecology reading on page 149.

7.1   Lymphatic System

The mammalian **lymphatic system** consists of lymphatic vessels and the lymphoid organs. This system, which is closely associated with the cardiovascular system, has three main functions: (1) lymphatic vessels take up excess tissue fluid and return it to the bloodstream; (2) lymphatic capillaries, called lacteals, absorb fats at the intestinal villi and transport them to the bloodstream; and (3) the lymphatic system helps to defend the body against disease.

Lymphatic Vessels Transport One Way

Lymphatic vessels are quite extensive; most regions of the body are richly supplied with lymphatic capillaries (Fig. 7.2). The construction of the larger lymphatic vessels is similar to that of cardiovascular veins, including the presence of valves. Also, the movement of lymph within these vessels is dependent upon skeletal muscle contraction. When the muscles contract, the lymph is squeezed past a valve that closes, preventing the lymph from flowing backwards.

The lymphatic system is a one-way system that begins with lymphatic capillaries in the skin. These capillaries take up fluid that has diffused from and has not been reabsorbed by the blood capillaries. **Edema** is swelling caused by the accumulation of tissue fluid. This can happen if too much tissue fluid is made and/or not enough of it is drained away. Once tissue fluid enters the lymphatic vessels, it is called **lymph.** The lymphatic capillaries join to form lymphatic vessels that merge before entering either the thoracic duct or the right lymphatic duct. The *thoracic duct* is much larger than the right lymphatic duct. It serves the lower extremities, the abdomen, the left arm, and the left side of both the head and the neck. The *right lymphatic duct* serves the right arm, the right side of the head and the neck, and the right thoracic area. The lymphatic ducts enter the subclavian veins, which are cardiovascular veins at the base of the neck.

Lymph flows one way from a capillary to ever-larger lymphatic vessels and finally to a lymphatic duct, which enters a subclavian vein.

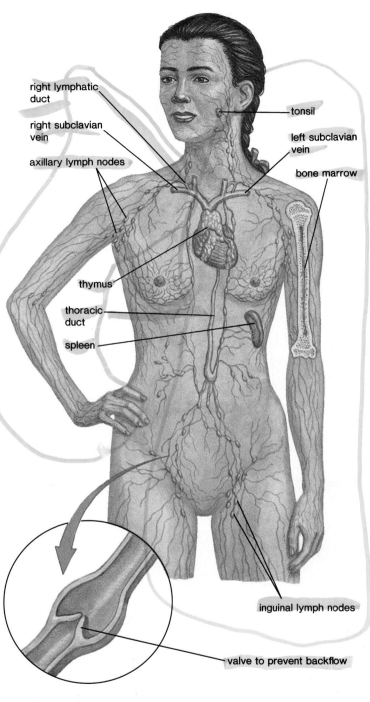

Figure 7.2   Lymphatic system.
The lymphatic vessels drain excess fluid from the tissues and return it to the cardiovascular system. The enlargement shows that lymphatic vessels have valves to prevent backward flow. The lymphoid organs are also shown.

 **Dynamic Human** | Lymphatic System

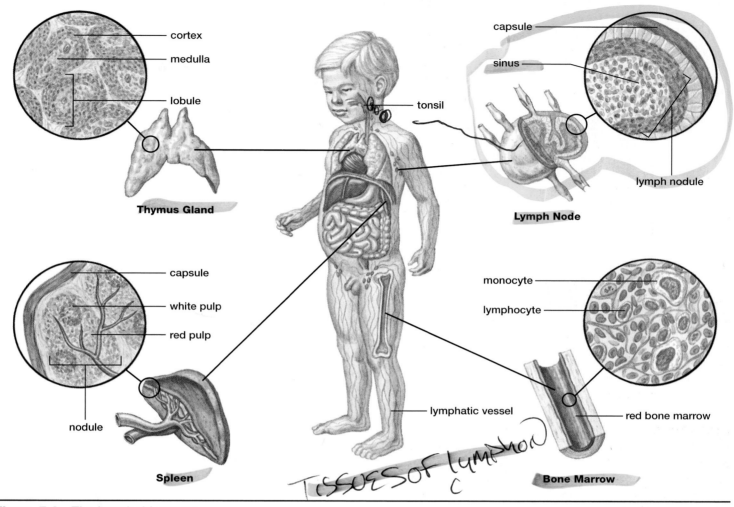

Figure 7.3 **The lymphoid organs.**
The lymphoid organs include the lymph nodes, the spleen, the thymus gland, and the red bone marrow, which all contain lymphocytes.

Lymphoid Organs Assist Immunity

The lymphoid organs of special interest are the lymph nodes, the spleen, the thymus gland, and the red bone marrow (Fig. 7.3).

Lymph nodes, which are small (about 1–25 mm) ovoid or round structures, are found at certain points along lymphatic vessels. A lymph node has a fibrous connective tissue capsule penetrated by incoming and outgoing lymphatic vessels. Connective tissue also divides a node into nodules. Each nodule contains a sinus (open space) lined with many lymphocytes and macrophages. As lymph passes through the sinuses, the macrophages purify it of infectious organisms and any other debris.

Nodules can occur singly or in groups. The *tonsils,* located in back of the mouth on either side of the tongue, and the *adenoids,* located on the posterior wall above the border of the soft palate, are composed of partly encapsulated lymph nodules. Also, nodules called *Peyer's patches* are found within the intestinal wall.

The lymph nodes occur in groups in certain regions of the body. For example, the inguinal nodes are in the groin and the axillary nodes are in the armpits.

The **spleen** is located in the upper left abdominal cavity just beneath the diaphragm. The construction of the spleen is similar to that of a lymph node. Outer connective tissue divides the organ into nodules, which contain sinuses. In the spleen, the sinuses are filled with blood instead of lymph. The blood vessels of the spleen can expand, and this enhances the carrying capacity of this organ as a blood reservoir and to make blood available in times of low blood pressure or when the body needs extra oxygen-carrying capacity.

A spleen nodule contains red pulp and white pulp. White pulp contains mostly lymphocytes. Red pulp contains red blood cells, lymphocytes, and macrophages. The red pulp helps to purify blood that passes through the spleen by removing bacteria and worn-out or damaged red blood cells. If the spleen ruptures due to injury, it can be removed.

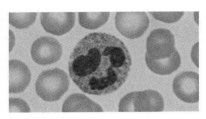

Neutrophil
40–70%
Phagocytizes
primarily bacteria

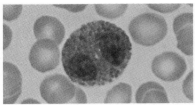

Eosinophil
1–4%
Phagocytizes and destroys
antigen-antibody
complexes

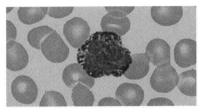

Basophil
0–1%
Releases histamine
when stimulated

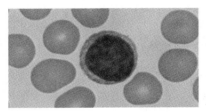

Lymphocyte
20–45%
B type produces antibodies
in blood and lymph;
T type kills virus-
containing cells.

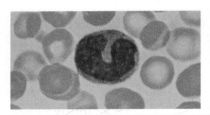

Monocyte
4–8%
Becomes macrophage—
phagocytizes bacteria and
viruses

Figure 7.4   Leukocytes.
There are five types of white blood cells, which differ according to
structure and function. The frequency of each type of cell is given
as a percentage of the total.

Although its functions are replaced by other organs, the individual is often slightly more susceptible to infections and may have to receive antibiotic therapy indefinitely.

The **thymus gland** is located along the trachea behind the sternum in the upper thoracic cavity. This gland varies in size, but it is larger in children than in adults and may disappear completely in old age. The thymus is divided into lobules by connective tissue. The T lymphocytes mature in the cortex of these lobules. The medulla, which consists mostly of epithelial cells, stains lighter than the cortex. It produces thymic hormones, such as thymosin, which are thought to aid in maturation of T lymphocytes. Thymosin may also have other functions in immunity.

Red bone marrow is the site of origination for all types of blood cells, including the five types of white blood cells that function in immunity (Fig. 7.4). The marrow contains stem cells that are ever capable of dividing and producing cells that go on to differentiate into the various types of blood cells. In a child, most bones have red bone marrow, but in an adult, it is present chiefly in the spongy bones of the skull, the sternum (breastbone), the ribs, the clavicle (collar bone), the pelvic bones, and the vertebral column. The red bone marrow consists of a network of connective tissue fibers, called reticular fibers, which are produced by cells called reticular cells. Reticular cells, the stem cells, and their progeny are packed about thin-walled sinuses filled with venous blood. Differentiated blood cells enter the bloodstream at these sinuses.

> The lymphoid organs have specific functions that assist immunity. Lymph is cleansed in lymph nodes; blood is cleansed in the spleen; T lymphocytes mature in the thymus although all white blood cells originate in the red bone marrow.

7.2   Nonspecific Defenses

Immunity is the ability of the body to defend itself against infectious agents, foreign cells, and even abnormal body cells, such as cancer cells. Natural immunity ordinarily keeps us free of diseases, including those caused by **pathogens,** and also cancer.

Immunity includes nonspecific and specific defenses. The three nonspecific defenses—barriers to entry, the inflammatory reaction, and protective proteins—are effective against many types of infectious agents.

Barring Entry

Skin and the mucous membranes lining the respiratory, digestive, and urinary tracts serve as mechanical barriers to entry by pathogens. Oil gland secretions contain chemicals that weaken or kill bacteria on the skin. The respiratory tract is lined by ciliated cells that sweep mucus and trapped particles up into the throat, where they can be swallowed, or expectorated (coughed out). The stomach has an acidic pH, which inhibits the growth of many types of bacteria. The various bacteria that normally reside in the intestine and other areas, such as the vagina, prevent pathogens from taking up residence.

Inflammatory Reaction

Whenever the skin is broken due to a minor injury, a series of events occurs that is known as the **inflammatory reaction.** The inflamed area has four symptoms: redness, pain, swelling, and heat. Figure 7.5 illustrates the participants in the inflammatory reaction. The mast cells, one type of participant, has the same stem cell as basophils (Fig. 7.4), but mast cells take up residence in the tissues.

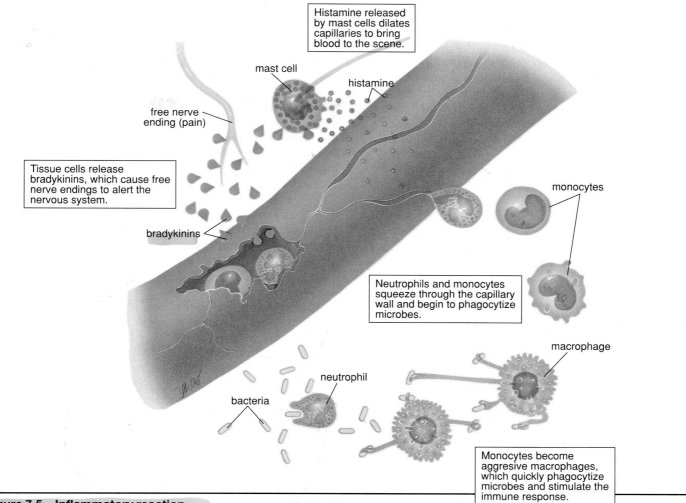

Histamine released by mast cells dilates capillaries to bring blood to the scene.

mast cell

histamine

free nerve ending (pain)

Tissue cells release bradykinins, which cause free nerve endings to alert the nervous system.

bradykinins

monocytes

Neutrophils and monocytes squeeze through the capillary wall and begin to phagocytize microbes.

macrophage

neutrophil

bacteria

Monocytes become aggresive macrophages, which quickly phagocytize microbes and stimulate the immune response.

Figure 7.5 Inflammatory reaction.
When a blood vessel is injured, bradykinin and then histamine are released. Bradykinin stimulates nerve endings and along with histamine causes blood vessels to dilate and become more permeable. Neutrophils and monocytes congregate at the injured site and squeeze through the capillary wall. The neutrophils begin to phagocytize bacteria. The monocytes become macrophages, large cells that are especially good at phagocytosis and also stimulate other white blood cells to action.

When an injury occurs, a capillary and several tissue cells are apt to rupture and to release **bradykinin.** This molecule initiates nerve impulses, resulting in the sensation of pain, and stimulates mast cells to release **histamine,** which, together with bradykinin, causes the capillaries to dilate and become more permeable. The enlarged capillaries causes the skin to redden, and the increased permeability allows proteins and fluids to escape so that swelling results.

The break in the skin allows pathogens to enter the body and triggers a migration of neutrophils and monocytes to the site of injury. Neutrophils and monocytes are amoeboid: they can change shape and squeeze through capillary walls to enter tissue fluid. Neutrophils phagocytize bacteria by forming endocytic vesicles. The engulfed bacteria are destroyed by hydrolytic enzymes when the vesicle combines with a lysosome, one of the cellular organelles. A rise in temperature due to the increased amount of blood in the capillaries increases phagocytosis by white blood cells.

When monocytes leave the bloodstream, they differentiate into **macrophages,** large phagocytic cells that are able to devour a hundred bacteria or viruses and still survive. Some tissues, particularly connective tissue, have resident macrophages, which routinely act as scavengers, devouring old blood cells, bits of dead tissue, and other debris. Macrophages can also bring about an explosive increase in the number of leukocytes by liberating a growth factor; this passes by way of blood to the red bone marrow, where it stimulates the production and the release of white blood cells, usually neutrophils.

As the infection is being overcome, some neutrophils die and become a part of pus, a whitish material that contains living and dead white blood cells, dead tissue, and bacteria.

The inflammatory reaction is a "call to arms"—it marshals phagocytic white blood cells to the site of bacterial invasion.

Chemo Warfare

The **complement system,** often simply called complement, is a group of plasma proteins, each designated by the letter C and a subscript. Once a complement protein is activated, it activates another protein, and the result is a set series of reactions. A limited amount of activated protein is needed because a domino effect occurs; each protein in the series is capable of activating many molecules of protein next in the series.

Complement is activated when pathogens enter the body. It "complements" certain immune responses, and this accounts for its name. For example, it is involved in and amplifies the inflammatory response because complement proteins attract phagocytes to the scene. Once complement proteins bind to the surface of pathogens, they will be phagocytized by a neutrophil or macrophage.

Another series of reactions is complete when complement proteins produce holes in bacterial cell walls and plasma membranes. When potassium ions leave, fluid and other ions enter the bacterial cell to the point that it undergoes lysis (bursts) (Fig. 7.6).

Interferon is a protein produced by virus-infected cells. Interferon binds to receptors of noninfected cells, causing the cells to prepare for possible viral attack by producing substances that interfere with viral replication. Interferon is specific to the species; therefore, only human interferon can be used in humans. Although it once was a problem to collect enough interferon for clinical and research purposes, interferon is now made as a biotechnology product.

Immunity includes these nonspecific defenses: barriers to entry, the inflammatory reaction, and protective proteins.

7.3   Specific Defenses

Sometimes, it is necessary to rely on a specific defense rather than a nonspecific defense against a particular antigen. An **antigen** is a protein or carbohydrate chain of a glycoprotein within the plasma membrane that the body recognizes as nonself. Pathogens have antigens, but antigens can also be part of a foreign human cell or a cancer cell.

Specific defense usually lasts for some time. For example, once we recover from the measles, we usually do not get the illness a second time. Specific defense is primarily the result of the action of the **B lymphocytes** and the **T lymphocytes,** which are produced in red *b*one marrow. B lymphocytes mature in the bone marrow, and T lymphocytes mature in the *t*hymus gland. B lymphocytes, also called B cells, give rise to plasma cells, which produce **antibodies,** proteins that are capable of combining with and neutralizing antigens. Antibodies are secreted into the blood and the lymph, where they are known as gamma globulins. T lymphocytes, also called T cells, do not produce antibodies; instead, certain T cells directly attack cells that bear antigens. Other T cells regulate the immune response.

Lymphocytes are capable of recognizing an antigen because they have receptor molecules for antigens on their surface. The shape of the receptors on any particular lymphocyte is complementary to a specific antigen. It is often said that the receptor and the antigen fit together like *a lock and a key.* It is estimated that during our lifetime, we encounter a million different antigens, so we need the same number of different lymphocytes for protection against those antigens. It is remarkable that diversification occurs to such an extent during the maturation process that there is a different lymphocyte type for each possible antigen. Despite this great diversity, none of the lymphocytes ordinarily attacks the body's own cells. It is believed when a lymphocyte arises that is equipped to respond to the body's own proteins, it is normally suppressed and develops no further.

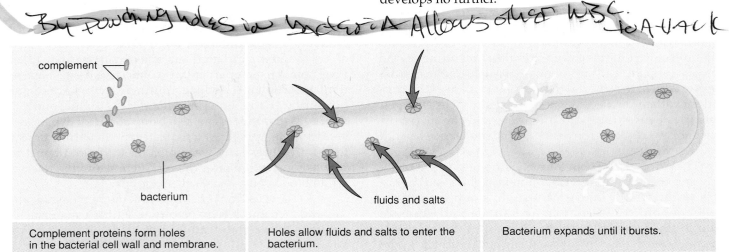

Figure 7.6   **Action of the complement system against a bacterium.**
When complement proteins in the plasma are activated by an immune reaction, they form holes in bacterial cell walls and plasma membranes, allowing fluids and salts to enter until the cell eventually bursts.

complement

bacterium

fluids and salts

Complement proteins form holes in the bacterial cell wall and membrane.

Holes allow fluids and salts to enter the bacterium.

Bacterium expands until it bursts.

B Cells Make Plasma and Memory Cells

Each type of B cell carries its specific antibody, as a membrane-bound receptor, on its surface. When a B cell in a lymph node or the spleen encounters an appropriate antigen, it becomes activated to divide many times. Most of the resulting cells are plasma cells, which secrete antibodies against this antigen. A **plasma cell** is a mature B cell that mass-produces antibodies in the blood or lymph.

The *clonal selection theory* states that the antigen selects which B cell will produce a clone of plasma cells (Fig. 7.7). Notice that a B cell does not clone until its antigen is present and binds to its receptors. B cells are then stimulated to clone by helper T cells, as is discussed in the next section.

Once antibody production is sufficient to agglutinate and/or neutralize the amount of antigen present in the body, the development of plasma cells ceases. Those members of a clone that do not participate in antibody production remain in the bloodstream as memory B cells. **Memory B cells** are the means by which long-term

immunity is possible. If the same antigen enters the system again, memory B cells quickly divide and give rise to new antibody-producing plasma cells.

Defense by B cells is called **antibody-mediated immunity** because the various types of B cells produce antibodies. It is also called *humoral immunity* because these antibodies are present in blood and lymph. A *humor* is any fluid normally occurring in the body.

Characteristics of B Cells

- Antibody-mediated immunity
- Produced and mature in bone marrow
- Undergo clonal selection in spleen and lymph nodes
- Direct recognition of antigen
- Clonal expansion produces antibody-secreting plasma cells as well as memory B cells

---

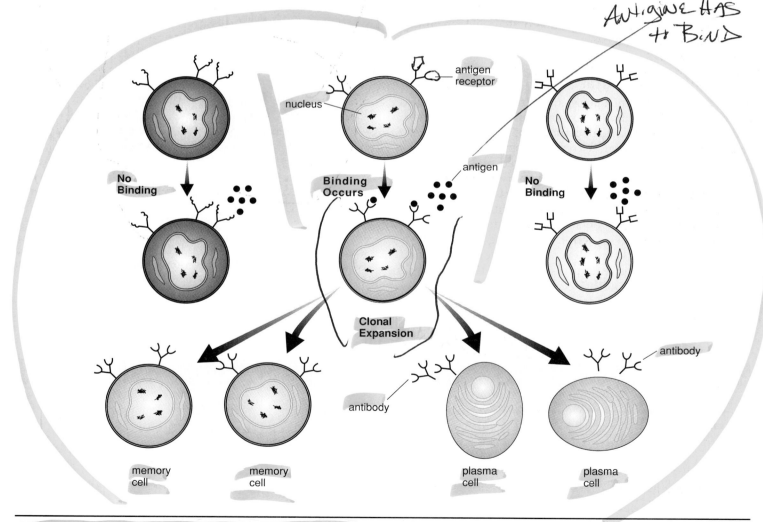

ANTIGENE HAS to BIND

Figure 7.7   Clonal selection theory as it applies to B cells.
In this diagram, color is used to signify that the membrane-bound antibody differs on three B cells. An antigen activates only the B cell with the appropriate antigen receptors; then the B cell undergoes clonal expansion. During the process, many plasma cells, which produce specific antibodies against this antigen, are produced. Memory cells, which retain the ability to recognize this antigen, are produced also.

How Antibodies Work

The most common type of antibody (IgG) is a Y-shaped protein molecule with two arms. Each arm has a "heavy" (long) polypeptide chain and a "light" (short) polypeptide chain of amino acids. These chains have *constant regions,* where the sequence of amino acids is set, and *variable regions,* where the sequence of amino acids varies (Fig. 7.8). The constant regions are not identical among all the antibodies. Instead, they are the same within a particular class of antibodies. The variable regions form an antigen-binding site with a shape that fits only a specific antigen. The antigen combines with the antibody at the antigen-binding site in a lock-and-key manner.

The antigen-antibody reaction can take several forms, but quite often the reaction produces complexes of antigens combined with antibodies. Such an antigen-antibody complex, sometimes called the immune complex, marks the antigen for destruction by other forces. For example, the complex may be engulfed by neutrophils or macrophages, or it may activate complement. Complement makes pathogens more susceptible to phagocytosis or can cause lysis, as discussed previously.

How Antibodies Differ

There are five different classes of circulating antibodies (Table 7.1). IgG antibodies are the major type in blood, and lesser amounts are also found in lymph and tissue fluid. IgG antibodies attack pathogens and their toxins. A toxin is a specific chemical (produced by bacteria, for example) that is poisonous to other living things. IgM antibodies are pentamers, meaning that they contain five of the Y-shaped structures shown in Figure 7.8a. These antibodies appear in blood soon after an infection begins and disappear before it is over. They are good activators of the complement system. IgA antibodies are dimers and contain two Y-shaped structures. They are the main type of antibody found in bodily secretions. They attack pathogens and their toxins before they reach the bloodstream. The role of IgD antibodies in immunity is uncertain, but

a very limited number is present in blood. IgE antibodies, which are responsible for allergic reactions, are discussed on page 156.

An antigen combines with an antibody at the antigen-binding site in a lock-and-key manner. The reaction can produce antigen-antibody complexes, which contain several molecules of antibody and antigen.

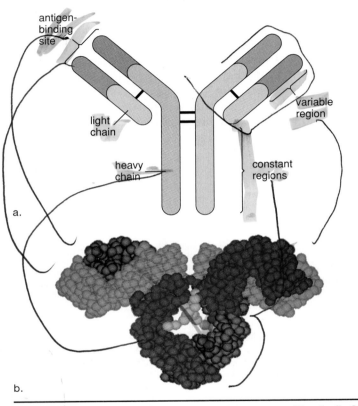

Figure 7.8 Structure of the most common antibody (IgG).
a. An IgG antibody contains two heavy (long) polypeptide chains and two light (short) chains arranged in two variable regions, where a particular antigen is capable of binding with the antibody.
b. Computer model of an antibody.

TABLE 7.1		
Antibodies		
Class	**Presence**	**Function**
IgG	Main antibody type in circulation	Attacks pathogens* and bacterial toxins; enhances phagocytosis
IgM	Antibody type found in circulation; largest antibody	Activates complement; clump cells
IgA	Main antibody type in secretions such as saliva and milk	Attacks pathogens and bacterial toxins
IgD	Antibody type found in circulation in extremely low quantity	Presence signifies maturity of B cell
IgE	Antibody type found as membrane-bound receptor on basophils in blood and on mast cells in tissues	Responsible for allergic reactions

*Viruses and bacteria

Ecology Focus

Pesticides: A Benefit Gone Wrong

A pesticide is any one of 55,000 chemical products used to kill insects, fungi, or rodents that interfere with human activities. The argument for pesticides is attractive. It is claimed they kill off disease-causing agents, increase yield and thus lower the cost of food, and work quickly with minimal risk. Over time it has been found that pesticides do not meet these claims. Instead, pests become resistant to pesticides, which kill off natural enemies in addition to the pest. Then the pest population explodes. At first DDT did a marvelous job killing off the mosquitos that carry malaria; now malaria is as big a problem as ever. In the meantime, DDT has accumulated in the tissues of wildlife and humans, causing harmful effects. The problem was made obvious when birds of prey became unable to reproduce due to weak eggshells. The use of DDT is now banned in the United States.

Today's crop losses due to insects are greater percentage-wise than they were in the 1940s before modern pesticides became available. In the meantime, farmers are sold ever-more expensive pesticides to control pests that are increasing in resistance and number. This is called the pesticide treadmill. At least 20 insect species are now apparently resistant to all known insecticides.

Increasingly, we have discovered that pesticides are harmful to the environment and humans (Fig. 7A). Death due to pesticide poisoning is a worldwide problem that involves mostly children who mistakenly ingest a large quantity. The worst industrial accident in the world occurred when a nerve gas used to produce an insecticide in India was released into the air; 3,700 people were killed, and 30,000 were injured. It is impossible to estimate how many people are eventually affected by exposure to pesticides at work or play. Pesticides are mostly used in farming, but 20% are used each year on lawns, gardens, parks, golf courses, and cemeteries. We even spray them in our houses! Vietnam veterans claim a variety of health effects due to contact with a herbicide called Agent Orange used as a defoliant during the Vietnam War. (The use of this herbicide is now also banned in the United States.)

The EPA says that pesticide residues on food pose a serious cancer risk to the U.S. population. Other possible effects are birth defects, disorders of the nervous system, and *suppression of the immune system.*

There is an alternative to the addictive use of pesticides. Integrated pest management uses a diversified environment, mechanical and physical means, natural

Figure 7A   Pesticide warning signs indicate that pesticides are harmful to our health.

enemies, disruption of reproduction, and resistant plants to control rather than eradicate pest populations. Chemicals are used only as a last resort. Maintaining hedgerows, weedy patches, and certain trees can provide diverse habitats and food for predators and parasites that help control pests. The use of strip farming and crop rotation by farmers denies pests a continuous food source.

Natural enemies abound in the environment. When lacewings were released in cotton fields, they reduced the boll weevil population by 96% and increased cotton yield threefold. A predatory moth was used to reclaim 60 million acres in Australia overrun by the prickly pear cactus, and another type of moth is now used to control alligator weed in the southern United States. Similarly, in this country the Chrysolina beetle controls Klamath weed, except in shady places where a root boring beetle is more effective.

The use of sex attractants and sterile insects is also a type of biological control. In Sweden and Norway, scientists synthesized the sex pheromone of the Ips bark beetle, which attacks spruce trees. Almost 100,000 baited traps collected

females normally attracted to males emitting this pheromone. Sterile males have also been used to reduce pest populations. The screwworm fly parasitizes cattle in the United States. Flies raised in a laboratory were made sterile by exposure to radiation. The entire Southeast was freed from this parasite when female flies mated with sterile males and then laid eggs that did not hatch.

In the past, only crossbreeding could produce resistant plants. Now genetic engineering, a new technique by which certain genes are introduced into organisms, including plants, may eventually make large-scale use of pesticides unnecessary. Already, 50 types of genetically engineered plants that resist insects, or viruses, have entered small-scale field trials.

In general, biological control is a more sophisticated method of controlling pests than the use of pesticides. It requires an in-depth knowledge of pests and/or their life cycles. Because it does not have an immediate effect on the pest population, the effects of biological control may not be apparent to the farmer or gardener who benefits from it.

It is important for citizens to do all they can to promote biological control of pests instead of relying on pesticides, which are unhealthy for the environment and humans.

- Urge elected officials to support legislation to protect humans and the environment against the use of pesticides.

- Allow native plants to grow on all or most of the land to give natural predators a place to live.

- Cut down on the use of pesticides, herbicides, and fertilizers for the lawn, garden, and house.

- Use alternative methods such as cleanliness and good sanitation to keep household pests under control.

- Keep homes in good repair so that pests are not encouraged to live there.

- Dispose of pesticides in a safe manner.

T Cells Become Cytotoxic, Helper, Memory, or Suppressor Cells

There are different types of T cells: cytotoxic T cells, helper T cells, memory T cells, and suppressor T cells. Cytotoxic and helper T cells differentiate in the thymus; memory T cells and suppressor T cells arise later when an activated T cell clones. All types of T cells look alike but can be distinguished by their functions.

Cytotoxic T cells attack and destroy antigen-bearing cells, such as virus-infected or cancer cells. Cytotoxic T cells have storage vacuoles containing perforin molecules. Perforin molecules perforate a plasma membrane, forming a hole that allows water and salts to enter. The cell under attack then swells and eventually bursts (Fig. 7.9). It often is said that T cells are responsible for **cell-mediated immunity**, characterized by destruction of antigen-bearing cells. Cytotoxic T cells and memory T cells are especially associated with cell-mediated immunity.

Helper T cells regulate immunity by enhancing the response of other immune cells. When exposed to an antigen, they enlarge and secrete **lymphokines**, stimulatory molecules released by helper T cells that cause helper T cells to clone and other immune cells to perform their functions. For example, lymphokines stimulate macrophages to phagocytize and stimulate B cells to become antibody-producing plasma cells. Because HIV that causes AIDS attacks helper T cells and certain other cells of the immune system, it inactivates the immune response.

Suppressor T cells regulate the immune response by suppressing further development of helper T cells. Since helper T cells also stimulate antibody production by B cells, the suppressor T cells help to prevent both the B cell and

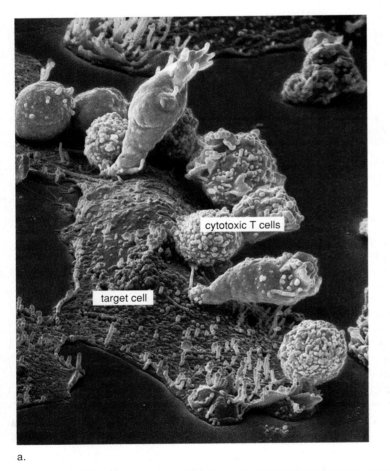

cytotoxic T cells

target cell

a.

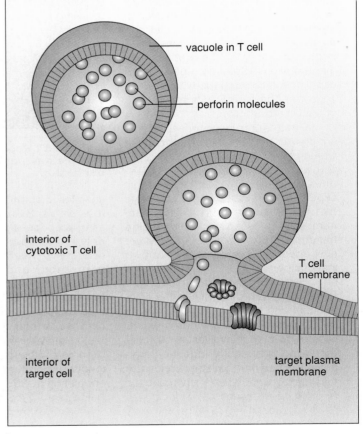

vacuole in T cell

perforin molecules

interior of
cytotoxic T cell

T cell
membrane

interior of
target cell

target plasma
membrane

b.

Figure 7.9   Cell-mediated immunity.

a. The scanning electron micrograph shows cytotoxic T cells attacking and destroying a cancer cell. **b.** During the killing process, the vacuoles in a cytotoxic T cell fuse with the cancer cell's plasma membrane and release units of the protein perforin. These units combine to form holes in the target cell's plasma membrane. Thereafter, fluid and salts enter so that the target cell eventually bursts.

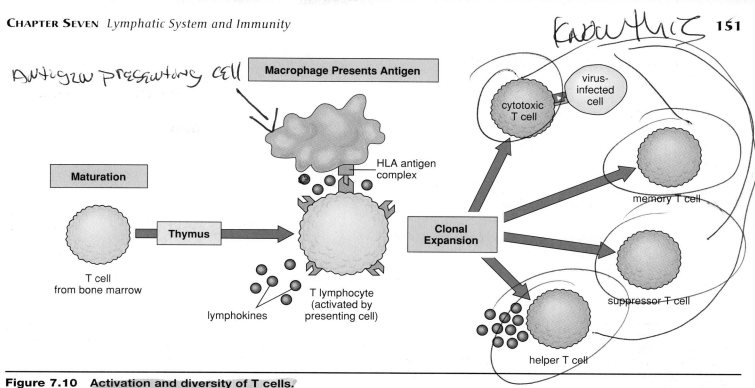

Figure 7.10 Activation and diversity of T cells.
T cells mature in the thymus and are activated when presented with an antigen by a macrophage. Clonal expansion follows with the production of the four types of T cells. Cytotoxic T cells are responsible for cell-mediated immunity.

T cell immune response from getting out of hand. Some doubt the existence of suppressor T cells and believe instead that suppression may depend upon the mixture of lymphokines released by T cells.

Following suppression, a population of **memory T cells** persists, perhaps for life. These cells are able to secrete lymphokines and to stimulate macrophages and B cells whenever the same antigen reenters the body. In this way, they contribute to active immunity.

Activating T Cells

T cells have receptors just as B cells do. Unlike B cells, however, cytotoxic T cells and helper T cells are unable to recognize an antigen that is simply present in lymph or blood. Instead, the antigen must be presented to them by an *antigen-presenting cell* (*APC*). When an APC, usually a macrophage, engulfs a pathogen, it is broken down to fragments within an endocytic vesicle. These fragments are antigenic, that is, they have the properties of an antigen. The fragments are linked to a major histocompatibility complex (MHC) protein in the plasma membrane so that they can be presented to a T cell.

Human MHC proteins are called **HLA (human leukocyte associated) proteins.** Because they mark the cell as belonging to a particular individual, HLA proteins are *self antigens*. The importance of self antigens in plasma membranes was first recognized when it was discovered that they contribute to the specificity of tissues and make it difficult to transplant tissue from one person to another. In other words, when the donor and the recipient are histo-(tissue) compatible (the same or nearly so), a transplant is more likely to be successful.

Figure 7.10 shows a macrophage presenting an antigen to a T cell. Once a T cell recognizes an antigen, it undergoes clonal expansion, producing other T cells that can also recognize this same antigen. Once a cytotoxic T cell recognizes an antigen, it attacks and destroys any cell that bears that antigen. In this way, T cells contribute to active immunity.

Characteristics of T Cells
- Cell-mediated immunity
- Produced in bone marrow; cytotoxic and helper T cells mature and differentiate in thymus
- Antigen must be presented, usually by macrophage
- Cytotoxic T cells search and destroy virus-infected, tumor, and transplanted cells
- Helper T cells secrete lymphokines and stimulate other immune cells

---

7.4   Induced Immunity

Immunity occurs naturally through infection or is brought about artificially by medical intervention. There are two types of induced immunity: active and passive. In active immunity, the individual alone produces antibodies against an antigen; in passive immunity, the individual is given prepared antibodies.

Active Immunity Is Long-Lived

Active immunity sometimes develops naturally after a person is infected with a pathogen. However, active immunity is often induced when a person is well so that possible future infection will not take place. To prevent infections, people can be artificially immunized against them.

Immunization involves the use of **vaccines,** substances that contain an antigen to which the immune system responds. Traditionally, vaccines are the pathogens themselves that have been treated so they are no longer virulent (able to cause disease). New methods of producing vaccines are being developed. Today, it is possible to genetically engineer bacteria to mass-produce a microbial surface protein, and this protein can be used as a vaccine. This method was used to prepare a vaccine against hepatitis B, a viral disease, and it is expected one day to allow production of a vaccine against malaria, a protozoan infection of the red blood cells. In order to reduce the number of injections required, perhaps it would be possible to produce a combination vaccine. First, a bacterium could be genetically engineered to produce several surface proteins at once. Then, these proteins could be packaged in plastic microspheres that would release them at different times so that booster shots would not be required.

After a vaccine is given, it is possible to follow an immune response by determining the amount of antibody present in a sample of plasma—this is called the *antibody titer*. A primary immune response occurs after the first exposure to a vaccine. For a period of several days, no antibodies are present; then, there is a slow rise in the titer, followed by first a plateau and then a gradual decline as the antibodies bind to the antigen or simply break down (Fig. 7.11). A secondary immune response occurs after a second exposure. The titer rises rapidly to a plateau level much greater than before. The second exposure is called a "booster" because it boosts the antibody titer to a high level. The high antibody titer now is expected to help prevent disease symptoms even if the individual is exposed to the disease-causing pathogen. Immunological memory causes an individual to be actively immune.

Memory Cells Provide a State of Readiness

Immunological memory is dependent upon the number of memory B and memory T cells capable of responding to a particular antigen. The receptors of memory B cells usually have a higher affinity for the antigen due to the

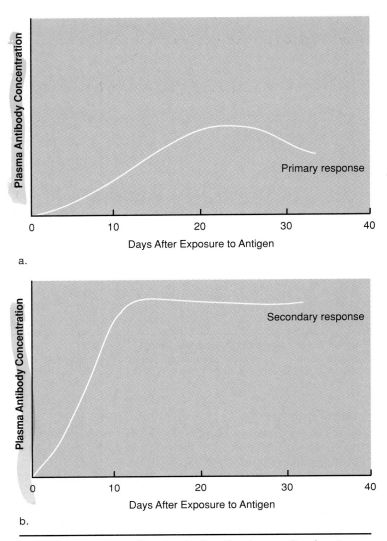

Figure 7.11   Development of active immunity due to immunization.
a. A primary immune response produces a lower concentration of plasma antibodies than does **(b)** a secondary immune response, which will occur following any subsequent booster shot.

selection process that occurred during the first exposure to the antigen, and, therefore, they are prone to make IgG earlier. Both memory B cells and memory T cells can respond to lower doses of antigen than they did during the first exposure to the antigen. Good active immunity lasts as long as clones of memory B and memory T cells are present in blood. Active immunity is usually long-lived.

Active (long-lived) immunity can be induced by the use of vaccines when a person is well and in no immediate danger of contracting an infectious disease. Active immunity is dependent upon the presence of memory B cells and memory T cells in the body.

Immunization of Children

Immunization protects children and adults from diseases. The success of immunization is witnessed by the fact that the vaccination against smallpox is no longer required because the disease has been eradicated. However, parents today often fail to get their children immunized because they don't realize the importance of it or can't bear the expense. Therefore, we might read in the newspapers about an outbreak of measles at a college or hospital in the United States because adults were not immunized as children. A recommended immunization schedule for children is given in the table (Fig. 7.12), and the United States is now committed to the goal of immunizing all children against the common types of childhood diseases listed. Diphtheria, whooping cough, and an Haemophilus influenza infection are all life-threatening respiratory diseases. Tetanus is characterized by muscular rigidity, including a locked jaw. These extremely serious infections are all caused by bacteria; the rest of the diseases listed are caused by viruses. Polio is a type of paralysis; measles and rubella, sometimes called German measles, are characterized by skin rashes; and mumps is characterized by enlarged parotid and other salivary glands. Although hepatitis B is not a childhood disease, it can be transmitted to the fetus by an infected mother. It is a serum-transmitted disease that can be acquired by transfusion, intravenous drug use, and sexual contact.

Bacterial diseases are subject to cure by antibiotic therapy such as penicillin, streptomycin, tetracycline, and erythromycin. Even so, it is better to rely on immunization rather than antibiotic therapy if possible. Some patients are allergic to antibiotics, and the reaction can be fatal. Antibiotics kill off not only disease-causing bacteria, but they also reduce the number of beneficial bacteria in the intestinal tract and elsewhere. These beneficial bacteria may have checked the spread of pathogens that now are free to multiply and to invade the body. This is the reason antibiotic therapy is often followed by a vaginal yeast infection in women. Antibiotic therapy also leads to strains that are resistant and therefore difficult to cure even with antibiotics. As mentioned previously, there are now resistant strains of bacteria that cause gonorrhea, a disease that has no vaccine.

Therefore, everyone should avail themselves of appropriate vaccinations. Preventing a disease by becoming actively immune to it is preferable to becoming ill and needing antibiotic therapy in order to be cured.

Figure 7.12 Suggested immunization schedule for infants and young children.
Children who are not immunized are subject to childhood diseases that can cause serious health consequences.

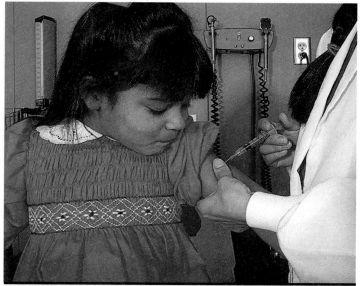

Vaccine	Age (Months)				Age (Years)
HepB* (hepatitis B)	Birth 1–2		12–15		4–6
DTP† (diphtheria, tetanus, whooping cough)	2	4	6	12–15	4–6
Td‡ (tetanus)					14–16
OPV§ (oral polio vaccine)	2	4	6		4–6
Hib§ (Haemophilus influenza, type b)	2	4	6	12–15	
MMR¶ (measles, mumps, rubella)	12–15 and 1 month later				

*Beginning in September 1996, 3 doses will be required for kindergarten entry.

†Five doses recommended for school entry.

‡First Td needed 10 years after last DTP.

§Doses 3 and 4 should be given according to manufacturer's guidelines.

¶A second dose given at least 1 month after first dose required for kindergarten entry.

Source: Immunization Program Guidelines, *Massachusetts Department of Public Health, June 1995.*

Passive Immunity Is Short-Lived

Passive immunity occurs artificially when an individual is given prepared antibodies (immunoglobulins) to combat a disease. Since these antibodies are not produced by the individual's B cells, passive immunity is short-lived. For example, newborn infants are passively immune to some diseases because IgG antibodies have crossed the placenta from the mother's blood. These antibodies soon disappear, however, so that within a few months, infants become more susceptible to infections. Breast-feeding prolongs the natural passive immunity an infant receives from the mother because antibodies are present in the mother's milk (Fig. 7.13).

Even though passive immunity does not last, it sometimes is used to prevent illness in a patient who has been unexpectedly exposed to an infectious disease. Usually, the patient receives a gamma globulin injection (serum that contains antibodies), perhaps taken from individuals who have recovered from the illness. In the past, horses were immunized, and serum was taken from them to provide the needed antibodies against such diseases as diphtheria, botulism, and tetanus. A patient who received these antibodies occasionally became ill because the serum contained proteins that the individual's immune system recognized as foreign. This was called serum sickness. But problems can still occur; more recently, an immunoglobulin intravenous product called Gammagard was withdrawn from the market because of possible implication in the transmission of hepatitis.

Passive immunity is needed when an individual is in immediate danger of succumbing to an infectious disease. Passive immunity is short-lived because the antibodies are administered to and not made by the individual.

Cytokines Boost White Blood Cells

Cytokines are messenger molecules produced by either lymphocytes or monocytes. They are called *lymphokines* when produced by lymphocytes and *monokines* when produced by monocytes. Because cytokines stimulate white blood cell formation and/or function, they are being investigated as possible adjunct therapy for cancer and AIDS. Both interferon and various other types of cytokines called interleukins have been used as immunotherapeutic drugs, particularly to enhance the ability of the individual's own T cells (and possibly B cells) to fight cancer.

Interferon, discussed previously on page 146, is a substance produced by leukocytes, fibroblasts, and probably most cells in response to a viral infection. When it is produced by T cells, interferon is called a lymphokine. Interferon still is being investigated as a possible anticancer drug, but so far it has proven to be effective only in certain patients, and the exact reasons for this as yet have not been determined.

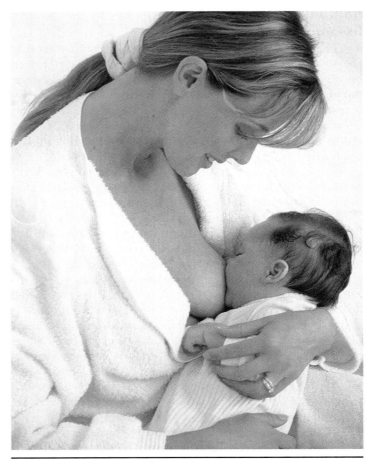

Figure 7.13   Example of passive immunity.
Breast-feeding is believed to prolong the passive immunity an infant receives from the mother because antibodies are present in the mother's milk.

When and if cancer cells carry an altered protein on their cell surface, they should be attacked and destroyed by cytotoxic T cells. Whenever cancer does develop, it is possible that the cytotoxic T cells have not been activated. In that case, lymphokines might awaken the immune system and lead to the destruction of the cancer. In one technique being investigated, researchers first withdraw T cells from the patient and activate the cells by culturing them in the presence of an interleukin. The cells then are reinjected into the patient, who is given doses of interleukin to maintain the killer activity of the T cells.

Those who are actively engaged in interleukin research believe that interleukins soon will be used as adjuncts for vaccines, for the treatment of chronic infectious diseases, and perhaps for the treatment of cancer. Interleukin antagonists also may prove helpful in preventing skin and organ rejection, autoimmune diseases, and allergies.

The interleukins and other lymphokines show some promise of enhancing the individual's own immune system.

Monoclonal Antibodies Have Same Specificity

As previously discussed, every plasma cell derived from the same B cell secretes antibodies against a specific antigen. These are **monoclonal antibodies** because all of them are one type and because they are produced by plasma cells derived from the same B cell. One method of producing monoclonal antibodies in vitro (outside the body in glassware) is depicted in Figure 7.14.

At present, monoclonal antibodies are being used for quick and certain diagnosis of various conditions. For example, HCG (human chorionic gonadotropin) is present in the urine of a pregnant woman. A monoclonal antibody can be used to detect this hormone, and if it is present, the woman is pregnant. Monoclonal antibodies also are used to identify infections. They are so accurate they can even sort out the different types of T cells in a blood sample. And because they can distinguish between cancer and normal tissue cells, they are used to carry radioactive isotopes or toxic drugs to tumors so that they can be selectively destroyed.

Monoclonal antibodies are considered to be a biotechnology product because a living system is used to mass-produce the product.

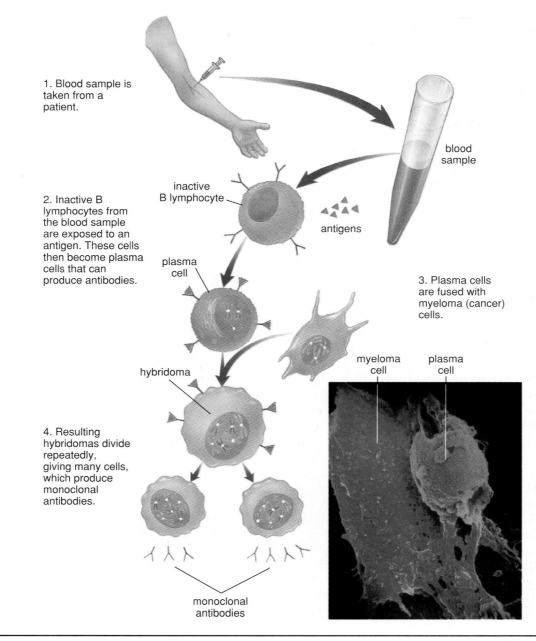

1. Blood sample is taken from a patient.

blood sample

2. Inactive B lymphocytes from the blood sample are exposed to an antigen. These cells then become plasma cells that can produce antibodies.

inactive B lymphocyte

antigens

plasma cell

3. Plasma cells are fused with myeloma (cancer) cells.

hybridoma

myeloma cell   plasma cell

4. Resulting hybridomas divide repeatedly, giving many cells, which produce monoclonal antibodies.

monoclonal antibodies

Figure 7.14 **Production of monoclonal antibodies.**
One possible method of producing human monoclonal antibodies.

Health Focus

Allergies

The runny nose and watery eyes of hay fever are often caused by an allergic reaction to the pollen of trees, grasses, and ragweed. Worse, the airways leading to the lungs constrict if one has asthma, resulting in difficult breathing characterized by wheezing.

Pollen grains, as you know, are produced by flowers within the anthers of stamens. When flowering plants reproduce, they release pollen, which becomes sexually mature microgametophytes carrying two sperm. Pollen grains are either windblown or carried by insects to the pistil. Once there, pollen grains produce a pollen tube through which the sperm travel on their way to an egg within a megagametophyte in an ovule. The base of the pistil is an ovary that contains the ovules.

Windblown pollen, particularly in the spring and fall, brings on the symptoms of hay fever. Most people can inhale pollen, eat peanuts, and take penicillin with no ill effects. But others have developed a hypersensitivity. Their immune systems have built up a type of antibody called immunoglobulin E that causes the release of histamine whenever they are exposed to an allergen. Histamine is a chemical that causes mucosal membranes of the nose and eyes to release fluid as a defense against pathogen invasion. But in the case of an allergy, copious fluid is released although no real danger is present.

Allergic individuals often respond to pollen, foods, animal fur and feathers, and even the feces of dust mites. They can reduce the chances of a reaction by avoiding those substances to which they are sensitive. The people in Figure 7B are trying to successfully use this approach. The taking of antihistamines can also be helpful. If these procedures are inadequate, patients can be tested to measure their susceptibility to any number of possible allergens. A small quantity of a suspected allergen is inserted just beneath the skin, and the strength of the subsequent reaction is noted. Then, ever-increasing doses of the allergen are injected subcutaneously with the hope that the body will build up a supply of IgG. IgG, in contrast to IgE, does not cause the release of histamine after it combines with the allergen. If IgG combines first upon exposure to the allergen, the allergic response does not occur. Patients know they are cured when the symptoms of hay fever and/or asthma no longer occur. Therapy may have to continue for as long as two to three years.

Allergic-type reactions can occur without involving the immune system. Wasp and bee stings contain substances that cause swellings, even in those whose immune system is not sensitized to substances in the sting. Also, jellyfish tentacles and foods such as fish that is not fresh and strawberries contain histamine or closely related substances that can cause a reaction. Immunotherapy is also not possible in those who are allergic to penicillin and bee stings. High sensitivity has built up upon the first exposure, and when reexposed, anaphylactic shock can occur. Among its many effects, histamine causes increased permeability of the smallest blood vessels, called capillaries. In these individuals, there is a drastic decrease in blood pressure that can be fatal within a few minutes. People who know they are allergic to bee stings can obtain a syringe of epinephrine to carry with them. This medication can delay the onset of anaphylactic shock until medical help is available.

Figure 7B   Protection against allergies.
The allergic reaction known as hay fever, and asthma attacks, can have many triggers, one of which is the pollen of a variety of plants. A dramatic solution to the problem has been found by this group of people.

7.5 Immunity Side Effects

The immune system protects us from disease because it can tell self from nonself. Sometimes, however, the immune system is underprotective, as when an individual develops cancer, or is overprotective, as when an individual develops allergies, suffers tissue rejection, or has an autoimmune disease.

Allergies: Overactive Immune System

Allergies are caused by an immune system that overreacts to an antigen or forms antibodies to substances that are usually not recognized as foreign. Unfortunately, as discussed in the Health Focus reading on page 156, allergies usually are accompanied by coldlike symptoms or, at times, by severe systemic reactions such as anaphylactic shock, which results in a sudden drop in blood pressure.

Of the five classes of antibodies—IgG, IgM, IgA, IgD, and IgE (see Table 7.1)—IgE antibodies are involved in allergic reactions. IgE antibodies are found in the bloodstream, but they, unlike the other types of antibodies, also reside in the membrane of mast cells found in the tissues. As mentioned, *mast cells* and basophils are thought to have a common stem cell. When the *allergen*, an antigen that provokes an allergic reaction, attaches to the IgE antibodies on mast cells or basophils, these cells release histamine and other substances.

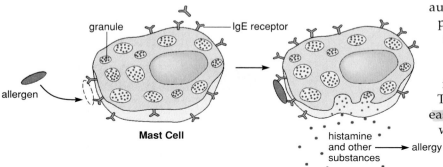

These substances cause mucous secretions and airway constriction, which are characteristic allergy symptoms. Mast cells and basophils produce histamine, resulting in an increased capillary permeability that can cause fluid loss and shock.

Allergy shots sometimes prevent the onset of allergic symptoms. Injections of the allergen cause the body to build up high quantities of IgG antibodies, and these combine with allergens received from the environment before they have a chance to reach the IgE antibodies located in the membrane of mast cells or basophils.

Histamine and other substances released by mast cells and basophils cause allergic symptoms.

Tissue Rejection: Foreign MHC Proteins

Certain organs, such as the skin, the heart, and the kidneys, could be successfully transplanted from one person to another without adjunct therapy if the body did not attempt to *reject* them. Rejection occurs because cytotoxic T cells bring about disintegration of foreign tissue in the body.

Organ rejection is presently controlled by careful selection of the organ to be transplanted and the administration of immunosuppressive drugs. It is best if the transplanted organ has the same type of MHC proteins as those of the recipient, because cytotoxic T cells recognize foreign MHC proteins. The immunosuppressive drug cyclosporine has been used for many years. A new drug, tacrolimus (formerly known as FK-506), shows some promise, especially in liver transplant patients. However, both drugs, which act by inhibiting the production of interleukin-2, are known to adversely affect the kidneys.

When an organ is rejected, the immune system is attacking cells that bear different MHC proteins from those of the individual.

Autoimmune Diseases: The Body Attacks Itself

Certain human illnesses are referred to as **autoimmune diseases** because they are due to an attack on tissues by the body's own antibodies and T cells. Exactly what causes autoimmune diseases is not known, but they seem to appear after the individual has recovered from an infection or traumatic injury. Some bacteria have been observed to produce toxic products that can cause T cells to bind prematurely to macrophages. Perhaps then the T cell reacts to the body's own tissues as if they were foreign. This might be the cause of at least some autoimmune diseases. In *myasthenia gravis*, neuromuscular junctions do not work properly, and muscular weakness results. In *multiple sclerosis* (*MS*), the myelin sheath of nerve fibers is attacked, and this causes various neuromuscular disorders. A person with *systemic lupus erythematosus* (*SLE*) has various symptoms, but death is usually due to kidney damage. In *rheumatoid arthritis*, the joints are affected. It is suspected that heart damage following rheumatic fever and type I diabetes are also autoimmune illnesses. There are no cures for autoimmune diseases.

Autoimmune diseases seem to be preceded by an infection that results in cytotoxic T cells or antibodies attacking the body's own organs.

7.6 Working Together

The Working Together box on page 158 shows how the lymphatic system works with the other systems of the body to maintain homeostasis.

HUMAN SYSTEMS WORK TOGETHER

Integumentary System

Lymphatic vessels pick up excess tissue fluid; immune system protects against skin infections.

Skin serves as a barrier to pathogen invasion; Langerhans' cells phagocytize pathogens; protects lymphatic vessels.

Skeletal System

Lymphatic vessels pick up excess tissue fluid; immune system protects against infections.

Red bone marrow produces leukocytes involved in immunity.

Muscular System

Lymphatic vessels pick up excess tissue fluid; immune system protects against infections.

Skeletal muscle contraction moves lymph; physical exercise enhances immunity.

Nervous System

Lymphatic vessels pick up excess tissue fluid; immune system protects against infections of nerves.

Microglial cells engulf and destroy pathogens.

Endocrine System

Lymphatic vessels pick up excess tissue fluid; immune system protects against infections.

Thymus is necessary to maturity of T lymphocytes.

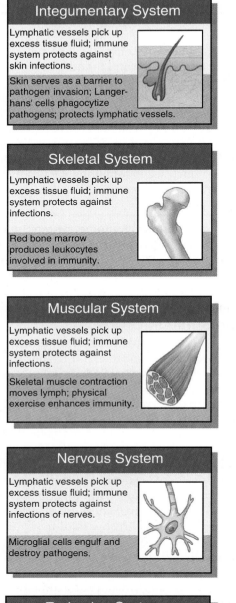
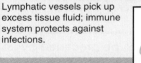

How the Lymphatic System works with other body systems

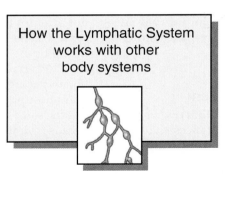

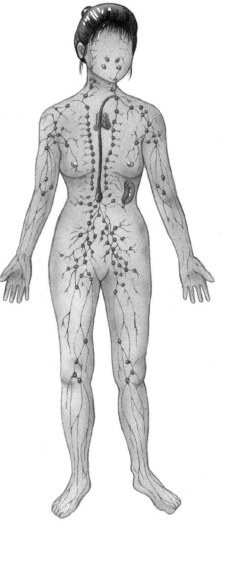

Circulatory System

Lymphoid organs produce and store formed elements; lymphatic vessels transport leukocytes and return tissue fluid to blood vessels; spleen serves as blood reservoir, filters blood. Blood vessels transport leukocytes and antibodies; blood services lymphoid organs and is source of tissue fluid that becomes lymph.

Respiratory System

Lymphatic vessels pick up excess tissue fluid; immune system protects against respiratory tract and lung infections.

Tonsils and adenoids occur along respiratory tract; breathing aids lymph flow; lungs carry out gas exchange.

Digestive System

Lacteals absorb fats; Peyer's patches prevent invasion of pathogens; appendix contains lymphoid tissue.

Digestive tract provides nutrients for lymphoid organs; stomach acidity prevents pathogen invasion of body.

Urinary System

Lymphatic system picks up excess tissue fluid, helping to maintain blood pressure for kidneys to function; immune system protects against infections.

Kidneys control volume of body fluids, including lymph.

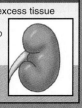

Reproductive System

Immune system does not attack sperm or fetus, even though they are foreign to the body.

Sex hormones influence immune functioning; acidity of vagina helps prevent pathogen invasion of body; milk passes antibodies to newborn.

Dynamic Human | Lymphatic System

SUMMARY

7.1 Lymphatic System

The lymphatic system consists of lymphatic vessels and lymphoid organs. The lymphatic vessels absorb fat molecules at intestinal villi and excess tissue fluid at the cardiovascular capillaries. Eventually, two main ducts empty into the subclavian veins.

Lymphocytes are produced and accumulate in the lymphoid organs (lymph nodes, spleen, thymus gland, and red bone marrow).

7.2 Nonspecific Defenses

Immunity involves nonspecific and specific defenses. Nonspecific defenses include barriers to entry, the inflammatory reaction, and protective proteins.

7.3 Specific Defenses

Specific defenses require lymphocytes, which are produced in the red bone marrow. B cells mature in the red bone marrow and undergo clonal selection in the lymph nodes and the spleen. Cytotoxic and helper T cells mature and differentiate in the thymus.

B cells are responsible for antibody-mediated immunity. An antibody is a Y-shaped molecule that has two binding sites. Each antibody is specific for a particular antigen. Activated B cells become antibody-secreting plasma cells and memory B cells. Memory B cells respond if the same antigen enters the body at a later date.

There are four types of T cells. Cytotoxic T cells kill cells on contact; helper T cells produce lymphokines and stimulate other immune cells; suppressor T cells suppress the immune response; and memory T cells remain in the body to provide long-lasting immunity.

7.4 Induced Immunity

Immunity can be induced in various ways. Vaccines are available to promote long-lived active immunity, and antibodies sometimes are able to provide an individual with short-lived passive immunity.

Cytokines, notably interferon and interleukins, are used in an attempt to promote the body's ability to recover from cancer and to treat AIDS.

7.5 Immunity Side Effects

Allergies result when an overactive immune system forms antibodies to substances not normally recognized as foreign. Cytotoxic T cells attack transplanted organs as nonself; therefore, immunosuppressive drugs must be administered. Autoimmune illnesses occur when antibodies and T cells attack the body's own tissues.

7.6 Working Together

The lymphatic system works with the other systems of the body in the ways described in the box on page 158.

STUDYING THE CONCEPTS

1. What is the lymphatic system, and what are its three functions?   142

2. Describe the structure and the function of lymph nodes, the spleen, the thymus gland, and bone marrow.   143–44

3. What are the body's nonspecific defense mechanisms?   144–46

4. Describe the inflammatory reaction, and give a role for each type of cell and molecule that participates in the reaction.   144–45

5. What is the clonal selection theory? B cells are responsible for which type of immunity?   147

6. Describe the structure of an antibody, and define the terms variable regions and constant regions.   148

7. Name the four types of T cells, and state their functions.   150–51

8. Explain the process by which a T cell is able to recognize an antigen.   151

9. How is active immunity achieved? How is passive immunity achieved?   152–54

10. What are cytokines, and how are they used in immunotherapy? 154

11. How are monoclonal antibodies produced, and what are their applications?   155

12. Discuss allergies, tissue rejection, and autoimmune diseases as they relate to the immune system.   157

APPLYING YOUR KNOWLEDGE

Concepts

1. Most individuals have experienced a swelling of lymph nodes or "glands" as they are often called. An infection of the throat often results in the "glands" in the neck area becoming swollen and sore. Explain what is happening to produce these symptoms.

2. What is the reason it is recommended that an individual receive a booster shot for tetanus every 10 years? Explain this on the basis of B cells.

3. Immunosuppresive drugs are often used in tissue transplant patients. What is the possible disadvantage to the use of these drugs?

Bioethical Issue

When AIDS first emerged in the 1980s, scientists had little information about the disease. One way they began to learn was to experiment with chimpanzees. Biologically speaking, these animals are similar to humans. So they make a good model for studying diseases that affect us. For example, researchers can study a chimp infected with HIV to learn just how the virus causes AIDS.

In the past, primate research has led to many therapies, including vaccines against polio and measles. Still, some animal rights advocates argue that it's unethical to infect a healthy animal in order to do research. What do you think? Is it fair to sacrifice an animal's life in an attempt to save human lives? Or should research be limited to nonanimal studies? Do some types of diseases warrant animal research? Defend your answers.

TESTING YOUR KNOWLEDGE

1. Lymphatic vessels take up excess _____ and return it to the _____ veins.

2. The function of lymph nodes is to _____ lymph.

3. T cells mature in the _____.

4. _____ is a group of proteins present in plasma that plays a role in destroying bacteria.

5. A stimulated B cell becomes antibody-secreting _____ cells and _____ cells, which are ready to produce the same type of antibody at a later time.

6. B cells are responsible for _____-mediated immunity.

7. T cells produce _____, which are stimulatory molecules for all types of immune cells.

8. In order for a T cell to recognize an antigen, it must be presented by a(n) _____ along with an MHC protein.

9. Give the function of the four types of T cells shown.

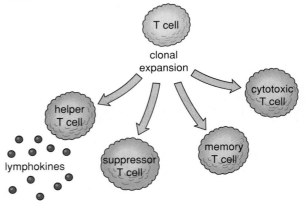

10. Immunization with _____ brings about active immunity.

11. Antibodies produced by the same plasma cell are _____ antibodies.

12. Allergic reactions are associated with the release of _____ and other substances from mast cells.

APPLYING TECHNOLOGY

Your study of the lymphatic system is supported by these available technologies:

Exploring the Internet

The Mader Home Page provides further resources for studying this chapter.

`http://www.mhhe.com/sciencemath/biology/mader/`

(Click on *Human Biology*.)

Dynamic Human: Lymphatic System CD-ROM

In *Anatomy*, major lymphatic organs are reviewed and macroscopic components are depicted in 3-dimension; in *Explorations*, lymph is shown flowing in lymphatic vessels, through lymph nodes, and into a cardiovascular vein; nonspecific defenses including fever, inflammatory response, interferon, and phagocytosis are depicted, and the function of B cells and T cells is explained; in *Histology*, microscopic slides are highlighted to reveal tissues; in *Clinical Concepts*, students have an opportunity to mix different types of blood, see how the AIDS virus invades and kills lymphocytes, and watch a patient being vaccinated.

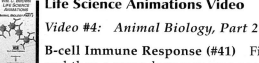

Explorations in Human Biology CD-ROM

Immune Response (#12) Macrophages, several types of T and B cells interact to destroy a bacterial, viral, or fungal pathogen as chosen by the student. (5)*

Life Science Animations Video

Video #4: Animal Biology, Part 2

B-cell Immune Response (#41) First a primary and then a secondary response occur in response to the same specific antigen. (2)

Structure and Function of Antibodies (#42) A review of antibody structure including constant and variable regions. (2)

Types of T cells (#43) Helper T cells are distinguished from cytotoxic (killer) T cells on the basis of activation and function. (2)

Relationship of Helper T cells and Killer T cells (#44) Both types of T cells respond to the same antigen in different ways. (3)

*Level of difficulty

SELECTED KEY TERMS

antibody Protein produced by plasma cells derived from B cells that binds with a specific antigen.   146

antibody-mediated immunity Specific mechanism of defense in which plasma cells derived from B cells produce antibodies that combine with antigens.   147

antigen Foreign substance, usually a protein or a polysaccharide, that stimulates the immune system to react, such as to produce antibodies.   146

autoimmune disease Disease that results when the immune system mistakenly attacks the body's own tissues.   157

B lymphocyte (LIM-fuh-syt) Lymphocyte that matures in the red bone marrow and, when stimulated by the presence of a specific antigen, gives rise to antibody-producing plasma cells.   146

bradykinin (bray-dih-KY-nen) Substance found in damaged tissue that initiates nerve impulses, resulting in the sensation of pain.   145

cell-mediated immunity Specific mechanism of defense in which T cells destroy antigen-bearing cells.   150

complement system Group of plasma proteins that form a nonspecific defense mechanism; its actions complement the antigen-antibody reaction.   146

cytotoxic T cell T lymphocyte that attacks and kills antigen-bearing cells.   150

edema Swelling due to tissue fluid accumulation in the intercellular spaces.   142

helper T cell T lymphocyte that releases lymphokines and stimulates certain other immune cells to perform their respective functions.   150

histamine Substance produced by basophils or mast cells that promotes some of the reactions associated with inflammation and allergies. 145

HLA (human lymphocyte associated) protein Protein in plasma membrane that identifies the cell as belonging to a particular individual and acts as a self-antigen.   151

immunity Ability of the body to protect itself from specific foreign substances and cells, including infectious pathogens.   144

inflammatory reaction Tissue response to injury that is characterized by redness, swelling, pain, and heat.   144

interferon (in-tur-FIR-ahn) Protein formed by a cell infected with a virus that can increase the resistance of other cells to the virus.   146

lymph Fluid, derived from tissue fluid, that is carried in lymphatic vessels.   142

lymph nodes Mass of lymphoid tissue located along the course of a lymphatic vessel.   143

lymphatic system Mammalian organ system consisting of lymphatic vessels and lymphoid organs.   142

lymphokine (LIM-fuh-kyn) Molecule secreted by lymphocytes that has the ability to affect the activity of all types of immune cells.   150

macrophage Large phagocytic cells derived from a monocyte that ingests pathogens and debris.   145

memory B cell Persistent population of B cells ready to produce antibodies specific to a particular antigen; accounts for the development of active immunity.   147

memory T cell Persistent population of T cells ready to recognize an antigen that previously invaded the body.   151

monoclonal antibody Antibody of one type produced by a single plasma cell.   155

pathogen Microscopic infectious agent, such as a bacterium or a virus.   144

plasma cell Cell derived from a B lymphocyte that is specialized to mass-produce antibodies.   147

red bone marrow Blood cell-forming tissue located in the spaces within spongy bone.   144

spleen Large, glandular organ located in the upper left region of the abdomen that stores and purifies blood.   143

suppressor T cell T lymphocyte that suppresses certain other T and B lymphocytes from continuing to divide and perform their respective functions.   150

T lymphocyte (LIM-fuh-syt) One of four types of T cells; cytotoxic T cells mature in the thymus and kill antigen-bearing cells outright.   146

thymus gland Organ that lies in the neck and chest area and is absolutely necessary to the development of immunity.   144

vaccine Antigens prepared in such a way that they can promote active immunity without causing disease.   152

Chapter 8

Respiratory System

Chapter Outline

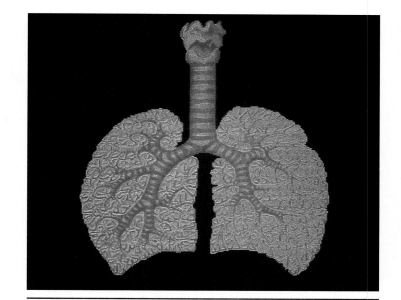

Figure 8.1 Vital organs.
Since structure suits function, gas-exchange surfaces tend to be permeable and moist, and this invites invasion by pathogens and pollutants.

Forget test scores or a winning football team. In 1994, La Quinta High School in Orange County, California, was known for one thing: tuberculosis (TB). An unsuspecting student carried TB into the school, infecting at least 12 people. By the time health officials caught the outbreak, 292 La Quinta students tested positive for TB. It was the worst TB incident at any U.S. high school in history.

TB is caused by a tiny bacterium that travels in coughs, shouts, and sneezes. When inhaled, TB bacteria are carried to the lungs by the downward path of air. Immune system cells race to destroy the invaders. If that doesn't work, the body traps the bacteria in tiny capsules called tubercles. But these capsules can crumble, unleashing the bacteria within. As more capsules form, the lungs may quickly become a scarred mass of lesions.

Beyond TB, a host of pathogens surround us, and we will consider several infections associated with the respiratory tract (Fig. 8.1). Aside from infections, two illnesses that have been attributed to breathing polluted air are emphysema and lung cancer. To a large degree, it is possible to protect oneself from these conditions because they may develop from smoking cigarettes. If infections or other illnesses interfere with breathing, death is likely. While it is possible to stop eating for several days, it is not possible to remain alive for longer than several minutes without breathing. Breathing is the first step of respiration, which also includes the movement of oxygen from the lungs to the tissues, and the process of cellular respiration, which produces ATP.

During breathing, air enters and exits the lungs by way of the respiratory tract, which consists of a series of tubes. Breathing is controlled by the brain, which automatically keeps the rib cage moving up and down. But the breathing rate can be modified, particularly by the amount of carbon dioxide in the blood, and we can voluntarily alter the quantity of air we take in and release. The respiratory system functions to have oxygen from the air enter the pulmonary capillaries and carbon dioxide from the tissues exit pulmonary capillaries. The manner in which oxygen and carbon dioxide is carried in the blood is of interest. The red blood cells and respiratory pigment hemoglobin play a role in both of these.

8.1 Respiratory Tract

During **inspiration** (breathing in) and **expiration** (breathing out), air is conducted toward or away from the lungs by a series of cavities, tubes, and openings, illustrated in Figure 8.2.

As air moves in along the air passages, it is filtered, warmed, and moistened. Filtering is accomplished by coarse hairs and cilia in the region of the nostrils and by cilia alone in the rest of the nose and the trachea. In the nose, the hairs and the cilia act as a screening device. In the trachea, cilia beat upward, carrying mucus, dust, and occasional bits of food that "went down the wrong way" into the pharynx, where the accumulation can be swallowed or expectorated. The air is warmed by heat given off by the blood vessels lying close to the surface of the lining of the air passages, and it is moistened by the wet surface of these passages.

Conversely, as air moves out during expiration, it cools and loses its moisture. As the air cools, it deposits its moisture on the lining of the windpipe and the nose, and the nose may even drip as a result of this condensation. The air still retains so much moisture, however, that upon expiration on a cold day, it condenses and forms a small cloud.

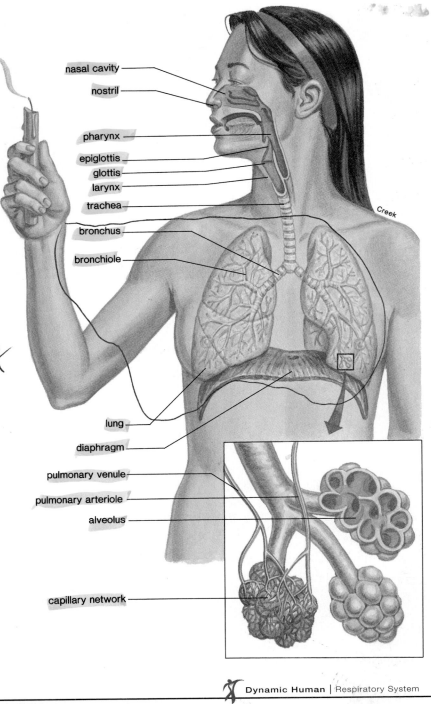

Figure 8.2   **The respiratory tract.**
The respiratory tract extends from the nose to the lungs, which are composed of air sacs called alveoli. Gas exchange occurs between air in the alveoli and blood within a capillary network that surrounds the alveoli.

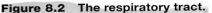

Air is filtered, warmed, and moistened as it moves from the nose toward the lungs.

TABLE 8.1
Path of Air

Structure	Description	Function
Nasal cavities	Hollow spaces in nose	Filter, warm, and moisten air
Pharynx	Chamber behind oral cavity and between nasal cavity and larynx	Connection to surrounding regions
Glottis	Opening into larynx	Passage of air into larynx
Larynx	Cartilaginous organ that contains vocal cords (voice box)	Sound production
Trachea	Flexible tube that connects larynx with bronchi (windpipe)	Passage of air to bronchi
Bronchi	Major divisions of the trachea that enter lungs	Passage of air to each lung
Bronchioles	Branched tubes that lead from the bronchi to the alveoli	Passage of air to each alveolus
Lungs	Soft, cone-shaped organs that occupy a large portion of the thoracic cavity	Gas exchange

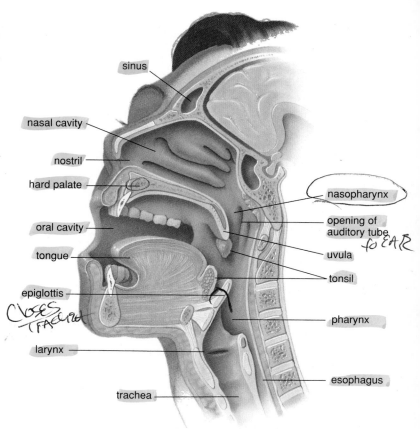

Figure 8.3   The upper respiratory tract.
The upper respiratory tract contains the nasal cavities, pharynx, larynx, and trachea.

The Nose Has Two Cavities

The nose contains two **nasal cavities** (Table 8.1), which are narrow canals with convoluted lateral walls separated from one another by a wall composed of bone and cartilage. Special ciliated cells in the narrow upper recesses of the nasal cavities (Fig. 8.3) act as odor receptors. Nerves lead from these cells to the brain, where the impulses generated by the odor receptors are interpreted as smell.

The tear (lacrimal) glands drain into the nasal cavities by way of tear ducts. For this reason, crying produces a runny nose. The nasal cavities also communicate with the cranial sinuses, air-filled, mucous membrane-lined spaces in the skull. If these membranes are inflamed due to a cold or an allergic reaction, mucus can accumulate in the sinuses, causing a sinus headache. The nasal cavities empty into the nasopharynx, the upper portion of the pharynx. The *eustachian tubes* lead from the nasopharynx to the middle ears.

The nasal cavities, which receive air, open into the nasopharynx.

The Pharynx Is a Crossroad

The **pharynx,** in the region of the throat, connects the nasal and oral cavities to the larynx. Therefore, the pharynx has three parts: the nasopharynx, where the nasal cavities open; the oropharynx, where the oral cavity opens; and the laryngopharynx, which opens into the larynx. In the pharynx, the air passage and the food passage cross because the larynx, which receives air, is superior to the esophagus, which receives food. The larynx lies at the top of the trachea. The larynx and trachea are normally open, allowing the passage of air, but the esophagus is normally closed and opens only when swallowing occurs.

Air from either the nose or the mouth enters the pharynx, as does food. The passage of air continues in the larynx and then the trachea proper.

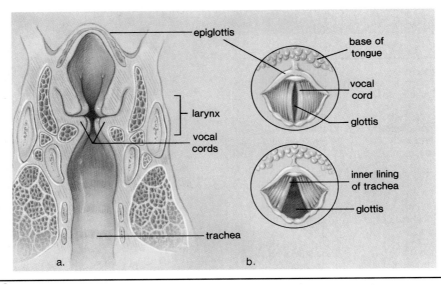

Figure 8.4   Placement of the vocal cords.

a. Longitudinal section of the larynx shows the anatomical location of the vocal cords inside the larynx. The vocal cords are stretched across the glottis. When air passes through the glottis, the vocal cords vibrate, producing sound. **b.** The glottis is narrow when we produce a high-pitched sound (*top*) and widens as the pitch deepens (*bottom*).

The Larynx Is the Voice Box

The **larynx** can be pictured as a triangular box whose apex, the Adam's apple, is located at the front of the neck. The Adam's apple is more prominent in men than women. At the top of the larynx is a variable-sized opening called the **glottis.** When food is swallowed, the glottis moves upward against the **epiglottis,** a flap of tissue that prevents food from passing into the larynx. You can detect this movement by placing your hand gently on your larynx and swallowing.

The larynx is called the voice box because the vocal cords are inside the larynx. The **vocal cords** are mucous membrane folds supported by elastic ligaments, which are stretched across the glottis (Fig. 8.4). When air passes through the glottis, the vocal cords vibrate, producing sound. At the time of puberty, the growth of the larynx and the vocal cords is much more rapid and accentuated in the male than in the female, causing the male to have a more prominent Adam's apple and a deeper voice. The voice "breaks" in the young male due to his inability to control the longer vocal cords. These changes cause the lower pitch of the voice in males.

The high or low pitch of the voice is regulated when speaking and singing by changing the tension on the vocal cords. The greater the tension, as when the glottis becomes more narrow, the higher the pitch. When the glottis is wider, the pitch is lower (Fig. 8.4). The loudness, or intensity, of the voice depends upon the amplitude of the vibrations, that is, the degree to which vocal cords vibrate.

The Trachea Is a Windpipe

The **trachea** is a tube held open by C-shaped cartilaginous rings. Cilia that project from the epithelium of the trachea keep the lungs clean by sweeping mucus and debris toward the throat. Smoking is known to destroy the cilia, and consequently the soot in cigarette smoke collects in the lungs. Smoking is discussed more fully at the end of this chapter.

If the trachea is blocked because of illness or the accidental swallowing of a foreign object, it is possible to insert a tube by way of an incision made in the trachea. This tube acts as an artificial air intake and exhaust duct. The operation is called a *tracheostomy.*

Bronchi Are Air Tubes

The trachea divides into two *bronchi* (sing., **bronchus**), which lead, respectively, into the right and left lungs (Fig. 8.5). The bronchi branch into a great number of smaller passages called **bronchioles.** The bronchi resemble the trachea in structure, but as the bronchiolar tubes divide and subdivide, their walls become thinner, and the small rings of cartilage are no longer present. During an asthma attack, smooth muscle cells lining the bronchioles constrict even to the point of closing, and movement of air through the narrowed tubes may result in the characteristic wheezing. Each bronchiole terminates in an elongated space enclosed by a multitude of air pockets, or sacs, called *alveoli* (sing., **alveolus**) (see Fig. 8.2). The alveoli make up the lungs.

Lungs Have Many Alveoli

The *lungs* are cone-shaped organs; the right lung has three lobes, and the left lung has two lobes, allowing room for the heart, which is on the left side of the body (Fig. 8.5). A lobe is further divided into lobules, each of which has a bronchiole serving many alveoli. The lungs lie on either side of the heart in the thoracic cavity. The base of each lung is broad and concave so that it fits the convex surface of the diaphragm. The other surfaces of the lungs follow the contours of the ribs and the diaphragm in the thoracic cavity.

Alveoli Are 300 Million Air Sacs

Each alveolar sac is made up of simple squamous epithelium surrounded by blood capillaries. Gas exchange occurs between air in the alveoli and blood in the capillaries (Fig. 8.6).

The alveoli of human lungs are lined with a surfactant, a film of lipoprotein that lowers the surface tension and prevents them from closing. The lungs collapse in some newborn babies, especially premature infants, who lack this film. The condition, called *infant respiratory distress syndrome,* is now treatable by surfactant replacement therapy.

There are approximately 300 million alveoli, with a total cross-sectional area of 50–70 m^2. This is at least 40 times the surface area of the skin. Because of their many air spaces, the lungs are very light; normally, a piece of lung tissue dropped in a glass of water floats. Such a test can be used to determine if a deceased baby has taken a breath or not.

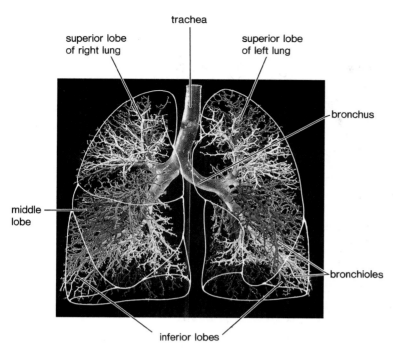

Figure 8.5   Airways to the lungs.
The trachea divides into the bronchi, which give rise to the bronchioles. The bronchioles have many branches and terminate at the alveoli. Each bronchiole and its branches are similarly colored.

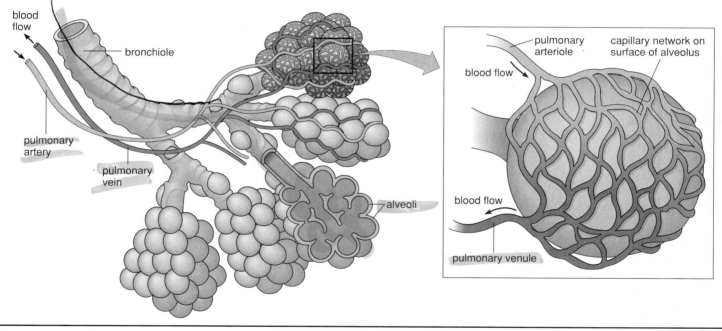

Figure 8.6   Gas exchange in the lungs.
The lungs consist of alveoli, surrounded by an extensive capillary network. Notice that the pulmonary arteriole carries deoxygenated blood (colored blue) and the pulmonary venule carries oxygenated blood (colored red).

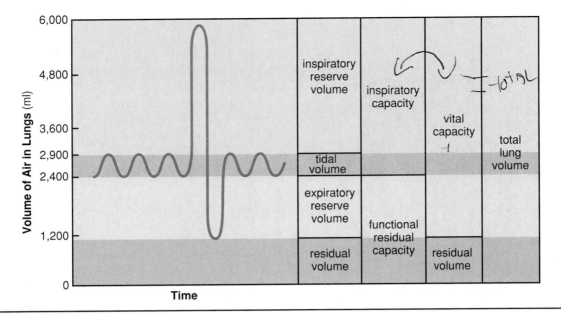

Figure 8.7 Vital capacity.
A spirometer measures the maximum amount of air that can be inhaled and exhaled when breathing by way of a tube connected to the instrument. During inspiration, a pen moves up, and during expiration, a pen moves down. The resulting pattern, such as the one shown here, is called a spirograph.

8.2   Mechanism of Breathing

The term respiration refers to the complete process of supplying oxygen to body cells for aerobic cellular respiration and the reverse process of ridding the body of carbon dioxide given off by cells. Respiration includes the following components:

1. Breathing: *inspiration* (entrance of air into the lungs) and *expiration* (exit of air from the lungs)

2. *External respiration:* exchange of the gases oxygen (O_2) and carbon dioxide (CO_2) between air and blood

3. *Internal respiration:* exchange of the gases O_2 and CO_2 between blood and tissue fluid

4. *Cellular respiration:* production of ATP in cells

Getting Air to the Lungs

When we breathe, the amount of air moved in and out with each breath is called the **tidal volume.** Normally, the tidal volume is about 500 ml, but we can increase the amount inhaled and exhaled by deep breathing. The maximum volume of air that can be moved in and out during a single breath is called the **vital capacity** (Fig. 8.7). First, we can increase inspiration by as much as 3,100 ml

of air by forced inspiration. This is called the *inspiratory reserve volume.* Similarly, we can increase expiration by contracting the abdominal and thoracic muscles. This is called the *expiratory reserve volume,* and it measures approximately 1,400 ml of air. Vital capacity is the sum of tidal, inspiratory reserve, and expiratory reserve volumes.

Note in Figure 8.7 that even after very deep breathing, some air (about 1,000 ml) remains in the lungs; this is called the **residual volume.** This air is no longer useful for gas exchange purposes. In some lung diseases, such as emphysema (p. 178), the residual volume builds up because the individual has difficulty emptying the lungs. This means that the lungs tend to be filled with useless air, and the vital capacity is thereby reduced.

Some of the inspired air never reaches the lungs; instead it fills the nose, trachea, bronchi, and bronchioles (see Fig. 8.2). These passages are not used for gas exchange, and therefore, they are said to contain *dead space.* To ensure that inspired air reaches the lungs, it is better to breathe slowly and deeply.

The air used for gas exchange excludes both the residual volume in the lungs and that in the dead space of the respiratory tract.

Ecology Focus

Photochemical Smog Can Kill

Most industrialized cities have photochemical smog at least occasionally. Photochemical smog arises when primary pollutants react with one another under the influence of sunlight to form a more deadly combination of chemicals. For example, the primary pollutants nitrogen oxides (NO_x) and hydrocarbons (HC) react with one another in the presence of sunlight to produce nitrogen dioxide (NO_2), ozone (O_3), and PAN (peroxyacetyl nitrate). Ozone and PAN are commonly referred to as oxidants. Breathing oxidants affects the respiratory and nervous systems, resulting in respiratory distress, headache, and exhaustion.

Cities with warm, sunny climates that are large and industrialized, such as Los Angeles, Denver, and Salt Lake City in the United States, Sydney in Australia, Mexico City in Mexico, and Buenos Aires in Argentina, are particularly susceptible to photochemical smog. If the city is surrounded by hills, a thermal inversion may aggravate the situation. Normally, warm air near the ground rises, so that pollutants are dispersed and carried away by air currents. But sometimes during a thermal inversion, smog gets trapped near the earth by a blanket of warm air (Fig. 8A). This may occur when a cold front brings in cold air, which settles beneath a warm layer. The trapped pollutants cannot disperse, and the results can be disastrous. In 1963, about 300 people died, and in 1966, about 168 people died in New York City when air pollutants accumulated over the city. Even worse were the events in London in 1957, when 700 to 800 people died, and in 1962, when 700 people died, due to the effects of air pollution.

Even though we have federal legislation to bring air pollution under control, more than half the people in the United States live in cities polluted by too much smog. We should place our emphasis on pollution prevention because, in the long run, prevention is usually easier and cheaper than pollution cleanup methods. Some prevention suggestions are as follows:

- Build more efficient automobiles or burn fuels that do not produce pollutants.
- Reduce the amount of waste to be incinerated by recycling materials.
- Reduce our energy use so that power plants need to provide less, and/or use renewable energy sources such as solar, wind, or water power.
- Require industries to meet clean air standards.

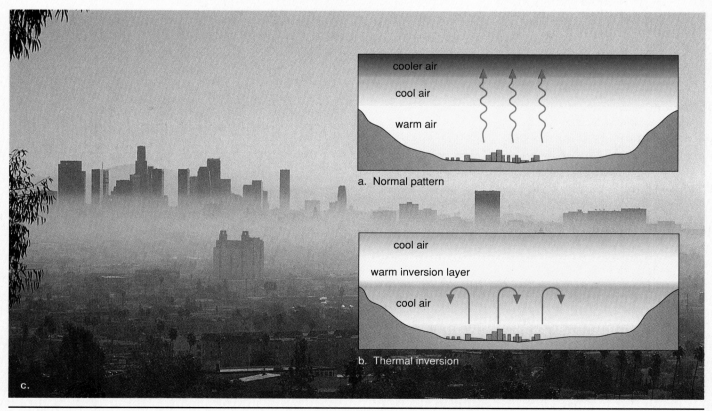

a. Normal pattern

b. Thermal inversion

c.

Figure 8A Thermal inversion.
a. Normally, pollutants escape into the atmosphere when warm air rises. **b.** During a thermal inversion, a layer of warm air (warm inversion layer) overlies and traps pollutants in cool air below. **c.** Los Angeles, a city of 8.5 million cars and thousands of factories, is particularly susceptible to thermal inversions, and this accounts for why this city is the "air pollution capital" of the United States.

Inspiration Precedes Expiration

To understand **ventilation**, the manner in which air enters and exits the lungs, it is necessary to remember first that normally there is a continuous column of air from the pharynx to the alveoli of the lungs.

Secondly, the lungs lie within the sealed-off thoracic cavity. The **rib cage** forms the top and sides of the thoracic cavity. It contains the ribs, hinged to the vertebral column at the back and to the sternum (breastbone) at the front, and the intercostal muscles that lie between the ribs. The **diaphragm,** a dome-shaped horizontal sheet of muscle and connective tissue, forms the floor of the thoracic cavity.

The lungs are enclosed by two serous membranes called **pleural membranes.** An infection of the pleural membranes is called pleurisy. The parietal pleura adheres to the rib cage and the diaphragm, and the visceral pleura is fused to the lungs. The two pleural layers lie very close to one another, separated only by a small amount of fluid. Normally, the intrapleural pressure is lower than atmospheric pressure by 4 mm Hg.

The importance of the reduced intrapleural pressure is demonstrated when, by design or accident, air enters the intrapleural space. The lungs collapse, and inspiration is impossible.

> The pleural membranes enclose the lungs and line the thoracic cavity. Intrapleural pressure is lower than atmospheric pressure.

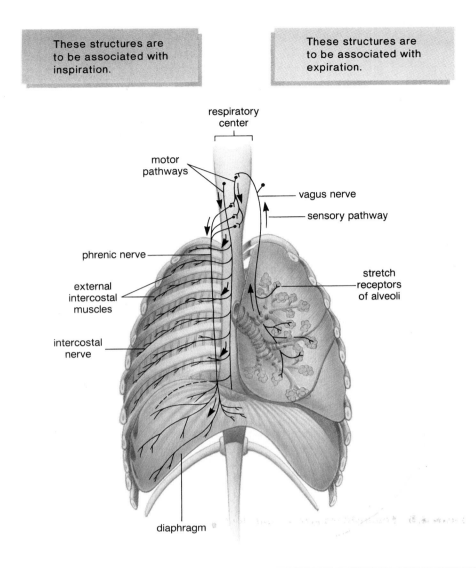

These structures are to be associated with inspiration.

These structures are to be associated with expiration.

respiratory center

motor pathways

vagus nerve

sensory pathway

phrenic nerve

stretch receptors of alveoli

external intercostal muscles

intercostal nerve

diaphragm

Figure 8.8   Nervous control of breathing.
During inspiration, the respiratory center stimulates the external intercostal (rib) muscles to contract via the intercostal nerves and the diaphragm to contract via the phrenic nerve. Should the lung volume increase above 1.5 liters, stretch receptors send inhibitory nerve impulses to the respiratory center via the vagus nerve. In either case, expiration occurs due to a lack of stimulation from the respiratory center to the diaphragm and intercostal muscles.

Inspiration Is Active

The **respiratory center,** located in the medulla oblongata, consists of a group of neurons that exhibit an automatic rhythmic discharge that triggers inspiration. Carbon dioxide (CO_2) and hydrogen ions (H^+) are the primary stimuli that directly cause changes in the activity of this center. This center is not affected by low oxygen (O_2) levels. There also are chemoreceptors in the *carotid bodies,* located in the carotid arteries, and in the *aortic bodies,* located in the aorta, that are sensitive to the level of hydrogen ions and also to the levels of carbon dioxide and oxygen in blood. When the concentration of hydrogen ions and carbon dioxide rise (and oxygen decreases), these bodies communicate with the respiratory center, and the rate and depth of breathing increase.

The respiratory center sends out impulses by way of nerves to the diaphragm and the muscles of the rib cage (Fig. 8.8). In its relaxed state, the diaphragm is dome-shaped, but upon stimulation, it contracts and lowers. Also, the external intercostal muscles contract, causing the rib cage to move upward and outward. Now the thoracic cavity increases in size, and the lungs expand. As the lungs expand, air pressure within the enlarged alveoli lowers and air enters through the nose or the mouth.

Controled By MecuA

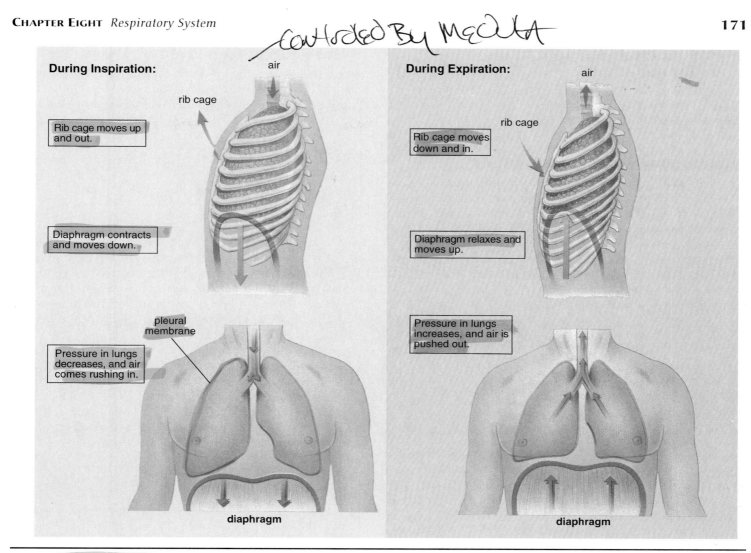

During Inspiration:

air

rib cage

Rib cage moves up and out.

Diaphragm contracts and moves down.

pleural membrane

Pressure in lungs decreases, and air comes rushing in.

diaphragm

During Expiration:

air

rib cage

Rib cage moves down and in.

Diaphragm relaxes and moves up.

Pressure in lungs increases, and air is pushed out.

diaphragm

Figure 8.9   Inspiration versus expiration.
During inspiration, the thoracic cavity and lungs expand so that air is drawn in. During expiration, the thoracic cavity and lungs resume their original positions and pressures. Now, air is forced out.

Inspiration is the active phase of breathing (Fig. 8.9). During this time, the diaphragm and the rib muscles contract, intrapleural pressure decreases, the lungs expand, and air comes rushing in. Note that air comes in because the lungs already have opened up; air does not force the lungs open. This is why it is sometimes said that *humans breathe by negative pressure.* The creation of a partial vacuum causes air to enter the lungs.

Expiration Is Usually Passive

When the respiratory center stops sending signals to the diaphragm and the rib cage, the diaphragm relaxes and it resumes its dome shape. The abdominal organs press up against the diaphragm, and the rib cage moves down and inward (Fig. 8.9). Now, the elastic lungs recoil, and air is pushed out. The respiratory center acts rhythmically to bring about breathing at a normal rate and volume. If by chance we inhale more deeply, the lungs are expanded and the alveoli stretch. This stimulates stretch receptors in the alveolar walls, and they initiate inhibi-

tory nerve impulses that travel from the inflated lungs to the respiratory center. This causes the respiratory center to stop sending out nerve impulses.

It is clear that while inspiration is the active phase of breathing, expiration is usually passive—the diaphragm and external intercostal muscles are relaxed when expiration occurs. In deeper and more rapid breathing, expiration can also be active. Contraction of internal intercostal muscles can force the rib cage to move downward and inward. Also, when the abdominal wall muscles are contracted, they push on the viscera, which push against the diaphragm, and the increased pressure in the thoracic cavity helps to expel air.

During inspiration, due to nervous stimulation, the diaphragm lowers and the rib cage lifts up and out. During expiration, due to a lack of nervous stimulation, the diaphragm rises and the rib cage lowers.

8.3 External and Internal Respiration

Figure 8.10 shows both external respiration and internal respiration. The principles of diffusion alone govern whether O_2 or CO_2 enters or leaves blood.

External Respiration Cleanses Blood

External respiration refers to the exchange of gases between air in the alveoli and blood in the pulmonary capillaries. Gases exert pressure, and the amount of pressure each gas exerts is its **partial pressure,** symbolized as P_{O_2} and P_{CO_2}. Blood flowing in the pulmonary capillaries has a higher P_{CO_2} than atmospheric air. Therefore, *CO_2 diffuses out of blood in the lungs.* Most of the CO_2 is being carried as bicarbonate ions (HCO_3^-). As the little remaining free CO_2 begins to diffuse out, the following reaction is driven to the right:

$$H^+ + HCO_3^- \longrightarrow H_2CO_3 \longrightarrow H_2O + CO_2 \uparrow$$

bicarbonate
ion

"Up" arrow indicates carbon
dioxide is leaving the body.

The enzyme *carbonic anhydrase,* present in red blood cells, speeds up the reaction. As the reaction proceeds, the respiratory pigment *hemoglobin* gives up the hydrogen ions (H^+) it has been carrying; HHb becomes Hb. Hb is called deoxyhemoglobin.

The pressure pattern is the reverse for O_2. Blood flowing in the pulmonary capillaries is low in oxygen, and alveolar air contains a much higher partial pressure of oxygen. Therefore, *O_2 diffuses into blood in the lungs.* Hemoglobin takes up this oxygen and becomes **oxyhemoglobin.**

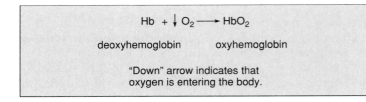

$$Hb + \downarrow O_2 \longrightarrow HbO_2$$

deoxyhemoglobin   oxyhemoglobin

"Down" arrow indicates that
oxygen is entering the body.

External respiration, the movement of O_2 and CO_2 between air within alveoli and blood in pulmonary capillaries, is dependent on the process of diffusion.

Internal Respiration Cleanses Tissue Fluid

Blood that enters the systemic capillaries is bright red in color because red blood cells contain oxyhemoglobin. Oxyhemoglobin gives up O_2, which diffuses out of blood into the tissues.

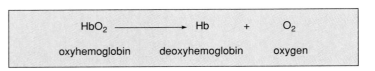

$$HbO_2 \longrightarrow Hb + O_2$$

oxyhemoglobin   deoxyhemoglobin   oxygen

Oxygen diffuses out of blood into the tissues because the P_{O_2} of tissue fluid is lower than that of blood. The lower P_{O_2} is due to cells continuously using up oxygen in aerobic cellular respiration. *Carbon dioxide diffuses into blood from the tissues* because the P_{CO_2} of tissue fluid is higher than that of blood. Carbon dioxide, produced continuously by cells, collects in tissue fluid.

After CO_2 diffuses into blood, it enters the red blood cells, where a small amount is taken up by hemoglobin, forming **carbaminohemoglobin.** Most of the CO_2 combines with water, forming carbonic acid (H_2CO_3), which dissociates to hydrogen ions (H^+) and bicarbonate ions (HCO_3^-). The increased concentration of CO_2 in the blood causes the reaction to proceed to the right.

$$CO_2 + H_2O \rightleftharpoons H_2CO_3 \rightleftharpoons H^+ + HCO_3^-$$

carbon
dioxide   water   carbonic
acid   hydrogen
ion   bicarbonate
ion

The enzyme carbonic anhydrase, present in red blood cells, speeds up the reaction. The globin portion of hemoglobin combines with excess hydrogen ions produced by the reaction, and Hb becomes HHb, called **reduced hemoglobin.** In this way, the pH of blood remains fairly constant. Bicarbonate ions diffuse out of red blood cells and are carried in the plasma. Blood that leaves the capillaries is deep purple in color because red blood cells contain reduced hemoglobin.

Internal respiration, the movement of O_2 and CO_2 between blood in the systemic capillaries and tissue fluid, is dependent on the process of diffusion.

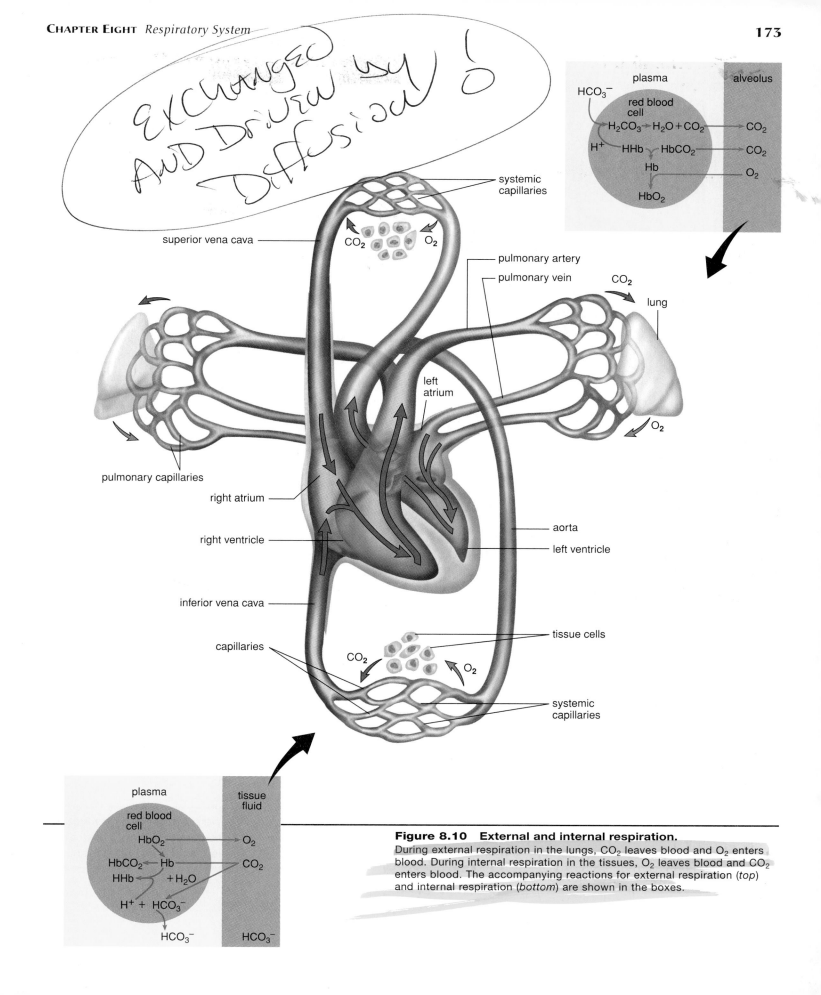

EXCHANGED by
AND DRIVEN by
Diffusion !

plasma   alveolus

HCO_3^-

red blood cell

$H_2CO_3 \rightarrow H_2O + CO_2 \rightarrow CO_2$

$H^+ \quad HHb \quad HbCO_2 \rightarrow CO_2$

Hb

O_2

HbO_2

systemic capillaries

superior vena cava

$CO_2 \qquad O_2$

pulmonary artery

pulmonary vein

CO_2

lung

left atrium

O_2

pulmonary capillaries

right atrium

right ventricle

aorta

left ventricle

inferior vena cava

tissue cells

capillaries

$CO_2 \qquad O_2$

systemic capillaries

plasma   tissue fluid

red blood cell

$HbO_2 \rightarrow O_2$

$HbCO_2 \leftarrow Hb \qquad CO_2$

$HHb \qquad + H_2O$

$H^+ + HCO_3^-$

$HCO_3^- \qquad\qquad HCO_3^-$

Figure 8.10 External and internal respiration.
During external respiration in the lungs, CO_2 leaves blood and O_2 enters blood. During internal respiration in the tissues, O_2 leaves blood and CO_2 enters blood. The accompanying reactions for external respiration (*top*) and internal respiration (*bottom*) are shown in the boxes.

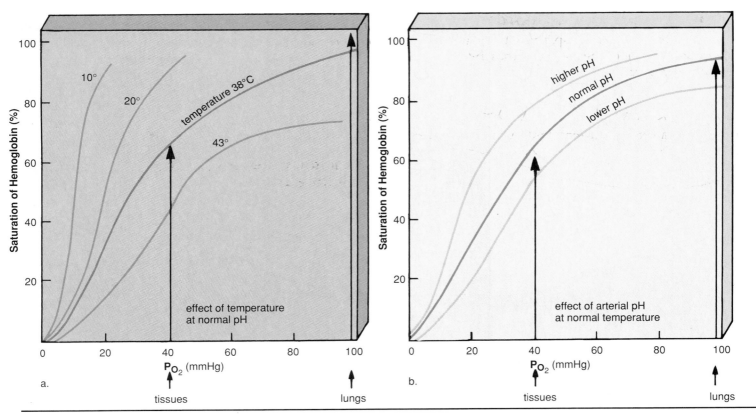

Figure 8.11   Effect of environmental conditions on hemoglobin saturation.
The partial pressure of oxygen (P_{O_2}) in pulmonary capillaries is about 98–100 mm Hg, but only about 40 mm Hg in tissue capillaries. Hemoglobin is about 98% saturated in the lungs because of P_{O_2}, and also because **(a)** the temperature is cooler and **(b)** the pH is higher in lungs. On the other hand, hemoglobin is only about 60% saturated in the tissues because of the P_{O_2} and also because **(a)** the temperature is warmer and the pH is lower in the tissues.

External and Internal Respiration Revisited

The binding capacity of hemoglobin is also affected by partial pressures. The P_{O_2} of air entering the lungs is about 100 mm Hg, and at this pressure the hemoglobin in the blood becomes saturated with O_2. This means that iron in hemoglobin molecules has combined with O_2. On the other hand, the P_{O_2} in the tissues is about 40 mm Hg, and then hemoglobin molecules release O_2, which diffuses into the tissue.

The lungs have a lower temperature and a higher pH than the tissues:

	pH	Temperature
Lungs	7.40	37°C
Tissues	7.38	38°C

How does this environmental difference between the lungs and tissues affect the combining properties of hemoglobin?

Both Figure 8.11*a* and *b* show that, as expected, hemoglobin is more saturated with O_2 in the lungs than in the tissues. This effect, which can be attributed to the difference in P_{O_2} between the lungs and tissues, is potentiated by the difference in temperature and pH between the lungs and tissues. Notice in Figure 8.11*a* that the saturation curve for hemoglobin is steeper at 10°C compared to 20°C, and so forth. Also, Figure 8.11*b* shows that the saturation curve for hemoglobin is steeper at higher pH than at lower pH.

This means that the environmental conditions in the lungs are favorable for the uptake of O_2 by hemoglobin, and the environmental conditions in the tissues are favorable for the release of O_2 by hemoglobin. Hemoglobin is about 98–100% saturated in the capillaries of the lungs and about 60–70% saturated in the tissues.

The difference in P_{O_2}, temperature, and pH between the lungs and tissues causes hemoglobin to take up oxygen in the lungs and release oxygen in the tissues.

8.4   Respiration and Health

We have seen that the entire respiratory tract has a warm, wet, mucous membrane lining, which is constantly exposed to environmental air.

Strep Throat: Risk of Rheumatic Fever

Strep throat is a very severe throat infection caused by the bacterium *Streptococcus pyogenes*. Swallowing may be difficult, and there is fever. Unlike a viral infection, strep throat should be treated with antibiotics. If not treated, it can lead to complications such as rheumatic fever, which can permanently damage the heart valves.

Bronchitis: Acute and Chronic

Viral infections can spread from the nasal cavities to the sinuses (sinusitis), to the middle ears (otitis media), to the larynx (laryngitis), and to the bronchi (bronchitis). Acute bronchitis (Fig. 8.12) usually is caused by a secondary bacterial infection of the bronchi, resulting in a heavy mucous discharge with much coughing. Acute bronchitis usually responds to antibiotic therapy. Chronic bronchitis, on the other hand, is not necessarily due to infection. It is often caused by constant irritation of the lining of the bronchi, which as a result undergo degenerative changes, including the loss of cilia and their normal cleansing action. There is frequent coughing, and the individual is more susceptible to respiratory infections. Chronic bronchitis is most often seen in cigarette smokers or those exposed to secondhand smoke or other types of polluted air.

Pneumonia: Lobules Fill and Breathing Ceases

Most forms of pneumonia are caused by either a bacterium or a virus that has infected the lungs. AIDS patients are subject to a particularly rare form of pneumonia caused by the protozoan *Pneumocystis carinii*. Sometimes, pneumonia is localized in specific lobules of the lungs. These lobules become nonfunctional as they fill with fluid. Obviously, the more lobules involved, the more serious the infection.

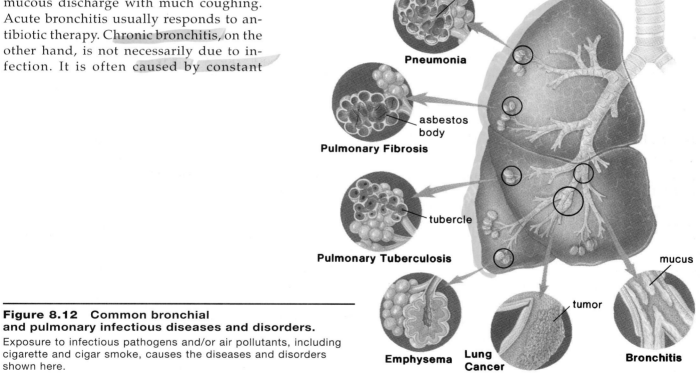

Figure 8.12   Common bronchial and pulmonary infectious diseases and disorders.
Exposure to infectious pathogens and/or air pollutants, including cigarette and cigar smoke, causes the diseases and disorders shown here.

The Most Often Asked Questions About Smoking, Tobacco, and Health, and . . . The Answers

Is there a safe way to smoke?

No. All cigarettes can cause damage and smoking even a small amount is dangerous. Cigarettes are perhaps the only legal product whose advertised and intended use—that is, smoking them—will hurt the body. Some people try to make smoking safer by smoking fewer cigarettes, but most smokers find this difficult. Some people think that switching from high tar/nicotine cigarettes to those with low tar/nicotine content makes smoking safer, but this does not always happen.... Even if smokers switch to lower tar brands and smoke fewer cigarettes, the health benefits are insignificant compared with the benefits of quitting altogether.

Does smoking cause cancer?

Yes, and not only lung cancer. Tobacco use is responsible for about 30% of all cancer deaths in the United States. Cigarette smoking causes about 87% of lung cancer deaths. Besides lung cancer, cigarette smoking is also a major cause of cancers of the mouth, larynx, and esophagus. In addition, smoking increases the risk of cancer of the bladder, kidney, pancreas, stomach, and the uterine cervix.

What are the chances of being cured of lung cancer?

Very low; the five-year survival rate is only 13%. Most forms of the disease start without producing any warning signs, so that it is rarely detected in the early stages when it is more likely to be cured. The past 15 years have brought little significant progress in the earlier diagnosis or treatment of lung cancer. Fortunately, lung cancer is a largely preventable disease. That is, by not smoking it can be prevented (Fig. 8B).

Do cigarettes cause other lung diseases?

Yes. Cigarette smoking causes other lung diseases which can be just as dangerous as lung cancer. It leads to chronic bronchitis—a disease where the airways produce excess mucus, which forces the smoker to cough frequently. Cigarette smoking is also the major cause of emphysema—a disease which slowly destroys a person's ability to breathe.... Chronic obstructive pulmonary disease (COPD), which includes chronic bronchitis and emphysema, kills about 81,000 people each year; cigarette smoking is responsible for 65,000 of these deaths.

Why do smokers have "smoker's cough"?

Cigarette smoke contains chemicals which irritate the air passages and lungs. When a smoker inhales these substances, the body tries to protect itself by coughing. The well-known "early morning" cough of smokers happens for a different reason. Normally, cilia (tiny hairlike formations which line the airways) beat outwards and "sweep" harmful material out of the lungs. Cigarette smoke, however, decreases this sweeping action, so some of the poisons in the smoke remain in the lungs and coughing is an attempt to remove them. Smokers are more likely to get pneumonia because damaged or destroyed cilia cannot protect the lungs from bacteria and viruses that float in the air.

If you smoke but don't inhale, is there any danger?

Yes. Wherever smoke touches living cells, it does harm. So, even if smokers don't inhale—including pipe and cigar smokers—they are at an increased risk for lip, mouth, and tongue cancer. Because it is virtually impossible to avoid inhaling tobacco smoke totally, these smokers also have an increased chance of getting lung cancer. Lung cancer is much more likely to occur in a person who has always smoked cigars or pipes than in a person who has never smoked at all.

Does cigarette smoking affect the heart?

Yes. Smoking cigarettes increases the risk of heart disease, which is America's number one killer. About 150,000 Americans die each year from heart attacks and other forms of heart disease caused by smoking. Smoking, high blood pressure, high blood cholesterol, and lack of exercise are all risk factors for heart disease. Smoking alone doubles the risk of heart disease. When a person smokes and has other risk factors, his chance of getting heart disease increases dramatically. For example, if smoking is combined with high blood pressure or high cholesterol, then the risk goes up four times. Put all three together—smoking, high blood pressure, and high cholesterol—and the risk goes up eight times. Smokers who have already had one heart attack are also more likely than nonsmokers to have another attack.

Is there any risk for pregnant women and their babies?

Pregnant women who smoke endanger the health and lives of their unborn babies. When a pregnant woman smokes, she really is smoking for two because the nicotine, carbon monoxide, and other dangerous chemicals in smoke enter the mother's bloodstream and then pass into the baby's body. Women who smoke during pregnancy risk having a miscarriage, or a premature or stillborn baby. Their babies are also more likely to be underweight, by an average of one-half pound.

Does smoking cause any special health problems for women?

Yes. Nonsmoking women who use oral contraceptives ("the Pill") double their chances of having a heart attack. However, when women use the Pill and smoke, they are 10 times more likely to suffer a heart attack than nonsmoking women who don't take the Pill. Women who smoke and use the Pill have an increased risk of stroke and blood clots in the legs as well. Women who smoke also run the risk of having trouble getting pregnant; the more they smoke, the more likely it is that they will have difficulty. Some studies show that female smokers,

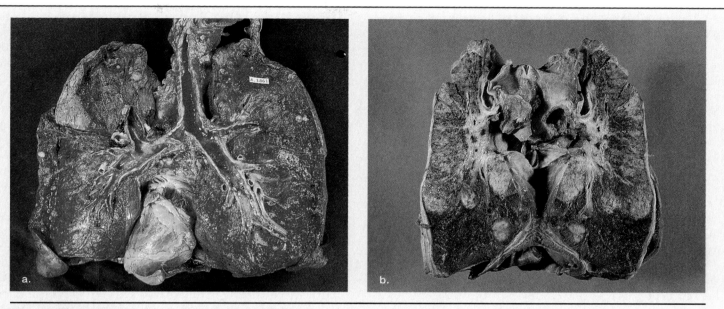

Figure 8B   Normal lung versus cancerous lung.
a. Normal lung with heart in place. Note the healthy red color. **b.** Lungs of a heavy smoker. Notice how black the lungs are except where cancerous tumors have formed.

especially the elderly, are at a higher risk for osteoporosis (a disease which weakens the bones and makes them more likely to break) than nonsmoking women. In addition, women who smoke increase their chances of getting cancer of the uterine cervix.

What are some of the short-term effects of smoking cigarettes?

Almost immediately, smoking can make it hard to breathe. Within a short time, it can also worsen asthma and allergies. Nicotine reaches the brain only seven seconds after taking a puff (faster than it takes heroin to reach the brain), where it produces a variety of effects.

Are there any other risks to the smoker?

Yes. There are many other risks. As we already mentioned briefly, smoking cigarettes causes stroke, which is the third leading cause of death in America. Smoking causes lung cancer, but if a person smokes and is exposed to radon or asbestos, the risk increases even more. Smokers are also more likely to have and die from stomach ulcers than nonsmokers. In addition, cigarettes can interact with medication the smoker is taking in unwanted ways—like preventing the drug from doing what it is supposed to do.

What are the dangers of passive smoking?

Passive smoking causes lung cancer in healthy nonsmokers. Children whose parents smoke are more likely to suffer from pneumonia or bronchitis in the first two years of life than children who come from smoke-free households. Nonsmokers who are married to smokers have a 30% greater risk for developing lung cancer than nonsmokers married to other nonsmokers.

Are chewing tobacco and snuff safe alternatives to cigarette smoking?

No, they are not. Many people who use chewing tobacco or snuff believe it can't harm them because there is no smoke. Wrong. Smokeless tobacco contains nicotine, the same addicting drug found in cigarettes. Snuff dippers also take in an average of over 10 times more cancer-causing substances (called nitrosamines) than cigarette smokers. In fact, some brands of smokeless tobacco contain as much as 20,000 times the legal limit of nitrosamines which are permitted in certain foods and consumer products (such as bacon, beer, and baby bottle nipples). While not inhaled through the lungs, the juice from smokeless tobacco is absorbed through the lining of the mouth. There it can cause sores and white patches which often lead to cancer of the mouth.

Excerpted from "The Most Often Asked Questions About Smoking, Tobacco, and Health, and... The Answers," revised July 1993. © American Cancer Society, Inc., Atlanta, GA. Used with permission.

Pulmonary Tuberculosis: Past and Recent Threat

Pulmonary tuberculosis is caused by the tubercle bacillus, a type of bacterium. When a person has tuberculosis, the alveoli burst and are replaced by inelastic connective tissue. It is possible to tell if a person has ever been exposed to tuberculosis with a skin test in which a highly diluted extract of the bacillus is injected into the skin of the patient. A person who has never been in contact with the bacillus shows no reaction, but one who has developed immunity to the organism shows an area of inflammation that peaks in about 48 hours. If these bacilli invade the lung tissue, the cells build a protective capsule about the foreigners and isolate them from the rest of the body. This tiny capsule is called a *tubercle*. If the resistance of the body is high, the imprisoned organisms die, but if the resistance is low, the organisms eventually can be liberated. If a chest X ray detects active tubercles, the individual is put on appropriate drug therapy to ensure the localization of the disease and the eventual destruction of any live bacterial organisms.

Tuberculosis was a major killer in the United States before the middle of this century, after which antibiotic therapy brought it largely under control. In recent years, however, the incidence of tuberculosis is on the rise, particularly among AIDS patients, the homeless, and the rural poor. Worse, the new strains are resistant to the usual antibiotic therapy. Therefore, some physicians would like to again quarantine patients in sanitoriums.

Emphysema: Bronchioles Collapse and Alveoli Burst

Emphysema refers to the destruction of lung tissue, with accompanying ballooning or inflation of the lungs due to trapped air. The trouble stems from the damage and the collapse of the bronchioles. When this occurs, the alveoli are cut off from renewed oxygen supply, and the air within them is trapped. The trapped air very often causes the alveolar walls to rupture (see Fig. 8.12), and a loss of elasticity makes breathing difficult. The victim is breathless and may have a cough. Since the surface area for gas exchange is reduced, not enough oxygen reaches the heart and the brain. Even so, the heart works furiously to force more blood through the lungs, which can lead to a heart condition. Lack of oxygen to the brain can make the person feel depressed, sluggish, and irritable.

Pulmonary Fibrosis: Inhaling Particles

Inhaling particles such as silica (sand), coal dust, asbestos (see Fig. 8.12), and now it seems fiberglass can lead to pulmonary fibrosis, a condition in which fibrous connective tissue builds up in the lungs. Breathing capacity can be seriously impaired, and the development of cancer is common. Since asbestos has been used so widely as a fireproofing and insulating agent, unwarranted exposure has occurred. It is projected that 2 million deaths could be caused by asbestos exposure—mostly in the workplace—between 1990 and 2020.

Lung Cancer: Women Catch Up

Lung cancer used to be more prevalent in men than in women, but recently it has surpassed breast cancer as a cause of death in women. This can be linked to an increase in the number of women who smoke today. Autopsies on smokers have revealed the progressive steps by which the most common form of lung cancer develops. The first event appears to be thickening and callusing of the cells lining the bronchi. (Callusing occurs whenever cells are exposed to irritants.) Then there is a loss of cilia so that it is impossible to prevent dust and dirt from settling in the lungs. Following this, cells with atypical nuclei appear in the callused lining. A tumor (see Fig. 8.12) consisting of disordered cells with atypical nuclei is considered to be cancer in situ (at one location). A final step occurs when some of these cells break loose and penetrate other tissues, a process called metastasis. Now the cancer has spread. The tumor may grow until the bronchus is blocked, cutting off the supply of air to that lung. The entire lung then collapses, the secretions trapped in the lung spaces become infected, and pneumonia or a lung abscess (localized area of pus) results. The only treatment that offers a possibility of cure is to remove a lobe or the lung completely before metastasis has had time to occur. This operation is called *pneumonectomy*.

Current research indicates that *involuntary smoking,* simply breathing in air filled with cigarette smoke, can also cause lung cancer and other illnesses that are apt to occur when a person smokes. The Health Focus reading on pages 176–77 lists the various illnesses associated with smoking. If a person stops both voluntary and involuntary smoking, and if the body tissues are not already cancerous, they usually return to normal over time.

8.5   Working Together

The Working Together box on page 179 shows how the respiratory system works with the other systems of the body to maintain homeostasis.

HUMAN SYSTEMS WORK TOGETHER

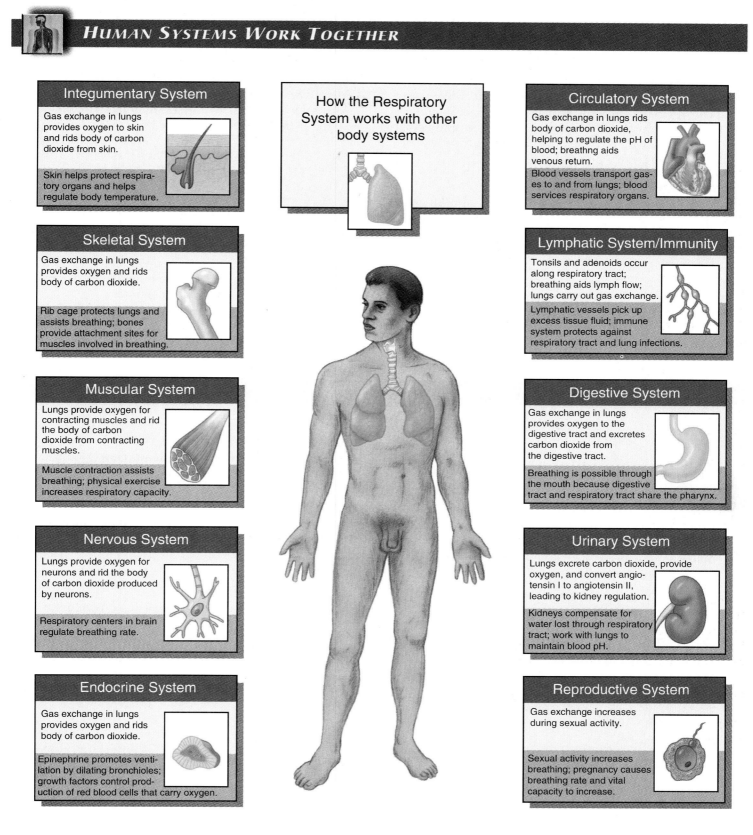

Integumentary System

Gas exchange in lungs provides oxygen to skin and rids body of carbon dioxide from skin.

Skin helps protect respiratory organs and helps regulate body temperature.

How the Respiratory System works with other body systems

Circulatory System

Gas exchange in lungs rids body of carbon dioxide, helping to regulate the pH of blood; breathng aids venous return.

Blood vessels transport gases to and from lungs; blood services respiratory organs.

Skeletal System

Gas exchange in lungs provides oxygen and rids body of carbon dioxide.

Rib cage protects lungs and assists breathing; bones provide attachment sites for muscles involved in breathing.

Lymphatic System/Immunity

Tonsils and adenoids occur along respiratory tract; breathing aids lymph flow; lungs carry out gas exchange.

Lymphatic vessels pick up excess tissue fluid; immune system protects against respiratory tract and lung infections.

Muscular System

Lungs provide oxygen for contracting muscles and rid the body of carbon dioxide from contracting muscles.

Muscle contraction assists breathing; physical exercise increases respiratory capacity.

Digestive System

Gas exchange in lungs provides oxygen to the digestive tract and excretes carbon dioxide from the digestive tract.

Breathing is possible through the mouth because digestive tract and respiratory tract share the pharynx.

Nervous System

Lungs provide oxygen for neurons and rid the body of carbon dioxide produced by neurons.

Respiratory centers in brain regulate breathing rate.

Urinary System

Lungs excrete carbon dioxide, provide oxygen, and convert angiotensin I to angiotensin II, leading to kidney regulation.

Kidneys compensate for water lost through respiratory tract; work with lungs to maintain blood pH.

Endocrine System

Gas exchange in lungs provides oxygen and rids body of carbon dioxide.

Epinephrine promotes ventilation by dilating bronchioles; growth factors control production of red blood cells that carry oxygen.

Reproductive System

Gas exchange increases during sexual activity.

Sexual activity increases breathing; pregnancy causes breathing rate and vital capacity to increase.

SUMMARY

8.1 Respiratory Tract

The respiratory tract consists of the nose (nasal cavities), the nasopharynx, the pharynx, the larynx (which contains the vocal cords), the trachea, the bronchi, and the bronchioles. The bronchi, along with the pulmonary arteries and veins, enter the lungs, which consist of the alveoli, air sacs surrounded by a capillary network.

8.2 Mechanism of Breathing

Inspiration begins when the respiratory center in the medulla oblongata sends excitatory nerve impulses to the diaphragm and the muscles of the rib cage. As they contract, the diaphragm lowers and the rib cage moves upward and outward; the lungs expand, creating a partial vacuum, which causes air to rush in. The respiratory center now stops sending impulses to the diaphragm and muscles of the rib cage. As the diaphragm relaxes, it resumes its dome shape, and as the rib cage retracts, air is pushed out of the lungs during expiration.

8.3 External and Internal Respiration

External respiration occurs when CO_2 leaves blood and O_2 enters blood at the alveoli. Oxygen is transported to the tissues in combination with hemoglobin as oxyhemoglobin (HbO_2). Internal respiration occurs when O_2 leaves blood and CO_2 enters blood at the tissues. Carbon dioxide is mainly carried to the lungs within the plasma as the bicarbonate ion (HCO_3^-). Hemoglobin combines with hydrogen ions and becomes reduced (HHb).

8.4 Respiration and Health

There are a number of illnesses associated with the respiratory tract. In addition to colds and flu, the lungs may be infected by the more serious pneumonia and tuberculosis. Two illnesses that have been attributed to breathing polluted air are emphysema and lung cancer.

8.5 Working Together

The respiratory system works with the other systems of the body in the ways described in the box on page 179.

STUDYING THE CONCEPTS

1. List the parts of the respiratory tract. What are the special functions of the nasal cavity, the larynx, and the alveoli?   165–67

2. Name and explain the four parts of respiration.   168

3. What is the difference between tidal volume and vital capacity? Of the air we breathe, what part is not used for gas exchange?   168

4. What are the steps in inspiration and expiration? How is breathing controlled?   170–71

5. Discuss the events of external respiration, including two pertinent equations in your discussion.   172

6. What two equations pertain to the exchange of gases during internal respiration?   172

7. State three factors that influence hemoglobin's O_2 binding capacity, and relate them to the environmental conditions in the lungs and tissues.   174

8. Name four lung infections other than cancer, and explain why breathing is difficult with these conditions.   175–78

9. What are emphysema and pulmonary fibrosis, and how do they affect a person's health?   178

10. List the steps by which lung cancer develops.   178

APPLYING YOUR KNOWLEDGE

Concepts

1. Although all of us breathe 14 to 20 times per minute and our muscles don't get tired and/or ache, breathing deeply for an extended period of time sometimes causes muscles in our chest and abdominal region to get sore. Explain the cause.

2. Based on our knowledge of breathing volumes, how is it possible to determine whether a baby was stillborn or whether the baby died shortly after birth?

3. With regard to the effect of temperature and pH on hemoglobin's affinity for oxygen, explain the body's ability to provide extra oxygen to the most active parts of the body.

Bioethical Issue

First it was the office. Then the supermarket. Now malls, airplanes, restaurants—virtually everywhere you go in the United States, establishments ban smoking. Amid renewed reports that smoking triples lung cancer risk—and even affects those who inhale secondhand smoke—people feel comfortable refusing smokers.

But some smokers say these bans infringe on their freedom. After all, they argue, these same places often sell alcohol or permit gambling, which can be unhealthy. Why, smokers ask, is *their* addiction getting the brunt of health regulations?

Do smoking bans treat smokers unfairly? Are establishments right to refuse smokers? How does the health impact on others play a role?

APPLYING TECHNOLOGY

Your study of the respiratory system is supported by these available technologies:

Exploring the Internet

The Mader Home Page provides further resources for studying this chapter.

`http://www.mhhe.com/sciencemath/biology/mader/`

(Click on *Human Biology.*)

Dynamic Human: Respiratory System CD-ROM

In *Gross Anatomy,* major organs are reviewed, and the thorax with lungs is presented in 3-dimension; in *Explorations,* Boyle's Law is explained as lungs expand; gas exchange and oxygen transport also occur; in *Histology,* microscopic slides are highlighted to reveal tissues; and in *Clinical Concepts,* bronchoscopy shows the passages of the bronchial tree and an interactive spirogram is given.

Explorations in Human Biology CD-ROM

Life Span and Life Style (#3) Life expectancy varies as students change the amount of smoking, drinking, fat in diet, and exercise. **(2)***

Smoking and Cancer (#6) The number of years it takes for cancer to develop decreases as students increase the years of smoking and the number of cigarettes smoked a day. **(2)**

*Level of difficulty

TESTING YOUR KNOWLEDGE

1. In tracing the path of air, the _____ immediately follows the pharynx.

2. The lungs contain air sacs called _____.

3. Label the diagram (right) of the human respiratory tract.

4. The breathing rate is primarily regulated by the amount of _____ and _____ in blood.

5. Air enters the lungs after they have _____.

6. Gas exchange is dependent on the physical process of _____.

7. Carbon dioxide is carried in blood as the _____ ion.

8. The hydrogen ions (H^+) given off when carbonic acid (H_2CO_3) dissociates are carried by _____.

9. Reduced hemoglobin becomes oxyhemoglobin in the _____ of the body.

10. Most cases of lung cancer actually begin in the _____.

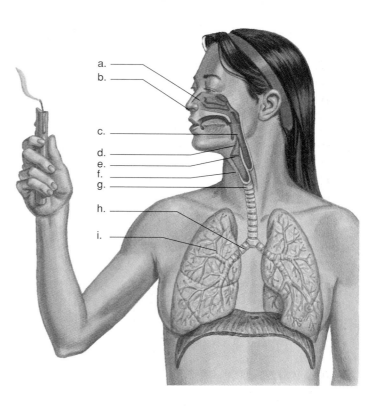

SELECTED KEY TERMS

alveolus (pl., alveoli) (al-VEE-uh-lus) Air sac of a lung.   166

bronchiole (BRAHNG-kee-ohl) Smaller air passages in the lungs that eventually terminate in alveoli.   166

bronchus (pl., bronchi) (BRAHNG-kus) One of two major divisions of the trachea leading to the lungs.   166

carbaminohemoglobin (kar-buh-MEE-noh-HEE-muh-gloh-bun) Hemoglobin carrying carbon dioxide.   172

diaphragm Dome-shaped horizontal sheet of muscle and connective tissue that separates the thoracic from the abdominal cavity. Its contraction helps expand the thoracic cavity during inspiration.   170

epiglottis (ep-uh-GLAHT-us) Structure that covers the glottis and closes off the air tract during the process of swallowing.   166

expiration Act of expelling air from the lungs; exhalation.   164

glottis (GLAHT-us) Opening for airflow into the larynx.   166

inspiration Act of taking air into the lungs; inhalation.   164

larynx (LAR-ingks) Cartilaginous organ located between pharynx and trachea that contains the vocal cords; voice box.   166

oxyhemoglobin (ahk-sih-HEE-muh-gloh-bun) Compound formed when oxygen combines with hemoglobin.   172

partial pressure Pressure produced by one gas in a mixture of gases.   172

pleural membrane (PLUR-al) Serous membrane that encloses the lungs.   170

reduced hemoglobin Hemoglobin that is carrying hydrogen ions.   172

residual volume Amount of air remaining in the lungs after a forceful expiration.   168

respiratory center Group of nerve cells in the medulla oblongata that sends out nerve impulses on a rhythmic basis, resulting in inspiration.   170

rib cage Top and sides of the thoracic cavity; contains ribs and intercostal muscles.   170

tidal volume Amount of air normally moved in the human body during an inspiration or expiration.   168

trachea (TRAY-kee-uh) Tube that is supported by C-shaped cartilaginous rings; lies between the larynx and the bronchi; also called the windpipe.   166

ventilation Breathing; the process of moving air into and out of the lungs.   170

vital capacity Maximum amount of air moved in or out of the human body with each breathing cycle.   168

vocal cord Fold of tissue within the larynx; creates vocal sounds when it vibrates.   166

Chapter 9

Urinary System and Excretion

Chapter Outline

Figure 9.1   Taking a drink of water.
Drinking water lowers the osmolarity of the blood. But when a drink is dehydrating, as is seawater, the osmolarity of blood and then urine increases.

O n a mild Saturday morning, Alex heads off in his dad's boat. Once out in the peaceful deep, he cuts the engine, turns on the little portable TV, and falls asleep. Three hours later, he wakes up. Cursing under his breath, Alex turns the key in the boat's ignition. It stalls. He tries again. It stalls again.

Alex decides to stay put and wait for help. Meanwhile, he searches the boat for something to drink. Staring at the ocean, Alex crouches down in the boat and scoops up seawater. He gulps it. That's better. He gets more. Fortunately for Alex, his dad came looking for him, or else he could have died from dehydration.

Ironically, the culprit would probably have been the water he kept drinking. Seawater is far too salty for the human body to process. In trying to flush away the excess salt, Alex's kidneys would have to excrete more liquid than consumed. If you're ever in Alex's position—or even if you just go on a long bike ride or other sports event—make sure you bring water. Your kidneys can't function without it.

When we are able to drink water (Fig. 9.1) as we should, the kidneys maintain the osmolarity and the pH of the blood within normal limits. And they are able to adjust the hydrogen ion concentration so that the pH of the blood remains about 7.4. As you may know, urine also contains the nitrogenous end products of amino acids, nucleic acids, and creatine phosphate metabolism, which are urea, uric acid, and creatinine, respectively. **Excretion** rids the body of metabolic wastes, which come from the breakdown of substances that have been metabolized in the body's cells.

Not Dedication

The correct functioning of the kidneys is essential to our good health, yet it is something that most of us take for granted until an illness strikes. Urinary tract infections and kidney stones cause much pain and may even result in permanent damage to the kidneys. Nationwide, there is a program of hemodialysis and transplantation for such patients.

The kidneys are a part of the urinary system because they produce urine, which is sent by way of the ureters to the bladder where it is stored before exiting the body through the urethra. Other organs aside from the kidneys excrete substances, and we shall discuss these after considering the urinary system. Excretion should not be confused with secretion, which is the release of a substance useful to the body. Also, defecation is not a form of excretion. Defecation refers only to the elimination of feces from the digestive tract.

9.1   Urinary System

The urinary system includes the kidneys and associated structures, which are illustrated in Figure 9.2.

Urinary Organs

The kidneys are found on either side of the vertebral column, just below the diaphragm. They lie in depressions against the deep muscles of the back beneath the peritoneum, the lining of the abdominal cavity, where they also receive some protection from the lower rib cage. But the kidneys can be damaged by blows on the back; kidney punches are not allowed in boxing.

The **kidneys** are bean-shaped, reddish brown organs, each about the size of a fist, which produce urine. They are covered by a tough capsule of fibrous connective tissue overlaid by adipose tissue. A depression (the hilum) on the concave side is where the renal blood vessels and the ureters enter or exit.

The **ureters** are muscular tubes about 25 cm long that convey the urine from the kidneys toward the bladder by peristalsis. Urine enters the bladder by peristaltic contractions, in jets that occur at the rate of five per minute.

The **urinary bladder,** which can hold up to 600 ml of urine, is a hollow, muscular organ that gradually expands as urine enters. In the male, the urinary bladder lies in front of the rectum, the seminal vesicles, and the vas deferens. In the female, the urinary bladder is in front of the uterus and the upper vagina. There are two sphincters in close proximity where the urethra joins the bladder.

The **urethra,** which extends from the urinary bladder to an external opening, differs in length in the female and the male. In the female, the urethra lies ventral to the vagina and is only about 4 cm long. The short length of the female urethra invites bacterial invasion and explains why the female is more prone to urinary tract infections. In the male, the urethra averages 20 cm when the penis is flaccid

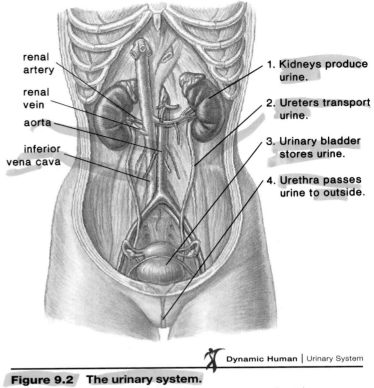

renal artery

renal vein

aorta

inferior vena cava

1. Kidneys produce urine.
2. Ureters transport urine.
3. Urinary bladder stores urine.
4. Urethra passes urine to outside.

Dynamic Human | Urinary System

Figure 9.2   The urinary system.
Urine is found only within the kidneys, the ureters, the urinary bladder, and the urethra.

(limp, nonerect). As the urethra leaves the male urinary bladder, it is encircled by the prostate gland. In older men, enlargement of the prostate gland can restrict urination, a condition that usually can be corrected surgically.

There is no connection between the genital (reproductive) and urinary systems in females; but there is a connection in males, where the urethra also carries sperm during ejaculation. This double function does not alter the path of urine, and it is important to realize that urine is found only in those structures noted in Figure 9.2.

Urination and the Nervous System

When the urinary bladder fills with urine to about 250 ml, stretch receptors send nerve impulses to the spinal cord. Urination occurs when nerve impulses leaving the spinal cord then cause the urinary bladder to contract and the sphincters to relax so that urination is possible. In older children and adults, the brain controls this reflex, delaying urination until a suitable time (Fig. 9.3).

Only the urinary system, consisting of the kidneys, the urinary bladder, the ureters, and the urethra, ever holds urine.

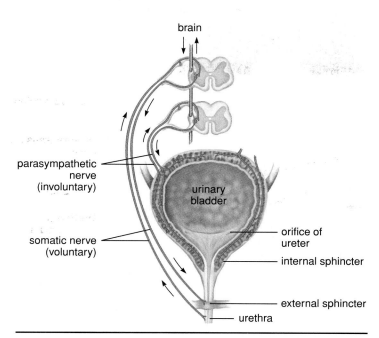

Figure 9.3 Urination.
As the bladder fills with urine, sensory impulses go to the spinal cord and then the brain. The brain can override the urge to void. When voiding occurs, excitatory impulses go by way of an involuntary nerve to the bladder, which contracts, and to an internal sphincter, which opens. Involuntary impulses reach the external sphincter, which also opens.

Water	**95%**
Solids	**5%**

TABLE 9.1
Composition of Urine

Nitrogenous wastes	(per 1,500 ml of urine)
Urea	30 grams
Creatinine	1–2 grams
Uric acid	1 gram
Salts	25 grams
Positive ions	
Sodium (Na^+)	
Potassium (K^+)	
Magnesium (Mg^{++})	
Ammonium (NH_4^+)	
Calcium (Ca^{++})	
Negative ions	
Chlorides (Cl^-)	
Sulfates ($SO_4^=$)	
Phosphates (PO_4^{-3})	
Bicarbonate (HCO_3^-)	

Functions of the Urinary System

The primary functions of the urinary system are carried out by the kidneys. The work of the kidneys is absolutely necessary to homeostasis; that is, maintaining the internal environment within normal limits.

The kidneys *secrete the hormone erythropoietin,* which stimulates red blood cell production, and they help activate the vitamin D precursor from the skin. Vitamin D promotes calcium (Ca^{++}) reabsorption from the digestive tract.

One of the most important functions of the kidneys is to maintain the water and salt balance of the body. Sometimes this is called regulating osmolarity because salts have the ability to cause osmosis. We tend to think of salt as sodium chloride, and indeed the blood sodium level is critical, but other ions are also regulated by the kidneys. Table 9.1 lists the various ions that can be found in urine. (A salt contains a negative and positive ion.)

The more water and salt there are in blood, the greater its volume and pressure; therefore, the kidneys *are involved in regulating blood volume and blood pressure.* The kidneys *secrete the enzyme renin,* which also helps maintain blood pressure in a manner to be described. Water and salt excretion are under hormonal control, and we will see how these hormones interact with the kidneys to regulate blood volume and pressure.

The kidneys also *regulate the pH of the blood.* In order for us to remain healthy, the blood pH should be just about 7.4. The kidneys monitor blood pH, mainly by excreting hydrogen ions (H^+). Urine usually has a pH of 6 or lower.

Finally, the kidneys remove metabolic wastes and foreign chemicals from the blood and excrete them in urine. Although this is a vital activity, it is not as immediately critical as maintaining the salt balance of the body. An ionic imbalance can lead to heart failure.

The kidneys are the primary organs to excrete the nitrogenous wastes generated by amino acid, nucleic acid, and creatinine metabolism. **Urea is the primary nitrogenous waste product of humans.** Amino acid breakdown results in ammonia, a highly toxic substance. The urea cycle in the liver consists of carrier molecules that combine two molecules of ammonia with carbon dioxide to form urea, which is far less toxic to cells.

$$2\,NH_3 \;+\; CO_2 \;\rightarrow\; H_2N-\overset{\displaystyle O}{\overset{\displaystyle \|}{C}}-NH_2 \;+\; H_2O$$
ammonia   carbon dioxide   urea

Creatine phosphate, which serves as a reservoir of high-energy phosphate in muscles, results in the end product *creatinine.* The breakdown of nucleotides produces **uric acid,** which is rather insoluble. If too much uric acid is present in blood, it precipitates out. Crystals of uric acid sometimes collect in the joints, producing a painful ailment called gout.

The kidneys are organs of homeostasis because they regulate the water and salt balance and the pH of the blood, and they excrete nitrogenous wastes.

Health Focus

Urinary Tract Infections Require Attention

Although males can get a urinary tract infection, the condition is 50 times more common in women. The explanation lies in a comparison of male and female anatomy (Fig. 9A). The female urethral and anal openings are closer together, and the shorter urethra makes it easier for bacteria from the bowels to enter and start an infection. Although it is possible to have no outward signs of an infection, usually urination is painful, and patients often describe a burning sensation. The urge to pass urine is frequent, but it may be difficult to start the stream. Chills with fever, nausea, and vomiting may be present.

Urinary tract infections can be confined to the urethra, in which case urethritis is present. If the bladder is involved, it is called cystitis, and should the infection reach the kidneys, the person has pyelonephritis. *Escherichia coli (E. coli)*, a normal bacterial resident of the large intestine, is usually the cause of infection. Since the infection is caused by a bacterium, it is curable by antibiotic therapy. The problem is, however, that reinfection is possible as soon as antibiotic therapy is finished.

It makes sense to try to prevent infection in the first place. These tips might help.

Men and women should drink lots of water. Try to drink from 2–2.5 liters of liquid a day. Try to avoid caffeinated drinks, which may be irritating. Cranberry juice is recommended because it contains a substance that stops bacteria from sticking to the bladder wall once an infection has set in. If an attack occurs, testing and antibiotic therapy may be in order. Keep in mind that sexually transmitted diseases such as gonorrhea, chlamydia, or herpes can cause urinary tract infections. All personal behaviors should be examined carefully, and suitable adjustments should be made to avoid urinary tract infections.

Most women have a urinary tract infection for the first time shortly after they become sexually active. Honeymoon cystitis was coined because of the common association of urinary tract infections with sexual intercourse. Washing the genitals before having sex and being careful not to introduce bacteria from the anus

into the urethra is recommended. Also, urinating immediately before and after sex will help to flush out any bacteria that are present. A diaphragm may press on the urethra and prevent adequate emptying of the bladder, and estrogen, such as in birth-control pills, can increase the risk of cystitis. A sex partner may have an asymptomatic (no symptoms) urinary infection that causes a woman to become infected repeatedly.

Women should wipe from the front to the back after using the toilet. Perfumed toilet paper and any other perfumed products that come in contact with the genitals may be irritating. Wearing loose clothing and cotton underwear discourages the growth of bacteria, while tight clothing, such as jeans and panty hose, provides an environment for the growth of bacteria.

Personal hygiene is especially important too at the time of menstruation. Hands should be washed before and after changing napkins and/or tampons. Superabsorbent tampons are not best if they are changed infrequently, as this may encourage the growth of bacteria. Also, sexual intercourse may cause menstrual flow to enter the urethra.

In males, the prostate is a gland that surrounds the urethra just below the bladder (Fig. 9A). The prostate contributes secretions to semen whenever semen enters the urethra prior to ejaculation. An infection of the prostate, called prostatitis, is often accompanied by a urinary tract infection. Fever is present and the prostate is tender and inflamed. The patient may have to be hospitalized and treated with a broad spectrum antibiotic. Prostatitis, which in a young person is often preceded by a sexually transmitted disease, can lead to a chronic condition. Chronic prostatitis may be asymptomatic or, as is more typical, there is irritation upon voiding and/or difficulty in voiding. The latter can lead to the need for surgery to remove the obstruction to urine flow.

In both males and females, it is wise to take all necessary steps to avoid urinary tract infections.

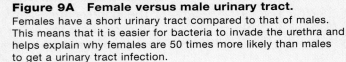

female   male

kidney

ureter

rectum   bladder

rectum

prostate
gland

pubic symphysis

vagina   urethra   urethra   penis

Figure 9A   Female versus male urinary tract.
Females have a short urinary tract compared to that of males. This means that it is easier for bacteria to invade the urethra and helps explain why females are 50 times more likely than males to get a urinary tract infection.

9.2 Kidneys

When a kidney is sliced lengthwise, it is possible to see that the renal artery and vein have many branches inside it (Fig. 9.4*a*). Without the presence of the blood vessels, it is easier to identify three regions of a kidney. The *renal cortex* is an outer granulated layer that dips down in between a radially striated, or lined, inner layer called the renal medulla. The *renal medulla* consists of cone-shaped tissue masses called renal pyramids. The *renal pelvis* is a central space, or cavity, that is continuous with the ureter (Fig. 9.4*b*).

Microscopically, the kidney is composed of over one million **nephrons,** sometimes called renal or kidney tubules (Fig. 9.4*c*). The nephrons produce urine and are positioned so that the urine flows into a collecting duct. Several nephrons enter the same collecting duct; the collecting ducts enter the renal pelvis.

Macroscopically, a kidney has three regions: renal cortex, renal medulla, and a renal pelvis that is continuous with the ureter. Microscopically, a kidney contains over one million nephrons.

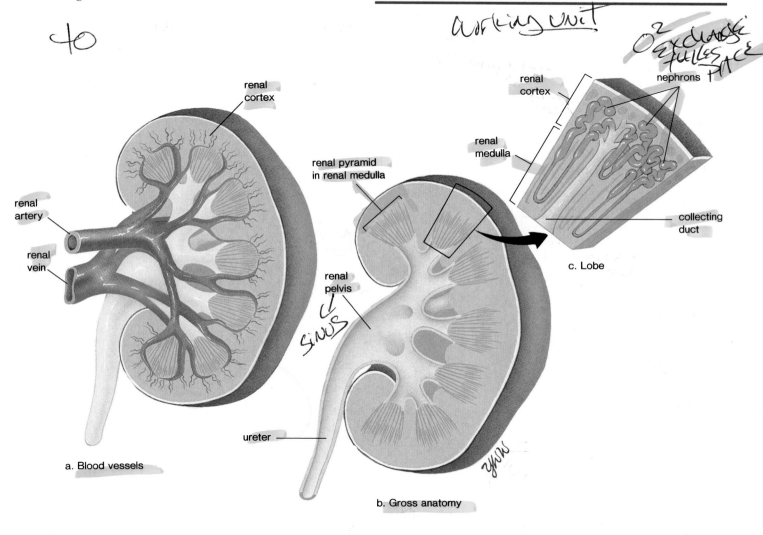

a. Blood vessels

b. Gross anatomy

c. Lobe

Figure 9.4 Gross anatomy of the kidney.
a. A longitudinal section of the kidney showing the blood supply. Note that the renal artery divides into smaller arteries, and these divide into arterioles. Venules join to form small veins, which join to form the renal vein. **b.** The same section without the blood supply. Now it is easier to distinguish the renal cortex, the renal medulla, and the renal pelvis, which connects with the ureter. The renal medulla consists of the renal pyramids. **c.** An enlargement of a lobe, showing the placement of nephrons in a renal pyramid.

Anatomy of a Nephron

Each nephron has its own blood supply, including two capillary regions. From the renal artery, an afferent arteriole leads to the **glomerulus,** a knot of capillaries inside the glomerular capsule. Blood leaving the glomerulus enters the efferent arteriole and then **peritubular capillaries,** which surround the rest of the nephron. From there the blood goes into a venule that joins the renal vein. The arrows in Figure 9.5 show the path of blood about a nephron.

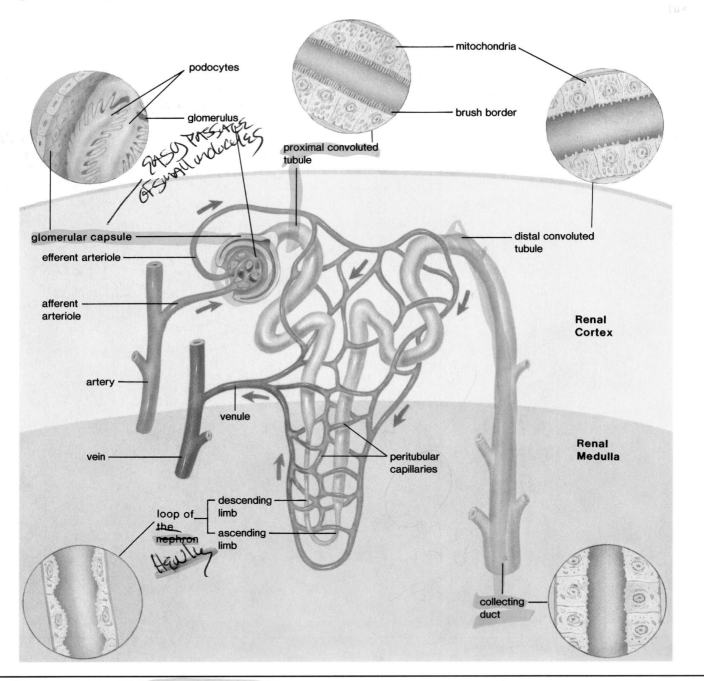

Figure 9.5   Nephron macroscopic and microscopic anatomy.
A nephron is made up of a glomerular capsule, the proximal convoluted tubule, the loop of the nephron, the distal convoluted tubule, and the collecting duct. The blowups show the types of tissue at these different locations. You can trace the path of blood about the nephron by following the arrows.

Nephron Has Several Parts

Each nephron is made up of several parts. The structure of each part suits its function.

First, the closed end of the nephron is pushed in on itself to form a cuplike structure called the **glomerular capsule** (Bowman's capsule). The outer layer of the glomerular capsule is composed of squamous epithelial cells; the inner layer is made up of *podocytes* that have long cytoplasmic processes. The podocytes cling to the capillary walls of the glomerulus and leave pores that allow easy passage of small molecules from the glomerulus to the inside of the glomerular capsule. This process, called *glomerular filtration*, produces a filtrate of blood.

Next, there is a **proximal** (meaning near the glomerular capsule) **convoluted tubule.** The cuboidal epithelial cells lining this part of the nephron have numerous microvilli, about one micrometer in length, that are tightly packed and form a brush border. A brush border greatly increases the surface area for the *tubular reabsorption* of filtrate components. Each cell also has many mitochondria, which can supply energy for active transport of molecules from the lumen to the peritubular capillary (Fig. 9.6).

Simple squamous epithelium appears as the tube narrows and makes a U-turn called the **loop of the nephron** (loop of Henle). Each loop consists of a descending limb that allows water to leave and an ascending limb that is impervious to water. Indeed, as we shall see, this arrangement facilitates the reabsorption of water by the nephron and collecting duct.

The cells of the **distal convoluted tubule** have numerous mitochondria, but they lack microvilli. This is consistent with the active role they play in moving molecules from the blood into the tubule, a step in urine formation called *tubular secretion*. The distal convoluted tubules of several nephrons enter one collecting duct. A kidney contains many collecting ducts, which carry urine to the renal pelvis.

As shown in Figure 9.5, the glomerular capsule and the convoluted tubules always lie within the renal cortex. The loop of the nephron dips down into the renal medulla; a few nephrons have a very long loop of the nephron, which penetrates deep into the renal medulla. **Collecting ducts** are also located in the renal medulla, and they give the renal pyramids their lined appearance.

Each part of a nephron is anatomically suited to its specific function in urine formation.

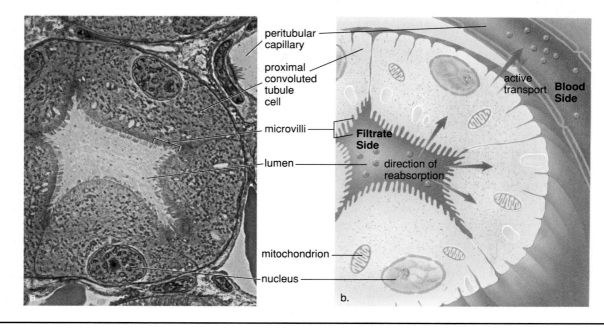

Figure 9.6 Proximal convoluted tubule.
a. This photomicrograph shows that the cells lining the proximal convoluted tubule have a brushlike border composed of microvilli, which greatly increases the surface area exposed to the lumen. The peritubular capillaries surround the cells. **b.** Diagrammatic representation of **(a)** shows that each cell has many mitochondria, which supply the energy needed for active transport, the process that moves molecules (green) from the lumen of the tubule to the capillary.

9.3   Urine Formation

Figure 9.7 gives an overview of urine formation, which is divided into these steps: (1) glomerular filtration, (2) tubular reabsorption, and (3) tubular secretion. Reabsorption of water is not an individual step because it occurs along the length of the nephron, but it is especially associated with the loop of the nephron and the collecting duct. Excretion is not a step in urine formation—it is the end result.

Glomerular Filtration Divides the Blood

Glomerular filtration occurs when whole blood enters the afferent arteriole and the glomerulus. Due to glomerular blood pressure, which is usually about 60 mm Hg, water and small molecules move from the glomerulus to the inside of the glomerular capsule. This is a filtration process because large molecules and formed elements are unable to pass through the capillary wall. In effect, then, blood in the glomerulus has two portions: the filterable components and the nonfilterable components.

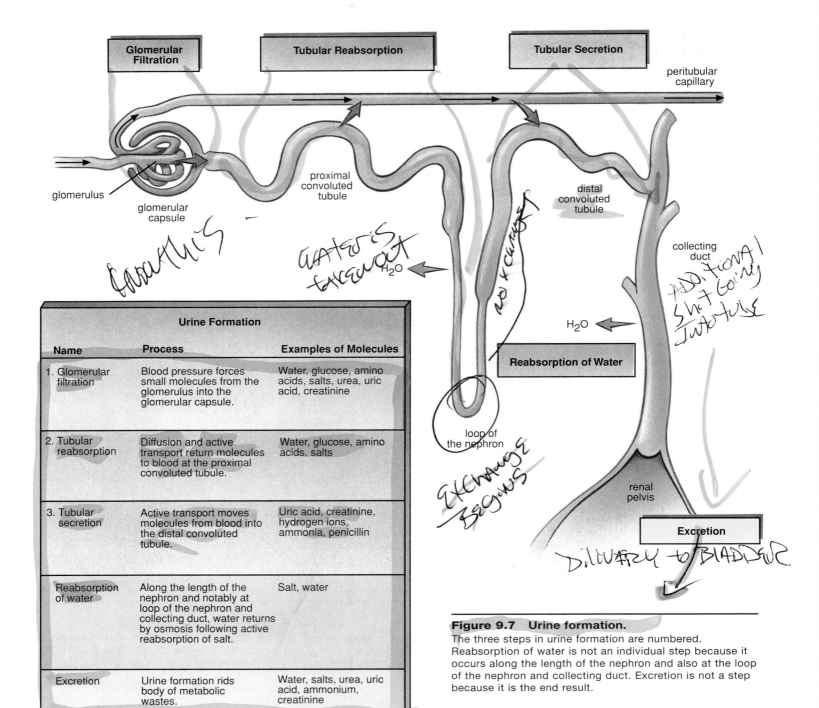

Urine Formation		
Name	**Process**	**Examples of Molecules**
1. Glomerular filtration	Blood pressure forces small molecules from the glomerulus into the glomerular capsule.	Water, glucose, amino acids, salts, urea, uric acid, creatinine
2. Tubular reabsorption	Diffusion and active transport return molecules to blood at the proximal convoluted tubule.	Water, glucose, amino acids, salts
3. Tubular secretion	Active transport moves molecules from blood into the distal convoluted tubule.	Uric acid, creatinine, hydrogen ions, ammonia, penicillin
Reabsorption of water	Along the length of the nephron and notably at loop of the nephron and collecting duct, water returns by osmosis following active reabsorption of salt.	Salt, water
Excretion	Urine formation rids body of metabolic wastes.	Water, salts, urea, uric acid, ammonium, creatinine

Figure 9.7   Urine formation.
The three steps in urine formation are numbered. Reabsorption of water is not an individual step because it occurs along the length of the nephron and also at the loop of the nephron and collecting duct. Excretion is not a step because it is the end result.

Filterable Blood Components	Nonfilterable Blood Components
Water	Formed elements (blood cells and platelets)
Nitrogenous wastes	Proteins
Nutrients	
Salts (ions)	

The **glomerular filtrate** contains small dissolved molecules in approximately the same concentration as plasma. Small molecules that escape being filtered and the nonfilterable components leave the glomerulus by way of the efferent arteriole.

As indicated in Table 9.2, 180 liters of water are filtered per day along with a considerable amount of small molecules, such as glucose and amino acids. If the composition of urine were the same as that of the glomerular filtrate, the body would continually lose water, salts, and nutrients. Death from dehydration, starvation, and low blood pressure would quickly follow. Therefore, we can conclude that the composition of the filtrate must be altered as this fluid passes through the remainder of the tubule.

During glomerular filtration, water, salts, nutrient molecules, and waste molecules move from the glomerulus to the inside of the glomerular capsule. The filtered substances are called the glomerular filtrate.

Tubular Reabsorption Is Both Passive and Active

Tubular reabsorption occurs as molecules and ions are both passively and actively reabsorbed from the nephron to the blood in the peritubular capillary. The osmolarity of the blood is maintained by the presence of plasma proteins and also by salt. When sodium ions (Na^+) are actively reabsorbed, chloride ions (Cl^-) follow passively. The reabsorption of salt (Na^+Cl^-) increases the osmolarity of the blood compared to the filtrate, and therefore, water moves passively from the tubule into the blood. About 60–70% of Na^+ are reabsorbed at the proximal convoluted tubule.

Nutrients such as glucose and amino acids also return to the blood at the proximal convoluted tubule. This is a selective process because only molecules recognized by carrier molecules are actively reabsorbed. Glucose is an example of a molecule that ordinarily is completely reabsorbed because there is a plentiful supply of carrier molecules for it. However, every substance has a maximum rate of transport, and after all its carriers are in use, any excess in the filtrate will appear in the urine. For example, as reabsorbed levels of glucose approach 180–200 mg/100 ml plasma, the rest will appear in the urine. In diabetes mellitus, excess glucose occurs in the blood, and then in the filtrate, and then in the urine, because the liver and muscles fail to store glucose as glycogen and the kidneys cannot reabsorb all of it. The

		TABLE 9.2		
		Reabsorption from Nephron		
Substance	**Amount Filtered (Per Day)**		**Amount Excreted (Per Day)**	**Reabsorption (%)**
Water, L	180		1.8	99.0
Sodium, g	630		3.2	99.5
Glucose, g	180		0.0	100.0
Urea, g	54		30.0	44.0

L = liters, g = grams

From A. J. Vander, et al. Human Physiology, 4th ed. © 1985 The McGraw-Hill Publishing Companies, Inc. All Rights Reserved. Reprinted by permission.

presence of glucose in the filtrate increases its osmolarity compared to blood, and therefore, less water is reabsorbed into the peritubular capillary. The frequent urination and increased thirst experienced by patients with uncontrolled diabetes mellitus is due to the fact that water is remaining in the filtrate and is not being reabsorbed.

We have seen that the filtrate that enters the proximal convoluted tubule is divided into two portions: the components that are reabsorbed from the tubule into blood, and the components that are nonreabsorbed and continue to pass through the nephron to be further processed into urine.

Reabsorbed Filtrate Components	Nonreabsorbed Filtrate Components
Most water	Some water
Nutrients	Much nitrogenous waste
Required salts (ions)	Excess salts (ions)

The substances that are not reabsorbed become the tubular fluid, which enters the loop of the nephron.

During tubular reabsorption, nutrient and salt molecules are actively reabsorbed from the proximal convoluted tubule into the peritubular capillaries, and water follows passively.

Tubular Secretion Adds Substances

Tubular secretion is a second way by which substances are removed from blood and added to tubular fluids. Hydrogen and ammonium ions, creatinine, and drugs such as penicillin are some of the substances that move from blood into the distal convoluted tubule. In the end, urine contains substances that underwent glomerular filtration and substances that underwent tubular secretion.

During tubular secretion, certain molecules are actively secreted from the peritubular capillaries into the distal convoluted tubule.

Health Focus

Spare Parts

Transplantation of the kidney, heart, liver, pancreas, lung, and other organs is now possible due to two major breakthroughs. First, solutions have been developed that preserve donor organs for several hours. This made it possible for one young boy to undergo surgery for 16 hours, during which time he received five different organs. Second, rejection of transplanted organs is now curtailed by immunosuppressive drugs; therefore, organs can be donated by unrelated individuals, living or dead. Living individuals can donate one kidney, or a portion of their liver or bone marrow, which quickly regenerate.

After death, it is still possible to give the "gift of life" to someone else—over 25 organs and tissues from the same person can be used for transplants at that time. A liver transplant, for example, can save the life of a child born with biliary atresia, a congenital defect in which the bile ducts do not form. Dr. Thomas Starzl, a pioneer in this field, reports there is a 90% chance of complete rehabilitation among children who survive a liver transplant (Fig. 9B). (He has also tried animal-to-human liver transplants, but so far these have not been successful.) So many heart recipients are now alive and healthy they have formed basketball and softball teams, demonstrating the normalcy of their lives after surgery.

One problem persists. Although the number of individuals waiting for organs is greater than ever, only a small percentage of people signify their willingness to donate organs at the time of their death. Organ and tissue donors must sign a donor card and carry it at all times (Fig. 9C). In many states, the back of the driver's license acts as a donor card. Age is no drawback, but the donor should have been in good health prior to death.

Organ and tissue donation will not interfere with funeral arrangements, and most religions do not object to the donation. Family members should know ahead of time about the desire to become a donor because they will be asked to sign permission papers at the time of death. No money is received for the gift organs, which are removed by a team of surgeons from the nearest organ procurement center.

The United Network for Organ Sharing (UNOS), based in Richmond, Va., which was established after the 1984 National Organ Transplant Act, has a computerized system for matching needy patients with available organs. The patients are ranked according to various medical criteria, and UNOS notifies the appropriate hospital of the availability of an organ. Donor and recipient identities are confidential.

Today, there are over 27,000 Americans waiting for a gift of life. Will they wait in vain?

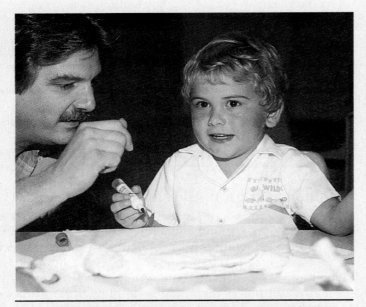

Figure 9B **Children receiving an organ transplant to cure a birth defect have a good chance of survival and complete recovery.**

ORGAN DONOR CARD

Print or type name of donor

In the hope that I may help others, I hereby make this anatomical gift, if medically acceptable, to take effect upon my death. The words and marks below indicate my desires.

I give: (a) _____ any needed organs or parts

        (b) _____ only the following organs or parts

Specify the organ(s) or part(s)

for the purposes of transplantation, therapy, medical research or education;

        (c) _____ my body for anatomical study if needed.

Limitations or special wishes, if any: _____

Signed by the donor and the following witnesses in the presence of each other:

Signature of Donor	Date of Birth of Donor
Date Signed	City & State
Witness	Witness

This is a legal document under the Uniform Anatomical Gift Act or similar laws. For further information consult your physician or

UNOS P.O. Box 13770 Richmond, Virginia 23225-8770

Figure 9C **An organ donor card gives the name of the donor and witnesses, one of whom should be a close family member.**

Source: Data from the Arkansas Regional Recovery Agency (ARORA) Little Rock, AR.

9.4 Regulatory Functions of the Kidneys

The kidneys help maintain the water and salt balance and the pH of the blood. When the kidneys regulate the water and salt balance, they also maintain the blood volume and blood pressure.

Reabsorbing Water

Usually, more than 99% of the water filtered at the glomerulus is returned to blood. Water is reabsorbed along the whole length of the nephron, but the excretion of a hypertonic urine (one that is more concentrated than blood) is dependent upon the reabsorption of water from the loop of the nephron (loop of Henle) and the collecting duct.

A *long loop of the nephron,* which typically penetrates deep into the renal medulla, is made up of a *descending* (going down) *limb* and an *ascending* (going up) *limb.* Salt (Na^+Cl^-) passively diffuses out of the lower portion of the ascending limb, but the upper, thick portion of the limb actively transports salt out into the tissue of the outer renal medulla (Fig. 9.8). Less and less salt is available for transport from the tubule as fluid moves up the thick portion of the ascending limb. Because of these circumstances, the loop of the nephron establishes an *osmotic gradient* within the tissues of the renal medulla: the concentration of salt is greater in the direction of the inner medulla. (Note that water cannot leave the ascending limb because the limb is impermeable to water.)

Also, if you examine Figure 9.8 carefully, you can see that the innermost portion of the inner medulla has the highest concentration of solutes. This cannot be due to salt because active transport of salt does not start until the thick portion of the ascending limb. Urea is believed to leak from the lower portion of the collecting duct, and it is this molecule that contributes to the high solute concentration of the inner medulla.

Because of the osmotic gradient within the renal medulla, water leaves the descending limb of the loop of the nephron along its length. This is a countercurrent mechanism: as water diffuses out of the descending limb, the remaining solution within the limb encounters an even greater osmotic concentration of solute; therefore, water will continue to leave the descending limb from the top to the bottom.

Fluid entering a collecting duct comes from the distal convoluted tubule. This fluid is now isotonic to the cells of the cortex. This means that to this point, the net effect of reabsorption of water and salt is the production of a fluid that has the same tonicity as blood. However, the urine within the collecting duct also encounters the same osmotic gradient mentioned earlier (Fig. 9.8). Therefore, water diffuses out of the collecting duct into the renal medulla, and the urine within the collecting duct becomes hypertonic to blood plasma.

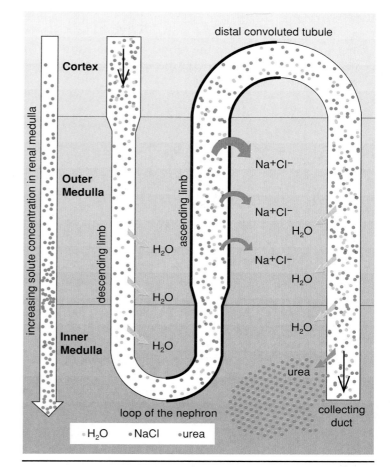

Figure 9.8 Reabsorption of water at the loop of the nephron and the collecting duct.

Salt (Na^+Cl^-) diffuses and is actively transported out of the ascending limb of the loop of the nephron into the renal medulla; also, urea is believed to leak from the collecting duct and to enter the tissues of the renal medulla. This creates a hypertonic environment, which draws water out of the descending limb and the collecting duct. This water is returned to the circulatory system. (The thick line means the ascending limb is impermeable to water.)

Antidiuretic hormone (ADH) released by the posterior lobe of the pituitary plays a role in water reabsorption at the collecting duct. In order to understand the action of this hormone, consider its name. Diuresis means increased amount of urine, and antidiuresis means decreased amount of urine. When ADH is present, more water is reabsorbed (blood volume and pressure rise), and a decreased amount of urine results. In practical terms, if an individual does not drink much water on a certain day, the posterior lobe of the pituitary releases ADH, causing more water to be reabsorbed and less urine to form. On the other hand, if an individual drinks a large amount of water and does not perspire much, ADH is not released. Now more water is excreted, and more urine forms.

Reabsorbing Salt

Usually, more than 99% of sodium (Na⁺) filtered at the glomerulus is returned to the blood. Most sodium (67%) is reabsorbed at the proximal tubule, and a sizable amount (25%) is extruded by the ascending limb of the loop of the nephron. The distal convoluted tubule and collecting duct reabsorb the rest.

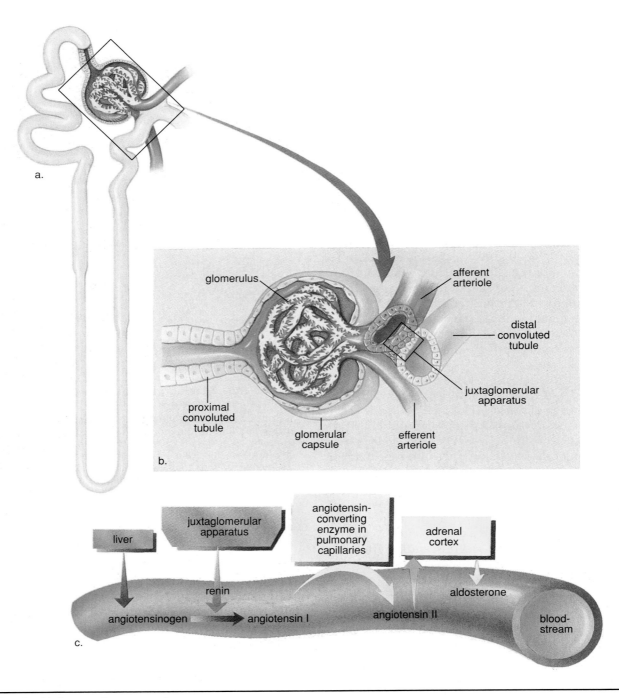

Figure 9.9   Juxtaglomerular apparatus.

a. This drawing shows how it is possible for the afferent arteriole and the distal convoluted tubule to lie next to one another. The juxtaglomerular apparatus occurs where they touch. **b.** Cross section shows exact location of the juxtaglomerular apparatus, which releases renin if the blood pressure in the afferent arteriole falls. **c.** Renin is an enzyme that changes angiotensinogen, a plasma protein made by the liver, to angiotensin I. Angiotensin I is changed to angiotensin II by a converting enzyme found in the lining of the pulmonary (lung) capillaries. Angiotensin II, a powerful vasoconstrictor, stimulates the adrenal cortex to release aldosterone into blood. Now the blood pressure rises as a result of vasoconstriction and reabsorption of sodium ions (Na⁺), followed by water.

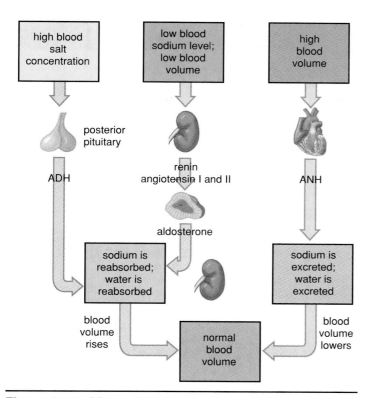

Figure 9.10 Maintaining blood volume.
Normal blood volume is maintained by ADH (antidiuretic hormone) and aldosterone, whose actions raise blood volume, and ANH (atrial natriuretic hormone), whose action lowers blood volume.

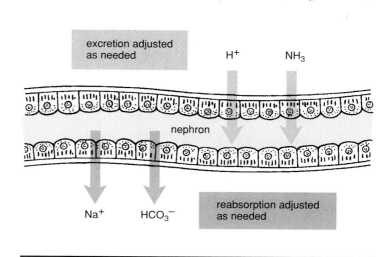

Figure 9.11 Maintaining blood pH.
The kidneys regulate blood pH largely by excreting hydrogen ions (H^+) in conjunction with ammonia (NH_3^+). They ordinarily reabsorb bicarbonate ions (HCO_3^-) but will excrete them if the blood is alkaline.

Hormones regulate the reabsorption of sodium by the kidneys. **Aldosterone** is a hormone secreted by the adrenal cortex, the outer portion of the adrenal glands, which lie atop the kidneys. Aldosterone promotes the excretion of potassium ions (K^+) and the reabsorption of sodium ions (Na^+). The secretion of aldosterone is set in motion by the kidneys themselves. The *juxtaglomerular apparatus* is a region of contact between the afferent arteriole and the distal convoluted tubule (Fig. 9.9). When blood volume, and therefore blood pressure, is not sufficient to promote glomerular filtration, the juxtaglomerular apparatus secretes renin. *Renin* is an enzyme that changes angiotensinogen (a large plasma protein produced by the liver) into angiotensin I. Later, angiotensin I is converted to angiotensin II, a powerful vasoconstrictor that also stimulates the adrenal cortex to release aldosterone. The reabsorption of sodium ions is followed by the reabsorption of water. Therefore, blood volume and blood pressure increase.

Atrial natriuretic hormone (ANH) is a hormone secreted by the atria of the heart when cardiac cells are stretched due to increased blood volume. ANH inhibits the secretion of renin by the juxtaglomerular apparatus and the secretion of aldosterone by the adrenal cortex. Its effect, therefore, is to promote the excretion of Na^+, that is, natriuresis. When Na^+ is excreted, so is water, and therefore, blood volume and blood pressure decrease (Fig. 9.10).

These examples also show that the kidneys regulate the salt balance in blood by controlling the excretion and the reabsorption of various ions. Sodium (Na^+) is an important ion in plasma that must be regulated, but the kidneys also excrete or reabsorb other ions, such as potassium ions, bicarbonate ions, and magnesium ions, as needed.

Diuretics

Diuretics are agents that increase the flow of urine. Drinking alcohol causes diuresis because it inhibits the secretion of ADH. The dehydration that follows is believed to contribute to the symptoms of a hangover. Caffeine is a diuretic because it increases the glomerular filtration rate and decreases the tubular reabsorption of Na^+. Diuretic drugs that have been developed to counteract high blood pressure in patients also inhibit the reabsorption of Na^+, leading to a decrease in water reabsorption and a decrease in blood volume.

Maintaining Blood pH

The kidneys help to maintain the pH level of blood around 7.4, and the whole nephron takes part in this process. The excretion of hydrogen ions (H^+) and ammonia (NH_3), together with the reabsorption of sodium ions (Na^+) and bicarbonate ions (HCO_3^-), is adjusted to keep the pH within normal bounds. If blood is acidic, hydrogen ions are excreted in combination with ammonia, while sodium ions and bicarbonate ions are reabsorbed. This restores the pH because $NaHCO_3$ is a base. If blood is basic, fewer hydrogen ions are excreted, and fewer sodium ions and bicarbonate ions are reabsorbed (Fig. 9.11).

HUMAN SYSTEMS WORK TOGETHER

Integumentary System

Kidneys compensate for water loss due to sweating; activate vitamin D precursor made by skin.

Skin helps regulate water loss; sweat glands carry on some excretion.

Skeletal System

Kidneys provide active vitamin D for Ca^{++} absorption and help maintain blood level of Ca^{++}, needed for bone growth and repair.

Bones provide support and protection.

Muscular System

Kidneys maintain blood levels of Na^+, K^+, and Ca^{++}, which are needed for muscle innervation, and eliminate creatinine, a muscle waste.

Smooth muscular contraction assists voiding of urine; skeletal muscles support and help protect urinary organs.

Nervous System

Kidneys maintain blood levels of Na^+, K^+, and Ca^{++}, which are needed for nerve conduction.

Brain controls nerves which innervate muscles that permit urination.

Endocrine System

Kidneys keep blood values within normal limits so that transport of hormones continues.

ADH and aldosterone, and atrial natriuretic hormone regulate reabsorption of Na^+ by kidneys.

How the Urinary System works with other body systems

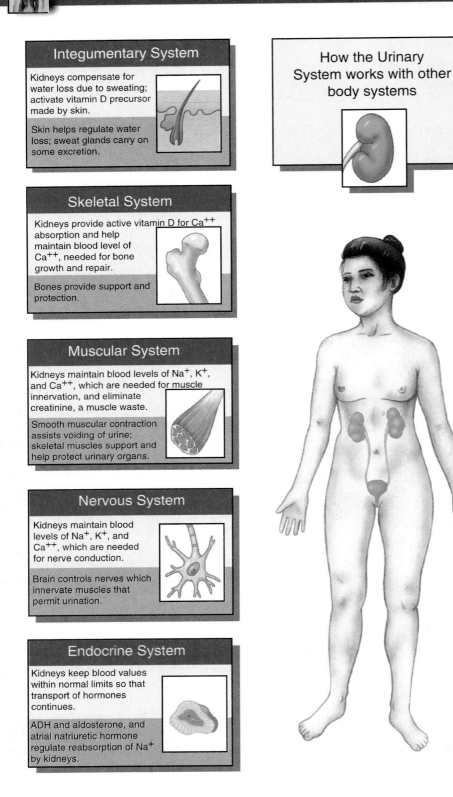

Circulatory System

Kidneys filter blood and excrete wastes; maintain blood volume, pressure, and pH; produce renin and erythropoietin.

Blood vessels deliver waste to be excreted; blood pressure aids kidney function; heart produces atrial natriuretic hormone.

Lymphatic System/Immunity

Kidneys control volume of body fluids, including lymph.

Lymphatic system picks up excess tissue fluid, helping to maintain blood pressure for kidneys to function; immune system protects against infections.

Respiratory System

Kidneys compensate for water lost through respiratory tract; work with lungs to maintain blood pH.

Lungs excrete carbon dioxide, provide oxygen, and convert angiotensin I to angiotensin II, leading to kidney regulation.

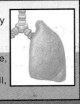

Digestive System

Kidneys convert vitamin D to active form needed for Ca^{++} absorption; compensate for any water loss by digestive tract.

Liver synthesizes urea; digestive tract excretes bile pigments from liver and provides nutrients.

Reproductive System

Semen is discharged through the urethra in males; kidneys excrete wastes and maintain electrolyte levels for mother and child.

Penis in males contains the urethra and performs urination; prostate enlargement hinders urination.

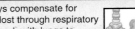

Dynamic Human | Urinary System

9.5 Working Together

The Working Together box on page 196 shows how the urinary system works with the other systems of the body to maintain homeostasis.

9.6 Other Excretory Organs

The kidneys are the primary excretory organs, but other organs also function in excretion (Fig. 9.12).

Lungs Remove CO_2

The process of expiration removes carbon dioxide (CO_2) and raises the pH of blood. The air we expire also contains moisture, as demonstrated by breathing onto a cool mirror.

Skin Perspires

The sweat glands in the skin excrete perspiration, which is a solution of water, salt, and some urea. This aids the overall process of excretion. However, we perspire not so much to rid the body of waste as to cool it; heat is lost as perspiration evaporates. In times of kidney failure, more urea is excreted by the sweat glands, even sometimes forming a so-called urea frost on the skin.

Liver Makes Bile Pigments

The liver excretes bile pigments, which are derived from the breakdown of hemoglobin. Bile is stored in the gallbladder before it passes into the small intestine by way of ducts. The yellow pigment found in urine, called urochrome, is also derived from the breakdown of heme, but this pigment is deposited in blood and is subsequently excreted by the kidneys.

9.7 Problems with Kidney Function

Because of the great importance of the kidneys to the maintenance of body fluid homeostasis, renal failure is a life-threatening event. There are many types of illnesses that cause progressive renal disease and renal failure.

Urinary tract infections are a fairly common occurrence, particularly in the female, since the shorter female urethra invites bacterial invasion more than the longer male urethra. Urethritis is an infection of the urethra; cystitis involves the bladder; and if the kidneys are infected, the infection is called pyelonephritis. The Health Focus reading on page 186 suggests ways to prevent urinary tract infections.

Glomerular damage sometimes leads to blockage of the glomeruli so that no fluid moves into the tubules, or damage can cause the glomeruli to become more permeable than usual. This is detected when a *urinalysis* is

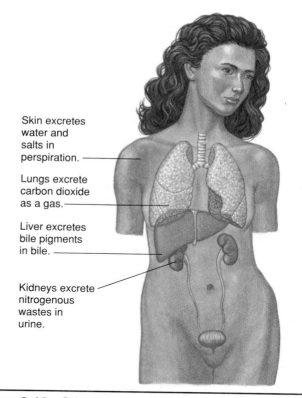

Figure 9.12 Other organs of excretion.
The organs of excretion rid the body of the waste products of metabolism.

Skin excretes water and salts in perspiration.

Lungs excrete carbon dioxide as a gas.

Liver excretes bile pigments in bile.

Kidneys excrete nitrogenous wastes in urine.

done. If the glomeruli are too permeable, albumin, white blood cells, or even red blood cells appear in the urine. A trace amount of protein in the urine is not a matter of concern, however.

When glomerular damage is so extensive that more than two-thirds of the nephrons are inoperative, waste substances accumulate in blood. This condition is called *uremia* because urea is one of the substances that accumulates. Although nitrogenous wastes can cause serious damage, the retention of water and salts (ions) is of even greater concern. The latter causes edema, fluid accumulation in the body tissues. Imbalance in the ionic composition of body fluids can even lead to loss of consciousness and to heart failure.

Replacing the Kidney

Patients with renal failure sometimes undergo a kidney transplant operation (see the Health Focus reading on page 192) during which a functioning kidney from a donor is received. As with all organ transplants, there is the possibility of organ rejection. Receiving a kidney from a close relative has the highest chance of success. The current one-year survival rate is 97% if the kidney is received from a relative and 90% if it is received from a nonrelative.

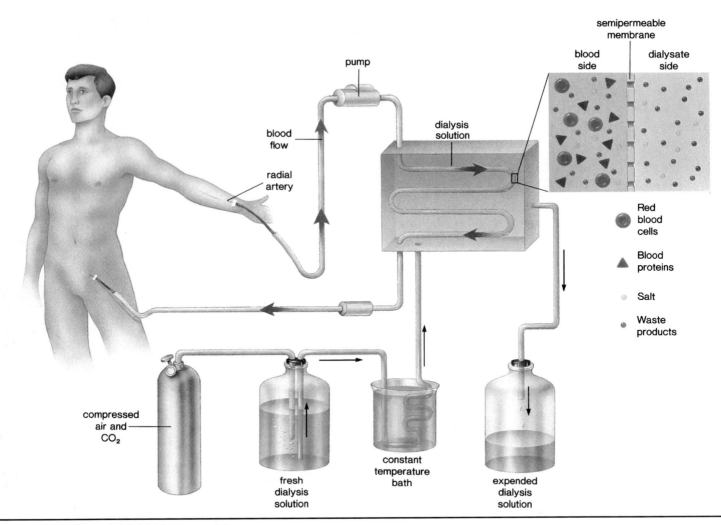

Figure 9.13   An artificial kidney machine.
As the patient's blood circulates through dialysis tubing, it is exposed to a dialysis solution (dialysate). Wastes exit from blood into the solution because of a preestablished concentration gradient. In this way, blood is not only cleansed, but its pH can also be adjusted.

Dialysis: Treating Blood

While waiting for a kidney transplant, it is probably necessary for the patient to undergo dialysis, utilizing either an artificial kidney machine or *continuous ambulatory peritoneal* (abdominal) *dialysis* (CAPD). **Dialysis** is defined as the diffusion of dissolved molecules through a semipermeable membrane (an artificial membrane with pore sizes that allow only small molecules to pass through). These molecules, of course, move across a membrane from the area of greater concentration to one of lesser concentration.

During *hemodialysis* (Fig. 9.13), the patient's blood is passed through a semipermeable membranous tube, which is in contact with a balanced salt (dialysis) solution. Substances more concentrated in blood diffuse into the dialysis solution, which is also called the dialysate, and substances more concentrated in the dialysate diffuse into blood. Accordingly, the artificial kidney can be utilized either to extract substances from blood, including waste products or toxic chemicals and drugs, or to add substances to blood—for example, bicarbonate ions (HCO_3^-) if blood is acidic. In the course of a six-hour hemodialysis, from 50–250 grams of urea can be removed from a patient, which greatly exceeds the amount excreted by normal kidneys. Therefore, a patient needs to undergo treatment only about twice a week.

In the case of CAPD, a fresh amount of dialysate is introduced directly into the abdominal cavity from a bag attached to a permanently implanted plastic tube. Waste and water molecules pass into the dialysate from the surrounding organs before the fluid is collected four or eight hours later. The individual can go about his or her normal activities during CAPD, unlike during hemodialysis.

SUMMARY

9.1 Urinary System

The path of urine is through the kidneys, ureters, urinary bladder, and finally, the urethra.

The kidneys produce urine, which is conducted by the ureters to the bladder where it is stored before being released by way of the urethra.

The kidneys maintain the water and salt balance of the body and help keep the blood pH within normal limits. They also excrete nitrogenous wastes, including urea, uric acid, and creatinine.

9.2 Kidneys

Macroscopically, the kidneys are divided into the renal cortex, renal medulla, and renal pelvis. Microscopically, they contain the nephrons.

Each nephron has its own blood supply; the afferent arteriole approaches the glomerular capsule and divides to become the glomerulus. The spaces between the podocytes of the glomerular capsule allow small molecules to enter the capsule from the glomerulus, a capillary tuft. The efferent arteriole leaves the capsule and immediately branches into peritubular capillaries.

Each region of the nephron is anatomically suited to its task in urine formation. The spaces between the podocytes of the glomerular capsule allow small molecules to enter the capsule from the glomerulus, a capillary knot. The cuboidal epithelial cells of the proximal convoluted tubule have many mitochondria and microvilli to carry out active transport (following passive transport) from the tubule to blood. In contrast, the cuboidal epithelial cells of the distal convoluted tubule have numerous mitochondria but lack microvilli. They carry out active transport from the blood to the tubule.

9.3 Urine Formation

Urine is composed primarily of nitrogenous waste products and salts in water.

The steps in urine formation are glomerular filtration, tubular reabsorption, and tubular secretion, as explained in Figure 9.7.

9.4 Regulatory Functions of the Kidneys

The kidneys regulate the water and salt balance of the body. Water is reabsorbed from all parts of the tubule, and the loop of the nephron establishes an osmotic gradient that draws water from the descending loop of the nephron and also the collecting duct. The permeability of the collecting duct is under the control of the hormone ADH.

The reabsorption of salt increases blood volume and pressure because more water is also reabsorbed. Two other hormones, aldosterone and ANH, control the kidneys' reabsorption of sodium (Na^+).

The kidneys keep blood pH within normal limits. They excrete H^+ together with NH_3^+ and reabsorb Na^+ and HCO_3^- as needed to maintain the pH at about 7.4.

9.5 Working Together

The urinary system works with the other systems of the body in the ways described in the box on page 196.

9.6 Other Excretory Organs

The lungs excrete carbon dioxide, and this raises blood pH; the liver excretes bile pigments. The skin excretes urea, salts, and water within perspiration.

9.7 Problems with Kidney Function

Various types of problems, including repeated urinary infections, can lead to kidney failure, which necessitates receiving a kidney from a donor or undergoing dialysis by utilizing a kidney machine or CAPD.

STUDYING THE CONCEPTS

1. State the path of urine and the function of each organ mentioned.   184

2. Describe the macroscopic anatomy of a kidney.   187

3. Trace the path of blood about a nephron.   188

4. Name the parts of a nephron, and tell how the structure of the convoluted tubules suits their function.   189

5. Describe the three steps of urine formation, and discuss how water is reabsorbed.   190

6. Tell how pH is regulated and how hormonal effects on the kidneys regulate blood volume.   193–95

7. Name four excretory organs as well as the substances they excrete. Include four nitrogenous waste products, and explain how each is formed in the body.   197

8. Explain how the artificial kidney machine works.   198

APPLYING YOUR KNOWLEDGE

Concepts

1. Although humans don't excrete ammonia, NH_3, anyone who has changed a diaper and placed it in a diaper pail has experienced the strong smell of ammonia. Knowing the chemical structure of urea, can you explain what has happened?

2. A common symptom experienced by heavy drinkers is a dry mouth and a feeling of thirst when they awaken the next morning. Is this just a psychological phenomenon? Explain.

3. During World War II, recruits who tested positive for diabetes were held overnight and retested the next day. The repeat test often showed no diabetes. How would it be possible to fake the condition of diabetes?

4. Materials can be extracted from a person's blood or can be added to the blood during dialysis. The functioning kidney can also extract or add substances to the blood. What is the major physiological difference between the artificial kidney and the human kidney?

Bioethical Issue

Across the United States, over 40,000 patients wait for a needed organ, be it a kidney, liver, or heart. Donor organs come from relatives or deceased or brain-dead donors. The problem is that patients outnumber donors. Each year, some 3,000 patients die waiting for a transplant.

In trying to get organs to patients who need them most, the federal government has established certain transplant guidelines. Chief among them is geographic limits. Before shipping, say, a kidney to a far-away hospital, an organ bank must first ensure that no local patients need that kidney.

This requirement protects local residents' rights to organs. The pitfall is that a patient in Dallas might die while the liver she desperately needs goes instead to a Houston resident who's only moderately ill. The law prevents organ banks from sending the liver across a given geographic boundary as long as anyone at home needs it.

Is this situation fair? Should the federal government prioritize organ transplants according to geographic boundaries? What other methods of allocation might work?

TESTING YOUR KNOWLEDGE

1. The liver excretes _____, which are derived from the breakdown of _____.

2. The primary nitrogenous end product from protein breakdown in humans is _____.

3. Urine leaves the urinary bladder by way of the _____.

4. Label this diagram of a nephron.

5. The cuboidal epithelial cells lining the proximal convoluted tubule have _____, and this is consistent with their function of _____ molecules from the tubule into blood.

6. _____ is a molecule that comprises most of the filtrate, is reabsorbed, and is still in urine.

7. _____ is a molecule that is found in the filtrate, is reabsorbed, and is concentrated in the urine.

8. Tubular secretion takes place at the _____, a portion of the nephron.

9. Reabsorption of water from the collecting duct is regulated by the hormone _____.

10. In addition to excreting nitrogenous wastes, the kidneys help adjust the _____ and _____ of blood.

11. Persons who have nonfunctioning kidneys often have their blood cleansed by _____ machines.

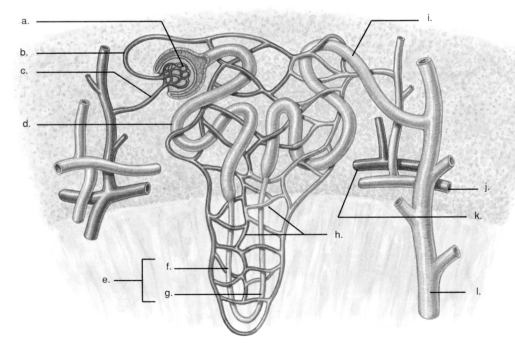

a.
b.
c.
d.
e.
f.
g.
h.
i.
j.
k.
l.

APPLYING TECHNOLOGY

Your study of the urinary system and excretion is supported by these available technologies:

Exploring the Internet

The Mader Home Page provides further resources for studying this chapter.

http://www.mhhe.com/sciencemath/biology/mader/

(Click on *Human Biology.*)

Dynamic Human: Urinary System CD-ROM

In *Anatomy,* the overall system, the kidney, and the nephron in 3-dimension are shown; in *Explorations,* zooming in leads to a depiction of urine formation; in *Histology,* microscopic slides are highlighted to reveal tissues; in *Clinical Concepts,* intravenous pyelography is used to make out the urinary system, and an ultrasound of the bladder shows the location of a kidney stone.

SELECTED KEY TERMS

aldosterone (al-DAHS-tuh-rohn) Hormone secreted by the adrenal cortex that regulates the sodium and potassium balance of the blood.   195

antidiuretic hormone (ADH) (ANT-ih-dy-yuu-RET-ik) Hormone secreted by the posterior pituitary that promotes the reabsorption of water by the collecting duct.   193

atrial natriuretic hormone (ANH) Substance secreted by the atria of the heart that promotes sodium excretion so that blood volume and blood pressure decreases.   195

collecting duct Tube that receives fluid from the distal convoluted tubules of several nephrons within a kidney.   189

dialysis Diffusion of dissolved molecules through a semipermeable membrane.   198

distal convoluted tubule Highly coiled region of a nephron that is distant from the glomerular capsule, where tubular secretion takes place.   189

excretion Removal of metabolic wastes from the body.   183

glomerular capsule (glu-MER-uh-lur) Double-walled cup that surrounds the glomerulus at the beginning of the nephron.   189

glomerular filtrate (glu-MER-uh-lur) Filtered portion of blood contained within the glomerular capsule.   191

glomerular filtration (glu-MER-uh-lur) Movement of small molecules from the glomerulus into the glomerular capsule due to the action of blood pressure.   190

glomerulus (glu-MER-uh-lus) Cluster; for example, the knot of capillaries surrounded by the glomerular capsule in a nephron, where glomerular filtration takes place.   188

kidney Organ in the urinary system that produces and excretes urine.   184

loop of the nephron (NEF-rahn) Portion of the nephron lying between the proximal convoluted tubule and the distal convoluted tubule that functions in water reabsorption.   189

nephron (NEF-rahn) Anatomical and functional unit of the kidney; kidney tubule.   187

peritubular capillary Capillary that surrounds a nephron and functions in reabsorption during urine formation.   188

proximal convoluted tubule Highly coiled region of a nephron near the glomerular capsule, where tubular reabsorption takes place.   189

tubular reabsorption Movement of molecules from the contents of the nephron into blood at the proximal convoluted tubule.   191

tubular secretion Movement of certain molecules from blood into the distal convoluted tubule so that they are added to urine.   191

urea Primary nitrogenous waste of humans derived from amino acid breakdown.   185

ureter (YUUR-ut-ur) One of two tubes that take urine from the kidneys to the urinary bladder.   184

urethra (yuu-REE-thruh) Tube that takes urine from the urinary bladder to outside.   184

uric acid Waste product of nucleotide metabolism.   185

urinary bladder Organ where urine is stored before being discharged by way of the urethra.   184

FURTHER READINGS FOR PART TWO

Becker, R. C. July/August 1996. Antiplatelet therapy. *Science & Medicine* 3(4):12. This article discusses the control of platelet aggregation.

Benjamini, E., and Leskowitz, S. 1996. *Immunology: A short course.* 3d ed. New York: John Wiley & Sons. Presents the essential principles of immunology.

Blaser, M. J. February 1996. The bacteria behind ulcers. *Scientific American* 274(2):104. Acid-loving microbes are linked to stomach ulcers and stomach cancer.

Brown, J. L., and Pollitt, E. February 1996. Malnutrition, poverty and intellectual development. *Scientific American* 274(2):38. Article discusses the complex role of essential nutrients in a child's mental development.

Capecchi, M. March 1994. Targeted gene replacement. *Scientific American* 270(3):52. Researchers are deciphering DNA segments that control development and immunity.

Gibbs, W. W. August 1996. Gaining on fat. *Scientific American* 275(2):88. Some weight problems are genetic or physiological in origin. New treatments might help.

Haen, P. J. 1995. *Principles of hematology.* Dubuque, Iowa: Wm. C. Brown Publishers. An introductory text for students planning a career in the medical sciences.

Hotez, P. J., and Pritchard, D. I. June 1995. Hookworm infection. *Scientific American* 272(6):68. Discusses how the biology of parasites offers clues to possible vaccines and also for new treatments for heart disease and immune disorders.

Johnson, H. M., et al. May 1994. How interferons fight disease. *Scientific American* 270(5):68. The properties of interferons are being utilized in treating a growing list of diseases.

Klatsky, A. L. March/April 1995. Cardiovascular effects of alcohol. *Scientific American Science & Medicine* 2(2):28. The effects of moderate and heavy alcohol use on the cardiovascular system is discussed.

Little, R. C., and Little, W. C. 1989. *Physiology of the heart and circulation.* 4th ed. Chicago: Year Book Medical Publishers, Inc. A good reference resource which gives an in-depth look at cardiovascular physiology.

Marieb, E. N. 1997. *Human anatomy and physiology.* 3d ed. Redwood City, Calif.: Benjamin/Cummings Publishing. A thorough anatomy and physiology text that can safely be used as a complete and accurate reference.

Marieb, E. N., and Mallatt, J. 1997. *Human anatomy.* 2d ed. Redwood City, Calif.: Benjamin/Cummings Publishing. An introductory anatomy text that covers gross, microscopic, developmental, and clinical anatomy.

Moon, R. E., et al. August 1995. The physiology of decompression illness. *Scientific American* 273(2):70. Discusses the physiological aspects of decompression illness.

Newman, J. December 1995. How breast milk protects newborns. *Scientific American* 273(6):76. Human milk contains special antibodies that boost the newborn's immune system.

Nowak, T. J., and Handford, A. G. 1994. *Essentials of pathophysiology.* Dubuque, Iowa: Wm. C. Brown Publishers. Introduces the concepts of pathophysiology for those in the medical fields.

Packer, L. March/April 1994. Vitamin E is nature's master antioxidant. *Scientific American Science & Medicine* 1(1):54. Evidence of the effectiveness of Vitamin E in disease prevention is accumulating.

Risher, C. E., and Easton, T. A. 1995. *Focus on human biology.* 2d. ed. New York: HarperCollins College Publishers. This comprehensive introductory textbook stresses basic human anatomy and physiology.

Science & Medicine. September/October 1996. 3(5). Issue contains articles on sleep apnea, and how fish oils reduce cardiovascular mortality. A&P: respiratory, cardiovascular/nutrition.

Sussman, N. L., and Kelly, J. H. May/June 1995. The artificial liver. *Scientific American Science & Medicine* 2(3):68. An artificial liver might assist temporarily while the natural liver regenerates, restoring normal function.

Thomas, E. D. September/October 1995. Hematopoietic stem cell transplantation. *Scientific American Science & Medicine* 2(5):38. Article discusses reconstituting marrow from cultured stem cells for bone marrow transplants.

Valtin, H. 1994. *Renal function.* 3d ed. Boston: Little, Brown and Company. A good reference resource which discusses renal mechanisms for preserving fluid and solute balance.

Weindruch, R. January 1996. Caloric restriction and aging. *Scientific American* 274(1):64. Consuming fewer calories may increase longevity.

West, J. B. 1990. *Respiratory physiology—the essentials.* 5th ed. Baltimore: Williams & Wilkins. A good reference resource which discusses all aspects of respiratory physiology including breathing, external, and internal respiration.

Yock, P., et al. September/October 1995. Intravascular ultrasound. *Scientific American Science & Medicine.* 2(3):68. Ultrasound images of coronary arteries helps diagnose atherosclerosis.

Part

Movement and Support

The skeletal and muscular systems give the body its shape and allow it to move. The skeletal system protects and supports other organs—the skull protects the brain, and the rib cage protects the lungs and heart; the pelvis supports the organs of the abdominal cavity. This system also contributes to homeostasis because it stores minerals and produces the blood cells.

Contraction of the skeletal muscles allows the body to move but also accounts for facial expressions and our ability to speak. Homeostasis would be impossible without movement of the rib cage up and down and contraction of the heart to keep the blood moving. Smooth muscle contraction allows the other internal organs to function; it helps move food along the digestive tract and urine along the urinary tract. Indeed, life as we know it is dependent on the skeletal and muscular systems.

Chapter 10

Skeletal System

Chapter Outline

10.1   TISSUES OF THE SKELETAL SYSTEM
- Various connective tissues are necessary to the anatomy of bones.   204

10.2   BONE GROWTH AND REPAIR
- Bone is a living tissue; therefore, it develops and undergoes repair.   206
- The fetal skeleton is cartilaginous, and then it is replaced by bone.   206
- The adult bones undergo remodeling—they are constantly being broken down and rebuilt.   207
- The mending of a fracture requires certain identifiable steps.   207

10.3   BONES OF THE SKELETON
- The bones of the skeleton are divided into those of the axial skeleton and those of the appendicular skeleton.   209

10.4   ARTICULATIONS
- Joints are classified according to their anatomy, and only one type is freely movable.   216

10.5   WORKING TOGETHER
- The skeletal system works with the other systems of the body to maintain homeostasis.   216

For years Ellen B. knew there was something different about the way she stood or walked. She leaned to one side, and her knees didn't line up properly. Other kids noticed this, but most of them did not understand that Ellen suffers from scoliosis, a sideways curvature of the spine that throws the rest of her body out of line.

Scoliosis, which strikes some 3% of adolescents, can be mild or serious. In Ellen's case, back pain was a nagging problem, and doctors worried that she might develop arthritis of the spine.

So, at 17, while most kids dwelled on high-school graduation, Ellen checked into the hospital. During surgery, doctors carefully hooked steel rods to vertebrae at the top and bottom of Ellen's spine, fusing the bones together with fragments taken from her hip. The fusions healed in a straightened position.

With her spine corrected, Ellen found herself two inches taller, virtually free of back pain, and physically similar to the other students in her class. In addition to being able to proceed with graduation plans, she now could go shopping for clothes without the concern that they wouldn't "hang" correctly. After the operation, Ellen could tell that our internal skeleton ordinarily permits flexible body movement without any pain. It serves as an attachment for muscles, whose contraction makes the bones move so that we can walk, play tennis, type papers, and do all manner of activities. The bones also support and protect. The large heavy bones of the legs support the entire body and its various organs against the pull of gravity. The skull protects our brain, and the rib cage protects the heart and lungs. Bones have other functions too: they are the site of blood cell formation, and they store inorganic calcium and phosphorus salts.

This chapter reviews the structure of bones and the manner in which they perform these numerous functions.

10.1   Tissues of the Skeletal System

The organs of the skeletal system are largely composed of connective tissue; particularly bone, cartilage, and fibrous connective tissue. Connective tissue contains cells that are separated by a matrix containing fibers.

Bone

Bones are strong because their matrix contains mineral salts, notably calcium phosphate. **Compact bone** is highly organized and composed of tubular units called osteons. In the cross section of an osteon, bone cells called osteocytes lie in lacunae, which are tiny chambers arranged in concentric circles around a central canal (Fig. 10.1). The mineralized matrix, which also contains protein fibers, fills the spaces between the lacunae. Tiny canals called *canaliculi* run through the matrix, connecting the lacunae with each other and with the central canal. Central canals contain blood vessels, lymphatic vessels, and nerves. Canaliculi bring nutrients from the blood vessel in the central canal to the cells in the lacunae.

Compared to compact bone, **spongy bone** has an unorganized appearance (Fig. 10.1). It contains numerous thin plates called *trabeculae* separated by unequal spaces. Although this makes spongy bone lighter than compact bone, spongy bone is still designed for strength. Just as braces are used for support in buildings, the trabeculae follow lines of stress. The spaces of spongy bone are often filled with **red bone marrow,** a specialized tissue that produces all types of blood cells. The osteocytes of spongy bone are irregularly placed within the trabeculae, and canaliculi bring them nutrients from the red bone marrow.

Cartilage

Cartilage is not as strong as bone, but it is more flexible because the matrix is gel-like and contains many collagenous and elastic fibers. The cells, called chondroblasts, lie within lacunae that are irregularly grouped. Cartilage has no blood vessels, and therefore, injured cartilage is slow to heal.

The three types of cartilage differ according to the type and arrangement of fibers in the matrix. *Hyaline cartilage* is firm and somewhat flexible. The matrix appears uniform and glassy, but actually it contains a generous supply of collagenous fibers. Hyaline cartilage is found at the ends of long bones and in the nose, at the ends of the ribs, and in the larynx and trachea.

Fibrocartilage is stronger than hyaline cartilage because the matrix contains wide rows of thick collagenous fibers. Fibrocartilage is able to withstand both tension and pressure, and this type of cartilage is found where support is of prime importance—in the disks located between the vertebrae and also in the cartilage of the knee.

Elastic cartilage is more flexible than hyaline cartilage because the matrix contains mostly elastin fibers. This type of cartilage is found in the ear flaps and epiglottis.

Fibrous Connective Tissue

Fibrous connective tissue contains rows of cells called fibroblasts separated by bundles of collagenous fibers that run in one direction. This tissue makes up the ligaments that connect bone to bone at a **joint.** A joint is also called an articulation.

Structure of a Long Bone

Figure 10.1 shows how the tissues we have been discussing are arranged in a long bone. Each end of a long bone is expanded into a region called an **epiphysis** (pl., epiphyses). The epiphyses are composed largely of spongy bone that contains red bone marrow where blood cells are made.

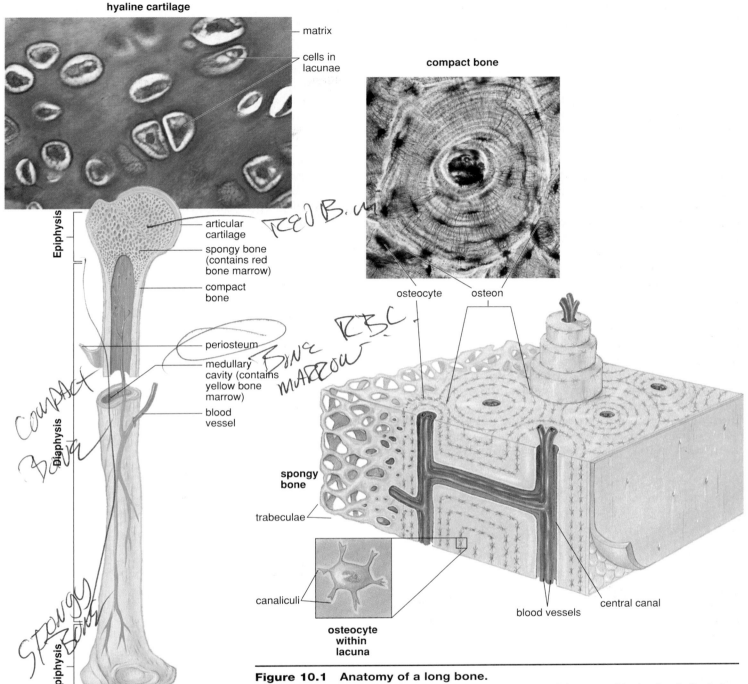

hyaline cartilage

matrix

cells in lacunae

compact bone

Epiphysis

articular cartilage

spongy bone (contains red bone marrow)

compact bone

periosteum

medullary cavity (contains yellow bone marrow)

blood vessel

Diaphysis

Epiphysis

osteocyte

osteon

spongy bone

trabeculae

canaliculi

osteocyte within lacuna

blood vessels

central canal

Figure 10.1   Anatomy of a long bone.
A long bone is encased by fibrous membrane except where it is covered by hyaline (articular) cartilage (see micrograph). Spongy bone located at each end may contain red bone marrow. The compact bone of the central shaft, which is shown in the enlargement and micrograph, contains yellow bone marrow.

The epiphyses are coated with a thin layer of hyaline cartilage, which is called **articular cartilage** because it occurs at a joint.

The shaft, or main portion of the bone, is called the **diaphysis.** The diaphysis has a large medullary that is filled with fatty *yellow bone marrow* in adults. A thin, vascular membrane called the *endosteum* lines the **medullary cavity,** whose walls are composed of compact bone.

Except for the articular cartilage on its ends, a long bone is completely covered by a layer of fibrous connective tissue called the *periosteum.* This covering contains blood vessels, lymphatic vessels, and nerves. Note in Figure 10.1, how a blood vessel penetrates the periosteum and enters the bone where it gives off branches that run within the central canals of compact bone. The periosteum is continuous with the ligaments that are connected to this bone.

10.2 Bone Growth and Repair

Bones are composed of living tissues as exemplified by their ability to grow and undergo repair. Several different types of cells are involved in bone growth and repair.

Osteogenitor cells are unspecialized cells present in the inner portion of the periosteum, in the endosteum, and in the central canal of compact bone.

Osteoblasts are bone-forming cells derived from osteogenitor cells. They are responsible for secreting the matrix, characteristic of bone.

Osteocytes are mature bone cells derived from osteoblasts. Once the osteoblasts are surrounded by matrix, they become the osteocytes found in bone.

Osteoclasts are thought to be derived from monocytes, a type of white blood cell present in red bone marrow. Osteoclasts perform bone resorption; that is, they break down bone and deposit calcium and phosphate in the blood. The work of osteoclasts is important to the growth and repair of bone.

Bone Development and Growth

The bones of the skeleton form during embryonic development in two distinctive ways. The term ossification refers to the formation of bone.

Some bones develop between sheets of fibrous connective tissue during a process called *intramembranous ossification*. The bones of the skull are examples of intramembranous bones. Osteogenitor cells derived from connective tissue cells become osteoblasts that begin to lay down matrix in various directions, forming the trabeculae of spongy bone. Other osteoblasts associated with the periosteum formed from the fibrous membrane lay down compact bone over the surface of the spongy bone. When the osteoblasts are surrounded by matrix, they become osteocytes.

Most of the bones of the human skeleton first appear as hyaline cartilage during prenatal development. Since these cartilaginous structures are shaped like the future bones, they provide "models" of these bones. The replacement of cartilaginous models by bone is called *endochondral ossification* (Fig. 10.2).

During endochondral ossification of a long bone, the cartilage begins to break down in the center of the diaphysis, which is now covered by a periosteum. Osteoblasts invade the region and begin to lay down spongy bone in what is called a primary ossification center. Other osteoblasts lay down compact bone beneath the periosteum. As the compact bone thickens, the spongy bone of the diaphysis is broken down by osteoclasts, and the cavity created becomes the medullary cavity.

The epiphyses (ends) of developing bone continue to grow, but soon secondary ossification centers form in these regions. Here spongy bone forms and does not break down. Also, a band of cartilage called the **epiphyseal plate** remains between the primary ossification center and each secondary center. The limbs keep increasing in length as long as the epiphyseal plates are still present. The rate of growth is controlled by hormones, such as growth hormones and the sex hormones. Eventually the plates become ossified, and the bone stops growing.

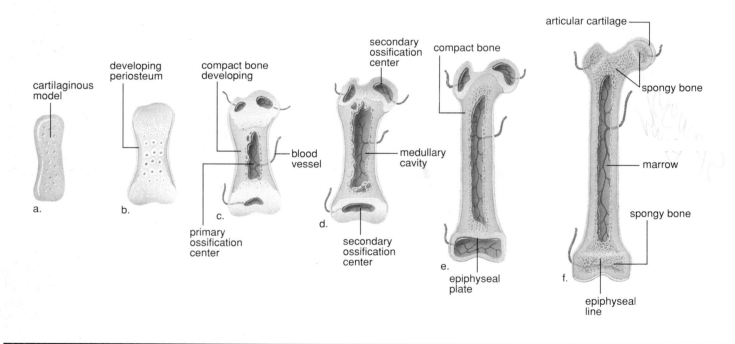

Figure 10.2 Endochondral bone formation.
a. A cartilaginous model develops during fetal development. **b.** A periosteum develops. **c.** A primary ossification center contains spongy bone surrounded by compact bone. **d.** Medullary cavity forms in diaphysis and secondary ossification centers develop in epiphyses. **e.** Growth is still possible as long as cartilage remains at epiphyseal plate. **f.** When the bone is fully formed, the epiphyseal plate becomes an epiphyseal line.

Remodeling of Bones

In the adult, bone is continually being broken down and built up again. Osteoclasts derived from monocytes in red bone marrow break down bone, remove worn cells, and deposit calcium in the blood. After a period of about three weeks, the osteoclasts disappear, and the bone is repaired by the work of osteoblasts. As they form new bone, osteoblasts take calcium from the blood. Eventually some of these cells get caught in the matrix they secrete and are converted to osteocytes, the cells found within the lacunae of osteons.

Thus, through this process of *remodeling*, old bone tissue is replaced by new bone tissue. Because of continual remodeling, the thickness of bones can change. Physical use and hormone balance affects the thickness of bones. Strange as it may seem, adults apparently require more calcium in the diet (about 1,000 to 1,500 mg daily) than do children in order to promote the work of osteoblasts. Otherwise, *osteoporosis*, a condition in which weak and thin bones easily fracture, may develop. Osteoporosis is discussed in the Health reading on page 208.

Bone Repair

Repair of a bone is required after it breaks or fractures. In some ways, bone repair parallels the development of a bone (Fig. 10.3) except that the first step mentioned here indicates that injury has occurred and fibrocartilage instead of hyaline cartilage precedes the production of compact bone.

1. *Hematoma forms* six to eight hours after fracture. Blood escapes from ruptured blood vessels and forms a hematoma (mass of clotted blood) in the space between the broken bones. The area is inflamed and swollen.
2. *Fibrocartilaginous callus* lasts about three weeks. Tissue repair begins, and fibrocartilage now fills the space between the ends of the broken bone.
3. *Bony callus* lasts about three to four months. Osteoblasts produce trabeculae of spongy bone and convert the fibrocartilage callus to a bony callus that joins the broken bones together.
4. *Remodeling.* Osteoblasts build new compact bone at the periphery, and osteoclasts reabsorb the spongy bone, creating a new medullary cavity.

The naming of fractures tells you what kind of break occurred. A fracture is complete if the bone is broken clear through and incomplete if the bone is not separated into two parts. A fracture is simple if it does not pierce the skin and compound if it does pierce the skin. Impacted means that the broken ends are wedged into each other, and a spiral fracture occurs when there is a ragged break due to twisting of a bone.

Bone is living tissue. It develops, grows, remodels, and repairs itself. In all these processes, osteoclasts break down bone, and osteoblasts build bone.

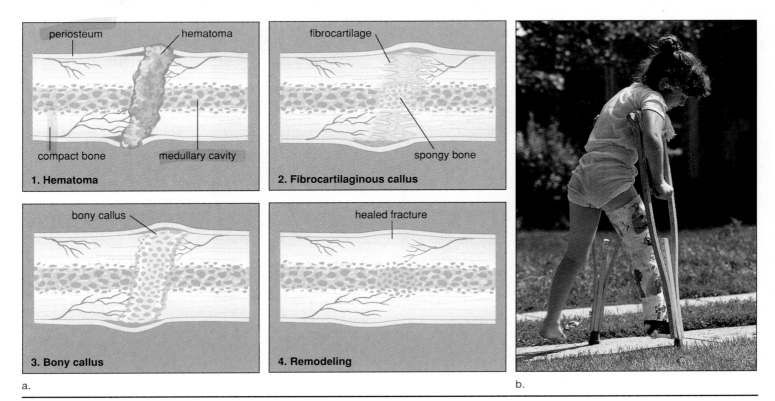

1. Hematoma — periosteum, hematoma, compact bone, medullary cavity

2. Fibrocartilaginous callus — fibrocartilage, spongy bone

3. Bony callus — bony callus

4. Remodeling — healed fracture

a.

b.

Figure 10.3 Bone fracture and repair.
a. Steps in the repair of a fracture. **b.** A plaster of Paris cast helps stabilize the bones while repair takes place.

Health Focus

You Can Avoid Osteoporosis

Osteoporosis is a condition in which the bones are weakened due to a decrease in the bone mass that makes up the skeleton. Throughout life, bones are continuously remodeled. While a child is growing, the rate of bone formation is greater than the rate of bone breakdown. The skeletal mass continues to increase until ages 20 to 30. After that, there is an equal rate of formation and breakdown of bone mass until ages 40 to 50. Then, reabsorption begins to exceed formation, and the total bone mass slowly decreases.

Over time, men are apt to lose 25% and women lose 35% of their bone mass. But we have to consider that men tend to have denser bones than women anyway, and their testosterone (male sex hormone) level generally does not begin to decline significantly until after age 65. In contrast, the estrogen (female sex hormone) level in women begins to decline at about age 45. Since sex hormones play an important role in maintaining bone strength, this difference means that women are more likely than men to suffer a higher incidence of fractures, involving especially the hip, vertebrae, long bones, and pelvis. Although osteoporosis may at times be the result of various disease processes, it is essentially a disease of aging.

There are measures that everyone can take to avoid osteoporosis when they get older. Adequate dietary calcium throughout life is an important protection against osteoporosis. The U.S. National Institutes of Health recommend a calcium intake of 1,200–1,500 mg per day during puberty. Males and females require 1,000 mg per day until the age 65 and 1,500 mg per day after age 65. In postmenopausal women not receiving estrogen replacement therapy, 1,500 milligrams per day is desirable.

A small daily amount of vitamin D is also necessary in order to use calcium correctly. Exposure to sunlight is required to allow skin to synthesize a precursor to vitamin D. If you reside on or north of a "line" drawn from Boston to Milwaukee, to Minneapolis, to Boise, chances are you're not getting enough vitamin D during the winter months. Therefore, you should avail yourself of vitamin D present in fortified foods such as low-fat milk and cereal.

Very inactive people, such as those confined to bed, lose bone mass 25 times faster than people who are moderately active. On the other hand, a regular, moderate, weight-bearing exercise like walking or jogging is another good way to maintain bone strength (Fig. 10A).

Postmenopausal women with any of these risk factors should have an evaluation of their bone density:

- white or Asian race
- thin body type
- family history of osteoporosis
- early menopause (before age 45)
- smoking
- a diet low in calcium, or excessive alcohol consumption and caffeine intake
- sedentary lifestyle

Presently bone density is measured by a method called dual energy X-ray absorptiometry (DEXA). This test measures bone density based on the absorption of photons generated by an X-ray tube. Soon there may be a blood and urine test to detect the biochemical markers of bone loss. Then it may be possible for physicians to screen all older women and at-risk men for osteoporosis.

If the bones are thin, it is worthwhile to take other measures to gain bone density because even a slight increase can significantly reduce fracture risk. Usually, experts believe that estrogen therapy in women is the treatment of choice, but there are other options. Evidence shows that estrogen replacement therapy will delay accelerated bone loss in postmenopausal women. Hormone replacement is most effective when begun at the start of menopause and continued over the long term. Some benefit, however, may still be obtained when treatment is begun later. A combination of hormone replacement and exercise apparently yields the best results. But there are other options, also. Calcitonin is a naturally occurring hormone whose main site of action is the skeleton where it inhibits the action of osteoclasts, the cells that break down bone. Also, alendronate is a drug that acts similarly to calcitonin. After three years of alendronate therapy, an increase in spinal density by about 8% and hip density by about 7% is obtained. Promising new drugs include slow-release fluoride therapy and certain growth hormones. These medications stimulate the formation of new bone.

Figure 10A   Preventing osteoporosis.
A dietary intake of calcium and vitamin D can help prevent osteoporosis. Exercise is also important. When playing golf, you should carry your own clubs and walk instead of using a golf cart.

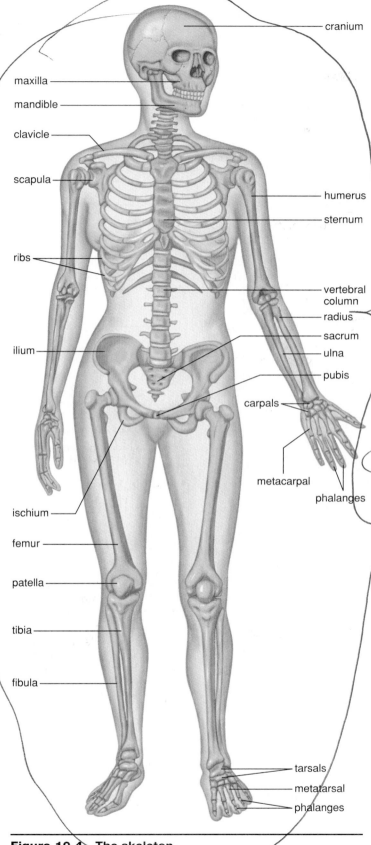

Figure 10.4 The skeleton.
The skeleton of a human adult contains bones that belong to the axial skeleton (at the midline) and those that belong to the appendicular skeleton (girdles and limbs).

Labels in figure: cranium, maxilla, mandible, clavicle, scapula, humerus, sternum, ribs, vertebral column, radius, sacrum, ulna, pubis, carpals, metacarpal, phalanges, ilium, ischium, femur, patella, tibia, fibula, tarsals, metatarsal, phalanges

10.3 Bones of the Skeleton

The 206 bones of the human skeleton are arranged as shown in Figure 10.4. The skeleton has two divisions: the axial skeleton and the appendicular skeleton. The axial skeleton is in the midline of the body, and the appendicular skeleton is the limbs along with their girdles.

Functions of the Skeleton

Let's discuss the functions of the skeleton in relation to particular bones.

The skeleton supports the body. The bones of the legs (the femur in particular and also the tibia) support the entire body when we are standing, and the coxal bones of the pelvis support the abdominal cavity.

The skeleton protects soft body parts. The bones of the skull protect the brain; and the rib cage, composed of the ribs, thoracic vertebrae, and sternum, protects the heart and lungs.

The skeleton produces blood cells. All bones in the fetus have spongy bone with red bone marrow that produces blood cells. In the adult, the flat bones of the skull, ribs, sternum, clavicles, and also the vertebrae and pelvis produce blood cells.

The skeleton stores calcium and phosphate. All bones have a matrix that contains calcium phosphate. When bones are remodeled, osteoclasts break down bone and return calcium ions and phosphorus ions to the bloodstream.

The skeleton, along with the muscles, permits flexible body movement. While articulations (joints) occur between all the bones, we can associate body movement in particular with the bones of the legs (especially the femur and tibia) and the feet (tarsals, metatarsals, and phalanges) because we use them when walking.

Classification of the Bones

The bones are classified according to their shape. Long bones, exemplified by the humerus and femur, are longer than they are wide. Short bones, such as the carpals and tarsals, are cube shaped—their lengths and widths are about equal. Flat bones, like those of the skull, are platelike with broad surfaces. Round bones, exemplified by the patella, are circular in shape. Irregular bones, such as the vertebrae and facial bones, have varied shapes that permit connections with other bones.

The bones of the skeleton are not smooth; they have articulating protuberances at various joints. And they have projections, often called processes, where the muscles attach. Also, there are depressions and openings for nerves and/or blood vessels.

The skeleton is divided into the axial and appendicular skeleton. Each has different types of bones with protuberances at joints and processes where the muscles attach.

Axial Skeleton

The **axial skeleton** lies in the midline of the body and consists of the skull, hyoid bone, vertebral column, and rib cage.

Skull

The **skull** is formed by the cranium, or braincase, and the facial bones. The categorization of these bones is somewhat arbitrary because, for example, the frontal bone forms the forehead of the face.

The *cranium* protects the brain and is composed of eight flat bones fitted tightly together in adults. In newborns, certain bones are not completely formed and instead are joined by membranous regions called **fontanels.** The fontanels usually close by the age of 16 months by the process of intramembranous ossification.

The bones of the cranium contain the **sinuses,** air spaces lined by mucous membrane, which reduce the weight of the skull and give a resonant sound to the voice. Two sinuses called the mastoid sinuses drain into the middle ear. *Mastoiditis,* a condition that can lead to deafness, is an inflammation of these sinuses.

The major bones of the cranium have the same names as the lobes of the brain: frontal, parietal, occipital, and temporal. On the top of the cranium (Fig. 10.5*a*), the **frontal bone** forms the forehead, the **parietal bones** extend to the sides, and the **occipital bone** curves to form the base of the skull. Here there is a large opening, the **foramen magnum** (Fig. 10.5*b*), through which the spinal cord passes and becomes the brain stem. Below the much larger parietal bones, each **temporal bone** has an opening that leads to the middle ear.

The **sphenoid bone,** which is shaped like a bat with wings outstretched, extends across the floor of the cranium from one side to the other. The sphenoid is considered to be the keystone bone of the cranium because all the other bones articulate with it. The sphenoid completes the sides of the skull and also contributes to forming the *orbits* (eye sockets). The **ethmoid bone,** which lies in front of the sphenoid, also helps form the orbits and the nasal septum. The orbits are completed by various facial bones. The term orbits refers to our ability to rotate the eyes.

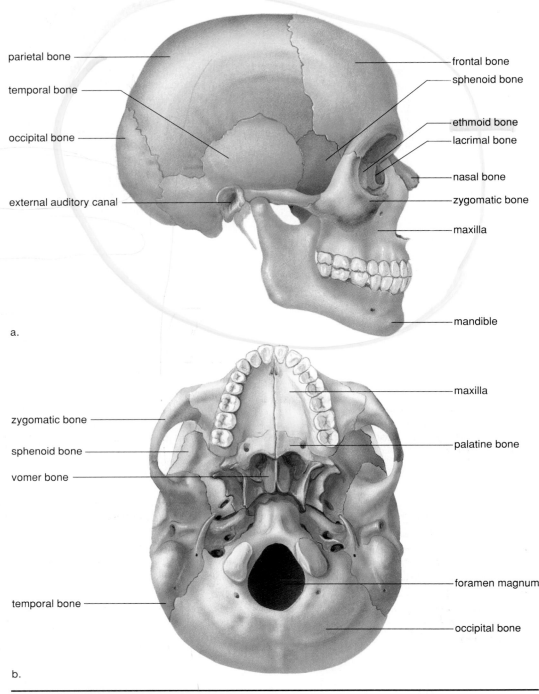

Figure 10.5   Bones of the skull.
a. Lateral view. **b.** Inferior view.

The cranium contains eight bones; the frontal, two parietal, the occipital, two temporal, the sphenoid, and the ethmoid.

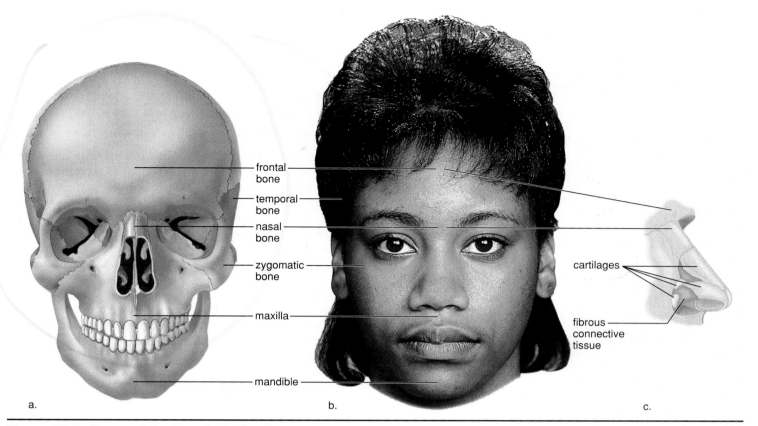

Figure 10.6   Bones of the face, including the nose.
a. The frontal bone forms the forehead and eyebrow ridges; the zygomatic bones form the cheekbones, and maxillae form the upper jaw. The maxillae are the most expansive facial bones, extending from the forehead to the lower jaw. The mandible has a projection we call the chin. **b.** The maxillae, frontal, and nasal bones help form the external nose. **c.** The rest of the nose is formed by cartilages and fibrous connective tissue.

Facial Bones

The most prominent of the *facial bones* are the mandible, maxillae (maxillary bones), the zygomatic bones, and the nasal bones.

The **mandible,** or lower jaw, is the only movable portion of the skull (Fig. 10.6*a*), and its action permits us to chew our food. It also forms the "chin." Tooth sockets are located on this bone and on the **maxillae** (maxillary bones), the upper jaw that also forms the anterior portion of the hard palate, the bony roof of the oral cavity. The palatine bones, which make up the posterior portion of the hard palate and the floor of the nasal cavity, cannot be seen externally.

The lips and cheeks have a core of skeletal muscle. The **zygomatic bones** are the cheekbone prominences, and the **nasal bones** form the bridge of the nose. The *lacrimal bone* contains the opening for the nasolacrimal canal, which passes from the orbits to the nasal cavity. The *vomer bone* is a part of the nasal septum.

Aside from the facial bones, the temporal and frontal bones, which are cranial bones, contribute to the face. The temples, which are the flattened spaces on each side of the forehead, are formed by the temporal bones. The frontal bone forms the forehead and has supraorbital ridges where the eyebrows are located. Glasses sit where the frontal bone joins the nasal bones.

While the ears are formed only by elastic cartilage and not by bone, the nose (Fig. 10.6*b*) is a mixture of bones and cartilages and fibrous connective tissue. The cartilages complete the tip of the nose and fibrous connective tissue forms the flared sides of the nose.

Among the facial bones, the mandible is the lower jaw where the chin is located, the two maxillae are the upper jaw, the two zygomatic bones are the cheekbones, and the two nasal bones form the bridge of the nose.

Hyoid Bone

Although the **hyoid bone** is not part of the skull, it will be mentioned here because it is a part of the axial skeleton. The larynx is the voice box at the top of the trachea in the neck region. The hyoid bone is located superior to the larynx, but it is the only bone in the body that does not articulate with another bone. It is attached to the temporal bones by muscles and ligaments. The hyoid bone anchors the tongue and serves as the site for the attachment of muscles associated with swallowing.

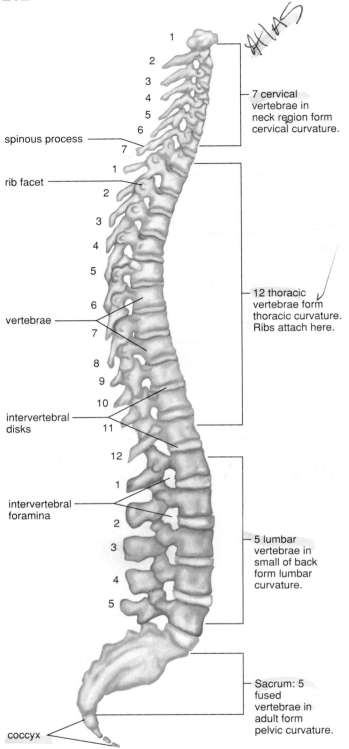

1

2

3

4

5

6

7   7 cervical vertebrae in neck region form cervical curvature.

spinous process

rib facet

vertebrae

1

2

3

4

5

6

7

8

9

10

11

12   12 thoracic vertebrae form thoracic curvature. Ribs attach here.

intervertebral disks

intervertebral foramina

1

2

3

4

5   5 lumbar vertebrae in small of back form lumbar curvature.

coccyx   Sacrum: 5 fused vertebrae in adult form pelvic curvature.

Figure 10.7   **The vertebral column.**
The vertebrae are named according to their location in the vertebral column, which is flexible due to the intervertebral disks. Note the presence of the coccyx, also called the tailbone.

Vertebral Column Supports

The **vertebral column** extends from the skull to the pelvis (the sacrum is a part of the pelvis) (Fig. 10.7). The vertebral column has many functions including:

- supports the head and trunk, allowing movement
- protects the spinal cord and nerves
- serves as a site for muscle attachment

Normally, the vertebral column has four curvatures that provide more resiliency and strength in an upright posture than a straight column could. As discussed in the introduction to this chapter, *scoliosis* is an abnormal lateral (sideways) curvature of the spine. There are two other well-known abnormal curvatures: *kyphosis* is an abnormal posterior curvature that often results in a hunchback, and *lordosis* is an abnormal anterior curvature resulting in a swayback.

The various vertebrae are named according to their location in the vertebral column (Fig. 10.7). The vertebrae join at articular *facets* and form a canal through which the spinal cord passes. The *spinous processes* of the vertebrae can be felt as bony projections along the midline of the back. The spinous processes and the transverse processes, which extend laterally, serve as attachment sites for the muscles that move the vertebral column.

The first *cervical vertebra*, called the *atlas*, holds up the head. It is so-named because Atlas, of Greek mythology, held up the world. Movement of the atlas permits the "yes" motion of the head. It also allows the head to tilt from side to side. The second cervical vertebra is called the *axis* because it allows a degree of rotation as when we shake the head "no." The bodies of the other cervical vertebrae are small compared to the thoracic and lumbar vertebrae (Fig. 10.8) The spinous process of the cervical vertebrae may be partially split, and arteries pass through the transverse foramen. The *thoracic vertebrae* have long, thin spinous processes, and they have an extra articular facet for the attachment of the ribs. *Lumbar vertebrae* have a large body and thick processes. The five *sacral vertebrae* are fused together in the sacrum, which is a part of the pelvic girdle. The *coccyx*, or tailbone, is also composed of fused vertebrae.

There are *intervertebral disks* composed of fibrocartilage between the vertebrae that act as a kind of padding. They prevent the vertebrae from grinding against one another and absorb shock caused by movements such as running, jumping, and even walking. The presence of the disks allows motion between vertebrae so that we can bend forward, backward, and from side to side. Unfortunately, these disks become weakened with age and can even slip and rupture. Pain will result if a slipped disk presses against the spinal cord and/or spinal nerves. If so, surgical removal of the disk may relieve the pain.

> The vertebral column consists of the vertebrae and serves as the backbone for the body. Disks between the vertebrae provide padding and account for flexibility of the column.

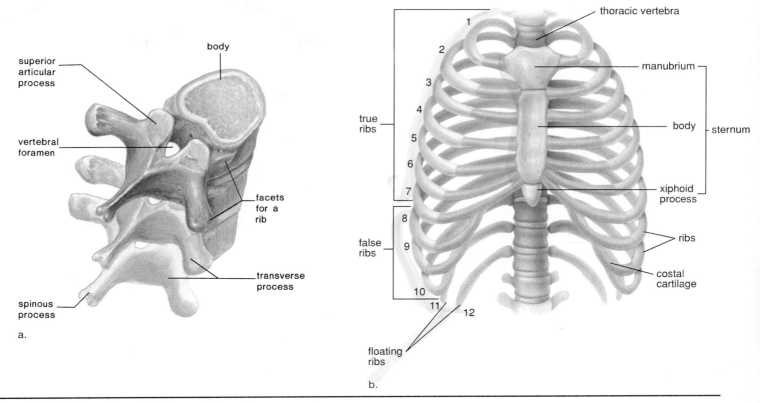

Figure 10.8 Thoracic vertebrae and the rib cage.
a. The thoracic vertebrae articulate with each other and with the ribs. A thoracic vertebrae has two facets for articulation with a rib; one is on the body, and the other is on the transverse process. **b.** The rib cage consists of the thoracic vertebrae, the ribs, the costal cartilages, and the sternum.

Rib Cage

The rib cage is composed of the thoracic vertebrae, the ribs and their associated cartilages, and the sternum (Fig. 10.8b).

The rib cage demonstrates how the skeleton is protective but also flexible. The rib cage protects the heart and lungs; yet it swings outward and upward upon inspiration and then downward and inward upon expiration.

Ribs

There are twelve pairs of **ribs.** All twelve pairs connect directly to the thoracic vertebrae in the back. A rib articulates with the body and transverse process of its corresponding thoracic vertebra. Each rib curves outward and then forward and downward.

In the front, ten pairs of ribs connect either directly or indirectly to the sternum via shafts of hyaline cartilages, called costal cartilages. The seven pairs of ribs that attach directly to the sternum are called true ribs. The three pairs of ribs that attach to the sternum by means of a common cartilage are called false ribs. The lower two pairs of ribs are known as "floating ribs" because they do not attach to the sternum. The lower ribs are not usually fractured because they are generally more flexible.

Sternum

The **sternum,** or breastbone, is a flat bone that has the shape of a blade. The sternum is attached to the ribs by coastal cartilages and helps them protect the heart and lungs. The sternum is actually composed of three bones that fuse during development. These bones are the manubrium, the body, and the xiphoid process.

An elevation called the *sternal angle* occurs where the *manubrium* joins with the body of the sternum. This is an important anatomical landmark because it occurs at the level of the second rib and therefore allows the ribs to be counted. Counting the ribs is sometimes used to determine where the apex of the heart is located. The apex of the heart is usually between the fifth and sixth ribs.

The *xiphoid process* is the third part of the sternum. Composed of hyaline cartilage in the child, it becomes ossified in the adult. The variably shaped xiphoid process serves as an attachment site for the diaphragm, the structure that divides the thoracic cavity from the abdominal cavity.

The rib cage, consisting of the thoracic vertebrae, the ribs, and the sternum, protects the heart and lungs in the thoracic cavity.

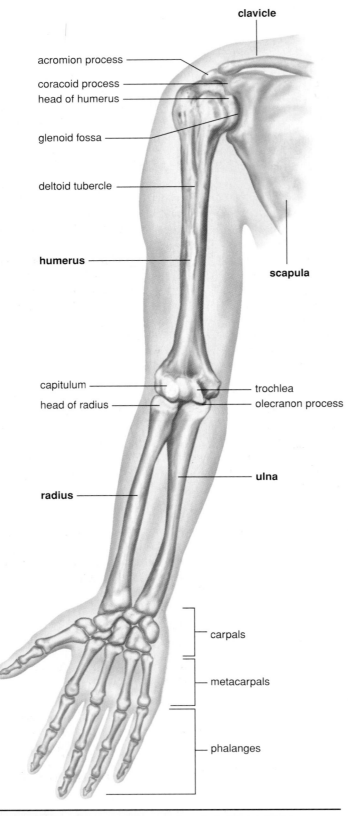

clavicle

acromion process

coracoid process

head of humerus

glenoid fossa

deltoid tubercle

humerus

scapula

capitulum

head of radius

trochlea

olecranon process

ulna

radius

carpals

metacarpals

phalanges

Figure 10.9   Bones of the pectoral girdle and arm.

Appendicular Skeleton

The **appendicular skeleton** consists of the bones within the pectoral and pelvic girdles and their attached limbs. The pectoral (shoulder) girdle and arm are specialized for flexibility; the pelvic girdle (hipbones) and legs are specialized for strength.

Pectoral Girdle and Arm

The **pectoral girdle** consists of the scapula (shoulder blade) and the clavicle (collarbone) (Fig. 10.9). The **clavicle** extends across the top of the thorax; it articulates with the *acromion process* of the scapula and also with the sternum. The **scapulae,** which are quite visible in the back, are held in place only by muscles. Muscles of the arm and chest attach to the *coracoid process* of the scapula. The **glenoid fossa** articulates with the head of the humerus. The glenoid fossa, a very shallow cavity, is much smaller than the *head of the humerus.* Although this means that the arm can move in almost any direction, there is little stability. Therefore, this is the joint that is most apt to dislocate. The components of the pectoral girdle are loosely linked together by ligaments, and this allows the girdle to follow freely the movements of the arm.

The **humerus,** the single long bone in the upper arm, has a smoothly rounded head that fits into the glenoid fossa of the scapula. The shaft of the humerus has a *tubercle* (protuberance) where the deltoid, the prominent muscle of the chest, attaches. Upon death, enlargement of this tubercle can be used as evidence that the person did a lot of heavy lifting.

The far end of the humerus has two protuberances called the *capitulum* and the *trochlea,* which articulate respectively with the **ulna** and the **radius,** at the elbow. The "funny bone" of the elbow is the *olecranon process* of the ulna.

When the arm is held so that the palm is turned frontward, the radius and ulna are about parallel to one another. When the arm is turned so that the palm is next to the body, the radius crosses in front of the ulna, a feature that contributes to the easy twisting motion of the forearm.

The hand has many bones, and this increases its flexibility. The wrist has eight **carpal** bones, which look like small pebbles. From these, five **metacarpal** bones fan out to form a framework for the palm. The metacarpal bone that leads to the thumb is placed in such a way that the thumb can reach out and touch the other digits. (**Digits** is a term that refers to either fingers or toes.) Your knuckles are the enlarged distal ends of the metacarpals. Beyond the metacarpals are the **phalanges,** the bones of the fingers and the thumb. The phalanges of the hand are long, slender, and lightweight.

The pectoral girdle and arm are specialized for flexibility of movement.

Pelvic Girdle and Leg

Figure 10.10 shows how the leg is attached to the pelvic girdle.

The **pelvic (hip) girdle** consists of two heavy, large coxal bones (hipbones) as well as the sacrum and coccyx. The strong bones of the pelvic girdle bear the weight of the body, protect the organs within the pelvic cavity, and serve as the place of attachment for the legs.

Each **coxal bone** has three parts: the ilium, the ischium, and the pubis, which are fused in the adult (Fig. 10.10). The hip socket, called the *acetabulum,* occurs where these three bones meet. The *ilium* is the largest part of the coxal bones, and our hips occur where it flares out. We sit on the *ischium,* which has a posterior spine called the ischial spine that projects into the pelvic cavity. The *pubis* (referring to pubic hair) is the anterior part of a coxal bone. The two pubic bones are joined together by a fibrocartilage disk at the pubic symphysis.

The male and female pelvis differ from one another. In the female, the iliac bones are more flared; the pelvic cavity is more shallow, but the outlet is wider. These adaptations facilitate giving birth.

The **femur** (thighbone) is the longest and strongest bone in the body. The head of the femur articulates with the coxal bones at the acetabulum, and the short *neck* better positions the legs for walking. The femur has two large processes, the *greater* and *lesser trochanter,* which are places of attachment for the muscles of the legs and buttocks. At its distal end, the femur has a *medial* and *lateral condyle* that articulate with the tibia of the lower leg. This is the region of the knee, and the **patella,** or kneecap, is held in place here by tendons that attach to the *tibial tuberosity.* At the distal end, the *medial malleolus* of the tibia causes the inner bulge of the ankle. The **fibula** is the more slender bone in the lower leg. The fibula has a head that articulates with the tibia and a distal *lateral malleolus* that forms the outer bulge of the ankle.

Each foot has an ankle, an instep, and five toes. The many bones of the foot give it considerably more flexibility, especially on rough surfaces. The ankle contains seven **tarsal** bones, one of which (the talus) can move freely where it joins the tibia and fibula. Strange to say, the *calcaneus,* or heel bone, is also considered to be part of the ankle. The talus and calcaneus support the weight of the body.

The instep has five elongated **metatarsal** bones. The distal end of the metatarsals form the ball of the foot. If the ligaments that bind the metatarsals together become weakened, flat feet are apt to result. The bones of the toes are called **phalanges,** just like those of the fingers, but in the foot, the phalanges are stout and extremely sturdy.

The pelvic girdle and leg are adapted to supporting the weight of the body. The femur is the longest and strongest bone in the body.

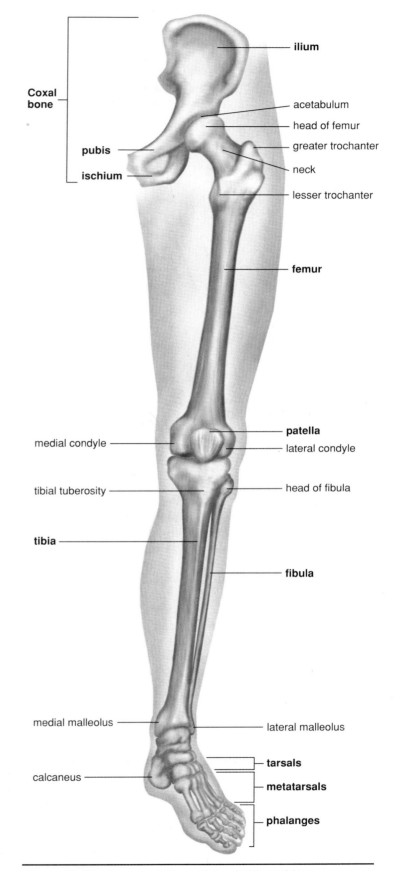

Figure 10.10   Bones of the pelvic girdle and leg.

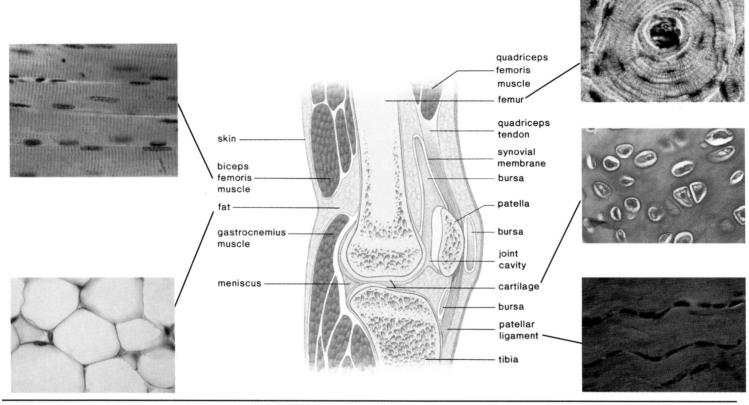

Figure 10.11 Knee joint.
The knee joint is a synovial joint. Notice the cavity between the bones, which is encased by ligaments and lined by synovial membrane. The patella (kneecap) serves to guide the quadriceps tendon over the joint when flexion or extension occurs.

10.4 Articulations

Bones are joined at the joints, which are classified as fibrous, cartilaginous, and synovial. Fibrous joints, such as the **sutures** between the cranial bones, are *immovable*. Cartilaginous joints are connected by hyaline cartilage as in the costal cartilages that join the ribs to the sternum or by fibrocartilage as seen in the intervertebral disks. Cartilaginous joints are *slightly movable*.

In *freely movable* **synovial joints,** the two bones are separated by a cavity. **Ligaments** composed of fibrous connective tissue hold the two bones in place as they form a capsule. **Tendons,** which bind muscles to bones, often help stabilize joints. The joint capsule is lined by a **synovial membrane,** which produces *synovial fluid,* a lubricant for the joint. The knee is an example of a synovial joint (Fig. 10.11). Aside from articular cartilage, the knee contains **menisci** (sing., meniscus), crescent-shaped pieces of hyaline cartilage between the bones. These give added stability and act as shock absorbers. Unfortunately, athletes often suffer injury of the menisci, known as torn cartilage. The knee joint also contains 13 fluid-filled sacs called bursae (sing., **bursa**), which ease friction between tendons and ligaments. Inflammation of the bursae is called *bursitis.* Tennis elbow is a form of bursitis.

There are different types of movable joints. The knee and elbow joints are **hinge joints** because, like a hinged door, they largely permit movement in one direction only. More movable are the **ball-and-socket joints;** for example, the ball of the femur fits into a socket on the hipbone. Ball-and-socket joints allow movement in all planes and even a rotational movement. The various movements of body parts at synovial joints are depicted in Figure 10.12.

Synovial joints are subject to *arthritis.* In *rheumatoid arthritis,* the synovial membrane becomes inflamed and grows thicker. Degenerative changes take place that make the joint almost immovable and painful to use. Evidence indicates that these effects are brought on by an autoimmune reaction. In *osteoarthritis,* the cartilage at the ends of the bones disintegrates so that the two bones become rough and irregular.

Joints are regions of articulations between bones. Synovial joints are freely movable and allow particular types of movements. Unfortunately, synovial joints are subject to various disorders.

10.5 Working Together

The Working Together box on page 218 shows how the skeletal system works with the other systems of the body to maintain homeostasis.

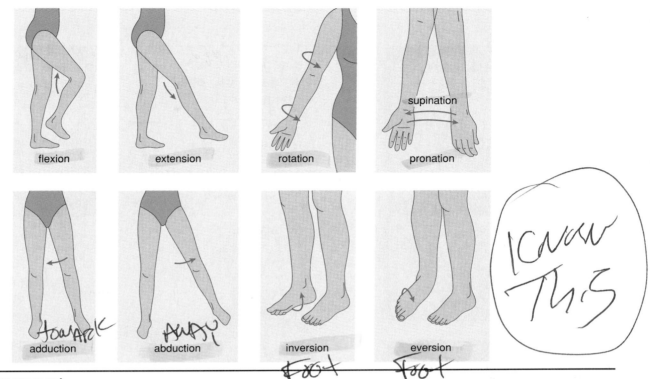

Figure 10.12   Joint movement.
Flexion decreases and extension increases the joint angle. Rotation is movement of a body part around its own axis. Supination is rotation of the lower arm so that the palm is upward; pronation is the opposite. Abduction is movement away from and adduction is movement toward the midline. Inversion is turning the foot so that the sole is inward, and eversion is the opposite.

SUMMARY

10.1 Tissues of the Skeletal System

The organs of the skeletal system are largely composed of connective tissue, particularly bone, cartilage, and fibrous connective tissue. Hyaline (articular) cartilage covers the ends while periosteum (fibrous connective tissue) covers the rest of a long bone. Spongy bone, which may contain red bone marrow, is in the epiphyses, and yellow bone marrow is in the medullary cavity of the diaphysis. The wall of the diaphysis is compact bone.

10.2 Bone Growth and Repair

Bone is a living tissue that can grow and repair. The prenatal human skeleton is at first cartilaginous, but it is later replaced by a bony skeleton. During adult life, bone is constantly being broken down by osteoclasts and then rebuilt by osteoblasts that become osteocytes in lacunae. Repair of a fracture requires four steps: (1) hematoma, (2) fibrocartilage callus, (3) bony callus, and (4) remodeling.

10.3 Bones of the Skeleton

The skeleton supports and protects the body, permits flexible movement, produces blood cells, and serves as a storehouse for mineral salts, particularly calcium phosphate.

The axial skeleton lies in the midline of the body and consists of the skull, the hyoid bone, the vertebral column, and the rib cage. The skull contains the cranium, which protects the brain and the facial bones. Considering the face, the frontal bone forms the forehead, the maxillae extends from the frontal bone to the upper row of teeth; the zygomatic bones are the cheekbones, and the nasal bones form the bridge of the nose. The rest of the nose is hyaline cartilage, except the flares are fibrocartilage. The ears are elastic cartilage. The mandible is the lower jaw and accounts for the chin.

The appendicular skeleton consists of the bones of the pectoral girdle, arms, pelvic girdle, and legs. The pectoral girdle and arms are adapted for flexibility. The glenoid fossa of the scapula is barely large enough to receive the head of the humerus, and this means that dislocated shoulders are more likely to happen than dislocated hips. The pelvic girdle and the legs are adapted for strength: the femur is the strongest bone in the body. However, the foot, like the hand, is flexible because it contains so many bones. Both fingers and toes are digits, and the term phalanges refers to the bones of each.

10.4 Articulations

There are three types of joints: fibrous, like the sutures of the cranium, are immovable; cartilaginous, like those between the ribs and sternum and the pubic symphysis, are slightly movable; and the synovial joints, consisting of a membrane-lined (synovial membrane) capsule, are freely movable. There are different kinds of synovial joints, and the movements they permit are varied.

HUMAN SYSTEMS WORK TOGETHER

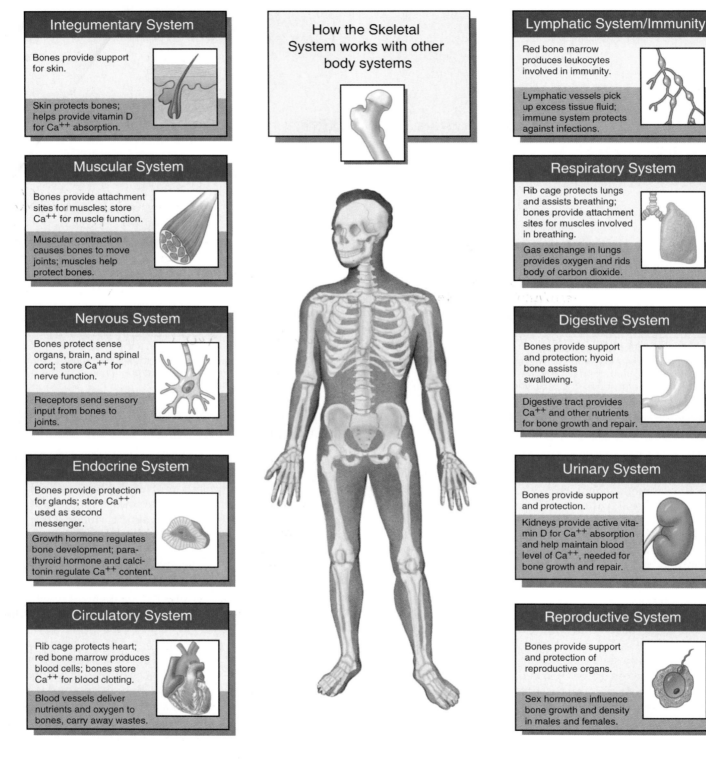

Integumentary System

Bones provide support for skin.

Skin protects bones; helps provide vitamin D for Ca^{++} absorption.

Muscular System

Bones provide attachment sites for muscles; store Ca^{++} for muscle function.

Muscular contraction causes bones to move joints; muscles help protect bones.

Nervous System

Bones protect sense organs, brain, and spinal cord; store Ca^{++} for nerve function.

Receptors send sensory input from bones to joints.

Endocrine System

Bones provide protection for glands; store Ca^{++} used as second messenger.

Growth hormone regulates bone development; parathyroid hormone and calcitonin regulate Ca^{++} content.

Circulatory System

Rib cage protects heart; red bone marrow produces blood cells; bones store Ca^{++} for blood clotting.

Blood vessels deliver nutrients and oxygen to bones, carry away wastes.

How the Skeletal System works with other body systems

Lymphatic System/Immunity

Red bone marrow produces leukocytes involved in immunity.

Lymphatic vessels pick up excess tissue fluid; immune system protects against infections.

Respiratory System

Rib cage protects lungs and assists breathing; bones provide attachment sites for muscles involved in breathing.

Gas exchange in lungs provides oxygen and rids body of carbon dioxide.

Digestive System

Bones provide support and protection; hyoid bone assists swallowing.

Digestive tract provides Ca^{++} and other nutrients for bone growth and repair.

Urinary System

Bones provide support and protection.

Kidneys provide active vitamin D for Ca^{++} absorption and help maintain blood level of Ca^{++}, needed for bone growth and repair.

Reproductive System

Bones provide support and protection of reproductive organs.

Sex hormones influence bone growth and density in males and females.

10.5 Working Together
The skeletal system works with the other systems of the body in the ways described in the box on page 218.

STUDYING THE CONCEPTS

1. Bones are organs composed of what types of tissues? What are some differences between compact bone and spongy bone?   204

2. Describe the makeup of long bone.   204–5

3. What are the two types of ossification? Describe endochondral ossification.   206

4. How does a long bone grow in children, and how is it remodeled in all age groups?   206–7

5. What are the four steps required for fracture repair?   207

6. What are the bones of the cranium and the face? Associate the parts of a face with particular bones or with cartilages.   210–11

7. What are the parts of the vertebral column, and what are its curvatures? Distinguish between the atlas, axis, sacrum, and coccyx.   212

8. What are the bones of the rib cage, and what are several of its functions?   213

9. What are the bones of the pectoral girdle? Give examples to demonstrate the flexibility of the pectoral girdle. Where does the scapula articulate with the clavicle?   214

10. What are the bones of the arm? Name several processes of the humerus; of the elbow.   214

11. What are the bones of the pelvic girdle, and what are their functions? Give examples to demonstrate the strength and stability of the pelvic girdle.   215

12. What are the bones of the leg? Name several processes of the femur; of the ankle.   215

13. How are joints classified? Give examples of the different types of synovial joints and the movements they permit.   216

TESTING YOUR KNOWLEDGE

Match the items in the key to the bones given in questions 1–6.

Key: a. forehead
   b. chin
   c. cheekbone
   d. elbow
   e. shoulder blade
   f. hip
   g. ankle

1. zygomatic
2. tibia and fibula
3. frontal bone
4. humerus
5. coxal bone
6. scapula

Match the items in the key to the bones listed in questions 7–12.

Key: a. glenoid fossa
   b. olecranon process
   c. acetabulum
   d. spinous process
   e. greater and lesser trochanter
   f. xiphoid process

7. femur
8. scapula
9. ulna
10. coxal bone
11. sternum
12. vertebra

Fill in the blanks.

13. Compact bone contains tubular units called _____.

14. The epiphysis of a long bone contains _____ bone, where blood cells are produced.

15. During bone remodeling, cells called _____ remove breakdown bone and return calcium to the blood.

16. The _____ are air-filled spaces in the cranium.

17. The sacrum is a part of the _____, and the sternum is a part of the _____.

18. The pectoral girdle is specialized for _____ while the pelvic girdle is specialized for _____.

19. The term phalanges refers to the bones in both the _____ and _____.

20. The knee is a freely movable joint of the _____ type.

APPLYING YOUR KNOWLEDGE

Concepts

1. Osteoporosis is a common disease, affecting 50% of women over the age of 45 and 90% of women over the age of 75. It is due to loss of bone mass. How is bone mass decreased?

2. A friend of mine was having severe neck and back pains. A surgeon at the Mayo Hospital inserted a metal rod alongside the cervical vertebrae and fused three cervical vertebrae. This took away the pain. What was her condition that warranted this operation?

3. The knee is a weak spot in many sports, but particularly in football. Explain the reason for this.

Bioethical issue

When a person constantly repeats tiny movements—say, typing on a keyboard, a painful condition called carpal tunnel syndrome (CTS) may develop. CTS occurs because tendons in the wrist's "carpal tunnel," a ring of bones, swell and pinch the nerve inside. As a result, the hand goes numb or prickly with pain.

Some people define CTS as an occupational hazard, arguing that employers should pay for doctors' visits or, in extreme cases, surgery to correct the condition. Some employers believe that CTS is not their responsibility because the condition can be avoided by resting the hands or frequently changing position. Should a company pay for an employee's CTS treatment? Or is it an employee's responsibility to avoid CTS?

APPLYING TECHNOLOGY

Your study of cell structure and function is supported by these available technologies:

Exploring the Internet

The Mader Home Page provides further resources for studying this chapter.

http://www.mhhe.com/sciencemath/biology/mader/

(Click on *Human Biology*.)

 ### Dynamic Human: Skeletal System CD-ROM

In *Anatomy*, the skeleton is reviewed with the cranium and thorax presented in 3-dimension; in *Explorations*, cross section of a long bone, types of joints, and a walking skeleton are shown; in *Histology*, movements of a synovial joint occur, and microscopic slides are highlighted to reveal various tissue; and in *Clinical Concepts*, arthroscopy, X rays of a dislocated shoulder and a fractured femur, an MRI of the knee, and a longitudinal section of a herniated disk are all included.

SELECTED KEY TERMS

appendicular skeleton (ap-un-DIK-yuh-ler) Portion of the skeleton forming the pectoral girdle and arms, and the pelvic girdle and legs.   214

articular cartilage Hyaline cartilaginous covering over the articulating surface of the bones of synovial joints.   205

axial skeleton (AK-see-ul) Portion of the skeleton that lies in the midline: the skull, the vertebral column, and the rib cage.   210

bursa Saclike, fluid-filled structure, lined with synovial membrane, that occurs near a joint.   216

compact bone Hard bone consisting of osteons (Haversian systems) cemented together by a matrix that contains calcium phosphate.   204

epiphyseal plate Cartilaginous layer within the epiphysis of a long bone that functions as a growing region.   206

fontanel Membranous region located between certain cranial bones in the skull of a fetus or infant.   210

joint Union of two or more bones; an articulation.   204

ligament Fibrous connective tissue that joins bone to bone at a joint.   216

medullary cavity Cavity within the diaphysis of a long bone occupied by yellow bone marrow.   205

menisci (sing., meniscus) (mun-NIS-ky) Cartilaginous wedges that separate the surfaces of bones in synovial joints.   216

osteoblast Bone-forming cell.   206

osteoclast Cell that breaks down and causes reabsorption of bone.   206

osteocyte Mature bone cell.   206

pectoral girdle Portion of the skeleton to which the arms are attached.   214

pelvic girdle Portion of the skeleton to which the legs are attached.   215

red bone marrow Blood-cell-forming tissue located in the spaces within spongy bone.   204

spongy bone Porous bone found at the ends of long bones where blood cells are formed.   204

synovial joint Freely movable joint.   216

vertebral column Backbone of spine through which the spinal cord passes.   212

Chapter 11

Muscular System

Chapter Outline

Figure 11.1   How important is a smile?
Facial expressions are dependent on muscles that allow us to open our mouths, wink our eyes, and of course, smile.

In December 1995, when most kids were dreaming of new toys, sugar cookies, and other holiday treats, 7-year-old Chelsey Thomas found herself hoping for just one thing—the chance to smile.

Chelsey was born with a rare disorder called Moebius syndrome, which causes paralysis of facial muscles. All through childhood—as her friends giggled and grinned—Chelsey could not force even the smallest of smiles. She stubbornly managed to stay happy. But it was hard. And Chelsey's parents worried that she would eventually become depressed and withdrawn.

So on the morning of December 15, after much preparation and talk with doctors, Chelsey and her parents drove to a medical center in California. There, a team of surgeons began a 10-hour operation designed to give Chelsey her smile.

Moving cautiously, doctors extracted a strip of muscle, vein, and artery from Chelsey's thigh. Making an incision along her cheek, the team carefully sewed the extractions into place along Chelsey's jaw. The tiny sutures, thinner than a human hair, were sewn while viewing them under a microscope.

The painstaking surgery allowed the transplanted muscle in the left side of Chelsey's face to work. She gave a beautiful, if lopsided, smile. Then, in April 1996, doctors performed the same surgery on the right side of her face. Finally, her long-awaited smile was complete. Today, Chelsey grins nonstop (Fig. 11.1). Her newfound smile dazzles her family and helps her fit in with kids at school.

Researchers are now learning to perform many kinds of muscle and nerve transplants. These operations can save or improve the use of a patient's arms or legs, for example. As knowledge of the muscular system improves, transplants will improve and become much easier and cheaper to perform. And that, say doctors, is something everyone can smile about.

All the activities of the body from smiling to balancing on tiptoe are dependent on muscles whose sole ability is to contract. That's why surgeons could take a muscle from the leg and transplant it to the face to allow Chelsey to smile. The location and function of certain well-known muscles are reviewed in this chapter, which concentrates on the structure of skeletal muscles from the macroscopic to the microscopic level. When muscles contract, two protein filaments slide past one another within individual muscle fibers. Much is known about the interaction of these filaments, and we will examine in detail how muscle contraction is activated.

Muscle contraction requires the presence of ATP, which is produced most efficiently when oxygen is present. Some muscle fibers, however, are adapted to anaerobic conditions, and these fibers may be associated with particular sports that require endurance, like weight lifting. Muscle anatomy and physiology are of extreme interest to sports enthusiasts, and this chapter emphasizes sports-related topics.

11.1 Skeletal Muscles

This chapter is primarily concerned with skeletal muscles—those muscles that make up the bulk of the human body. Muscles are covered by several layers of fibrous connective tissue called fascia, which extends beyond the muscle to become its **tendon** (Fig. 11.2). Fascia also surrounds fascicles, which are bundles of muscle fibers, and separates the muscle fibers within a fascicle. A muscle fiber contains many contractile elements called myofibrils. Myofibrils run the length of a muscle fiber, and the striations seen in observations of muscular tissue are due to the placement of myofilaments within sarcomeres, the units of a myofibril. There are thin (actin) filaments and thick (myosin) filaments in the myofibril.

Skeletal muscles are attached to the skeleton, and their contraction causes the movement of bones at a joint. When muscles contract, one bone remains fairly stationary, and the other one moves. The **origin** of a muscle is on the stationary bone, and the **insertion** of a muscle is on the bone that moves.

Frequently, a body part is moved by a group of muscles working together. Even so, one muscle does most of the work and is called the *prime mover*. The assisting muscles are called the *synergists*. When muscles contract, they shorten; therefore, muscles can only pull; they cannot push.

Muscles have antagonists, and *antagonistic pairs* work opposite to one another to bring about movement in op-

posite directions. For example, the biceps brachii and the triceps brachii are antagonists; one flexes the forearm, and the other extends the forearm (Fig. 11.3).

When muscles cooperate to achieve movement, some act as prime movers, others are synergists, and still others are antagonists.

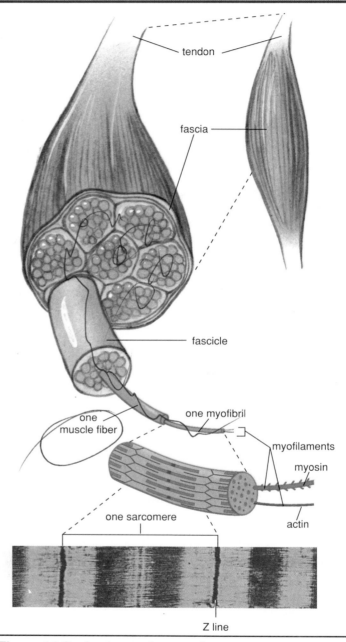

Figure 11.2 Anatomy of a muscle.
A fascia covers the surface of a muscle and contributes to the tendon, which attaches the muscle to a bone. Fascia also separates bundles of muscle fibers that contain contractile elements called myofibrils. The arrangement of myofilaments within a myofibril accounts for skeletal muscle striations. A sarcomere is a contractile unit of a myofibril.

TABLE 11.1
Muscles (anterior view)

Name	Function
Head and neck	
Frontalis	Wrinkles forehead and lifts eyebrows
Orbicularis oculi	Closes eye (winking)
Zygomaticus	Raises corner of mouth (smiling)
Masseter	Closes jaw
Orbicularis oris	Closes and protrudes lips (kissing)
Arms and trunk	
External oblique	Compresses abdomen; rotates trunk
Rectus abdominis	Flexes spine
Pectoralis major	Flexes and adducts shoulder and arm ventrally (pulls arm across chest)
Deltoid	Abducts and raises arm at shoulder joint
Biceps brachii	Flexes forearm and supinates hand
Leg	
Adductor longus	Adducts thigh
Iliopsoas	Flexes thigh
Sartorius	Rotates thigh (sitting cross-legged)
Quadriceps femoris group	Extends lower leg
Peroneus longus	Everts foot
Tibialis anterior	Dorsiflexes and inverts foot
Flexor and extensor digitorum longus	Flex and extend toes

TABLE 11.2
Muscles (posterior view)

Name	Function
Head and neck	
Occipitalis	Moves scalp backward
Sternocleidomastoid	Turns head to side; flexes neck and head
Trapezius	Extends head; raises and adducts shoulders dorsally (shrugging shoulders)
Arm and trunk	
Latissimus dorsi	Extends and adducts shoulder and arm dorsally (pulls arm across back)
Deltoid	Abducts and raises arm at shoulder joint
External oblique	Rotates trunk
Triceps brachii	Extends forearm
Flexor and extensor carpi group	Flex and extend hand
Flexor and extensor digitorum	Flex and extend fingers
Buttocks and legs	
Gluteus medius	Abducts thigh
Gluteus maximus	Extends thigh (forms buttocks)
Hamstring group	Flexes lower leg
Gastrocnemius	Extends foot (tiptoeing)

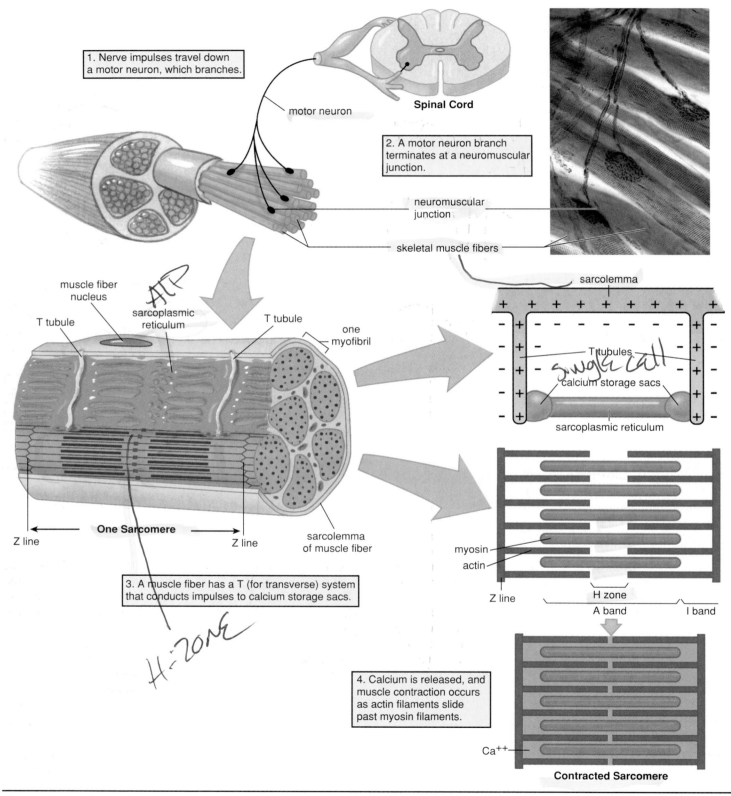

1. Nerve impulses travel down a motor neuron, which branches.

motor neuron

Spinal Cord

2. A motor neuron branch terminates at a neuromuscular junction.

neuromuscular junction

skeletal muscle fibers

sarcolemma

T tubules

calcium storage sacs

sarcoplasmic reticulum

muscle fiber nucleus

ATP

sarcoplasmic reticulum

T tubule

T tubule

one myofibril

single cell

Z line

One Sarcomere

Z line

sarcolemma of muscle fiber

myosin

actin

Z line

H zone

A band

I band

3. A muscle fiber has a T (for transverse) system that conducts impulses to calcium storage sacs.

H-ZONE

4. Calcium is released, and muscle contraction occurs as actin filaments slide past myosin filaments.

Ca^{++}

Contracted Sarcomere

Figure 11.5   Contraction of a muscle.
The boxed statements explain the sequence of events that lead up to contraction of sarcomeres within myofibrils.

11.2 Mechanism of Muscle Fiber Contraction

A series of events leads up to muscle fiber contraction. After discussing these events, we will examine a neuromuscular junction and filament sliding in detail.

Overview of Muscular Contraction

Figure 11.5 shows the steps that lead to muscle contraction. Muscles are stimulated to contract by nerve impulses that begin in the brain or spinal cord. These nerve impulses travel down a motor neuron to a neuromuscular junction, a region where a motor neuron fiber meets a muscle fiber.

Each *muscle fiber* is a cell containing the usual cellular components, but special names have been assigned to some of these components. The plasma membrane is called the **sarcolemma,** the cytoplasm is the *sarcoplasm*, and the endoplasmic reticulum is the *sarcoplasmic reticulum*. A muscle fiber also has some unique anatomical characteristics. For one thing, it has a *T* (for transverse) system; the sarcolemma forms *T tubules* that penetrate, or dip down, into the cell so that they come into contact—but do not fuse—with expanded portions of the sarcoplasmic reticulum. The expanded portions of the sarcoplasmic reticulum, called *calcium-storage sacs,* contain calcium ions (Ca^{++}), which are essential for muscle contraction. The sarcoplasmic reticulum encases hundreds and sometimes even thousands of **myofibrils,** which are the contractile portions of the fibers.

Myofibrils and Sarcomeres

Myofibrils are cylindrical in shape and run the length of the muscle fiber. The light microscope shows that muscle fibers have light and dark bands called *striations* (Fig. 11.6). The electron microscope shows that the striations of myofibrils are formed by the placement of myofilaments within contractile units called **sarcomeres.** A sarcomere extends between two dark lines called the Z lines. A sarcomere contains two types of protein myofilaments. The thick filaments are made up of a protein called **myosin,** and the thin filaments are made up of a protein called **actin.** The I band is light colored because it contains only actin filaments attached to a Z line. The dark regions of the A band contain overlapping actin and myosin filaments, and its H zone has only myosin filaments.

Sliding Filaments

Impulses generated at a neuromuscular junction travel down a T tubule to the calcium storage sacs, and calcium is released into the muscle fiber. Now the muscle fiber contracts as the sarcomeres within the myofibrils shorten. When a sarcomere shortens, the actin (thin) filaments slide past the myosin (thick) filaments and approach one another. This causes the I band to shorten and the H zone to almost or

TABLE 11.3	
Muscle Contraction	
Name	**Function**
Actin filaments	Slide past myosin, causing contraction
Ca^{++}	Needed for myosin to bind the actin
Myosin filaments	Pull actin filaments by means of cross-bridges; are enzymatic and split ATP
ATP	Supplies energy for muscle contraction

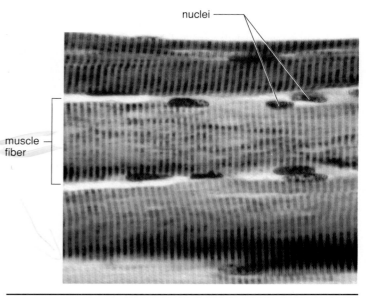

Figure 11.6   Light micrograph of skeletal muscle.
The striations of skeletal muscle tissue are produced by alternating dark A bands and light I bands. See the sarcomere in Figure 11.5.

completely disappear. The movement of actin filaments in relation to myosin filaments is called the **sliding filament theory** of muscle contraction. During the sliding process, the sarcomere shortens even though the filaments themselves remain the same length.

The participants in muscle contraction have the functions listed in Table 11.3. ATP supplies the energy for muscle contraction. Although the actin filaments slide past the myosin filaments, it is the myosin filaments that do the work. Myosin filaments break down ATP and have cross-bridges that pull the actin filaments toward the center of the sarcomere.

Muscle fibers are innervated, and when they are stimulated to contract, myofilaments slide past one another causing sarcomeres to shorten.

At the Neuromuscular Junction

Muscle fibers are stimulated to contract by motor neurons. Each motor nerve fiber has several branches, and each branch ends in a synaptic bulb. The region where a synaptic bulb lies in close proximity to the sarcolemma of a muscle fiber is called a **neuromuscular junction** (Fig. 11.7).

A synaptic bulb contains synaptic vesicles that are filled with a substance called a neurotransmitter. There is a small gap called the synaptic cleft between the synaptic bulb and the sarcolemma of a muscle fiber. The neurotransmitter plays a role in transmitting a signal to contract across this cleft between motor neuron and muscle fiber.

When nerve impulses traveling down a motor neuron arrive at the synaptic bulb, the synaptic vesicles release a neurotransmitter called acetylcholine (ACh) into the synaptic cleft. ACh quickly diffuses to and binds with their receptor sites on the sarcolemma. Now the sarcolemma generates impulses that spread over the sarcolemma and down the T system to the region of calcium (Ca^{++}) storage sacs. The release of calcium causes the myofibrils within a muscle fiber to contract as discussed previously.

> At a neuromuscular junction, nerve impulses bring about the release of a neurotransmitter substance that signals a muscle fiber to contract.

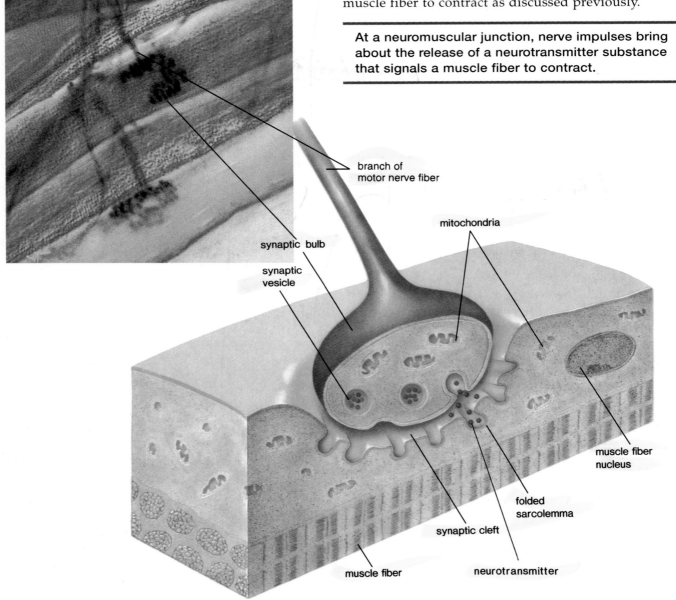

Figure 11.7   Neuromuscular junction.
The branch of a motor nerve fiber terminates in a synaptic bulb that meets but does not touch a muscle fiber. A synaptic cleft separates the synaptic bulb from the sarcolemma of the muscle fiber. Nerve impulses traveling down a motor neuron cause synaptic vesicles to discharge a neurotransmitter, which diffuses across the synaptic cleft. When the neurotransmitter is received by the sarcolemma of a muscle fiber, contraction follows.

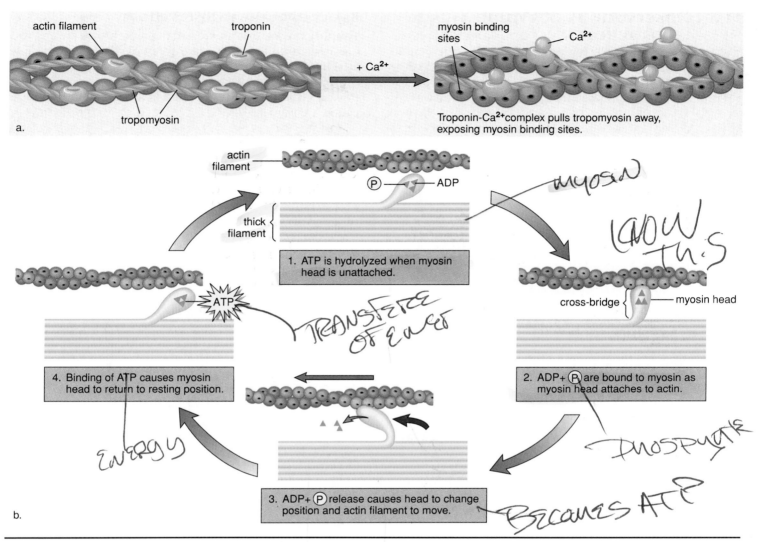

Figure 11.8 The role of calcium and myosin in muscle contraction.
a. Upon release, calcium binds to troponin, exposing myosin binding sites. **b.** After breaking down ATP, myosin heads bind to an actin filament, and later, a power stroke causes the actin filament to move.

As Contraction Occurs

Figure 11.8 shows the placement of two other proteins associated with a thin filament, which is composed of a double row of twisted actin molecules. Threads of *tropomyosin* wind about an actin filament, and *troponin* occurs at intervals along the threads. Calcium ions that have been released from their storage sac combine with troponin. After binding occurs, the tropomyosin threads shift their position, and myosin binding sites are exposed.

The thick filament is actually a bundle of myosin molecules, each having a double globular head with an ATP binding site. The heads function as ATPase enzymes, splitting ATP into ADP and $\textcircled{P}$. This reaction activates the head so that it will bind to actin. The ADP and $\textcircled{P}$ remain on the myosin heads until the heads attach to actin, forming a *cross-bridge*. Now, ADP and $\textcircled{P}$ are released, and this

causes the cross-bridges to change their positions. This is the *power stroke*, that pulls the thin filaments toward the middle of the sarcomere. When another ATP molecule binds to a myosin head, the cross-bridge is broken as the head detaches from actin. The cycle begins again; the actin filaments move nearer the center of the sarcomere each time the cycle is repeated.

Contraction continues until nerve impulses cease and calcium ions are returned to their storage sacs. The membranes of the sarcoplasmic reticulum contain active transport proteins that pump calcium ions back into the storage sacs.

Myosin filament heads break down ATP and then attach to an actin filament, forming cross-bridges that pull the actin filament to the center of a sarcomere.

11.3 Observations of Whole Muscle Contraction

It is possible to study the contraction of individual whole muscles in the laboratory, and certain fundamental principles have been discovered. These principles can be applied to the contraction of muscles in the human body.

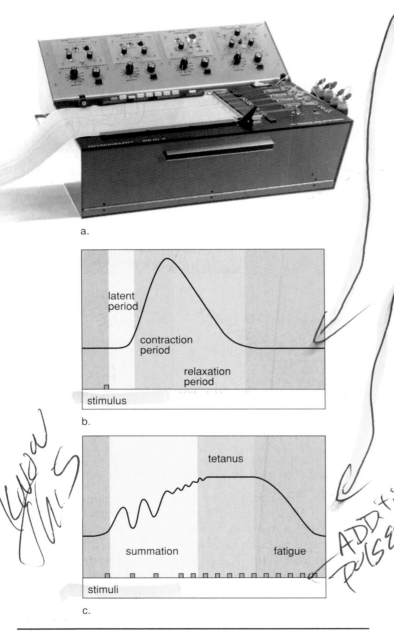

Figure 11.9   Physiology of skeletal muscle contraction.
a. A Physiograph is an apparatus used to record a myogram, a visual representation of the contraction of a muscle that has been dissected from an animal. **b.** Simple muscle twitch is composed of three periods: latent, contraction, and relaxation. **c.** Summation and tetanus. When stimulation frequency increases, the muscle does not relax completely between stimuli, and the contraction gradually increases in intensity. The muscle becomes maximally contracted until it fatigues.

Basic Laboratory Observations

To study muscle contraction in the laboratory, the gastrocnemius (calf muscle) is usually removed from a frog and attached to a movable lever. The muscle is stimulated, and the mechanical force of contraction is recorded by a Physiograph® (Fig. 11.9a). The resulting visual pattern is called a *myogram.*

Muscle Twitch

At first, the stimulus may be too weak to cause a contraction, but as soon as the strength of the stimulus reaches a threshold stimulus, the muscle contracts and then relaxes. This action—a single contraction that lasts only a fraction of a second—is called a *muscle twitch.* Figure 11.9b is a myogram of a twitch, which is customarily divided into the latent period, or the period of time between stimulation and initiation of contraction; the contraction period, when the muscle shortens; and the relaxation period, when the muscle returns to its former length.

Stimulation of an individual fiber within a muscle usually results in a maximal, *all-or-none contraction.* But the contraction of a whole muscle, as evidenced by the size of a muscle twitch, can vary in strength depending on the number of fibers contracting.

Summation and Tetanus

If a muscle is given a rapid series of threshold stimuli, it can respond to the next stimulus without relaxing completely. In this way, muscle contraction *summates* until maximal sustained contraction, called **tetanus,** is achieved (Fig. 11.9c). The myogram no longer shows individual twitches; rather, they are fused and blended completely into a straight line. Tetanus continues until the muscle fatigues due to depletion of energy reserves. *Fatigue* is apparent when a muscle relaxes even though stimulation continues.

Muscle Tone in the Body

Tetanic contractions ordinarily occur in the body's muscles whenever skeletal muscles are actively used. But while some fibers are contracting, others are relaxing. Because of this, intact muscles rarely fatigue completely. Even when muscles appear to be at rest, they exhibit **tone** in which some of their fibers are contracting. Muscle tone is particularly important in maintaining posture. If all the fibers within the muscles of the neck, trunk, and legs suddenly relax, the body collapses.

Maintenance of the right amount of tone requires the use of special receptors called muscle spindles. A muscle spindle consists of a bundle of modified muscle fibers, with sensory nerve fibers wrapped around a short, specialized region. A muscle spindle contracts along with muscle fibers, but thereafter, it sends stimuli to the central nervous system, which then regulates muscle contraction so that tone is maintained (Fig. 11.10).

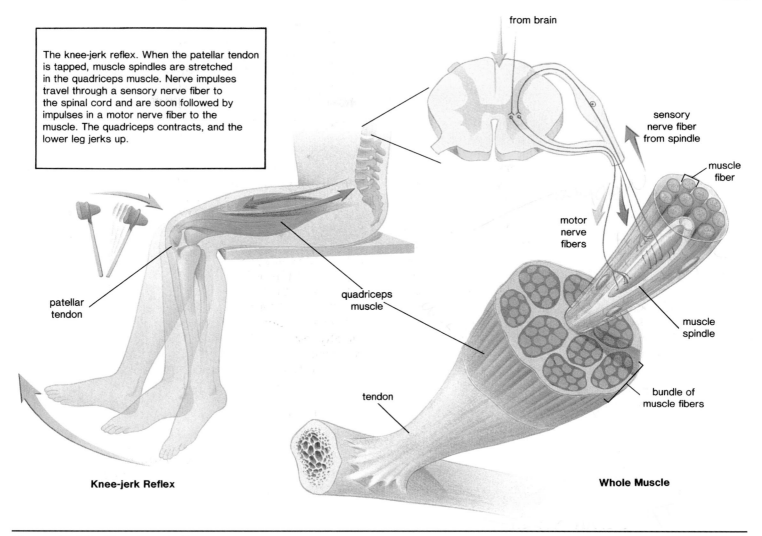

The knee-jerk reflex. When the patellar tendon is tapped, muscle spindles are stretched in the quadriceps muscle. Nerve impulses travel through a sensory nerve fiber to the spinal cord and are soon followed by impulses in a motor nerve fiber to the muscle. The quadriceps contracts, and the lower leg jerks up.

Figure 11.10 Muscle spindles.

Muscle spindles are stretch receptors that supply information about muscle fiber contraction. Muscle spindle input is also responsible for the knee-jerk reflex.

Recruitment and the Strength of Contraction

Each motor neuron innervates several muscle fibers at a time. A motor neuron together with all of the muscle fibers that it innervates is called a **motor unit.** As the intensity of nervous stimulation increases, more and more motor units are activated. This phenomenon, known as recruitment, results in stronger and stronger contractions.

Another variable of importance is the number of fibers within a motor unit. In the ocular muscles that move the eyes, the innervation ratio is one neuron per 23 muscle fibers, while in the gastrocnemius muscle of the buttocks, the ratio is about one neuron per 1,000 muscle fibers. Moving the eyes requires finer control than moving the legs.

When a muscle contracts maximally, it is under maximal tension. *Tension* is the force exerted on an object by a contracting muscle.

Isotonic Versus Isometric Contraction

Ordinarily, muscle tension is exhibited when a muscle shortens and movement occurs. This is called an **isotonic contraction.** For example, the biceps brachii contracts isotonically when a weight is lifted. On occasion, a muscle contracts but does not shorten or produce movement. This is called an **isometric contraction.** When a stationary bar is pulled on, the biceps brachii contracts isometrically, and no movement occurs.

Observation of muscle contraction in the laboratory applies to muscle contraction in the body. A muscle at rest exhibits tone dependent on tetanic contractions; a contracting muscle exhibits degrees of tension dependent on recruitment; and only isotonic contraction results in movement.

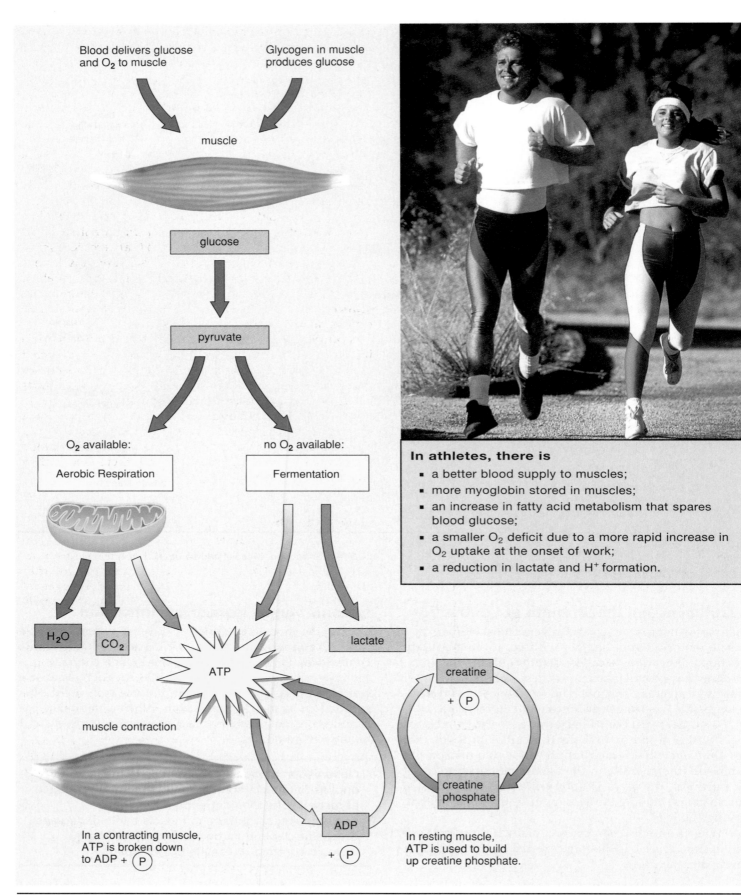

Figure 11.11   Energy sources for muscle contraction.

11.4 Energy for Muscle Contraction

Muscle contraction requires a plentiful supply of ATP, which can be supplied in four different ways. Previously produced ATP lasts a few seconds during strenuous exercise, and then the mechanisms described in Figure 11.11 produce new ATP. Aerobic respiration occurring in mitochondria usually provides most of a muscle's ATP. A muscle cell can make use of glycogen and fatty acids stored in the muscle as fuel, but still also needs a supply of oxygen in order to produce ATP:

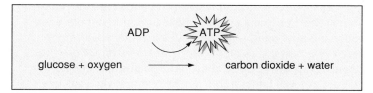

Aerobic respiration is not an immediate way of supplying an active muscle's increased need for ATP but has an advantage in that the end products (carbon dioxide and water), can be rapidly disposed of. A by-product of aerobic respiration is heat, which is used to keep the body warm.

There are two anaerobic ways by which muscles can acquire ATP immediately. The first method depends on **creatine phosphate,** a high-energy compound built up when a muscle is resting. Creatine phosphate cannot participate directly in muscle contraction. Instead, it is used to regenerate ATP by the following reaction:

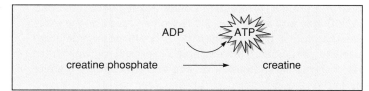

This reaction occurs in the midst of sliding filaments, and therefore, this method of supplying ATP is the speediest energy source available to muscles. For short bursts of intense activity, say sprinting, creatine phosphate provides enough energy for only about eight seconds, and then it is spent. Creatine phosphate is rebuilt when a muscle is resting by transferring a phosphate group from ATP to creatinine.

Fermentation is the second way muscles have of supplying ATP without consuming oxygen. Fermentation, which depends solely on glycolysis (the breakdown of glucose), occurs quickly, but the end product is lactate (lactic acid):

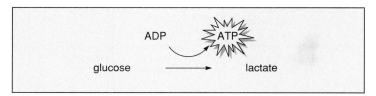

The accumulation of lactate in a muscle fiber makes the cell's internal environment more acidic, and eventually the glycolytic enzymes will cease to function well.

Altogether, creatine phosphate and fermentation provide a muscle with only enough energy for about two or three minutes of intense exercise. If continued beyond this amount of time, cramping and fatigue will set in. Cramping seems to be due to lack of ATP needed to pump calcium ions back into the sarcoplasmic reticulum and to break the linkages between the actin and myosin filaments so that muscle fibers can relax.

For almost any exercise, the three pathways work together, if the activity is not too strenuous. Usually, the anaerobic pathways are no longer needed as the body achieves an aerobic steady state. At this point some lactate has accumulated but not enough to bring on exhaustion. **Myoglobin,** an muscle oxygen carrier similar to blood hemoglobin, helps supply oxygen so that aerobic respiration is possible. Myoglobin is synthesized in muscle cells and is responsible for the reddish brown color of skeletal muscle fibers. This ability to store oxygen temporarily reduces a muscle's need for oxygen from the bloodstream during muscle contraction.

In persons who train, the number of mitochondria increases, and there is a greater reliance on them rather than fermentation to produce ATP. The mitochondria start consuming O_2 as soon as the ADP concentration starts rising due to muscle contraction and subsequent breakdown of ATP. Mitochondria can break down fatty acid instead of glucose, and this spares the blood glucose content. It also means that less lactate is produced, and the pH of the blood remains steady. And, there is less of an oxygen debt.

Oxygen Debt

When a muscle uses the anaerobic means of supplying energy needs, it incurs an **oxygen debt.** Oxygen debt is obvious when a person continues to breathe heavily after exercising. The ability to run up an oxygen debt is one of muscle tissue's greatest assets. Brain tissue cannot last nearly as long as muscles can without oxygen.

Repaying an oxygen debt requires replenishing creatine phosphate supplies and disposing of lactic acid. Lactic acid can be changed back to pyruvic acid and metabolized completely in mitochondria, or it can be sent to the liver to reconstruct glycogen. A marathon runner who has just crossed the finish line is not exhausted due to oxygen debt. Instead the runner has used up all the muscles' and probably the liver's glycogen supply. It takes about two days to replace glycogen stores on a high-carbohydrate diet and more than five days on a protein and lipid diet.

Working muscles require a supply of ATP. Anaerobic creatine phosphate breakdown and glycolysis can quickly generate ATP. Aerobic respiration in mitochondria is best for sustained exercise.

Health Focus

Exercise, Exercise, Exercise

Exercise programs improve muscular strength, muscular endurance, and flexibility. Muscular strength is the force a muscle group (or muscle) can exert against a resistance in one maximal effort. Muscular endurance is judged by the ability of a muscle to contract repeatedly or sustain a contraction for an extended period. Flexibility is tested by observing the range of motion about a joint.

Exercise also improves cardiorespiratory endurance. The heart rate and capacity increase, and the air passages dilate so that the heart and lungs are able to support prolonged muscular activity. The blood level of high-density lipoprotein (HDL), the molecule that prevents the development of plaque in blood vessels, increases. Also, body composition, the proportion of protein to fat, changes favorably when you exercise.

Exercise also seems to help prevent certain kinds of cancer. Cancer prevention involves eating properly, not smoking, avoiding cancer-causing chemicals and radiation, undergoing appropriate medical screening tests, and knowing the early warning signs of cancer. However, studies show that people who exercise are less likely to develop colon, breast, cervical, uterine, and ovarian cancers.

Physical training with weights can improve bone density and strength and muscular strength and endurance in all adults regardless of age. Even men and women in their eighties and nineties make substantial gains in bone and muscle strength, which can help them lead more independent lives. Exercise helps prevent osteoporosis, a condition in which the bones are weak and tend to break. Exercise promotes the activity of osteoblasts in young people as well as older people.

The stronger the bones when a person is young, the less chance of osteoporosis as a person ages. Exercise helps prevent weight gain not only because the level of activity increases but also because muscles metabolize faster than other tissues. As a person becomes more muscular, it is less likely that fat will accumulate.

Exercise relieves depression and enhances the mood. Some report exercise actually makes them feel more energetic, and after exercising, particularly in the late afternoon, they sleep better that night. Self-esteem rises not only because of improved appearance, but also due to other factors that are not well understood. It is known that vigorous exercise releases endorphins, hormone-like chemicals that are known to alleviate pain and provide a feeling of tranquility.

A sensible exercise program is one that provides all the benefits without the detriments of a too strenuous program. Overexertion can actually be harmful to the body and might result in sports injuries such as a bad back or bad knees. The programs suggested in Table 11A are tailored according to age and, if followed, are beneficial.

Dr. Arthur Leon at the University of Minnesota performed a study involving 12,000 men, and the results showed that only moderate exercise is needed to lower the risk of a heart attack by one-third. In another study conducted by the Institute for Aerobics Research in Dallas, Texas, which included 10,000 men and more than 3,000 women, even a little exercise was found to lower the risk of death from circulatory diseases and cancer. Increasing daily activity by walking to the corner store instead of driving and by taking the stairs instead of the elevator can improve your health.

TABLE 11A

A Checklist for Staying Fit

Children, 7–12	Teenagers, 13–18	Adults, 19–55	Seniors, 55 and up
Vigorous activity 1–2 hours daily	Vigorous activity 3–5 times a week	Vigorous activity for 1/2 hour, 3 times a week	Moderate exercise 3 times a week
Free play	Build muscle with calisthenics	Exercise to prevent lower back pain: aerobics, stretching, yoga	Plan a daily walk
Build motor skills through team sports, dance, swimming	Plan aerobic exercise to control buildup of fat cells	Take active vacations: hike, bicycle, cross-country ski	Daily stretching exercise
Encourage more exercise outside of physical education classes	Pursue tennis, swimming, horseback riding—sports that can be enjoyed for a lifetime	Find exercise partners: join a running club, bicycle club, outing group	Learn a new sport or activity: golf, fishing, ballroom dancing
Initiate family outings: bowling, boating, camping, hiking	Continue team sports, dancing, hiking, swimming		Try low-impact aerobics. Before undertaking new exercises, consult your doctor

11.5 Exercise and Muscle Contraction

Aside from athletes who excel in a particular sport, there is a general interest today in staying fit by exercising.

Exercise and Size of Muscles

Muscles that are not used or that are used for only very weak contractions decrease in size, or atrophy. **Atrophy** can occur when a limb is placed in a cast or when the nerve serving a muscle is damaged. If nerve stimulation is not restored, muscle fibers gradually are replaced by fat and fibrous tissue. Unfortunately, atrophy can cause muscle fibers to shorten progressively, leaving body parts contracted in contorted positions.

Forceful muscular activity over a prolonged period causes muscle to increase in size as the number of myofibrils within muscle fibers increase. Increase in muscle size, called **hypertrophy,** occurs only if the muscle contracts to at least 75% of its maximum tension.

Some athletes take anabolic steroids, either testosterone or related chemicals, to promote muscle growth. This practice has many undesirable side effects such as cardiovascular disease, liver and kidney dysfunction, impotency and sterility, and even an increase in rash behavior called "roid mania."

Slow-Twitch and Fast-Twitch Muscle Fibers

We have seen that all muscle fibers metabolize both aerobically and anaerobically. Some fibers, however, utilize one method more than the other to provide myofibrils with ATP. Slow-twitch fibers tend to be aerobic, and fast-twitch fibers tend to be anaerobic (Fig. 11.12).

Slow-Twitch Fibers

Slow-twitch fibers have a steadier tug and have more endurance despite motor units with a smaller number of fibers. These fibers are most helpful in sports like long-distance running, biking, jogging, and swimming. Because they produce most of their energy aerobically, they tire only when their fuel supply is gone. Slow-twitch fibers have many mitochondria and are dark in color because they contain myoglobin. They are also surrounded by dense capillary beds and draw more blood and oxygen than fast-twitch fibers. Slow-twitch fibers have a low maximum tension, which develops slowly, but these fibers are highly resistant to fatigue. Because slow-twitch fibers have a substantial reserve of glycogen and fat, their abundant mitochondria can maintain a steady, prolonged production of ATP when oxygen is available.

Fast-Twitch Fibers

Fast-twitch fibers are those that tend to be anaerobic and seem to be designed for strength because their motor units contain many fibers. They provide explosions of energy and are most helpful in sports like sprinting, weight lifting, swinging a golf club, or putting a shot. Fast-twitch fibers are white in color because they have fewer mitochondria, little or no myoglobin, and fewer blood vessels than slow-twitch fibers do. Fast-twitch fibers can develop maximum tension more rapidly than slow-twitch fibers can, and their maximum tension is greater. However, their dependence on anaerobic energy leaves them vulnerable to an accumulation of lactic acid that causes them to fatigue quickly.

11.6 Working Together

The Working Together box on page 236 shows how the muscular system works with other systems of the body to maintain homeostasis.

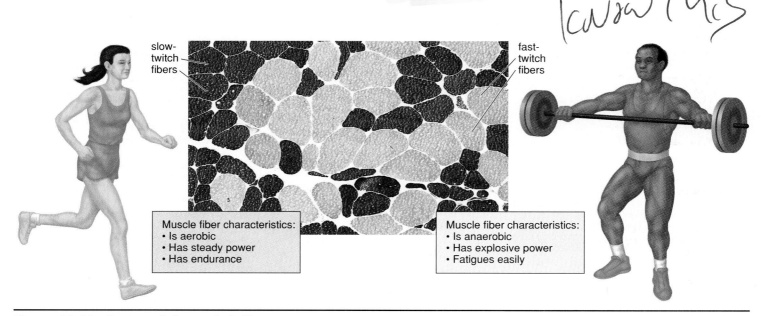

slow-twitch fibers

fast-twitch fibers

Muscle fiber characteristics:
• Is aerobic
• Has steady power
• Has endurance

Muscle fiber characteristics:
• Is anaerobic
• Has explosive power
• Fatigues easily

Figure 11.12   Fast- and slow-twitch fibers.
If your muscles contain many slow-twitch fibers (dark color), you would probably do better at a sport like cross-country running. But if your muscles contain many fast-twitch fibers (light color), you would probably do better at a sport like weight lifting.

HUMAN SYSTEMS WORK TOGETHER

Integumentary System

Muscle contraction provides heat to warm skin.

Skin protects muscles; rids the body of heat produced by muscle contraction.

How the Muscular System works with other body systems

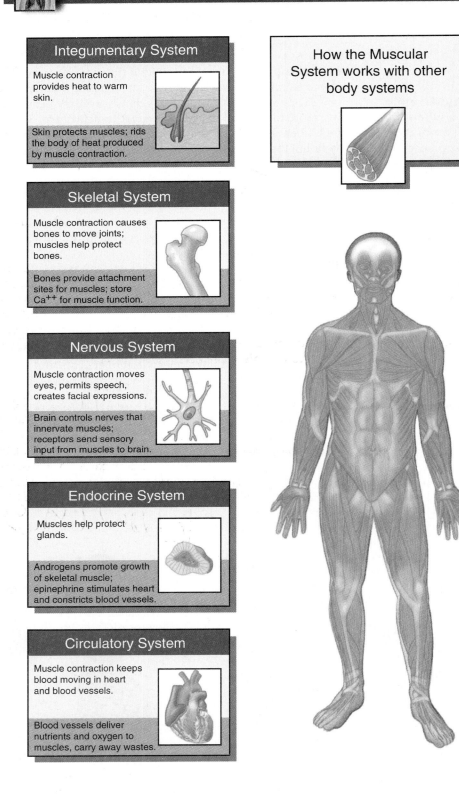

Lymphatic System/Immunity

Skeletal muscle contraction moves lymph; physical exercise enhances immunity.

Lymphatic vessels pick up excess tissue fluid; immune system protects against infections.

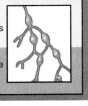

Skeletal System

Muscle contraction causes bones to move joints; muscles help protect bones.

Bones provide attachment sites for muscles; store Ca^{++} for muscle function.

Respiratory System

Muscle contraction assists breathing; physical exercise increases respiratory capacity.

Lungs provide oxygen for, and rid the body of carbon dioxide from, contracting muscles.

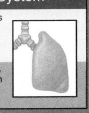

Nervous System

Muscle contraction moves eyes, permits speech, creates facial expressions.

Brain controls nerves that innervate muscles; receptors send sensory input from muscles to brain.

Digestive System

Smooth muscle contraction accounts for peristalsis; skeletal muscles support and help protect abdominal organs.

Digestive tract provides glucose for muscle activity; liver metabolizes lactic acid following anaerobic muscle activity.

Endocrine System

Muscles help protect glands.

Androgens promote growth of skeletal muscle; epinephrine stimulates heart and constricts blood vessels.

Urinary System

Smooth muscle contraction assists voiding of urine; skeletal muscles support and help protect urinary organs.

Kidneys maintain blood levels of Na^+, K^+, and Ca^{++}, which are needed for muscle innervation, and eliminate creatinine, a muscle waste.

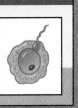

Circulatory System

Muscle contraction keeps blood moving in heart and blood vessels.

Blood vessels deliver nutrients and oxygen to muscles, carry away wastes.

Reproductive System

Muscle contraction occurs during orgasm and moves gametes; abdominal and uterine muscle contraction occurs during childbirth.

Androgens promote growth of skeletal muscle.

SUMMARY

11.1 Skeletal Muscles

Skeletal muscles have levels of organization. A whole muscle contains bundles of muscle fibers; muscle fibers contain myofibrils; and myofibrils contain actin and myosin filaments. Skeletal muscles are usually attached to the skeleton where some are prime movers, some are synergists, and others are antagonists. Muscles have various functions; they provide movement and heat, help maintain posture, and protect underlying organs.

Muscles are named for their size, shape, direction of fibers, location, number of attachments, and action.

11.2 Mechanism of Muscle Fiber Contraction

Nerve impulses travel down motor neurons and meet muscle fibers at neuromuscular junctions. The sarcolemma of a muscle fiber forms T tubules that extend into the fiber and almost touch the calcium storage sacs of the sarcoplasmic reticulum. When calcium ions are released into muscle fibers, actin filaments slide past myosin filaments within the sarcomeres of a myofibril.

At a neuromuscular junction, synaptic vesicles release acetylcholine (ACh), which binds to receptor sites on the sarcolemma, causing impulses to travel down the T system and calcium to leave the calcium storage sacs. Myofibril contraction follows.

Calcium ions bind to troponin and cause tropomyosin threads winding around actin filaments to shift their position, revealing myosin binding sites. The myosin filament is composed of many myosin molecules, each containing a head with an ATP binding site. Myosin is an ATPase, and once it breaks down ATP, the myosin head is ready to attach to actin. The release of ATP + Ⓟ causes the head to change its position. This is the power stroke that causes the actin filament to slide toward the center of a sarcomere. When myosin catalyzes another ATP, the head detaches from actin, and the cycle begins again.

11.3 Observations of Whole Muscle Contraction

In the laboratory, muscle contraction is described in terms of a muscle twitch, summation, and tetanus. In the body, muscles exhibit tone, in which tetanic contraction involving a number of fibers is the rule. The strength of muscle contraction varies according to recruitment of motor units. In an isotonic contraction, muscle fibers exhibit tension when movement occurs; in an isometric contraction, the muscle has tension, but movement does not occur.

11.4 Energy for Muscle Contraction

A muscle fiber has four ways to acquire ATP: (1) ATP is built up by aerobic respiration when the muscle is resting; (2) ATP is produced by aerobic respiration when the muscle is contracting; (3) creatine phosphate, also built up when a muscle is resting, can donate a phosphate to ADP, forming ATP when a muscle is contracting; and (4) fermentation with the concomitant accumulation of lactic acid can occur when a muscle is contracting. Two of these ways (1 and 2) are aerobic (they require oxygen), and two of these ways (3 and 4) are anaerobic (they do not require oxygen). Fermentation can result in oxygen debt because oxygen is needed to complete the metabolism of lactic acid.

11.5 Exercise and Muscle Contraction

Exercise results in muscle fiber increase called hypertrophy. Certain sports like running and swimming can be associated with slow-twitch fibers, which rely on aerobic respiration to acquire ATP. They have a plentiful supply of mitochondria and myoglobin, which gives them a dark color. Other sports like weight lifting can be associated with fast-twitch fibers, which rely on an anaerobic means of acquiring ATP. They have few mitochondria and myoglobin, and their motor units contain more muscle fibers. Fast-twitching fibers are known for their explosive power, but they fatigue quickly.

11.6 Working Together

The skeletal system works with the other systems of the body in the ways described in the box on page 236.

STUDYING THE CONCEPTS

1. Describe the levels of structure of a muscle from the whole muscle level to the myofilaments within a myofibril.   222

2. List and discuss the functions of muscles.   223

3. Describe the steps resulting in muscle contraction by starting with the motor neuron and ending with the sliding of actin filaments.   227

4. Describe the structure and function of a neuromuscular junction.   228

5. Describe the cyclical events as myosin pulls actin toward the center of a sarcomere.   229

6. Contrast a muscle twitch with summation and tetanus.   230

7. What is tone, and how is it maintained?   230

8. By what mechanism does the strength of muscle contraction vary?   231

9. Contrast isotonic with isometric contraction.   231

10. What are the four ways a muscle fiber can acquire ATP for muscle contraction? How are the four ways interrelated?   232

11. What is atrophy? Hypertrophy?   235

12. Contrast slow-twitch and fast-twitch fibers in as many ways as possible.   235

APPLYING YOUR KNOWLEDGE

Concepts

1. After giving birth, a cow collapses on the floor of the barn but recovers and can stand up after a veterinarian injects it with a liter of calcium salt solution. Explain the ability of the cow to now stand.

2. Under laboratory conditions, nervous stimulation no longer causes a muscle to contract, but direct stimulation of the muscle does produce contraction. Explain.

3. Muscles store glucose as glycogen and use it as a source of glucose for muscle contraction. The liver stores glucose as glycogen and then returns glucose to the bloodstream. Explain the advantage of this difference in the two organs.

Bioethical Issue

If there's anyone who knows about building muscles, it's an athlete. Daily workouts literally enlarge muscles, increasing tissue size and density. Such intense experiences can leave an athlete inspired, invigorated—or exhausted.

In recent years, some members of the sports community have worried that the combination of extreme physical fitness and intense competition might prove too much for young athletes. And in sports like tennis and skating, athletes are getting younger all the time.

Is it ethical for parents and coaches to expect 13- or 14-year-old athletes to endure grueling workouts, day in and day out? When combined with competition, is the pressure inappropriate for preteen sports stars? Who has the right to decide what's "too much" for these athletes? The athletes themselves? Parents? Coaches? Defend your answers.

TESTING YOUR KNOWLEDGE

1. Skeletal muscles contain bundles of _____ _____.

2. When muscles contract, body parts _____.

3. Impulses traveling down the T tubules cause _____ ions to leave their storage sacs.

4. _____ is the neurotransmitter released by a motor neuron.

5. Myosin does the work of muscle contraction. It breaks down _____, and it pulls the actin filament.

6. Most muscles in the body exhibit tone; they always have some fibers that are in _____ contraction.

7. When a muscle contracts isometrically, it (does or does not) shorten.

8. _____ stores high-energy phosphate bonds in muscle fibers.

9. Muscles placed in a cast are apt to _____.

10. _____-twitch muscle fibers are associated with swimming.

11. Muscle cramping appears to be due to a depletion of _____.

12. Label this diagram of a muscle fiber, using these terms: myofibril, mitochondrion, T tubules, sarcomere, sarcolemma, endoplasmic reticulum.

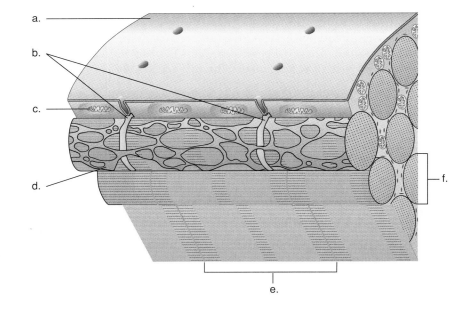

APPLYING TECHNOLOGY

Your study of the muscular system is supported by these available technologies:

Exploring the Internet

The Mader Home Page provides further resources for studying this chapter.

`http://www.mhhe.com/sciencemath/biology/mader/`

(Click on *Human Biology*.)

Dynamic Human: Muscular System CD-ROM

In *Anatomy,* location of muscle types are shown; in *Explorations,* isometric is contrasted with isotonic contraction; the work of the biceps is contrasted with the triceps; there is a depiction of the sliding filament theory and zooming in occurs until a neuromuscular junction appears; in *Histology,* microscopic slides are highlighted to reveal tissues.

Explorations in Human Biology CD-ROM

Muscle Contraction (#4) The degree of filament sliding varies as students change the rate of stimulation and the amount of ATP available in a sarcomere. **(4)***

Life Science Animations Video

Video #1: Chemistry, The Cell, and Energetics

Levels of Muscle Structure (#29) Skeletal muscle structure is reviewed, and the filament sliding occurs in a sarcomere. **(2)**

Sliding Filament Model of Muscle Contraction (#30) ATP attaches to myosin heads, which then bind to actin. **(1)**

Regulation of Muscle Contraction (#31) Due to nerve impulses, calcium leaves the sarcoplasmic reticulum and attaches to troponin so that tropomyosin shifts its position, exposing myosin binding sites. **(3)**

*Level of difficulty

SELECTED KEY TERMS

actin One of two major proteins of muscle; makes up thin filaments in myofibrils of muscle fibers. See myosin.   227

atrophy Wasting of a muscle due to lack of use.   235

creatine phosphate Compound unique to muscles that contains a high-energy phosphate bond.   232

hypertrophy Increase in muscle size following long-term exercise.   235

insertion End of a muscle that is attached to a movable bone.   222

isometric contraction Muscular contraction in which the muscle fails to shorten.   231

isotonic contraction Muscular contraction in which the muscle shortens.   231

motor unit Motor neuron and the muscle fibers associated with it.   231

myofibril Contractile portion of a muscle fiber.   227

myoglobin Pigmented compound in muscle tissue that stores oxygen.   232

myosin (MY-uh-sun) One of two major proteins of muscle; makes up thick filaments in myofibrils and is capable of breaking down ATP. See actin.   227

neuromuscular junction Point of communication between a neuron and a muscle fiber.   228

origin End of a muscle that is attached to a relatively immovable bone.   222

oxygen debt Oxygen that is needed to metabolize lactate, a compound that accumulates during vigorous exercise.   232

sarcolemma (sahr-kuh-LEM-uh) Plasma membrane of a muscle fiber.   227

sarcomere (SAHR-kuh-mir) Structural and functional unit of a myofibril; contains actin and myosin filaments.   227

sliding filament theory Movement of actin in relation to myosin; accounts for muscle contraction.   227

tendon Fibrous connective tissue that joins muscle to bone.   222

tetanus Sustained maximal muscle contraction.   230

tone Tension that is maintained even when a muscle is at rest.   230

FURTHER READINGS FOR PART THREE

Clemente, C. D. 1996. *Anatomy: A regional atlas of the human body.* 4th ed. Philadelphia: Lea and Febiger. This atlas contains both drawings and photographs of anatomical regions of the human body.

Fox, S. I. 1996. *Human physiology.* 5th ed. Dubuque, Iowa: Wm. C. Brown Publishers. This is an introductory physiology text.

Grillner, S. January 1996. Neural networks for vertebrate locomotion. *Scientific American* 274(1):64. Discoveries about how the brain coordinates muscle movement raise hopes for restoration of mobility for some accident victims.

Gunstream, S. E. 1995. *Anatomy and physiology.* 2d ed. Dubuque, Iowa: Wm. C. Brown Publishers. This is an introductory anatomy and physiology text-workbook.

Guyton, A. C., and Hall, J. E. 1996. *Textbook of medical physiology.* Philadelphia: Saunders College Publishing. Presents physiological principles for those in the medical fields.

Hole, J. W., Jr. 1995. *Essentials of human anatomy and physiology.* 5th ed. Dubuque, Iowa: Wm. C. Brown Publishers. An introductory text that is written in an easily understood manner.

Marieb, E. N. 1995. *Human anatomy and physiology.* 3d ed. Redwood City, Calif.: Benjamin/Cummings Publishing. A thorough anatomy and physiology text that can safely be used as a complete and accurate reference.

Mellion, M. 1994. *Sports medicine secrets.* St. Louis: Mosby-Year Book, Inc. This book covers frequently asked questions in sports medicine.

National Geographic Society. 1994. *The incredible machine.* Washington, D.C.: National Geographic Society. A colorfully illustrated reference of the structures and functions of the human body.

Nowak, T. J., and Handford, A. G. 1994. *Essentials of pathophysiology.* Dubuque, Iowa: Wm. C. Brown Publishers. Introduces the concepts of pathophysiology for those in the medical fields.

Risher, C. E., and Easton, T. A. 1995. *Focus on human biology.* 2d ed. New York: HarperCollins College Publishers. This comprehensive introductory textbook stresses basic human anatomy and physiology.

Shier, D., et al. 1996. *Hole's human anatomy & physiology.* 7th ed. Dubuque, Iowa: Wm. C. Brown Publishers. An introductory anatomy and physiology text that has proved useful to beginning students.

Tate, P., et al. 1994. *Understanding the human body.* St. Louis: Mosby-Year Book, Inc. The relationship between structure and function is stressed in this introductory text.

Tortora, G. J. 1996. *Atlas of the human skeleton.* New York: Harper-Collins Publishers. This supplement contains carefully selected and labeled photographs of the human skeleton.

Van De Graaff, K. M. 1995. *Human anatomy.* 4th ed. Dubuque, Iowa: Wm. C. Brown Publishers. This introductory text provides a balanced presentation of anatomy at various levels.

Van De Graaff, K. M., and Fox, S. I. 1995. *Concepts of human anatomy & physiology.* 4th ed. Dubuque, Iowa: Wm. C. Brown Publishers. An introductory anatomy and physiology text that comprehensively presents basic principles.

Vander, A. J., et al. 1994. *Human physiology: The mechanisms of body function.* 6th ed. New York: McGraw-Hill, Inc. Presents the principles of human physiology with an emphasis on physiological mechanisms.

Part

Integration and Coordination in Humans

The nervous system is the ultimate coordinator of homeostasis, that is, an internal environment that is in dynamic equilibrium. Nerves bring information to the brain and spinal cord from receptors that are responding to changes both inside and outside the body. Then, the nerves take the commands given by the brain and spinal cord to effectors, allowing the body to respond quickly to these changes.

The endocrine system, like the nervous system, regulates other organs, but it acts more slowly and brings about a response that lasts longer. The endocrine organs secrete chemical messengers called hormones into the bloodstream. After arriving at their target organs, hormones alter cellular metabolism.

We now know that the nervous system and the endocrine system are joined in numerous ways despite their very different modes of operation.

Chapter 12

Nervous System

Chapter Outline

12.1 NEURON STRUCTURE
- The nervous system is made up of cells called neurons, which are specialized to carry nerve impulses. 242

12.2 NERVE IMPULSE
- A nerve impulse is an electrochemical change that travels along the length of a neuron. 244

12.3 TRANSMISSION ACROSS A SYNAPSE
- Transmission of impulses between neurons is accomplished by means of chemicals called neurotransmitters. 246

12.4 PERIPHERAL NERVOUS SYSTEM
- The peripheral nervous system contains nerves that conduct nerve impulses between body parts and the central nervous system. 247

12.5 CENTRAL NERVOUS SYSTEM
- The central nervous system, made up of the spinal cord and the brain, is highly organized. 252
- In the brain, consciousness is associated with the cerebrum, which is more highly developed in humans than in other animals. 255

12.6 DRUG ABUSE
- The use of psychoactive drugs such as alcohol, nicotine, marijuana, cocaine, and heroin is detrimental to the body. 257

12.7 WORKING TOGETHER
- The nervous system works with the other systems of the body to maintain homeostasis. 260

I t was a warm spring Saturday in 1995 when a crowd gathered in Culpeper, Virginia, to enjoy a horse riding competition. Everything was fine until—suddenly—one horse stopped dead in its tracks. The horse's rider, Superman star Christopher Reeve, went tumbling through the air.

Hitting the ground, Reeve crushed the top two vertebrae in his neck, damaging the sensitive spinal cord underneath. In a split second, he was paralyzed. Doctors say he is lucky to be alive.

The spinal cord is a ropelike bundle of long nerve tracts that shuttle messages between the brain and the rest of the body. Together the brain and spinal cord comprise the **central nervous system (CNS),** which receives information from external sources and coordinates the body's response. Nerves, located outside the central nervous system in the **peripheral nervous system (PNS),** gather and transmit sensory information to the central nervous system. Then nerves cause the body to act on the orders of the CNS.

When Reeve hurt his spinal cord, his brain lost the ability to communicate with that portion of his body located below the site of damage (Fig. 12.1). He receives no sensation nor can he command his limb muscles. But that has not stopped this superhero. His cranial nerves still give him sensory input, and his brain still enables him to have emotions, remember, and reason. Reeve is working on a book about his experience.

Of all the body's splendid components, biologists still are trying to understand how the nervous system functions, especially since this is fundamentally dependent upon the structure of just one type of cell, the neuron.

12.1 Neuron Structure

The nervous tissue of the human body contains many billions of **neurons,** large complex cells that conduct nerve impulses from one part of the body to another part. Neurons are quite complex, but each one has just three parts: dendrite(s), cell body, and an axon (Fig. 12.2). A **dendrite** conducts signals toward the cell body. The **cell body** is the part of a neuron that contains the nucleus and other organelles. An **axon** conducts nerve impulses away from the cell body. Dendrites and axons collectively are called *neuron fibers.*

There are three types of neurons: sensory neurons, motor neurons, and interneurons, whose functions are best described in relation to the CNS. A sensory neuron is sometimes referred to as the *afferent neuron,* and the motor neuron is sometimes called the *efferent neuron.* These words, which are derived from Latin, mean running to and running away from, respectively. A **sensory neuron** takes nerve impulses to the CNS, and a **motor neuron** takes nerve impulses away from the CNS. Notice that a motor neuron has short dendrites and a long axon.

An **interneuron** (also called an associated neuron) is always found completely within the CNS and conveys nerve impulses between parts of that system. An interneuron has short dendrites and either a long or a short axon.

> There are three types of neurons: sensory neurons take nerve impulses to the CNS; motor neurons take nerve impulses away from the CNS; and interneurons take nerve impulses within the CNS.

Central Nervous System   Peripheral Nervous System

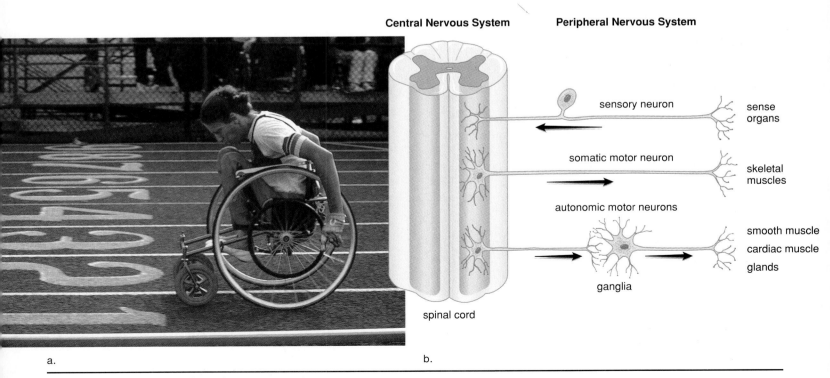

a.   b.

Figure 12.1   Organization of nervous system.

a. In paraplegics, messages no longer flow between the legs and the central nervous system (the brain and spinal cord). **b.** The sensory neurons of the peripheral nervous system take nerve impulses from the sense organs to the central nervous system (CNS), and motor neurons (both somatic to skeletal muscles and autonomic to internal organs) take nerve impulses from the CNS to the organs mentioned.

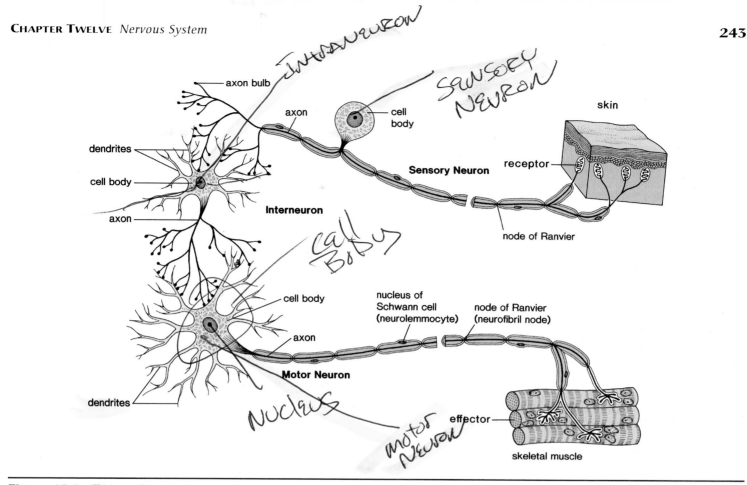

Figure 12.2   Types of neurons.
A sensory neuron, an interneuron, and a motor neuron are drawn here to show their arrangement in the body. (The breaks indicate that the fibers are much longer than shown.) How does this arrangement correlate with the function of each neuron?

Most long fibers are covered by tightly packed spirals of **neuroglial cells.** Neuroglial cells service the neurons—they have supportive, nutritive, and perhaps some communicative functions. *Schwann cells* (neurolemmocytes) are neuroglial cells found in the PNS that encircle a fiber, leaving gaps called the **nodes of Ranvier** (neurofibril nodes). Schwann cells wrap themselves around the axon many times and in this way lay down several layers of plasma membrane containing myelin, a lipid substance that is an excellent insulator. Myelin, which gives nerve fibers their white, glistening appearance, forms a **myelin sheath** (Fig. 12.3). Scarlike patches begin to replace myelin in multiple sclerosis, and the motor difficulties result. The outer membrane of Schwann cells plays an important role in nerve regeneration. If a nerve fiber is accidentally severed, the part distant from the nerve fiber degenerates, except for this membrane, which serves as a passageway for new axonal growth.

Long nerve fibers are covered by a myelin sheath formed by the plasma membrane of Schwann cells.

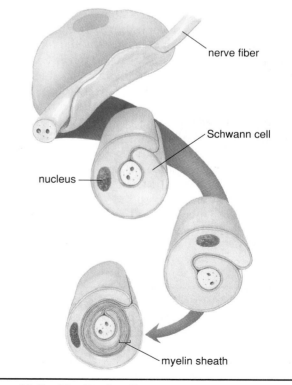

Figure 12.3   Myelin sheath.
A myelin sheath forms when Schwann cells wrap themselves around a nerve fiber in the manner shown.

12.2 Nerve Impulse

A **nerve impulse** is the way a neuron transmits information. The nature of a nerve impulse has been studied by using excised axons and an instrument called a voltmeter. *Voltage* (mV=millivolts) is a measure of the electrical potential difference between two points, which in this case are the inside and the outside of the axon. The change in voltage is displayed on an *oscilloscope,* an instrument with a screen that shows a trace, or pattern, indicating a change in voltage with time.

In the long fibers of human nerves, the speed of conduction is quite rapid (200 meters per second) because of the large diameter of the fibers and because the nerve impulse jumps from node of Ranvier to node of Ranvier. This is called *saltatory* (saltatory means jumping) *conduction.*

Resting Potential: Inside Is Negative

In the experimental setup shown in Figure 12.4, an oscilloscope is wired to two electrodes, one placed inside and the other placed outside an axon. The axon is essentially a membranous tube filled with axoplasm (cytoplasm of the axon). When the axon is not conducting an impulse, the oscilloscope records a *membrane potential* (potential difference across a membrane) equal to about −65 mV. This reading indicates that the inside of the neuron is negative compared to the outside. This is called the **resting potential** because the axon is not conducting an impulse.

The existence of this polarity (charge difference) can be correlated with a difference in ion distribution on either side of the axomembrane (plasma membrane of the axon). As Figure 12.4a shows, the concentration of sodium ions (Na^+) is greater outside the axon than inside, and the concentration of potassium ions (K^+) is greater inside the axon than outside. The unequal distribution of these ions is due to the action of the **sodium-potassium pump.** This is a transport protein in the membrane that pumps Na^+ out of and K^+ into the axon. The work of the pump maintains the unequal distribution of Na^+ and K^+ across the membrane.

The pump is always working because the membrane is somewhat permeable to these ions, and they tend to diffuse toward their lesser concentration. Since the membrane is more permeable to potassium than to sodium, there are always more positive ions outside the membrane than inside; this accounts in part for the polarity recorded by the oscilloscope. There are also large, negatively charged proteins in the axoplasm that are too large to cross the membrane.

> The resting potential indicates that the inside of a fiber is negative compared to the outside. Because of the sodium-potassium pump, there is a concentration of Na^+ outside a fiber and K^+ inside a fiber.

Action Potential: Upswing and Downswing

When an axon is stimulated by an electric shock, a sudden difference in pH, or a pinch, *threshold* may be reached for an action potential. An **action potential** is obvious because a rapidly moving pattern appears on the oscilloscope screen. This pattern, which represents rapid polarity changes, has an upswing and a downswing (Fig. 12.4b).

The action potential requires two types of special protein-lined channels in the membrane. There is a channel that allows sodium (Na^+) to pass through the membrane and another that allows potassium (K^+) to pass through the membrane. Each of these types of channels has a gate: the sodium channel has a gate called the sodium gate, and the potassium channel has a gate called the potassium gate.

An action potential is an electrochemical change because it consists of (1) polarity changes brought about by (2) the movement of Na^+ and K^+.

Sodium Gates Open

Stimulation of an axon causes the gates of the sodium channels to open temporarily, allowing Na^+ to flow into the axon. As Na^+ rapidly moves across the membrane to the inside of the axon, the action potential swings up from −65 mV to +40 mV. The sudden permeability of the membrane causes the oscilloscope to record a *depolarization*: the charge inside the fiber changes from negative to positive as Na^+ enters the axoplasm of the axon.

Potassium Gates Open

Now the potassium gates open, and K^+ moves from inside the axon to the outside of the axon. As K^+ leaves, the action potential swings down from +40 mV to −65 mV. The membrane is suddenly permeable to K^+ because the potassium gates of the potassium channels temporarily open, allowing K^+ to flow out of the axon. The oscilloscope records a *repolarization*: the inside of the axon resumes a negative charge.

Membrane Rests During a Refractory Period

A fiber can conduct a volley of nerve impulses because only a small number of ions are exchanged with each impulse. As soon as an impulse has passed by each successive portion of a fiber, it undergoes a refractory period during which it is unable to conduct an impulse. This ensures a one-way direction of the impulse. During a refractory period, the sodium gate cannot yet open.

> The nerve impulse is the action potential (an electrochemical change) traveling along a fiber. Depolarization occurs when Na^+ moves to the inside and repolarization occurs when K^+ moves to the outside of a fiber.

a. Resting Potential

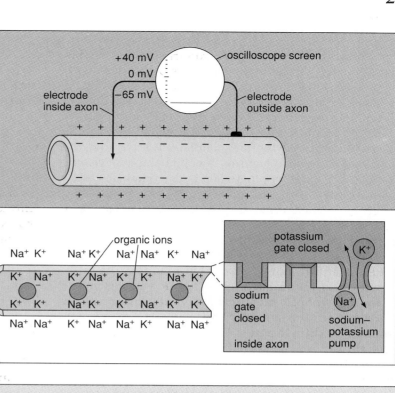

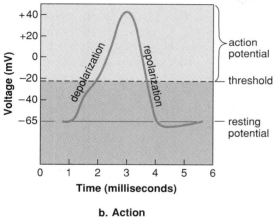

b. Action Potential

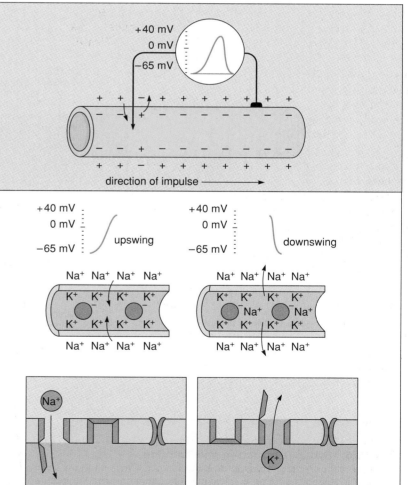

Figure 12.4 Resting and action potential.
a. Resting potential. The oscilloscope (shown in photograph) records a resting potential of -65 mV due to the presence of large organic ions inside a fiber. Note also the unequal distribution of Na^+ and K^+ across the membrane due to the work of the sodium-potassium pump. **b.** Action potential (shown enlarged). A depolarization (upswing) is due to the movement of Na^+ to the inside of a fiber and a repolarization (downswing) is due to the movement of K^+ to the outside of a fiber.

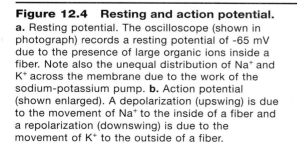

12.3 Transmission Across a Synapse

Every axon branches into many fine terminal branches, each of which is tipped by a small swelling, called a synaptic bulb, or more specifically, an axon bulb (Fig. 12.5*a, b*). Each bulb lies very close to the dendrite (or the cell body) of another neuron. This region of close proximity is called a **synapse.** At a synapse, the membrane of the first neuron is called the *presynaptic membrane,* and the membrane of the next neuron is called the *postsynaptic membrane.* The small gap between is the **synaptic cleft.**

In humans, transmission of information across a synaptic cleft is usually carried out by chemicals known as **neurotransmitters,** which are stored in synaptic vesicles (Fig. 12.5*b, c*). When nerve impulses traveling along an axon reach an axon bulb, gated channels for calcium ions (Ca^{++}) open and calcium enters the bulb. Following a series of reactions, synaptic vesicles merge with the presynaptic membrane, and neurotransmitter is released into the synaptic cleft. The neurotransmitter molecules diffuse across the cleft to the postsynaptic membrane, where they bind with a receptor in a lock-and-key manner (Fig. 12.5*c*).

Depending on the type of neurotransmitter and/or the type of receptor, the response can be excitation or inhibition. Neurotransmitters that utilize gated ion channels are fast acting. If excitation occurs, the Na$^+$ channels open, and the likelihood of the neuron transmitting a nerve impulse increases. Other neurotransmitters affect the metabolism of the postsynaptic cell, and therefore, they are slower acting.

Neurotransmitters Are Varied

At least 25 different types of neurotransmitters are known. Some of these are gases like CO and NO, and some are neuropeptide or amino acids or derivatives of these. Two extremely well-known transmitters are **acetylcholine (ACh)** and **norepinephrine (NE).** ACh and NE can be excitatory or inhibitory according to the type of receptor at the postsynaptic membrane.

Once a neurotransmitter has been released into a synaptic cleft and has initiated a response, it is removed. In some synapses, the postsynaptic membrane contains enzymes that rapidly inactivate the neurotransmitter. For example, the enzyme **acetylcholinesterase (AChE)** breaks

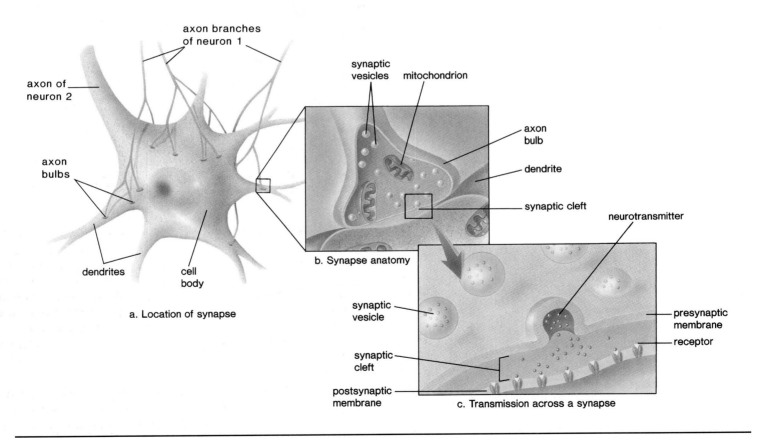

Figure 12.5 Synapse structure and function.
a. A synapse occurs where an axon bulb comes close to a dendrite or to a cell body. **b.** An axon bulb contains synaptic vesicles, which in turn contain a neurotransmitter. **c.** Transmission across a synapse occurs when synaptic vesicles release a neurotransmitter that binds to receptors in the postsynaptic membrane.

down acetylcholine. In other synapses, the synaptic ending rapidly reabsorbs the neurotransmitter, possibly for repackaging in synaptic vesicles or for chemical breakdown. The short existence of neurotransmitters in the synapse prevents continuous stimulation (or inhibition) of postsynaptic membranes.

Transmission across a synapse is dependent on the release of neurotransmitters, which diffuse across the synaptic cleft.

12.4 Peripheral Nervous System

The nervous system is divided into the central nervous system and the peripheral nervous system (Fig. 12.6). The *peripheral nervous system (PNS)* lies outside the central nervous system and contains both cranial nerves and spinal nerves. The paired cranial nerves connect to the brain, and the paired spinal nerves lie on either side of the spinal cord. In the PNS, the somatic system controls the skeletal muscles, and the autonomic system controls the smooth muscles, cardiac muscles, and glands. There are two parts to the autonomic system: the sympathetic system and the parasympathetic system.

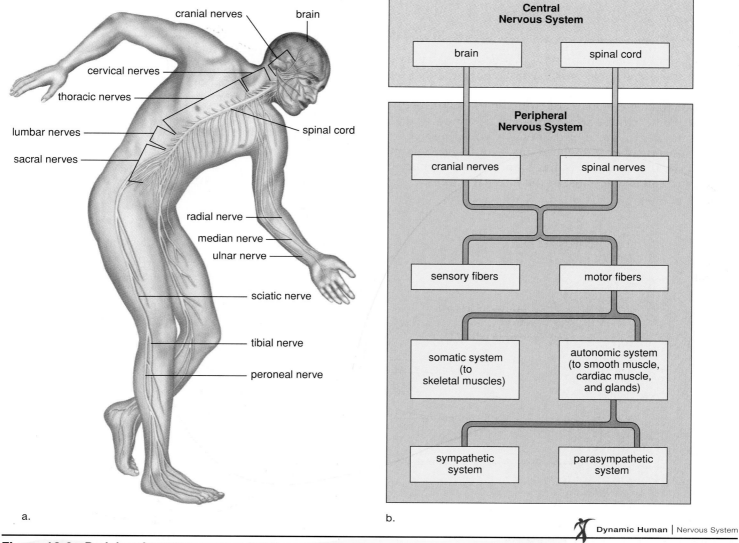

a.

b.

Dynamic Human | Nervous System

Figure 12.6 Peripheral nervous system (PNS) compared with central nervous system (CNS).
a. The CNS consists of the brain and the spinal cord, and the PNS consists of the nerves. **b.** In the somatic system, nerves send sensory impulses from receptors to the skeletal muscles. In the autonomic system, consisting of the sympathetic and parasympathetic systems, motor impulses travel to smooth muscle, cardiac muscle, and the glands.

Cranial and Spinal Nerves

The peripheral nervous system is made up of nerves. **Nerves** are bundles of fibers, which are the processes of neurons. The cell bodies of neurons are found in the CNS—that is, the brain and spinal cord—or in ganglia. Ganglia (sing., **ganglion**) are collections of cell bodies within the PNS. Only the fibers of neurons are found in the nerves of the PNS.

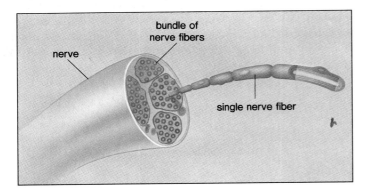

Sensory nerves contain sensory fibers, motor nerves contain motor fibers, and mixed nerves contain both types of fibers.

Humans have 12 pairs of cranial nerves attached to the brain. Some of these are sensory nerves, some are motor nerves, and others are mixed nerves. **Cranial nerves** are largely concerned with the head, neck, and facial regions of the body; the vagus nerve has branches to the pharynx and larynx and also to most of the internal organs.

Humans have 31 pairs of **spinal nerves.** Each spinal nerve emerges from the spinal cord by two short branches, or roots, which lie within the vertebral column (Fig. 12.7). The *dorsal root* contains the fibers of sensory neurons, which conduct impulses to the spinal cord. The *ventral root* contains the fibers of motor neurons, which conduct impulses away from the cord. These two roots join just before a spinal nerve leaves the vertebral column. Therefore, all spinal nerves are mixed nerves that contain many sensory fibers and motor fibers. Each spinal nerve serves the particular region of the body in which it is located.

An individual nerve fiber obeys an all-or-none law, meaning that it fires maximally or it does not fire. A nerve does not obey the all-or-none law—a nerve can have a number of degrees of performance because it contains many fibers, any number of which can be carrying nerve impulses.

In the PNS, cranial nerves take impulses to and/or from the brain, and spinal nerves take impulses to and from the spinal cord.

Somatic System Serves Skin and Muscles

The **somatic system** includes the mixed nerves that serve the musculoskeletal system and the exterior sense organs, including those in the skin.

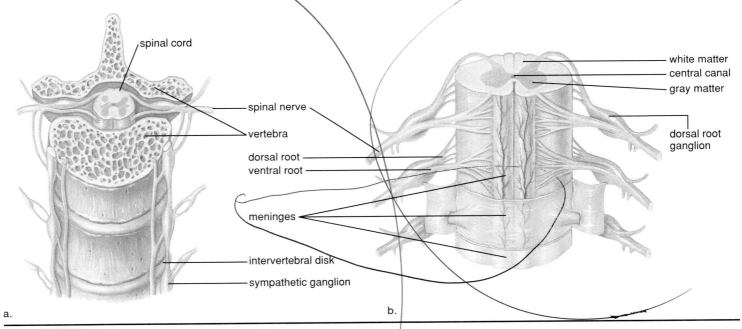

Figure 12.7   Spinal cord anatomy.
a. Cross section of the spine, showing spinal nerves. The human body has 31 pairs of spinal nerves. **b.** This cross section of the spinal cord shows that a spinal nerve has a dorsal root and a ventral root. Also, the cord is protected by three layers of tissue called the meninges.

Reflexes Are Automatic

A **reflex** is an automatic, involuntary response to changes occurring inside or outside the body. In the somatic system, outside stimuli often initiate a reflex. Some reflexes, such as blinking your eye, involve the brain, while others, such as withdrawing your hand from a hot object, do not necessarily involve the brain. Figure 12.8 illustrates the path of the second type of reflex involving the spinal cord and a spinal nerve, which is called a *spinal reflex,* or *reflex arc.*

If you touch a sharp pin, a **receptor,** which is a receiver of environmental stimuli, generates nerve impulses that move along a *sensory neuron* toward the cell body and the CNS. The cell body of a sensory neuron is located in the *dorsal-root ganglion,* just outside the spinal cord. From the cell body, the impulses travel along the sensory neuron and enter the cord dorsally. The impulses then pass to many interneurons, one of which connects with a *motor neuron.* The short

dendrites and the cell body of the motor neuron lead to the axon, which leaves the cord ventrally. The nerve impulses travel along the axon to an **effector,** which brings about a response to a stimulus. The effector in this case is muscle that contracts so that you withdraw your hand from the pin. There are various other reactions—the person will most likely look at the pin, wince, and cry out in pain. This whole series of responses is explained by the fact that the sensory neuron stimulates several interneurons. They take impulses to all parts of the CNS, including the cerebrum, which in turn makes the person conscious of the stimulus and directs his or her reaction to it.

The reflex arc is a major functional unit of the nervous system. It allows us to react rapidly to internal and external stimuli.

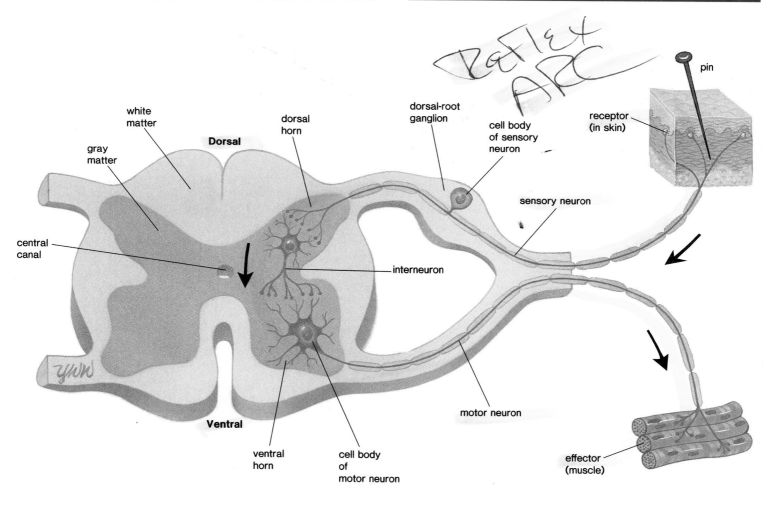

Figure 12.8   A reflex arc showing the path of a spinal reflex.
When a receptor in the skin is stimulated, nerve impulses (see arrows) move along a sensory neuron to the spinal cord. (Note that the cell body of a sensory neuron is in a ganglion outside the cord.) The nerve impulses are picked up by an interneuron, which lies completely within the cord, and passed to the dendrites and the cell body of a motor neuron that lies ventrally within the cord. The nerve impulses then move along the motor neuron to an effector, such as a muscle, which contracts. The brain receives information concerning sensory stimuli by way of other interneurons, with long fibers in tracts that run up and down the cord within the white matter.

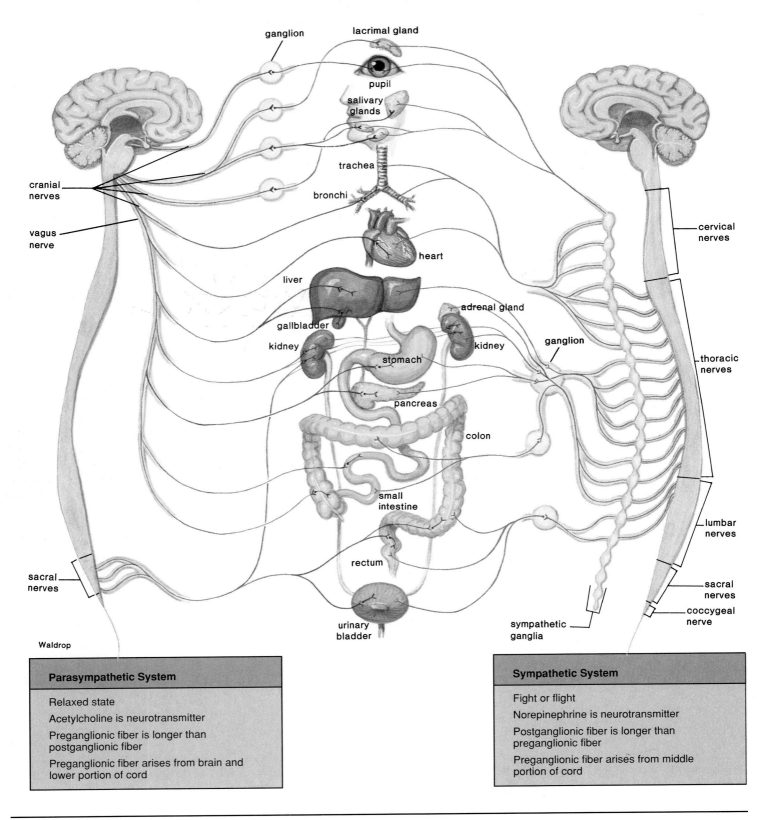

Figure 12.9   Autonomic system structure and function.
The sympathetic fibers arise from the thoracic and lumbar portions of the spinal cord; the parasympathetic fibers arise from the brain and the sacral portion of the spinal cord. Each system innervates the same organs but has contrary effects. For example, the sympathetic system speeds up the beat of the heart, while the parasympathetic system slows it down.

Autonomic System Serves Internal Organs

The sensory neurons that come from the internal organs receive impulses from receptors, which they pass on to the CNS. Reflex actions, such as those that regulate the blood pressure and breathing rate, are especially important to the maintenance of homeostasis. The motor neurons that control the internal organs function automatically and usually without need for conscious intervention.

The motor neurons of the **autonomic system** are divided into the sympathetic and parasympathetic systems (Fig. 12.9 and Table 12.1). Both of these systems (1) function automatically and usually subconsciously in an involuntary manner; (2) innervate all internal organs; and (3) utilize two neurons and one ganglion for each impulse. The first of these two neurons has a cell body within the CNS and a *preganglionic fiber.* The second neuron has a cell body within the ganglion and a *postganglionic fiber.*

The autonomic system serves the internal organs, without conscious control.

Sympathetic System: Fight or Flight

Most preganglionic fibers of the **sympathetic system** arise from the middle, or *thoracic-lumbar,* portion of the spinal cord and almost immediately terminate in ganglia that lie near the cord. Therefore, in this system, the preganglionic fiber is short, but the postganglionic fiber that makes contact with an organ is long.

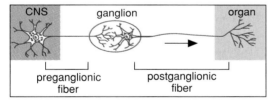

The sympathetic system is especially important during emergency situations and is associated with "fight or flight." If you need to fend off a foe or flee from danger,

active muscles require a ready supply of glucose and oxygen. The sympathetic system accelerates the heartbeat and dilates the bronchi. On the other hand, the sympathetic system inhibits the digestive tract—digestion is not an immediate necessity if you are under attack. The neurotransmitter released by the postganglionic axon is primarily norepinephrine (NE). The structure of NE is like that of epinephrine (adrenaline), an adrenal medulla hormone that is sometimes used as a heart stimulant.

The sympathetic system brings about those responses we associate with "fight or flight."

Parasympathetic System: Relaxed State

A few cranial nerves, including the vagus nerve, together with fibers that arise from the sacral (bottom) portion of the spinal cord, form the **parasympathetic system.** Therefore, this system often is referred to as the *craniosacral portion* of the autonomic system. In the parasympathetic system, the preganglionic fiber is long, and the postganglionic fiber is short because the ganglia lie near or within the organ.

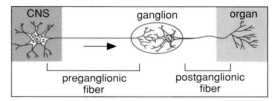

The parasympathetic system, sometimes called the "housekeeper system," promotes all the internal responses we associate with a relaxed state; for example, it causes the pupil of the eye to contract, promotes digestion of food, and retards the heartbeat. The neurotransmitter utilized by the parasympathetic system is acetylcholine (ACh).

The parasympathetic system brings about the responses we associate with a relaxed state.

TABLE 12.1			
Comparison of Somatic Motor and Autonomic Motor Pathways			
Items	**Somatic Motor Pathway**	**Autonomic Motor Pathways**	
		Sympathetic	**Parasympathetic**
Level of consciousness	Voluntary	Involuntary	Involuntary
Number of neurons per message	One	Two (preganglionic shorter than postganglionic)	Two (preganglionic longer than postganglionic)
Location of motor fiber	Most cranial nerves and all spinal nerves	Spinal nerves	Cranial (vagus) and sacral spinal nerves
Neurotransmitter	Acetylcholine	Norepinephrine	Acetylcholine
Effectors	Skeletal muscles	Smooth and cardiac muscle, glands	Smooth and cardiac muscle, glands

12.5 Central Nervous System

The *central nervous system* (*CNS*) consists of the spinal cord and the brain where nerve impulses are received, coordinated, and interpreted. The spinal cord is surrounded by vertebrae, and the brain is enclosed by the skull. Also, both the spinal cord and the brain are wrapped in three protective membranes known as **meninges** (sing., meninx); meningitis is an infection of these coverings (see Fig. 12.7). The spaces between the meninges are filled with **cerebrospinal fluid**, which cushions and protects the CNS. Cerebrospinal fluid is contained within the *ventricles* of the brain, which are interconnecting spaces that produce and serve as a reservoir for cerebrospinal fluid, and in the central canal of the spinal cord. A small amount of this fluid sometimes is withdrawn for laboratory testing when a spinal tap (i.e., lumbar puncture) is performed.

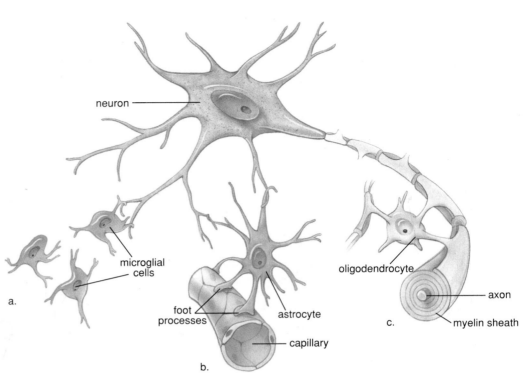

Figure 12.10 Neuroglial cells in the central nervous system.
a. Microglial cells are phagocytes that clean up debris. **b.** Astrocytes lie between neurons and a capillary; therefore, substances entering neurons from the blood must first pass through astrocytes. **c.** Oligodendrocytes form the myelin sheaths around fibers in the CNS.

Neuroglial Cells Are Versatile

There are several different types of neuroglial cells in the CNS (Fig. 12.10), and much research is currently being conducted to determine how much "glial" cells contribute to the functioning of the central nervous system. Neuroglial cells outnumber neurons nine to one and take up more than half the volume of the brain, but until recently, they were thought to merely support and nourish neurons. There are no Schwann cells in the CNS; instead oligodendrocytes form myelin. Microglial cells, in addition to supporting neurons, also phagocytize bacterial and cellular debris. Aside from providing nutrients to neurons, astrocytes absorb the neurotransmitter glutamate, and they produce a growth factor known as *glial-derived growth factor* (*GDGF*) that someday might be used as a cure for Parkinson disease and other diseases caused by neuron degeneration. Neuroglial cells don't have long processes, but even so, researchers are now beginning to gather evidence that they do communicate among themselves and with neurons!

Spinal Cord Communicates

The spinal cord has two main functions: (1) it is the center for many reflex actions, and (2) it provides a means of communication between the brain and the spinal nerves, which leave the spinal cord.

The spinal cord has white matter and gray matter (see Fig. 12.7). Unmyelinated cell bodies and short fibers give *gray matter* its color. In cross section, the gray matter looks like a butterfly or the letter H. Portions of sensory neurons and motor neurons are found here, as are short interneurons that connect these two types of neurons.

Myelinated long fibers of interneurons that run together in bundles called *tracts* give *white matter* its color. These tracts connect the spinal cord to the brain. Dorsally, there are primarily ascending tracts taking information to the brain, and ventrally, there are primarily descending tracts carrying information from the brain. Because the tracts at one point cross over, the left side of the brain controls the right side of the body, and the right side of the brain controls the left side of the body.

The CNS, which lies in the midline of the body and consists of the brain and the spinal cord, receives sensory information and initiates motor control.

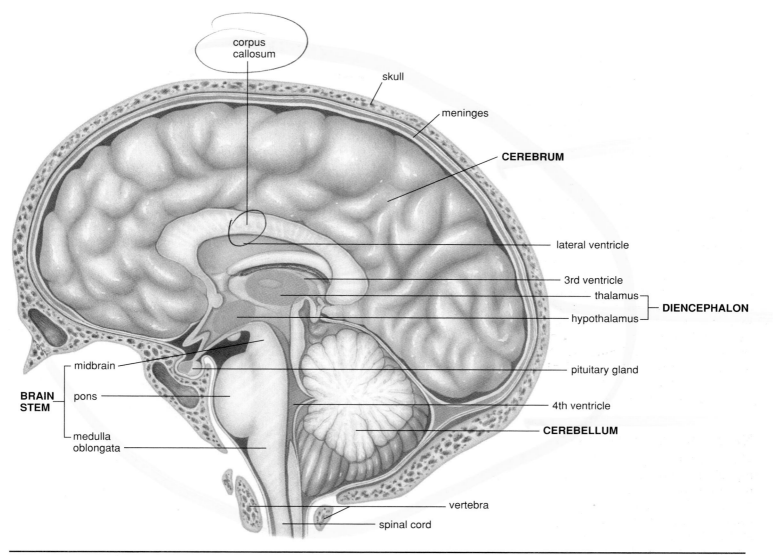

Figure 12.11　The human brain.
Note how large the cerebrum is compared to the rest of the brain.

The Brain Commands

The human brain is divided into these parts: medulla oblongata, cerebellum, pons, midbrain, hypothalamus, thalamus, and cerebrum. The brain has four cavities, called **ventricles:** two lateral ventricles, the third ventricle, and the fourth ventricle.

Brain Stem Is Closest to the Cord

The *brain stem*, which maintains life support systems, consists of the medulla oblongata, pons, and midbrain (Fig. 12.11). The **medulla oblongata** lies between the spinal cord and the pons and is anterior to the cerebellum. It contains a number of *vital centers* for regulating heartbeat, breathing, and vasoconstriction (blood pressure). It also contains

the reflex centers for vomiting, coughing, sneezing, hiccuping, and swallowing. The medulla contains tracts that ascend or descend between the spinal cord and the brain's higher centers.

The word *pons* means bridge in Latin, and true to its name, the pons contains bundles of axons traveling between the cerebellum and the rest of the CNS. In addition, the pons functions with the medulla to regulate breathing rate and has reflex centers concerned with head movements in response to visual and auditory stimuli.

The **midbrain** acts as a relay station for tracts passing between the cerebrum and the spinal cord or cerebellum, and it also has reflex centers for visual, auditory, and tactile responses.

Diencephalon Surrounds the Third Ventricle

The *diencephalon*, which consists of the hypothalamus and thalamus, facilitates emotions and regulates body functions. The **hypothalamus** forms the floor of the third ventricle. It maintains homeostasis, or the constancy of the internal environment, and contains centers for regulating hunger, sleep, thirst, body temperature, water balance, and blood pressure. The hypothalamus also controls the pituitary gland and thereby serves as a link between the nervous and endocrine systems.

The **thalamus** consists of two masses of gray matter located in the sides and roof of the third ventricle. It is the last portion of the brain for sensory input before the cerebrum. It processes this information and serves as a central relay station for sensory impulses traveling upward from other parts of the body and brain to the cerebrum. It receives all sensory impulses (except those associated with the sense of smell) and channels them to appropriate regions of the cortex for interpretation.

Cerebellum Coordinates Muscle Activity

The **cerebellum,** which lies below and behind the cerebrum, is separated from the brain stem by the fourth ventricle. It has two portions that are joined by a narrow median portion. The surface of the cerebellum is gray matter, and the interior is largely white matter. The cerebellum functions in muscle coordination, integrating impulses received from higher centers to ensure that all of the skeletal muscles work together to produce smooth and coordinated motions. The cerebellum is also responsible for maintaining normal muscle tone and transmitting impulses that maintain posture. It receives information from the inner ear indicating the position of the body and then sends impulses to the muscles, whose contractions maintain or restore balance.

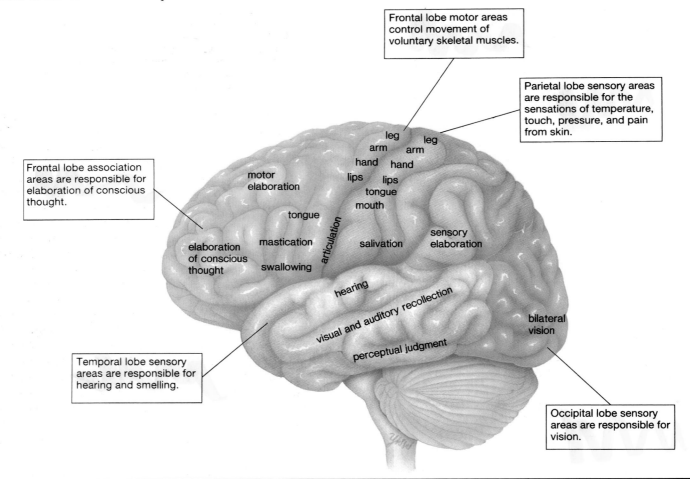

Figure 12.12   The cerebral cortex.
The cortex of the cerebrum is divided into four surface lobes: frontal, parietal, temporal, and occipital. Physiologists have mapped the cerebral cortex and report that each area has these functions.

Cerebrum Is Paramount

The **cerebrum,** the foremost part of the brain, is the primary area responsible for consciousness. It is the largest portion of the brain in humans. The outer layer of the cerebrum, called the *cerebral cortex,* is gray in color and contains cell bodies and unmyelinated short fibers. The cerebrum is divided into halves, known as the right and left **cerebral hemispheres,** which are connected by a bridge of nerve fibers called the *corpus callosum.* Each hemisphere contains four surface lobes: *frontal, parietal, temporal,* and *occipital* (Fig. 12.12). Little is known about the functions of a fifth lobe, the insula, which lies beneath the surface.

Certain areas of the cerebral cortex have been "mapped" in great detail. Physiologists have identified motor areas of the frontal lobe, which initiate contraction of various skeletal (voluntary) muscles, and sensory areas of the parietal lobe, which receive impulses from certain sense organs. The association areas, which are in communication with all the other lobes, integrate information into higher, more complex levels of consciousness. Association areas are concerned with intellect, artistic and creative abilities, learning, and memory. Their presence suggests that the brain has plasticity, that is, certain areas can take over the function of other areas if the need arises.

The *basal nuclei* constitute the central gray matter of the cerebrum. The precise functions of the basal nuclei are not known, but they may have some control over voluntary muscle action, because when they are diseased, Parkinson disease and Sydenham chorea may develop. Both of these are neuromuscular disturbances.

Consciousness is the province of the cerebrum, the most developed portion of the human brain. The cerebrum is responsible for higher mental processes, including the interpretation of sensory input and the initiation of voluntary muscular movements.

The Limbic System Controls Memory and Emotions

The **limbic system** involves portions of both the subconscious and the conscious brain (Fig. 12.13). It lies just beneath the cerebral cortex and contains neural pathways that connect portions of the frontal lobes, the temporal lobes, the thalamus, and the hypothalamus. The *basal nuclei* are also a part of the limbic system.

Stimulation of different areas of the limbic system causes the subject to experience pain, pleasure, rage, affection, sexual interest, fear, or sorrow. By causing pleasant or unpleasant feelings about experiences, the limbic system influences how the individual will behave in the future.

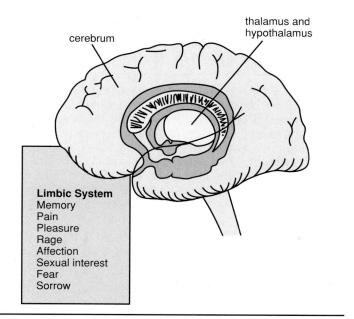

Limbic System
Memory
Pain
Pleasure
Rage
Affection
Sexual interest
Fear
Sorrow

Figure 12.13   The limbic system.
The limbic system (colored orange), which includes portions of the cerebrum, the thalamus, and the hypothalamus, is sometimes called the emotional brain because it seems to control the memory and emotions listed.

The limbic system also is involved in the processes of learning and memory. Learning requires memory, but just what permits memory development is not definitely known. Experimentation with invertebrates such as slugs and snails indicates that learning is accompanied by an increased use of some synapses in the brain, while forgetting involves a decrease in the use of synapses.

Experiments with monkeys have led to the conclusion that the limbic system is absolutely essential to both short-term and long-term memory. An example of short-term memory in humans is the ability to recall a telephone number long enough to dial it; an example of long-term memory is the ability to recall the events of the day. It is believed that at first, impulses move only within the limbic circuit, but eventually the basal nuclei transmit the neurotransmitter acetylcholine (ACh) to the sensory areas, where memories are stored. The involvement of the limbic system certainly explains why emotionally charged events result in our most vivid memories. The fact that the limbic system communicates with the sensory areas for touch, smell, vision, hearing, and taste accounts for the ability of any particular sensory stimulus to awaken a complex memory.

The limbic system is particularly involved in emotions and in memory and learning.

Health Focus

A Cure for Alzheimer and Parkinson Disease

It appears that Alzheimer and Parkinson disease might someday be treatable, according to recent research findings. Alzheimer disease (AD) is a disorder characterized by a gradual loss of reason that begins with memory lapses and ends with an inability to perform any type of daily activity. Personality changes signal the onset of AD. A normal 50- to 60-year-old adult might forget the name of a friend not seen for years. However, people with AD forget the name of a neighbor who visits daily. With time, they have trouble finding their way and cannot perform simple errands. People afflicted with AD become confused and tend to repeat the same question over and over. Other signs of mental disturbances eventually appear, and patients gradually become bedridden and die of a complication such as pneumonia.

A neuron damaged by Alzheimer disease (AD) is shown in Figure 12A. The AD neuron has two abnormalities not seen in the normal neuron. Bundles of fibrous protein, called neurofibrillary tangles, surround the nucleus in the cell, and protein-rich accumulations, called amyloid plaques, envelop the axon branches. These abnormal neurons are especially seen in the portions of the brain that are involved in reason and memory (frontal lobe and limbic system). In order to see the abnormal brain neurons, brain tissue must be examined microscopically after the patient dies.

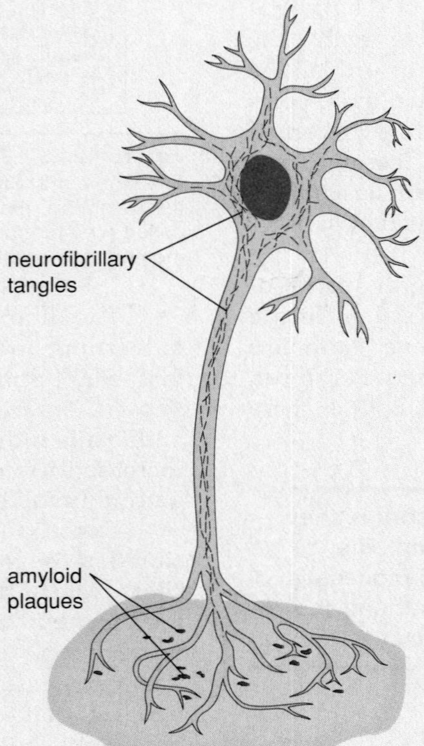

The likelihood of developing AD is higher if there is a family history of AD. Researchers have discovered that in some families whose members have a 50% chance of AD, a genetic defect is present on chromosome 21, the same chromosome associated with Down syndrome. Further, the genetic defect affects the normal production of amyloid precursor protein (APP), which may be the cause of the amyloid plaques.

Nevertheless, it appears that the neurotransmitter acetylcholine may be in short supply in the brains of patients with AD. Drugs that enhance acetylcholine production are now available for AD patients.

Parkinson is a disease of the basal nuclei that is usually not seen until after age 60. The three obvious signs of Parkinson disease are slowness of movement, tremor, and rigidity. The person walks with a small-stepped shuffle, and there is no swinging of the arms. An involuntary shaking of the hands occurs even when they are at rest. Rigidity of facial muscles gives the appearance of a masklike face although there is frequent eye blinking. All these symptoms are due to the degeneration of neurons in the basal nuclei that normally produce the neurotransmitter dopamine. Therefore, we know that dopamine is essential for normal coordinated movements.

The cause of Parkinson disease is not identifiable in most cases. Some families have a high incidence of the disease, but Parkinson can strike one twin and not the other. Therefore, it appears that Parkinson is not inherited. Parkinson-like symptoms are known to develop after carbon monoxide poisoning, encephalitis (inflammation of the brain), or the intake of certain drugs.

Dopamine cannot be administered as a drug because it cannot cross the *blood-brain barrier*—the neuroglial cells associated with the capillaries in the brain prevent many substances from entering the brain. But L-dopa, which can be converted to dopamine by brain neurons, does cross the blood-brain barrier, and it is presently given to Parkinson patients. Unfortunately, L-dopa begins to lose its therapeutic effectiveness as more dopamine-secreting neurons die off. Presently, some Parkinson patients are receiving neuron transplants, but this requires a major operation in which cells are implanted in the brain. A significant number of individuals are opposed to the use of fetal brain cells for these operations. As an alternative, researchers are experimenting with cultured cells from a number of sources, even from pigs.

Medical treatment for Alzheimer and Parkinson disease is most desirable, and another may have been found. It's been discovered that neuroglial cells produce neurotropic proteins that can be mass produced through genetic engineering of bacteria. Nerve growth factor (NGF) is being considered for treatment of Alzheimer disease, and glial-derived growth factor (GDGF) is being considered for treatment of Parkinson disease. If the administration of these drugs can stop the deterioration of brain cells, new medical treatment will be made available to all.

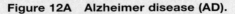

neurofibrillary tangles

amyloid plaques

Figure 12A   Alzheimer disease (AD).
An AD neuron has neurofibrillary tangles and amyloid plaques. AD neurons are particularly present in the frontal lobe and limbic system. This accounts for the development of symptoms of Alzheimer disease.

12.6 Drug Abuse

A wide variety of drugs can be used to alter the mood and/or emotional state (see Appendix D). Drugs that affect the nervous system have two general effects: (1) they impact the limbic system, and (2) they either promote or decrease the action of a particular neurotransmitter (Fig. 12.14). Stimulants increase the likelihood of excitation of the receiving neurons and depressants decrease the likelihood. Increasingly, researchers believe that dopamine, a neurotransmitter in the brain, is primarily responsible for mood. Cocaine is known to potentiate the effects of dopamine by interfering with its uptake from synaptic clefts. Many new medications developed to counter drug addiction and mental illness affect the release, reception, or breakdown of dopamine.

Drug abuse is apparent when a person takes a drug at a dose level and under circumstances that increase the potential for a harmful effect. Drug abusers are apt to display either a psychological and/or a physical dependence on the drug. Dependence has developed when the person spends much time thinking about the drug or arranging to get it and often takes more of the drug than was intended. With physical dependence, formerly called an addiction to the drug, the person is tolerant to the drug—that is, must increase the amount of the drug to get the same effect and has withdrawal symptoms when he or she stops taking the drug.

Drugs that affect the nervous system can cause physical dependence and withdrawal symptoms.

Alcohol

Wine, beer, and whiskey production require yeast fermentation of grape juice, barley, and usually corn or rye, respectively. It is possible that *alcohol* might influence the action of transmitters like GABA, an inhibiting transmitter, or glutamate, an excitatory neurotransmitter. Once imbibed, alcohol is primarily metabolized in the liver, where it disrupts the normal workings of glycolysis and the Krebs cycle. The cell begins to carry on fermentation, and lactic acid builds up. The pH of the blood decreases and becomes acidic.

Since the Krebs cycle is not working, fat cannot be broken down, and the liver turns fatty. Fat accumulation, the first stage of liver deterioration, begins after only a single night of heavy drinking. If heavy drinking continues, fibrous scar tissue appears during a second stage of deterioration. If heavy drinking stops, the liver can still recover and become normal once again. If not, the final and irrevocable stage, cirrhosis of the liver, occurs: liver cells die, harden, and turn orange (cirrhosis means orange).

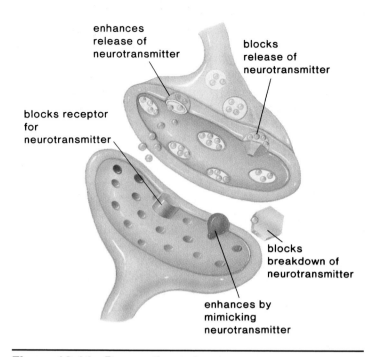

Figure 12.14 Drug actions at a synapse.
Each drug typically has one of these specific functions.

enhances release of neurotransmitter

blocks release of neurotransmitter

blocks receptor for neurotransmitter

blocks breakdown of neurotransmitter

enhances by mimicking neurotransmitter

The surgeon general recommends that pregnant women drink no alcohol at all. Alcohol crosses the placenta freely and causes *fetal alcohol syndrome*, which is characterized by mental retardation and various physical defects.

Good nutrition is difficult when a person is a heavy drinker. Alcohol contains considerable energy, and its breakdown can be coupled to the production of ATP molecules. However, alcohol, unlike many foods, contains little but calories; it does not supply amino acids, vitamins, or minerals. Without adequate vitamins, red blood cells and white blood cells cannot be formed in the bone marrow. The immune system is depressed, and the chances of stomach, liver, lung, pancreas, colon, and tongue cancer are increased. Protein digestion and amino acid metabolism is so upset that even adequate protein intake will not prevent amino acid deficiencies. Muscles atrophy and weakness results. Fat deposits occur in the heart wall and hypertension develops. There is an increased risk of cardiac arrhythmias and stroke.

Alcohol, which is the most abused drug in the United States, can lead to serious health consequences.

Nicotine

Nicotine, an alkaloid derived from tobacco, is a widely used neurological agent. When smoking a cigarette, nicotine is quickly distributed to all body organs, including the central and peripheral nervous systems. In the central nervous system, nicotine causes neurons to release dopamine, a neurotransmitter mentioned earlier. The excess of dopamine has a reinforcing effect that leads to dependence on the drug. In the peripheral nervous system, nicotine stimulates the same postsynaptic receptors as acetylcholine and leads to increased skeletal muscular activity. It also increases the heartbeat rate and blood pressure, and digestive tract mobility. Nicotine may even occasionally induce vomiting and/or diarrhea. It also causes water retention by the kidneys.

Many cigarette smokers find it difficult to give up the habit because nicotine induces both physiological and psychological dependence. Withdrawal symptoms include headache, stomach pain, irritability, and insomnia. Tobacco contains not only nicotine, it also contains many other harmful substances. Cigarette smoking contributes to early death from cancer, including not only lung cancer but also cancer of the larynx, mouth, throat, pancreas, and urinary bladder. Chronic diseases like bronchitis and emphysema are likely to develop, and there is increased risk of heart attack due to cardiovascular disease.

Now that women are as apt to smoke as men, lung cancer has surpassed breast cancer as a cause of death. Cigarette smoking in young women who are sexually active is most unfortunate because if they become pregnant, nicotine, like other psychoactive drugs, adversely affects a developing embryo and fetus.

Marijuana

The dried flowering tops, leaves, and stems of the Indian hemp plant *Cannabis sativa* contain and are covered by a resin that is rich in THC (tetrahydrocannabinol). The names *cannabis* and *marijuana* apply to either the plant or THC.

The effects of marijuana differ depending upon the strength and the amount consumed, the expertise of the user, and the setting in which it is taken. Usually, the user reports experiencing a mild euphoria along with alterations in vision and judgment, which result in distortions of space and time. Motor incoordination occurs, as well as the inability to concentrate and to speak coherently.

Intermittent use of low-potency marijuana generally is not associated with obvious symptoms of toxicity, but heavy use can produce chronic intoxication. Intoxication is recognized by the presence of hallucinations, anxiety, depression, rapid flow of ideas, body image distortions, paranoid reactions, and similar psychotic symptoms. The terms *cannabis psychosis* and *cannabis delirium* refer to such reactions.

Marijuana is classified as a hallucinogen. It is possible that, like LSD (lysergic acid diethylamide), it has an effect on the action of serotonin, an excitatory neurotransmitter in the brain.

The use of marijuana does not seem to produce physical dependence, but a psychological dependence on the euphoric and sedative effects can develop. Craving or difficulty in stopping use also can occur as a result of regular heavy use.

Marijuana has been called a *gateway drug* because adolescents who have used marijuana also tend to try other drugs. For example, in a study of 100 cocaine abusers, 60% had smoked marijuana for more than 10 years.

Usually, marijuana is smoked in a cigarette form called a joint. Since this allows toxic substances, including carcinogens, to enter the lungs, chronic respiratory disease and lung cancer are considered dangers of long-term, heavy use. Some researchers claim that marijuana use leads to long-term brain impairment. Others report that males and females suffer reproductive dysfunctions. *Fetal cannabis syndrome,* which resembles fetal alcohol syndrome, has been reported.

Some psychologists are very concerned about the use of marijuana among adolescents. Marijuana can be used to avoid dealing with the personal problems that often develop during this maturational phase.

Although marijuana does not produce physical dependence, it does produce psychological dependence.

Cocaine

Cocaine is an alkaloid derived from the shrub *Erythroxylum cocoa.* Cocaine is sold in powder form and as *crack,* a more potent extract (Fig. 12.15). Users often describe the feeling of euphoria that follows intake of the drug as a *rush.* Snorting (inhaling) produces this effect in a few minutes, injection within 30 seconds, and smoking in less than 10 seconds. The rush only lasts a few seconds and then is replaced by a state of arousal, which lasts from 5 minutes to 30 minutes. Then the user begins to feel restless, irritable, and depressed. To overcome these symptoms, the user is apt to take more of the drug, repeating the cycle until there is no more drug left. A binge of this sort can go on for days, after which the individual suffers a crash. During the binge period, the user is hyperactive and has little desire for food or sleep but has an increased sex drive. During the crash period, the user is fatigued, depressed, and irritable, has memory and concentration problems, and displays no interest in sex. Indeed, men are often impotent. Other drugs, such as marijuana, alcohol, or heroin, often are taken to ease the symptoms of the crash.

Like nicotine, cocaine increases the concentration of dopamine in synaptic clefts. However, cocaine affects the reuptake of dopamine. After release into a synapse, dopamine ordinarily is withdrawn into the presynaptic cell for recycling and reuse. Because cocaine prevents this reuptake, there is an excess of dopamine in the synaptic cleft, and the user experiences the sensation of a rush.

The epinephrine-like effects of dopamine account for the state of arousal that lasts for some minutes after the rush experience.

With continued cocaine use, the body begins to make less dopamine to compensate for a seemingly excess supply. The user, therefore, now experiences *tolerance, withdrawal symptoms,* and an intense *craving* for cocaine. These are indications that the person is highly dependent upon the drug or, in other words, that cocaine is extremely addictive. Overdosing on cocaine is a real possibility. The number of deaths from cocaine and the number of emergency-room admissions for drug reactions involving cocaine have increased greatly. High doses can cause seizures and cardiac and respiratory arrest.

Individuals who snort the drug can suffer damage to the nasal tissues and even perforation of the septum between the nostrils. It is possible that long-term cocaine abuse causes brain damage; babies born to addicts suffer withdrawal symptoms and may suffer neurological and developmental problems.

Heroin

Heroin is derived from morphine, an alkaloid of *opium.* Heroin usually is injected. After intravenous injection, the onset of action is noticeable within one minute and reaches its peak in 3–6 minutes. There is a feeling of euphoria along with relief of pain. Side effects can include nausea, vomiting, dysphoria, and respiratory and circulatory depression leading to death.

Heroin binds to receptors meant for the **endorphins,** the special neurotransmitters that kill pain and produce a feeling of tranquility. They are believed to alleviate pain by preventing the release of a neurotransmitter termed substance P from certain sensory neurons in the region of the spinal cord. When substance P is released, pain is felt, and when substance P is not released, pain is not felt. Endorphins and heroin also bind to receptors on neurons that travel from the spinal cord to the limbic system. Stimulation of these can cause a feeling of pleasure.

Individuals who inject heroin become physically dependent on the drug. With time, the body's production of endorphins decreases. *Tolerance* develops so that the user needs to take more of the drug just to prevent *withdrawal* symptoms. The euphoria originally experienced upon injection is no longer felt.

Heroin withdrawal symptoms include perspiration, dilation of pupils, tremors, restlessness, abdominal cramps, gooseflesh, defecation, vomiting, and increase in systolic pressure and respiratory rate. Those who are excessively dependent may experience convulsions, respiratory failure, and death. Infants born to women who are physically dependent also experience these withdrawal symptoms.

Cocaine and heroin produce a very strong physical dependence. An overdose of these drugs can cause death.

a.

b.

Figure 12.15 Cocaine use.
a. Crack is the ready-to-smoke form of cocaine. It is a more potent and a more deadly form than the powder. **b.** Users often smoke crack in a glass water pipe. The high produced consists of a "rush" lasting a few seconds, followed by a few minutes of euphoria. Continuous use makes the user extremely dependent on the drug.

HUMAN SYSTEMS WORK TOGETHER

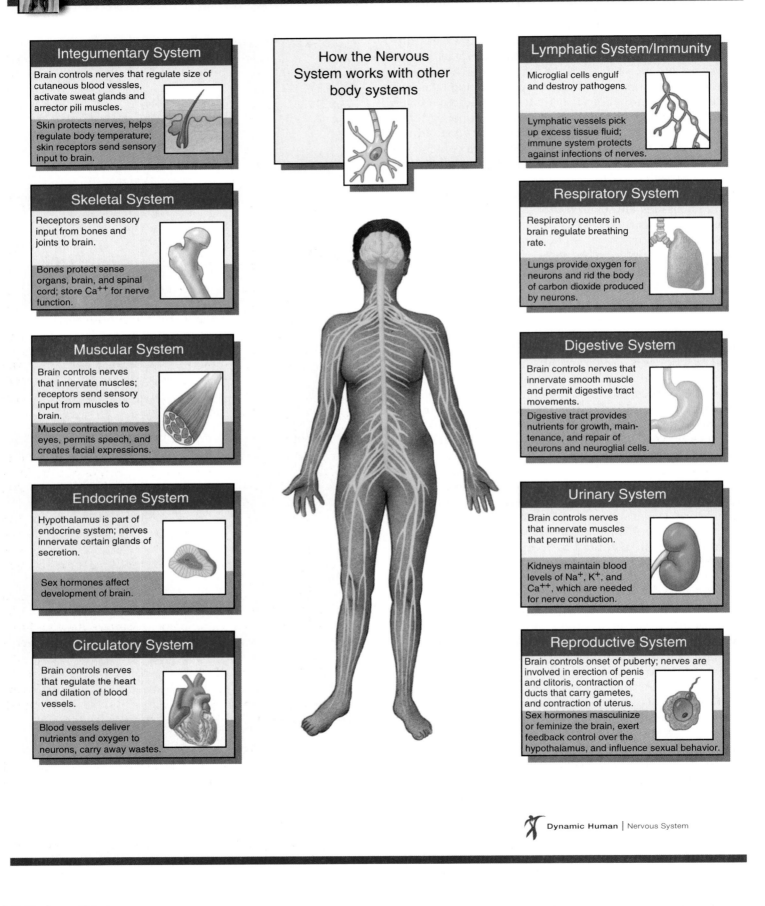

Integumentary System

Brain controls nerves that regulate size of cutaneous blood vessels, activate sweat glands and arrector pili muscles.

Skin protects nerves, helps regulate body temperature; skin receptors send sensory input to brain.

Skeletal System

Receptors send sensory input from bones and joints to brain.

Bones protect sense organs, brain, and spinal cord; store Ca^{++} for nerve function.

Muscular System

Brain controls nerves that innervate muscles; receptors send sensory input from muscles to brain.

Muscle contraction moves eyes, permits speech, and creates facial expressions.

Endocrine System

Hypothalamus is part of endocrine system; nerves innervate certain glands of secretion.

Sex hormones affect development of brain.

Circulatory System

Brain controls nerves that regulate the heart and dilation of blood vessels.

Blood vessels deliver nutrients and oxygen to neurons, carry away wastes.

How the Nervous System works with other body systems

Lymphatic System/Immunity

Microglial cells engulf and destroy pathogens.

Lymphatic vessels pick up excess tissue fluid; immune system protects against infections of nerves.

Respiratory System

Respiratory centers in brain regulate breathing rate.

Lungs provide oxygen for neurons and rid the body of carbon dioxide produced by neurons.

Digestive System

Brain controls nerves that innervate smooth muscle and permit digestive tract movements.

Digestive tract provides nutrients for growth, maintenance, and repair of neurons and neuroglial cells.

Urinary System

Brain controls nerves that innervate muscles that permit urination.

Kidneys maintain blood levels of Na^+, K^+, and Ca^{++}, which are needed for nerve conduction.

Reproductive System

Brain controls onset of puberty; nerves are involved in erection of penis and clitoris, contraction of ducts that carry gametes, and contraction of uterus.

Sex hormones masculinize or feminize the brain, exert feedback control over the hypothalamus, and influence sexual behavior.

Dynamic Human | Nervous System

Methamphetamine (Ice)

Methamphetamine is related to amphetamine, a well-known stimulant. Both methamphetamine and amphetamine have been drugs of abuse for some time, but a new form of methamphetamine known as "ice" is now used as an alternative to cocaine. Ice is a pure, crystalline hydrochloride salt that has the appearance of sheetlike crystals. Unlike cocaine, ice can be illegally produced in this country in laboratories and does not need to be imported.

Ice, like crack, will vaporize in a pipe, so it can be smoked, avoiding the complications of intravenous injections. After rapid absorption into the bloodstream, the drug moves quickly to the brain. It has the same stimulatory effect as cocaine, and subjects report they cannot distinguish between the two drugs after intravenous administration. Methamphetamine effects, however, persist for hours instead of a few seconds. Therefore, it is the preferred drug of abuse by many.

Designer Drugs

Designer drugs are analogs; that is, they are chemical compounds of controlled substances slightly altered in molecular structure. One such drug is MPPP (1-methyl-4-phenylprionoxy-piperidine), an analog of the narcotic fentanyl. Even small doses of the drug are very toxic and can cause death.

12.7 Working Together

The Working Together box on page 260 shows how the nervous system works with the other systems of the body to maintain homeostasis.

SUMMARY

12.1 Neuron Structure

The anatomical unit of the nervous system is the neuron, of which there are three types: sensory, motor, and interneuron. Each of these is made up of dendrites, a cell body, and an axon. Long fibers (dendrites and axons) are covered by a myelin sheath.

12.2 Nerve Impulse

When an axon is not conducting a nerve impulse, the resting potential indicates that the inside of an axon is negative compared to the outside. Because of the sodium-potassium pump, there is a concentration of Na^+ outside an axon and K^+ inside an axon.

When an axon is conducting a nerve impulse, an action potential (i.e., electrochemical change) travels along a neuron. Depolarization (inside becomes positive) is due to the movement of Na^+ to the inside, and then repolarization (inside becomes negative again) is due to the movement of K^+ to the outside of an axon.

12.3 Transmission Across a Synapse

Transmission of the nerve impulse from one neuron to another takes place across a synapse. In humans, synaptic vesicles release a chemical, known as a neurotransmitter, into the synaptic cleft. The binding of the neurotransmitter to receptors in the postsynaptic membrane increases the chance of a nerve impulse (stimulation) or decreases the chance of a nerve impulse (inhibition) in the next neuron.

12.4 Peripheral Nervous System

The peripheral nervous system contains the somatic system and the autonomic system. Nerves, such as cranial and spinal nerves, take impulses to and from the CNS.

Reflexes are automatic, and some do not require involvement of the brain. A simple reflex requires the use of neurons that make up a reflex arc. In the somatic system, a sensory neuron conducts the nerve impulse from a receptor to an interneuron, which in turn transmits the impulse to a motor neuron, which conducts it to an effector.

While the motor portion of the somatic (voluntary) division of the PNS controls skeletal muscle, the motor portion of the autonomic (involuntary) division controls smooth muscle of the internal organs and glands. The sympathetic system, which is often associated with those reactions that occur during times of stress, and the parasympathetic system, which is often associated with those activities that occur during times of relaxation, are both parts of the autonomic system.

12.5 Central Nervous System

Most of the cells in the CNS are neuroglial cells, which recently have been shown to participate actively in the functioning of the system and to produce growth factors that can possibly be used to cure neurological diseases.

The CNS consists of the spinal cord and brain. The gray matter of the cord contains neuron cell bodies; the white matter contains tracts that consist of the long axons of interneurons. These run from all parts of the cord, even up to the cerebrum.

The brain integrates all nervous system activity and commands all voluntary activities. In the brain stem, the medulla oblongata and pons have centers for vital functions, like breathing and the heartbeat. The cerebellum coordinates muscle contractions. In the diencephalon, the hypothalamus, in particular, controls homeostasis, and the thalamus specializes in sense reception. The cerebrum, which is responsible for consciousness, can be mapped, and each lobe seems to have particular functions.

12.6 Drug Abuse

Neurological drugs, although quite varied, have been found to affect the limbic system by either promoting or preventing the action of neurotransmitters.

12.7 Working Together

The nervous system works with the other systems of the body in the ways described in the box on page 260.

STUDYING THE CONCEPTS

1. What are the three types of neurons? How are they similar, and how are they different?   242
2. What does the term resting potential mean, and how is it brought about? Describe the two parts of an action potential and the changes that can be associated with each part.   244
3. What is the sodium-potassium pump, and when is it active?   244
4. What is a neurotransmitter, where is it stored, how does it function, and how is it destroyed? Name two well-known neurotransmitters.   246–47
5. What are the two main divisions of the nervous system? Explain why these names are appropriate.   247
6. Distinguish between cranial and spinal nerves.   248
7. Trace the path of a reflex action after discussing the structure and the function of the spinal cord and the spinal nerve.   249
8. What is the autonomic system, and what are its two major divisions? Give several similarities and differences between these divisions.   251
9. Name the major parts of the brain, and give a function for each.   253–55
10. Describe the physiological effects and mode of action of alcohol, marijuana, cocaine, and heroin.   257–59

APPLYING YOUR KNOWLEDGE

Concepts

1. Explain the reason that drinking an alcoholic beverage has an almost immediate effect on the body.
2. What is an advantage of a reflex action? What is a disadvantage of a reflex action?
3. Poliomyelitis, polio, is a viral disease in which the virus destroys the motor nerve cell bodies in the anterior horns of the spinal cord. What are the effects of this disease?
4. What is the reason that being angry or excited during the eating of a meal often results in an uneasy feeling associated with the digestive tract and food remaining in the stomach for a longer than normal period?

Bioethical Issue

Some people argue that mind-altering drugs should be legalized. Marijuana enthusiasts, for example, point out that doctors sometimes prescribe the drug to help terminally ill patients deal with pain. Therefore, it should be available to everyone. They also note that alcohol, a legal depressant, may do as much damage to a person's health as illegal substances.

Still, many are strongly opposed to legalizing drugs. They argue that people who take drugs such as LSD are not just risking their own lives; they also risk the lives of those around them. Psychoactive drugs impair judgment and perception, and a hallucinating person may grow violent or act irresponsibly.

Should psychoactive drugs be legalized? If not, who should be allowed access to these drugs and under what conditions? Offer reasons for your answers.

APPLYING TECHNOLOGY

Your study of the nervous system is supported by these available technologies:

Exploring the Internet

The Mader Home Page provides further resources for studying this chapter.

http://www.mhhe.com/sciencemath/biology/mader/

(Click on *Human Biology*.)

Dynamic Human: Nervous System CD-ROM

In *Anatomy,* the brain is shown in 3-dimension, the lobes of the cerebral cortex are reviewed, and a cross section of the spinal cord is explained; four *Explorations* pertain to the nervous system: Innervation of the Tongue, Motor and Sensory Pathways, Neural Network, and the Reflex Arc.; in *Histology,* microscopic slides of neurons and the spinal cord are shown; in *Clinical Concepts,* three different types of strokes are emphasized.

Explorations in Human Biology CD-ROM

Nerve Conduction (#8) The speed of the nerve impulse varies as students change the degree of myelination and the size of the axon. **(2)***

Synaptic Transmission (#9) The occurrence of a nerve impulse in a neuron varies as students change the amount of excitatory and inhibitory neurotransmitter released by the presynaptic membrane and the receptor composition of the postsynaptic membrane. **(3)**

Life Science Animations Video

Video #3: Animal Biology, Part 1

Formation of Myelin Sheath (#22) Schwann cells are seen wrapping around an axon to form the myelin sheath of a nerve fiber. The similar role of oligodendrocytes in the central nervous system is reviewed. **(1)**

Saltatory Nerve Conduction (#23) Sodium and potassium ions are seen moving across an axomembrane as the mechanics of the nerve impulse are explained. Ion movement only at nodes is shown in a myelinated fiber. **(3)**

Signal Integration (#24) Inhibitory and excitatory signals are seen arriving at a dendrite, and the proportion of each influences whether the axon fires or not. **(2)**

Reflex Arc (#25) The knee jerk is used as an example of the reflex arc. **(1)**

*Level of difficulty

TESTING YOUR KNOWLEDGE

1. Label this diagram.

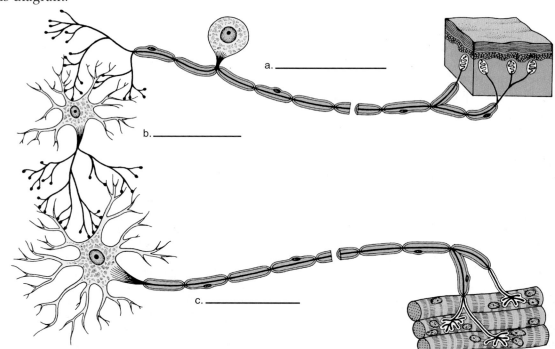

a. _____

b. _____

c. _____

2. A(n) _____ carries nerve impulses away from the cell body.

3. During the upswing of the action potential, _____ ions are moving to the _____ of the nerve fiber.

4. The space between the axon of one neuron and the dendrite of another is called the _____.

5. ACh is broken down by the enzyme _____ after ACh has altered the permeability of the postsynaptic membrane.

6. Motor nerves innervate _____ and _____.

7. In a reflex arc, only the neuron called the _____ is completely within the CNS.

8. The vagus nerve is a(n) _____ nerve that controls _____.

9. The brain and the spinal cord are covered by protective layers called _____.

10. The _____ is the major part of the brain that allows us to be conscious.

11. The _____ is the part of the brain responsible for coordination of body movements.

12. Many cigarette smokers are dependent upon the drug _____, and this is why they have difficulty giving up their habit.

SELECTED KEY TERMS

acetylcholine (ACh) (uh-set-ul-KOH-leen) Neurotransmitter active in both the peripheral and central nervous systems.   246

acetylcholinesterase (AChE) (uh-set-ul-koh-luh-NES-tuh-rays) Enzyme that breaks down acetylcholine bound to postsynaptic receptors within a synapse.   246

action potential Polarity changes due to the movement of ions across the plasma membrane of an active neuron.   244

autonomic system (awt-uh-NAHM-ik) Branch of the peripheral nervous system that has involuntary control over the internal organs; consists of the sympathetic and parasympathetic systems.   251

axon Fiber of a neuron that conducts nerve impulses away from the cell body.   242

central nervous system (CNS) Portion of the nervous system consisting of the brain and spinal cord.   242

cerebellum (ser-uh-BEL-um) Part of the brain located inferior to the cerebrum that coordinates skeletal muscles to produce smooth, coordinated motions.   254

cerebral hemisphere One of the large, paired structures that together constitute the cerebrum of the brain.   255

cerebrospinal fluid (suh-ree-broh-SPYN-ul) Fluid found in the ventricles of the brain, in the central canal of the spinal cord, and in association with the meninges.   252

cerebrum (suh-REE-brum) Part of the brain that provides higher mental function and consists of two large masses, or cerebral hemispheres.   255

cranial nerve Nerve that arises from the brain.   248

dendrite Part of a neuron, typically branched, that conducts signals toward the cell body.   242

effector Structure such as a muscle or a gland that allows an organism to respond to internal and external stimuli.   249

endorphin Neuropeptide synthesized in the pituitary gland that suppresses pain.   259

ganglion (GANG-glee-un) Collection of neuron cell bodies within the peripheral nervous system.   248

hypothalamus (hy-poh-THAL-uh-mus) Part of the brain located below the thalamus that helps regulate the internal environment of the body.   254

limbic system Portion of the brain concerned with memory and emotions.   255

limbic system Portion of the brain concerned with memory and emotions.   255

medulla oblongata A portion of the brain located between the pons and the spinal cord that is concerned with the control of internal organs.   253

midbrain Most superior part of the brain stem that contains traits and reflex centers.   253

myelin sheath (MY-uh-lun) Schwann plasma membranes that cover long neuron fibers, giving them a white, glistening appearance.   243

neuroglial cell (nuh-ROH-glee-uhl) One of several types of cells found in nervous tissue that supports, protects, and nourishes neurons.   243

neuron Nerve cell that characteristically has three parts: dendrites, cell body, and axon; occurs as sensory neuron, motor neuron, and interneuron.   242

neurotransmitter Chemical stored at the ends of axons that is responsible for signal transmission across a synapse.   246

nerve Bundle of fibers outside the central nervous system.   248

nerve impulse Action potential (electrochemical change) traveling along a neuron.   244

node of Ranvier Gap in the myelin sheath around a nerve fiber.   243

norepinephrine (NE) (nor-eh-puh-NEH-frun) Neurotransmitter active in the peripheral and central nervous systems; also a hormone secreted by the adrenal medulla in times of stress.   246

parasympathetic system Part of the autonomic system that usually promotes activities associated with a restful state.   251

peripheral nervous system (PNS) Nerves and ganglia that lie outside the central nervous system.   242

receptor Sensory structure specialized to receive information from the environment and to generate nerve impulses.   249

reflex Automatic, involuntary response of an organism to a stimulus.   249

resting potential Polarity across the plasma membrane of a resting fiber due to an unequal distribution of ions.   244

sodium-potassium pump Plasma membrane transport protein that moves sodium ions out of and potassium ions into cells; important in nerve and muscle cells.   244

somatic system Portion of the peripheral nervous system containing motor neurons that control skeletal muscles.   248

spinal nerve Nerve that arises from the spinal cord.   248

sympathetic system Part of the autonomic system that usually promotes activities associated with emergency (fight-or-flight) situations.   251

synapse (SIN-aps) Region between two neurons where information is transmitted from one to the other, usually from the axon to the dendrite or cell body of the next neuron.   246

synaptic cleft Small gap between presynaptic and postsynaptic membranes of a synapse.   246

thalamus Part of the brain located in the sides and roof of the third ventricle that serves as the integrating center for sensory input; it plays a role in arousing the cerebral cortex.   254

ventricle Cavity in an organ, such as the ventricles of the brain.   253

Chapter 13

Senses

Chapter Outline

Figure 13.1   Eating habits.
Do we like certain foods because we associate their taste and smell with fun?

After a bad day, what's your favorite thing to eat? Chocolate? Mom's homemade pasta? So-called "comfort food" soothes our spirits and—if only for a minute—makes the world seem okay again.

That's because we learn to link certain tastes and smells with emotion (Fig. 13.1). Sensory cells, like those found in the nose, send messages to the parts of our brain that control emotion and memory. So we remember those freshly baked cookies that once brightened a depressing day. We may then reach for a cookie when we feel down.

Not only do we depend on taste and smell to experience pleasure, they also help ensure our very survival. In nature, animals learn to avoid tempting substances that lead to illness. Likewise, humans shy away from foods linked to negative experiences. A person who comes down with food poisoning at a Japanese restaurant, for example, is unlikely to crave the taste of sushi in the near future. Smell alone can sometimes be protective. Many lives have been saved by a sensitive nose picking up a whiff of smoke or poisonous vapor or detecting spoilage in food.

The senses work together. Once an animal has experienced a noxious substance, it may recognize it by sight alone in the future. Eyes watch out for imminent dangers and ears hear warning sounds or we would never learn to drive in traffic. These same organs allow us to communicate with others and enrich our lives as when we see a play or go to the ballet.

Sense organs are described as the "windows of the brain" because they generate nerve impulses that travel to the central nervous system. Those impulses reaching the cerebrum of the brain result in conscious sensation.

13.1 Receptors and Sensations

Receptors are structures specialized to receive certain environmental stimuli and generate nerve impulses.

Receptors According to Stimuli

Receptors in humans can be classified into just six types.

Chemoreceptors are sensitive to chemical substances in the immediate vicinity. The sense of taste and smell are well known to utilize this type of receptor, but there are also chemoreceptors in various other organs that are sensitive to internal conditions. One of these monitors the hydrogen ion concentration $[H^+]$ in the blood, and if the pH lowers, the breathing rate increases. As more carbon dioxide is expired, the blood pH will rise.

Mechanoreceptors are stimulated by mechanical forces, which are most often pressure of some sort. The sense of touch is dependent on pressure receptors that are sensitive to either strong or slight pressures. Baroreceptors located in the wall of the aorta sense changes in blood pressure, and stretch receptors in the lungs sense the degree of lung inflation. Even hearing is dependent on mechanoreceptors. In this case, the receptors are sensitive to pressure waves in inner ear fluid. Pressure receptors that provide information regarding equilibrium are also located in the inner ear.

Proprioceptors sense the degree of muscle contraction, the stretch of tendons, and the movement of ligaments. The information sent by muscle spindles to the central nervous system is used to maintain the body's posture despite the force of gravity always acting upon the skeleton and muscles. Golgi tendon organs act to decrease the degree of muscle contraction. The information they provide helps ensure that muscles are not pulled away from their attachments by the degree of muscle contraction.

Thermoreceptors are stimulated by changes in temperature. Those that respond when temperatures rise are called heat receptors, and those that respond when temperatures lower are called cold receptors. There are internal thermoreceptors in the hypothalamus and surface thermoreceptors in the skin.

Pain receptors are technically called nociceptors. They are naked dendrites that respond to chemicals released by damaged tissues or to excess stimuli of heat or pressure. It is very protective for us to have pain receptors because pain receptors alert us to possible danger. Without the pain of appendicitis, we may never seek the medical help that is needed to avoid a ruptured appendix.

Photoreceptors detect light. Eyes contain photoreceptors, which are responsible for the sense of vision because they are sensitive to light.

Receptors initiate nerve impulses, which are conducted to the CNS. Sensation and the interpretation of stimuli is dependent upon the brain.

Senses According to Receptors

These various types of receptors are responsible for the senses we have.

Somatic Senses

Senses that are associated with the skin, muscles, joints, and internal organs are called *somatic senses*. Mechanoreceptors in the skin give us a sense of touch and throughout the body a sense of pressure. A sense of temperature is due to receptors located in the skin and the brain. A sense of pain is due to pain receptors located in the skin and also among internal organs. Proprioception, which is a sense of knowing the position of the limbs, is due to proprioceptors like muscle spindles.

Special Senses

The sense organs for taste, smell, vision, equilibrium, and hearing contain a number of receptors all specialized to give us a particular sense. Table 13.1 lists the special senses and their associated sense organs.

TABLE 13.1			
Special Sense Organs			
Sense	**Type of Receptor**	**Specific Receptors**	**Sense Organ**
Taste	Chemoreceptor	Taste cells	Taste buds
Smell	Chemoreceptor	Olfactory cells	Olfactory epithelium
Vision	Photoreceptor	Rods and cones in retina	Eye
Equilibrium	Mechanoreceptor	Hair cells in utricle, saccule, and semicircular canals	Ear
Hearing	Mechanoreceptor	Hair cells in organ of Corti	Ear

13.2 Skin

The dermis of skin contains receptors for touch, pressure, pain, and temperature (Fig. 13.2). It is a mosaic of these tiny receptors, as you can determine by slowly passing a metal probe over your skin. At certain points, there is a feeling of pressure, and at others, there is a feeling of hot or cold (depending on the temperature of the probe). Certain parts of the skin contain more receptors for a particular sensation; for example, the fingertips have an abundance of touch receptors.

The sense of touch is dependent on mechanoreceptors in the skin. *Pacinian corpuscles* are pressure receptors located deep in the dermis. *Meissner corpuscles* and *Merkel disks* are touch receptors located closer to the surface in the skin. Some think that *Ruffini end organs* may be heat receptors and *end-bulbs of Krause* may be cold receptors, but there are others who believe that these are modified pressure receptors. They think that the dendrites of certain sensory neurons are the actual thermoreceptors. Certainly, naked dendrites are responsible for the sensation of pain. The skin has pain receptors, but so do internal organs.

Adaptation occurs when a receptor becomes so accustomed to stimulation that it stops generating impulses, even though the stimulus is still present. The touch receptors adapt: soon after we put on an article of clothing, we are no longer aware of the feel of clothes against our skin.

Specialized receptors in the human skin respond to touch, pressure, pain, and temperature (hot and cold).

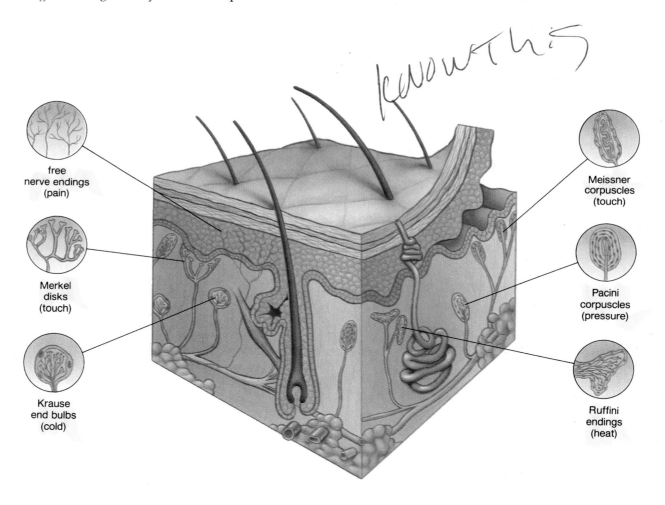

free nerve endings (pain)

Merkel disks (touch)

Krause end bulbs (cold)

Meissner corpuscles (touch)

Pacini corpuscles (pressure)

Ruffini endings (heat)

Figure 13.2 Receptors in human skin.
The classical view shown here is that each receptor has the main function indicated. However, investigations in this century indicate that matters are not so clear-cut. For example, microscopic examination of the skin of the ear shows only free nerve endings (pain receptors), and yet the skin of the ear is sensitive to all sensations. Therefore, it appears that the receptors of the skin are somewhat, but not completely, specialized.

13.3 Chemoreceptors

There are chemoreceptors located in the aortic arch that are sensitive to the pH of the blood. When the pH drops, they signal the brain, and immediately thereafter, the breathing rate increases. The expiration of CO_2 will raise the pH of the blood.

Taste and smell are called the chemical senses because the receptors for these senses are sensitive to chemical substances in the food we eat and air we breathe.

Taste Buds in the Mouth

Taste buds are located primarily on the tongue (Fig. 13.3). Many lie along the walls of the papillae, the small elevations on the tongue that are visible to the naked eye. Isolated ones are also present on the hard palate, the pharynx, and the epiglottis.

Taste buds are pockets of cells that extend through the tongue epithelium and open at a taste pore. Taste buds have supporting cells and a number of elongated taste cells that end in *microvilli*. A plasma membrane receptor is a protein, or more likely a glycoprotein, that projects from the surface of the cell and binds to a particular molecule. The microvilli bear plasma membrane receptors for certain chemicals. The binding of a molecule to a plasma membrane receptor causes electrochemical changes that lead to the generation of nerve impulses in associated sensory nerve fibers. These nerve impulses go to the parietal lobe of the cerebrum.

It is believed that there are four types of tastes (bitter, sour, salty, sweet) and that taste buds for each are concentrated on the tongue in particular regions (Fig. 13.3a). Sweet receptors are most plentiful near the tip of the tongue. Sour receptors occur primarily along the margins of the tongue. Salty receptors are most common on the tip and the upper front portion of the tongue. Bitter receptors are located toward the back of the tongue. Associated sensory fibers have graded, rather than all-or-nothing, sensitivities to the four taste types. The brain appears to survey the overall pattern of incoming sensory impulses and to take a "weighted average" of their taste messages as the perceived taste. Information goes directly to the cerebrum and also to the brain stem. This may have survival value because it suggests that babies will continue to nurse even in the absence of conscious control.

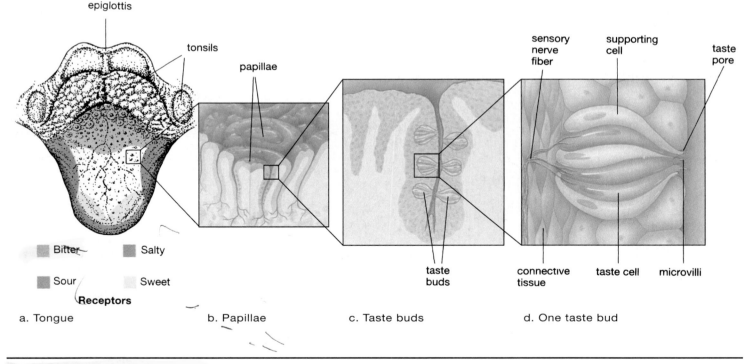

Figure 13.3   Taste buds.
a. Elevations on the tongue are called papillae. Shown here are locations of taste buds responsive to sweet, sour, salty, and bitter. **b.** Enlargement of papillae. **c.** The taste buds occur along the walls of the papillae. **d.** The various cells that make up a taste bud. Taste cells in a bud end in microvilli that are sensitive to the chemicals exhibiting the tastes noted in **(a).** When the chemicals combine with membrane-bound receptors, nerve impulses are generated.

Olfactory Cells in the Nose

Our sense of smell is dependent on **olfactory cells** located high in the roof of the nasal cavity (Fig. 13.4). Olfactory cells are modified neurons. Each cell ends in a tuft of about five olfactory cilia, which bear plasma membrane receptors for various chemicals. When odorous molecules bind to the plasma membrane receptors, nerve impulses pass along nerve fibers to the olfactory bulb, which is located in the front of the brain. Some processing probably occurs here before olfactory information is sent primarily to the temporal lobe of the cerebrum, which produces the sensation of smell.

Recent work has determined that there are around 1,000 different odor receptors. Many olfactory cells have the same specific type of receptors. A given odor activates a characteristic combination of cells, and this information is pooled in the olfactory bulb. The brain then determines the precise pattern of the types of receptors activated.

The olfactory receptors, like touch and temperature receptors, adapt to outside stimuli. In other words, after a while, the presence of a particular chemical no longer causes the olfactory cells to generate nerve impulses, and we are no longer aware of a particular smell.

The sense of taste and the sense of smell supplement each other, creating a combined effect when interpreted by the cerebrum. For example, when you have a cold, you think that food has lost its taste, but actually you have lost the ability to sense its smell. This method works in reverse also. When you smell something, some of the molecules move from the nose down into the mouth region and stimulate the taste buds there. Therefore, part of what we refer to as smell may actually be taste (Fig. 13.5).

The receptors for taste (taste cells) and the receptors for smell (olfactory cells) work together to give us our sense of taste and our sense of smell.

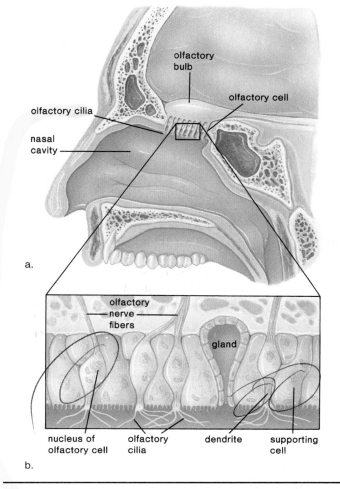

Figure 13.4   Olfactory cell location and anatomy.
a. The olfactory epithelium in humans is located high in the nasal cavity. **b.** Enlargement of the olfactory cells shows they are modified neurons located between supporting cells. When olfactory cells are stimulated by chemicals, olfactory nerve fibers conduct nerve impulses to the olfactory bulb. An olfactory tract within the bulb takes the nerve impulses to the brain.

Figure 13.5   Sensation and diet.
If you are on a diet, it is safer to eat the same monotonous food each day instead of tempting the palate. A variety of interesting tastes and smells can cause you to eat more than is warranted.

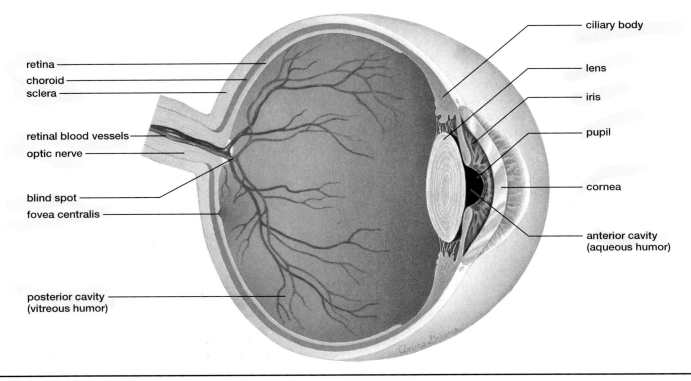

Figure 13.6   **Anatomy of the human eye.**
Notice that the sclera, the outer layer of the eye, becomes the cornea and that the choroid, the middle layer, is connected to the ciliary body and the iris. The retina, the inner layer, contains the receptors for vision; the fovea centralis is the region where vision is most acute. A blind spot occurs where the optic nerve leaves the retina; there are no receptors for light in this location.

13.4 Eyes

The anatomy and physiology of the eye are complex, and we will consider each topic separately.

How the Eye Looks

The eyeball is an elongated sphere about 2.5 cm in diameter and has three layers, or coats: the sclera, the choroid, and the retina (Fig. 13.6 and Table 13.2). The outer layer, the **sclera,** is white and fibrous except for the transparent cornea, the window of the eye. The middle, thin, dark-brown layer, the **choroid,** is vascular and absorbs stray light rays. Toward the front, the choroid thickens and forms the ring-shaped ciliary body. The **ciliary body** contains the *ciliary muscle,* which controls the shape of the lens for near and far vision. Also, the choroid becomes a thin, circular, muscular, pigmented diaphragm, the **iris,** which regulates the size of the **pupil,** a hole in the center through which light enters the eyeball. The **lens,** attached to the ciliary body by ligaments, divides the cavity of the eye into two smaller cavities. A clear, gelatinous material, the **vitreous humor,** fills the posterior cavity behind the lens.

TABLE 13.2	
Function of Parts of the Eye	
Part	**Function**
Lens	Refracts and focuses light rays
Iris	Regulates light entrance
Pupil	Admits light
Choroid	Absorbs stray light
Sclera	Protects eyeball
Cornea	Refracts light rays
Humors	Refracts light rays
Ciliary body	Holds lens in place, accommodation
Retina	Contains receptors for sight
Rods	Make black-and-white vision possible
Cones	Make color vision possible
Optic nerve	Transmits impulse to brain
Fovea centralis	Makes acute vision possible

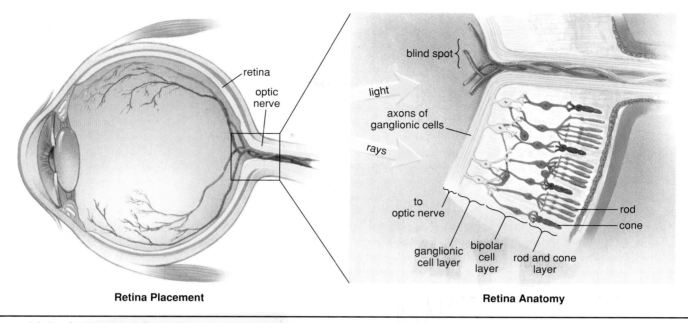

Figure 13.7   Anatomy of the retina.
The retina is the inner layer of the eye. Rods and cones are located at the back of the retina, followed by the bipolar cells and the ganglionic cells, whose fibers become the optic nerve. (Notice that rods share bipolar cells but cones do not. Cones, therefore, distinguish more detail.) The optic nerve carries impulses to the occipital lobe of the cerebrum.

The anterior cavity between the cornea and the lens is filled with a clear, watery fluid secreted by the ciliary body and called the **aqueous humor.**

A small amount of aqueous humor is continually produced each day. Normally, it leaves the anterior cavity by way of tiny ducts located where the iris meets the cornea. When a person has *glaucoma,* these drainage ducts are blocked, and aqueous humor builds up. If glaucoma is not treated, the resulting pressure compresses the arteries that serve the nerve fibers of the retina, where the receptors for sight are located. The nerve fibers begin to die due to lack of nutrients, and the person becomes partially blind. Eventually, total blindness can result.

How the Retina Is Structured

The inner layer of the eye, the **retina,** has three layers of cells (Fig. 13.7). The layer closest to the choroid contains the sense receptors for vision, the **rods** and the **cones;** the middle layer contains bipolar cells; and the innermost layer contains ganglionic cells, whose fibers become the **optic nerve.** Only the rods and the cones contain light-sensitive pigments, and therefore, light must penetrate to the back of the retina before nerve impulses are generated.

The rods and the cones synapse with the bipolar cells, which in turn synapse with ganglionic cells that initiate nerve impulses. The fibers of the ganglionic cells pass in front of the retina, forming the optic nerve, which turns to pierce the layers of the eye. Notice in Figure 13.7 that there are many more rods and cones than ganglionic cells. In fact, the retina has as many as 150 million rods but only one million ganglionic cells and optic nerve fibers. This means that there is considerable mixing of messages and a certain amount of integration before nerve impulses are sent to the thalamus and then to the occipital lobe of the cerebrum. There are no rods or cones where the optic nerve passes through the retina; therefore, this is a **blind spot,** where vision is impossible.

The retina contains a very special region called the **fovea centralis** (Fig. 13.6), an oval, yellowish area with a depression where there are only cones. Vision is most acute in the fovea centralis. The placement of the fovea centralis means that we can see an object better if we don't look directly at it.

The eye has three layers: the outer sclera, the middle choroid, and the inner retina. Only the retina contains receptors for light energy.

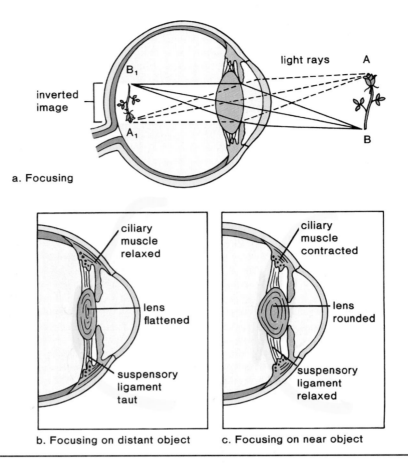

Figure 13.8   Focusing.
a. Light rays from each point on an object are bent by the cornea and the lens in such a way that they are directed to a single point after emerging from the lens. By this process, an inverted image of the object forms on the retina. **b.** When focusing on a distant object, the lens is flat because the ciliary muscle is relaxed and the suspensory ligament is taut. **c.** When focusing on a near object, the lens accommodates; it becomes rounded because the ciliary muscle contracts, causing the suspensory ligament to relax.

How the Eyelids and Eyelashes Work

The thin skin of an eyelid covers muscle and connective tissue. Lining the inner surface is a transparent mucous membrane called the **conjunctiva.** The conjunctiva folds back and covers the anterior of the eye except for the cornea and prevents tears from entering the orbits. The eyelids have eyelashes that can trap a tiny particle of grit, and when they do, the eyes close immediately. Sebaceous glands associated with each eyelash produce an oily secretion that lubricates the eye. Inflammation of any of the glands is called a *sty.*

The eyelids are operated by muscles that close and raise the lid. When a person suffers from *myasthenia gravis,* muscle weakness occurs due to an inability to respond to ACh, and the eyelids often have to be taped open.

The lacrimal gland, which lies in the orbit above the eye, produces tears that flow over the eye when the eyelids are blinked. The tears, collected by two small ducts, pass into a sac before draining into the nasopharynx.

Focusing Uses the Lens

When we look at an object, light rays are **focused** on the retina (Fig. 13.8*a*). In this way, an *image* of the object appears on the retina. The image on the retina occurs when the rods and the cones in a particular region are excited. Obviously, the image is much smaller than the object. To produce this small image, light rays must be bent (refracted) and brought into focus. They are initially bent as they pass through the cornea, and further bending occurs as the rays pass through the lens and the humors.

Light rays are reflected from an object in all directions. For distant objects, only nearly parallel rays enter the eye, and the cornea alone is needed for focusing. For close objects, many of the rays are at sharp angles to one another, and additional focusing is required. The lens provides this additional focusing power as **accommodation** occurs: the lens remains flat when we view distant objects, but it rounds up when we view near objects.

The shape of the lens is controlled by the ciliary muscle within the ciliary body. When we view a distant object, the ciliary muscle is relaxed, causing the suspensory ligaments attached to the ciliary body to be taut; therefore, the lens remains relatively flat (Fig. 13.8b). When we view a near object, the ciliary muscle contracts, releasing the tension on the suspensory ligaments, and the lens rounds up due to its natural elasticity (Fig. 13.8c). Because close work requires contraction of the ciliary muscle, it very often causes muscle fatigue known as eyestrain.

Usually after the age of 40, the lens loses some of its elasticity and is unable to accommodate. Corrective lenses are then usually necessary, as is discussed on page 274. Also with aging or possibly exposure to the sun (see the Health reading on page 278), the lens is subject to *cataracts*; the lens can become opaque and therefore incapable of transmitting rays of light. Recent research suggests that cataracts develop when crystalline lens proteins contained by special cells within the interior oxidize, causing the three-dimensional shape to change. If so, researchers believe that eventually they may be able to find ways to restore the normal configuration of the crystalline lens so that cataracts can be treated medically instead of surgically.

For the present, however, surgery is the only viable treatment for cataracts. First, a surgeon opens the eye near the rim of the cornea. Zonulysin, an enzyme, may be used to digest away the ligaments holding the lens in place. Most surgeons then use a cryoprobe, which freezes the lens for easy removal. An intraocular lens attached to the iris can then be implanted so that the patient does not need to wear thick glasses or contact lenses.

The lens, assisted by the cornea and the humors, focuses images on the retina.

The Image Is Upside Down

The image formed on the retina is inverted (Fig. 13.8a), and it is thought that perhaps this image is righted in the brain by experience. In one experiment, scientists wore glasses that inverted the field of vision. At first, they had difficulty adjusting to the placement of the objects, but they soon became accustomed to their inverted world. Experiments such as these suggest that if we see the world upside down, the brain learns to see it right side up.

The Image Is 3-D

We can see well with either eye alone, but the two eyes functioning together provide us with *stereoscopic vision.* Normally, the two eyes are directed by the eye muscles toward the same object, and therefore, the object is focused on corresponding points of the two retinas. Each eye, however, sends its own data to the brain about the placement

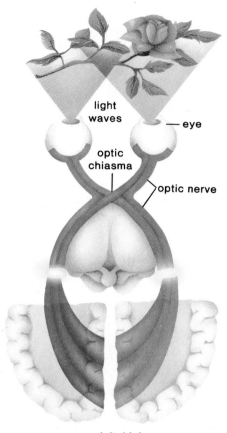

Figure 13.9 Optic chiasma.
Both eyes "see" the entire object, but data from the right half of each retina go to the right occipital lobe of the cerebrum, and data from the left half of the retina go to the left occipital lobe because of the optic chiasma. When the data are combined, the brain "sees" the entire object in depth.

of the object because each forms an image from a slightly different angle. These data are pooled to produce depth perception by a two-step process. First, because some optic nerves cross at the optic chiasma (Fig. 13.9), each half of the brain receives information from both eyes about the same part of an object. Later, the two halves of the brain communicate to arrive at a three-dimensional interpretation of the whole object.

The eyes and the brain working together allow us to see the world right side up and in three dimensions.

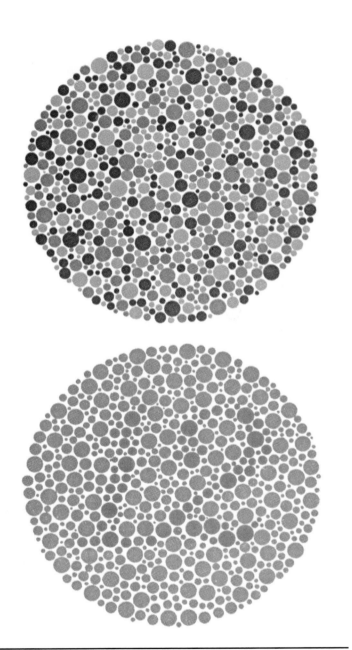

Figure 13.10   Test plates for color blindness.
When looking at the top plate, a person with normal color vision sees the number 8, and when looking at the bottom plate, a person with normal color vision sees the number 12. The most common form of color blindness involves an inability to distinguish reds from greens.

The above has been reproduced from Ishihara's Tests for Colour Blindness published by KANEHARA & CO., LTD., Tokyo, Japan, but tests for color blindness cannot be conducted with this material. For accurate testing, the original plate should be used.

Inability to See Color

Complete color blindness is extremely rare. In most instances, a particular type of cone is lacking or deficient in number. The lack of red or green cones is the most common, affecting about 5–8% of human males. If the eye lacks red cones, the green colors are accentuated, and vice versa (Fig. 13.10).

Inability to See Clearly

The majority of people can see what is designated as a size 20 letter 20 feet away, and so are said to have 20/20 vision. Persons who can see close objects but cannot see the letters from this distance are said to be nearsighted. Nearsighted people can see close objects better than they can see objects at a distance. These individuals often have an elongated eyeball, and when they attempt to look at a distant object, the image is brought to focus in front of the retina (Fig. 13.11). They can see close objects because they can adjust the lens to allow the image to focus on the retina, but to see distant objects, these people must wear concave lenses, which diverge the light rays so that the image can be focused on the retina. There is a treatment for nearsightedness called radial keratotomy, or RK. From four to eight cuts are made in the cornea so that they radiate out from the center like spokes in a wheel. When the cuts heal, the cornea is flattened. Lasers are now also used to flatten or change the shape of the cornea. Although some patients are satisfied with the result, others complain of glare and varying visual acuity.

Persons who can easily see the optometrist's chart but cannot see close objects well are farsighted; these individuals can see distant objects better than they can see close objects. They often have a shortened eyeball, and when they try to see close objects, the image is focused behind the retina. When the object is distant, the lens can compensate for the short eyeball, but when the object is close, these persons must wear a convex lens to increase the bending of light rays so that the image can be focused on the retina.

When the cornea or lens is uneven, the image is fuzzy. The light rays cannot be evenly focused on the retina. This condition, called **astigmatism,** can be corrected by an unevenly ground lens to compensate for the uneven cornea.

As We Age

With normal aging, the lens loses some of its ability to change shape and focus on close objects. Because nearsighted individuals still have difficulty seeing objects clearly in the distance, they must wear bifocals, which means that the upper part of the lens is for distant vision and the remainder is for near vision.

The shape of the eyeball determines the need for corrective lenses; the inability of the lens to accommodate as we age also requires corrective lenses for close vision.

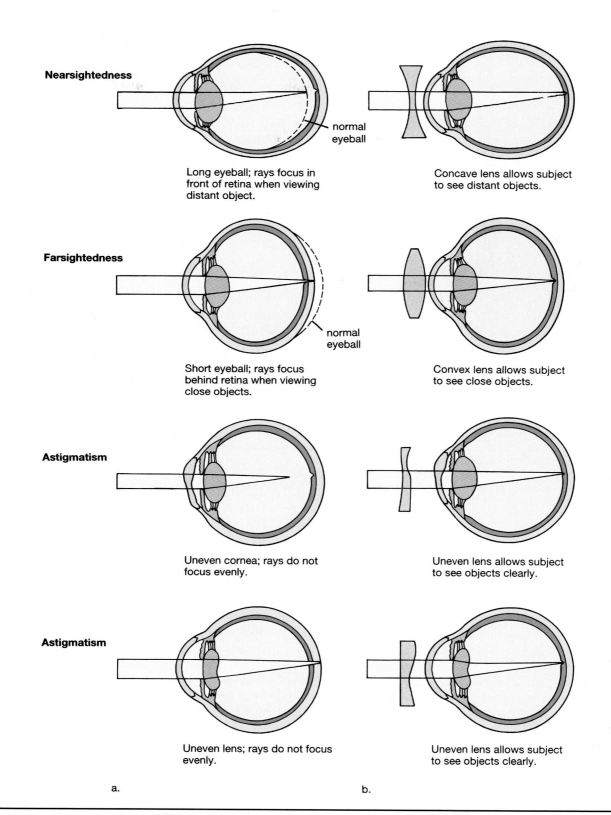

Nearsightedness

Long eyeball; rays focus in front of retina when viewing distant object.

normal eyeball

Concave lens allows subject to see distant objects.

Farsightedness

Short eyeball; rays focus behind retina when viewing close objects.

normal eyeball

Convex lens allows subject to see close objects.

Astigmatism

Uneven cornea; rays do not focus evenly.

Uneven lens allows subject to see objects clearly.

Astigmatism

Uneven lens; rays do not focus evenly.

Uneven lens allows subject to see objects clearly.

a.

b.

Figure 13.11 Common abnormalities of the eye, with possible corrective lenses.
a. The cornea and the lens function in bringing light rays (lines) to focus, but sometimes they are unable to compensate for the shape of the eyeball or for an uneven cornea. **b.** In these instances, corrective lenses can allow the individual to see normally.

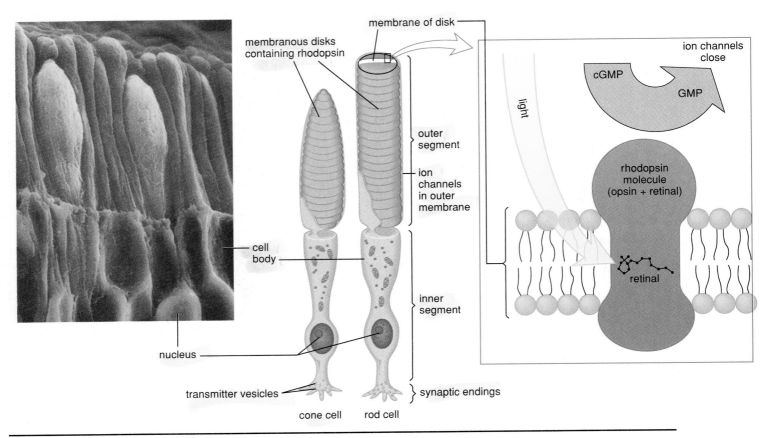

Figure 13.12   Structure and function of rods and cones.
The outer segment of rods and cones contains stacks of membranous disks, which contain visual pigments. In rods, the membrane of each disk contains rhodopsin, a complex molecule containing the protein opsin and the pigment retinal. When retinal absorbs light energy, it changes shape, activating rhodopsin to begin a series of reactions that end when cGMP (cyclic guanosine monophosphate) is converted to GMP. Thereafter, ion channels close.

Seeing Uses Chemistry

Only dim light is required to stimulate rods; therefore, they are responsible for *night vision*. The numerous rods are also better than cones at detecting motion, but they cannot provide distinct and/or color vision. This causes objects to appear blurred and look gray in dim light, as in the early morning or at dusk. Because the rods are more densely located in the periphery of the retina, you see dim objects better at night if you look at them a little to one side.

What is it that permits the rods to detect light? Many molecules of **rhodopsin** are located within the membrane of the disks (lamellae) found in the outer segment of the rods (Fig. 13.12). Rhodopsin is a complex molecule that contains the protein opsin and a pigment molecule called *retinal*, which is a derivative of vitamin A. When light strikes retinal, it changes shape and rhodopsin is activated. The membrane of the outer segment of a rod cell has many ion channels that are held open in the dark by cyclic guanosine monophosphate (cGMP). Activation of rhodopsin leads to a conversion of cGMP to GMP and closure of some sodium ion channels. The resulting increased negativity of the rod interior causes electrochemical changes in bipolar and then ganglionic cell layers, which send nerve impulses to the brain. Rod cells are sensitive to dim light because one molecule of rhodopsin leads to the conversion of many cGMP to GMP molecules. There has been an *amplification* of the original stimulus.

The cones, located primarily in the fovea and activated by bright light, detect the fine detail and the color of an object. *Color vision* depends on three different kinds of cones, which contain pigments called the B (blue), G (green), and R (red) pigments. Each pigment is made up of retinal and opsin, but there is a slight difference in the opsin structure of each, which accounts for their individual absorption patterns. Various combinations of cones are believed to be stimulated by in-between shades of color, and the combined nerve impulses are interpreted in the brain as a particular color.

> The receptors for sight are the rods and the cones. The rods are responsible for vision in dim light, and the cones are responsible for vision in bright light and for color vision.

13.5 Ears

The ear accomplishes two sensory functions: equilibrium (balance) and hearing. The receptors for both of these are located in the inner ear and consist of hair cells with stereocilia that respond to mechanical stimulation. Each hair cell has from 30 to 150 stereocilia. When the stereocilia of any particular hair cell are displaced in a certain direction, the cell generates nerve impulses, which are sent along a cranial nerve to the brain.

How the Ear Appears

Figure 13.13 is a drawing of the ear, and Table 13.3 lists the parts of the ear. The ear has three divisions: outer, middle, and inner. The **outer ear** consists of the pinna (external flap) and the auditory canal. The opening of the auditory canal

is lined with fine hairs and sweat glands. Modified sweat glands are located in the upper wall of the canal; they secrete earwax, a substance that helps to guard the ear against the entrance of foreign materials, such as air pollutants.

The **middle ear** begins at the **tympanic membrane** (eardrum) and ends at a bony wall containing two small openings covered by membranes. These openings are called the *oval* and *round windows*. Three small bones are found between the tympanic membrane and the oval window. Collectively called the **ossicles**, individually they are the **malleus** (hammer), the **incus** (anvil), and the **stapes** (stirrup) because their shapes resemble these objects (Fig. 13.13). The malleus adheres to the tympanic membrane, and the stapes touches the oval window. The posterior wall of the middle ear has an opening that leads to mastoid sinuses of the skull.

TABLE 13.3

The Ear

	Outer Ear	Middle Ear	Inner Ear	
			Cochlea	Sacs and Semicircular Canals
Function	Directs sound waves to tympanic membrane	Picks up and amplifies sound waves	Hearing	Maintains equilibrium
Anatomy	Pinna; auditory canal	Tympanic membrane; ossicles	Organ of Corti; auditory nerve	Saccule and utricle; semicircular canals
Medium	Air	Air (auditory tube)	Fluid	Fluid

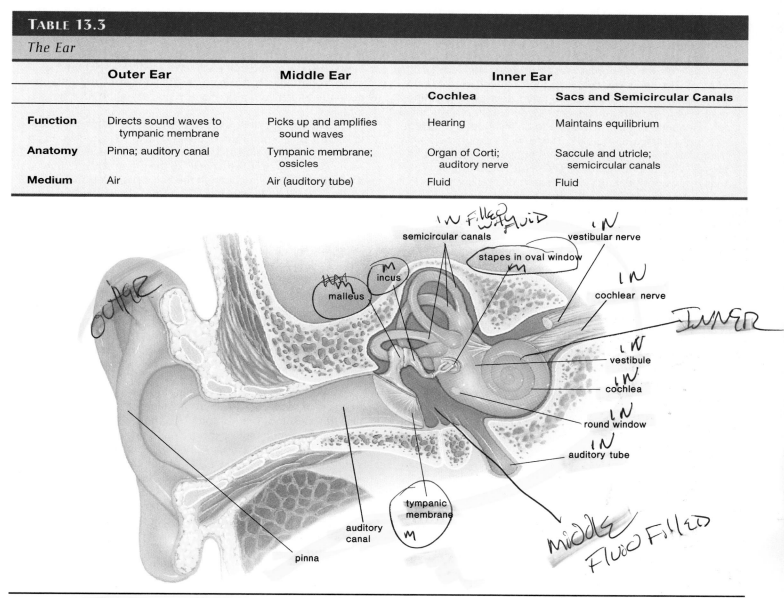

Figure 13.13 Anatomy of the human ear.
In the middle ear, the malleus (hammer), the incus (anvil), and the stapes (stirrup) amplify sound waves. Otosclerosis is a condition in which the stapes becomes attached to the inner ear and is unable to carry out its normal function. It can be replaced by a plastic piston, and thereafter the individual hears normally because sound waves are transmitted as usual to the cochlea, which contains the receptors for hearing.

Health Focus

Protecting Vision and Hearing

Older age can be accompanied by a serious loss of vision and hearing. The time to start preventive measures for such problems however is when we are younger.

PREVENTING A LOSS OF VISION

The eye is subject to both injuries and disorders. Although flying objects sometimes penetrate the cornea and damage the iris, lens, or retina, careless use of contact lenses is the most common cause of injuries to the eye. Injuries cause only 4% of all cases of blindness; the most frequent causes are retinal disorders, glaucoma, and cataracts, in that order. Retinal disorders are varied. In diabetic retinopathy, which blinds many people between the ages of 20 and 74, capillaries to the retina burst and blood spills into the vitreous fluid. Careful regulation of blood glucose levels in these patients may be protective. In macular degeneration, the cones are destroyed because thickened choroid vessels no longer function as they should. Glaucoma occurs when the drainage system of the eyes fails, so that fluid builds up and destroys nerve fibers responsible for peripheral vision. Eye doctors always check for glaucoma, but it is advisable to be aware of the disorder in case it comes on quickly. Those who have experienced acute glaucoma report that the eyeball feels as heavy as a stone. In cataracts, cloudy spots on the lens of the eye eventually pervade the whole lens. The milky yellow-white lens scatters incoming light and blocks vision.

Regular visits to an eye-care specialist, especially in the elderly, are a necessity in order to catch conditions, such as glaucoma, early enough to allow effective treatment. It has not been proven, however, that consuming particular foods or vitamin supplements will reduce the risk of cataracts. Even so, it's possible that estrogen replacement therapy may be somewhat protective in postmenopausal women.

There are preventive measures that we can take to reduce the chance of defective vision as we age. Accumulating evidence suggests that both macular degeneration and cataracts, which tend to occur in the elderly, are caused by long-term exposure to the ultraviolet rays of the sun. It is recommended, therefore, that everyone, especially those who live in sunny climates or work outdoors, wear glass and not plastic sunglasses that absorb ultraviolet light. Large lenses worn close to the eyes offer further protection. The Sunglass Association of America has devised the following system for categorizing sunglasses:

- Cosmetic lenses absorb at least 70% of UV-B and 20% UV-A and 60% of visible light. Such lenses are worn for comfort rather than protection.
- General purpose lenses absorb at least 95% of UV-B and 60% of UV-A and 60–92% of visible light. They are good for outdoor activities in temperate regions.
- Special purpose lenses block at least 99% of UV-B and 60% UV-A, and 20–97% of visible light. They are good for bright sun combined with sand, snow, or water.

Health care providers have found an increased incidence of cataracts in heavy cigarette smokers. In men, smoking 20 cigarettes or more a day, and in women, smoking more than 35 cigarettes a day doubles the risk of cataracts. It is possible that smoking reduces the delivery of blood and therefore nutrients to the lens.

PREVENTING A LOSS OF HEARING

Especially when we are young, the middle is ear is subject to infections that can lead to hearing impairments if they are not treated promptly by a physician. The mobility of ossicles decreases with age, and in otosclerosis, new filamentous bone grows over the stirrup impeding its movement. Surgical treatment is the only remedy for this type of conduction deafness. However, age-associated nerve deafness due to stereocilia damage from exposure to loud noises is preventable. Hospitals are now aware that even the ears of the newborn need to be protected from noise and are taking steps to make sure neonatal intensive care units and nurseries are as quiet as possible.

In today's society, exposure to the types of noises listed in Table 13A is a common occurrence. Noise is measured in decibels, and any noise above a level of 80 decibels could result in damage to the hair cells of the organ of Corti. Eventually, the stereocilia and then the hair cells disappear completely (Fig. 13A). If listening to city traffic for extended periods can damage hearing, it stands to reason that frequent attendance at rock concerts, constantly playing a stereo loudly, or using earphones at high volume is also damaging to hearing. The first hint of danger could be temporary hearing loss, a "full" feeling in the ears, muffled hearing, or tinnitus (e.g., ringing in the ears). If you have any of these symptoms, modify your listening habits immediately to prevent further damage. If exposure to noise is unavoidable, specially designed noise-reduction earmuffs are available, and it is also possible to purchase earplugs made from a compressible, spongelike material at the drugstore or sporting-goods store. These earplugs are not the same as those worn for swimming, and they should not be used interchangeably.

Aside from loud music, noisy indoor or outdoor equipment, such as a rug-cleaning machine or a chain saw, are also troublesome. Even motorcycles and recreational vehicles such as snowmobiles and motocross bikes can contribute to a gradual loss of hearing. Exposure to intense sounds of short duration, such as a burst of gunfire, can result in an immediate hearing loss. Hunters may have a significant hearing reduction in the ear opposite to the shoulder where the gun is carried. The butt of the rifle offers some protection to the ear nearest the gun when it is shot.

Finally, people need to be aware that some medicines are ototoxic. Anticancer drugs, most notably cisplatin, and certain antibiotics (e.g., streptomycin, kanamycin, gentamicin) make ears especially susceptible to a hearing loss. Anyone taking such medications needs to be especially careful to protect the ears from any loud noises.

TABLE 13A

Noises That Affect Hearing

Type of Noise	Sound Level (decibels)	Effect
"Boom car," jet engine, shotgun, rock concert	over 125	Beyond threshold of pain; potential for hearing loss high
Discotheque, "boom box," thunderclap	over 120	Hearing loss likely
Chain saw, pneumatic drill, jackhammer, symphony orchestra, snowmobile, garbage truck, cement mixer	100–200	Regular exposure of more than one minute risks permanent hearing loss
Farm tractor, newspaper press, subway, motorcycle	90–100	Fifteen minutes of unprotected exposure potentially harmful
Lawn mower, food blender	85–90	Continuous daily exposure for more than eight hours can cause hearing damage
Diesel truck, average city traffic noise	80–85	Annoying; constant exposure may cause hearing damage

Source: National Institute on Deafness and Other Communication Disorders, January 1990, National Institute of Health.

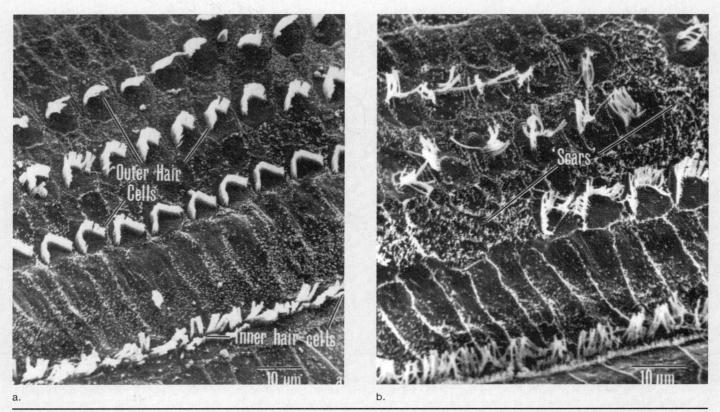

a.   b.

Figure 13A **The higher the decibel reading noted in the table, the more likely a noise will damage hearing.**
a. Normal hair cells in the organ of Corti of a guinea pig. **b.** Damaged cells. This damage occurred after 24-hour exposure to a noise level like that at heavy-metal rock concerts. Hearing is permanently impaired because lost cells will not be replaced, and damaged cells may also die.

An **auditory** (eustachian) **tube,** which extends from each middle ear to the nasopharynx, permits equalization of air pressure. Chewing gum, yawning, and swallowing in elevators and airplanes help to move air through the auditory tubes upon ascent and descent. As this occurs we often hear the ears "pop."

Whereas the outer ear and the middle ear contain air, the inner ear is filled with fluid. The **inner ear** (Fig. 13.14*a*), anatomically speaking, has three areas: the semicircular canals and the vestibule are concerned with equilibrium; the cochlea is concerned with hearing.

The **semicircular canals** are arranged so that there is one in each dimension of space. The base of each of the three canals, called the **ampulla** (Fig. 13.14*b*), is slightly enlarged. Little hair cells whose stereocilia are embedded within a gelatinous material called a *cupula* are found within the ampullae.

A vestibule, or chamber, lies between the semicircular canals and the cochlea. It contains two small membranous sacs called the **utricle** and the **saccule** (Fig. 13.14*c*). Both of these sacs contain little hair cells, whose stereocilia are embedded within a gelatinous material called an *otolithic membrane* (Fig. 13.14*c*). Calcium carbonate (CaCO$_3$) granules, or **otoliths,** rest on this membrane.

The **cochlea** resembles the shell of a snail because it spirals. Three canals are located within the tubular cochlea: the vestibular canal, the **cochlear canal,** and the tympanic canal. Along the length of the basilar membrane, which forms the lower wall of the cochlear canal, are little hair cells whose stereocilia are embedded within a gelatinous material called the *tectorial membrane.* The hair cells of the cochlear canal, called the **organ of Corti** (Fig. 13.14*d*), synapse with nerve fibers of the cochlear (auditory) nerve. The **cochlear nerve** carries nerve impulses to the brain, which interprets them as sound.

> The ear has three major divisions: outer ear, middle ear, and inner ear. The outer ear contains the auditory canal; the middle ear contains the ossicles; and the inner ear contains the semicircular canals, a vestibule, and the cochlea.

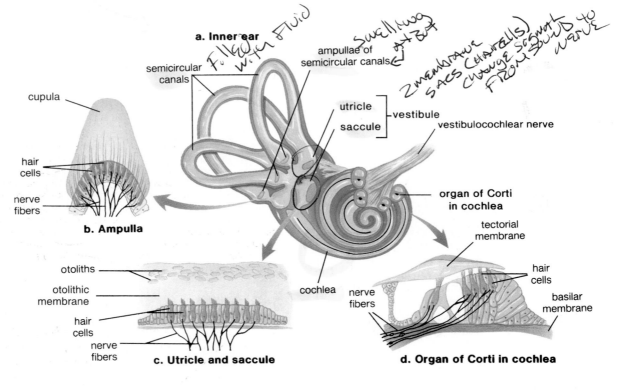

Figure 13.14   Anatomy of the inner ear.
a. The inner ear contains the semicircular canals, the utricle and the saccule within a vestibule, and the cochlea. The cochlea has been cut to show the location of the organ of Corti. **b.** An ampulla at the base of each semicircular canal contains the receptors (hair cells) for dynamic equilibrium. **c.** The utricle and the saccule are small sacs that contain the receptors (hair cells) for static equilibrium. **d.** The receptors for hearing (hair cells) are in the organ of Corti.

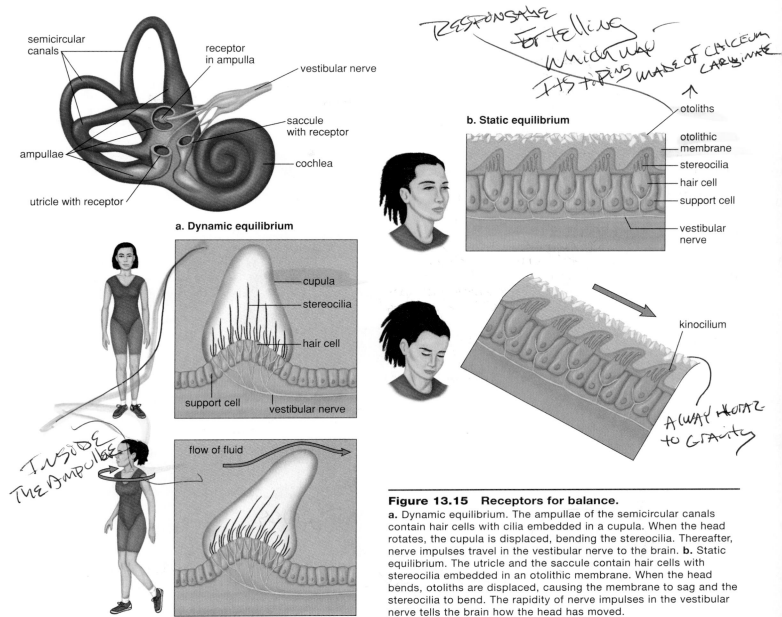

a. Dynamic equilibrium

cupula

stereocilia

hair cell

support cell

vestibular nerve

flow of fluid

INSIDE THE AMPULLAE

b. Static equilibrium

RESPONSABLE for telling which way its topins MADE OF CALCEUM CARBONATE

otoliths

otolithic membrane

stereocilia

hair cell

support cell

vestibular nerve

kinocilium

ALWAY HORIZ to GRAVITY

Figure 13.15 Receptors for balance.
a. Dynamic equilibrium. The ampullae of the semicircular canals contain hair cells with cilia embedded in a cupula. When the head rotates, the cupula is displaced, bending the stereocilia. Thereafter, nerve impulses travel in the vestibular nerve to the brain. **b.** Static equilibrium. The utricle and the saccule contain hair cells with stereocilia embedded in an otolithic membrane. When the head bends, otoliths are displaced, causing the membrane to sag and the stereocilia to bend. The rapidity of nerve impulses in the vestibular nerve tells the brain how the head has moved.

Hair Cells Keep Us Balanced

The sense of balance has been subdivided into two senses: *dynamic equilibrium,* involving angular and/or rotational movement of the head, and *static equilibrium,* involving movement of the head in one plane, either vertical or horizontal.

Because there are three semicircular canals, each ampulla responds to head rotation in a different plane of space. As fluid within a semicircular canal flows over and displaces a cupula, the stereocilia of the hair cells bend, and the pattern of impulses carried by the vestibular nerve to the brain changes (Fig. 13.15*a*). Continuous movement of fluid in the semicircular canals causes one form of motion sickness.

The utricle is especially sensitive to horizontal movements and the bending of the head, while the saccule responds best to vertical (up-down) movements. When the body is still, the otoliths in the utricle and the saccule rest on the otolithic membrane above the hair cells (Fig. 13.15*b*). When the head bends or the body moves in the horizontal and vertical planes, the otoliths are displaced and the otolithic membrane sags, bending the stereocilia of the hair cells beneath. If the stereocilia move toward the larger kinocilium, nerve impulses in the vestibular nerve increase, and if the stereocilia move away from the larger kinocilium, nerve impulses in the vestibular nerve decrease. These data tell the brain the direction of the movement of the head.

Movement of cupula within the semicircular canals contributes to the sense of dynamic equilibrium. Movement of the otolithic membrane within the utricle and the saccule accounts for static equilibrium.

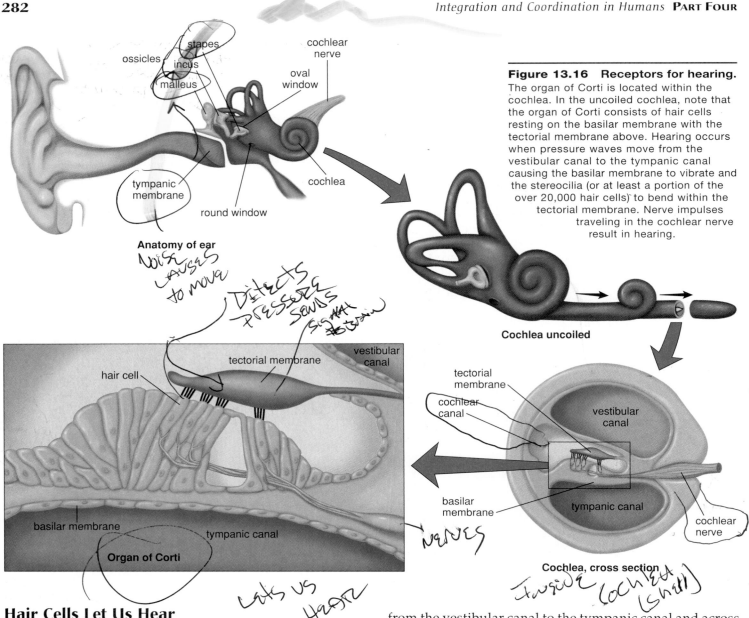

Figure 13.16   Receptors for hearing.
The organ of Corti is located within the cochlea. In the uncoiled cochlea, note that the organ of Corti consists of hair cells resting on the basilar membrane with the tectorial membrane above. Hearing occurs when pressure waves move from the vestibular canal to the tympanic canal causing the basilar membrane to vibrate and the stereocilia (or at least a portion of the over 20,000 hair cells) to bend within the tectorial membrane. Nerve impulses traveling in the cochlear nerve result in hearing.

Anatomy of ear

Cochlea uncoiled

Organ of Corti

Cochlea, cross section

Hair Cells Let Us Hear

The process of hearing begins when sound waves enter the auditory canal (Fig. 13.16). Just as ripples travel across the surface of a pond, sound travels by the successive vibrations of molecules. Ordinarily, sound waves do not carry much energy, but when a large number of waves strike the tympanic membrane, it moves back and forth (vibrates) ever so slightly. The malleus then takes the pressure from the inner surface of the tympanic membrane and passes it by means of the incus to the stapes in such a way that the pressure is multiplied about 20 times as it moves from the tympanic membrane to the stapes. The stapes strikes the membrane of the oval window, causing it to vibrate, and in this way, the pressure is passed to the fluid within the cochlea.

If the cochlea is unwound, you can see that the vestibular canal connects with the tympanic canal, which leads to the round window membrane. When the stapes strikes the membrane of the oval window, pressure waves move from the vestibular canal to the tympanic canal and across the basilar membrane, and the round window bulges. As the basilar membrane moves up and down, the stereocilia of the hair cells embedded in the tectorial membrane bend. Then, nerve impulses begin in the cochlear nerve and travel to the brain stem. Eventually they reach the temporal lobe of the cerebrum where they are interpreted as a sound.

The organ of Corti is narrow at its base but widens as it approaches the tip of the cochlear canal. Each part of the organ is sensitive to different wave frequencies, or pitch. Near the tip, the organ of Corti responds to low pitches, such as a tuba, and near the base, it responds to higher pitches, such as a bell or a whistle. The nerve fibers from each region along the length of the organ of Corti lead to slightly different areas in the brain. The pitch sensation we experience depends upon which region of the basilar membrane vibrates and which area of the brain is stimulated.

Volume is a function of the amplitude of sound waves. Loud noises cause the fluid of the cochlea to vibrate to a greater degree, and this, in turn, causes the basilar membrane to move up and down to a greater extent. The resulting increased stimulation is interpreted by the brain as volume. It is believed that the tone of a sound is an interpretation of the brain based on the distribution of hair cells stimulated.

The sense receptors for sound are hair cells on the basilar membrane (the organ of Corti). When the basilar membrane vibrates, the stereocilia of the hair cells bend, and nerve impulses are transmitted to the brain.

There are two major types of deafness: *conduction deafness* and *nerve deafness*. Conduction deafness can be due to a congenital defect, and it can also be due to infections that have caused the ossicles to fuse, restricting their ability to magnify sound waves.

Nerve deafness most often occurs when cilia on the sense receptors within the cochlea have worn away. Since this can happen with normal aging, elderly people are more likely than younger persons to have trouble hearing. However, as discussed in the Health Focus reading on page 278, nerve deafness also occurs due to noise pollution. Hearing aids are not helpful for nerve deafness. Costly cochlear implants, which stimulate the auditory nerve directly, are available, but those who have these electronic devices report that the speech they hear is like that of a robot.

SUMMARY

13.1 Receptors and Sensations

Each type of receptor is sensitive to a particular kind of stimulus. When stimulation occurs, receptors initiate nerve impulses that are transmitted to the spinal cord and/or brain. Only when nerve impulses reach the brain are we conscious of sensation.

13.2 Skin

The skin contains receptors for touch, pressure, pain, and temperature (hot and cold).

13.3 Chemoreceptors

Taste and smell are due to chemoreceptors that are stimulated by chemicals in the environment. The taste buds contain taste cells that communicate with nerve fibers, while the receptors for smell are neurons.

After molecules bind to plasma membrane receptors on the microvilli of taste cells and the cilia of olfactory cells, nerve impulses eventually reach the brain, which determines the taste and odor according to the pattern of receptors stimulated.

13.4 Eyes

Vision is dependent on the eye, the optic nerve, and the visual cortex of the cerebrum. The eye has three layers. The outer layer, the sclera, can be seen as the white of the eye; it also becomes the transparent bulge in the front of the eye called the cornea. The middle layer, called the choroid, absorbs stray light rays. The rods, receptors for vision in dim light, and the cones, receptors that depend on bright light and provide color and detailed vision, are located in the retina, the inner layer of the eyeball. The cornea, the humors, and especially the lens bring the light

rays to focus on the retina. To see a close object, accommodation occurs as the lens rounds up. Due to the optic chiasma, both sides of the brain must function together to give three-dimensional vision.

When light strikes rhodopsin, a molecule composed of opsin and retinal, retinal changes shape and rhodopsin is activated. Chemical reactions that produce electrochemical changes eventually result in nerve impulses that are carried in the optic nerve to the brain.

13.5 Ears

Hearing is a specialized sense dependent on the ear, the cochlear nerve, and the temporal lobe of the cerebrum. The ear is divided into three parts: outer, middle, and inner. The outer ear consists of the pinna and the auditory canal, which direct sound waves to the middle ear. The middle ear begins with the tympanic membrane and contains the ossicles (malleus, incus, stapes). The malleus is attached to the tympanic membrane, and the stapes is attached to the oval window, which is covered by membrane. The inner ear contains the cochlea and the semicircular canals, plus the utricle and saccule. The outer and middle portions of the ear simply convey and magnify the sound waves that strike the oval window. Its vibrations set up pressure waves within the cochlea, which contains the organ of Corti, consisting of hair cells whose stereocilia are embedded within the tectorial membrane. When the stereocilia of the hair cells bend, nerve impulses begin in the cochlear nerve and are carried to the brain.

The ear also contains receptors for our sense of balance. Dynamic equilibrium is dependent on the stimulation of hair cells within the ampullae of the semicircular canals. Static equilibrium relies on the stimulation of hair cells within the utricle and the saccule.

STUDYING THE CONCEPTS

1. What are the six types of receptors in the human body? Relate each type of somatic and special sense with one of these types of receptors.   266

2. List and discuss the receptors of the skin.   267

3. Describe the structure of a taste bud, and tell how a taste cell functions.   268

4. Describe the structure of an olfactory cell, and tell how it functions.   269

5. Describe the anatomy of the eye, and explain focusing and accommodation.   270–73

6. Relate the need for corrective lenses to three possible eye shapes. Discuss bifocals.   274

7. Describe sight in dim light. What chemical reaction is responsible for vision in dim light? Discuss color vision.   276

8. Describe the anatomy of the ear and how we hear.   277–83

9. Describe the role of the semicircular canals, the utricle, and the saccule in balance.   281

10. Discuss the two causes of deafness, including why younger people frequently suffer a hearing loss.   283

APPLYING YOUR KNOWLEDGE

Concepts

1. A blow to the eye produces a sensation of flashing lights ("seeing stars"). What does this tell you about the receptors in an eye?

2. A friend of yours is red-green color blind. Will eating carrots help your friend? Why or why not?

3. The inner ear is involved in balance; why does it help to open the eyes when standing on one foot?

4. As people age, they can hear male voices better than female voices. What does that tell you about loss of hearing?

Bioethical Issue

Walking out of a concert, everything sounds muffled. The cars, for example, sound quieter. And when you talk, your voice booms, even though you think you're speaking at a normal level.

Attend enough concerts with music playing at 120 decibels or louder, and this reaction could become permanent. Such loud noise damages cells in the ear that help interpret sound. That's why performers often wear earplugs while on stage.

Given the potential damage, should concert arenas and clubs be required to keep band music below a certain decibel level? Or is each consumer responsible for his or her own hearing? Defend your answer.

APPLYING TECHNOLOGY

Your study of the senses is supported by these available technologies:

Exploring the Internet

The Mader Home Page provides further resources for studying this chapter.

`http://www.mhhe.com/sciencemath/biology/mader/`

(Click on *Human Biology.*)

Life Science Animations Video

Video #31:   Animal Biology, Part 1

Organs of Static Equilibrium (#26)   Inner ear anatomy is explained, and the movement of otoliths is seen as the head tilts back and forth.   (3)*

Organ of Corti (#27)   Sound waves amplified by the ossicles cause the organ of Corti to respond, and nerve impulses are generated in the cochlear nerve that travel to the brain.   (2)

*Level of difficulty

TESTING YOUR KNOWLEDGE

1. Vision, hearing, taste, and smell do not occur unless nerve impulses reach the proper portion of the _____.

2. The _____ of skin contains _____ for touch, pressure, pain, and temperature.

3. _____ occurs when a receptor becomes accustomed to stimulation and stops generating impulses.

4. Taste cells and olfactory cells are _____ because they are sensitive to chemicals in the air and in food.

5. The receptors for vision, the _____ and the _____, are located in the _____, the inner layer of the eye.

6. Label this diagram of an eye. State a function for each structure labeled.

7. The cones give us _____ vision and work best in _____ light.

8. Vision in dim light is dependent on the presence of _____ in the rods.

9. The lens _____ for viewing close objects.

10. People who are nearsighted cannot see objects that are _____. A _____ lens restores this ability.

11. The ossicles are the _____, the _____, and the _____.

12. The semicircular canals are involved in our sense of _____.

13. The organ of Corti is located in the _____ canal of the _____.

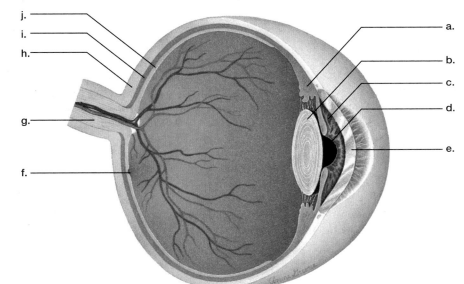

SELECTED KEY TERMS

accommodation Lens adjustment to see near objects. 272

adaptation Decrease in the excitability of a receptor in response to continuous constant-intensity stimulation. 267

ampulla Base of a semicircular canal in the inner ear. 280

aqueous humor Clear, watery fluid between the cornea and lens of the eye. 271

astigmatism Blurred vision due to an irregular curvature of the cornea or the lens. 274

auditory tube Extension from the middle ear to the nasopharynx for equalization of air pressure on the eardrum. 280

blind spot Area of the eye containing no rods or cones and where the optic nerve passes through the retina. 271

chemoreceptor Sensory receptor that is sensitive to chemical stimulation—for example, receptors for taste and smell. 266

choroid (KOR-oyd) Vascular, pigmented middle layer of the eyeball. 270

ciliary body Structure associated with the choroid layer that contains ciliary muscle, which controls the shape of the lens of the eye. 270

cochlea (KOH-klee-uh) Portion of the inner ear that resembles a snail's shell and contains the organ of Corti, the sense organ for hearing. 280

cochlear canal Canal within the cochlea that contains the organ of Corti with hair cells that function as hearing receptors. 280

cochlear nerve Cranial nerve that carries nerve impulses from the organ of Corti to the brain, thereby contributing to the sense of hearing; auditory nerve. 280

cone Bright-light receptor in the retina of the eye that detects color and provides visual acuity. 271

conjunctiva Membranous covering on the anterior surface of the eye. 272

focus Manner by which light rays are bent by the cornea and lens, creating an image on the retina. 272

fovea centralis (FOH-vee-uh sen-TRA-lus) Region of the retina consisting of densely packed cones that is responsible for the greatest visual acuity. 271

incus Middle of three ossicles of the ear that serves to conduct vibrations from the tympanic membrane to the oval window of the inner ear. 277

inner ear Portion of the ear consisting of a vestibule, semicircular canals, and the cochlea where balance is maintained and sound is transmitted. 280

iris Muscular ring that surrounds the pupil and regulates the passage of light through this opening. 270

lens Clear membranelike structure found in the eye behind the iris; brings light rays to focus on the retina. 270

malleus (MA-lee-us) First of three ossicles of the ear that serves to conduct vibrations from the tympanic membrane to the oval window of the inner ear. 277

mechanoreceptor Sensory receptor that is sensitive to mechanical stimulation, such as that from pressure, sound waves, and gravity. 266

middle ear Portion of the ear consisting of the tympanic membrane and the ossicles; where sound is amplified. 277

olfactory cell (ahl-FAK-tuh-ree) A neuron modified as a receptor for the sense of smell. 269

optic nerve Either of two cranial nerves that carry nerve impulses from the retina of the eye to the brain, thereby contributing to the sense of sight. 271

organ of Corti Hair cells that function as hearing receptors within the inner ear. 280

ossicle One of the small bones of the middle ear—malleus, incus, stapes. 277

otolith (OH-tul-ith) Calcium carbonate granule associated with stereociliated cells in the utricle and the saccule. 280

outer ear Portion of ear consisting of the pinna and auditory canal. 277

pain receptor Sensory receptor that is sensitive to chemicals released by damaged tissues or excess stimuli of heat or pressure. 266

photoreceptor Light-sensitive receptor. 266

proprioceptor (proh-pree-oh-SEP-tur) Sensory receptor in the muscles that assists the brain in knowing the position of the limbs. 266

pupil Opening in the center of the iris of the eye. 270

receptor Sensory structure specialized to receive information from the environment and to generate nerve impulses; also a protein located in the plasma membrane or within the cell that binds to a substance that alters some aspect of the cell. 266

retina Innermost layer of the eyeball, which contains the rods and the cones. 271

rhodopsin (roh-DAHP-sun) Visual pigment found in the rods whose activation by light energy leads to vision. 276

rod Dim-light receptor in the retina of the eye that detects motion but no color. 271

saccule Saclike cavity in the vestibule of the inner ear; contains receptors for static equilibrium. 280

sclera (SKLER-uh) White, fibrous, outer layer of the eyeball. 270

semicircular canal One of three tubular structures within the inner ear that contains receptors responsible for the sense of dynamic equilibrium. 280

stapes (STAY-peez) Last of three ossicles of the ear that serves to conduct vibrations from the tympanic membrane to the oval window of the inner ear. 277

taste bud Sense organ containing the receptors associated with the sense of taste. 268

thermoreceptor Sensory receptor that is sensitive to changes in temperature. 266

tympanic membrane (tim-PAN-ik) Eardrum that receives sound waves and is located at the start of the middle ear. 277

utricle (YOO-trih-kul) Saclike cavity in the vestibule of the inner ear that contains receptors for static equilibrium. 280

vitreous humor (VIH-tree-us) Clear, gelatinous material between the lens of the eye and the retina. 270

Chapter 14

Endocrine System

Chapter Outline

Figure 14.1 Hormone regulation.
Coordination by the nervous and endocrine systems never ceases, even when we sleep. Melatonin, the sleep hormone, and many other hormones are now coursing through this person's body even as she sleeps.

It was another sleepless night of TV reruns for Joanne. For the past month, Joanne had tossed and turned, occasionally peering out the window at darkened houses down her street. She was exhausted.

Finally, Joanne decided to call a friend who was a budding neurobiologist. That's when she learned about a natural hormone called melatonin. Produced by the brain's pea-sized pineal gland, melatonin lulls us to sleep. At night, our eyes signal a biological clock located in the brain that darkness is upon us, and the clock directs the pineal gland to produce melatonin. Our metabolism slows. Our body temperature falls. Blissfully, we fall asleep (Fig. 14.1).

Some researchers believe that a melatonin supplement—often taken about an hour before bedtime—can launch such nighttime reactions in restless bodies. Joanne, like many other Americans, began taking over-the-counter melatonin pills. Sure enough, she did fall asleep faster.

A nation of well-rested people sounds ideal, but melatonin is useful only for those with delayed sleep syndrome, and there are many other types of sleep disorders. And the longtime effects of melatonin on other hormonal levels is not yet known. Therefore, it's important to learn about the possible side effects of melatonin before becoming a true believer and taking the medication.

Melatonin is only one of the many hormones produced by the body. The presence or absence of hormones affects our metabolism, our appearance, and our behavior. Hormones are usually produced by endocrine glands, so called because they secrete their products internally, placing them directly in the blood. Although their secretions are carried throughout the body by the blood, each one affects only particular organs, appropriately called target organs. Their cells have the necessary receptors in the plasma membranes.

The study of hormones is one of the most exciting disciplines within modern biology; it has produced many useful advances such as the synthesis of melatonin and other hormones, contraceptive methods, treatment of infertility and disorders such as diabetes, which was once fatal but now is controlled by injections of the hormone insulin.

The endocrine glands are a part of the *endocrine system*, which, like the nervous system, coordinates the functioning of body parts. The nervous system is fast acting and utilizes nerve impulses that travel rapidly along nerve fibers. The endocrine system is slower acting because **hormones** are chemical messengers that affect the metabolism of cells.

14.1 How Hormones Work

Hormones fall into two basic categories: (1) peptides (used here to include amino acid, polypeptide, and protein hormones) and (2) steroid hormones. All steroids are a complex of four carbon rings, but each type has different side chains. Peptide hormones activate existing enzymes in the cell, and steroid hormones bring about the synthesis of new proteins.

A hormone does not seek out a particular organ; rather the organ is awaiting the arrival of the hormone. Cells that can react to a hormone have hormone receptors that combine with the hormone in a lock-and-key manner.

Steroid Hormones Activate DNA

Steroid hormones do not bind to hormone receptors; they can enter the cell and the nucleus freely (Fig. 14.2). Once inside the nucleus, steroid hormones, such as estrogen and progesterone, bind to hormone receptors. The hormone-receptor complex then binds to DNA. Particular genes are activated; transcription of DNA and translation of messenger ribonucleic acid (mRNA) follow. In this manner, steroid hormones lead to protein synthesis.

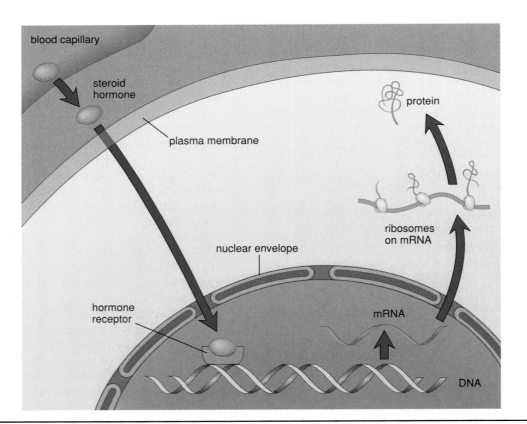

Figure 14.2   Cellular activity of steroid hormones.
After passing through the plasma membrane, a steroid hormone binds to a receptor inside the cell. Recent evidence indicates that the receptors reside in the nucleus rather than the cytoplasm. The hormone-receptor complex then binds to DNA, and this leads to activation of certain genes and protein synthesis.

Indirect [handwritten]

Peptide Hormones Activate Enzymes

Epinephrine is a peptide hormone secreted by the adrenal medulla, an endocrine gland. Among its many effects, epinephrine causes muscle cells to break down glycogen so that glucose is produced. In Figure 14.3, epinephrine is about to bind to a hormone receptor. This will lead to the activation of an enzyme that produces *cyclic AMP* (cyclic adenosine monophosphate). Cyclic AMP (cAMP) is made from ATP, but it contains only one phosphate group, which is attached to the adenosine portion of the molecule at two spots. Thus the molecule is cyclic.

Notice that epinephrine never enters the cell. Since peptide hormones do not enter the cell, they are sometimes called the *first messengers,* while the cAMP, which often sets the metabolic machinery in motion, is called the *second messenger.* In other cells, the binding of a peptide hormone leads to a different type of second messenger. Calcium is a common second messenger, and this helps explain why calcium regulation in the body is so important.

G proteins are plasma membrane proteins that are named for their ability to break down GTP (guanosine triphosphate), an energy carrier similar to ATP. A G protein lies between the hormone receptor and the enzyme (adenylase cyclase) that changes ATP to cAMP. In this instance, the G protein activates adenylate cyclase, and cAMP results. Other G proteins regulate the action of other membrane enzymes.

cAMP is a second messenger that sets an *enzyme cascade* into motion. It's called an enzyme cascade because cAMP activates only one particular enzyme in the cell, but this enzyme in turn activates another, and so forth. The activated enzymes, of course, can be used repeatedly. Therefore, at every step in the enzyme cascade, more and more reactions take place—the binding of a single hormone molecule eventually results in a thousandfold response. But peptide hormones tend to act relatively quickly for only a short period of time.

Hormones are chemical messengers that influence the metabolism of the cell either indirectly by causing the production of a particular protein in the cell (steroid hormone) or directly by increasing enzyme activity (peptide hormone).

---

2 levels of regulation [handwritten]

Genes are DNA. [handwritten]

1 Gene = 1 protein [handwritten]

± 125,000 [handwritten]

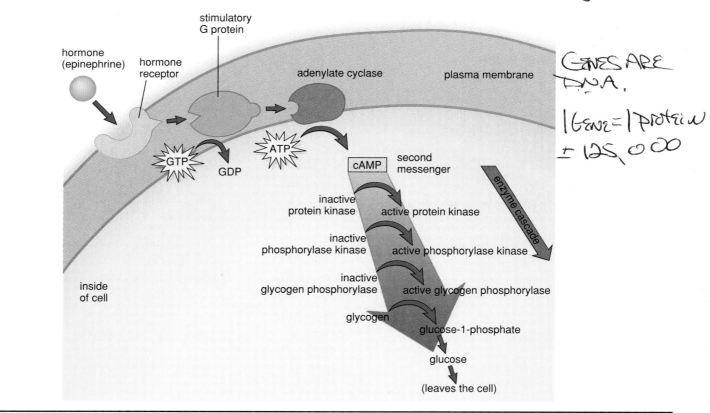

Figure 14.3   **Cellular activity of peptide hormones.**
Peptide hormones, which are called the first messengers, bind to a specific receptor protein in the plasma membrane. The hormone-receptor complex activates a G protein, which binds to GTP; a portion of the complex splits off and activates adenylate cyclase. Adenylate cyclase acts on ATP to produce cAMP, the second messenger, which activates an enzyme cascade. There are also inhibitory G proteins.

TABLE 14.1

Principal Endocrine Glands and Hormones

Endocrine Gland	Hormone Released	Target Tissues/Organ	Chief Function(s) of Hormone
Hypothalamus	Hypothalamic-releasing and release-inhibiting hormones	Anterior pituitary	Regulate anterior pituitary hormones
Anterior pituitary	Thyroid-stimulating (TSH, thyrotropic)	Thyroid	Stimulates thyroid
	Adrenocorticotropic (ACTH)	Adrenal cortex	Stimulates adrenal cortex
	Gonadotropic [follicle-stimulating (FSH), luteinizing (LH)]	Gonads	Egg and sperm production, and sex hormone production
	Prolactin (PRL)	Mammary glands	Milk production
	Growth (GH, somatotropic)	Soft tissues, bones	Cell division, protein synthesis, and bone growth
	Melanocyte-stimulating (MSH)	Melanocytes in skin	Unknown function in humans; regulates skin color in lower vertebrates
Posterior pituitary	Antidiuretic (ADH, vasopressin)	Kidneys	Stimulates water reabsorption by kidneys
	Oxytocin	Uterus, mammary glands	Stimulates uterine muscle contraction and release of milk by mammary glands
Thyroid	Thyroxin	All tissues	Increases metabolic rate; regulates growth and development
	Calcitonin	Bones, kidneys, intestine	Lowers blood calcium level
Parathyroids	Parathyroid (PTH)	Bones, kidneys, intestine	Raises blood calcium level
Adrenal cortex	Glucocorticoids (cortisol)	All tissues	Raise blood glucose level; stimulate breakdown of protein
	Mineralocorticoids (aldosterone)	Kidneys	Reabsorb sodium and excrete potassium
	Sex hormones	Gonads, skin, muscles, bones	Stimulate sex characteristics
Adrenal medulla	Epinephrine and norepinephrine	Cardiac and other muscles	Emergency situations; raise blood glucose level
Pancreas	Insulin	Liver, muscles, adipose tissue	Lowers blood glucose level; promotes formation of glycogen
	Glucagon	Liver, muscles, adipose tissue	Raises blood glucose level
Gonads			
Testes	Androgens (testosterone)	Gonads, skin, muscles, bones	Stimulate secondary male sex characteristics
Ovaries	Estrogens and progesterone	Gonads, skin, muscles, bones	Stimulate female sex characteristics
Thymus	Thymosins	T lymphocytes	Production and maturation of T lymphocytes
Pineal gland	Melatonin	Brain	Circadian and circannual rhythms; possibly involved in maturation of sex organs

Endocrine Glands

Table 14.1 lists the hormones released by the principal endocrine glands, which are depicted in Figure 14.4. The hypothalamus and pituitary gland are located in the brain, the thyroid and parathyroids are located in the neck, while the adrenal glands and pancreas are located in the abdominal cavity. The gonads include the ovaries, located in the abdominal cavity, and the testes, located outside this cavity in the scrotum. Also listed in Table 14.1 are the pineal gland, located in the brain, and the thymus, which lies ventral to the thorax.

Like the nervous system, the endocrine system is especially involved in homeostasis; that is, the dynamic equilibrium of the internal environment. The internal environment is the blood and tissue fluid that surrounds the body's cells. Notice that several hormones directly affect the blood water, glucose, calcium, and sodium levels. Other hormones are involved in the maturation and function of the reproductive organs. In fact, many people are most familiar with the effect of hormones on sexual functions.

There are two mechanisms that control the effect of endocrine glands. Quite often a negative feedback mechanism controls the secretion of hormones. An endocrine gland can be sensitive to either the condition it is regulating or to the blood level of the hormone it is producing. For example, when the blood glucose level rises, the pancreas produces insulin, which causes the liver to store glucose. The stimulus for the production of insulin has thereby been dampened, and therefore, the pancreas stops producing insulin. The effect of insulin, however, can also be offset by the production of glucagon by the pancreas. In other words, contrary actions of hormones is the second way their effects can be controlled.

Notice there are other examples of contrary hormonal actions in Table 14.1. The thyroid lowers the blood calcium level, but the parathyroids raise the blood calcium level. We will also have the opportunity to point out other instances in which hormones work opposite to one another and thereby bring about the regulation of a substance in the blood.

The effect of hormones is controlled in two ways:
(1) negative feedback opposes their release, and
(2) contrary hormones oppose each other's actions.
The end result is regulation of a bodily substance or function within normal limits.

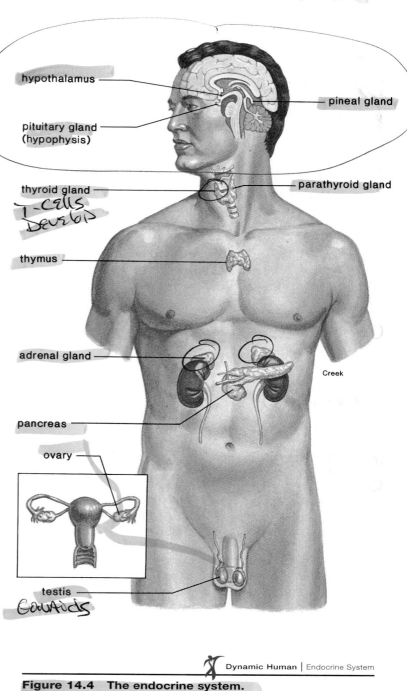

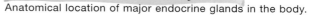

Figure 14.4 The endocrine system.
Anatomical location of major endocrine glands in the body.

14.2 Hypothalamus and Pituitary Gland

The *hypothalamus*, a portion of the brain, regulates the internal environment. For example, it helps to control heart rate, body temperature, and water balance, as well as the glandular secretions of the **pituitary gland.** The pituitary, a small gland about 1 cm in diameter, is connected to the hypothalamus by a stalklike structure. The pituitary has two portions: the posterior pituitary and the anterior pituitary (see Table 14.1).

Posterior Pituitary Stores Two Hormones

The structural and functional relation between the hypothalamus and the **posterior pituitary** illustrates the overlap between the nervous and endocrine systems. There are neurons in the hypothalamus that are called neurosecretory cells because they respond both to neurotransmitters and produce hormones. The hormones pass through axons that run between the hypothalamus and the posterior pituitary. There the hormones are stored in axon endings (Fig. 14.5).

The axon endings in the posterior pituitary store **antidiuretic hormone (ADH)**—sometimes called vasopressin—and oxytocin. ADH promotes the reabsorption of water from the collecting ducts, which receive urine produced by nephrons within the kidneys. There are neurons in the hypothalamus that act as a sensor because they are sensitive to the tonicity of the blood. When these cells determine that the blood is too concentrated, ADH is released into the bloodstream from the axon endings in the posterior pituitary, and upon reaching the kidneys, water is reabsorbed. As the blood becomes dilute, ADH is no longer released. This is an example of control by negative feedback because the effect of the hormone (to dilute blood) acts to shut down the release of the hormone.

Inability to produce ADH causes diabetes insipidus (watery urine), in which a person produces copious amounts of urine with a resultant loss of ions from the blood. The condition can be corrected by the administration of ADH.

Oxytocin is the other hormone that is made in the hypothalamus and stored in the posterior pituitary. Oxytocin causes the uterus to contract and is used to artificially induce labor. It also stimulates the release of milk from the mammary glands when a baby is nursing.

The posterior pituitary stores two hormones, ADH and oxytocin, both of which are produced by and released from neurosecretory cells in the hypothalamus.

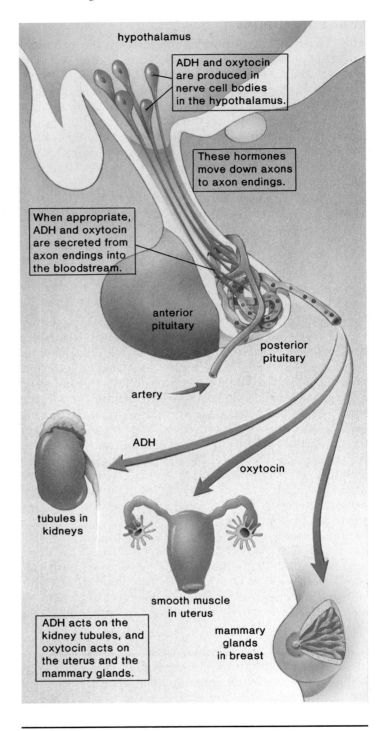

Figure 14.5 **Hypothalamus and posterior pituitary.**
The hypothalamus produces two hormones, ADH and oxytocin, which are stored and secreted by the posterior pituitary.

Anterior Pituitary Is the Master Gland

A *portal system* consists of two capillary systems connected by a vein. Detailed anatomical studies show that there is a portal system lying between the hypothalamus and the anterior pituitary (Fig. 14.6). The hypothalamus controls the anterior pituitary by producing *hypothalamic-releasing* and *release-inhibiting hormones*. For example, there is a thyroid-releasing hormone (TRH) and a thyroid-release-inhibiting hormone (TRIH). TRH stimulates the pituitary to release thyroid-stimulating hormone, and TRIH inhibits the pituitary from releasing thyroid-stimulating hormone.

The anterior pituitary produces at least six different types of hormones, each by a distinct cell type. **Growth hormone (GH),** or somatotropic hormone, promotes cell division, protein synthesis, and bone growth. It stimulates the transport of amino acids into cells and increases the activity of ribosomes, both of which are essential to protein synthesis. In bones, it stimulates cartilaginous plates in long bones and the replacement of these plates into bone by osteoblasts. Evidence suggests that the effects on cartilage and bone may actually be due to hormones called *somatomedins,* which are released by the liver in response to GH.

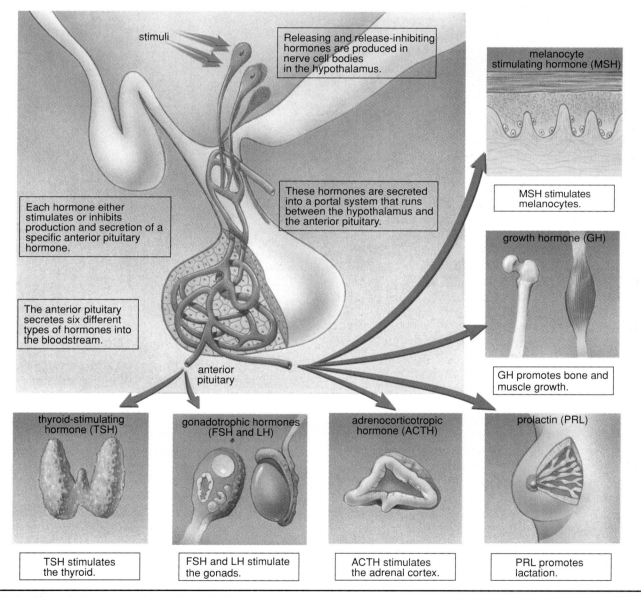

stimuli

Releasing and release-inhibiting hormones are produced in nerve cell bodies in the hypothalamus.

Each hormone either stimulates or inhibits production and secretion of a specific anterior pituitary hormone.

These hormones are secreted into a portal system that runs between the hypothalamus and the anterior pituitary.

The anterior pituitary secretes six different types of hormones into the bloodstream.

anterior pituitary

melanocyte stimulating hormone (MSH)

MSH stimulates melanocytes.

growth hormone (GH)

GH promotes bone and muscle growth.

thyroid-stimulating hormone (TSH)

TSH stimulates the thyroid.

gonadotrophic hormones (FSH and LH)

FSH and LH stimulate the gonads.

adrenocorticotropic hormone (ACTH)

ACTH stimulates the adrenal cortex.

prolactin (PRL)

PRL promotes lactation.

Figure 14.6 Hypothalamus and anterior pituitary.
The hypothalamus controls the secretions of the anterior pituitary, and the anterior pituitary controls the secretions of the thyroid, adrenal cortex, and gonads, which are also endocrine glands.

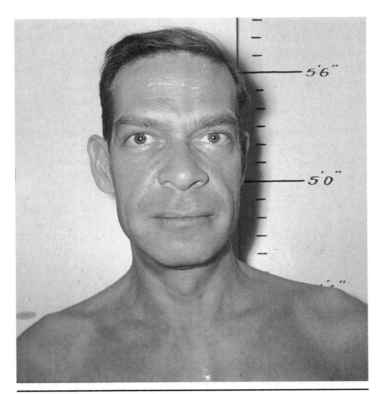

Figure 14.7   Acromegaly.
Acromegaly is caused by overproduction of GH in the adult. It is
characterized by an enlargement of the bones in the face, the
fingers, and the toes of an adult.

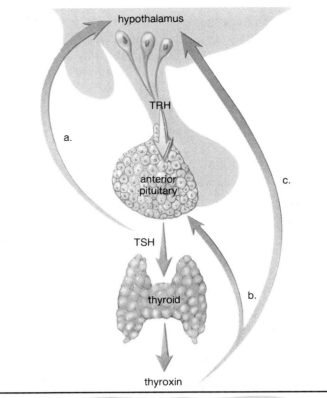

Figure 14.8   Hypothalamus-anterior pituitary-thyroid axis.
Negative feedback in the regulation of the hypothalamus, anterior
pituitary, and thyroid. **a.** The level of TSH exerts feedback control over
the hypothalamus. **b.** The level of thyroxin exerts feedback control over
the anterior pituitary. **c.** The level of thyroxin exerts feedback control
over the hypothalamus. In this way, thyroxin controls its own secretion.
Cortisol and sex hormone levels are controlled in similar ways.

GH is produced in greatest quantities during childhood and adolescence, when most body growth is occurring. If too little GH is produced during childhood, the
individual becomes a pituitary dwarf, characterized by
perfect proportions but small stature. If too much GH is
secreted, a person can become a giant. Giants usually have
poor health, primarily because GH has a secondary effect
on the blood sugar level, promoting an illness called diabetes mellitus.

On occasion, there is overproduction of growth hormone in the adult, and a condition called *acromegaly*
results. Long bone growth is no longer possible in adults.
Only the feet, hands, and face (particularly the chin, nose,
and eyebrow ridges) can respond, and these portions of
the body become overly large (Fig. 14.7).

Prolactin (PRL) is produced in quantity only after
childbirth. It causes the mammary glands in the breasts to
develop and produce milk. It also plays a role in carbohydrate and fat metabolism.

Melanocyte-stimulating hormone (MSH) causes skin-
color changes in many fishes, amphibians, and reptiles
that have melanophores, special skin cells that produce
color variations. The concentration of this hormone in
humans is very low.

The **anterior pituitary** (hypophysis) is sometimes
called the *master gland* because it controls the secretion of
some other endocrine glands (see Fig. 14.4 and Table 14.1).
The anterior pituitary secretes the following hormones,
which have an effect on other glands:

1. **Thyroid-stimulating hormone (TSH)**
2. **Adrenocorticotropic hormone (ACTH),** which
   stimulates the adrenal cortex
3. **Gonadotropic hormones (FSH and LH),** which
   stimulate the gonads—the testes in males and the
   ovaries in females

TSH causes the thyroid to produce thyroxin; ACTH
causes the adrenal cortex to produce cortisol; and gonadotropic hormones cause the gonads to secrete sex hormones
and stimulate gamete (sperm and egg) production. A three-
tiered relationship exists among the hypothalamus, the
anterior pituitary, and the other endocrine glands: the hypothalamus produces hormones, which control the anterior pituitary; and the anterior pituitary produces
hormones that control the thyroid, the adrenal cortex, and
the gonads. In turn, these glands produce hormones that,
through a negative feedback mechanism, regulate the
secretion of the appropriate hypothalamic-releasing hormone and pituitary hormone (Fig. 14.8).

The hypothalamus, the anterior pituitary, and
other glands controlled by the anterior pituitary are
all involved in a self-regulating negative feedback
mechanism.

14.3 Thyroid and Parathyroid Glands

The **thyroid gland** is a large gland located in the neck, where it is attached to the trachea just below the larynx (see Fig. 14.4). The parathyroid glands are imbedded in the posterior surface of the thyroid gland.

Thyroid Gland

The thyroid gland is composed of a large number of follicles, each a small spherical structure made of thyroid cells and filled with stored thyroid hormone.

Thyroxin Speeds Metabolic Rate

Thyroxin, the hormone produced by the thyroid gland, occurs in two forms. Thyroxin is usually secreted as T_4 (tetraiodothyronine), which contains four iodine atoms, but eventually this form is converted to T_3 (triiodothyronine), the active form of the hormone. Iodine, derived from food, is actively transported into the thyroid gland, where its concentration can increase to as much as 25 times that of the blood. If iodine is lacking in the diet, the thyroid gland is unable to produce thyroxin and it enlarges, resulting in a **simple goiter** (Fig. 14.9).

The cause of a simple goiter becomes clear if we refer to Figure 14.8. The anterior pituitary produces TSH, which stimulates the thyroid to secrete thyroxin. The increased level of thyroxin ordinarily exerts feedback control over the anterior pituitary, which ceases to produce TSH. When there is a low level of thyroxin in the blood (called hypothyroidism), the anterior pituitary produces TSH, which stimulates the thyroid. Because the level of thyroxin continues to be the same, TSH continues to stimulate the thyroid. The result is enlargement of the gland, or a simple goiter. Some years ago it was discovered that the use of iodized salt allows the thyroid to produce thyroxin and therefore helps prevent simple goiter.

Thyroxin increases the metabolic rate. It does not have one target organ; instead, it stimulates all organs of the body to metabolize at a faster rate. More glucose is broken down and more energy is utilized.

If the thyroid fails to develop properly, a condition called **cretinism** results. Individuals with this condition are short and stocky and have had extreme hypothyroidism since childhood or infancy. Thyroxin therapy can initiate growth, but unless treatment is begun within the first two months, mental retardation results. The occurrence of hypothyroidism in adults produces the condition known as **myxedema,** which is characterized by lethargy, weight gain, loss of hair, slower pulse rate, lowered body temperature, and thickness and puffiness of the skin (Fig. 14.10). The administration of adequate doses of thyroxin restores normal function and appearance.

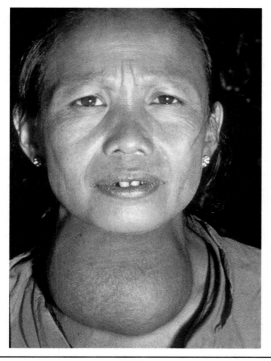

Figure 14.9   Simple goiter.
An enlarged thyroid gland often is caused by a lack of iodine in the diet. Without iodine, the thyroid is unable to produce thyroxin, and continued anterior pituitary stimulation causes the gland to enlarge.

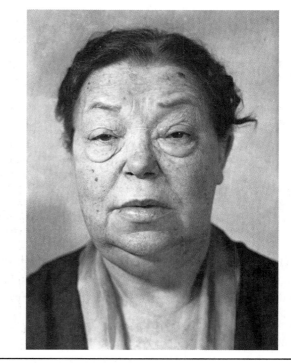

Figure 14.10   Myxedema.
This condition is caused by thyroxin insufficiency in the older adult. An unusual type of edema leads to swelling of the face and bagginess under the eyes.

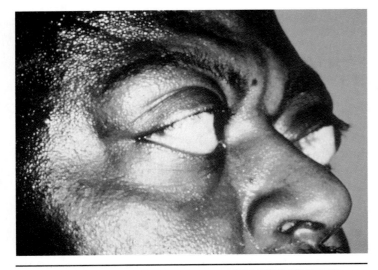

Figure 14.11   Exophthalmic goiter (Graves' disease).
In hyperthyroidism, the eyes protrude because of edema in the
tissues of the eye sockets.

In the case of hyperthyroidism, or *Graves' disease,* the
thyroid gland is enlarged and overactive, causing a goiter
to form. The eyes protrude because of edema in eye socket
tissues and swelling of muscles that move the eyes. This
type of goiter is called **exophthalmic goiter** (Fig. 14.11). The
patient usually becomes hyperactive, nervous, irritable, and
suffers from insomnia. Removal or destruction of a portion
of the thyroid by means of radioactive iodine is sometimes
effective in curing the condition. Hyperthyroidism can also
be caused by a thyroid tumor, which is usually detected as
a lump during physical examination. Again, the treatment
is surgery in combination with administration of radioactive iodine. The prognosis for most patients is excellent.

Calcitonin Lowers Blood Calcium

In addition to thyroxin, the thyroid gland also produces
the hormone **calcitonin,** which helps regulate the blood
calcium level in the blood and opposes the action of parathyroid hormone. Calcitonin lowers blood calcium (Ca^{++})
by increasing the buildup of bone (Fig. 14.12). The interaction of these two hormones is also mentioned below.

Parathyroid Glands Are Embedded in Thyroid

Many years ago, the four parathyroid glands were sometimes mistakenly removed during thyroid surgery because
they are so small.

Parathyroid Hormone Raises Blood Calcium

Parathyroid hormone (PTH), the hormone produced by
the **parathyroid glands,** causes the blood phosphate ($HPO_4^=$)
level to decrease and the blood calcium (Ca^{++}) level to increase. PTH inhibits the activity of osteoblasts and promotes the activity of *osteoclasts* in bone. Calcitonin has the
opposite effect, and therefore, the homeostatic blood

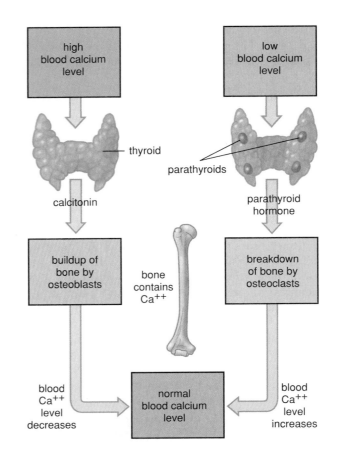

Figure 14.12   Regulation of blood calcium level.
The opposing actions of parathyroid hormone and calcitonin on
bone breakdown maintains the homeostatic level of blood calcium.

calcium balance is achieved through the action of both PTH
and calcitonin. How PTH and calcitonin work together is
described in Figure 14.12.

Aside from promoting the *breakdown* of bone, PTH
causes the *retention* of calcium by the kidneys and activates
vitamin D, which, in turn, stimulates the *absorption* of calcium from the intestine. In addition, PTH promotes the excretion of phosphate in urine by the kidneys (Fig. 14.13).

If insufficient parathyroid hormone production causes
a significant drop in the blood calcium level, tetany results.
In **tetany,** the body shakes from continuous muscle contraction. The effect is brought about by increased excitability
of the nerves, which initiate nerve impulses spontaneously
and without rest. Calcium plays an important role in both
nervous conduction and muscle contraction. It is also necessary to blood clotting.

The level of PTH secretion is controlled by a negative
feedback mechanism involving calcium. When the blood
calcium level rises, PTH secretion is inhibited, and when
the blood calcium level lowers, PTH secretion is stimulated.

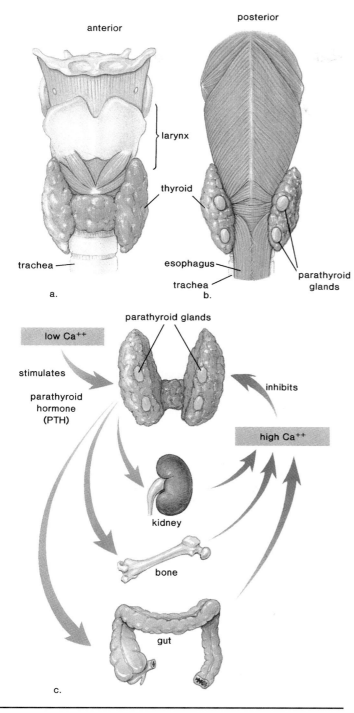

Figure 14.13 Thyroid and parathyroid glands.
a. The thyroid gland is located in the neck in front of the trachea.
b. The four parathyroid glands are embedded in the posterior surface of the thyroid gland. Yet the parathyroid and thyroid glands have no anatomical or physiological connection with one another.
c. Regulation of parathyroid hormone (PTH) secretion. A low blood calcium level causes the parathyroids to secrete PTH, which causes the kidneys and the intestine to retain calcium and osteoclasts to break down bone. The result is a higher blood calcium level. A high blood calcium level then inhibits secretion of PTH.

14.4 Adrenal Glands Have Two Parts

Each of the two **adrenal glands** lies atop a kidney (Fig. 14.4). Each consists of an inner portion called the *medulla* and an outer portion called the *cortex*. These portions, like the anterior pituitary and the posterior pituitary, have no physiological connection with one another.

The adrenal hormones increase during times of stress. Those secreted by the adrenal medulla allow us to respond to emergency situations, and those from the adrenal cortex help us to recover from stress.

The hypothalamus exerts control over the activity of both portions of the adrenal glands. It can initiate nerve impulses that travel by way of the brain stem, spinal cord, and sympathetic nerve fibers to the adrenal medulla, which then secretes its hormones. The hypothalamus, by means of ACTH-releasing hormone, controls the anterior pituitary's secretion of ACTH, which, in turn, stimulates the adrenal cortex. Stress of all types, including both emotional and physical trauma, prompts the hypothalamus to stimulate the adrenal glands.

The adrenal glands have two parts, an outer cortex and an inner medulla. The adrenal medulla is under nervous control, and the cortex is under the control of ACTH, a hormone of the anterior pituitary.

Adrenal Medulla Responds to Stress

Epinephrine (adrenaline) and **norepinephrine** (noradrenaline) are produced by the *adrenal medulla*. The postganglionic fibers of the sympathetic system, which controls the adrenal medulla, also secrete norepinephrine.

Epinephrine and norepinephrine are involved in the body's immediate response to stress. They bring about all the bodily changes that occur when an individual reacts to an emergency:

- blood glucose level rises, and the metabolic rate increases
- bronchioles dilate, and breathing rate increases
- blood vessels to the digestive tract and skin constrict; those to skeletal muscles dilate
- cardiac muscle contracts more forcefully, and heart rate increases

---

The adrenal medulla releases epinephrine and norepinephrine into the bloodstream, helping us cope with emergency situations.

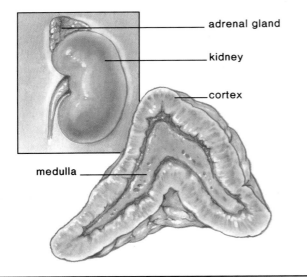

Figure 14.14   Adrenal glands.
Location and structure of adrenal glands.

Adrenal Cortex Also Responds to Stress

The two major types of hormones produced by the *adrenal cortex* are the *glucocorticoids,* which help regulate the blood glucose level, and *mineralocorticoids,* which help regulate the level of minerals in the blood. It also secretes a small amount of male sex hormones and a small amount of female sex hormones in both sexes—that is, in the male, both male and female sex hormones are produced by the adrenal cortex, and in the female, both male and female sex hormones are also produced by the adrenal cortex (Fig. 14.14).

Glucocorticoids Raise Blood Glucose

Cortisol is the glucocorticoid responsible for the greatest amount of activity. Cortisol promotes the hydrolysis of muscle protein to amino acids, which enter blood. This leads to a higher blood glucose level when the liver converts these amino acids to glucose. Cortisol also favors metabolism of fatty acids rather than carbohydrate. In opposition to insulin, therefore, cortisol raises the blood glucose level. Cortisol also counteracts the inflammatory response that leads to the pain and the swelling of joints in arthritis and bursitis. The administration of cortisol aids these conditions because it reduces inflammation.

Mineralocorticoids Regulate Blood Na⁺ and K⁺ Levels

Aldosterone is the most important of the mineralocorticoids. The primary target organ of aldosterone is the kidney, where it promotes renal absorption of sodium (Na^+) and renal excretion of potassium (K^+).

The secretion of mineralocorticoids is not under the control of the anterior pituitary. When the blood volume and blood sodium level is low, the kidneys secrete **renin** (Fig. 14.15). Renin is an enzyme that converts the plasma

protein angiotensinogen to angiotensin I, which is changed to angiotensin II by a converting enzyme found in the lungs. Angiotensin II stimulates the adrenal cortex to release aldosterone. The effect of this system, called the renin-angiotensin-aldosterone system, is to raise the blood volume and pressure in two ways. First, angiotensin II constricts the arterioles directly, and secondly, aldosterone causes the kidneys to reabsorb sodium. When the blood sodium level rises, water is reabsorbed, and blood volume and pressure are maintained.

Two other hormones play a role in the homeostatic maintenance of blood volume. We have already discussed that the hormone ADH helps increase blood volume by causing the kidney to reabsorb water. When the atria of the heart are stretched due to increased blood volume, cardiac cells release a hormone called *atrial natriuretic hormone (ANH),* which inhibits the secretion of renin by the kidneys and the secretion of aldosterone from the adrenal cortex. The effect of this hormone is, therefore, to cause the excretion of sodium, that is, *natriuresis.* When sodium is excreted, so is water, and therefore, blood volume and blood pressure decrease.

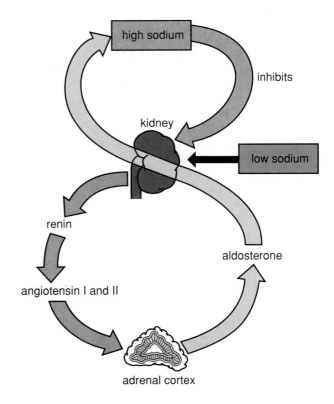

Figure 14.15   Renin-angiotensin-aldosterone homeostatic system.
If the blood sodium level is low, the kidneys secrete renin. The increased renin acts via the increased production of angiotensin I and II to stimulate aldosterone secretion. Aldosterone promotes reabsorption of sodium by the kidneys; when the blood sodium level rises, water is reabsorbed, and blood pressure rises.

Adrenal Cortex Can Malfunction

When there is a low level of adrenal cortex hormones due to hyposecretion, a person develops Addison disease. Typically, there is a peculiar bronzing of the skin in a person with **Addison disease** (Fig. 14.16). Because the lack of cortisol results in a drop in blood glucose level that can be severe, there is high susceptibility to any kind of stress due to an insufficient energy supply. Even a mild infection can cause death. Due to the lack of aldosterone, the blood sodium level is low, and the person experiences low blood pressure and possibly severe dehydration. Left untreated, Addison disease can be fatal.

When there is a high level of glucocorticoids in the body due to hypersecretion, a person develops **Cushing** syndrome (Fig. 14.17). Excess cortisol causes a tendency toward diabetes mellitus, a decrease in muscular protein, and an increase in subcutaneous fat. Because of these effects, the person usually has an obese trunk, while the arms and legs remain normal. Due to the high blood sodium level, the blood is basic (pH greater than normal), hypertension occurs, and there is edema of the face, which gives it a moonlike shape. Masculinization may occur in women due to oversecretion of adrenal male sex hormone.

Addison disease is due to adrenal cortex hyposecretion, and Cushing syndrome is due to adrenal cortex hypersecretion.

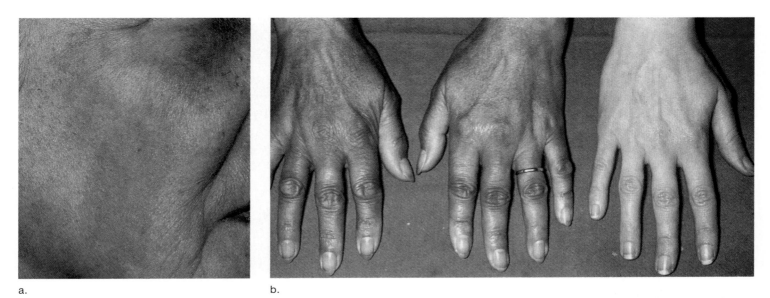

a.   b.

Figure 14.16 Addison disease.
This condition is characterized by a peculiar bronzing of the skin, particularly noticeable in these light-skinned individuals. Note the color of **(a)** the face and **(b)** the hands compared to the hand of an individual without the disease.

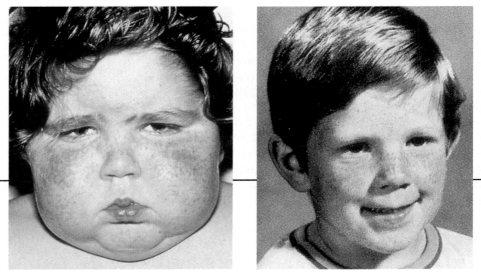

Figure 14.17 Cushing syndrome.
This condition results from hypersecretion of glucocorticoids by an adrenal cortex tumor.
a. First diagnosed with Cushing syndrome.
b. Four months later, after therapy.

a.   b.

14.5 Pancreas Produces Two Hormones

The **pancreas** is a long organ that lies transversely in the abdomen between the kidneys and near the duodenum of the small intestine. It is composed of two types of tissue. Exocrine tissue produces and secretes *digestive juices* that go by way of ducts to the small intestine. Endocrine tissue, called the **pancreatic islets** (of Langerhans), produces and secretes the hormones **insulin** and **glucagon** directly into the blood.

Insulin is secreted when there is a high blood glucose level, which usually occurs just after eating. Insulin has three different actions: (1) it stimulates liver, fat, and muscle cells to take up and metabolize glucose; (2) it stimulates the liver and the muscles to store glucose as glycogen; and (3) it promotes the buildup of fats and proteins and inhibits their use as an energy source. Therefore, insulin is a hormone that promotes storage of nutrients so that they are on hand during leaner times. It also helps to lower the blood glucose level.

Glucagon is secreted from the pancreas in between meals, and its effects are opposite to those of insulin. Glucagon stimulates the breakdown of stored nutrients and causes the blood glucose level to rise (Fig. 14.18).

Diabetes Mellitus: Two Types

The symptoms of **diabetes mellitus** include the following:

  Sugar in the urine

  Frequent, copious urination

  Abnormal thirst

  Rapid weight loss

  General weakness

  Drowsiness and fatigue

  Itching of the genitals and the skin

  Visual disturbances, blurring

  Skin disorders, such as boils, carbuncles, and infection

Many of these symptoms develop because sugar is not being metabolized by the cells. The liver fails to store glucose as glycogen, and all the cells have a reduced ability to take up and utilize glucose as an energy source. This means that the blood glucose level rises very high after eating, causing glucose to be excreted in the urine. More water than usual is therefore excreted so that the diabetic is extremely thirsty.

Since carbohydrate is not being metabolized, the body turns to the breakdown of protein and fat for energy. The metabolism of fat leads to the buildup of ketones in the blood and acidosis (acid blood) that can eventually cause coma and death of the diabetic. The symptoms of hyperglycemia (high blood sugar) develop slowly (Table 14.2), and there is time for intervention and reversal of symptoms.

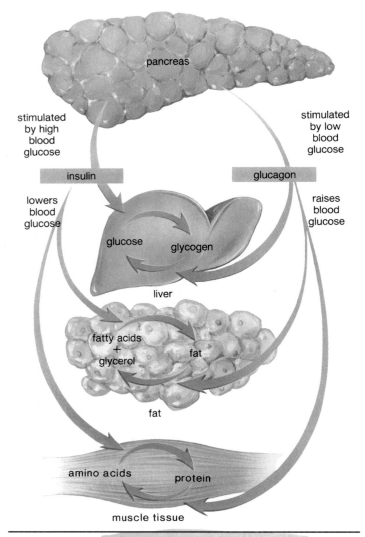

Figure 14.18   Insulin and glucagon homeostatic system.
When the blood glucose level is high, the pancreas secretes insulin. Insulin promotes the storage of glucose as glycogen and the synthesis of proteins and fats (as opposed to their use as energy sources). Therefore, insulin lowers the blood glucose level. When the blood glucose level is low, the pancreas secretes glucagon. Glucagon acts in opposition to insulin in all respects; therefore, glucagon raises the blood glucose level.

There are two types of diabetes mellitus. In *type I (insulin-dependent) diabetes*, the pancreas is not producing insulin. The condition is believed to be brought on by exposure to an environmental agent, most likely a virus, whose presence causes cytotoxic T cells to destroy the pancreatic islets (of Langerhans). As a result, the individual must have daily insulin injections. These injections control the diabetic symptoms but still can cause inconveniences, since either an overdose of insulin or the absence of regular eating can bring on the symptoms of hypoglycemia (low blood sugar) (Table 14.2). These symptoms appear when the blood glucose level falls below the normal range. Since the brain requires a constant supply of sugar, unconsciousness can result. The cure is quite simple: an immediate source of sugar, such as a sugar cube or fruit juice, can very quickly counteract hypoglycemia.

Obviously, insulin injections are not the same as a fully functioning pancreas that responds on demand to a high glucose level by supplying insulin. For this reason, some doctors advocate an islet transplant for type I diabetes.

Of the 16 million people who now have diabetes in the United States, most have *type II (noninsulin-dependent) diabetes*. This type of diabetes mellitus usually occurs in people of any age who are obese and inactive. The pancreas produces insulin, but the cells do not respond to it. The cells increasingly lack the receptors necessary to detect the presence of insulin, and therefore, the liver and muscles in particular are incapable of taking up glucose. If type II diabetes is untreated, the results can be as serious as type I diabetes. (Diabetics are prone to blindness, kidney disease, and circulatory disorders. Pregnancy carries an increased risk of diabetic coma, and the child of a diabetic is somewhat more likely to be stillborn or to die shortly after birth.) It is important, therefore, to prevent or at least to control type II diabetes. The best defense is a low-fat diet and regular exercise. If this fails, oral drugs that stimulate the pancreas to secrete more insulin and enhance the action of insulin on the liver and muscles are available.

Diabetes mellitus is caused by the lack of insulin or the insensitivity of cells to insulin. Insulin lowers blood glucose levels by causing the cells, particularly those of the liver and muscles, to take up glucose and convert it to glycogen.

TABLE 14.2

Symptoms of Hyperglycemia and Hypoglycemia

Hyperglycemia	Hypoglycemia
Slow, gradual onset	Sudden onset
Dry, hot skin	Perspiration, pale skin
No dizziness	Dizziness
No palpitation	Palpitation
No hunger	Hunger
Excessive urination	Normal urination
Excessive thirst	Normal thirst
Deep, labored breathing	Shallow breathing
Fruity breath odor	Normal breath odor
Large amounts of urinary sugar	Urinary sugar absent or slight
Ketones in urine	No ketones in urine
Drowsiness and great lethargy leading to stupor	Confusion, disorientation, strange behavior

Data from Henry Dolger and Bernard Seeman, How to Live with Diabetes, *1986.*

14.6 Other Endocrine Glands

The testes and ovaries are endocrine glands. There are also lesser known glands and some tissues that produce hormones.

Testes Are in Males and Ovaries Are in Females

The gonads are the testes in the male and the ovaries in the female. The *testes* are located in the scrotum, and the ovaries are located in the abdominal cavity. The testes produce *androgens* (e.g., *testosterone*), which are the male sex hormones, and the *ovaries* produce estrogen and progesterone, the female sex hormones. The hypothalamus and the pituitary gland control the hormonal secretions of these organs in the same manner that was previously described for the thyroid gland.

The male sex hormone, testosterone, has many functions. It is essential for the normal development and functioning of the sex organs in males. It is also necessary for the maturation of sperm.

Greatly increased testosterone secretion at the time of puberty stimulates the growth of the penis and the testes. Testosterone also brings about and maintains the secondary sex characteristics in males that develop at the time of puberty. Testosterone causes growth of a beard, axillary (underarm) hair, and pubic hair. It prompts the larynx and the vocal cords to enlarge, causing the voice to change. It is partially responsible for the muscular strength of males, and this is the reason some athletes take supplemental amounts of **anabolic steroids,** which are either testosterone or related chemicals. The contraindications of taking anabolic steroids are discussed in the reading on the next page. Testosterone also stimulates oil and sweat glands in the skin; therefore, it is largely responsible for acne and body odor. Another side effect of testosterone is baldness. Genes for baldness probably are inherited by both sexes, but baldness is seen more often in males because of the presence of testosterone.

Testosterone is believed to be largely responsible for the sex drive. It may even contribute to the suggested aggressiveness of males.

The female sex hormones, *estrogens* and *progesterone*, have many effects on the body. In particular, estrogens secreted at the time of puberty stimulate the growth of the uterus and the vagina. Estrogen is necessary for egg maturation and is largely responsible for the secondary sex characteristics in females. It is responsible for female body hair and fat distribution. In general, females have a more rounded appearance than males because of a greater accumulation of fat beneath the skin. Also, the pelvic girdle extends more in females than in males, resulting in females having a larger pelvic cavity and wider hips. Both estrogen and progesterone are required for breast development and regulation of the uterine cycle, which includes monthly menstruation (discharge of blood and mucosal tissues from the uterus).

Health Focus

Dangers of Anabolic Steroids

Anabolic steroids are synthetic forms of the male sex hormone testosterone. Trainers may have been the first to acquire anabolic steroids for weight lifters, bodybuilders, and other athletes such as professional football players. When taken in large doses (10 to 100 times the amount prescribed by doctors for illnesses) and accompanied by exercise, anabolic steroids promote larger muscles. Occasionally, steroid abuse makes the news because an Olympic winner tests positive for the drug and must relinquish a medal. Steroid use has been outlawed by the International Olympic Committee.

The U.S. Food and Drug Administration bans the importation of most steroids, but they are brought into the country illegally and sold through the mail or in gyms and health clubs. According to federal officials, 1 to 3 million Americans now take anabolic steroids. Their increased use by teenagers wishing to build bulk quickly is of special concern. Some attribute this to society's emphasis on physical appearance and the need of insecure youngsters to feel better about how they look.

Physicians, teachers, and parents are quite alarmed about anabolic steroid abuse. It's even predicted that two or three months of high-dosage use in a youngster can cause death two or three decades later. The many harmful effects of anabolic steroids on the body are listed in Figure 14A. In addition, these drugs increase aggression and make a person feel invincible. One abuser even had his friend videotape him as he drove his car at 40 miles per hour into a tree!

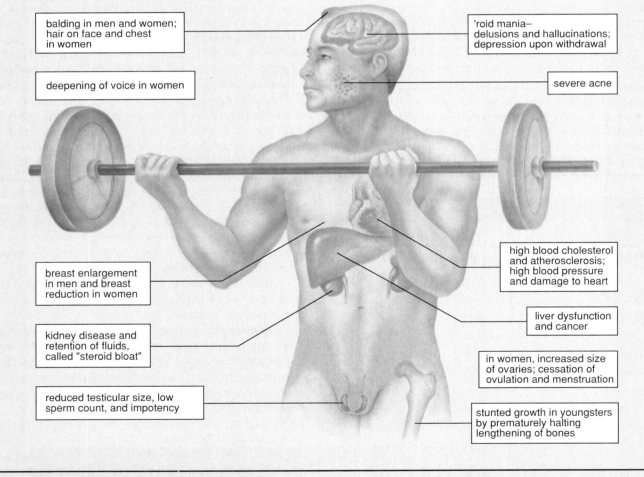

balding in men and women; hair on face and chest in women

deepening of voice in women

breast enlargement in men and breast reduction in women

kidney disease and retention of fluids, called "steroid bloat"

reduced testicular size, low sperm count, and impotency

'roid mania— delusions and hallucinations; depression upon withdrawal

severe acne

high blood cholesterol and atherosclerosis; high blood pressure and damage to heart

liver dysfunction and cancer

in women, increased size of ovaries; cessation of ovulation and menstruation

stunted growth in youngsters by prematurely halting lengthening of bones

Figure 14A   The effects of anabolic steroid use.

Thymus: Most Active in Children

The *thymus* is a lobular gland that lies inside the thorax on the ventral side (see Fig. 14.4). This organ reaches its largest size and is most active during childhood. With aging, the organ gets smaller and becomes fatty. Certain lymphocytes that originate in the bone marrow and then pass through the thymus are transformed into T lymphocytes. The thymus produces various hormones called *thymosins*, which aid in the differentiation of T lymphocytes and may stimulate immune cells in general. There is hope that these hormones can be used in conjunction with lymphokine therapy to restore or to stimulate T lymphocyte function in patients suffering from AIDS or cancer.

Pineal Gland: Hormone at Night

The **pineal gland,** which is located in the brain (see Fig. 14.4), produces the hormone called melatonin, primarily at night. Melatonin is involved in a daily cycle called a **circadian rhythm.** Normally we grow sleepy at night when melatonin levels increase and awaken once daylight returns and melatonin levels are low. Shift work is usually troublesome because it upsets this normal daily rhythm. Similarly, travel to another time zone, as when going to Europe from the United States, results in jet lag because the body is still producing melatonin according to the old schedule. Some people even have Seasonal Affective Disorder (SAD); they become depressed and have an uncontrollable desire to sleep with the onset of winter. Giving melatonin makes their symptoms worse, but exposure to a bright light improves them.

Many animals go through a yearly cycle that includes enlargement of reproductive organs during the summer when melatonin levels are low. Mating occurs in the fall and young are born in the spring. It is of interest that children with a brain tumor that destroys the pineal gland experience early puberty. It's possible that the pineal gland is involved in human sexual development.

Nontraditional Sources

Even organs that are usually not considered endocrine glands do secrete hormones. The heart produces *atrial natriuretic hormone,* which helps to regulate the sodium and water balance of the body. It promotes the renal excretion of sodium, and when water follows passively, blood pressure lowers. It also inhibits the release of renin and the hormones aldosterone and ADH. Atrial natriuretic hormone is a peptide that is released not only by the atria but also by the aortic arch, ventricles, lungs, and pituitary gland in response to increases in blood pressure. The stomach and the small intestine produce peptide hormones that regulate digestive secretions.

A number of different types of organs and cells produce peptide *growth factors,* which stimulate cell division and mitosis. They are like hormones in that they act on cell types with specific receptors to receive them. Some, including lymphokines and blood cell growth factors, are released into the blood; others diffuse to nearby cells. Other growth factors are described in the following listing:

Platelet-derived growth factor is released from platelets and from many other cell types. It helps in wound healing and causes an increase in the number of fibroblasts, smooth muscle cells, and certain cells of the nervous system.

Epidermal growth factor and nerve growth factor stimulate the cells indicated by their names, as well as many others. These growth factors are also important in wound healing.

Tumor angiogenesis factor stimulates the formation of capillary networks and is released by tumor cells. One treatment for cancer is to prevent the activity of this growth factor.

Prostaglandins (PG) are another class of chemical messengers that are produced and act locally. They are derived from fatty acids stored in plasma membranes as phospholipids. When a cell is stimulated by reception of a hormone or even by trauma, a series of synthetic reactions takes place in the plasma membrane, and PG is first released into the cytoplasm and then secreted from the cell. There are many different types of prostaglandins produced by many different tissues. In the uterus, certain prostaglandins cause muscles to contract; therefore, they are implicated in the pain and discomfort of menstruation in some women. Aspirin helps reduce pain because it inhibits the synthesis of prostaglandins. Also, prostaglandins mediate the effects of pyrogens, chemicals that are believed to reset the temperature regulatory center in the brain. Aspirin reduces the body temperature of a person with a fever because of its effect on prostaglandins.

On the other hand, certain prostaglandins are used to treat ulcers because they reduce gastric secretion. Others are used to treat hypertension because they lower blood pressure and to prevent thrombosis because they inhibit platelet aggregation. Because the different prostaglandins can have contrary effects, it has been very difficult to standardize their use, and in most instances, prostaglandin therapy is still considered experimental.

14.7 Working Together

The Working Together box on page 304 shows how the endocrine system works with the other systems of the body to maintain homeostasis.

HUMAN SYSTEMS WORK TOGETHER

Integumentary System

Androgens activate sebaceous glands and help regulate hair growth.

Skin provides sensory input that results in action by certain endocrine glands.

Skeletal System

Growth hormone regulates bone development; parathyroid hormone and calcitonin regulate Ca^{++} content.

Bones provide protection for glands; store Ca^{++} used as second messenger.

Muscular System

Androgens promote growth of skeletal muscle; epinephrine stimulates heart and constricts blood vessels.

Muscles help protect glands.

Nervous System

Sex hormones affect development of brain.

Hypothalamus is part of endocrine system; nerves innervate glands of secretion.

Circulatory System

Epinephrine increases blood pressure; ADH, aldosterone, and atrial natriuretic hormone help regulate blood volume; growth factors control blood cell formation.

Blood vessels transport hormones from glands; blood services glands; heart produces atrial natriuretic hormones.

How the Endocrine System works with other body systems

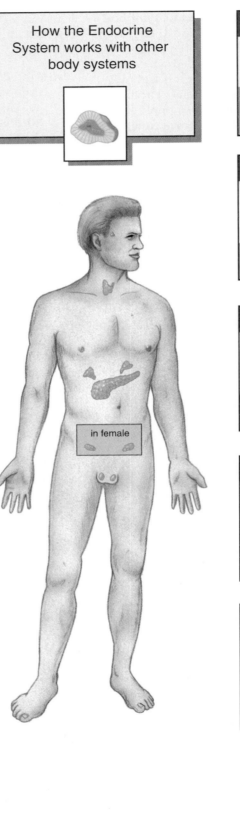

in female

Lymphatic System/Immunity

Thymus is necessary for maturity of T lymphocytes.

Lymphatic vessels pick up excess tissue fluid; immune system protects against infections.

Respiratory System

Epinephrine promotes ventilation by dilating bronchioles; growth factors control production of red blood cells that carry oxygen.

Gas exchange in lungs provides oxygen and rids body of carbon dioxide.

Digestive System

Hormones help control secretion of digestive glands and accessory organs; insulin and glucagon regulate glucose storage in liver.

Stomach and small intestine produce hormones.

Urinary System

ADH, aldosterone, and atrial natriuretic hormone regulate reabsorption of water and Na^+ by kidneys.

Kidneys keep blood values within normal limits so that transport of hormones continues.

Reproductive System

Hypothalamic, pituitary, and sex hormones control sex characteristics and regulate reproductive processes.

Gonads produce sex hormones.

Dynamic Human | Endocrine System

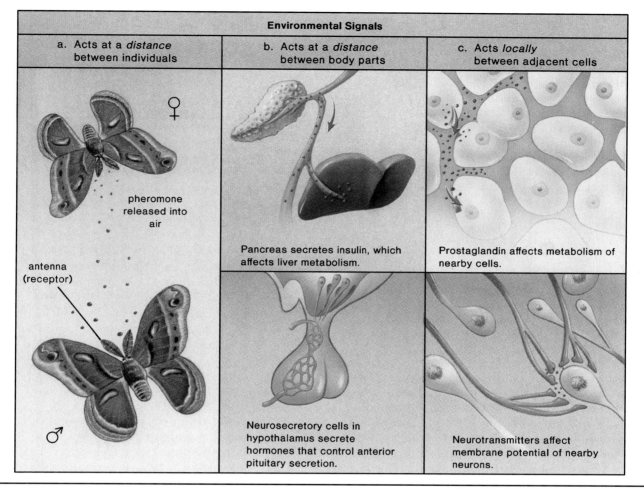

Environmental Signals

a. Acts at a *distance* between individuals

pheromone released into air

antenna (receptor)

♀

♂

b. Acts at a *distance* between body parts

Pancreas secretes insulin, which affects liver metabolism.

Neurosecretory cells in hypothalamus secrete hormones that control anterior pituitary secretion.

c. Acts *locally* between adjacent cells

Prostaglandin affects metabolism of nearby cells.

Neurotransmitters affect membrane potential of nearby neurons.

Figure 14.19   The three categories of environmental signals.
a. Pheromones are chemical messengers that act at a distance between individuals. **b.** Endocrine hormones and neurosecretions typically are carried in the bloodstream and act at a distance within the body of a single organism. **c.** Some chemical messengers have local effects only; they pass between cells that are adjacent to one another. This, of course, includes neurotransmitters.

14.8 Environmental Signals

In this chapter, we concentrated on the functions of the human endocrine glands and their hormonal secretions. We already know that hormones are only one type of chemical messenger, or environmental signal, between cells. In fact, the concept of chemical messengers has now been broadened to include at least the following three categories of environmental signals (Fig. 14.19).

Environmental signals that act at a distance between organisms. Many organisms release chemical messengers called **pheromones** into the air or in externally deposited body fluids. These are intended to be messages for other members of the species. For example, ants lay down a pheromone trail to direct other ants to food, and the female silkworm moth releases bombykol, a sex attractant that is received by male moth antennae even several kilometers away. This chemical is so potent that it has been estimated that only 40 out of 40,000 receptors on the male antennae need to be activated for the male to respond. Mammals also release pheromones; for example, the urine of dogs

serves as a territorial marker. Some studies are being conducted to determine if humans have pheromones, and other studies have suggested that humans respond to pheromones. For example, investigators have observed that women who live in close quarters tend to have coinciding menstrual cycles. They reason that this might be caused by a pheromone, but they don't know what the pheromone might be or how it might exert its effect.

Environmental signals that act at a distance between body parts. This category includes the endocrine secretions, which traditionally have been called hormones. It also includes the secretions of the neurosecretory cells in the hypothalamus—the production and action of ADH and oxytocin illustrate the close relationship between the nervous system and the endocrine system. Neurosecretory cells produce these hormones, which are released when these cells receive nerve impulses. As another example of the overlap between the nervous and endocrine systems, consider that endorphins on occasion travel in the bloodstream, but they act on nerve cells to alter their membrane potential. Also, norepinephrine is secreted by the adrenal medulla but is also a neurotransmitter in the sympathetic nervous system.

Environmental signals that act locally between adjacent cells. Neurotransmitters belong in this category, as do substances that are sometimes called local hormones. For example, when the skin is cut, histamine, released by mast cells, promotes the inflammatory response.

> Today, hormones are categorized as one type of environmental signal. There are environmental signals that work at a distance between individuals (e.g., pheromones), at a distance between body parts (e.g., hormones), and locally between adjacent cells (e.g., neurotransmitters).

SUMMARY

14.1 How Hormones Work

Steroid hormones enter the nucleus and combine with a receptor hormone, and the complex attaches to and activates DNA. Transcription and translation lead to protein synthesis.

The peptide hormones are usually received by a hormone receptor located in the plasma membrane. Most often their reception leads to activation of an enzyme that changes ATP to cyclic AMP (cAMP). cAMP then activates another enzyme, which activates another, and so forth.

14.2 Hypothalamus and Pituitary Gland

Neurosecretory cells in the hypothalamus produce antidiuretic hormone (ADH) and oxytocin, which are stored in axon endings in the posterior pituitary until they are released.

The hypothalamus produces hypothalamic-releasing and hypothalamic-release-inhibiting hormones, which pass to the anterior pituitary by way of a portal system. The anterior pituitary produces at least six types of hormones, and some of these stimulate other hormonal glands to secrete hormones. Therefore, the anterior pituitary is sometimes called the master gland.

14.3 Thyroid and Parathyroid Glands

The thyroid gland produces thyroxin and triiodothyronine, hormones that play a role in growth and development of immature forms; in mature individuals, they increase the metabolic rate. The thyroid gland also produces calcitonin, which helps lower the blood calcium level. The parathyroid glands raise the blood calcium and decrease the blood phosphate level.

14.4 Adrenal Glands Have Two Parts

The adrenal medulla secretes epinephrine and norepinephrine, which bring about responses we associate with emergency situations.

The adrenal cortex primarily produces the glucocorticoids (cortisol) and the mineralocorticoids (aldosterone). Cortisol stimulates hydrolysis of proteins to amino acids that are converted to glucose; in this way, it raises the blood glucose level. Aldosterone causes the kidneys to reabsorb sodium ions (Na^+) and excrete potassium ions (K^+).

14.5 Pancreas Produces Two Hormones

The pancreatic islets secrete insulin, which lowers the blood glucose level, and glucagon, which has the opposite effect. The most common illness due to hormonal imbalance is diabetes mellitus, which is due to the failure of the pancreas to produce insulin or the cells to take it up.

14.6 Other Endocrine Glands

The gonads produce the sex hormones; the thymus secretes thymosins, which stimulate T lymphocyte production and maturation; the pineal gland produces melatonin, which may be involved in circadian rhythms and the development of the reproductive organs.

14.7 Working Together

The endocrine system works with the other systems of the body in the ways described on page 304.

14.8 Environmental Signals

There are three categories of environmental signals; those that act at a distance between individuals (pheromones); those that act at a distance within the individual (traditional endocrine hormones and secretions of neurosecretory cells); and local messengers (such as prostaglandins, growth factors, and neurotransmitters). Since there is great overlap between these categories, perhaps the definition of a hormone should be expanded to include all of them.

STUDYING THE CONCEPTS

1. Give a definition of a hormone that includes how they are transported in the body and how they are received.   288

2. Categorize endocrine hormones according to their chemical composition.   288

3. Explain how steroid hormones and peptide hormones affect the metabolism of the cell.   288–89

4. Explain the relationship of the hypothalamus to the posterior pituitary gland and to the anterior pituitary gland.   292–94

5. Discuss two hormones secreted by the anterior pituitary that have an effect on the body proper rather than on other glands. What medical conditions, if any, are associated with each hormone?   293–94

6. Give an example of the three-tiered relationship among the hypothalamus, the anterior pituitary, and other endocrine glands. Explain why the anterior pituitary can be called the master gland.   293–94

7. List the hormones secreted by the anterior pituitary that affect other endocrine glands.   294

8. What types of goiters are associated with a malfunctioning thyroid? Explain each type.   295–96

9. How do the thyroid and the parathyroid work together to control the blood calcium level?   296

10. What hormones are secreted by the adrenal medulla, and what effects do these hormones have?   297

11. Name the most significant glucocorticoid and mineralocorticoid, and discuss their functions. Explain the symptoms of Addison disease and Cushing syndrome.   298–99

12. Draw a diagram to explain the contrary actions of insulin and glucagon. Use your diagram to explain three major symptoms of type I diabetes mellitus.   300

13. Name the other endocrine glands discussed in this chapter, and discuss the function of the hormones they secrete.   301–3

14. Categorize chemical messengers into three groups based upon the distance between site of secretion and receptor site, and give examples of each group.   305–6

15. Give examples to show that there is an overlap between the mode of operation of the nervous system and that of the endocrine system. Explain why the traditional definition of a hormone may need to be expanded.   305–6

APPLYING YOUR KNOWLEDGE

Concepts

1. Explain the fact that an injection of epinephrine (adrenaline) causes almost an immediate effect on the body but an injection of testosterone takes a much longer time to produce an effect.

2. The anterior pituitary is often referred to as the master gland. Offer an argument that this term should be reserved for the hypothalamus.

3. In the early twentieth century, thyroxin was included in diet pills. Since then this practice has been banned. What was the reason for including thyroxin in diet pills?

4. Overseas travelers are encouraged to take melatonin to enable them to get on the sleep schedule of the foreign country. Explain the reason for this treatment and the time of day that one would take the hormone.

Bioethical Issue

In the past few years, some high schools and universities have begun requiring students on sports teams to undergo testing for steroids and other drugs. The concern is real. Public health experts estimate that thousands of professional athletes take steroids. Staying off the steroid track can spare athletes damage to their bodies and psyches.

Still, some athletes argue that drug testing invades their privacy and infringes on their personal freedom. What do you think? Should college athletes be required to take tests proving they are drug-free? Should all students? What about professional athletes? Do the health benefits of drug testing outweigh its invasiveness? Defend your answers.

TESTING YOUR KNOWLEDGE

1. Peptide hormones are received by a receptor located in the _____.

2. The hypothalamus produces the hormones _____ and _____, released by the posterior pituitary.

3. _____ secreted by the hypothalamus control the anterior pituitary.

4. Hormone production is self-regulated by a _____ mechanism.

5. Growth hormone (GH) is produced by the _____ pituitary.

6. Simple goiter occurs when the thyroid produces _____ (too much or too little) _____.

7. Parathyroid hormone increases the level of _____ in blood.

8. ACTH, produced by the anterior pituitary, stimulates the adrenal _____.

9. An overproductive adrenal cortex results in the condition called _____.

10. Label this diagram and explain how a negative feedback system keeps the hormone level constant in the body.

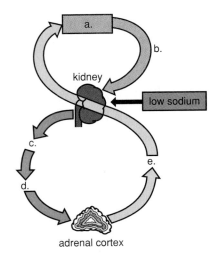

kidney

low sodium

adrenal cortex

11. Type I diabetes mellitus is due to a malfunctioning _____, while type II diabetes mellitus is due to an insensitivity to _____.

12. Prostaglandins are not carried in _____, as are hormones secreted by the endocrine glands.

13. Environmental signals are _____ that influence the behavior of organisms, organs, or adjacent cells.

APPLYING TECHNOLOGY

Your study of the endocrine system is supported by these available technologies:

Exploring the Internet

The Mader Home Page provides further resources for studying this chapter.

`http://www.mhhe.com/sciencemath/biology/mader/`

(Click on *Human Biology*.)

Dynamic Human: Endocrine System CD-ROM

In *Gross Anatomy*, major organs are reviewed; in *Explorations*, secretion of insulin is related to liver function, and zooming in occurs as the hypothalamus-anterior pituitary-thyroid axis is explained; in *Histology*, microscopic slides are highlighted to review parts; and in *Clinical Concepts*, zooming in occurs as the two types of diabetes are explained.

Explorations in Cell Biology & Genetics CD-ROM

Cell-Cell Interactions (#4) Glucose production varies as students make changes in a receptor's intracellular amplifying system. **(5)***

Explorations in Human Biology CD-ROM

Hormone Action (#11) Blood glucose levels vary due to insulin and glucagon secretion as students change their caloric intake and exercise levels. **(4)**

Life Science Animations Video

Video #3: Animal Biology, Part 1

Peptide Hormone Action (#28) Reception of a hormone causes cAMP to activate a kinase; cAMP is deactivated by a phosphodiesterase. **(2)**

_____
*Level of difficulty

SELECTED KEY TERMS

adrenal gland (uh-DREEN-ul) Gland that lies atop a kidney; the adrenal medulla produces the hormones epinephrine and norepinephrine, and the adrenal cortex produces the glucocorticoid and mineralocorticoid hormones.   297

adrenocorticotropic hormone (ACTH) (uh-DREEN-noh-kawrt-ih-koh-TROH-pik) Hormone secreted by the anterior lobe of the pituitary gland that stimulates activity in the adrenal cortex.   294

aldosterone (al-DAHS-tuh-rohn) Hormone secreted by the adrenal cortex that regulates the sodium and potassium balance of the blood.   298

anterior pituitary Portion of the pituitary gland that produces six types of hormones and is controlled by hypothalamic-releasing and release-inhibiting hormones.   294

antidiuretic hormone (ADH) (ANT-ih-dy-yuu-RET-ik) Hormone secreted by the posterior pituitary that promotes the reabsorption of water by the kidneys.   292

calcitonin Hormone secreted by the thyroid gland that helps regulate the blood calcium level.   296

circadian rhythm Regular physiological or behavioral event that occurs on an approximately 24-hour cycle.   303

cortisol Glucocorticoid secreted by the adrenal cortex that affects metabolism and has an anti-inflammatory and immunosuppressive effect.   298

cretinism Condition resulting from improper development of the thyroid in an infant.   295

diabetes mellitus Condition characterized by a high blood glucose level and the appearance of glucose in the urine, due to a deficiency of insulin secretion or decreased sensitivity of target cells to insulin.   300

epinephrine (eh-puh-NEH-frun) Hormone secreted by the adrenal medulla in times of stress; also called adrenaline.   297

exophthalmic goiter (ek-sahf-THAL-mik) Enlargement of the thyroid gland accompanied by secretion of too much thyroxin.   296

glucagon Hormone secreted by the pancreas that causes the liver to break down glycogen and raises the blood glucose level.   300

gonadotropic hormone (goh-nad-oh-TROH-pik) Substance secreted by anterior pituitary that regulates the activity of the ovaries and testes; principally, follicle-stimulating hormone **(FSH)** and luteinizing hormone **(LH).**   294

growth hormone (GH) Hormone secreted by the anterior pituitary that promotes cell division, protein synthesis, and bone growth.   293

hormone Chemical messenger produced in low concentrations that has physiological and/or developmental effects, usually in another part of the organism.   288

insulin Hormone secreted by the pancreas that lowers the blood glucose level by promoting the uptake of glucose by cells and the conversion of glucose to glycogen by the liver and skeletal muscles.   300

myxedema Condition resulting from a deficiency of thyroid hormone in an adult.   295

norepinephrine (NE) (nor-eh-puh-NEH-frun) Hormone secreted by the adrenal medulla in times of stress; also a neurotransmitter active in the peripheral and central nervous systems.   297

oxytocin (ahk-sih-TO-sin) Hormone released by the posterior pituitary that causes contraction of uterus and milk letdown.   292

pancreas Elongated, flattened organ in the abdominal cavity that secretes digestive enzymes into the small intestine (exocrine function) and hormones into the blood (endocrine function).   300

pancreatic islets (of Langerhans) Distinctive group of cells within the pancreas that secretes insulin and glucagon.   300

parathyroid gland (par-uh-THY-royd) One of four glands embedded in the posterior surface of the thyroid gland; produces parathyroid hormone.   296

parathyroid hormone (PTH) Hormone secreted by parathyroid glands that increases the blood calcium level and decreases the blood phosphate level.   296

pheromone Chemical substance secreted by one organism that influences the behavior of another.   305

pineal gland (PY-nee-ul) Gland located near the surface of the body (fish, amphibians) or in the third ventricle of the brain (mammals) that produces melatonin.   303

pituitary gland Small gland that lies just inferior to the hypothalamus; the anterior pituitary secretes several hormones, some of which control other endocrine glands; the posterior pituitary stores and secretes oxytocin and antidiuretic hormone.   292

posterior pituitary Portion of the pituitary gland that stores and secretes oxytocin and antidiuretic hormone, which are produced by the hypothalamus.   292

prolactin (PRL) Hormone secreted by the anterior pituitary that stimulates the production of milk from the mammary glands.   294

prostaglandins (PG) Hormones that have various and powerful local effects.   303

simple goiter Condition in which an enlarged thyroid produces low levels of thyroxin.   295

thyroid gland Organ that is in the neck and secretes several important hormones, including thyroxin and calcitonin.   295

thyroid-stimulating hormone (TSH) Hormone produced by the anterior pituitary that causes the thyroid to secrete thyroxin.   294

thyroxin (thy-RAHK-sin) Hormone secreted from the thyroid gland that promotes growth and development; in general, it increases the metabolic rate in cells.   295

FURTHER READINGS FOR PART FOUR

Achord, J. L. March/April 1995. Alcohol and the liver. *Scientific American Science & Medicine* 2(2):16. Article discusses the high incidence of hepatitis C virus in alcoholics.

Axel, R. October 1995. The molecular logic of smell. *Scientific American* 273(4):154. Article discusses how the brain identifies a scent by neuron activation.

Black, P. H. November/December 1995. Psychoneuroimmunology: Brain and immunity. *Scientific American Science & Medicine* 2(6):16. Article discusses the role of stress in susceptibility to infections, and cancer and HIV progression.

Carey, J., editor. 1990. *Brain facts: A primer on the brain and nervous system.* Washington, D.C.: Society for Neuroscience. Written for the general public, this booklet provides an introduction to the brain and nervous system and their disorders.

Davis, D. L., and Bradlow, H. L. April 1995. Can environmental estrogens cause breast cancer? *Scientific American* 273(4):166. Estrogenlike compounds found in the environment may contribute to breast cancer.

Fox, S. I. 1996. *Human physiology.* 5th ed. Dubuque, Iowa: Wm. C. Brown Publishers. This is an introductory physiology text.

Fuller, R. W. July/August 1995. Neurologic effects of serotonin. *Scientific American Science & Medicine* 2(4):48. Neurons communicate by means of serotonin, reaching into parts of the brain that control many functions.

Guyton, A. C., and Hall, J. E. 1996. *Textbook of medical physiology.* Presents physiological principles for those in the medical fields.

Hadley, M. E. 1992. *Endocrinology.* 3d ed. Englewood Cliffs, New Jersey: Prentice Hall. This advanced text provides a modern and complete coverage of the field of neuroendocrine physiology.

Hales, C. N. July/August 1994. Fetal nutrition and adult diabetes. *Scientific American Science & Medicine* 1(3):54. Poor fetal nutrition could be a link to noninsulin dependent diabetes.

Hole, J. W., Jr. 1995. *Essentials of human anatomy and physiology.* 5th ed. Dubuque, Iowa: Wm. C. Brown Publishers. An introductory text that is written in an easily understood manner.

Julien, R. M. 1995. *A primer of drug action.* 7th ed. New York: W. H. Freeman and Company. A concise, nontechnical guide to the actions, uses, and side effects of psychoactive drugs.

Klatsky, A. L. March/April 1995. Cardiovascular effects of alcohol. *Scientific American Science & Medicine* 2(2):28. The effects of moderate and heavy alcohol use on the cardiovascular system is discussed.

Lacy, P. E. July 1995. Treating diabetes with transplanted cells. *Scientific American* 273(1):50. New technology may allow replacement of pancreatic cells in diabetes patients.

Lowe, W. L., Jr. March/April 1996. Insulin-like Growth Factors. *Scientific American Science & Medicine* 3(2):62. Article discusses the role of IGF-I and IGF-II in carrying out the effects of growth hormone.

Marieb, E. N. 1995. *Human anatomy and physiology.* 4th ed. Redwood City, Calif.: Benjamin/Cummings Publishing. This introductory text presents anatomy and physiology to students in the allied health fields.

Moore, P. S., and Broome, C. V. November 1994. Cerebrospinal meningitis epidemics. *Scientific American* 271(5):38. Research has provided new information which may eventually allow the prediction and control of meningitis outbreaks.

National Geographic Society. 1994. *The incredible machine.* Washington, D.C.: National Geographic Society. A colorfully illustrated reference of the structures and functions of the human body.

Nolte, J. 1993. *The human brain.* 3d ed. St. Louis: Mosby-Year Book, Inc. Beginners are guided through the basic aspects of brain structure and function.

Nowak, T. J., and Handford, A. G. 1994. *Essentials of pathophysiology.* Dubuque, Iowa: Wm. C. Brown Publishers. Introduces the concepts of pathophysiology for those in the medical fields.

Ray, O., and Ksir, C. 1993. *Drugs, society, & human behavior.* 6th ed. St. Louis: Mosby-Year Book, Inc. This textbook addresses drugs from psychological, pharmacological, historical, and legal perspectives.

Risher, C. E., and Easton, T. A. 1995. *Focus on human biology.* 2d. ed. New York: HarperCollins College Publishers. This comprehensive introductory textbook stresses basic human anatomy and physiology.

Roses, A. D. September/October 1995. Apolipoprotein E and Alzheimer disease. *Scientific American Science & Medicine* 2(3):16. The cause of structural changes and symptoms in Alzheimer disease is being reexamined.

Schwartz, W. J. May/June 1996. Internal timekeeping. *Science & Medicine* 3(3):44. Article discusses circadian rhythm mechanisms.

Science & Medicine. July/August 1994. 3(4). This issue includes articles on migranes, fetal alcohol syndrome, and targeting drugs to the brain.

Scientific American Science & Medicine. July/August 1994. 1(3). This issue includes information on the actions of estrogens.

Shier, D., et al. 1996. *Hole's human anatomy & physiology.* 7th ed. Dubuque, Iowa: Wm. C. Brown Publishers. An introductory anatomy and physiology text that has proved useful to beginning students.

Springer, S. P., and Deutsch, G. 1993. *Left brain, right brain.* 4th ed. New York: W. H. Freeman and Company. This text provides information at several levels covering current research into cerebral hemisphere function, and left brain versus right brain function.

Streit, W. J., and Kincaid-Colton, C. A. November 1995. The brain's immune system. *Scientific American* 273(5):54. Microganglia may be responsible for damage to tissue caused by strokes and Alzheimer disease.

Swerdlow, J. L. June 1995. The brain. *National Geographic* 187(6):2. New research leads to treatments for many age-old disorders.

Tortora, G. J., and Grabowski, S. R. 1996. *Principles of anatomy and physiology.* 8th ed. New York: HarperCollins College Publishers. An introductory anatomy and physiology text that presents basic information in an easy to understand manner.

Van De Graaff, K. M., and Fox, S. I. 1995. *Concepts of human anatomy & physiology.* 4th ed. Dubuque, Iowa: Wm. C. Brown Publishers. An introductory anatomy and physiology text that comprehensively presents basic principles.

Vander, A. J., et al. 1994. *Human physiology: The mechanisms of body function.* 6th ed. New York: McGraw-Hill, Inc. Presents the principles of human physiology with an emphasis on physiological mechanisms.

Part

5

Human Reproduction

Human beings are one of two sexes, male and female. The reproductive organs of each sex function to produce the sex cells that join prior to the development of a new individual. The embryo develops into a fetus within the body of the female, and birth occurs when there is a reasonable chance for independent existence. The steps of human development can be outlined from the fertilized egg to the birth of the child.

We are in the midst of a sexual revolution. We have the freedom to engage in varied sexual practices and to reproduce by alternative methods of conception, such as in vitro fertilization. With freedom comes a responsibility to be familiar with the biology of reproduction and health-related issues, such as sexually transmitted diseases, not only for ourselves but for our potential offspring.

Chapter 15

Reproductive System

Chapter Outline

15.1   MALE REPRODUCTIVE SYSTEM
- The male reproductive system is designed for continuous sperm production and delivery within a fluid medium.   312

15.2   FEMALE REPRODUCTIVE SYSTEM
- The female reproductive system is designed for the monthly production of an egg and preparation of the uterus to house the developing fetus.   316

15.3   FEMALE HORMONE LEVELS
- Hormones control the monthly reproductive cycle in females and play a significant role in maintaining pregnancy, should it occur.   318

15.4   DEVELOPMENT OF MALE AND FEMALE SEX ORGANS
    Male or female sex organs develop from the same indifferent tissue.   322

15.5   CONTROL OF REPRODUCTION
- Birth-control measures vary in effectiveness from those that are very effective to those that are minimally effective.   324
- There are alternative methods of reproduction today, including in vitro fertilization followed by introduction to the uterus.   328

15.6   WORKING TOGETHER
- The reproductive system works with the other systems of the body to maintain homeostasis.   328

I t had seemed simple enough. Leigh Anne and Joe graduated from college, launched their careers, and got married. Some years later, they bought a house. Then, they decided to have a baby.

That's when things got complicated.

For some reason, Leigh Anne just didn't get pregnant. After two years of trying to conceive, the couple headed to a well-known fertility specialist. After some tests, the doctor explained a variety of fertility treatments and drugs designed to help couples conceive.

Leigh Anne and Joe weighed their options and decided to try in vitro fertilization. During a series of visits to the clinic, a doctor removed eggs from Leigh Anne and combined them with sperm from Joe. Nurtured in the lab, the mixture formed fertilized eggs, which the doctor then placed back into Leigh Anne's uterus.

Fortunately, the procedure worked. Leigh Anne's pregnancy was normal and healthy. Today, the couple's three year old races around the house, plays leapfrog over the family dog and eats everything in sight. At some point, Leigh Anne and Joe might try in vitro fertilization again. For now, though, they'd just like their toddler to try a nap.

15.1   Male Reproductive System

The male reproductive system includes the organs pictured in Figure 15.1 and listed in Table 15.1. The male gonads are paired testes (sing., **testis**), which are suspended within the *scrotal sacs* of the **scrotum**. The testes begin their development inside the abdominal cavity, but they descend into the scrotal sacs as development proceeds. *Cryptorchidism,*

TABLE 15.1	
Male Reproductive System	
Organ	**Function**
Testes	Produce sperm and sex hormones
Epididymides	Maturation and some storage of sperm
Vasa deferentia	Conduct and store sperm
Seminal vesicles	Contribute nutrients and fluid to semen
Prostate gland	Contributes basic fluid to semen
Urethra	Conducts sperm
Bulbourethral glands	Contribute mucoid fluid to semen
Penis	Organ of copulation

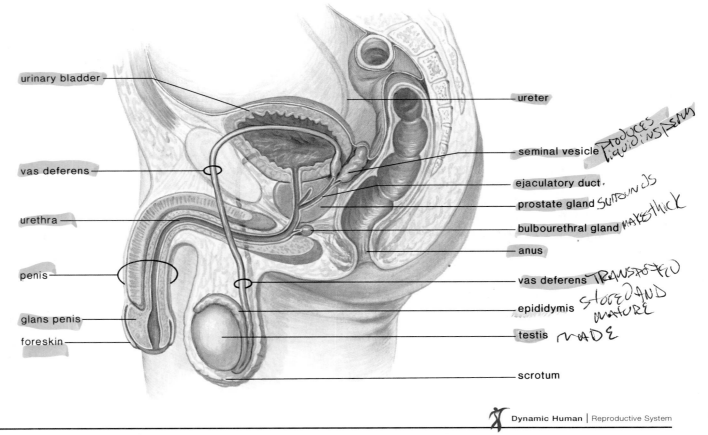

Dynamic Human | Reproductive System

Figure 15.1   The male reproductive system.
The testes produce sperm. The seminal vesicles, the prostate gland, and the bulbourethral gland provide a fluid medium for the sperm. Circumcision is the removal of the foreskin. Notice that the penis in this drawing is not circumcised because the foreskin is present.

or nondescent of the testes, sometimes occurs. If cryptorchidism is not treated by hormone therapy or surgery to place the testes in the scrotum, *sterility* (the inability to produce offspring) results. Sterility occurs because normal sperm production is inhibited at body temperature; a cooler temperature is required.

Sperm produced by the testes mature within the epididymides (sing., **epididymis**), which are tightly coiled tubules lying just outside the testes. Maturation seems to be required for the sperm to swim to the egg. Once the sperm have matured, they are propelled into the vasa deferentia (sing., **vas deferens**) by muscular contractions. Sperm are stored in both the epididymides and the vasa deferentia. When a male becomes sexually aroused, sperm enter the urethra, part of which is located within the penis.

The **penis** is a cylindrical organ that hangs in front of the scrotum (Fig. 15.2). Spongy, erectile tissue containing distensible blood spaces extends through the shaft of the penis. During sexual arousal, nervous reflexes cause an increase in arterial blood flow to the penis. This increased blood flow fills the blood space in the erectile tissue, and the penis, which is normally limp (flaccid), stiffens and increases in size. These changes are called **erection.** If the penis fails to become erect, the condition is called *impotency.*

Semen (seminal fluid) is a thick, whitish fluid that contains sperm and fluids. As Table 15.1 indicates, three types of glands contribute fluids to semen. The **seminal vesicles** lie at the base of the bladder. Each joins a vas deferens to form an *ejaculatory duct* that enters the urethra. As sperm pass from the vasa deferentia into the urethra, these vesicles secrete a thick, viscous fluid containing nutrients for possible use by the sperm. Just below the bladder is the **prostate gland,** which secretes a milky alkaline fluid believed to activate or increase the motility of the sperm. In older men, the prostate gland may become enlarged, thereby constricting the urethra and making urination difficult. Slightly below the prostate gland, on either side of the urethra, is a pair of small glands called **bulbourethral glands,** which have mucous secretions with a lubricating effect. Notice in Figure 15.1 that the urethra also carries urine from the bladder during urination.

Sexual arousal can cause an erection, and if arousal reaches its peak, **ejaculation** follows. After sperm enter the ejaculatory duct, seminal vesicles, prostate gland, and bulbourethral glands release their secretions. As orgasm occurs, rhythmical contractions of the urethra expels semen from the penis. An erection lasts for only a limited amount of time, and after ejaculation, the penis returns to its normal flaccid state. Following ejaculation, a male may typically experience a period of time, called *refractory period,* during which stimulation does not bring about an erection. The contractions that expel semen from the penis are a part of male **orgasm,** which are the physiological and psychological sensations that occur at the climax of sexual stimulation.

Sperm produced by the testes mature in the epididymides and pass from the vasa deferentia to the urethra, where certain glands add fluid to sperm, forming semen.

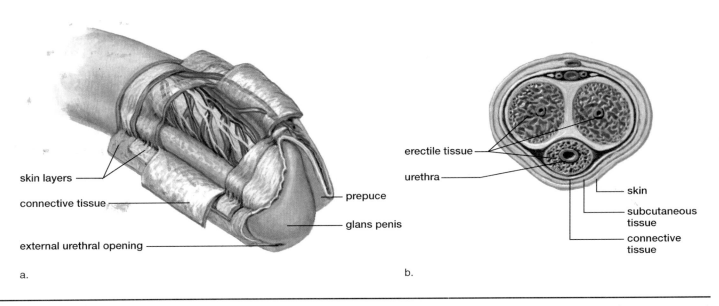

Figure 15.2   Penis anatomy.
a. Beneath the skin and the connective tissue lies the urethra, surrounded by erectile tissue. This tissue expands to form the glans penis, which in uncircumcised males is partially covered by the foreskin. **b.** Two other columns of erectile tissue in the penis are located dorsally.

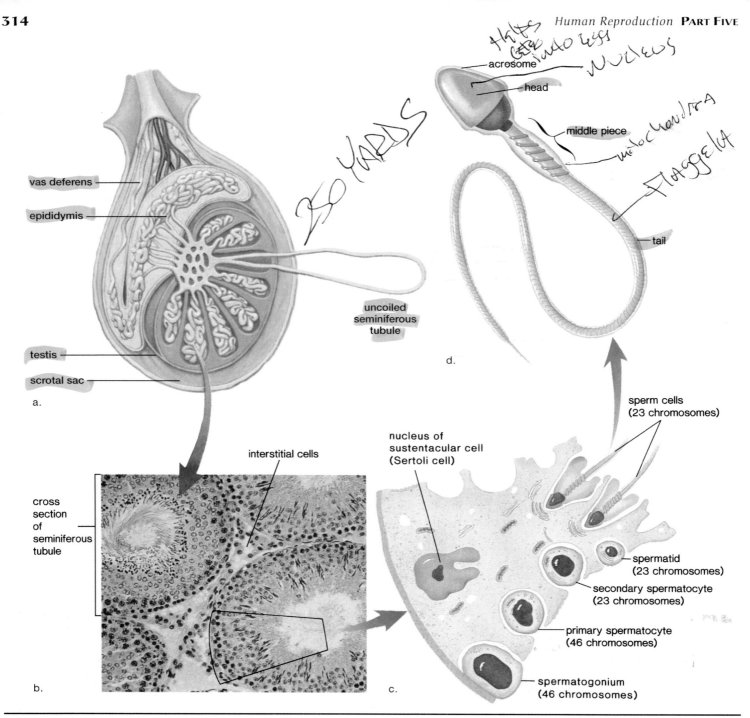

Figure 15.3   Testis and sperm.
a. The lobules of a testis contain seminiferous tubules. **b.** Light micrograph of cross section of seminiferous tubules where spermatogenesis occurs. **c.** Diagrammatic representation of spermatogenesis, which occurs in the wall of the tubules. **d.** A sperm has a head, a middle piece, and a tail. The nucleus, containing 23 chromosomes, is in the head, capped by the enzyme-containing acrosome.

Testes Produce Sperm and Hormones

A longitudinal section of a testis shows that it is composed of compartments called lobules, each of which contains one to three tightly coiled *seminiferous tubules* (Fig. 15.3). Altogether, these tubules have a combined length of approximately 250 meters. A microscopic cross section of a seminiferous tubule shows that it is packed with cells undergoing *spermatogenesis*, which involves meiosis. Also present are sustentacular (Sertoli) cells, which support, nourish, and regulate the spermatogenic cells.

Mature **sperm,** or spermatozoa, have three distinct parts: a head, a middle piece, and a tail. The middle piece and the tail contain microtubules in the characteristic 9 + 2 pattern of cilia and flagella. In the middle piece, mitochondria are wrapped around the microtubules and provide the energy for movement. The head contains a nucleus covered by a cap called the *acrosome,* which stores enzymes needed to penetrate the egg. The human egg is surrounded by several layers of cells and a thick noncellular membrane—these acrosomal enzymes play a role in allowing a sperm to reach the surface of the egg. The normal human male usually releases several hundred million sperm per ejaculation, assuring an adequate number for fertilization to take place. Fewer than 100 ever reach the vicinity of the egg, however, and only one sperm normally enters an egg.

Hormonal Regulation in Males

The hypothalamus has ultimate control of the testes' sexual function because it secretes a hormone called gonadotropic-releasing hormone, or GnRH, that stimulates the anterior pituitary to secrete the gonadotropic hormones. There are two gonadotropic hormones—*follicle-stimulating hormone (FSH)* and *luteinizing hormone (LH)*—in both males and females. In males, FSH promotes spermatogenesis in the seminiferous tubules, which also release the hormone inhibin.

LH in males is sometimes given the name *interstitial cell-stimulating hormone (ICSH)* because it controls the production of testosterone by the interstitial cells, which are scattered in the spaces between the seminiferous tubules. All these hormones are involved in a negative feedback relationship that maintains the fairly constant production of sperm and testosterone (Fig. 15.4).

Testosterone Is the Male Sex Hormone

Testosterone is the main sex hormone in males. It is essential for the normal development and functioning of the organs listed in Table 15.1. Testosterone is also necessary for the maturation of sperm.

Testosterone brings about and maintains the male secondary sex characteristics that develop at the time of puberty. Testosterone causes growth of a beard, axillary (underarm) hair, and pubic hair. It prompts the larynx and vocal cords to enlarge, causing the voice to change.

Males commonly experience a growth spurt later than females, and they grow for a longer period of time. This means that males are generally taller than females and have broader shoulders and longer legs relative to trunk length. Testosterone is responsible for the greater muscle strength of males, and this is the reason some athletes take supplemental amounts of anabolic steroids, which are either testosterone or related steroid hormones resembling testosterone. The practice can lead to health problems in-

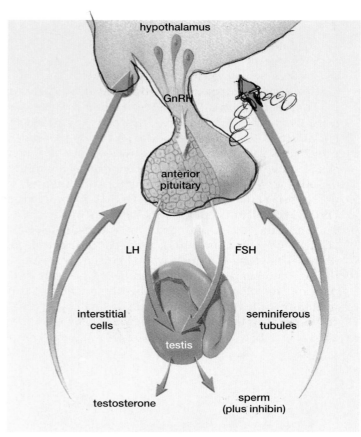

Figure 15.4 Hormonal control of testes.
GnRH (gonadotropic-releasing hormone) stimulates the anterior pituitary to secrete the gonadotropic hormones FSH and LH. FSH stimulates the testes to produce sperm, and LH stimulates the testes to produce testosterone. Testosterone and inhibin exert negative feedback control over the hypothalamus and the anterior pituitary, and this ultimately regulates the level of testosterone in the blood.

volving the kidneys, the circulatory system, and hormonal imbalances. The testes shrink in size, and feminization in regard to other male traits occurs.

Testosterone has various side effects. It causes oil and sweat glands in the skin to secrete; therefore, it is largely responsible for acne and body odor. Its presence is associated with the pattern baldness seen more often in males than females. Testosterone is believed to be largely responsible for the sex drive, and some who attribute aggressiveness to males think that testosterone may play a role in this characteristic.

FSH promotes spermatogenesis, and LH promotes testosterone production in testes. Testosterone maintains the primary and secondary sex characteristics and promotes maturation of sperm.

15.2   Female Reproductive System

The female reproductive system includes the ovaries, the oviducts, the uterus, and the vagina (Fig. 15.5 and Table 15.2). The *ovaries*, which release an egg each month, lie in shallow depressions, one on each side of the upper pelvic cavity. The *oviducts*, also called uterine or fallopian tubes, extend from the ovaries to the uterus; however, the oviducts are not attached to the ovaries. Instead, they have fingerlike projections called fimbriae that sweep over the ovaries. When an egg bursts from an ovary during ovulation, it usually is swept into an oviduct by the combined action of the fimbriae and the beating of cilia that line the oviducts. Fertilization, if it occurs, normally takes place in an oviduct, and the developing embryo is propelled slowly by ciliary movement and tubular muscle contraction to the uterus. The *uterus* is a thick-walled, muscular organ about the size and shape of an inverted pear. The embryo completes its development after embedding itself in the uterine lining, called the endometrium. A small opening at the cervix leads to the vaginal canal. The *vagina* is a tube at a 45° angle with the small of the back. The mucosal lining of the vagina lies in folds and can extend. This is especially important when the vagina serves as the birth canal, and it also can facilitate sexual intercourse, when the vagina receives the penis.

The external genital organs of the female include two folds of skin, called labia minora and labia majora, on either side of the urethral and vaginal orifices (openings).

TABLE 15.2
Female Reproductive Organs

Organ	Function
Ovaries	Produce egg and sex hormones
Oviducts (fallopian tubes)	Conduct egg; location of fertilization
Uterus (womb)	Houses developing fetus
Cervix	Contains opening to uterus
Vagina	Receives penis during sexual intercourse and serves as birth canal

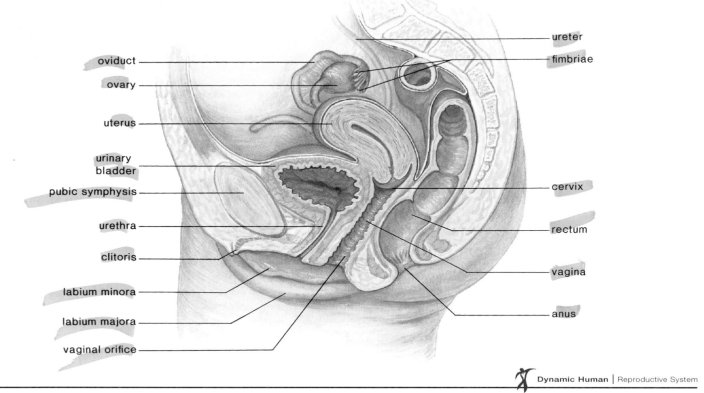

Dynamic Human | Reproductive System

Figure 15.5   The female reproductive system.
The ovaries release one egg a month; fertilization occurs in the oviduct, and development occurs in the uterus. The vagina is the birth canal and the organ of sexual intercourse.

Ovaries and Oviducts

The female gonads are called **ovaries.** The ovaries are small—approximately 3 cm by 1 cm—ovoid structures located one on either side of the pelvic cavity. Each is held in place by several ligaments, the largest of which is attached to an oviduct and the uterus. Each month one of the ovaries releases an *egg,* which bursts from the ovary during **ovulation.**

The Oviducts Are Tubes to the Uterus

The **oviducts** are also called uterine or fallopian tubes. These muscular tubes, which extend from the uterus to the ovaries, are lined by cells possessing cilia that beat toward the uterus. The tubes are not attached to the ovaries and instead have fingerlike projections called **fimbriae** that sweep over the ovary at the time of ovulation. Once in the oviduct, the egg is at first propelled rapidly by cilia movement and tubular muscle contraction. Soon the contractions diminish, and the egg moves more slowly toward the uterus. Fertilization, the completion of oogenesis, and zygote formation usually occur in an oviduct. Occasionally, the embryo becomes embedded in the wall of an oviduct, where it begins to develop. These *"tubular pregnancies"* cannot succeed because the oviducts are not anatomically capable of allowing full development to occur. An *ectopic pregnancy* is one that begins anywhere outside the uterus.

The Uterus and Vagina

The **uterus** is held in place near the bladder by uterine ligaments. Its position changes according to the fullness of the urinary bladder and rectum. If the bladder is full, the uterus is moved backward; if the rectum is full, the uterus is moved more forward in the pelvic cavity.

The muscular uterus has three portions: the fundus, the body, and the **cervix.** The oviducts join the uterus just below the fundus, and the opening of the cervix, called the os, leads to the vaginal canal.

The uterus, sometimes called the womb, is approximately 5 cm wide in its usual state but is capable of stretching to over 30 cm to accommodate the growing baby. The lining of the uterus, called the **endometrium,** participates in the formation of the placenta, which supplies nutrients needed for development. The endometrium has two layers: a basal layer and a functional layer next to the lumen of the uterus. In the nonpregnant female, the functional layer of the endometrium varies in thickness according to a monthly reproductive cycle, called the uterine cycle (p. 321).

Cancer of the cervix is a common form of cancer in women. Early detection is sometimes possible by means of a **Pap smear,** which requires the removal of a few cells from the region of the cervix for microscopic examination. If the cells are cancerous, a hysterectomy may be recommended. A *hysterectomy* is the removal of the uterus.

The **vagina** is a 5- to 6-inch muscular tube that makes a 45° angle with the small of the back. The lining of the vagina lies in folds called rugae, which are capable of extension as the muscular wall stretches. Substances produced by cells in the lining of the vagina are metabolized to acids inside the vagina. These acids are believed to prevent bacterial infections of the genital tract. But they also make the vagina hostile to sperm, because sperm prefer a basic rather than an acidic environment. Since seminal fluid is basic, it is capable of neutralizing the vagina.

Females Have External Genitals

The external genitals of the female are known collectively as the **vulva** (Fig. 15.6). The **labia majora** extend backward from the *mons pubis,* a fatty prominence underlying the pubic hair. The **labia minora** lie just inside the labia majora. They extend forward from the vaginal orifice (opening) to encircle and form a foreskin for the clitoris, an organ that is homologous to the penis. Although quite small, the clitoris has a shaft of erectile tissue and is capped by a pea-shaped glans. The glans clitoris has sensory receptors, which allow it to function as a sexually sensitive organ. The vagina may be partially closed by a ring of tissue called the hymen. If unbroken, the hymen is ruptured by initial sexual intercourse; but often it is disrupted prior to this by vigorous physical activity or even by the use of tampons.

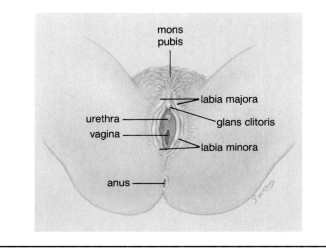

Figure 15.6 External genitals of the female.
At birth, the opening of the vagina is partially blocked by a membrane called the hymen. Physical activities and sexual intercourse disrupt the hymen.

15.3 Female Hormone Levels

The following glands and hormones are involved in regulating female hormone levels.

Hypothalamus: secretes *GnRH (gonadotropic-releasing hormone)*

Anterior pituitary: secretes the gonadotropic hormones **FSH (follicle-stimulating hormone)** and **LH (luteinizing hormone)**

Ovaries: secrete the female sex hormones **estrogen** and **progesterone**

The Ovarian Cycle

A longitudinal section through an ovary shows many **follicles,** each containing an oocyte (Fig. 15.7). A female is born with as many as 2 million follicles, but the number is reduced to 300,000–400,000 by the time of puberty. Only a small number of follicles (about 400) ever mature because a female usually releases only one egg per month during her reproductive years.

As the follicle in the ovary undergoes maturation, it develops from a primary follicle to a secondary follicle to a Graafian follicle. During *oogenesis*, the chromosome

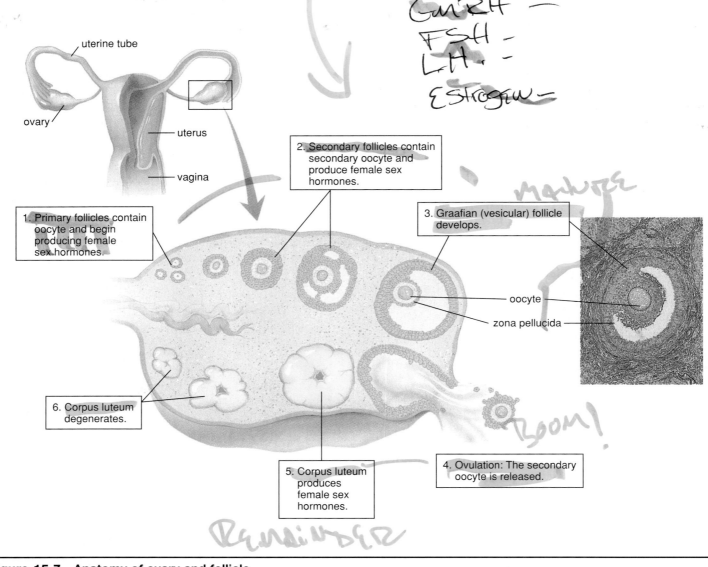

Figure 15.7   Anatomy of ovary and follicle.
As a follicle matures, the oocyte enlarges and is surrounded by layers of follicular cells and fluid. Eventually, ovulation occurs, the mature follicle ruptures, and the secondary oocyte is released. A single follicle actually goes through all stages in one place within the ovary.

number is reduced, and a secondary oocyte is formed and pushed to the side of the fluid-filled cavity within the secondary follicle. In a *Graafian follicle,* the fluid-filled cavity increases to the point that the follicle wall balloons out on the surface of the ovary and bursts, releasing the secondary oocyte surrounded by a clear membrane and follicular cells. As mentioned, this is referred to as *ovulation.* Actually, the second meiotic division is not completed unless fertilization occurs. In the meantime, the follicle in the ovary is developing into the **corpus luteum.** If pregnancy does not occur, the corpus luteum in the ovary begins to degenerate after about ten days.

These events, called the **ovarian cycle,** are under the control of the gonadotropic hormones, *follicle-stimulating hormone (FSH)* and *luteinizing hormone (LH)* (Fig. 15.8.) The gonadotropic hormones are not present in constant amounts but instead are secreted at different rates during the cycle. For simplicity's sake, it can be emphasized that during the first half, or *follicular phase,* of the cycle, FSH promotes the development of a follicle in the ovary, which secretes estrogen. As the estrogen level in the blood rises, it exerts

feedback control over the anterior pituitary secretion of FSH so that the follicular phase comes to an end.

Presumably, the high level of estrogen in the blood also causes a sudden secretion of a large amount of GnRH from the hypothalamus. This leads to a surge of LH production by the anterior pituitary and to ovulation at about the fourteenth day of a twenty-eight-day cycle.

During the second half, or *luteal phase,* of the ovarian cycle, it can be emphasized that LH promotes the development of the corpus luteum, which secretes progesterone. Progesterone causes the uterine lining to build up. As the blood level of progesterone rises, it exerts feedback control over the anterior pituitary secretion of LH so that the corpus luteum in the ovary begins to degenerate. As the luteal phase comes to an end, menstruation occurs.

One ovarian follicle per month produces a secondary oocyte. Following ovulation, the follicle develops into the corpus luteum.

| The hypothalamus produces GnRH (gonadotropic-releasing hormone). |

| GnRH stimulates the anterior pituitary to produce FSH (follicle-stimulating hormone) and LH (luteinizing hormone). |

| FSH stimulates the follicle to produce estrogen, and LH stimulates the corpus luteum to produce progesterone. |

| Estrogen and progesterone affect the sex organs (e.g., uterus) and the secondary sex characteristics and exert feedback control over the hypothalamus and the anterior pituitary. |

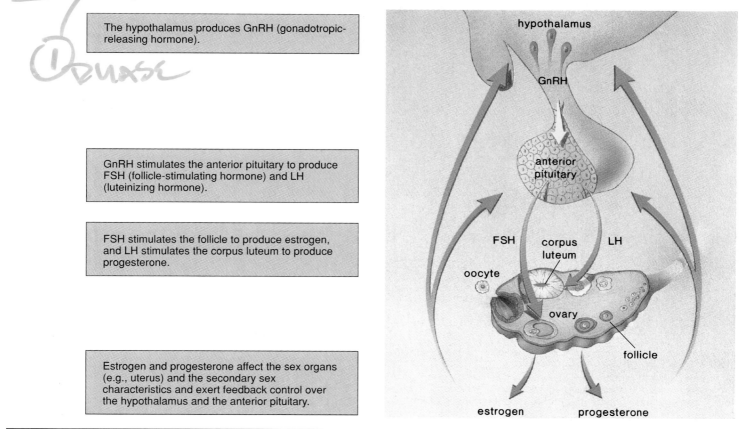

Figure 15.8 Hormonal control of ovaries.

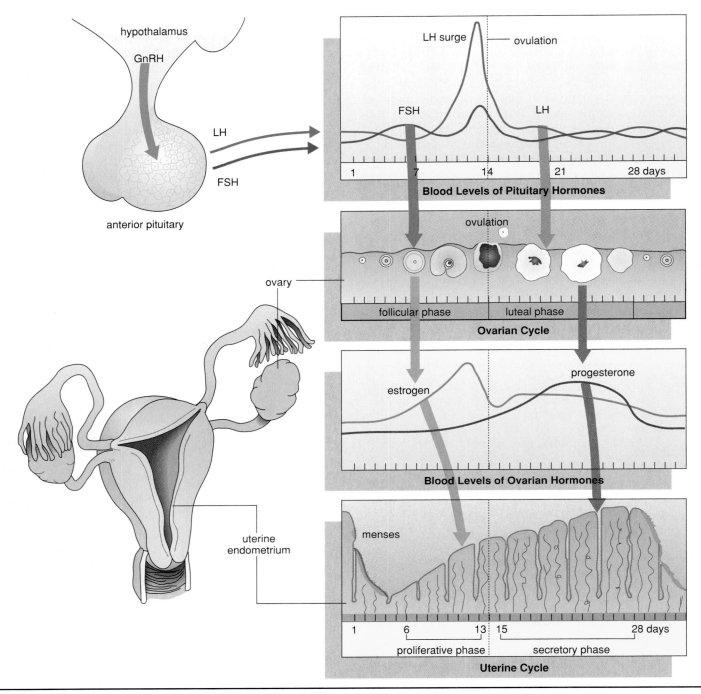

Figure 15.9 Female hormone levels.

During the follicular phase of the ovarian cycle, FSH released by the anterior pituitary promotes the maturation of a follicle in the ovary. The ovarian follicle produces increasing levels of estrogen, which causes the endometrium to thicken during the proliferative phase of the uterine cycle. After ovulation and during the luteal phase of the ovarian cycle, LH promotes the development of the corpus luteum. This structure produces increasing levels of progesterone, which causes the endometrial lining to become secretory. Menstruation begins when progesterone production declines to a low level.

TABLE 15.3

Ovarian and Uterine Cycles (Simplified)

Ovarian Cycle	Events	Uterine Cycle	Events
Follicular phase—Days 1–13	FSH	Menstruation—Days 1–5	Endometrium breaks down
	Follicle maturation	Proliferative phase—Days 6–13	Endometrium rebuilds
	Estrogen		
Ovulation—Day 14*	LH spike		
Luteal phase—Days 15–28	LH	Secretory phase—Days 15–28	Endometrium thickens and glands are secretory
	Corpus luteum		
	Progesterone		

*Assuming 28-day cycle.

The Uterine Cycle

The female sex hormones, estrogen and progesterone, have numerous functions. The effects of these hormones on the endometrium of the uterus cause the uterus to undergo a cyclical series of events known as the **uterine cycle** (Table 15.3 and Fig. 15.9). Twenty-eight-day cycles are divided as follows.

During *days 1–5,* there is a low level of female sex hormones in the body, causing the uterine lining to disintegrate and its blood vessels to rupture. A flow of blood and tissues, known as the menses, passes out of the vagina during **menstruation,** also known as the menstrual period.

During *days 6–13,* increased production of estrogen by an ovarian follicle in the ovary causes the endometrium of the uterus to thicken and to become vascular and glandular. This is called the proliferative phase of the uterine cycle.

Ovulation usually occurs on the fourteenth day of the twenty-eight-day cycle.

During *days 15–28,* increased production of progesterone by the corpus luteum in the ovary causes the endometrium of the uterus to double or triple in thickness (from 1 mm to 2–3 mm) and the uterine glands to mature, producing a thick mucoid secretion. This is called the secretory phase of the uterine cycle. The endometrium now is prepared to receive the developing embryo. If pregnancy does not occur, the corpus luteum in the ovary degenerates, and the low level of sex hormones in the female body results in the uterine lining breaking down. The menstrual discharge begins at this time. Even while menstruation is occurring, the anterior pituitary begins to increase its production of FSH, and a new follicle begins to mature in the ovary.

During the uterine cycle, the endometrium of the uterus builds up and then is broken down during menstruation.

Pregnancy Follows Fertilization

If fertilization does occur, an embryo begins development even as it travels down the oviduct to the uterus. The endometrium is now prepared to receive the developing embryo, which becomes embedded in the lining several days following fertilization (Fig. 15.10). The *placenta* originates from both maternal and fetal tissues. It is the region of exchange of molecules between fetal and maternal blood, although there is rarely any mixing of the two. At first the placenta produces human chorionic gonadotropin (HCG), which maintains the corpus luteum in the ovary until the placenta begins its own production of progesterone and estrogen. Progesterone and estrogen have two effects: They shut down the anterior pituitary so that no new follicle in the ovaries matures, and they maintain the lining of the uterus so that the corpus luteum in the ovary is not needed. Usually, there is no menstruation during pregnancy.

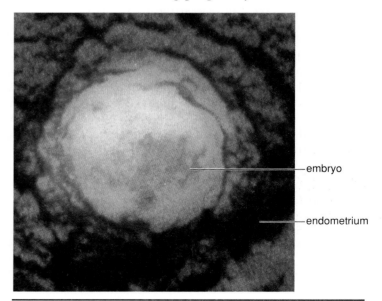

Figure 15.10   Implantation.
A scanning electron micrograph showing embryo implanted on day 12 following fertilization.

Estrogen and Progesterone Are Female Sex Hormones

Estrogen and progesterone not only affect the uterus; they affect other parts of the body as well. Estrogen is largely responsible for the secondary sex characteristics in females, including body hair and fat distribution. In general, females have a more rounded appearance than males because of a greater accumulation of fat beneath the skin. Like males, females develop axillary and pubic hair during puberty. In females, the upper border of pubic hair is horizontal, but in males, it tapers toward the navel. Both estrogen and progesterone are also required for breast development. Other hormones are involved in milk production and letdown following a pregnancy.

The pelvic girdle is wider and deeper in females so that the pelvic cavity usually has a larger relative size compared to males. This means that females have wider hips than males and that the thighs converge at a greater angle toward the knees. Because the female pelvis tilts forward, females tend to have protruding buttocks, more of a lower back curve than males, an abdominal bulge, and a tendency to be somewhat knock-kneed.

After Menopause the Ovaries Don't Respond

Menopause, the period in a woman's life during which the ovarian and uterine cycles cease, is likely to occur between ages 45 and 55. The ovaries are no longer responsive to the gonadotropic hormones produced by the anterior pituitary, and the ovaries no longer secrete estrogen or progesterone. At the onset of menopause, the uterine cycle becomes irregular, but as long as menstruation occurs, it is still possible for a woman to conceive. Therefore, a woman usually is not considered to have completed menopause until there has been no menstruation for a year.

The hormonal changes during menopause often produce physical symptoms, such as "hot flashes," which are caused by circulatory irregularities, dizziness, headaches, insomnia, sleepiness, and depression. These symptoms may be mild or even absent. If they are severe, medical attention should be sought. Women sometimes report an increased sex drive following menopause. It has been suggested that this may be due to androgen production by the adrenal cortex.

Estrogen and progesterone produced by the ovaries are the female sex hormones responsible for the primary sex characteristics, the uterine cycle, and the secondary sex characteristics of females.

15.4 Development of Male and Female Sex Organs

The sex of an individual is determined at the moment of fertilization. Males have the chromosomes X and Y, while girls have two X chromosomes. During the first few months of development, it is impossible to tell whether the unborn child is a boy or girl by external inspection. Gonads don't start developing until the seventh week of development. The tissue that gives rise to the gonads is called indifferent because it can become testes or ovaries depending on the action of hormones. Genes on the Y chromosome cause testes to develop and produce androgenic hormones, and these determine the course of development.

In Figure 15.11*a* notice that at six weeks both males and females have the same type tissues and ducts. During this indifferent stage an embryo has the potential to develop into a male or female. If a Y chromosome is present, androgenic (male) hormones stimulate the mesonephric ducts to become male genital ducts. The mesonephric ducts enter the urethra, which belongs to both the urinary and reproductive systems in males. Androgenic hormones suppress development of paramesonephric ducts in males.

In the absence of a Y chromosome and in the presence of two X chromosomes, ovaries develop instead of testes from the same indifferent tissue. Now the mesonephric ducts regress, and the paramesonephric ducts develop into the uterus and uterine tubes. A developing vagina also extends from the uterus. There is no connection between the urinary and genital system in females.

At fourteen weeks, both the primitive testes and ovaries are located deep inside the abdominal cavity. An inspection of the interior of the testes would show that sperm are even now starting to develop, and similarly, the ovaries already contain large numbers of tiny follicles, each containing an ovum. Toward the end of development, the testes descend into the scrotal sac; the ovaries remain in the abdominal cavity.

Figure 15.11*b* shows the development of the external genitals. These tissues are also indifferent at first—they can develop into either male or female genitals. At six weeks, a small bud appears between the legs that can develop into the male penis or the female clitoris, depending on the presence or absence of the Y chromosome and androgenic hormones. At nine weeks, there is a groove called the urogenital groove bordered by two swellings. By fourteen weeks, in males, this groove has disappeared, and the scrotum has formed from the original swellings. In females, the groove persists and becomes the vaginal opening. Labia majora and labia minora are present instead of a scrotum.

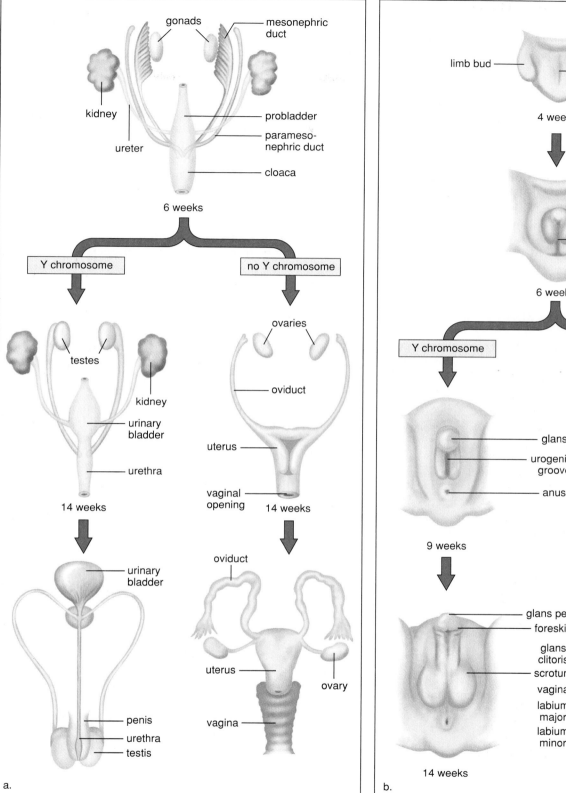

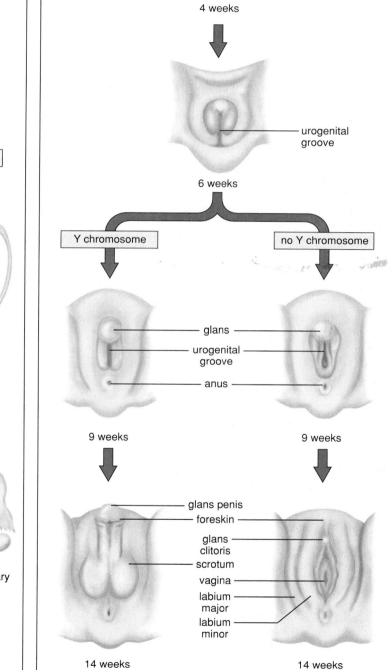

Figure 15.11   Male and female organs.
Development of gonads and ducts.

15.5 Control of Reproduction

Several means are available to dampen or enhance our reproductive potential.

Birth Control Is Varied

The most reliable method of birth control is abstinence, that is, the absence of sexual intercourse. This form of birth control has the added advantage of preventing transmission of a sexually transmitted disease. Other, perhaps more common, means of birth control used in this country are given in Table 15.4. The table gives the effectiveness for the various birth-control methods listed. For example, with the least effective method given in the table, we expect that within a year, 70 out of 100 women, or 70%, of sexually active women will not get pregnant, while 30 women will get pregnant.

Figure 15.12 features some of the most effective and commonly used means of birth control. *Oral contraception* (birth-control pills) usually involves taking a combination of estrogen and progesterone for 21 days of a 28-day cycle (beginning at the end of menstruation). The estrogen and progesterone in the birth-control pill effectively shut down the pituitary production of both FSH and LH so that no follicle in the ovary begins to develop in the ovary; and since ovulation does not occur, pregnancy cannot take place. Since there are possible side effects, those taking birth-control pills should see a physician regularly.

An *intrauterine device (IUD)* is a small piece of molded plastic that is inserted into the uterus by a physician. IUDs are believed to alter the environment of the uterus and oviducts so that fertilization probably does not occur—but if fertilization should occur, implantation cannot take place. The type of IUD featured in Figure 15.12 has copper wire wrapped around the plastic.

The *diaphragm* is a soft latex cup with a flexible rim that lodges behind the pubic bone and fits over the cervix. Each woman must be properly fitted by a physician, and the diaphragm can be inserted into the vagina two hours at most before sexual relations. It must be used with spermicidal jelly or cream and should be left in place at least six hours after sexual relations. The cervical cap is a minidiaphragm.

TABLE 15.4

Common Birth-Control Methods

Name	Procedure	Methodology	Effectiveness	Risk
Abstinence	Refrain from sexual intercourse	No sperm in vagina	100%	None
Vasectomy	Vasa deferentia cut and tied	No sperm in seminal fluid	Almost 100%	Irreversible sterility
Tubal ligation	Oviducts cut and tied	No eggs in oviduct	Almost 100%	Irreversible sterility
Oral contraception	Hormone medication is taken daily	Anterior pituitary does not release FSH and LH	Almost 100%	Thromboembolism, especially in smokers
Depo-Provera injection	Four injections of progesterone-like steroid given per year	Anterior pituitary does not release FSH and LH	About 99%	Breast cancer? Osteoporosis?
Contraceptive implants	Tubes of progestin (form of progesterone) implanted under skin	Anterior pituitary does not release FSH and LH	More than 90%	Presently none known
Intrauterine device (IUD)	Plastic coil inserted into uterus by physician	Prevents implantation	More than 90%	Infection (pelvic inflammatory disease, PID)
Diaphragm	Latex cup inserted into vagina to cover cervix before intercourse	Blocks entrance of sperm to uterus	With jelly, about 90%	Presently none known
Cervical cap over cervix	Latex cap held by suction	Delivers spermicide near cervix	Almost 85%	Cancer of cervix
Male condom	Latex sheath fitted over erect penis	Traps sperm and prevents STDs	About 85%	Presently none known
Female condom	Polyurethane liner fitted inside vagina	Blocks entrance of sperm to uterus and prevents STDs	About 85%	Presently none known
Coitus interruptus	Penis withdrawn before ejaculation	Prevents sperm from entering vagina	About 75%	Presently none known
Jellies, creams, foams	These spermicidal products inserted before intercourse	Kills a large number of sperm	About 75%	Presently none known
Natural family planning	Day of ovulation determined by record keeping; various methods of testing	Intercourse avoided on certain days of the month	About 70%	Presently none known
Douche	Vagina and uterus cleansed after intercourse	Washes out sperm	Less than 70%	Presently none known

A *male condom* is a thin skin (lambskin) or latex sheath that fits over the erect penis. The ejaculate is trapped inside the sheath and, thus, does not enter the vagina. When used in conjunction with a spermicide, the protection is better than with the condom alone. The condom is generally recognized as giving protection against sexually transmitted diseases.

Contraceptive implants utilize a synthetic progesterone to prevent ovulation by disrupting the ovarian cycle. Six match-sized, time-release capsules are surgically inserted under the skin of a woman's upper arm. The effectiveness of this system may last five years. *Depo-Provera* injections, which change the endometrium, utilize a synthetic progesterone that must be administered every three months. Changes occur in the endometrium that make it less likely that pregnancy will occur.

Searching for Other Means of Birth Control

There has been a revival of interest in *barrier methods* of birth control, including the male condom, because these methods offer some protection against sexually transmit-ted diseases. A female condom, now available, consists of a large polyurethane tube with a flexible ring that fits onto the cervix. The open end of the tube has a ring that covers the external genitals.

Investigators have long searched for a *"male pill."* Analogues of gonadotropic-releasing hormone have been used to prevent the hypothalamus from stimulating the anterior pituitary. Inhibin has also been used to prevent the anterior pituitary from producing FSH. Testosterone and/or related chemicals have been used to inhibit spermatogenesis in males, but this hormone must be administered by injection.

Contraceptive vaccines are now being developed. For example, a vaccine developed to immunize women against HCG, the hormone so necessary to **implantation** of the embryo, was successful in a limited clinical trial. Since HCG is not normally present in the body, no untoward autoimmune reaction is expected, and the immunization does wear off with time. Others believe that it would also be possible to develop a safe antisperm vaccine that would be used in women.

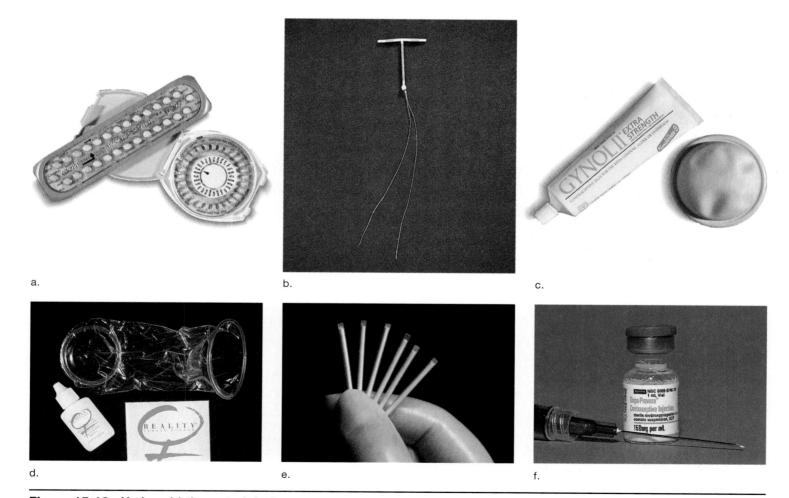

a. b. c.

d. e. f.

Figure 15.12 Various birth-control devices.
a. Oral contraception (birth-control pills). **b.** Intrauterine device. **c.** Spermicidal jelly and diaphragm. **d.** Male condom. **e.** Contraceptive implants. **f.** Depo-Provera injection.

Health Focus

Endometriosis

In the female reproductive tract, there is a small space between an ovary and the oviduct that leads to the uterus. The lining of the uterus, called the endometrium, loses its outer layer during menstruation, and sometimes a portion of the menstrual discharge is carried backwards up the oviduct and into the abdominal cavity instead of being discharged through the opening in the cervix (Fig. 15A). There, endometrial tissue can become attached to and implanted in various organs such as the ovaries; the wall of the vagina, bowel, or bladder; or even on the nerves that serve the lower back or legs. This painful condition is called endometriosis, and it affects 1–3% of women of reproductive age.

Women with uterine cycles of less than 27 days and with a menstrual flow lasting longer than one week have an increased chance of endometriosis. Women who have taken the birth-control pill for a long time or have had several pregnancies have a decreased chance of endometriosis.

When a woman has endometriosis, the displaced endometrial tissue reacts as if it were still in the uterus—thickening, becoming secretory, and then breaking down. The discomfort of menstruation is then felt in other organs of the abdominal cavity, resulting in pain. An area of endometriosis can degenerate and become a scar. Scars that hold two organs together are called adhesions, which can distort organs and lead to infertility.

Only direct observation of the abdominal organs can confirm endometriosis. First, a half-inch incision is made near the navel, and the gas carbon dioxide is injected into the abdominal cavity to separate the organs. Then, a laparoscope (optical telescope) is inserted into the abdomen, allowing the physician to see the organs. Patches of endometriosis show up as purple, blue, or red spots, and there may be dark brown cysts filled with blood. When these rupture, there is a great deal of pain. In the severest cases, there is scarring and the formation of adhesions and abnormal masses around the pelvic organs. A second incision, usually at the pubic hairline, allows the insertion of other instruments that can be used to remove endometrial implants.

Further treatment can take one of two courses. The drug nafarelin can be administered as a nasal spray, which acts through hormonal controls to stop the production of estrogen and, therefore, to stop the uterine cycle. Or, the ovaries can be removed, stopping the uterine cycle. In either case, the woman will suffer symptoms of menopause that can be relieved in the first instance by withdrawing the drug nafarelin, or in the second instance by giving the woman estrogen in doses that do not reactivate the uterine cycle.

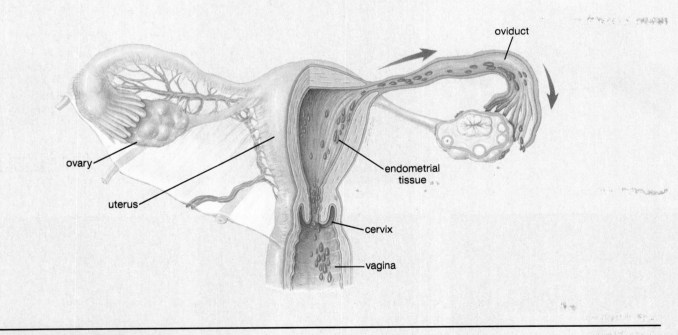

Figure 15A Endometriosis.
It has been suggested that endometriosis is caused by a backward menstrual flow, as represented by the arrows in this drawing. This backward flow allows endometrial cells to enter the abdominal cavity, where they take up residence and respond to monthly cyclic changes in hormonal levels, and the result is extreme discomfort.

Morning-after Pills

There are morning-after regimens available that, depending on when the woman begins medication, either prevent fertilization altogether or stop the fertilized egg from ever implanting. These regimens involve taking pills containing synthetic progesterone and/or estrogen in a manner prescribed by a physician. Many women do not realize that this method of birth control is available, and yet it is estimated that use of these regimens could greatly reduce the number of unintended pregnancies. Effective treatment sometimes causes nausea and vomiting, which can be severe.

Mifepristone, better known as RU-486, is a pill that causes the loss of an implanted embryo by blocking the progesterone receptors of the cells in the uterine lining. Without functioning receptors for progesterone, the uterine lining sloughs off, carrying the embryo with it. When taken in conjunction with a prostaglandin to induce uterine contractions, RU-486 is 95% effective. It is possible that some day the medication will be used by women who are experiencing delayed menstruation without knowing if they are actually pregnant.

There are numerous well-known birth-control methods and devices available to those who wish to prevent pregnancy. Their effectiveness varies. In addition, new methods are expected to be developed.

Some Couples Are Infertile

Sometimes couples do not need to prevent pregnancy; conception or fertilization does not occur despite frequent intercourse. The American Medical Association estimates that 15% of all couples in this country are unable to have any children and therefore are properly termed *sterile;* another 10% have fewer children than they wish and therefore are termed *infertile*. The latter assumes that the couple has been unsuccessfully trying to become pregnant for at least one year.

What Causes Infertility?

The two major causes of infertility in females are blocked oviducts, possibly due to pelvic inflammatory disease (PID), discussed in a following section, and failure to ovulate due to low body weight. *Endometriosis* is the presence of uterine tissue outside the uterus, particularly in the oviducts and on the abdominal organs. As discussed in the Health reading on page 326, endometriosis can contribute to infertility. Endometriosis occurs when the menstrual discharge flows up into the oviducts and out into the abdominal cavity. This backward flow allows living uterine cells to establish themselves in the abdominal cavity where they go through the usual uterine cycle, causing pain and structural abnormalities that make it more difficult for a woman to conceive.

Figure 15.13 Mother and child.
Sometimes couples utilize alternative methods of reproduction in order to experience parenthood.

Sometimes the causes of infertility can be corrected surgically and/or medically so that couples can have children (Fig. 15.13). If no obstruction is apparent and body weight is normal, it is possible to give females HCG, extracted from the urine of pregnant women, along with gonadotropins extracted from the urine of postmenopausal women. This treatment may cause multiple ovulations and sometimes multiple pregnancies.

The most frequent cause of infertility in males is low sperm count and/or a large proportion of abnormal sperm. Disease, radiation, chemical mutagens, high testes temperature, and the use of psychoactive drugs can contribute to this condition.

When reproduction does not occur in the usual manner, many couples adopt a child. Others sometimes first try one of the alternative reproductive methods discussed in the following paragraphs. If all the alternative methods discussed are considered, it is possible for a baby to have five parents: (1) sperm donor, (2) egg donor, (3) surrogate mother, and (4) and (5) adoptive mother and father.

Ecology Focus

Fertility and the Environment

The last half century has seen an explosion in the number of manufactured chemicals. Worldwide, about 70,000 chemicals are in everyday use, and about 1,000 new ones are added each year. Most of us are well aware that some of these chemicals can contribute to the development of cancer; but now it appears that they may have more subtle effects, such as affecting human fertility (Fig. 15B).

Researchers at the National University Hospital in Copenhagen, Denmark, as well as in other countries, tell us that sperm counts and the quality of sperm—whether they are vigorous and healthy as opposed to sluggish and malformed—have declined significantly in the past 50 years. Stress, smoking, drug use, and perhaps the use of brief underwear may all be involved, but the suspicion is that the exposure to certain chemicals such as DDT, TCDD (the most toxic form of dioxin), PCBs (polychlorinated biphenyls), and BPA (bisphenol-A), a compound used in the plastic coating of food cans, may be at fault. These chemicals have hormonelike properties. In laboratory animal experiments, a single oral dose of TCDD fed to pregnant females causes abnormal sperm counts and other reproductive malfunctions in male offspring. They note that besides a decline in human sperm counts, there has been a rise in testicular cancer. Such a combination of effects is known to occur when male fetuses are exposed to too high a ratio of female-to-male hormones in the womb. Researchers speculate that the chemicals mentioned first accumulate in the mother's body and then are passed to the fetus by way of the placenta. The result is reduced fertility in male offspring. In another laboratory study, dioxin was found to increase the chance of endometriosis several years later. Endometriosis is becoming more common in the United States and seems to be occurring in ever-younger women.

Figure 15B   Crop duster.
Some investigators believe that pesticides sprayed on fields cause widespread infertility.

Not all scientists accept the link between environmental estrogens and reproductive ills. Those that do point out that just as laboratory studies of chlorofluorocarbons (CFCs) allowed us to forecast the existence of ozone holes, laboratory work on environmental toxins allows us to explain the decline in sperm count.

Alternative Methods of Reproducing Are Available

Artificial Insemination by Donor (AID). During artificial insemination, sperm are placed in the vagina by a physician. Sometimes, a woman is artificially inseminated by her husband's sperm. This is especially helpful if the husband has a low sperm count—the sperm can be collected over a period of time and concentrated so that the sperm count is sufficient to result in fertilization. Often, however, a woman is inseminated by sperm acquired from a donor who is a complete stranger to her. At times, a mixture of husband and donor sperm are used.

A variation of AID is intrauterine insemination (IUI). IUI involves hormonal stimulation of the ovaries, followed by placement of the donor's sperm in the uterus rather than in the vagina.

In Vitro Fertilization (IVF). During IVF, conception occurs in laboratory glassware. The newer ultrasound machines can spot follicles in the ovaries that hold immature eggs; therefore, the latest method is to forego the administration of fertility drugs and retrieve immature eggs by using a needle. The immature eggs are then brought to maturity in glassware before concentrated sperm from the male are added. After about two to four days, the embryos are inserted into the uterus of the woman, who is now in the secretory phase of her uterine cycle. If implantation is successful, development is normal and continues to term.

Gamete Intrafallopian Transfer (GIFT). Gamete intrafallopian transfer (GIFT) was devised as a means to overcome the low success rate (15–20%) of in vitro fertilization. The method is exactly the same as in vitro fertilization, except the eggs and the sperm are placed in the oviducts immediately after they have been brought together. GIFT has an advantage in that it is a one-step procedure for the woman—the eggs are removed and are reintroduced all in the same time period. For this reason, it is less expensive—approximately $1,500 compared to $3,000 and up for in vitro fertilization.

Surrogate Mothers. In some instances, women are paid to have babies. These women are called surrogate mothers. Other individuals contribute sperm (or eggs) to the fertilization process in such cases.

When corrective procedures fail to reverse infertility, it is possible to consider an alternative method of reproduction.

15.6   Working Together

The Working Together box on page 329 shows how the reproductive system works with the other systems of the body to maintain homeostasis.

HUMAN SYSTEMS WORK TOGETHER

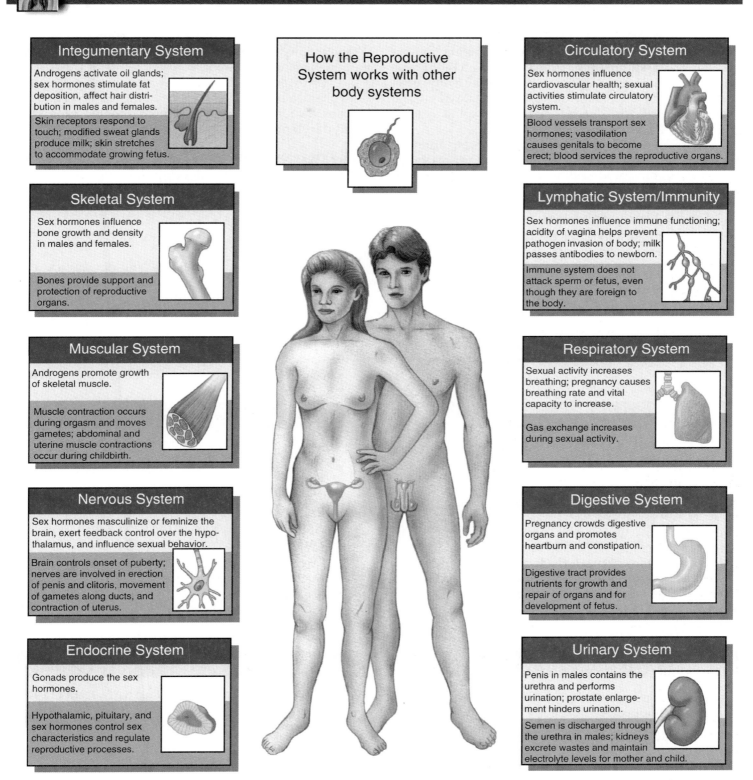

Integumentary System

Androgens activate oil glands; sex hormones stimulate fat deposition, affect hair distribution in males and females.

Skin receptors respond to touch; modified sweat glands produce milk; skin stretches to accommodate growing fetus.

Skeletal System

Sex hormones influence bone growth and density in males and females.

Bones provide support and protection of reproductive organs.

Muscular System

Androgens promote growth of skeletal muscle.

Muscle contraction occurs during orgasm and moves gametes; abdominal and uterine muscle contractions occur during childbirth.

Nervous System

Sex hormones masculinize or feminize the brain, exert feedback control over the hypothalamus, and influence sexual behavior.

Brain controls onset of puberty; nerves are involved in erection of penis and clitoris, movement of gametes along ducts, and contraction of uterus.

Endocrine System

Gonads produce the sex hormones.

Hypothalamic, pituitary, and sex hormones control sex characteristics and regulate reproductive processes.

How the Reproductive System works with other body systems

Circulatory System

Sex hormones influence cardiovascular health; sexual activities stimulate circulatory system.

Blood vessels transport sex hormones; vasodilation causes genitals to become erect; blood services the reproductive organs.

Lymphatic System/Immunity

Sex hormones influence immune functioning; acidity of vagina helps prevent pathogen invasion of body; milk passes antibodies to newborn.

Immune system does not attack sperm or fetus, even though they are foreign to the body.

Respiratory System

Sexual activity increases breathing; pregnancy causes breathing rate and vital capacity to increase.

Gas exchange increases during sexual activity.

Digestive System

Pregnancy crowds digestive organs and promotes heartburn and constipation.

Digestive tract provides nutrients for growth and repair of organs and for development of fetus.

Urinary System

Penis in males contains the urethra and performs urination; prostate enlargement hinders urination.

Semen is discharged through the urethra in males; kidneys excrete wastes and maintain electrolyte levels for mother and child.

SUMMARY

15.1 Male Reproductive System

In males, spermatogenesis, occurring in seminiferous tubules of the testes, produces sperm that mature and are stored in the epididymides and may be stored in the vasa deferentia before entering the urethra, along with secretions produced by seminal vesicles, the prostate gland, and bulbourethral glands. Semen is ejaculated during male orgasm, when the penis is erect.

Hormonal regulation, involving secretions from the hypothalamus, the anterior pituitary, and the testes, maintains testosterone, produced by the interstitial cells of the testes, at a fairly constant level.

15.2 Female Reproductive System

In females, an egg produced by an ovary enters an oviduct, which leads to the uterus. The uterus opens into the vagina. The external genital area includes the vaginal opening, the clitoris, the labia minora, and the labia majora.

15.3 Female Hormone Levels

In the nonpregnant female, the ovarian and uterine cycles are under hormonal control of the hypothalamus, anterior pituitary, and the female sex hormones estrogen and progesterone.

If fertilization occurs, the corpus luteum in the ovary is maintained because of HCG production. Progesterone production does not cease, and the embryo implants itself in the thick uterine lining.

Estrogen and progesterone maintain the secondary sex characteristics of females, including less body hair than males, a wider pelvic girdle, a more rounded appearance, and development of breasts.

15.4 Development of Male and Female Sex Organs

Male and female organs develop from the same indifferent tissue depending on whether the chromosomes are XY or XX. External examination does not reveal the sex of the fetus until after the third month.

15.5 Control of Reproduction

Numerous birth-control methods and devices are available for those who wish to prevent pregnancy. Infertile couples are increasingly resorting to alternative methods of reproduction.

15.6 Working Together

The reproductive system works with the other systems of the body in the ways described in the box on page 329.

STUDYING THE CONCEPTS

1. Outline the path of sperm. What glands contribute fluids to semen?   312–13

2. Discuss the anatomy and physiology of the testes. Describe the structure of sperm.   314–15

3. Name the endocrine glands involved in maintaining the sex characteristics of males and the hormones produced by each.   315

4. Describe the organs of the female genital tract. Where do fertilization and implantation occur? Name two functions of the vagina.   316

5. Name and describe the external genitals in females.   317

6. Discuss the anatomy and the physiology of the ovaries. Describe the ovarian cycle.   318–19

7. Describe the uterine cycle and relate it to the ovarian cycle. In what way is menstruation prevented if pregnancy occurs?   321

8. Name three functions of the female sex hormones.   322

9. Discuss the various means of birth control and their relative effectiveness in preventing pregnancy.   324–27

10. Describe how in vitro fertilization is carried out.   328

APPLYING TECHNOLOGY

Your study of the reproductive system is supported by these available technologies:

Exploring the Internet

The Mader Home Page provides further resources for studying this chapter.

http://www.mhhe.com/sciencemath/biology/mader/

(Click on *Human Biology.*)

Dynamic Human: Reproductive System CD-ROM

In *Anatomy,* male and female reproductive anatomies and a 3-dimensional view of both are shown; in *Explorations,* oogenesis and spermatogenesis occur, the menstrual cycle is related to the ovarian cycle, and an erection occurs as the erectile tissue fills with blood; in *Histology,* microscopic slides are highlighted to reveal tissues; and in *Clinical Concepts,* tubal ligation and vasectomy are depicted.

APPLYING YOUR KNOWLEDGE

Concepts

1. Vasectomy is one method of contraception. Many males have the concern that they will not be able to function sexually after the operation. What advice could you offer them about this concern?

2. If one is practicing the rhythm method of birth control, what is the reason there is a period of five to seven days that are considered to be "unsafe" for unprotected intercourse, even though the egg is discharged only on one day each month?

3. Men whose partners are having difficulty conceiving are sometimes told to switch from jockey shorts to boxer shorts. Is there any justification in doing so? Explain.

Bioethical Issue

Of all the medical issues in the past decade, nothing has caused more uproar than abortion. Political campaigns and passionate debates on the topic continue.

That's because the basic questions involved are largely subjective, ethical—and difficult to answer. Try to answer these seemingly simple queries, and you'll understand:

At what stage in an embryo's development does that embryo become a child? Are there certain circumstances—and time frames—in which abortion is reasonable? Who should control human reproduction? Do would-be parents have the right to make decisions about the continuation of a pregnancy, or should this be decided by doctors, judges, ministers, etc.? Explain your reason(s) for your answers.

TESTING YOUR KNOWLEDGE

1. In tracing the path of sperm, the structure that follows the epididymis is the _____.

2. The prostate gland, the bulbourethral glands, and the _____ all contribute to seminal fluid.

3. The main male sex hormone is _____.

4. An erection is caused by the entrance of _____ into sinuses within the penis.

5. Label this diagram of the male reproductive system, and trace the path of sperm.

6. In the female reproductive system, the uterus lies between the oviducts and the _____.

7. In the ovarian cycle, once each month a(n) _____ releases an egg. In the uterine cycle, the _____ lining of the uterus is prepared to receive the embryo.

8. The female sex hormones are _____ and _____.

9. Pregnancy in the female is detected by the presence of _____ in blood or urine.

10. In vitro fertilization occurs in _____.

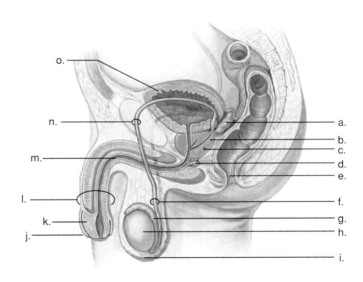

SELECTED KEY TERMS

bulbourethral gland Either of two small structures located below the prostate gland in males; adds secretions to semen. 313

cervix Narrow end of the uterus, which leads into the vagina. 317

corpus luteum (KOR-pus LOOT-ee-um) Yellow body that forms in the ovary from a follicle that has discharged its egg; it secretes progesterone and estrogen. 319

ejaculation Ejection of seminal fluid. 313

endometrium (en-doh-MEE-tree-um) Lining of the uterus, which becomes thickened and vascular during the uterine cycle. 317

epididymis Coiled tubule next to the testes where sperm mature and may be stored for a short time. 313

erection Condition of the penis when it is turgid and erect, instead of being flaccid or lacking turgidity. 313

estrogen Female sex hormone, which, along with progesterone, maintains the primary sex organs and stimulates development of the female secondary sex characteristics. 318

fimbria (FIM-bree-uh) Fingerlike extension from the oviduct near the ovary. 317

follicle Structure in the ovary that produces the egg and, in particular, the female sex hormones, estrogen and progesterone. 318

follicle-stimulating hormone (FSH) Hormone secreted by the anterior pituitary gland that stimulates the development of an ovarian follicle in a female or the production of sperm in a male.   318

implantation Attachment and penetration of the embryo into the lining of the uterus (endometrium).   325

labia majora (LAY-bee-uh) Outer folds of the vulva.   317

labia minora Inner folds of the vulva.   317

luteinizing hormone (LH) (LOOT-ee-nyzing) Hormone produced by the anterior pituitary gland that stimulates the development of the corpus luteum in the ovary in females and the production of testosterone in males.   318

menopause Termination of the ovarian and uterine cycles in older women.   322

menstruation Loss of blood and tissue from the uterus at the end of a uterine cycle.   321

orgasm Physical and emotional climax during sexual intercourse; results in ejaculation in the male.   313

ovarian cycle Monthly changes occurring in the ovary that determine the level of sex hormones in the blood.   319

ovary Female gonad, the organ that produces eggs, estrogen, and progesterone.   317

oviduct Tube that transports eggs to the uterus; also called uterine tube.   317

ovulation Discharge of a mature egg from the follicle within the ovary.   317

Pap smear Analysis done on cervical cells for detection of cancer.   317

penis External organ in males through which the urethra passes and that serves as the organ of sexual intercourse.   313

progesterone (proh-JES-tuh-rohn) Female sex hormone secreted by the corpus luteum of the ovary and by the placenta.   318

prostate gland Gland located around the male urethra below the urinary bladder; adds secretions to semen.   313

scrotum Pouch of skin that encloses the testes.   312

semen Thick, whitish fluid consisting of sperm and secretions from several glands of the male reproductive tract.   313

seminal vesicle Convoluted, saclike structure attached to the vas deferens near the base of the urinary bladder in males; adds secretions to semen.   313

sperm Male sex cell with three distinct parts at maturity: head, middle piece, and tail.   315

testis Male gonad, the organ that produces sperm and testosterone.   312

uterine cycle Monthly occurring changes in the characteristics of the uterine lining (endometrium).   321

uterus Organ located in the female pelvis where the fetus develops; the womb.   317

vagina Organ that leads from the uterus to the vestibule and serves as the birth canal and organ of sexual intercourse in females.   317

vas deferens Tube that leads from the epididymis to the urethra in males.   313

vulva External genitals of the female that surround the opening of the vagina.   317

Chapter 16

Sexually Transmitted Diseases

Chapter Outline

Figure 16.1   Sexual relationships.
Everyone who has sexual relationships should be aware of the possibility of acquiring a sexually transmitted disease and take all necessary precautions.

By the time Jennifer W. realizes what has happened, it's too late. She is sitting in a doctor's office, quietly explaining that every time she goes to the bathroom, she has this terrible burning pain. She doesn't want to tell her friends. She can't tell her parents.

At least she is telling the doctor.

The source of Jennifer's agony is genital herpes, a very common—and very embarrassing—condition. Genital herpes is a sexually transmitted disease (STD) caused by a virus. As in Jennifer's case, most people catch the disease by having sex with someone who's infected but has no symptoms (Fig. 16.1). That's because herpes viruses—as well as AIDS and a host of other STDs—can lie latent in the body, hiding from sight. Even if you can't see the disease, you should try to prevent passage by *(1) practicing abstinence; or (2) having a monogamous (always the same partner) sexual relationship with someone who does not have an STD, and in the case of AIDS and hepatitis B, is not an intravenous drug user; or (3) always using a female or male latex condom in the proper manner with a water-based vaginal spermicide containing nonoxynol-9. You should also avoid oral/genital contact—just touching the genitals can transfer an STD in some cases.*

With a prescription drug, Jennifer recovers from her present symptoms of a herpes infection. But she'll never be free of the virus, a possible recurrence of symptoms, or the knowledge that her pain could have been avoided. Perhaps if Jennifer had taken Human Biology she would have known about STDs and how to protect herself.

STDs are now a major worldwide health problem especially because many young people are sexually active and the age of sexual intercourse has been declining. This chapter discusses pathogens such as viruses, bacteria, and fungi and the STDs they cause. In addition, more information about a human immunodeficiency viral (HIV) infection and AIDS can be found in the supplement that follows this chapter.

16.1 Viral in Origin

Sexually transmitted diseases (STDs) are contagious diseases caused by pathogens that are passed from one human to another by sexual contact. Viruses cause numerous diseases in humans (Table 16.1), including AIDS, herpes, genital warts, and hepatitis B, four sexually transmitted diseases of great concern today.

Viruses are incapable of independent reproduction and reproduce only inside a living **host** cell. For this reason, viruses are called *obligate intracellular parasites*. To maintain viruses in the laboratory, they are injected into laboratory-bred animals, live chick embryos, or animal cells maintained in tissue culture. Viruses infect all sorts of cells—from bacterial cells to human cells—but they are very specific. For example, viruses called bacteriophages infect only bacteria, the tobacco mosaic virus infects only plants, and the rabies virus infects only mammals. Human viruses even specialize in a particular tissue. Human immunodeficiency virus (HIV) enters certain blood cells, the polio virus reproduces in spinal nerve cells, the hepatitis viruses infect only liver cells.

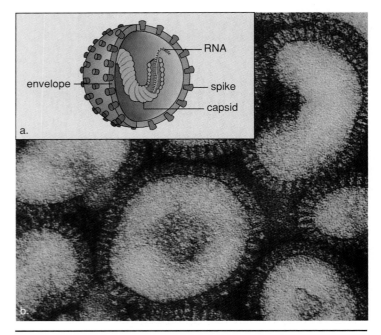

Figure 16.2   Influenza virus.
This RNA virus causes influenza in humans. **a.** Diagram of virus. **b.** Electron micrograph of actual virus.

Viruses are noncellular, and they have a unique construction. These tiny particles always have at least two parts: an outer capsid composed of protein subunits and an inner core of nucleic acid—either DNA or RNA (Fig. 16.2). The capsid of an animal virus is often surrounded by an outer envelope, which is derived from the host-cell plasma membrane, that also contains viral glycoprotein spikes. The glycoprotein spikes within the envelope allow the virus to adhere to plasma membrane receptors. Then the entire virus enters the host cell by endocytosis. When the virus is uncoated, the envelope and capsid are removed. Once the viral genes are free inside the cell, viral components are synthesized and assembled into new viruses (Fig. 16.3). Release of the virus from the host cell occurs by *budding*. During budding, the virus gets its envelope, which consists of host plasma membrane components and glycoproteins that were coded for by viral genes. These glycoproteins are the spikes that allow the virus to attach to the plasma membrane of a host cell. The process of budding does not necessarily kill the host cell.

Some DNA viruses that enter human cells—for example, the papillomaviruses, the herpes viruses, the hepatitis viruses, and the adenoviruses—can undergo a period of latency. While latent, the viral genes are integrated into the host cell chromosome DNA, and they are copied whenever the host cell reproduces. Certain environmental factors, such as ultraviolet radiation, can induce the virus to replicate and bud from the cell. Latent viruses are of special concern because their presence in the host-cell chromosome can alter the cell and make it become cancerous. Some viruses are cancer-producing because they bring with them *oncogenes*, cancer-causing genes.

TABLE 16.1	
Infectious Diseases Caused by Viruses	
Respiratory Tract	**Nervous System**
Common colds	Encephalitis
Flu*	Polio*
Viral pneumonia	Rabies*
Skin Reactions	**Liver**
Measles*	Yellow fever*
German measles*	Hepatitis A, C, and D
Chicken pox*	**Other**
Shingles	Mumps*
Warts	Cancer
Sexually Transmitted	
AIDS	
Genital herpes	
Genital warts	
Hepatitis B*	

*Vaccines available. Yellow fever, rabies, and flu vaccines are given only if the situation requires them. Smallpox vaccinations are no longer required.

Retroviruses are RNA viruses that have a DNA stage in their replication, and these viruses can also become latent. A retrovirus contains a special enzyme called reverse transcriptase that carries out RNA → cDNA transcription. This DNA is called cDNA because it is a DNA copy of the viral RNA genes. During latency, the cDNA is integrated into host DNA and then it leaves when viral reproduction and budding occur. Retroviruses are of interest because human immunodeficiency virus (HIV), which causes AIDS, is a retrovirus. Retroviruses also cause certain forms of cancer.

Viral diseases are controlled by preventing transmission, by administering vaccines, and only recently by administering antiviral drugs as discussed in the Health reading on page 339. **Pathogens** are viruses, bacteria, and other organisms that cause diseases in humans. Knowing how a particular pathogen is transmitted can help prevent its spread. Covering the mouth and nose when one coughs or sneezes helps prevent the spread of a cold, and use of a condom during intercourse helps prevent the transmission of sexually transmitted diseases. Vaccines, which are drugs administered to stimulate immunity to a pathogen so that it cannot later cause disease, are available for some viral diseases such as polio, measles, mumps, and hepatitis B, a sexually transmitted disease (Table 16.1). Antibiotics which are designed to interfere with bacterial metabolism, have no effect on viral illnesses. Instead, drugs must be designed that interfere with either entry of the virus into the host cell, replication of its DNA when copies are made, or exit of the virus from the cell.

A number of viruses cause diseases in humans. Some of these are significant sexually transmitted diseases (STDs).

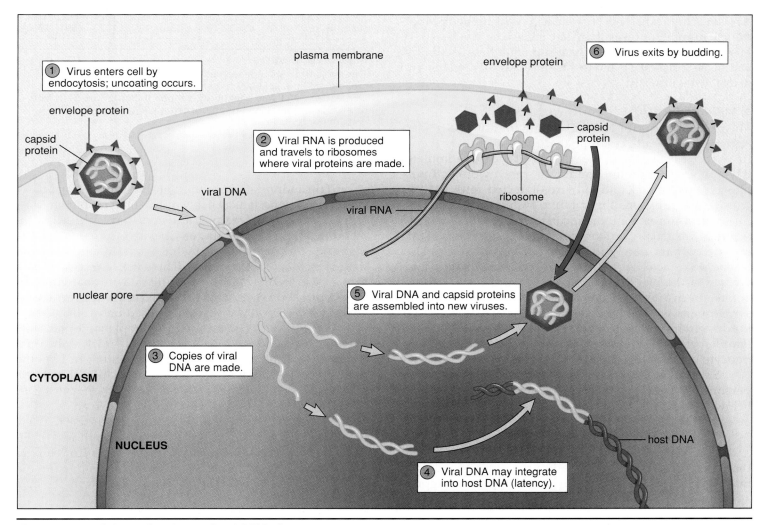

Figure 16.3   Life cycle of an animal DNA virus.
DNA viruses reproduce within a cell and become latent by inserting their DNA into host DNA.

HIV Infection

AIDS (acquired immunodeficiency syndrome) is caused by a **human immunodeficiency virus (HIV)** that infects specific blood cells. The brief discussion here may be supplemented by the discussion beginning on page 349.

Prevalence of an HIV infection and AIDS

A new HIV infection is believed to occur every 15 seconds, the majority in heterosexuals. Estimates vary, but as many as 20 million people worldwide may now be infected with HIV, and there could be as many as 40 million individuals infected by the end of the century. More than 74 thousand new cases were reported in the United States during 1995 (Fig. 16.4). AIDS is not distributed equally throughout the world. Most infected people live in Africa (66%) where it is believed HIV infections first began, but new infections are now occurring at the fastest rate in Southeast Asia and the Indian subcontinent.

In the United States, AIDS is concentrated in large cities but is now spreading to small towns and rural areas. HIV infection is more prevalent among particular groups; African Americans and Hispanics have a higher proportionate number of cases than do Caucasians. Even so, HIV poses a threat to sexually active adults regardless of ethnicity and sexual orientation and to all who inject themselves with drugs intravenously.

Symptoms of AIDS

The primary hosts for HIV are certain immune cells: macrophages, the white cell type that first encounters the virus; *helper T lymphocytes,* the white cell type that stimulates B lymphocytes to produce antibodies; and cytotoxic T lymphocytes that attack and kill virus-infected cells. In other words, HIV threatens the immune system.

During an asymptomatic carrier stage (called category A), there are usually no symptoms, yet the person is highly infectious. Immediately after infection and before the blood test becomes positive, there is a large number of infectious viruses in the blood that could be passed on to another person. Even after the blood test becomes positive, the person remains well as long as the body stays ahead of the hordes of viruses entering the blood. Millions of helper T lymphocytes are produced each day, and the helper T lymphocyte count is higher than 500 per mm^3.

Several months to several years after infection in persons not treated, the helper T lymphocyte count falls below 500 per mm^3, and the symptoms of pre-AIDS (called category B) begin to appear. Lymph nodes are swollen, severe fatigue is common, and fever with night sweats and diarrhea will be present. There may be indications that the virus has entered the brain as loss of memory, inability to think clearly, loss of judgment, and/or depression become apparent.

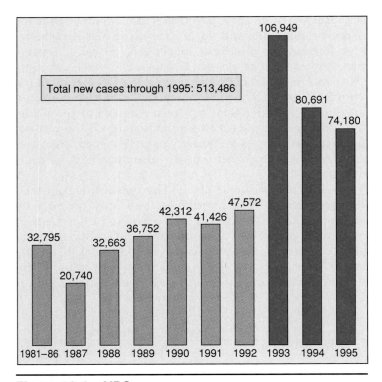

Figure 16.4   AIDS.
The number of AIDS cases has risen since AIDS was first recognized as a sexually transmitted disease. The dramatic increase in 1993 was caused by a modification of the definition of AIDS. Now anyone with an opportunistic disease is considered to have AIDS.
Source: HIV/AIDS Surveillance Report (7)2:5,19. Atlanta: Centers for Disease Control and Prevention.

If the individual develops non-life-threatening but recurrent infections, it most likely means that full-blown AIDS will occur shortly. One possible infection is thrush, a infection caused by the fungus *Candida albicans,* which is identified by the presence of white spots and ulcers on the tongue and inside the mouth. The *Candida* infection may also spread to the vagina, resulting in a chronic infection there. Another frequent infection that often appears is oral and genital herpes, a disease discussed on page 338.

Until recently, the majority of people with pre-AIDS eventually developed full-blown AIDS, which is the final stage (called category C) of an HIV infection. In 1993 the definition of AIDS was broadened to include those persons with a severe depletion of helper T lymphocytes (less than 200 per mm^3 of blood) and/or who have an opportunistic infection. An **opportunistic infection** is one that has the opportunity to occur only because the immune system is severely weakened. Persons with AIDS die from one or more opportunistic diseases, such as *P. carinii* pneumonia and Kaposi's sarcoma, a form of cancer—not from the HIV infection itself.

Treatment for an HIV Infection

Most of the drugs available for treatment of an HIV infection, including the well-known drug called AZT, interfere with the ability of reverse transcriptase to produce viral DNA. Another class of drugs called proteinase inhibitors block the action of another viral enzyme called proteinase. When HIV proteinase is blocked, the resulting viruses lack the capacity to cause infection. Although intensive multidrug therapy will lower the blood level of the virus even to zero, it is possible the virus has not been wiped out, since it may remain in other tissues such as the spleen and lymph nodes. Therefore, therapy must be continued indefinitely.

Drugs have also been developed to deal with opportunistic diseases in AIDS patients. These drugs help people with AIDS lead a fairly normal life for some months until they finally weaken and cannot recover.

Many investigators are working on an AIDS vaccine. Some are trying to develop a vaccine in the traditional way. Traditionally, vaccines are made by treating a pathogen chemically, thereby weakening it so that it can be injected into the body without causing disease. Giving new impetus to this approach is the recent finding that certain long-term HIV survivors are infected with mutated viruses that are unable to reproduce. It is possible to produce such viruses in the laboratory, and perhaps they will be safe enough to use as a vaccine for everyone. Others are working on subunit vaccines that utilize just a single HIV protein as the vaccine. The various vaccines are at different stages in their development. Thus far, certain vaccines have caused antibodies to appear, but none of the vaccines has prevented future infection.

Transmission of HIV

The largest proportion of people with AIDS in the United States are homosexual men, but the proportions attributed to intravenous drug users and heterosexuals are rising. Women now account for 19% of all newly diagnosed cases of AIDS. An infected woman can pass HIV to her unborn children by way of the placenta or to a newborn through breast milk.

It is clear that sexual behaviors and drug-related activities are the major means by which AIDS is transmitted in the United States. Essentially, HIV is spread by passing virus-infected T lymphocytes found in body fluids, such as semen or blood, from one person to another. (Blood and blood products are now tested for the presence of HIV, so the risk of contracting an infection in this manner is now considered very unlikely.) In order to prevent transmission of HIV, you should follow the general advice given in the introduction to this chapter (see p. 333).

More people, many of whom are heterosexuals, are infected with the HIV virus each year. All possible steps should be taken to avoid transmission because treatment is costly and there is no cure.

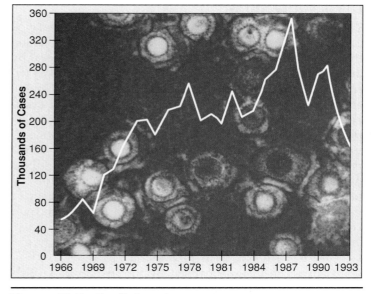

Figure 16.5 Genital warts.
A graph depicting the incidence of new cases of genital warts in the United States from 1966 to 1993 is superimposed on a micrograph of human papillomaviruses.
Source: National Disease and Therapeutic Index (IMS America, Ltd.).

Genital Warts

Human papillomaviruses (HPVs) cause warts, including common warts, plantar warts, and also genital warts, which are sexually transmitted. Over one million persons become infected each year with a form of HPV, which causes **genital warts** (Fig. 16.5).

Transmission and Symptoms

Quite often, carriers of genital warts do not detect any sign of warts, although flat lesions may be present. When present, the warts commonly are seen on the penis and foreskin of men and near the vaginal opening in women. A newborn can become infected by passage through the birth canal.

Genital warts are associated with cancer of the cervix, as well as tumors of the vulva, the vagina, the anus, the penis, and the mouth. Some researchers believe that HPVs are involved in 90 to 95% of all cases of cancer of the cervix. Teenagers with multiple sex partners seem to be particularly susceptible to HPV infections. More cases of cancer of the cervix are being seen among this age group.

Treatment

Presently, there is no cure for an HPV infection, but it can be treated effectively by surgery, freezing, application of an acid, or laser burning, depending on severity. If the warts are removed, they may recur. Also, even after treatment, the virus can be transmitted. Therefore, abstinence or use of a condom with vaginal spermicide containing nonoxynol-9 is necessary to prevent the spread of genital warts.

Genital warts is a prevalent STD associated with cervical cancer in women.

Herpes Infections

The herpes viruses cause various illnesses. Chicken pox and shingles are caused by a type of herpes virus called the varicella-zoster virus, and mononucleosis is caused by another type of herpes virus, called the Epstein-Barr virus. These conditions do not appear to be spread by sexual contact. The **herpes simplex viruses** that do cause sexually transmitted disease are large DNA viruses that infect the mucosal linings of the body such as the mouth and vagina. There are two types of herpes simplex viruses: type 1 usually causes cold sores and fever blisters, while type 2 more often causes genital herpes. Cross-over infections do occur, however. That is, type 1 has been known to cause a genital infection, while type 2 has been known to cause cold sores and fever blisters.

Cold Sores and Fever Blisters

Cold sores are usually caused by herpes simplex virus type 1. Infection with this virus usually occurs during childhood with symptoms of a mild upper respiratory illness. In some cases, however, painful blisters may appear in the throat, on or below the tongue, and on other mouth parts, including the lips. The sores heal, but the virus remains latent in ganglia outside the central nervous system and spinal cord. Thereafter, a recurrence of symptoms takes the form of a cold sore around the lips. A high fever, stress, colds, menstruation, or exposure to sunlight seems to trigger a reactivation of the virus. It is important to realize that herpes lesions are infectious for at least three to four days until the sores begin to heal. Contact with the sores or any contaminated object can cause the virus to be transmitted. Oral/genital contact can lead to a genital herpes infection.

Genital Herpes

Genital herpes, usually caused by the herpes simplex virus type 2, is as prevalent as genital warts with over a million or more new cases appearing each year (Fig. 16.6). At any one time, millions of persons could be having recurring symptoms.

Persons usually get infected with herpes simplex virus type 2 when they are adults. In some people there are no symptoms. Or there may be a tingling or itching sensation before blisters appear on the genitals (within 2–20 days). Once the blisters rupture, they leave painful ulcers that may take as long as three weeks or as little as five days to heal. The blisters may be accompanied by fever, pain on urination, swollen lymph nodes in the groin, and in women, a copious discharge.

After the ulcers heal, the disease is only latent, and blisters can recur although usually at less frequent intervals and with milder symptoms. Again, fever, stress, sunlight, and menstruation are associated with a recurrence of symptoms. When there are no symptoms, the virus primarily resides in the ganglia of sensory nerves associated with the affected skin. Although herpes simplex type 2 was formerly thought to be a cause of cervical cancer, this is no longer believed to be the case.

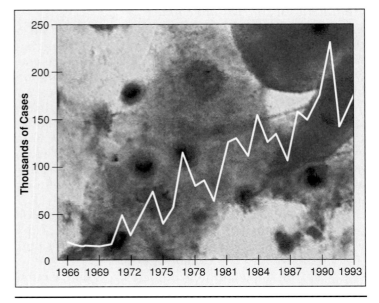

Figure 16.6   Genital herpes.
A graph depicting the incidence of new cases of genital herpes in the United States from 1966 to 1993 is superimposed on a micrograph of cells infected with the herpes virus.
Source: National Disease and Therapeutic Index (IMS America, Ltd.).

Infection of the newborn can occur if the child comes in contact with a lesion in the birth canal. At worst, the infant can be gravely ill, become blind, have neurological disorders including brain damage, or die. Birth by cesarean section prevents these occurrences; therefore, all pregnant women infected with the virus should be sure to tell their health-care provider.

Transmission

Genital herpes lesions shed infective viruses. Live viruses have occasionally been cultured from the skin of persons with no lesions; therefore, it is believed that the virus can be spread even when there are no visible lesions. Persons who are infected should take extreme care to prevent the possibility of spreading the infection to other parts of their own body, such as the eyes. Certainly, sexual contact should be avoided until all lesions are completely healed, and then the general directions given in the introduction for avoiding STD transfer should be followed (see p. 333).

Treatment

Presently, there is no cure for genital herpes. The drugs acyclovir and vidarabine disrupt viral reproduction. The ointment form of acyclovir relieves initial symptoms, and the oral form prevents the recurrence of symptoms as long as it is being taken. Research is being conducted in an attempt to develop a vaccine.

Herpes simplex viruses cause genital herpes, an extremely infectious disease whose symptoms recur and for which there is no cure.

Health Focus

STDs and Medical Treatment

Treatment of STDs is troublesome at best. Presently, there are no vaccines available except for a hepatitis B infection, and to date, not many persons have been inoculated. Unfortunately, there is no treatment once a hepatitis B infection has occured. The antiviral drugs acyclovir and vidarabine are helpful against genital herpes, but they do not cure herpes. Once an individual stops the therapy, the lesions are apt to recur. The current recommended therapy of AIDS includes an expensive combination of drugs that prevent HIV reproduction and curtail its ability to infect other cells.

Antibiotics are used to treat sexually transmitted diseases caused by bacteria. Antibiotics such as penicillin, streptomycin, and tetracycline interfere with metabolic pathways unique to bacteria, and therefore, they are not expected to harm host cells. Still, there are problems associated with antibiotic therapy. Some individuals are allergic to antibiotics, and the reaction may even be fatal. Antibiotics not only kill off disease-causing bacteria, they also reduce the number of beneficial bacteria in the intestinal tract. The use of antibiotics sometimes prevents natural immunity from occurring, leading to the necessity for recurring antibiotic therapy.

Most important, perhaps, is the growing resistance of certain strains of bacteria to a particular antibiotic. Antibiotics were introduced in the 1940s, and for several decades they worked so well it appeared that infectious diseases had been brought under control. However, we now know that bacterial strains can mutate and become resistant to a par-ticular antibiotic. Worse yet, bacteria can swap bits of DNA, and in this way, resistance can pass to other strains of bacteria. Penicillin and tetracycline, long used to cure gonorrhea, now have a failure rate of more than 20% against certain strains of *Gonococcus*. To help prevent resistant bacteria, antibiotics should be administered only when absolutely necessary, and the prescribed therapy should be finished. Also, as a society we have to continue to develop new antibiotics to which resistance has not yet occurred.

An alarming new concern has been the upsurge of tuberculosis in persons with sexually transmitted diseases. Tuberculosis is a chronic lung infection caused by the bacterium *Mycobacterium tuberculosis*. This bacteria is often called the tubercle bacillus (TB) because usually the bacilli are walled off within tubercles that can sometimes be seen in X rays of the lungs. TB can pass from an HIV-infected person to a healthy person more easily than HIV. Sexual contact is not needed—only close personal contact, and therefore, it is possible that TB will now spread from those with STDs to the general populace. Some of the modern strains of TB are resistant to antibiotic therapy; therefore, just as with HIV, research is needed for the development of an effective vaccine.

Prevention is still the best way to manage STDs. The use of a condom and the avoidance of oral/genital contact is essential unless you have a monogamous relationship (always the same partner) with someone who is free of STDs and is not an intravenous drug user.

Hepatitis Infections

There are several types of **hepatitis.** Hepatitis A, caused by HAV (hepatitis A virus), is usually acquired from sewage-contaminated drinking water. This infection can also be sexually transmitted through oral/anal contact.

Hepatitis B, caused by HBV, is usually spread by sexual contact. HBV, which is more infectious than the AIDS virus, can also be spread by blood transfusions or contaminated needles. Fortunately, a vaccine is now available. An HDV infection sometimes accompanies an HBV infection, and if so, liver function is expected to deteriorate more rapidly.

Hepatitis C, caused by HCV, is called the post-transfusion form of hepatitis. This type of hepatitis, which is usually acquired by contact with infected blood, is also of great concern. Infection can lead to chronic hepatitis, liver cancer, and death.

Hepatitis E, caused by HEV, is usually seen in developing countries. Only imported cases, in travelers to the country, or visitors to endemic regions, have been reported in the United States.

Still other types of hepatitis are now under investigation. An international team reported finding a hepatitis F virus, but this has not yet been confirmed. Hepatitis G-1 and G-2 have been discovered in animals, while hepatitis G-3 viruses are now believed to cause long-term liver ailments not explainable before.

Hepatitis B

HBV is a DNA virus that is spread in the same way as HIV, the cause of AIDS. Sharing needles by drug abusers and sexual contact between either heterosexuals or homosexual men transmits the disease. Over 90% of persons with AIDS also have an HBV infection. Also, like HIV, HBV can be passed from mother to child by way of the placenta.

Only about 50% of infected persons have flu-like symptoms, including fatigue, fever, headache, nausea, vomiting, muscle aches, and dull pain in the upper right of the abdomen. Jaundice, a yellowish cast to the skin, can also be present. Some persons have an acute infection that lasts only three to four weeks. Others have a chronic form of the disease that leads to liver failure and a need for a liver transplant.

Since there is no treatment for an HBV infection, prevention is imperative. The general directions given on page 333 should be followed, but inoculation with the HBV vaccine is the best protection. The vaccine, which is safe and does not have any major side effects, is now on the list of recommended immunizations for children.

Hepatitis B is an infection that can lead to liver failure. Because it is spread in the same way as AIDS, many persons are infected with both viruses at the same time.

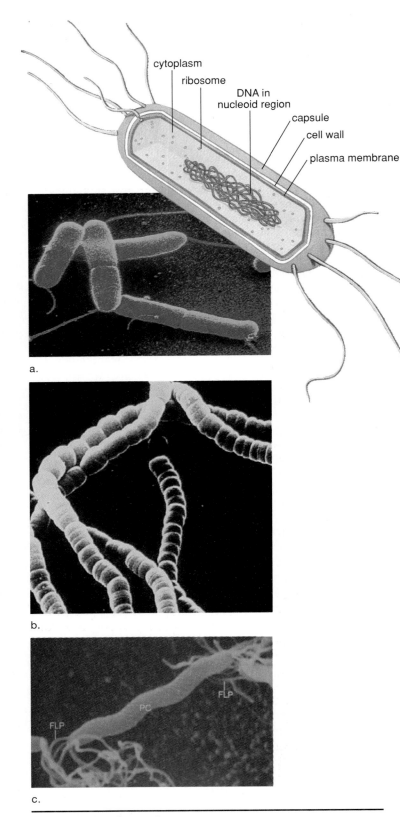

cytoplasm
ribosome
DNA in
nucleoid region
capsule
cell wall
plasma membrane

a.

b.

c.

Figure 16.7   Bacteria.
Bacteria occur as **(a)** a bacillus (rod shape), **(b)** a coccus (round
shape), and **(c)** a spirillum. The structure of a bacterium is also
shown in **(a)**.

16.2 Bacterial in Origin

Although **bacteria** are generally larger than viruses, they are still microscopic. As a result, it is not always obvious that they are abundant in the air, water, and soil and on most objects. It has even been suggested that the combined weight of all bacteria would exceed that of any other type of organism on earth. Bacteria have long been used by humans to produce various products commercially. Chemicals, such as ethyl alcohol, acetic acid, butyl alcohol, and acetone, are produced by bacteria. Bacterial action is involved in the production of butter, cheese, sauerkraut, rubber, cotton, silk, coffee, wine, and cocoa. By means of genetic engineering, bacteria are now used to produce human insulin and interferon, as well as other types of proteins. Even certain antibiotics are produced by bacteria.

Bacteria occur in three basic shapes: rod (bacillus); round, or spherical (coccus); and spiral (a curved shape called a spirillum) (Fig. 16.7). Human cells contain a membrane-bound nucleus and several kinds of membranous organelles; therefore, they are called eukaryotic (true nucleus) cells. Bacterial cells lack a membrane-bound nucleus and the other membranous organelles typical of human cells; therefore, they are called prokaryotic (before nucleus) cells. A nucleoid region contains their DNA, and a cytoplasm contains the enzymes of their many metabolic pathways. Bacteria are enclosed in a cell wall, and some are also surrounded by a polysaccharide or polypeptide capsule that inhibits destruction by the host. Motile bacteria often have flagella (sing., flagellum).

Bacteria usually reproduce asexually by **binary fission.** First, the single chromosome duplicates, and then the two chromosomes move apart into separate areas. Next, the plasma membrane and cell wall grow inward and partition the cell into two daughter cells, each of which has its own chromosome (Fig. 16.8). Under favorable conditions, growth may be very rapid, with cell division occurring as often as every 20 minutes. When faced with unfavorable environmental conditions, some bacteria can form **endospores.** During spore formation, the cell shrinks, rounds up within the former plasma membrane, and secretes a new, thicker wall inside the old one. Endospores are amazingly resistant to extreme temperatures, drying out, and harsh chemicals, including acids and bases. When conditions are again suitable for growth, the spore absorbs water, breaks out of the inner shell, and becomes a typical bacterial cell capable of reproducing.

Most bacteria are free-living organisms that perform many useful services in the environment. They are able to live in just about any habitat and use just about any substance as a food source because of their wide metabolic ability. Most bacteria are decomposers—they break down dead organic matter in the environment by secreting digestive enzymes, and then they absorb the nutrient molecules.

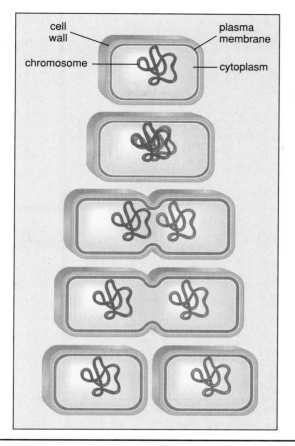

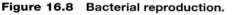

Figure 16.8 Bacterial reproduction.
When bacteria reproduce, the single chromosome is attached to the plasma membrane, where it is replicating. As the plasma membrane and the cell wall lengthen, the two chromosomes separate. Once fission has taken place, each bacterium has its own chromosome.

Many bacteria are *symbiotic,* that is, they live in close association with another species. There are a number of bacteria that are normal inhabitants of our bodies. These bacteria, called the *normal microbial flora,* cause disease only when the opportunity arises. The bacterium *Escherichia coli,* which lives in our intestines, breaks down remains that we have not digested. The microflora of our intestines, including *E. coli,* provide us with the vitamins thiamine, riboflavin, pyridoxine, B₁₂, and K. The last two vitamins are necessary for the production of blood components. *E. coli* causes cystitis if it gets into the urinary tract. And *Staphylococcus aureus,* which lives on our skin, causes an infection if it gets into a wound.

The diseases listed in Table 16.2 are caused by bacteria that are ordinarily parasitic; that is, they usually do cause an illness. Most bacteria are aerobic, and they require a supply of oxygen just as we do, but several human diseases are caused by anaerobes. Among them are botulism, gas gangrene, and tetanus, which are listed in Table 16.2.

Diseases caused by pathogens are termed communicable because the pathogen must be transmitted to the new

TABLE 16.2

Some Infectious Diseases Caused by Bacteria

Respiratory Tract	**Nervous System**
Strep throat	Tetanus*
Pneumonia*	Botulism
Whooping cough*	Meningitis
Tuberculosis*	**Digestive Tract**
Skin Reactions	Food poisoning
Staph (pimples and boils)	Dysentery
Other	Cholera*
Gas gangrene* (wound infections)	**Sexually Transmitted**
Diphtheria*	Gonorrhea
Typhoid fever*	Chlamydia
	Syphilis

*Vaccines are available. Tuberculosis vaccine is not used in this country. Typhoid fever, cholera, and gas gangrene vaccines are given if the situation requires it. Others are routinely given.

host. Pathogens are transferred by a variety of mechanisms, including food and water, human or animal bites, contact, and aerosols (droplets in the air). Sexually transmitted diseases require intimate contact between persons for their successful transmission.

To be a pathogen, a bacterium must

- have an ability to pass from one host to the next
- penetrate into the host's tissues
- withstand the host's defense mechanisms
- induce illness in the host

Bacterial diseases are controlled by preventing transmission, by administering vaccines, and by antibiotic therapy. **Antibiotics** are medications that kill bacteria by interfering with one of their unique metabolic pathways. Antibiotic therapy is highly successful if it is carried out in the recommended manner. Otherwise, as discussed in the Health reading on page 339, resistant strains can develop. Unfortunately, there are no vaccines for STDs caused by bacteria; therefore, preventing transmission becomes all the more important. Abstinence or monogamous relations (always the same partner) with someone who is free of an STD will prevent transmission. Otherwise, the use of a condom with a vaginal spermicide containing nonoxynol-9 and avoidance of oral/genital contact is recommended.

Bacteria lack a membrane-bound nucleus and other membranous organelles found in eukaryotic cells, but they are capable of an independent existence. Most are free-living, but a few cause human diseases that can often be cured by antibiotic therapy.

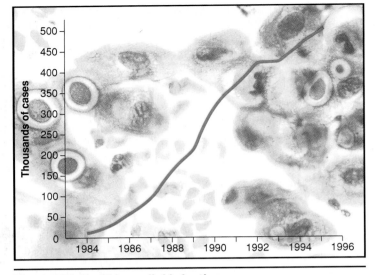

Figure 16.9   Chlamydial infection.
A graph depicting the incidence of new cases of chlamydia in the United States from 1984 to 1995 is superimposed on a micrograph of a cell containing different stages of the organism.

Source: Sexually Transmitted Disease Surveillance, 1995. Atlanta: Centers for Disease Control and Prevention, September 1996.

Chlamydia

Chlamydia is named for the tiny bacterium that causes it (*Chlamydia trachomatis*). For years, chlamydiae were considered to be more closely related to viruses than to bacteria, but today it is known that these organisms are cellular in nature. Even so, they are obligate parasites due to their inability to produce ATP molecules. After chlamydia enters a cell by endocytosis, the life cycle occurs inside the endocytic vacuole, which eventually bursts and liberates many new infective chlamydiae.

New chlamydial infections are higher than any other sexually transmitted disease. From 1984 through 1994, reported incidences of chlamydia increased dramatically, from less than fifty thousand cases per year to about five hundred thousand cases. Some estimate that the actual incidence could be as high as six million new cases per year. For every reported case in men, more than five cases are detected in women (Fig. 16.9). This is mainly due to increased detection of asymptomatic infections through screening. The low rates in men suggest that many of the sex partners of women with chlamydia are not diagnosed or reported.

Symptoms

Chlamydial infections of the lower reproductive tract usually are mild or asymptomatic, especially in women. About 8–21 days after infection, men may experience a mild burning sensation on urination and a mucoid discharge. Women may have a vaginal discharge along with the symptoms of a urinary tract infection. Unfortunately, a physician mistakenly may diagnose it as a gonorrheal or urinary infection and prescribe the wrong type of antibiotic, or the person may never seek medical help. In these instances, there is a particular risk of the infection spreading from the cervix to the oviducts so that **pelvic inflammatory disease (PID)** results. This very painful condition can result in a block-

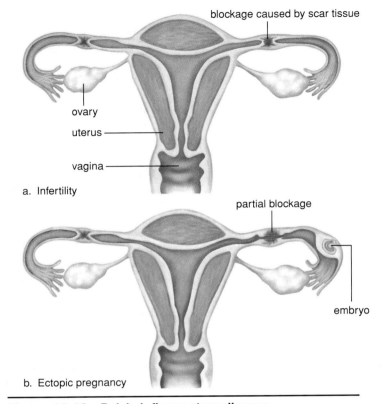

Figure 16.10   Pelvic inflammatory disease.
Following a chlamydial infection, the oviducts can be **(a)** completely blocked by scar tissue so that infertility results or **(b)** partially blocked so that fertilization occurs but the embryo is unable to pass to the uterus. The growing embryo can cause the oviduct to burst.

age of the oviducts with the possibility of sterility or ectopic pregnancy (Fig. 16.10).

Some believe that chlamydial infections increase the possibility of premature and stillborn births. If a newborn comes in contact with chlamydia during delivery, pneumonia or inflammation of the eyes can result. Erythromycin eyedrops at birth prevent this occurrence. If a chlamydial infection is detected in a pregnant woman, treatment with erythromycin should be used during pregnancy.

Diagnosis and Treatment

New and faster laboratory tests are now available for detection of a chlamydial infection. Their expense sometimes prevents public clinics from using them, however. It's been suggested that these criteria could help physicians decide which women should be tested: no more than 24 years old; having had a new sex partner within the preceding two months; having a cervical discharge; bleeding during parts of the vaginal exam; and using a nonbarrier method of contraception. A chlamydial infection is cured with the antibiotics tetracycline, doxycycline, and azithromycin; therefore, treatment should begin immediately.

PID and sterility are possible effects of a chlamydial infection in women. This condition may accompany a gonorrheal infection discussed next.

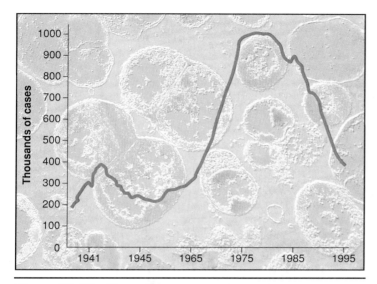

Figure 16.11 Gonorrhea.
A graph depicting the incidence of new cases of gonorrhea in the United States from 1941 to 1995 is superimposed on a micrograph of a urethral discharge from an infected male. Gonorrheal bacteria (*Neisseria gonorrhoeae*) occur in pairs; for this reason, they are called diplococci.

Source: Sexually Transmitted Disease Surveillance, 1995. Atlanta: Centers for Disease Control and Prevention, September 1996.

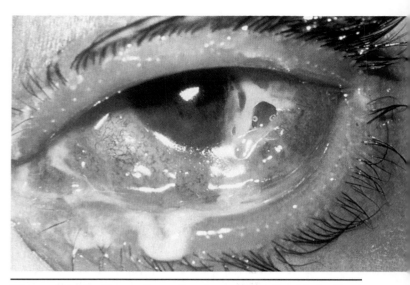

Figure 16.12 Secondary sites for a gonorrheal infection.
Gonorrhea infection of the eye is possible whenever the bacterium comes in contact with the eyes. This can happen when the newborn passes through the birth canal. Manual transfer from the genitals to the eyes is also possible.

Gonorrhea

Gonorrhea is caused by the bacterium *Neisseria gonorrhoeae*, which is a diplococcus, meaning that generally there are two spherical cells in close proximity. The incidence of gonorrhea has been declining since an all-time high in 1978 (Fig. 16.11). However, it can be noted that gonorrhea rates among African Americans is 28 times greater than the rate among Caucasians. Also, women using the birth-control pill have a greater risk of gonorrhea because the vagina is receptive to this pathogen, and the male does not need to wear a condom.

Symptoms

The diagnosis of gonorrhea in men is not difficult as long as they display typical symptoms (as many as 20% of men may be asymptomatic). The patient complains of pain on urination and has a milky urethral discharge three to five days after contact. In women, the bacteria may first settle within the vagina or near the cervix, from which they may spread to the oviducts. Unfortunately, the majority of women are asymptomatic until they develop severe pains in the abdominal region due to *PID* (*pelvic inflammatory disease*) (Fig. 16.10). PID from gonorrhea is especially apt to occur in women using an IUD (intrauterine device) as a birth-control measure.

PID from either a chlamydial or gonorrheal infection affects about one million women a year in the United States. The total number of ectopic pregnancies in the United States in 1992 was estimated at about 20 cases per 1,000 pregnancies, and many of these were most likely due to PID.

Gonorrhea proctitis, an infection of the anus, with symptoms including anal pain and blood or pus in the feces also occurs in patients. Oral/genital contact can cause infection of the mouth, throat, and the tonsils. Gonorrhea can spread to internal parts of the body, causing heart damage or arthritis. If, by chance, the person touches infected genitals and then his or her eyes, a severe eye infection can result (Fig. 16.12).

Eye infection leading to blindness can occur as a baby passes through the birth canal. Because of this, all newborn infants receive erythromycin eyedrops as a protective measure.

Transmission and Treatment

The chances of getting a gonorrheal infection are good. Women have 50–60% risk while men have a 20% risk of contracting the disease after even a single exposure to an infected partner. Therefore, the preventive measures listed in the introduction on page 333 should be followed.

Blood tests for gonorrhea are being developed, but in the meantime, it is necessary to diagnose the condition by microscopically examining the discharge of men or by growing a culture of the bacterium from either the male or the female to positively identify the organism. Because there is no blood test, it is very difficult to recognize asymptomatic carriers, who are capable of passing on the condition without realizing it. If the infection is diagnosed, gonorrhea can usually be cured using the antibiotics penicillin or tetracycline; however, resistance to antibiotic therapy is becoming more common. In 1994, resistance was noted in as many as 40% of strains tested.

Gonorrheal infections are presently on the decline; however, resistance to antibiotic therapy is becoming increasingly prevalent.

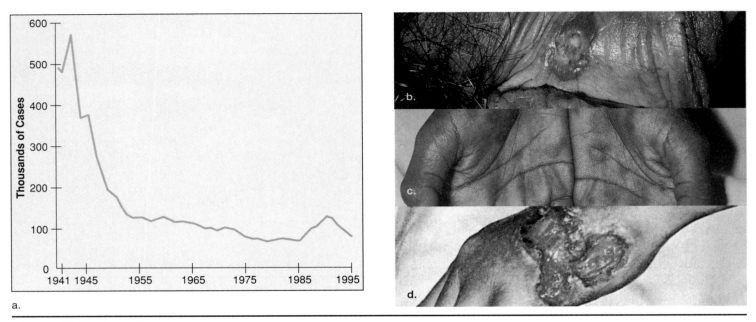

Figure 16.13   Syphilis.
a. A graph depicting the incidence of total cases of syphilis in the United States from 1941 to 1994. **b.** First stage of syphilis is a chancre where the bacterium enters the body. **c.** Second stage is a body rash that occurs even on the palms of the hands and soles of the feet. **d.** In the tertiary stage, gummas may appear on skin or internal organs.
Source: Sexually Transmitted Disease Surveillance, 1994. Atlanta: Centers for Disease Control and Prevention, September 1995.

Syphilis

Syphilis is caused by a bacterium called *Treponema pallidum*, an actively motile, corkscrewlike organism that is classified as a spirochete. The number of new cases of syphilis in 1995 were the fewest reported in the United States since 1977 (Fig. 16.13).

Syphilis has three stages, which can be separated by latent stages in which the bacteria are not multiplying. During the primary stage, a hard chancre (ulcerated sore with hard edges) indicates the site of infection. The chancre can go unnoticed, especially since it usually heals spontaneously, leaving little scarring. During the secondary stage, proof that bacteria have invaded and spread throughout the body is evident when the individual breaks out in a rash. Curiously, the rash does not itch and is seen even on the palms of the hands and the soles of the feet. There can be hair loss and infectious gray patches on the mucous membranes, including the mouth. These symptoms disappear of their own accord.

Not all cases of secondary syphilis go on to the tertiary stage. Some spontaneously resolve the infection and some do not progress beyond the secondary stage. During a tertiary stage, which lasts until the patient dies, syphilis may affect the cardiovascular system, and weakened arterial walls (aneurysms) are seen, particularly in the aorta. In other instances, the disease may affect the nervous system. The patient may become mentally impaired, blind, walk with a shuffle, or show signs of insanity. **Gummas,** large destructive ulcers, may develop on the skin or within the internal organs.

Congenital syphilis is caused by syphilitic bacteria crossing the placenta. The child is stillborn, or born blind, with many other possible anatomical malformations.

Diagnosis and Treatment

Diagnosis of syphilis can be made by blood tests or by microscopic examination of fluids from lesions. One blood test is based on the presence of reagin, an antibody that appears during the course of the disease. Currently, the most used test is Rapid Plasma Reagin (RPR), which is used for screening large numbers of test serums. Because the blood tests can give a false positive, they are followed up by microscopic detection. Dark-field microscopic examination of fluids from lesions can detect the living organism. Alternately, test serum is added to a freeze-dried *T. pallidum* on a slide. Any antibodies present that react to *T. pallidum* are detected by fluorescent-labeled antibodies to human gamma globulin. The organism will now fluoresce when examined under a fluorescence microscope.

Syphilis is a very devastating disease. Control of syphilis depends on prompt and adequate treatment of all new cases; therefore, it is very important for all sexual contacts to be traced so that they can be treated. The cure for all stages of syphilis is some form of penicillin. To prevent transmission, the general instructions given in the introduction (see p. 333) should be faithfully followed.

Syphilis is a very devastating sexually transmitted disease, which is curable by antibiotic therapy. The chance of it developing resistance is always a threat.

STUDYING THE CONCEPTS

1. Describe the structure and life cycle of a DNA virus.   334

2. Describe the cause and symptoms of an HIV infection.   336–37

3. Among which groups of society is AIDS now increasing most rapidly? How might transmission be prevented?   336

4. Give the cause and symptoms for genital warts, genital herpes, and a heptitis B infection.   337–39

5. Discuss the treatment for STDs caused by viruses. How might these viruses be prevented from spreading?   339

6. State the three shapes of bacteria, and describe the structure of a prokaryotic cell.   340

7. Describe the symptoms and results of a chlamydial infection and gonorrhea in men and in women. What is PID, and how does it affect reproduction?   342–43

8. Describe the three stages of syphilis.   344

9. How does the newborn acquire an infection of AIDS, herpes, chlamydia, gonorrhea, or syphilis? What effects do these infections have on infants?   337, 338, 342, 343, 344

10. Describe the symptoms of vaginitis and pubic lice.   345

APPLYING YOUR KNOWLEDGE

Concepts

1. In the United States, the total number of STDs of bacterial origin remains about the same each year, but the total number of STDs of viral origin increases each year. Explain the reason for this.

2. Some animal viruses are said to have developed disguises that make it more difficult for the host immune system to recognize them as foreign objects. How is this done?

3. How may the advent of the birth-control pill have resulted in an increase in STDs?

4. If a young female has contracted genital warts, why should she be sure to have a yearly pap test?

Bioethical Issue

When it comes to sexually transmitted diseases (STDs), information is critical. In order to avoid these diseases, you've got to know the methods of prevention. Sex education classes can explain, for example, how well condoms prevent disease. Classes also can detail how each disease is carried, how to avoid it, etc.

But some people worry that sex education doesn't belong in the school. Parents may not want teachers discussing private details of sex-related diseases with their children. That, they argue, is a parent's job.

What do you think? Are schools responsible for educating kids about sexually transmitted diseases? Or is that something parents should do? Is it fair to ask students to attend courses on private matters like sex? Explain the reasoning for your answers.

TESTING YOUR KNOWLEDGE

1. All viruses have an inner core of _____ and a capsid of _____ .

2. AIDS is caused by a type of RNA virus known as a _____ .

3. Although it is a sexually transmitted disease, intravenous drug users who share needles get AIDS because the virus lives in _____ cells.

4. Herpes simplex virus type 1 causes _____ , and type 2 causes _____ .

5. Bacterial cells have one of three shapes: _____ , _____ , and _____ .

6. Women are often asymptomatic for a chlamydial infection until they develop _____ .

7. The use of a _____ can help reduce the risk of acquiring an STD.

8. In the tertiary stage of syphilis, there may be large sores called _____ .

9. These three sexually transmitted diseases are usually curable by antibiotic therapy: _____ , _____ , and _____ .

10. Women who take the birth-control pill are more likely to acquire vaginitis due to a _____ infection.

11. Identify the STD by these symptoms:

    a. _____ milky discharge

    b. _____ swollen lymph glands and night sweats

    c. _____ chancre, followed by nonitching rash

SELECTED KEY TERMS

acquired immunodeficiency syndrome (AIDS) Disease caused by HIV that is characterized by failure of the immune system.   336

antibiotic Medicine that specifically interferes with bacterial metabolism and in that way cures humans of a bacterial disease.   341

bacterium Unicellular organism lacking a membrane-bound nucleus and the other membranous organelles typical of eukaryotic cells.   340

binary fission Reproduction by division into two daughter cells without the utilization of a mitotic spindle.   340

chlamydia (klah-MID-ee-ah) A sexually transmitted disease caused by a tiny bacterium of the same name and particularly characterized by urethritis.   342

endospore Resistant body formed by bacteria when environmental conditions worsen.   340

genital herpes Sexually transmitted disease characterized by open sores on the external genitalia.   338

genital warts Sexually transmitted disease caused by papillomavirus and characterized by warts on the genitals.   337

gonorrhea (gahn-ah-REE-ah) Contagious sexually transmitted disease caused by bacteria and leading to inflammation of the urogenital tract.   343

gumma (GUM-ah) Large unpleasant sore that may occur during the tertiary stage of syphilis.   344

hepatitis Inflammation of the liver.   339

herpes simplex virus Cold sore virus (type 1) and genital herpes virus (type 2).   338

host Organism a parasite lives on or in, to obtain nutrients.   334

human immunodeficiency virus (HIV) Virus responsible for AIDS.   336

human papillomavirus (HPV) (pap-ih-LOH-mah-vy-rus) Human genital wart virus.   337

opportunistic infection Infection that has an opportunity to occur because the immune system has been weakened.   336

pathogen Disease-causing agent.   335

pelvic inflammatory disease (PID) Disease state of the reproductive organs usually caused by a chlamydial infection or gonorrhea.   342

retrovirus Virus that contains only RNA and carries out RNA to DNA transcription, called reverse transcription.   335

syphilis (SIF-ih-lus) Chronic, contagious sexually transmitted disease caused by a bacterium that is a spirochete.   344

virus Nonliving, obligate, intracellular parasite consisting of an outer capsid and an inner core of nucleic acid.   334

yeast Small unicellular fungus.   345

AIDS
Supplement

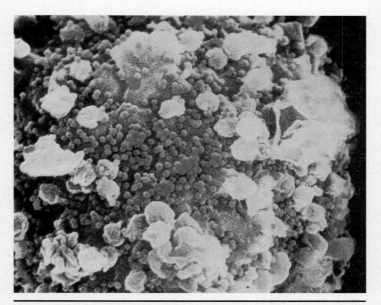

Figure S.1 HIV (green) budding from an infected T lymphocyte.

Figure S.1 shows HIV (human immunodeficiency virus) budding from a T lymphocyte, the primary host for these viruses. No wonder the immune system falters in a person with an HIV infection. The very cells that orchestrate the immune response are under viral attack. As a society, can we afford expensive multidrug therapy for all those expected to become infected? Will there instead be a vaccine? If so, B lymphocytes would prepare a supply of antibodies to disarm HIV before the viruses have a chance to enter T lymphocytes. It would be a feat. There might be something about the virus—mode of transmission or the course of the disease—that will make any type of vaccine ineffective. The burden it seems, is on the individual. We all must come to realize the importance of our T lymphocytes and take measures to protect them from possible destruction by HIV. The various ways to prevent infection discussed on page 356 should be followed faithfully.

An HIV infection leads to AIDS (acquired immunodeficiency syndrome). The full name of AIDS can be explained in this way: *acquired* means that the condition is caught rather than inherited; *immune deficiency* means that the virus attacks the immune system so there is greater susceptibility to certain opportunistic infections and cancer; and *syndrome* means that some fairly typical infections and cancers usually occur in the infected person. These are the conditions that cause AIDS patients to die.

An HIV infection precedes the occurrence of AIDS by several years. At first, the body is able to deal effectively with the infection, and only later do symptoms arise that permit a diagnosis of AIDS. Therefore, the number of persons diagnosed as having AIDS is much smaller than those that are actually infected with HIV. In the United States, over one million (1/300 people) are believed to be infected with the virus, but only about 515,000 individuals have been diagnosed as having AIDS. The *U.S. Center for Disease Control and Prevention (CDC)* is now using a new definition for AIDS that allows more persons to be counted as having AIDS. The new definition includes persons with severely impaired immune systems without the necessity of their having come down with a life-threatening disease.

An HIV infection precedes the condition of AIDS by several years. AIDS is characterized by a severely impaired immune system and very serious illnesses.

S.1   Origin of AIDS

It's generally accepted that HIV originated in Africa and then spread to the United States and Europe by way of the Caribbean. Even today, there are monkeys in Africa infected with immunosuppressive viruses that could have mutated to HIV after humans ate monkey meat.

Most likely, HIV entered the United States on numerous occasions as early as the 1950s. Presently, the first documented case is a 15-year-old male who died in Missouri in 1969 with skin lesions now known to be characteristic of an AIDS-related cancer. Doctors froze some of his tissues because they could not identify the cause of death. Recently, these tissues were examined and found to be infected with HIV. Researchers also want to test the preserved tissue samples of a 49-year-old Haitian who died in New York during 1959 of the type of pneumonia now known to be AIDS related.

British scientists were able to show by examining preserved tissues that a Manchester seaman most probably died in 1959 of AIDS. This may be one of the first cases of AIDS because scientists believe the immunosuppressive monkey virus may have evolved into HIV during the late 1950s.

During the 1960s, it was the custom to list leukemia as the cause of death in immunodeficient patients. Most likely some of these people actually died of AIDS. Since HIV is not extremely infectious, it took several decades for the number of AIDS cases to increase to the point that AIDS

TABLE S.1
Transmission of HIV-1

Possible Routes

Homosexual and heterosexual contact

Intravenous drug use

Transfusion (unlikely in U.S.)

Crossing placenta during pregnancy; breast-feeding

Increasing the Risk

Promiscuous behavior (large number of partners, sex with a prostitute)

Drug abuse with needle sharing

Presence of another sexually transmitted disease

---

became recognizable as a specific and separate disease. The name AIDS was coined in 1982, and HIV was found to be the cause of AIDS in 1983–84.

HIV-1, the virus that causes most cases of AIDS in the United States, spread rapidly through the population during the early to mid-1980s. During this early period, all age groups were equally at risk, but now the incidence of new infections among older individuals has stabilized while the number among young people is still increasing.

The virus (HIV-1) that causes most cases of AIDS in the United States took hold in the early 1980s.

Prevalence of AIDS Today

Estimates vary, but as many as 20 million people worldwide may now be infected with HIV, and there could be as many as 40 million individuals infected by the end of the century. A new infection is believed to occur every 15 seconds, the majority from heterosexual contact. HIV infections are not distributed equally throughout the world. Since AIDS arose in Africa, it is not surprising that most infected people live in Africa (66%), but new infections are now occurring at the fastest rate in Southeast Asia and the Indian subcontinent, areas which now account for 16% of all those infected. The explosion of the epidemic in Southeast Asia and India is attributed to a commercial sex industry. A study conducted in northern Thailand estimated that of newly infected military recruits, most have a history of sexual contact with female prostitutes.

As many as 40 million persons worldwide may acquire an HIV infection by the end of the century.

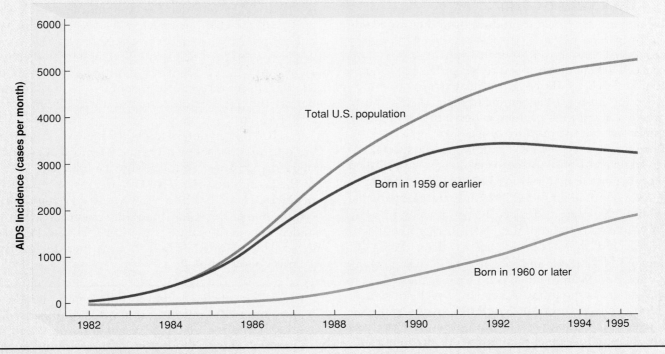

Figure S.2   Monthly AIDS incidence per age.
The rate of AIDS incidence is slowing down in the overall U.S. population and for individuals born in or before 1959. The AIDS incidence for individuals born after 1960 continues to rise.

In the United States, AIDS cases are concentrated in certain large cities such as New York City, Los Angeles, Chicago, and Washington, D.C., but AIDS is spreading to smaller cities and rural areas as well. Caucasians and African Americans each have about 40% of all AIDS cases. But because African Americans are a minority, they have a higher prevalence rate compared to Caucasians. One of every 50 African-American males may now be infected with HIV. Hispanics represent only about 15% of all AIDS cases but again have a higher prevalence rate compared to Caucasians.

In the United States, one-quarter of all new AIDS cases are now occurring among people born after 1960, and the incidence is still rising, whereas the incidence has stabilized in those born before 1960 (Fig. S.2). All young people should be aware that their age group will account for most of new HIV infections. One infected person can spread the disease to others, who then spread it to others, and soon a large number of persons have been infected. As of 1992, AIDS had become the leading cause of death for U.S. men aged 25–44 years and the third leading cause of death among women in this age group. However, a newly infected patient is most apt to be a young African-American woman.

At Risk Behaviors

Table S.1 lists the most frequent ways by which HIV is transmitted. HIV is transmitted by sexual contact with an infected person, including vaginal or rectal intercourse and oral/genital contact. Also, needle-sharing among intravenous drug users is high-risk behavior. A less common mode of transmission (and now rarely in countries where blood is screened for HIV) is through transfusions of infected blood or blood-clotting factors. Babies born to HIV-infected women may become infected before or during birth, or through breast-feeding after birth.

HIV first spread through the homosexual community, and male-to-male sexual contact still accounts for the largest percentage of new AIDS cases in the United States. But the largest increases of HIV infections are occurring through heterosexual contact or by intravenous drug use. Now, women account for 19% of all newly diagnosed cases of AIDS. The rise in the incidence of AIDS among women of reproductive age is paralleled by a rise in the incidence of AIDS in children younger than 13. The chance of an infected mother passing HIV to her unborn child is thought to be about 15–25%.

Behaviors that help prevent transmission are discussed later in this supplement. These behaviors should be followed by everyone in order to prevent the spread of AIDS.

Certain large cities and minorities have a higher prevalence rate of AIDS. Young adults have an increasing incidence of AIDS compared to older adults.

New HIV infections are increasing faster among heterosexuals than homosexuals; women now account for nearly 20% of new AIDS cases.

S.2 Life Cycle of HIV

HIV, like all viruses, do not reproduce until they are inside a host cell (Fig. S.3). HIV are *retroviruses*, meaning that their genetic material consists of RNA instead of DNA. For HIV-1 to enter a cell, its envelope spike, called gp120 (gp for glycoprotein and 120 for its weight), must bind with a CD4 receptor in the host-cell plasma membrane ①. It appears that a plasma membrane protein, called a fusion cofactor, is also necessary to the binding process.

Once inside the host cell ②, HIV uses a special enzyme called reverse transcriptase to make a DNA copy, called cDNA, of its genetic material ③. This single strand replicates ④. Now double-stranded DNA integrates into a host chromosome ⑤ where it directs the production of more viral RNA ⑥. Each strand of viral RNA brings about synthesis of protein, most of which are capsid proteins ⑦. Assembly of RNA strands and capsids now occurs ⑧. Many viruses then bud from the host cell before it dies ⑨.

HIV destroys the immune system, which consists of white blood cells and certain lymphoid organs and usually protects us from disease. The immune system protects us by producing *antibodies,* proteins that attack foreign substances called antigens, and by attacking infected cells outright. HIV-1 is an antigen, and a cell that harbors the virus is an infected cell to which the immune system should react.

How does HIV destroy the mmune system? The specific hosts for HIV-1 are *helper T lymphocytes*, which stimulate both B lymphocytes to produce antibodies and *cytotoxic T lymphocytes* to attack and kill virus-infected cells. Macrophages, which present *antigens* to helper T lymphocytes, are also under attack. CD4 receptors are found on a number of other cells involved in immunity and on some epithelial cells.

HIV-1 often enters the body by way of vaginal, rectal, or oral mucosal membranes. Macrophages present in these membranes are the first to be infected. They move to the lymph nodes where HIV begins to replicate itself inside of CD4 T lymphocytes. Many HIV remain in local lymph nodes for some time; but eventually, the lymph nodes degenerate, and large numbers of HIV enter the general bloodstream, where they are then detected as CD4 T lymphocytes die off.

> **HIV-1 is a retrovirus that infects immune cells carrying a CD4 receptor and a fusin cofactor. The body fights the disease for several years and then eventually loses the battle as the immune system collapses, judged by decreasing blood concentration of CD4 T lymphocytes.**

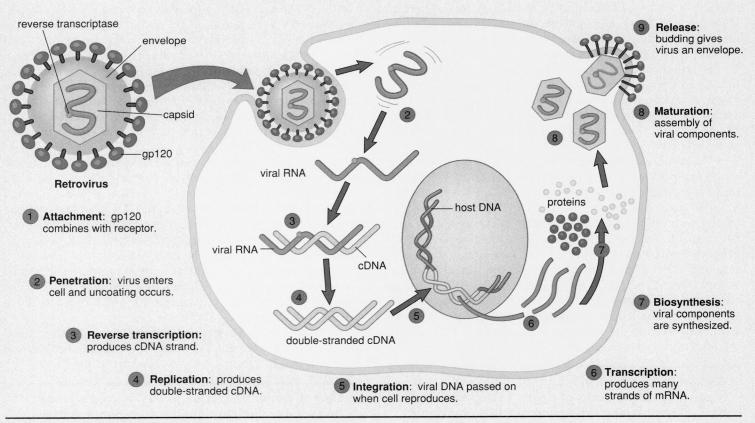

Figure S.3   Reproduction of HIV-1.
HIV-1 is a retrovirus that utilizes reverse transcription to produce cDNA (DNA copy of RNA genes). cDNA integrates into the cell's chromosomes before it reproduces and buds from the cell.

Health Focus

The HIV Blood Test

It is now well recognized that acquired immunodeficiency syndrome (AIDS) is caused primarily by a human immunodeficiency virus (HIV-1) that infects T lymphocytes. Even though the virus attacks the immune system, B cells do begin to produce antibodies within weeks or months after the infection begins. All of the commonly used tests detect the presence of these antibodies by having them react to an antigen. The tests use plastic beads that are coated with either inactivated virus or just specific viral envelope proteins (Fig. S.A).

At the start of the test, an antigen-coated bead is placed in a reaction tube. The serum sample to be tested is added. If the specimen contains HIV-1 antibodies, they bind to the bead and remain bound when the bead is washed and rinsed. These antibodies are designated as Ab_1 because they are the first antibodies attached to the bead.

The next step is to get a color reaction that can be read photometrically by a laboratory instrument. First, a conjugate designated as Ab_2-Enz is added to the tube. The conjugate is made up of an antibody called antihuman globulin attached to an enzyme. Ab_1 is a human globulin; therefore, the Ab_2-Enz will bind to Ab_1 on the bead. After washing and rinsing, a substrate is added that will react with the enzyme and cause a color change. The intensity of the color reaction is compared to a known negative test result (no HIV-1 antibody is present in the specimen) and a known positive test result (HIV-1 antibody is present in the specimen).

The test we have described is called an HIV-1 antibody *Enzyme-Linked-ImmunoSorbent Assay* (ELISA). Note that since this is a test for HIV-1 antibody, it cannot detect persons who are infectious but have not yet produced HIV-1 antibodies. Nor can it reliably detect persons infected with HIV-2, another HIV that can cause AIDS. On the other hand, false positive results are possible in persons suffering from diseases other than AIDS. To guard against false positive results, it is a common practice to do a more specialized and expensive test on specimens that give a positive ELISA test. Patients who give a positive result with both HIV-1 antibody ELISA and with the confirmatory test are told they have an HIV infection and that they should seek immediate treatment.

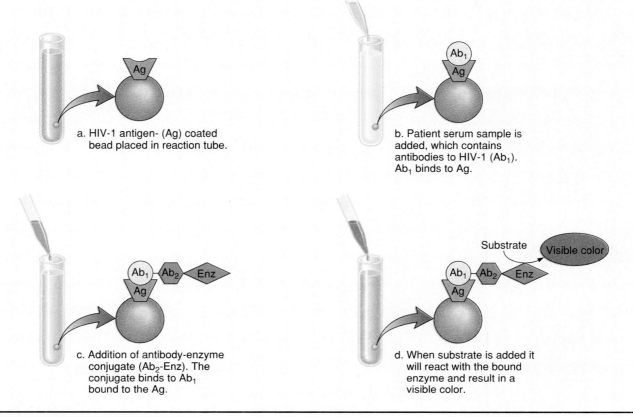

a. HIV-1 antigen- (Ag) coated bead placed in reaction tube.

b. Patient serum sample is added, which contains antibodies to HIV-1 (Ab_1). Ab_1 binds to Ag.

c. Addition of antibody-enzyme conjugate (Ab_2-Enz). The conjugate binds to Ab_1 bound to the Ag.

d. When substrate is added it will react with the bound enzyme and result in a visible color.

Figure S.A   A positive HIV blood test.
a. An antigen-coated bead is placed in a reaction tube. **b.** HIV-1 antibodies (Ab_1) in the specimen bind to HIV-1 antigens on the bead. **c.** Antibody (Ab_2)-enzyme conjugate binds to Ab_1. **d.** A substrate for the enzyme is added. When the enzyme acts on the substrate, a color change occurs that signifies the test is positive. Washing and rinsing must take place between *b* and *c* and between *c* and *d* to rid the tube of any unbound antibodies. If the test is negative instead of positive, all antibodies are washed away and there is no color change.

S.3 Stages of an HIV Infection

The Centers for Disease Control and Prevention now recognize three categories of an HIV-1 infection. AIDS is the end stage of the infection.

Category A

During this stage, the CD4 T lymphocyte count is 500 per mm³ or greater (Fig. S.4). For a period of time after the initial infection with HIV, people don't usually have any symptoms at all. A few (1–2%) do have mononucleosis-like symptoms that may include fever, chills, aches, swollen lymph nodes, and an itchy rash. These symptoms disappear, however, and there are no other symptoms for quite some time. Although there are no symptoms, the person is highly infectious. Large numbers of infectious viruses may be present in the plasma, but the HIV blood test, which is described in the Health Focus reading on page 353, is not yet positive. It tests for the presence of antibodies and not for the presence of HIV itself. This means that HIV can still be transmitted even when the HIV blood test is negative.

As mentioned, HIV is at first sequestered within the lymph nodes, which may appear swollen in some individuals. The immune system is producing antibodies, the blood test is positive, but the number of viruses in circulation is quite low. Although it may seem as if the HIV infection is under control, actually there is a great unseen battle going on. The body is staying ahead of the hordes of viruses entering the blood by producing as many as one to two billion new helper T lymphocytes each day. As long as the body is able to keep pace with the virus, the number of detected viruses in the blood and the number of infected helper T lymphocytes stays low and the individual stays healthy.

Category B

Several months to several years after infection, the infected individual will probably progress to category B. During this stage, the T lymphocyte count is 200 to 499 per mm³ and symptoms most likely will begin to appear. There will be swollen lymph nodes in the neck, armpits, or groin that persist for three months or more. Other symptoms that indicate category B are severe fatigue not related to exercise or drug use; unexplained persistent or recurrent fevers, often with night sweats; persistent cough not associated with smoking, a cold, or the flu; and persistent diarrhea. Also possible are signs of nervous system impairment, including loss of memory, inability to think clearly, loss of judgment, and/or depression.

When the individual develops non-life-threatening but recurrent infections, it is a signal that full-blown AIDS will occur shortly. One possible infection is thrush, a *Candida albicans* infection that is identified by the presence of white spots and ulcers on the tongue and inside the mouth. The fungus may also spread to the vagina, resulting in a chronic infection there. Another frequent infection is herpes simplex, with painful and persistent sores on the skin surrounding the anus, the genital area, and/or the mouth.

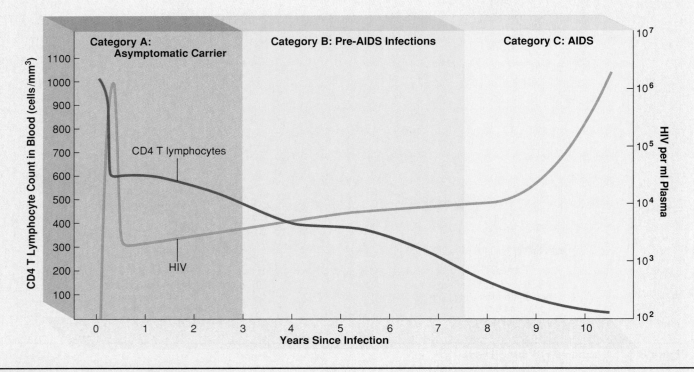

Figure S.4 Stages of an HIV infection.
In category A individuals, the number of HIV in plasma rises upon infection and then falls. The number of CD4 T lymphocytes falls, but stays above 400 per mm³. In category B individuals, the number of HIV in plasma is slowly rising and the number of T lymphocytes is decreasing. In category C individuals, the number of HIV in plasma rises dramatically as the number of T lymphocytes falls below 200 per mm³.

Category C

Until recently, the majority of infected persons have proceeded to category C. During this stage, the CD4 T lymphocyte count falls below 200 per mm^3 and the lymph nodes have degenerated. The patient, who is now suffering from "slim disease" (as AIDS is called in Africa), experiences severe weight loss and weakness due to persistent diarrhea and coughing, and will most likely succumb to one of the opportunistic infections. An **opportunistic infection** is one that has the *opportunity* to occur only because the immune system is severely weakened. Persons with AIDS die from one or more of these diseases and not from the HIV infection itself. Some of the opportunistic infections are the following:

- *Pneumocystis carinii* pneumonia. The *lungs* become useless as they fill with fluid and debris due to an infection with this organism. There is not a single documented case of *P. carinii* pneumonia in a person with normal immunity.

- *Mycobacterium tuberculosis.* This bacterial infection, usually of the *lungs*, is seen more often as an infection of lymph nodes and other organs in patients with AIDS. Of special concern, tuberculosis is spreading into the general populace and is multidrug resistant.

- Toxoplasmic encephalitis is caused by a one-cell parasite that lives in cats and other animals as well as humans. Many persons harbor a latent infection in the *brain* or muscle, but in AIDS patients, the infection leads to loss of brain cells, seizures, weakness, or decreased sensation on one side of the body.

- Kaposi's sarcoma is an unusual *cancer* of blood vessels, which gives rise to reddish purple, coin-size spots and lesions on the skin.

- Invasive cervical cancer. This *cancer* of the cervix spreads to nearby tissues. This condition has been added to the list because the incidence of AIDS has now increased in women.

Drugs have been developed to deal with opportunistic diseases in AIDS patients. These drugs help people with AIDS lead a fairly normal life for some months, but eventually, patients are repeatedly hospitalized due to weight loss, constant fatigue, and multiple infections. Death usually follows in 2–4 years.

During the asymptomatic stage of an HIV infection, the person is highly infectious although there may be no symptoms; during pre-AIDS, there may be swollen lymph nodes and various infections; during the last stage, which is called AIDS, the patient usually succumbs to an opportunistic infection.

S.4 Treatment for AIDS

There is no cure for AIDS, but there are two types of drugs presently available for treatment. The well-known drug called AZT and several others are analogs of nucleotides. Analogs are chemically altered forms so that when reverse transcriptase chooses them instead of a normal nucleotide, reverse transcription stops and viral DNA is not produced. The other type of drug, called a proteinase inhibitor, blocks the action of the viral enzyme called proteinase, which is required for normal viral assembly. When HIV proteinase is blocked, the resulting virus lacks the capacity to cause infection.

Now that it is apparent that HIV-1 does not have a latent stage, physicians believe that drug therapy should begin as soon as an HIV infection is detected. Recently, multidrug therapy with the analogs AZT and 3TC and a proteinase inhibitor has met with encouraging success. The use of two analogs at the same time seems to retain the sensitivity of HIV to both of them. Otherwise, the virus will probably mutate to a resistant strain. It is important to realize that these drugs are very expensive and they must be taken indefinitely or else the virus will most likely rebound. Also, although HIV may apparently disappear, the immune system is still impaired.

In addition to interfering with viral reproduction, it may be possible to help the immune system counter the virus by injecting the patient with a stimulatory cytokine like interferon or interleukin-2. Thus far, this approach has not met with great success, but researchers are still hopeful. A novel idea is to administer antibodies against the CD4 receptor. Perhaps these antibodies will attach to CD4 receptors in helper T lymphocyte plasma membranes, and then HIV would not be able to gain entry. Chemicals called chemokines are now known to attach to the fusin receptors, and perhaps they or some other agent that blocks these receptors could also be used as a medication to prevent HIV infection of susceptible cells.

Many investigators are working on a vaccine for AIDS. Some are trying to develop a vaccine in the traditional way. Traditionally, vaccines are made by treating a pathogen chemically, thereby weakening it so that it can be injected into the body without causing disease. Others are working on subunit vaccines that utilize just a single HIV protein, such as gp120, as the vaccine. Recently, researchers were heartened to find that an injection of the DNA for gp120 alone can act as a vaccine. After entering a cell, the DNA causes the viral protein to appear in the plasma membrane. Chimpanzees are showing a strong immune reaction, and human trials have begun.

Combination drug therapy has met with encouraging success against an HIV infection. Immunotherapy and the possibility of a vaccine are still being pursued.

a. AIDS patient, Tom Moran, July 1987

c. AIDS patient, Tom Moran, late January 1988

b. AIDS patient, Tom Moran, early January 1988

Figure S.5   The course of an AIDS infection.
These photos show the effect of an HIV infection in one individual who progressed through all the stages of AIDS. All possible means should be taken to prevent an HIV infection in the first place.

S.5   Preventing Transmission of HIV

Sexual behaviors and drug-related activities are the major means by which AIDS is transmitted in the United States (see Table S.1). HIV is spread by passing virus-infected macrophages and T lymphocytes found in body secretions or in blood from one person to another.

Sexual activities transmit HIV. Therefore,

1. Abstain from sexual intercourse or develop a long-term monogamous (always the same partner) sexual relationship with a partner who is free of HIV.

2. Refrain from multiple sex partners or having relations with someone who has multiple sex partners. If you have sex with two other people and each of these has sex with two people and so forth, the number of people who are relating is quite large.

3. Remember that the prevalence of AIDS is presently higher among homosexuals and bisexuals than among those that are heterosexual. Also, having relations with an intravenous drug user is risky because the behavior of this group risks AIDS. In addition, anyone who already has another sexually transmitted disease is more likely to contract AIDS.

4. Avoid anal-rectal intercourse during which the penis is inserted into the rectum because this behavior increases the risk of infection. The lining of the rectum is thin, and infected T lymphocytes can easily enter the body here. Also, the rectum is supplied with many blood vessels, and insertion of the penis into the rectum is likely to cause tearing and bleeding that facilitate the entrance of HIV. The vaginal lining is thick and difficult to penetrate, but the lining of the uterus is only one cell thick and

does allow infected T lymphocytes to enter. Uncircumcised males are more likely than circumcised males to become infected because vaginal secretions can remain under the foreskin for a long period of time.

5. Practice safe sex. If you do not know for certain that your partner has been free of HIV for the past five years, always use a latex condom during sexual intercourse. Be sure to follow the directions supplied by the manufacturer. Use of a water-based spermicide containing nonoxynol-9 in addition to the condom can offer further protection because nonoxynol-9 also kills the virus and virus-infected lymphocytes.

6. Avoid fellatio (kissing and insertion of the penis into a partner's mouth) and cunnilingus (kissing and insertion of the tongue into the vagina) because they may be a means of transmission. The mouth and gums often have cuts and sores that facilitate the entrance of infected T lymphocytes.

Drug use transmits HIV. Therefore,

1. Stop, if necessary, or do not start the habit of injecting drugs into your veins.

2. If you are a drug user and cannot stop your behavior, then always use a new sterile needle for injection or one that has been cleaned in bleach.

3. Aside from intravenous drug use, do not use alcohol or any drugs in a way that may prevent you from being able to control your behavior.

At present, there is no way to stop the AIDS epidemic (Fig. S.5). But these behaviors will help prevent the spread of AIDS and decrease the projected number of new cases.

Chapter 17

Development and Aging

Chapter Outline

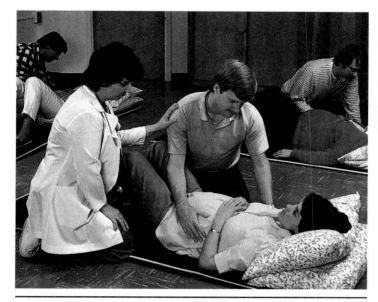

Figure 17.1 Childbirth classes.
At childbirth classes, both prospective parents learn how to facilitate the birthing process.

"At last," thought Mary Jean as she left the doctor's office. "I'm going to have a baby." Bill was sure to be overjoyed and so would the prospective grandparents. Bill and Mary Jean had been trying to have a baby for three years with no success. Recently, they had been talking to their doctor about artificial insemination, in vitro fertilization, and other methods of getting pregnant. It was a wonderful surprise and a relief to find that the pregnancy had happened naturally.

Now there are many decisions for Mary Jean. "Where should I have the baby—at a birthing center or hospital? Should I breast- or bottle-feed? How much weight can I gain?" She would discuss all these concerns with Bill and have them sign up soon for childbirth classes (Fig. 17.1). She wanted Bill to be with her every step of the way.

Thinking back over her Human Biology course, Mary Jean knew that the embryo had implanted itself by now in the uterine wall. At two months it would have a human appearance and become a fetus floating within a watery sac. While there, the fetus would be completely dependent upon the placenta, a special structure through which nourishment and oxygen are supplied and wastes are removed. But only at eight months or more would her child have an excellent chance of becoming an independently functioning human being. She was glad to have a doctor who would oversee her health while she waited for that all important day.

This chapter is about the events and the processes that occur as a single cell becomes a complex organism. The reproductive systems were discussed in a previous chapter. A male continually produces sperm in the testes, a female releases one egg a month from the ovaries. The release of the egg, called ovulation, usually occurs on about day 14 of an ovarian cycle that lasts 28 days. The hormones produced by the ovary control the uterine cycle, which consists of menstruation, a proliferative phase during which the endometrium (uterine lining) thickens due to the increased production of glands and blood vessels, and a secretory phase during which the endometrium secretes nutrient substances to sustain a developing embryo. These two phases prepare the uterus to accept the fertilized egg (zygote).

This time for Mary Jean and Bill, everything went perfectly. The 200 to 600 million sperm released by Bill during sexual intercourse were ejaculated against the opening of the cervix at the far end of the vagina. The semigelatinous seminal fluid protected the sperm from the acid of the vagina for several minutes, and many managed to enter the uterus. Whether or not sperm enter the uterus depends in part on the consistency of the cervical mucus. Three to four days prior to ovulation and on the day of ovulation, the mucus is watery, and the sperm can penetrate it easily. During the other days of the uterine cycle, the mucus is thicker and has a sticky consistency, and the sperm can rarely penetrate it. Some high-speed sperm reach the oviducts in only 30 minutes, but generally, the journey from vagina to oviduct takes several hours. Many sperm get lost along the way, and only several hundred sperm ever reach the oviducts.

17.1 Fertilization

Fertilization (Fig. 17.2) normally occurs in the upper third of the oviduct, where the sperm encounter an egg if ovulation has occurred. The sperm must swim against the downward current created by the ciliary action of the epithelium lining the oviducts. However, it is believed that uterine and oviduct contractions help transport the sperm and that prostaglandins within seminal fluid promote these contractions.

Fertilization occurs after a sperm and egg have interacted. A sperm has three distinct parts: a head, a middle piece, and a tail. The tail is a flagellum, which allows the sperm to swim toward the egg, and the middle piece contains energy-producing mitochondria. The head contains a haploid nucleus capped by a membrane-bounded acrosome containing enzymes that allow the sperm to penetrate the egg.

Once sperm have made their way past the corona radiata, layers of nutrient cells around the egg, several mechanisms assure fertilization takes place in a species-specific manner (Fig. 17.2). The egg has a plasma membrane, a vitelline membrane, and a jelly coat called the *zona pellucida*. The acrosomal enzymes digest a hole in the zona pellucida as the acrosome extrudes a filament that attaches to receptors located in the vitelline membrane. This is a lock-and-key reaction that is species-specific. Then, the egg

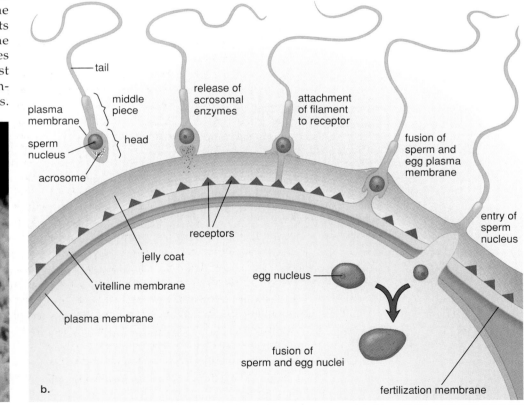

Figure 17.2 Fertilization.

a. During fertilization, a single sperm enters the egg. **b.** A head of a sperm has a membrane-bound acrosome filled with enzymes. When released, these enzymes digest away the jelly coat around the egg, and the acrosome extrudes a filament that attaches to a receptor on the vitelline membrane. Now the sperm nucleus enters and fuses with the egg nucleus, and the resulting zygote begins to divide. The vitelline membrane becomes the fertilization membrane, which prohibits any more sperm from entering the egg.

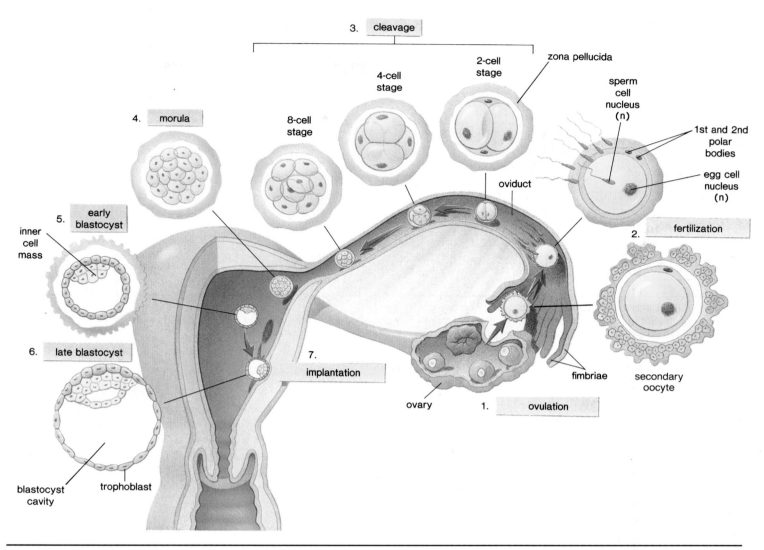

Figure 17.3 Human development before implantation.
Structures and events proceed counterclockwise. At ovulation, the secondary oocyte leaves the ovary. A single sperm penetrates the zona pellucida, and fertilization occurs in the oviduct. As the zygote moves along the oviduct, it undergoes cleavage to produce an embryo that implants itself in the uterine lining.

plasma membrane and sperm plasma membrane fuse, allowing the sperm nucleus to enter. Fusion of the nuclei takes place and the zygote begins development.

As soon as the plasma membrane of sperm and egg fuse, the plasma membrane and the vitelline membrane undergo changes that prevent the entrance of any other sperm. The vitelline membrane is now called the fertilization membrane.

Following fertilization, the fertilized egg, more properly called the zygote, begins dividing. The developing embryo travels very slowly down the oviduct to the uterus, where after two to three days it implants itself in the prepared uterine lining (Fig. 17.3). Upon implantation, the woman is pregnant. If implantation takes place, the uterine lining is maintained, because the membrane surrounding the embryo begins to produce a hormone called HCG (human chorionic gonadotropin) that prevents degeneration

of the corpus luteum. Pregnancy tests, which are readily available in hospitals, clinics, and now even drug and grocery stores, are based on the fact that HCG is present in the blood and urine of a pregnant woman. There is also an early detection test that can be done in a doctor's office, and the results are available within the hour. The signs that often prompt a woman to have a pregnancy test are cessation of menstruation, increased frequency of urination, morning sickness, and increase in the size and fullness of the breasts, as well as darkening of the areolae, a ring of pigmented epithelium surrounding the nipples.

Following fertilization of an egg by a sperm, the zygote begins developing, and the embryo implants itself in the uterine lining. Pregnancy has now taken place.

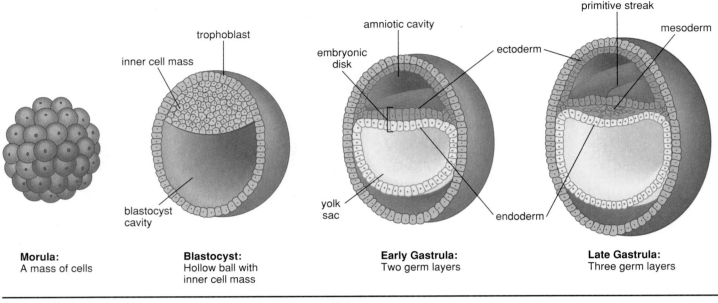

Morula:
A mass of cells

Blastocyst:
Hollow ball with
inner cell mass

Early Gastrula:
Two germ layers

Late Gastrula:
Three germ layers

Figure 17.4   Early developmental stages in cross section.
Cleavage results in the inner cell mass. Morphogenesis occurs as cells rearrange themselves, and differentiation is exemplified first by the formation of three different germ layers and then by the formation of organs such as the neural tube.

17.2   Human Development Before Birth

This section considers the major processes and events that take place from the time of fertilization to the time of birth.

Processes of Development

Embryonic development includes these processes:

Cleavage   Immediately after fertilization, the zygote begins to divide so that at first there are 2, then 4, 8, 16, and 32 cells, and so forth. Increase in size does not accompany these divisions.

Morphogenesis   *Morphogenesis* refers to the shaping of the embryo and is first evident when certain cells are seen to move, or migrate, in relation to other cells. By these movements, the embryo begins to assume various shapes.

Differentiation   *Differentiation* occurs as cells take on a specific structure and function. Nerve cells have long processes that conduct nerve impulses, and muscle cells contain contractile elements.

Growth   During embryonic development, cell division is accompanied by an increase in size of the daughter cells, and *growth* (in the true sense of the term) takes place.

These processes can be observed in the early developmental stages, which humans share with all animals (Fig. 17.4).

Morula

Cleavage is a process that occurs during the first stage of development. During cleavage, cell division without growth results in a mass of tiny cells. The cells are uniform in size because the cytoplasm has been equally divided among them. This solid mass of cells is called a *morula,* which means a bunch of berries.

Blastula

Morphogenesis begins as the cells of the morula position themselves to create a cavity. All animal blastulas have an empty cavity, but since the human blastula is called a **blastocyst,** the cavity is called the blastocyst cavity. The inner cell mass is at one end of the blastocyst.

Gastrula

In humans, a space called the amniotic cavity appears above the inner cell mass. The inner cell mass becomes an embryonic disk composed of two layers: the upper layer is called ectoderm, and the lower layer is called the endoderm.

Now, the bilayered disk elongates to form a primitive streak, at the midline region of the embryo. Morphogenesis continues as some of the upper cells within the primitive streak invaginate and spread out between the ectoderm and endoderm. The invaginating cells are the mesoderm layer.

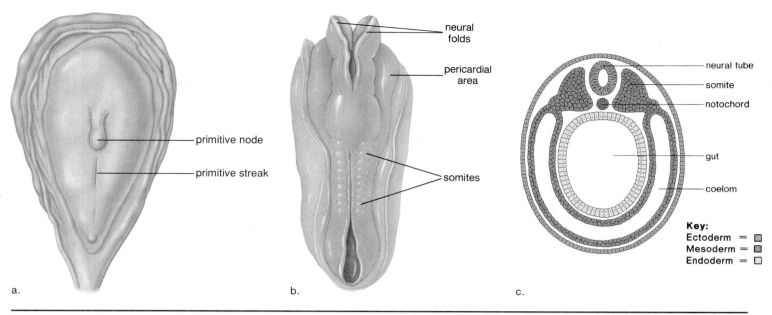

Figure 17.5   Primitive streak and neurula.
a. At 16 days, the primitive node marks the extent of the primitive streak where invagination results in a three-layered embryo. **b.** At 21 days, the neural tube is forming along the midline of the body. The pericardial area contains the primitive heart, and the somites give rise to the muscles and the vertebrae, which replace the notochord. **c.** Generalized cross section of a human neurula. Each of the germ layers can be associated with the later development of particular organs as listed in Table 17.1.

Differentiation is in progress because **ectoderm, mesoderm,** and **endoderm** are the germ layers that give rise to all the other tissues and organs of the body (Table 17.1). Gastrulation is the movement of cells that results in a *gastrula,* an embryo composed of three germ layers.

TABLE 17.1

Germ Layers and Organ Development

Ectoderm	Mesoderm	Endoderm
Skin epidermis, including hair, nails, and sweat glands	All muscles	Lining of digestive tract, trachea, bronchi, lungs, gallbladder, and urethra
Nervous system, including brain, spinal cord, ganglia, and nerves	Dermis of skin	Liver
Retina, lens, and cornea of eye	All connective tissue, including bone, cartilage, and blood	Pancreas
Inner ear	Blood vessels	Thyroid, parathyroid, and thymus glands
Lining of nose, mouth, and anus	Kidneys	Urinary bladder
Tooth enamel	Reproductive organs	

Neurula

Newly formed mesoderm cells that lie along the main axis become the notochord, a dorsal supporting rod. (In humans, the notochord is later replaced by the vertebral column.) The nervous system develops from ectoderm located just above the notochord. A neural plate thickens into neural folds that fuse, forming a neural tube. The neural tube develops into the spinal cord and the brain.

Neurulation involves **induction,** the ability of one tissue to influence the development of another tissue. Experiments have shown that the nervous system does not form unless there is a notochord present. Today, investigators believe that induction explains the process of differentiation. Induction requires direct contact or the production of a chemical by one tissue that most likely activates certain genes in the cells of the other tissue. These genes then direct how differentiation is to occur.

Midline mesoderm not contributing to the formation of the notochord becomes two longitudinal masses of tissue. From these blocklike portions of mesoderm, called somites, the muscles of the body and the vertebrae of the spine develop. The coelom, an embryonic body cavity that forms at this time, is completely lined by mesoderm. In humans, the coelom becomes the thoracic and abdominal cavities. Figure 17.5 shows an intact human neurula and a generalized cross section of the embryo indicating the location of the three germ layers.

Development involves certain processes that can be related to particular stages of early development.

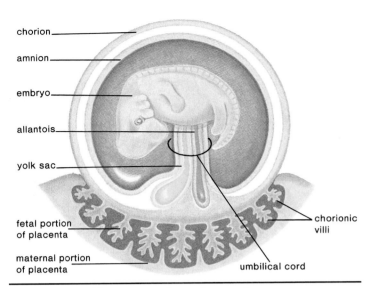

Figure 17.6   **The extraembryonic membranes.**
The chorion and amnion surround the embryo. The two other extraembryonic membranes, the yolk sac and allantois, contribute to the umbilical cord.

Extraembryonic Membranes

One of the major events in early development is the establishment of the extraembryonic membranes (Fig. 17.6). The term **extraembryonic membranes** is apt because these membranes extend out beyond the embryo. One of the membranes, the **amnion,** provides a fluid environment for the developing embryo and fetus. It is a remarkable fact that all animals, even land-dwelling humans, develop in a watery medium. Amniocentesis is a process by which amniotic fluid and the fetal cells floating in it are withdrawn for examination. One authority describes the functions of amniotic fluid in this way:

> The colorless amniotic fluid by which the fetus is surrounded serves many purposes. It prevents the walls of the uterus from cramping the fetus and allows it unhampered growth and movement. It encompasses the fetus with a fluid of constant temperature which is a marvelous insulator against cold and heat. Above all, it acts as an excellent shock absorber. A blow on the mother's abdomen merely jolts the fetus, and it floats away.[1]

The **yolk sac** is another extraembryonic membrane. Yolk is a nutrient material utilized by other animal embryos—the yellow of a chick's egg is yolk. However, in humans, the yolk sac contains no yolk and is the first site of red blood cell formation. Part of this membrane becomes incorporated into the umbilical cord. Another extraembryonic membrane, the **allantois,** contributes to the circulatory system: its blood vessels become umbilical blood vessels that transport fetal blood to and from the placenta. The **chorion,** the outer extraembryonic membrane, becomes part of the **placenta** (Fig. 17.7), where the fetal blood exchanges gases, nutrients, and wastes with maternal blood.

Placenta: For Exchange

The placenta begins formation once the embryo is implanted fully. Treelike extensions of the chorion called **chorionic villi** project into the maternal tissues. Later, these disappear in all areas except where the placenta develops. By the tenth week, the placenta is formed fully and begins to produce progesterone and estrogen. These hormones have two effects: due to their negative feedback effect on the hypothalamus and the anterior pituitary, they prevent any new follicles from maturing, and they maintain the lining of the uterus—the corpus luteum is not needed now. There is usually no menstruation during pregnancy.

The placenta has a fetal side contributed by the chorion and a maternal side consisting of uterine tissues. Notice in Figure 17.7 how the chorionic villi meet maternal blood vessels and blood of the mother and the fetus never mix since exchange always takes place across plasma membranes. Carbon dioxide and other wastes move from the fetal side to the maternal side, and nutrients and oxygen move from the maternal side to the fetal side of the placenta by diffusion. The **umbilical cord** stretches between the placenta and the fetus. Although it may seem that the umbilical cord travels from the placenta to the intestine, actually it simply contains the blood vessels taking fetal blood to and from the placenta. The umbilical cord is the lifeline of the fetus because it contains the umbilical arteries and vein, which transport waste molecules (carbon dioxide and urea) to the placenta for disposal and take oxygen and nutrient molecules from the placenta to the rest of the fetal circulatory system.

Harmful chemicals also can cross the placenta, and this is of particular concern during the embryonic period, when various structures are first forming. Each organ or part seems to have a sensitive period during which a substance can alter its normal function. The first Health reading for this chapter (see pp. 364–365) concerns the origination of birth defects and explains ways to detect genetic defects before birth.

Fetal Circulation

The fetus has circulatory features that are not present in the adult circulation (Fig. 17.7). All of these features can be related to the fact that the fetus does not use its lungs for gas exchange. For example, much of the blood entering the right atrium is shunted into the left atrium through the *oval opening* (foramen ovale) between the two atria. Also, any blood that does enter the right ventricle and is pumped into the pulmonary trunk is shunted into the aorta by way of the *arterial duct* (ductus arteriosus).

Blood within the aorta travels to the various branches, including the iliac arteries, which connect to the *umbilical arteries* leading to the placenta. Exchange between maternal blood and fetal blood takes place at the placenta. The *umbilical vein* carries blood rich in nutrients and oxygen to the fetus. The umbilical vein enters the liver and then joins the venous duct, which merges with the inferior vena cava, a vessel that returns blood to the heart. It is interesting to note that the umbilical arteries and vein run along side one another in the umbilical cord, which is cut at birth, leaving only the umbilicus (navel).

1. A. F. Guttmacher. *Pregnancy, Birth and Family Planning* (New York: New American Library, 1974), p. 74.

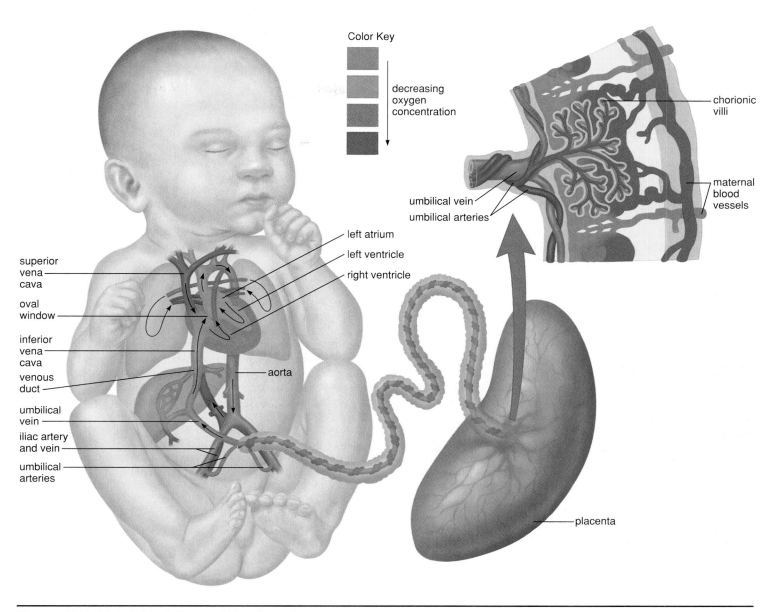

Color Key

decreasing
oxygen
concentration

chorionic
villi

umbilical vein
umbilical arteries

maternal
blood
vessels

left atrium
left ventricle
right ventricle

superior
vena
cava

oval
window

inferior
vena
cava

venous
duct

umbilical
vein

iliac artery
and vein

umbilical
arteries

aorta

placenta

Figure 17.7   Fetal circulation and the placenta.
The lungs are not functional in the fetus, and the blood passes directly from the right atrium to the left atrium or from the right ventricle to the aorta. The umbilical arteries take fetal blood to the placenta where exchange of molecules between fetal and maternal blood takes place across the walls of the chorionic villi. Oxygen and nutrient molecules diffuse into the fetal blood, and carbon dioxide and urea diffuse from fetal blood. The umbilical vein returns blood from the placenta to the fetus.

The most common of all cardiac defects in the newborn is the persistence of the oval opening. With the tying of the cord and the expansion of the lungs, blood enters the lungs in quantity. Return of this blood to the left side of the heart usually causes a flap to cover the opening. Incomplete closure occurs in nearly one out of four individuals, but even so, passage of the blood from the right atrium to the left atrium rarely occurs because either the opening is small or it closes when the atria contract. In a small number of cases, the passage of impure blood from the right side to the left side of the heart is sufficient to cause a "blue baby." Such a condition now can be corrected by open heart surgery.

The arterial duct closes because endothelial cells divide and block off the duct. Remains of the arterial duct and parts of the umbilical arteries and vein later are transformed into connective tissue.

Fetal circulation shunts blood away from the lungs, toward and away from the placenta within the umbilical blood vessels located within the umbilical cord. Exchange of substances between fetal blood and maternal blood takes place at the placenta, which forms from the chorion, an extraembryonic membrane, and uterine tissue.

Health Focus

Preventing Birth Defects

It is believed that at least one in 16 newborns has a birth defect, either minor or serious, and the actual percentage may be even higher. It is estimated that only 20% of all birth defects are due to heredity. Those that are hereditary can sometimes be detected before birth. **Amniocentesis** allows the fetus to be tested for abnormalities of development; chorionic villi sampling allows the embryo to be tested; and a new method has been developed for screening eggs to be used for in vitro fertilization (Fig. 17A).

It is recommended that all females take everyday precautions to protect any future and/or presently developing embryos and fetuses from defects. Proper nutrition is a must (deficiency in folic acid causes neural tube defects). X-ray diagnostic therapy should be avoided during pregnancy because X rays cause mutations in the developing embryo or fetus. Children born to women who received X-ray treatment are apt to have birth defects and/or to develop leukemia later. Toxic chemicals, such as pesticides and many organic industrial chemicals, which are also mutagenic, can cross the placenta. Cigarette smoke not only contains carbon monoxide but also other fetotoxic chemicals. Babies born to smokers are often underweight and subject to convulsions.

Pregnant Rh⁻ women should receive an Rh immunoglobulin injection to prevent the production of Rh antibodies. These antibodies can cause nervous system and heart defects.

Sometimes, birth defects are caused by pathogens. Females can be immunized before the childbearing years for rubella (German measles), which in particular causes birth defects such as deafness. Unfortunately, immunization for sexually transmitted diseases is not possible. The AIDS virus can cross the placenta, and over 1,500 babies who contracted AIDS while in their mother's womb are now mentally retarded. When a mother has herpes, gonorrhea, or chlamydia, newborns can become infected as they pass through the birth canal. Blindness and other physical and mental defects may develop. Birth by cesarean section could prevent these occurrences.

Pregnant women should not take any type of drug except with a doctor's prescription. Certainly illegal drugs, such as marijuana, cocaine, and heroin, should be completely avoided. "Cocaine babies" now make up 60% of drug-affected babies. Severe fluctuations in blood pressure that are produced by the use of cocaine temporarily deprive the developing brain of oxygen. Cocaine babies have visual problems, lack coordination, and are mentally retarded. The drugs aspirin, caffeine (present in coffee, tea, and cola), and alcohol should be severely limited. It is not unusual for babies of drug addicts and alcoholics to display withdrawal symptoms and to have various abnormalities. Babies born to women who drink while pregnant are apt to have fetal alcohol syndrome (FAS). These babies have decreased weight, height, and head size, with malformation of the head and face. Mental retardation is common in FAS infants.

Medications can also cause problems. When the synthetic hormone DES was given to pregnant women to prevent miscarriage, their daughters showed various abnormalities of the reproductive organs and an increased tendency toward cervical cancer. Other sex hormones, including birth-control pills, can possibly cause abnormal fetal development, including abnormalities of the sex organs. The tranquilizer thalidomide is well known for having caused deformities of the arms and legs in children born to women who took the drug. Therefore, a woman has to be very careful about taking medications while pregnant.

Now that physicians and lay people are aware of the various ways in which birth defects can be prevented, it is hoped that the incidence of birth defects will decrease in the future.

Figure 17A   Three methods for genetic-defect testing before birth.

a. Amniocentesis is usually performed from the fifteenth to the seventeenth week of pregnancy. A long needle is passed through the abdominal wall to withdraw a small amount of amniotic fluid, along with fetal cells. Since there are only a few cells in the amniotic fluid, testing may be delayed as long as four weeks until cell culture produces enough cells for testing purposes. About 40 tests are available for different defects.

b. Chorionic villi sampling is usually performed from the fifth to the twelfth week of pregnancy. The doctor inserts a long, thin tube through the vagina into the uterus. With the help of ultrasound, which gives a picture of the uterine contents, the tube is placed between the lining of the uterus and the chorion. Then a sampling of the chorionic villi cells is obtained by suction. Chromosome analysis and biochemical tests for genetic defects can be done immediately on these cells.

c. Screening eggs for genetic defects is a new technique. Preovulatory eggs are removed by aspiration after a laparoscope (optical telescope) is inserted into the abdominal cavity through a small incision in the region of the navel. Only the first polar body is tested because if the woman is heterozygous for a genetic defect and it is found in the polar body, then the egg must be normal. Normal eggs then undergo in vitro fertilization and are placed in the prepared uterus. At present, only one in ten attempts results in a birth, but it is known ahead of time that the child will be normal for the genetic traits tested.

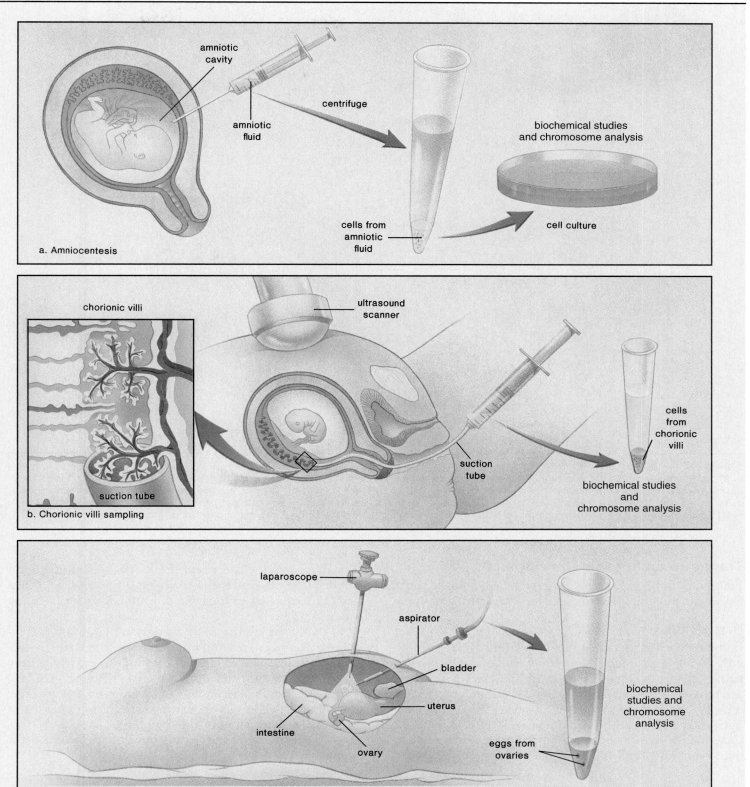

a. Amniocentesis

amniotic cavity

amniotic fluid

centrifuge

cells from amniotic fluid

biochemical studies and chromosome analysis

cell culture

b. Chorionic villi sampling

chorionic villi

suction tube

ultrasound scanner

suction tube

cells from chorionic villi

biochemical studies and chromosome analysis

c. Obtaining eggs for screening

laparoscope

aspirator

bladder

uterus

intestine

ovary

eggs from ovaries

biochemical studies and chromosome analysis

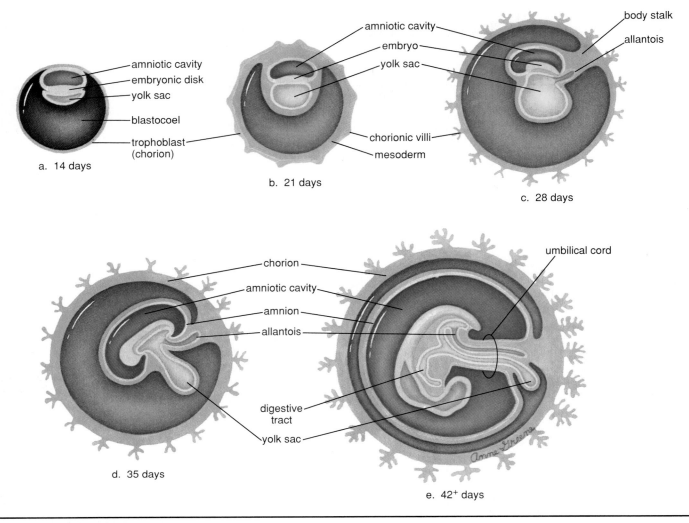

Figure 17.8 Embryonic development.
a. At first there are only tissues present in the embryo. The amniotic cavity is above the embryo, and the yolk sac is below. **b.** The chorion is developing villi, so important to exchange between mother and child. **c.** The allantois and yolk sac are two more extraembryonic membranes. **d.** These extraembryonic membranes are positioned inside the body stalk as it becomes the umbilical cord. **e.** At 42 days, the embryo has a head region and a tail region. The umbilical cord takes blood vessels between the embryo and the chorion (placenta).

Timing of Embryonic Development

Embryonic development lasts from fertilization to the end of the second month. All major organs develop during this time.

First Month

Immediately after fertilization, the zygote divides repeatedly as it passes down the oviduct to the uterus. The resulting morula becomes the hollow blastocyst with an inner cell mass to one side. The blastocyst is bounded by a layer of cells that becomes the chorion (see Fig. 17.3). The early appearance of the chorion emphasizes the complete dependence of the developing embryo on this extraembryonic membrane. The inner cell mass is the embryo. Once it

has arrived in the uterus on the fourth or fifth day after fertilization, it spends about two or three days, and then the blastocyst begins to implant itself in the uterine lining (see Fig. 17.3).

By the end of the second week, implantation is completed. The ever-growing number of cells becomes a gastrula with three germ layers. The amniotic cavity is seen above the embryo, and the yolk sac is below (Fig. 17.8a) During the third week, another extraembryonic membrane, the allantois, makes its appearance briefly, but later it and the yolk sac become part of the umbilical cord as it forms (Fig. 17.8c–d). Organs are already developed, including the spinal cord and heart.

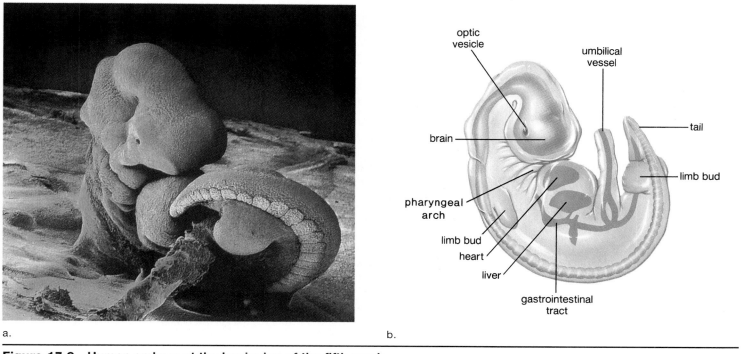

a.

b.

Figure 17.9 Human embryo at the beginning of the fifth week.
a. Scanning electron micrograph. **b.** The embryo is curled so that the head touches the heart. The organs of the gastrointestinal tract are forming. The bones in the tail will regress and become those of the coccyx. The arms and legs will develop from the bulges that are called limb buds.

By the end of the first month, the placenta is forming. The embryo has a nonhuman appearance largely due to the presence of a tail, but also because the arms and legs, which begin as limb buds, resemble paddles. The head is much larger than the rest of the embryo, and the whole embryo bends under its weight (Fig. 17.9). The eyes, ears, and nose are just appearing. The enlarged heart beats and the bulging liver takes over the production of blood cells for blood, which will carry nutrients to the developing organs and wastes from the developing organs.

Second Month

At the end of two months, the embryo's tail has disappeared, and the arms and legs are more developed with fingers and toes apparent (Fig. 17.10). The head is very large, the nose is flat, the eyes are far apart, and the ears are distinctively present. Internally, all major organs have appeared. Embryonic development is now finished; Table 17.2 outlines the main events.

At the end of the embryonic period, all organ systems are established, and there is a mature and functioning placenta. The embryo is only about 38 mm (1 1/2 in) long.

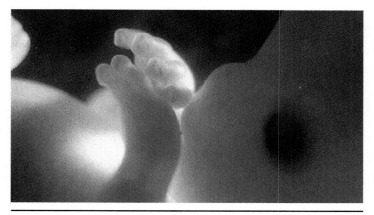

Figure 17.10 Two-month-old embryo.
The embryonic period is over, and from now on a more human appearance takes shape.

Table 17.2 also outlines the events for the mother. When first pregnant, the mother may experience nausea and vomiting, loss of appetite, and fatigue. Other changes are swelling and tenderness of the breasts, increased urination, and irregular bowel movements. Some women, however, report increased energy levels and a general sense of well-being during this time.

TABLE 17.2

Human Development

Time	Events for Mother	Events for Baby
Embryonic Development		
First week	Ovulation occurs.	Fertilization occurs. Cell division begins and continues. Chorion appears.
Second week	Symptoms of early pregnancy (nausea, breast swelling and tenderness, fatigue) are present.	Implantation. Amnion and yolk sac appear. Embryo has tissues. Placenta begins to form.
Third week	First missed menstruation. Blood pregnancy test is positive.	Nervous system begins development. Allantois and blood vessels are present. Placenta is well formed.
Fourth week	Urine pregnancy test is positive.	Limb buds form. Heart is noticeable and beating. Nervous system is prominent. Embryo has tail. Other systems form.
Fifth week	Uterus is the size of hen's egg. Frequent need to urinate due to pressure of growing uterus on bladder.	Embryo is curved. Head is large. Limb buds show divisions. Nose, eyes, and ears are noticeable.
Sixth week	Uterus is the size of an orange.	Fingers and toes are present. Cartilaginous skeleton.
Two months	Uterus can be felt above the pubic bone.	All systems are developing. Bone is replacing cartilage. Refinement of facial features. 38 mm (1 1/2 in).
Fetal Development		
Third month	Uterus is the size of a grapefruit.	Possible to distinguish sex. Fingernails appear.
Fourth month	Fetal movement is felt by those who have been pregnant before.	Skeleton visible. Hair begins to appear. 150 mm (6 in), 170 g (6 oz).
Fifth month	Fetal movement is felt by those who have not been pregnant before. Uterus reaches up to level of umbilicus and pregnancy is obvious.	Protective cheesy coating begins to be deposited. Heartbeat can be heard.
Sixth month	Doctor can tell where baby's head, back, and limbs are. Breasts have enlarged, nipples and areolae are darkly pigmented, and colostrum is produced.	Body is covered with fine hair. Skin is wrinkled and reddish.
Seventh month	Uterus reaches halfway between umbilicus and rib cage.	Testes descend into scrotum. Eyes are open. 300 mm (12 in), 1,350 g (3 lb).
Eighth month	Weight gain is averaging about a pound a week. Difficulty in standing and walking because center of gravity is thrown forward.	Body hair begins to disappear. Subcutaneous fat begins to be deposited.
Ninth month	Uterus is up to rib cage, causing shortness of breath and heartburn. Sleeping becomes difficult.	Ready for birth. 530 mm (20 1/2 in), 3,400 g (7 1/2 lb).

Timing of Fetal Development

Fetal development extends from the third to the ninth month as shown in Table 17.2. The fetus has a human appearance, but refinements are still taking place.

Third and Fourth Months

At the beginning of the third month (Fig. 17.11), head growth begins to slow down as the rest of the body increases in length. Epidermal refinements, such as eyelashes, eyebrows, hair on the head, fingernails, and nipples, appear.

Cartilage is replaced by bone as ossification centers appear in the bones. The skull has six large fontanels (membranous areas or soft spots), which permit a certain amount of flexibility as the head passes through the birth canal and allow rapid growth of the brain during infancy. The fontanels disappear by two years of age.

Sometime during the third month, it is possible to distinguish males from females. Once the testes differentiate, they produce androgens, the male sex hormones. The androgens, especially testosterone, stimulate the growth and differentiation of the male external genitals. In the absence of androgens, female genitals form. The ovaries do not produce estrogen because there is plenty of it circulating in the mother's bloodstream.

At this time, the testes or ovaries are located within the abdominal cavity. Later, in the last trimester of male fetal development, the testes descend into the scrotal sacs of the scrotum. Sometimes the testes fail to descend, in which case an operation can be performed to place them in their proper location.

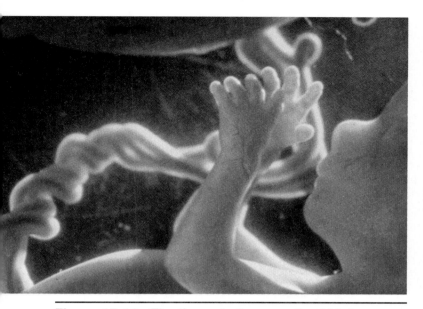

Figure 17.11 The three- to four-month-old fetus looks human.
Face, hands, and fingers are well defined.

By the end of the fourth month, the fetus is less than 150 mm (6 in) in length and weighs a little more than 170 grams (6 oz).

> During the third and fourth months, the skeleton is becoming ossified. The sex of the individual is now distinguishable.

Fifth through Seventh Months

During the fifth through seventh months (Fig. 17.12), the mother begins to feel fetal movement. At first, there is only a fluttering sensation, but as the legs grow and develop, kicks and jabs are felt. The fetal heartbeat is loud enough to be heard when a physician applies a stethoscope to the mother's abdomen. The fetus is in the fetal position with the head bent down and in contact with the flexed knees.

The wrinkled skin is covered by a fine down called **lanugo.** The lanugo is coated with a white, greasy, cheeselike substance called **vernix caseosa,** which is believed to protect the delicate skin from the amniotic fluid. During these months, the eyelids open fully.

At the end of this period, the fetus is almost 300 mm (12 in), and the weight has increased to almost 1,350 grams (3 lb). It is possible that if born now, the baby will survive; however, the lungs lack surfactant, which reduces surface tension within the alveoli (air sacs). Babies born without surfactant risk respiratory distress syndrome or collapsed lungs.

Eighth and Ninth Months

As the end of development approaches, the fetus usually rotates so that the head is pointed toward the cervix. However, if the fetus does not turn, then the likelihood of a breech birth (rump first) may prescribe a cesarean section. It is very difficult for the cervix to expand enough

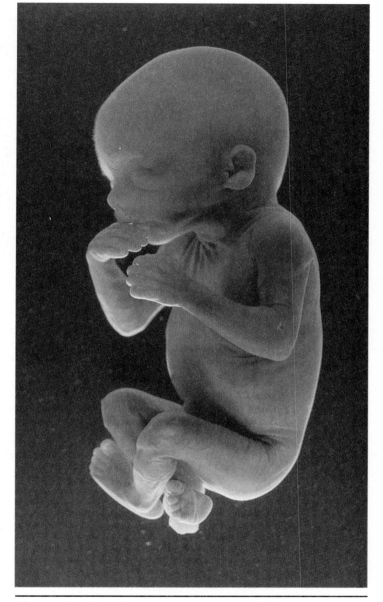

Figure 17.12 A five- to seven-month-old fetus.

to accommodate this form of birth, and asphyxiation of the baby is more likely to occur.

At the end of nine months, the fetus is about 530 mm (20 1/2 in) long and weighs about 3,400 grams (7 1/2 lb) (Fig. 17.13). Weight gain is due largely to an accumulation of fat beneath the skin. Fullterm babies have a better chance of survival.

> From the fifth to the ninth month, the fetus continues to grow and to gain weight. Babies born after six or seven months may survive, but fullterm babies have a better chance of survival.

Table 17.2 continues with the effects of pregnancy on the mother.

17.3   Birth

Investigation and experimentation show that when the fetal brain is sufficiently mature, the hypothalamus causes the pituitary to stimulate the adrenal cortex so that androgens are released into the bloodstream. The placenta utilizes androgens as a precursor for estrogen, a molecule that stimulates the local production of prostaglandins and oxytocin. All three of these molecules cause the uterus to contract and expel the fetus.

The uterus has contractions throughout pregnancy. At first, these are light, lasting about 20–30 seconds and occurring every 15–20 minutes. Near the end of pregnancy, the contractions may become stronger and more frequent so that the woman may think that she is in labor. However, the onset of true labor is marked by uterine contractions that occur regularly every 15–20 minutes and last for 40 seconds or more.

Stage 1

Prior to or at the first stage of **parturition**, which includes labor and birth of the fetus, there can be a "bloody show" caused by expulsion of a mucous plug from the cervical canal. This plug prevents bacteria and sperm from entering the uterus during pregnancy.

Uterine contractions during the first stage of labor occur in such a way that the cervical canal slowly disappears as the lower part of the uterus is pulled upward toward the baby's head (Fig. 17.13b). This process is called *effacement*, or "taking up the cervix." With further contractions, the baby's head acts as a wedge to assist cervical dilation. If it has not occurred already, the amniotic membrane is apt to rupture during this stage, releasing the amniotic fluid, which leaks out the vagina (sometimes referred to as breaking water). The first stage of labor ends once the cervix is dilated completely.

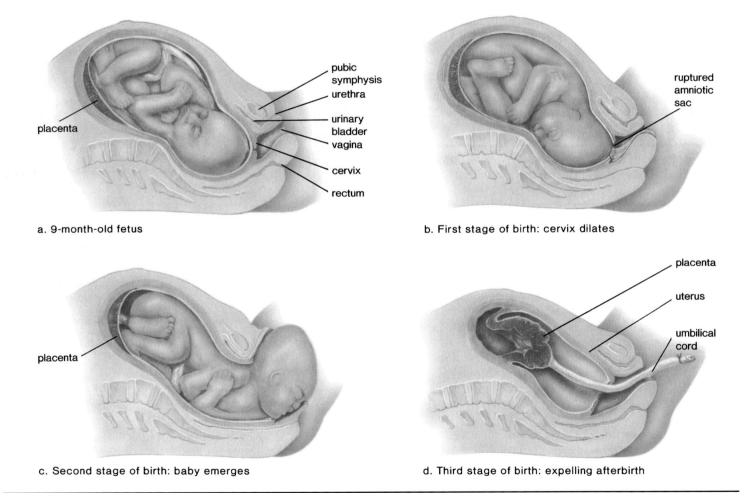

a. 9-month-old fetus

b. First stage of birth: cervix dilates

c. Second stage of birth: baby emerges

d. Third stage of birth: expelling afterbirth

Figure 17.13   Three stages of parturition.
a. Position of fetus just before birth begins. **b.** Dilation of cervix. **c.** Birth of baby. **d.** Expulsion of afterbirth.

Stage 2

During the second stage of parturition, the uterine contractions occur every 1–2 minutes and last about one minute each. They are accompanied by a desire to push, or bear down. As the baby's head gradually descends into the vagina, the desire to push becomes greater. When the baby's head reaches the exterior, it turns so that the back of the head is uppermost (Fig. 17.13c). Since the vaginal orifice may not expand enough to allow passage of the head without tearing, an **episiotomy** often is performed. This incision, which enlarges the opening, is sewn together later and heals more perfectly than a tear. As soon as the head is delivered, the baby's shoulders rotate so that the baby faces either to the right or the left. At this time, the physician may hold the head and guide it downward, while one shoulder and then the other emerges. The rest of the baby follows easily.

Once the baby is breathing normally, the umbilical cord is cut and tied, severing the child from the placenta. The stump of the cord shrivels and leaves a scar, which is the **umbilicus.**

Stage 3

The placenta, or *afterbirth,* is delivered during the third stage of labor (Fig. 17.13d). About 15 minutes after delivery of the baby, uterine muscular contractions shrink the uterus and dislodge the placenta. The placenta then is expelled into the vagina. As soon as the placenta and its membranes are delivered, the third stage of labor is complete.

During the first stage of birth, the cervix dilates; during the second stage, the child is born; and during the third stage, the afterbirth is expelled.

Female Breast and Lactation

A female breast contains 15 to 25 lobules, each with its own milk duct, which begins at the nipple and divides into numerous other ducts that end in blind sacs called *alveoli* (Fig. 17.14).

During pregnancy, the breasts enlarge as the ducts and alveoli increase in number and size. The same hormones that affect the mother's breast can also affect the child's. Some newborns, including males, even secrete a small amount of milk for a few days.

Usually, there is no production of milk during pregnancy. The hormone *prolactin* is needed for lactation to begin, and the production of this hormone is suppressed because of the feedback control that the increased amount of estro-

gen and progesterone during pregnancy has on the pituitary. Once the baby is delivered, however, the pituitary begins secreting prolactin. It takes a couple of days for milk production to begin, and, in the meantime, the breasts produce **colostrum,** a thin, yellow, milky fluid rich in protein, including antibodies.

The continued production of milk requires a suckling child. When a breast is suckled, the nerve endings in the areola are stimulated, and a nerve impulse travels along neural pathways from the nipples to the hypothalamus, which directs the pituitary gland to release the hormone *oxytocin.* When this hormone arrives at the breast, it causes contraction of the lobules so that milk flows into the ducts (called milk letdown), where it may be drawn out of the nipple by the suckling child. Some women choose to breast-feed and some do not.

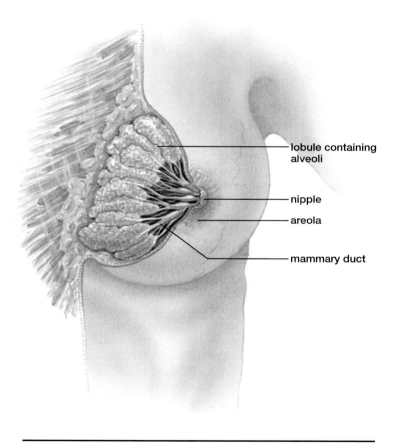

lobule containing alveoli

nipple

areola

mammary duct

Figure 17.14   Female breast anatomy.
The female breast contains lobules consisting of ducts and alveoli. The alveoli are lined by milk-producing cells in the lactating (milk-producing) breast.

17.4 Human Development After Birth

Development does not cease once birth has occurred but continues throughout the stages of life: infancy, childhood, adolescence, and adulthood. **Aging** encompasses these progressive changes that contribute to an increased risk of infirmity, disease, and death (Fig. 17.15).

Today, there is great interest in **gerontology,** the study of aging, because there are now more older individuals in our society than ever before, and the number is expected to rise dramatically. In the next half-century, those over age 75 will rise from the present 13 million to 34–45 million, and those over age 80 will rise from three million to 6 million individuals. The human life span is judged to be a maximum of 110–115 years. The present goal of gerontology is not necessarily to increase the life span, but to increase the health span, the number of years that an individual enjoys the full functions of all body parts and processes.

Why We Age

There are many theories about what causes aging. Three of these are considered here.

Genetic in Origin

Several lines of evidence indicate that aging has a genetic basis. (1) The number of times a cell divides is species-specific. The maximum number of times human cells divide is around 50. Perhaps as we grow older, more and more cells are unable to divide, and instead, they undergo degenerative changes and die. (2) Some cell lines may become nonfunctional long before the maximum number of divisions has occurred. Whenever DNA replicates, mutations can occur, and this can lead to the production of nonfunctional proteins. Eventually, the number of inadequately functioning cells can build up, which contributes to the aging process. (3) The children of long-lived parents tend to live longer than those of short-lived parents. Recent work suggests that when an animal produces fewer free radicals, it lives longer. Free radicals are unstable molecules that carry an extra electron. In order to stabilize themselves, free radicals donate an electron to another molecule like DNA or proteins (e.g., enzymes) or lipids found in plasma membranes. Eventually these molecules are unable to function, and the cell is destroyed. There are genes that code for antioxidant enzymes that detoxify free radicals. This research suggests that animals with particular forms of these genes—and therefore more efficient antioxidant enzymes—live longer.

Whole-Body Process

A decline in the hormonal system can affect many different organs of the body. For example, type II diabetes is common in older individuals. The pancreas makes insulin, but the cells lack the receptors that enable them to respond. Menopause in women occurs for a similar reason.

Figure 17.15 Aging.
Aging is a slow process during which the body undergoes changes that eventually bring about death, even if no marked disease or disorder is present. Medical science is trying to extend the human life span and the health span, the length of time the body functions normally.

There is plenty of follicle-stimulating hormone in the bloodstream, but the ovaries do not respond. Perhaps aging results from the loss of hormonal activities and a decline in the functions they control.

The immune system, too, no longer performs as it once did, and this can affect the body as a whole. The thymus gland gradually decreases in size, and eventually most of it is replaced by fat and connective tissue. The incidence of cancer increases among the elderly, which may signify that the immune system is no longer functioning as it should. This idea is substantiated, too, by the increased incidence of autoimmune diseases in older individuals.

It is possible, though, that aging is not due to the failure of a particular system that can affect the body as a whole, but to a specific type of tissue change that affects all organs and even the genes. It has been noticed for some time that proteins—such as collagen fibers which are present in many support tissues—become increasingly cross-linked as people age. Undoubtedly, this cross-linking contributes to the stiffening and the loss of elasticity characteristic of aging tendons and ligaments. It may also account for the inability of such organs as the blood vessels, the heart, and the lungs to function as they once did. Some researchers have now found that glucose has the tendency to attach to any type of protein, which is the first step in a cross-linking process. They are presently experimenting with drugs that can prevent cross-linking.

Extrinsic Factors

The current data about the effects of aging are often based on comparisons of the characteristics of the elderly to younger age groups; but perhaps today's elderly were not as aware when they were younger of the importance of, for example, diet and exercise to general health. It is possible, then, that much of what we attribute to aging is instead due to years of poor health habits.

Consider, for example, osteoporosis. This condition is associated with a progressive decline in bone density in both males and females so that fractures are more likely to occur after only minimal trauma. Osteoporosis is common in the elderly—by age 65, one-third of women will have vertebral fractures, and by age 81, one-third of women and one-sixth of men will have suffered a hip fracture. While there is no denying that a decline in bone mass occurs as a result of aging, certain extrinsic factors are also important. The occurrence of osteoporosis itself is associated with cigarette smoking, heavy alcohol intake, and perhaps inadequate calcium intake. Not only is it possible to eliminate these negative factors by personal choice, it is also possible to add a positive factor. A moderate exercise program has been found to slow down the progressive loss of bone mass.

Even more important, a proper diet that includes at least five servings of fruits and vegetables a day and a sensible exercise program will most likely help eliminate cardiovascular disease, the leading cause of death today. Experts no longer believe that the cardiovascular system necessarily suffers a large decrease in functioning ability with age. Persons 65 years of age and older can have well-functioning hearts and open coronary arteries if their health habits are good and they continue to exercise regularly.

How Aging Affects Body Systems

Data about how aging affects body systems should be accepted with reservations.

Skin

As aging occurs, skin becomes thinner and less elastic because the number of elastic fibers decreases and the collagen fibers undergo cross-linking, as discussed previously. Also, there is less adipose tissue in the subcutaneous layer; therefore, older people are more likely to feel cold. The loss of thickness partially accounts for sagging and wrinkling of the skin.

Homeostatic adjustment to heat is also limited because there are fewer sweat glands for sweating to occur. There are fewer hair follicles, so the hair on the scalp and the extremities thins out. The number of oil (sebaceous) glands is reduced, and the skin tends to crack. Older people also experience a decrease in the number of melanocytes, making hair gray and skin pale. In contrast, some of the remaining pigment cells are larger, and pigmented blotches appear in skin.

Processing and Transporting

Cardiovascular disorders are the leading cause of death among the elderly. The heart shrinks because there is a reduction in cardiac muscle cell size. This leads to loss of cardiac muscle strength and reduced cardiac output. Still, it is observed that the heart, in the absence of disease, is able to meet the demands of increased activity. It can increase its rate to double or triple the amount of blood pumped each minute even though the maximum possible output declines.

Because the middle coat of arteries contains elastic fibers, which most likely are subject to cross-linking, the arteries become more rigid with time, and their size is further reduced by plaque, a buildup of fatty material. Therefore, blood pressure readings gradually rise. Such changes are common in individuals living in Western industrialized countries but not in agricultural societies. A low cholesterol and saturated fatty acid diet has been suggested as a way to control degenerative changes in the cardiovascular system.

There is reduced blood flow to the liver, and this organ does not metabolize drugs as efficiently as before. This means that as a person gets older, less medication is needed to maintain the same level in the bloodstream.

Circulatory problems often are accompanied by respiratory disorders, and vice versa. Growing inelasticity of lung tissue means that ventilation is reduced. Because we rarely use the entire vital capacity, these effects are not noticed unless there is increased demand for oxygen.

There is also reduced blood supply to the kidneys. The kidneys become smaller and less efficient at filtering wastes. Salt and water balance are difficult to maintain, and the elderly dehydrate faster than young people. Difficulties involving urination include incontinence (lack of bladder control) and the inability to urinate. In men, the prostate gland may enlarge and reduce the diameter of the urethra, making urination so difficult that surgery is often needed.

The loss of teeth, which is frequently seen in elderly people, is more apt to be the result of long-term neglect than aging. The digestive tract loses tone, and secretion of saliva and gastric juice is reduced, but there is no indication of reduced absorption. Therefore, an adequate diet, rather than vitamin and mineral supplements, is recommended. There are common complaints of constipation, increased amount of gas, and heartburn, but gastritis, ulcers, and cancer can also occur.

Integration and Coordination

It is often mentioned that while most tissues of the body regularly replace their cells, some at a faster rate than others, the brain and the muscles ordinarily do not. However, contrary to previous opinion, recent studies show that few neural cells of the cerebral cortex are lost during the normal aging process. This means that cognitive skills remain unchanged even though there is characteristically a loss in short-term memory. Although the elderly learn more slowly than the young, they can acquire and remember new material. It is noted that when more time is given for the subject to respond, age differences in learning decrease.

Neurons are extremely sensitive to oxygen deficiency, and if neuron death does occur, it may not be due to aging itself but to reduced blood flow in narrowed blood vessels. Specific disorders, such as depression, Parkinson disease, and Alzheimer disease, are sometimes seen, but they are not common. Reaction time, however, does slow, and more stimulation is needed for hearing, taste, and smell receptors to function as before. After age 50, there is a gradual reduction in the ability to hear tones at higher frequencies, and this can make it difficult to identify individual voices and to understand conversation in a group. The lens of the eye does not accommodate as well and also may develop a cataract. Glaucoma, the buildup of pressure due to increased fluid, is more likely to develop because of a reduction in the size of the anterior cavity of the eye.

Loss of skeletal muscle mass is not uncommon, but it can be controlled by a regular exercise program. There is a reduced capacity to do heavy labor, but routine physical work should be no problem. A decrease in the strength of the respiratory muscles and inflexibility of the rib cage contribute to the inability of the lungs to expand as before, and reduced muscularity of the urinary bladder contributes to difficulties with urination.

As noted before, aging is accompanied by a decline in bone density. Osteoporosis, characterized by a loss of calcium and mineral from bone, is not uncommon, but there is evidence that proper health habits can prevent its occurrence. Arthritis, which causes pain upon movement of the joint, is also seen.

Weight gain occurs because the basal metabolism decreases and inactivity increases. Muscle mass is replaced by stored fat and retained water.

The Reproductive System

Females undergo menopause, and thereafter the level of female sex hormones in blood falls markedly. The uterus and the cervix are reduced in size, and there is a thinning of the walls of the oviducts and the vagina. The external genitals become less pronounced. In males, the level of androgens falls gradually over the age span of 50–90, but sperm production continues until death.

It is of interest that as a group, females live longer than males. Although their health habits may be better, it

Figure 17.16 Remaining active.
The aim of gerontology is to allow the elderly to enjoy living.

is also possible that the female sex hormone estrogen offers women some protection against circulatory disorders when they are younger. Males suffer a marked increase in heart disease in their forties, but an increase is not noted in females until after menopause. Then women lead men in the incidence of stroke. Men are still more likely than women to have a heart attack, however.

How To Age Well

We have listed many adverse effects due to aging, but it is important to emphasize that while such effects are seen, they are not a necessary occurrence (Fig. 17.16). We must discover any extrinsic factors that precipitate these adverse effects and guard against them. Just as it is wise to make the proper preparations to remain financially independent when older, it is also wise to realize that biologically successful old age begins with the health habits developed when we are younger.

SUMMARY

17.1 Fertilization

Only one sperm head actually enters the egg, and this sperm's nucleus fuses with the egg nucleus. The zygote begins to develop into an embryo, which travels down the oviduct and embeds itself in the uterine lining. Cells surrounding the embryo produce HCG, and the presence of this hormone indicates that the female is pregnant.

17.2 Human Development Before Birth

The processes of development (cleavage, morphogenesis, differentiation, and growth) occur during the early developmental stages, which consist of the morula, blastocyst, gastrula, and neurula. The extraembryonic membranes, including the placenta, are special features of human development. Fetal lungs do not operate, and the fetal circulation takes blood to the placenta, where exchange takes place.

Human development consists of embryonic (first two months) and fetal (third to ninth month) development. During the embryonic period, the extraembryonic membranes appear and serve important functions: the embryo acquires organ systems. During the fetal period, there is a refinement of these systems.

17.3 Birth

Birth, or parturition, has three phases. During the first stage, the cervix dilates to allow passage of the baby's head and body. The amnion usually bursts sometime during this stage. During the second stage, the baby is born and the umbilical cord is cut. As the baby takes his or her first breath, anatomical changes convert fetal circulation to adult circulation. During the third stage, the placenta is delivered.

Milk is not produced during pregnancy because of hormonal suppression, but once the child is born, milk production begins. Prolactin promotes the production of milk, and oxytocin allows milk letdown.

17.4 Human Development After Birth

Development after birth consists of infancy, childhood, adolescence, and adulthood. Young adults are at their prime, and then the aging process begins. Aging encompasses progressive changes from about age 20 on that contribute to an increased risk of infirmity, disease, and death. Perhaps aging is genetic in origin, perhaps it is due to a change that affects the whole body, or perhaps it is due to extrinsic factors.

STUDYING THE CONCEPTS

1. Describe the process of fertilization and the events immediately following it.   358–59

2. What is the basis of the pregnancy test?   359

3. What are the processes of development? Relate these processes to the early development stages.   360–61

4. Name the four extraembryonic membranes, and give a function for each.   362

5. Describe the structure and function of the placenta.   362

6. During which period of pregnancy does a woman have to be the most careful about the intake of medications and other drugs?   362

7. Describe the vascular components of the umbilical cord and how they relate to fetal circulation.   362

8. Specifically, what events normally occur during the first, second, third, and fourth weeks of development?   366–67, 368   What events normally happen during the second through the ninth months?   367–69

9. In general, describe the physical changes in the mother during pregnancy.   368

10. What are the three stages of birth? Describe the events of each stage.   370–71

11. Describe the suckling reflex.   371

12. Discuss three theories of aging. What are the major changes in body systems that have been observed as adults age?   372–74

APPLYING YOUR KNOWLEDGE

Concepts

1. Fertilization of an egg that has a chance of becoming an embryo normally takes place in the upper one-third of the oviduct. Explain the benefit for this.

2. Fetal hemoglobin is different than adult hemoglobin. It has a greater affinity for oxygen. What advantage is there to this difference?

3. Osteoporosis is very common in older people, particularly women. What factors can you control that will decrease your chances of suffering from osteoporosis?

4. Rarely more than one sperm enters an egg. If any egg were fertilized by more than one sperm, what would be the probable result?

Bioethical Issue

No one is more dependent on a parent than an unborn child. Nestled in the womb, babies simply absorb the environment around them. They eat what the mother eats, drink what the mother drinks, hear (to some extent) what she hears.

That's why it's so critical that mothers take care of themselves during pregnancy. Substances like caffeine, alcohol, or nicotine cross the placenta and enter the baby's bloodstream. Tiny amounts of some chemicals may do little harm. Moderate amounts of others—like alcohol—can cause brain damage.

For some women, life-style changes required during pregnancy are difficult. How responsible is a mother for ensuring the health of her baby? Is it reasonable to expect a mother to always eat right, exercise, and avoid all potentially harmful substances when pregnant? And if she doesn't, how accountable should she be for the effects on her baby?

TESTING YOUR KNOWLEDGE

1. Fertilization occurs when the _____ nucleus fuses with the _____ nucleus.

2. The zygote divides as it passes down an oviduct. This process is called _____.

3. Once the embryo arrives at the uterus, it begins to _____ itself in the uterine lining.

4. When cells take on a specific structure and function, _____ is occuring.

5. The _____ membranes include the chorion, the _____ , the yolk sac, and the allantois.

6. During embryonic development, all major _____ form.

7. Fetal development begins at the end of the _____ month.

8. During development, the nutrient needs of the developing embryo (fetus) are served by the _____.

9. In most deliveries, the _____ appears before the rest of the body.

10. The hormone _____ is required for milk letdown during the suckling reflex.

11. As we age, the proteins in the body undergo _____, a process that causes body parts to become stiff and rigid.

12. Label this diagram.

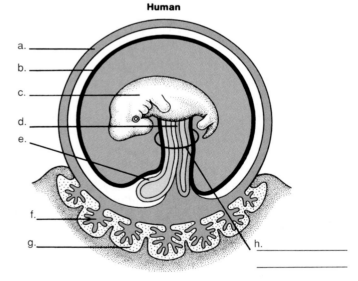

Human

a. _____

b. _____

c. _____

d. _____

e. _____

f. _____

g. _____

h. _____

_____

APPLYING TECHNOLOGY

Your study of development and aging is supported by these available technologies:

Exploring the Internet

The Mader Home Page provides further resources for studying this chapter.

http://www.mhhe.com/sciencemath/biology/mader/

(Click on *Human Biology*.)

Explorations in Human Biology CD-ROM

Life Span and Life-Style (#3)   Life expectancy varies as students change the amount of smoking, drinking, fat in diet, and exercise.   (2)*

Life Science Animations Video

Video #2:   Cell Division/Heredity/Genetics/ Reproduction and Development
Human Embryonic Development (#21)
Developmental changes are observed from germ layer formation to a human form.

Video #4:   Animal Biology II
Common Congenital Defects of the Heart (#39)
Circulatory problems due to failure of the oval opening and the arterial duct to close, as well as other defects, are illustrated.   (2)

*Level of difficulty

SELECTED KEY TERMS

aging Progressive changes over time, leading to loss of physiological function and eventual death.   372

allantois (ahl-un-TOE-us) Extraembryonic membrane that contributes to the formation of blood vessels.   362

amniocentesis (AM-nee-oh-sen-TEE-sis) Removal of a small amount of amniotic fluid to examine the chromosomes and the enzymatic potential of fetal cells.   364

amnion (AM-nee-ahn) Extraembryonic membrane that forms a fluid-filled sac around the embryo.   362

blastocyst Early stage of embryonic development that consists of a hollow ball of cells.   360

chorion (KOR-ee-ahn) Extraembryonic membrane that forms an outer covering around the embryo and contributes to the formation of the placenta.   362

chorionic villi Treelike extensions of the chorion of the embryo, projecting into the maternal tissues.   362

colostrum (kul-LAHS-trum) Thin, milky fluid rich in proteins, including antibodies, that is secreted by the mammary glands a few days prior to or after delivery before true milk is secreted.   371

differentiation Process and the developmental stages by which a cell becomes specialized for a particular function.   360

ectoderm Outer germ layer of the embryonic gastrula; it gives rise to the nervous system and skin.   361

embryo Organism in its early stages of development; first week to two months.   359

endoderm Inner germ layer that lines the archenteron/gut of the gastrula; it becomes the lining of the digestive and respiratory tracts and associated organs.   361

episiotomy (eh-peez-ee-AHT-oh-mee) Surgical procedure performed during childbirth in which the opening of the vagina is enlarged to avoid tearing.   371

extraembryonic membranes Membranes that are not a part of the embryo but are necessary to the continued existence and health of the embryo.   362

fertilization Union of a sperm nucleus and an egg nucleus, which creates a zygote with the diploid number of chromosomes.   358

gerontology Study of aging.   372

implantation Attachment and penetration of the embryo to the lining of the uterus (endometrium).   359

induction Ability of a chemical or a tissue to influence the development of another tissue.   361

lanugo (lah-NOO-goh) Short, fine hair that is present during the later portion of fetal development.   369

mesoderm Middle germ layer of the embryonic gastrula; gives rise to the muscles, the connective tissue, and the circulatory system.   361

parturition (par-too-RISH-un) Birth of a human and the expulsion of the extraembryonic membranes through the terminal portion of the female reproductive tract.   370

placenta Structure formed from the chorion and uterine tissue through which nutrient and waste exchange occur for the embryo and later the fetus.   362

umbilical cord Cord through which blood vessels pass, connecting the fetus to the placenta.   362

umbilicus Navel where the umbilical blood vessels enter the umbilical cord; remains after the umbilical cord has been cut off following birth.   371

vernix caseosa (VER-niks kah-see-OH-sah) Cheeselike substance covering the skin of the fetus.   369

yolk sac Extraembryonic membrane that serves as the first site for blood cell formation.   362

zygote Diploid cell formed by the union of two gametes; the product of fertilization.   359

FURTHER READINGS FOR PART FIVE

Alcamo, I. E. 1996. *AIDS: The biological basis.* Dubuque, Iowa: Wm. C. Brown Publishers. This easily understood book focuses on the biology of AIDS.

Alexander, N. J. March/April 1996. Barriers to sexually transmitted diseases. *Scientific American Science & Medicine* 3(2):32. Article discusses the effectiveness of certain contraceptives in protecting women against STDs.

Alexander, N. J. September 1995. Future contraceptives. *Scientific American* 273(3):136. Article discusses the possibility of contraceptive vaccines for both men and women.

Anderson, R. M., and May, R. M. Understanding the AIDS pandemic. *Scientific American Special Issue.* 1993, p. 86. Mathematical models are used to explore the biology of AIDS and its transmission.

Black, P. H. November/December 1995. Psychoneuroimmunology: Brain and immunity. *Scientific American Science & Medicine.* The role of stress in susceptibility to infections, and cancer and HIV progression is discussed.

Caldwell, J. C., and Caldwell, P. March 1996. The African AIDS epidemic. *Scientific American* 273(3):62. Article discusses a factor most likely responsible for causing the high rate of AIDS transmission in Africa.

Capecchi, M. March 1994. Targeted gene replacement. *Scientific American* 270(3):52. Researchers are deciphering DNA segments that control development and immunity.

Carlson, B. M. 1994. *Human embryology and developmental biology.* St. Louis: Mosby-Year Book, Inc. This text for students in the medical field features contemporary research and new technologies in the field of embryology.

Cohen, P. T., et al. 1994. *The AIDS knowledge base.* Boston: Little, Brown and Company. A comprehensive text on HIV disease.

Crooks, R., and Baur, K. 1996. *Our sexuality.* 6th ed. Redwood City, Calif.: Benjamin/Cummings Publishing. Introduction to the biological, psychosocial, behavioral, and cultural aspects of sexuality.

Duan, L., and Pomerantz, R. J. May/June 1996. Intracellular antibodies for HIV-1 gene therapy. *Science & Medicine* 3(3):24. Article discusses the cloning of synthetic antibody fragments that can inhibit the function of viral proteins.

Dusenbery, D. B. 1996. *Life at small scale: The behavior of microbes.* New York: Scientific American Library. This easy-to-read, well-illustrated text describes how microbes respond to the physical demands of their environment.

Emini, E. A. May/June 1995. Hurdles in the path to an HIV-1 vaccine. *Scientific American Science & Medicine* 2(3):38. This article discusses the obstacles involved in finding a vaccine for HIV-1.

Fan, H., et al. 1994. *The biology of AIDS.* 3d ed. Boston: Jones and Bartlett Publishers. Provides a firm scientific overview of AIDS to the nonspecialized student.

Gilbert, S. F. 1994. *Developmental biology.* 4th ed. Sunderland, Mass.: Sinauer Associates. This text emphasizes modern developmental biology and contains current information in this expanding field.

Jensen, M. M., and Wright, D. N. 1992. *Introduction to microbiology for the health sciences.* 3d ed. Englewood Cliffs, New Jersey: Prentice Hall. This self-instructive microbiology text is adaptable to nonclassroom as well as classroom settings; its emphasis is on the role of microorganisms in disease processes.

MacDonald, P. C., and Casey, M. L. March/April 1996. Preterm birth. *Scientific American Science & Medicine* 3(2):42. Article discusses the role of oxytocin, prostaglandins, and infections in the initiation of human labor.

Mader, S. S. 1990. *Human reproductive biology.* 2d ed. Dubuque, Iowa: Wm. C. Brown Publishers. An introductory text covering human reproduction in a clear, easily understood manner.

Newman, J. December 1995. How breast milk protects newborns. *Scientific American* 273(6):76. Human milk contains special antibodies that boost the newborn's immune system.

Nowak, M. A., and McMichael, A. J. August 1995. How HIV defeats the immune system. *Scientific American* 273(2):58. Full-blown AIDS results when the proliferating virus finally overwhelms the body's defenses, a process that may take years.

Nusslein-Volhard, C. August 1996. Gradients that organize embryo development. *Scientific American* 275(2):54. Nobel Prize-winning researcher describes how chemical gradients of substances called morphogens give an evolving embryo its shape.

Perls, T. T. January 1995. The oldest old. *Scientific American* 272(1):70. A survey of the very elderly often finds them in good physical condition.

Prescott, L. M., et al. 1996. *Microbiology.* 3d ed. Dubuque, Iowa: Wm. C. Brown Publishers. This introductory text covers all major areas of microbiology.

Ricklefs, R. E., and Finch, C. E. 1995. *Aging: A natural history.* New York: Scientific American Library. This text emphasizes the nature of aging and the mechanisms of physiological deterioration.

Risher, C. E., and Easton, T. A. 1995. *Focus on human biology.* 2d ed. New York: HarperCollins College Publishers. This comprehensive introductory textbook stresses basic human anatomy and physiology.

Ross, I. K. 1995. *Aging of cells, humans and societies.* Dubuque, Iowa: Wm. C. Brown Publishers. Presents current concepts on aging.

Science & Medicine. July/August 1994. 3(4). This issue includes an article on fetal alcohol syndrome.

Selkoe, D. J. September 1992. Aging brain, aging mind. *Scientific American* 267(3):134. Discusses how health affects changes in learning, memory, and reasoning as aging progresses.

Spicer, D. V., and Pike, M. C. July/August 1995. Hormonal manipulation to prevent breast cancer. *Scientific American Science & Medicine* 2(4):58. This contraceptive regimen could reduce the risk of breast cancer.

Stolley, P. D., and Lasky, T. 1995. *Investigating disease patterns: The science of epidemiology.* New York: Scientific American Library. The process of epidemiology and its contribution to the understanding of disease is covered in this interesting, easy-to-read book.

Tortora, G. J., et al. 1995. *Microbiology: An introduction.* Redwood City, Calif.: Benjamin/Cummings Publishing. This introductory microbiology text presents the diversity of microbial life and the roles of microbes in nature and in our daily lives.

Van De Graaff, K. M. 1995. *Survey of infectious and parasitic diseases.* Dubuque, Iowa: Wm. C. Brown Publishers.

Weindruch, R. January 1996. Caloric restriction and aging. *Scientific American* 274(1):64. Consuming fewer calories may increase longevity.

Part

6

Human Genetics

Human beings practice sexual reproduction, which requires gamete production. Gametes carry half the total number and various combinations of chromosomes and genes. It is sometimes possible to determine the chances of an offspring receiving a particular parental gene and, therefore, inheriting a genetic disorder.

Genes, now known to be constructed of DNA, control not only the metabolism of the cell but also, ultimately, the characteristics of the individual. The step-by-step procedure by which DNA specifies protein synthesis has been discovered. Biotechnology is a new and burgeoning field that permits the extraction of DNA from one organism and its insertion in a different organism for a purpose useful to human beings.

Cancer is a cellular disease brought on by DNA mutations that transform a normal cell into a cancer cell. Cancer-causing genes are derived from normal genes, which keep cell division under control. Because environmental factors play a major role in causing or promoting cancer, the possibility exists of reducing its incidence. Knowledge about the detection and treatment of cancer is improving daily.

Chapter 18

Chromosomal Inheritance

Chapter Outline

18.1 CHROMOSOMAL INHERITANCE
- Normally, humans inherit 22 pairs of autosomes and one pair of sex chromosomes for a total of 46 chromosomes.   380
- Normally, males have the sex chromosomes XY and females have the sex chromosomes XX.   380
- Abnormalities arise when humans inherit an abnormal number or type of chromosome.   381–383

18.2 HUMAN LIFE CYCLE
- The human life cycle involves two types of cell divisions: mitosis and meiosis.   384

18.3 MITOSIS
- Mitosis, cell division in which the chromosome number remains constant, is involved in growth.   385

18.4 MEIOSIS
- Meiosis, cell division in which the chromosome number is reduced by one-half, is involved in gamete production.   388

Maria and Tom wanted to have their son, Tommy, who has Down syndrome, be in classes with "normal" students. Tommy's school, however, was reluctant. Administrators expressed concern that the couple's son would draw too much energy, time, and attention away from other students. What's more, they argued, he would learn more in a specially designed environment.

But Maria and Tom prevailed. Tommy went to the local elementary school. With tutoring and lots of parental attention, he graduated from high school with his peers. Today, he lives in an apartment close to his parents and works at a nearby library.

Physicians knew from his chromosome makeup that Tommy had Down syndrome. This chapter describes how it is possible to examine the chromosomes of a body cell and describes various chromosome abnormalities. Chromosome abnormalities are apt to occur during meiosis, the type of cell division needed for gamete production. During meiosis, the chromosome number is reduced so that egg and sperm usually carry only one-half the full number of chromosomes. The chromosome number is restored when a sperm fertilizes an egg.

18.1 Chromosomal Inheritance

In a nondividing cell, the nucleus contains indistinct and diffuse *chromatin*, but in a dividing cell, chromatin becomes the short and thick *chromosomes*. A cell may be photographed just prior to division so that a picture of the chromosomes is obtained. The picture may be entered into a computer and the chromosomes electronically arranged by pairs (Fig. 18.1). Individual chromosomes are recognized by their size, location of the centromere (a constriction), and characteristic banding due to staining. The resulting display of pairs of chromosomes is called a **karyotype.** Although both males and females have 23 pairs of chromosomes, one of these pairs is of unequal length in males. The larger chromosome of this pair is called the X and the smaller is called the Y. Females have two X chromosomes in their karyotype. The X and Y chromosomes are called the **sex chromosomes** because they contain the genes that determine sex. The other chromosomes, known as **autosomes,** include all of the pairs of chromosomes except the X and Y chromosomes. Each pair of autosomes in the human karyotype is numbered.

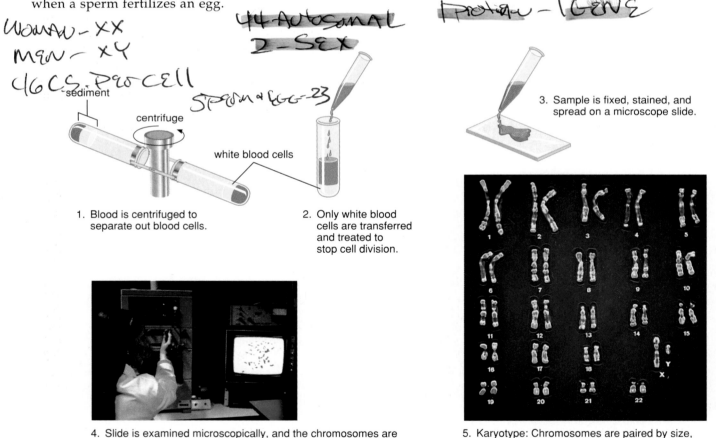

1. Blood is centrifuged to separate out blood cells.

2. Only white blood cells are transferred and treated to stop cell division.

3. Sample is fixed, stained, and spread on a microscope slide.

4. Slide is examined microscopically, and the chromosomes are photographed. Computer arranges the chromosomes into pairs.

5. Karyotype: Chromosomes are paired by size, centromere location, and banding patterns.

Figure 18.1 Human karyotype preparation.
As illustrated here, the stain used can result in chromosomes with a banded appearance. The bands help researchers identify and analyze the chromosomes.

Down Syndrome and Cri du Chat

Two chromosome disorders of interest are Down syndrome and cri du chat.

Down Syndrome.

Down syndrome (Fig. 18.2) is easily recognized by these characteristics: short stature; an eyelid fold; stubby fingers; a wide gap between the first and second toes; a large, fissured tongue; a round head; a palm crease, the so-called simian line; and, unfortunately, mental retardation, which can sometimes be severe.

Down syndrome is also called trisomy 21 because the individual usually has three copies of chromosome 21. In most instances, the egg had two copies instead of one of this chromosome. (In 23% of the cases studied, however, the sperm had the extra chromosome 21.) The chances of a woman having a Down syndrome child increase rapidly with age, starting at about age 35. The frequency of Down syndrome is 1 in 800 births for mothers under 40 years of age and 1 in 80 for mothers over 40 years of age.

Although an older woman is more likely to have a Down syndrome child, most babies with Down syndrome are born to women younger than age 40 because this is the age group having the most babies. Amniocentesis (removing fluid and cells from the amnionic sac surrounding the fetus) followed by karyotyping can detect a Down syndrome child. However, young women are not usually encouraged to undergo this procedure because the risk of complications from the procedure is greater than the risk of having a Down syndrome child. It has been observed that a woman carrying a Down syndrome child sometimes has a lower than usual amount of a substance called α-fetoprotein (AFP) in her blood. These results should certainly be followed up with amniocentesis and karyotyping.

It is known that the genes that cause Down syndrome are located on the bottom third of chromosome 21 (Fig. 18.2b), and extensive investigative work has been directed toward discovering the specific genes responsible for the characteristics of the syndrome. Thus far, investigators have discovered several genes that may account for various conditions seen in persons with Down syndrome. For example, they have located genes most likely responsible for the increased tendency toward leukemia, cataracts, accelerated rate of aging, and mental retardation. The gene for mental retardation, dubbed the *Gart* gene, causes an increased level of purines in the blood, a finding associated with mental retardation. It is hoped that someday it will be possible to control the expression of the *Gart* gene even before birth so that at least this symptom of Down syndrome does not appear.

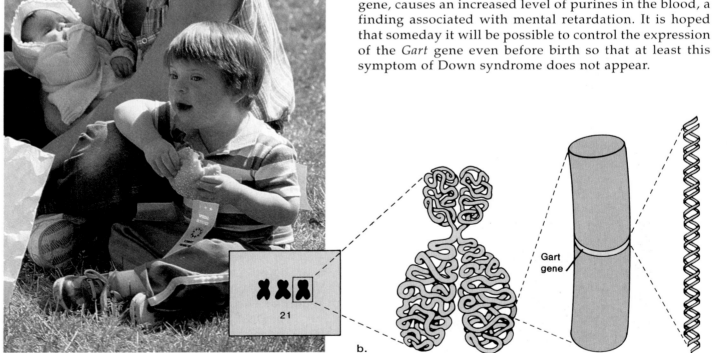

a.

b.

Figure 18.2   Down syndrome.

a. Common characteristics of the syndrome include a wide, rounded face and a fold of the upper eyelids. Mental retardation, along with an enlarged tongue, makes it difficult for a person with Down syndrome to speak distinctly. **b.** Karyotype of an individual with Down syndrome shows an extra chromosome 21. More sophisticated technologies allow investigators to pinpoint the location of specific genes associated with the syndrome. An extra copy of the *Gart* gene, which leads to a high level of purines, may account for the mental retardation seen in persons with Down syndrome.

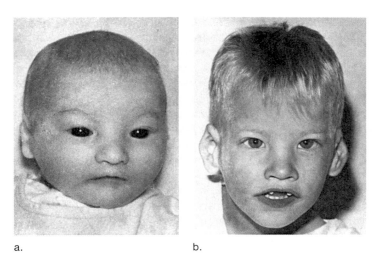

a.   b.

Figure 18.3   Cri du chat syndrome.
a. An infant and **(b)** an older child with this syndrome.

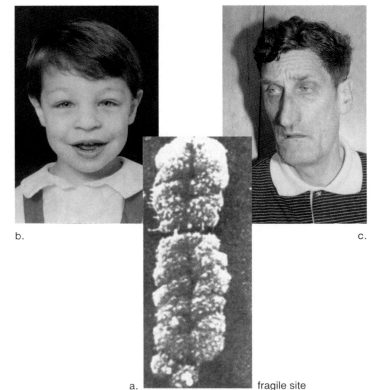

b.   c.

a.   fragile site

Figure 18.4   Fragile X syndrome.
a. An arrow points out the fragile site of this fragile X chromosome.
b. A young person with the syndrome appears normal but **(c)** with age, the elongated face has a prominent jaw, and the ears protrude.

Cri du Chat Syndrome

A chromosome deletion is responsible for *cri du chat* (cat's cry) *syndrome,* which has a frequency of 1 in 50,000 live births (Fig. 18.3). An infant with this syndrome has a moon face, small head, and a cry that sounds like the meow of a cat because of a malformed larynx. An older child has an eyelid fold and misshapen ears that are placed low on the head. Severe mental retardation becomes evident as the child matures. A karyotype shows that a portion of one chromosome 5 is missing (deleted), while the other chromosome 5 is normal, as are all other chromosomes.

Sex Chromosomal Inheritance

The sex chromosomes in humans are called X and Y. Since women are XX, an egg always bears an X, but since males are XY, a sperm can bear an X or a Y. Therefore, the sex of the newborn child is determined by the father. If a Y-bearing sperm fertilizes the egg, then the XY combination results in a male. On the other hand, if an X-bearing sperm fertilizes the egg, the XX combination results in a female. All factors being equal, there is a 50% chance of having a girl or a boy.

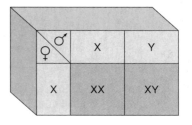

However, for reasons that are not clear, more males than females are conceived, but from then on, the death rate among males is higher than for females. By age 85, there are twice as many females as males.

Fragile X Syndrome

Fragile X syndrome occurs in one in 1,000 male births and one in 2,500 female births. In these individuals, an X chromosome is nearly broken, leaving the tip hanging by a flimsy thread (Fig. 18.4). As children, fragile X syndrome individuals appear to be normal except they may be hyperactive or autistic. Their speech is delayed in development and is often repetitive in nature. As adults, they are short in stature with a long face. The jaw is prominent, and there are big, usually protruding ears (Fig. 18.4*b* and *c*). Males also have large testicles. Stubby hands, lax joints, and a heart defect may also be seen. Mental impairment varies but is more apt to occur in males who have inherited an abnormal X chromosome.

The DNA sequence at the fragile site was isolated and found to have trinucleotide repeats. The base triplet CGG was repeated over and over again. There are about 6 to 50 copies of this repeat in normal persons but over 230 copies in persons with fragile X syndrome. A so-called permutation—between 50 and 230 repeats—is sometimes inherited in both males and females. But only females with a permutation have fragile X children. The reason(s) why a male with a permutation does not have a fragile X child is still being investigated. Nucleotide repeats have been found in several other inherited conditions and is an active area of research today.

Too Many/Too Few Sex Chromosomes

Turner syndrome occurs in one in 6,000 births. The individual is XO with one sex chromosome, an X; the O signifies the absence of a second sex chromosome. These females are short, have a broad chest, and webbed neck. The ovaries, oviducts, and uterus are very small and nonfunctional. Turner females do not undergo puberty or menstruate, and there is a lack of breast development (Fig. 18.5a). They are usually of normal intelligence and can lead fairly normal lives, but they are infertile even if they receive hormone supplements.

Klinefelter syndrome occurs in one in 1,500 births. These males with two or more X chromosomes in addition to a Y chromosome are sterile. The testes and prostate gland are underdeveloped, and there is no facial hair. Also, there may be some breast development (Fig. 18.5b). Affected individuals have large hands and feet and very long arms and legs. They are usually slow to learn but not mentally retarded unless they inherit more than two X chromosomes.

The *triplo-X syndrome* occurs in one in 1,500 births. A triplo-X female has three or more X chromosomes. It might be supposed that the XXX female is especially feminine, but this is not the case. Although in some cases there is a tendency toward learning disabilities, most triplo-X females have no apparent physical abnormalities except that they may have menstrual irregularities, including early onset of menopause.

Jacob syndrome occurs in one in 1,000 births. These XYY males are usually taller than average, suffer from persistent acne, and tend to have speech and reading problems. At one time, it was suggested that these men were likely to be criminally aggressive, but it has since been shown that the incidence of such behavior among them may be no greater than among XY males.

Individuals sometimes are born with the sex chromosomes XO (Turner syndrome), XXY (Klinefelter syndrome), XXX (triplo-X syndrome), and XYY (Jacob syndrome). No matter how many X chromosomes there are, an individual with a Y chromosome is usually a male.

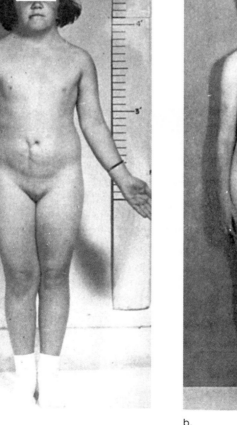

a.   b.

Figure 18.5 Abnormal sex chromosome inheritance.
a. Female with Turner (XO) syndrome, which includes a web neck, short stature, and immature sexual features. **b.** A male with Klinefelter (XXY) syndrome, which is marked by small testes and development of the breasts in some cases.

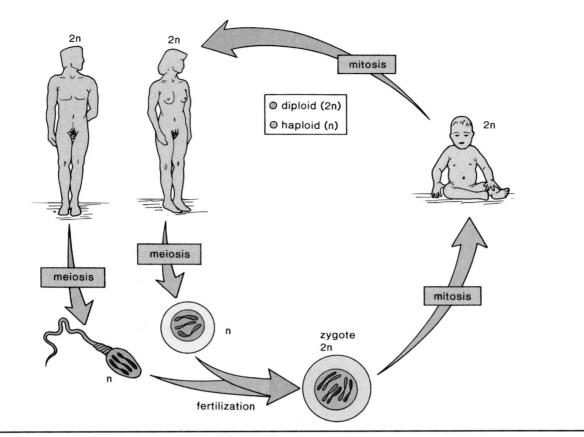

Figure 18.6   Life cycle of humans.
Meiosis in males is a part of sperm production, and meiosis in females is a part of egg production. When a haploid sperm fertilizes a haploid egg, the zygote is diploid. The zygote undergoes mitosis as it develops into a newborn child. Mitosis continues after birth until the individual reaches maturity, and then the life cycle begins again.

TABLE 18.1			
Mitosis Versus Meiosis			
Location	**Cell Division**	**Description**	**Result**
Somatic (body) cells	Mitosis	2n (diploid) → 2n (diploid)	Growth and repair
Sex organs	Meiosis	2n (diploid) → n (haploid)	Gamete production

18.2   Human Life Cycle

The human life cycle involves growth and sexual reproduction (Fig. 18.6). During growth, cells divide by a process called *mitosis,* which ensures that each and every cell has a complete number of chromosomes. Sexual reproduction requires the production of sex cells, which have half the number of chromosomes. A type of cell division called *meiosis* reduces the chromosome number by one-half.

Meiosis occurs in the sex organs. In males, it produces the cells that become sperm; in females, it produces the cells that become eggs. The sperm and the egg are the sex cells, or **gametes.** Gametes contain half the number of chromosomes compared to **somatic** (body) **cells**—one chromosome from each of the pairs of chromosomes. This is called the **haploid (n)** number of chromosomes; the haploid number of chromosomes in humans is 23.

A new individual comes into existence when a sperm fertilizes an egg. The resulting zygote has the **diploid (2n)** number of chromosomes. Each parent contributes one chromosome to each of the pairs of chromosomes present. As the individual develops, *mitosis* occurs and each somatic (body) cell has the diploid number of chromosomes. In humans, the diploid number is 46 because there are 23 pairs of chromosomes.

Table 18.1 summarizes the major differences between mitosis and meiosis in multicellular animals.

The life cycle of humans requires two types of cell division: mitosis and meiosis.

18.3 Mitosis

Mitosis is cell division that produces *two daughter cells, each with the same number and kinds of chromosomes as the parent cell, the cell that divides.*[1] Therefore, following mitotic cell division, the parent cell and the daughter cells are genetically identical. Mitosis occurs as part of the cell cycle.

Cell Cycle

The **cell cycle** consists of mitosis and interphase. The cell divides and then it enters interphase before dividing again. Therefore, **interphase** is the interval of time between cell divisions. The length of time required for the entire cell cycle varies according to the organism and even the type of cell within the organism, but 18–24 hours is typical for animal cells. Mitosis lasts less than an hour to slightly more than 2 hours; for the rest of the time, the cell is in interphase.

It used to be said that interphase was a resting stage, but we now know that this is not the case. The organelles are metabolically active and are carrying on their normal functions. If the cell is going to divide, *DNA replication* occurs. During replication, DNA is copied and each chromosome becomes duplicated. A duplicated chromosome is composed of two sister chromatids. *Sister chromatids* are genetically identical—they contain the same genes. Also, organelles, including the *centrioles*, duplicate. A nondividing cell has one pair of centrioles, but in a cell that is going to divide, this pair duplicates, and there are two pairs of centrioles outside the nucleus.

The cell cycle includes mitosis and interphase. During interphase, DNA replication results in each chromosome having sister chromatids. The organelles, including centrioles, also duplicate during interphase.

Overview of Mitosis: 2n → 2n

During interphase, the chromosomes are indistinct chromatin, but when mitosis is going to occur, chromatin becomes condensed, and the chromosomes become visible. Before mitosis begins, the parental cell is 2n, and the sister chromatids are held together in a region called the **centromere.** At the completion of mitosis, each chromosome consists of a single **chromatid.**

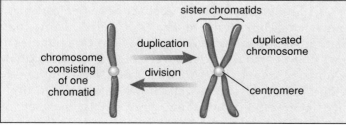

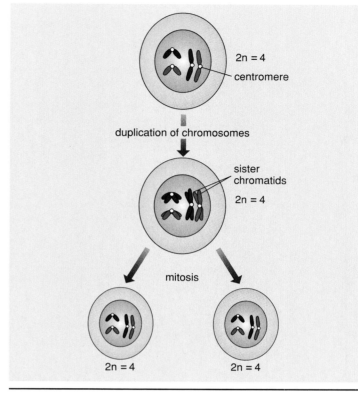

Figure 18.7 Overview of mitosis.
The blue chromosomes were inherited from one parent and the red chromosomes were inherited from the other parent.

Figure 18.7 gives an overview of mitosis; for simplicity, only four chromosomes are depicted. (In determining the number of chromosomes, it is necessary to count only the number of independent centromeres.) During mitosis, the centromeres divide, the sister chromatids separate, and one of each kind goes into each daughter cell. Therefore, each daughter cell gets a complete set of chromosomes and is 2n. (Following separation, each chromatid is called a chromosome.) Since each daughter cell receives the same number and kinds of chromosomes as the parent cell, each is genetically identical to each other and to the parent cell.

Mitosis occurs in humans when tissues grow or when repair occurs. Following fertilization, the zygote begins to divide mitotically, and mitosis continues during development and the life span of the individual. In the adult, tissues differ as to their ability to divide; nervous and muscle tissue cells seem to lose the ability to divide. Epidermal cells, which line the respiratory tract and the digestive tract and form the outer layer of the skin, divide continuously. Stem cells in the red bone marrow divide to produce millions of blood cells every day. Whenever repair takes place, as when a broken bone is mended, mitosis has occurred.

Following mitosis, each of two daughter cells has the same number and kinds of chromosomes as the parental cell. Body (somatic) cells undergo mitosis, a process that is necessary to the growth and repair of tissues.

---

1. The terms *mitosis* and *meiosis* technically refer only to nuclear division, but for convenience, they are used here to refer to the division of the entire cell.

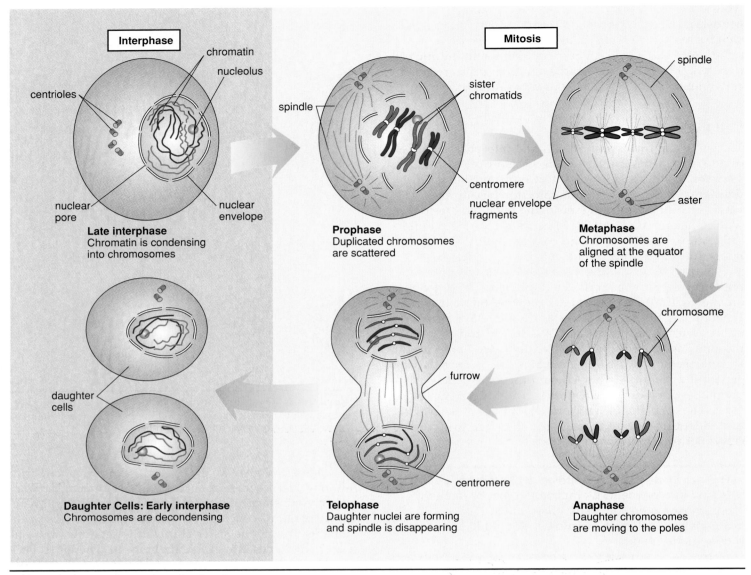

Figure 18.8 Interphase and mitosis.
The blue chromosomes were inherited from one parent and the red chromosomes were inherited from the other parent.

Stages of Mitosis

As an aid in describing the events of mitosis, the process is divided into four phases: prophase, metaphase, anaphase, and telophase (Fig. 18.8). Although it is helpful to depict the stages of mitosis as if they were separate, they are actually continuous and flow from one stage to another with no noticeable interruption.

Prophase

It is apparent during **prophase** that cell division is about to occur. The two pairs of centrioles outside the nucleus begin moving away from each other toward opposite ends of the nucleus. **Spindle fibers** appear between the separating centriole pairs, the nuclear envelope begins to fragment, and the nucleolus begins to disappear.

The chromosomes are now visible. Each is composed of sister chromatids held together at a centromere. Spindle

fibers attach to the centromeres as the chromosomes continue to shorten and to thicken. At prophase, chromosomes are randomly placed in the nucleus and have not yet aligned at the equator of the spindle.

Structure of the Spindle   At the end of prophase, a cell has a fully formed spindle. A **spindle** has poles, asters, and fibers. The **asters** are arrays of short microtubules that radiate from the poles, and the fibers are bundles of microtubules that stretch between the poles. Microtubule organizing centers (MTOC) are associated with the centrioles at the poles. It is well known that a MTOC organizes microtubules including, presumably, those of the spindle. It is possible that the centrioles assist in this function, but it could also be that their location at the poles of a spindle simply ensures that each daughter cell receives a pair of centrioles.

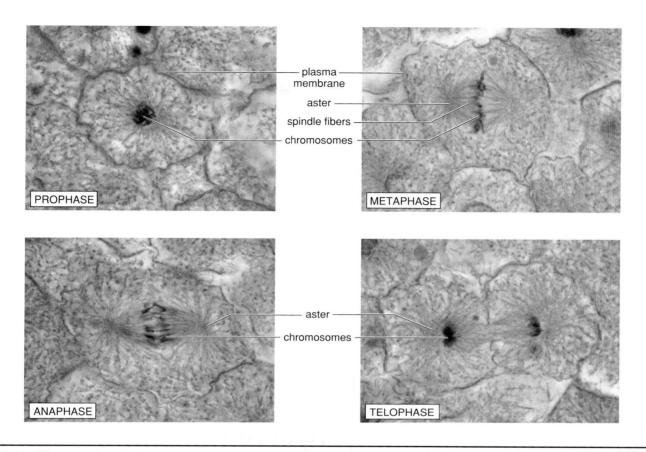

Figure 18.9   Micrographs of mitosis occurring in a whitefish embryo.

Metaphase

During **metaphase,** the nuclear envelope is fragmented, and the spindle occupies the region formerly occupied by the nucleus. The chromosomes are now at the *equator* (center) of the spindle. Metaphase is characterized by a fully formed spindle, with the chromosomes, each having two sister chromatids, aligned at the equator (Fig. 18.9). At the close of metaphase, the centromeres uniting the chromatids split.

Anaphase

At the start of **anaphase,** the sister chromatids separate. *Once separated, the chromatids are called chromosomes.* Separation of the sister chromatids ensures that each cell receives a copy of each type of chromosome and thereby has a full complement of genes. During anaphase, the daughter chromosomes move to the poles of the spindle. Anaphase is characterized by the diploid number of chromosomes moving toward each pole.

Function of the Spindle   The spindle brings about chromosome movement. Two types of spindle fibers are involved in the movement of chromosomes during anaphase. One type extends from the poles to the equator of the spindle; there they overlap. As mitosis proceeds, these fibers increase in length, and this helps push the chromosomes apart. The chromosomes themselves are attached to other spindle fibers that simply extend from their centromeres to the poles.

These fibers get shorter and shorter as the chromosomes move toward the poles, and eventually disappear. These fibers pull the chromosomes apart.

Spindle fibers, as stated, are composed of microtubules. Microtubules can assemble and disassemble by the addition or subtraction of tubulin (protein) subunits. This is what enables spindle fibers to lengthen and shorten and what ultimately causes the movement of the chromosomes.

Telophase

Telophase begins when the chromosomes arrive at the poles. During telophase, the chromosomes become indistinct chromatin again. The spindle disappears as nucleoli appear, and nuclear envelopes form in each cell. Telophase is characterized by the presence of two daughter nuclei.

In animal cells, a slight indentation called a **cleavage furrow** passes around the circumference of the cell. Actin filaments form a contractile ring, and as the ring gets smaller and smaller, the cleavage furrow pinches the cell in half. As a result, each cell becomes enclosed by its own plasma membrane.

Following mitosis, each daughter cell is 2n. When the sister chromatids separate during anaphase, each newly forming cell receives the same number and kinds of chromosomes as the parental cell.

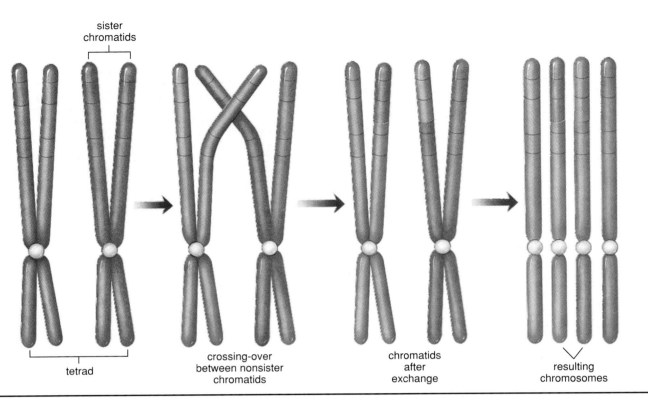

Figure 18.10 Crossing-over.
When homologous chromosomes are in synapsis, the nonsister chromatids exchange genetic material. Following crossing-over, there is a different combination of genes on each chromatid.

18.4 Meiosis   *Germ cells eggs + sperm*

Meiosis, which requires two cell divisions, results in *four daughter cells, each having one of each kind of chromosome and therefore half the number of chromosomes as the parent cell.*[2] The parent cell has the 2n number of chromosomes, while the daughter cells have the n number of chromosomes. Therefore, meiosis is often called reduction division. Following meiotic cell division, the daughter cells are not genetically identical.

Overview of Meiosis: 2n → n

Meiosis results in four daughter cells because it consists of two divisions called **meiosis I** and **meiosis II.** Before meiosis I begins, each chromosome has duplicated and is composed of two sister chromatids. The parental cell is 2n. Recall that when a cell is 2n, the chromosomes occur in pairs. For example, the 46 chromosomes of humans occur in 23 pairs of chromosomes. These pairs are called **homologous chromosomes.**

During meiosis I, the homologous chromosomes of each pair come together and line up side by side due to a means of attraction still unknown. This so-called synapsis results in a tetrad, an association of four chromatids that stay in close proximity until they separate. During synapsis, nonsister chromatids may exchange genetic material. The exchange of genetic material between chromatids is

called **crossing-over.** Crossing-over recombines the genes of the parental cell without the loss or gain of genetic material (Fig. 18.10).

Following synapsis during meiosis I, the homologous chromosomes of each pair separate. This separation means that one chromosome from each homologous pair will be found in each daughter cell. There are no restrictions as to which chromosome goes to each daughter cell, and therefore, all possible combinations of chromosomes occur within the daughter cells.

Notice that following meiosis I, the daughter cells have half the number of chromosomes and the chromosomes are still duplicated (Fig. 18.11). Again, counting the number of centromeres tells the number of chromosomes in each daughter cell.

During meiosis I, homologous chromosomes separate, and the daughter cells receive one of each pair. The daughter cells are not genetically identical. The chromosomes are still duplicated.

When meiosis II begins, the chromosomes are still duplicated. Therefore, no duplication of chromosomes is needed between meiosis I and meiosis II. The chromosomes are *dyads* because each one is composed of two sister chromatids. During meiosis II, the sister chromatids sepa-

---

2. See footnote 1.

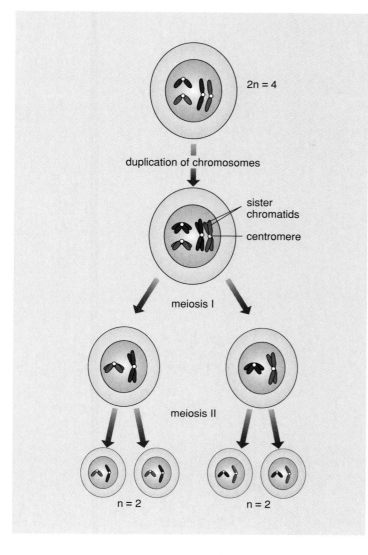

duplication of chromosomes

sister chromatids

centromere

meiosis I

meiosis II

2n = 4

n = 2   n = 2

Figure 18.11   Overview of meiosis.
Following duplication of chromosomes, the parent cell undergoes two divisions, meiosis I and meiosis II. During meiosis I, homologous chromosomes separate, and during meiosis II, chromatids separate. The final daughter cells are haploid. (The blue chromosomes were inherited from one parent, and the red chromosomes were inherited from the other parent.)

rate in each of the cells from meiosis I. Each of the resulting four daughter cells has the haploid number of chromosomes (Fig. 18.11).

During meiosis II, the sister chromatids separate, and the resulting four daughter cells are each haploid.

The Importance of Meiosis

The *gametes* are the sperm in males and eggs in females. In humans, meiosis occurs in the testes and ovaries during the production of the gametes. Meiosis is not complete in egg production until fertilization occurs. Because of meiosis, the chromosome number stays constant in each generation of humans. When a haploid sperm fertilizes a haploid egg, the new individual has the diploid number of chromosomes. There are three ways in which the new

individual is assured a different combination of genes than either parent:

1. Crossing-over recombines the genes on the sister chromatids of homologous pairs.
2. Following meiosis, each gamete has a different combination of chromosomes.
3. Upon fertilization, recombination of chromosomes occurs.

Following meiosis, each of four daughter cells has the haploid number of chromosomes, whereas the parental cell was diploid. Meiosis takes place in the testes and ovaries during the production of the gametes. It ensures that the chromosome number stays constant in each generation and that the new individual has a different combination of genes than either parent.

Stages of Meiosis

The same four stages in mitosis—prophase, metaphase, anaphase, and telophase—occur during both meiosis I and meiosis II.

The First Division

The stages of meiosis I are diagrammed in Figure 18.12*a*. During *prophase I*, the spindle appears while the nuclear envelope fragments and the nucleolus disappears. The homologous chromosomes, each having two sister chromatids, undergo **synapsis,** forming tetrads. Crossing-over occurs now, but for simplicity, this event has been omitted from Figure 18.12. In *metaphase I*, tetrads line up at the equator of the spindle. During *anaphase I*, homologous chromosomes of each pair separate and move to opposite poles of the spindle. During *telophase I*, nucleoli appear and nuclear envelopes form as the spindle disappears. In certain species, the plasma membrane furrows to give two cells, and in others, the second division begins without benefit of complete furrowing. Regardless, each daughter cell contains only one chromosome from each homologous pair. The chromosomes are dyads, and each has two sister chromatids. No replication of DNA occurs during a period of time called *interkinesis.*

The Second Division

The stages of meiosis II for an animal cell are diagrammed in Figure 18.12*b*. At the beginning of *prophase II*, a spindle appears while the nuclear envelope fragments and the nucleolus disappears. Dyads (one dyad from each pair of homologous chromosomes) are present, and each attaches to the spindle independently. During *metaphase II*, the dyads are lined up at the equator. At the close of metaphase, the centromeres split. During *anaphase II*, the sister chromatids of each dyad separate and move toward the poles. Each pole receives the same number of chromosomes. In *telophase II*, the spindle disappears as nuclear envelopes form. The plasma membrane furrows to give two complete cells, each of which has the haploid, or n, number of chromosomes. Since each cell from meiosis I undergoes meiosis II, there are four daughter cells altogether.

Meiosis involves two cell divisions. During meiosis I, tetrads form and crossing-over occurs. Homologous chromosomes separate, and each daughter cell receives a pair of sister chromatids. During meiosis II, separation of chromatids in daughter cells from meiosis I results in four daughter cells, each with the haploid number of chromosomes.

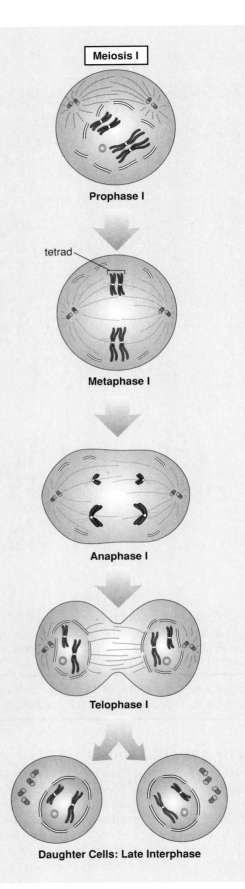

Meiosis I

Prophase I

tetrad

Metaphase I

Anaphase I

Telophase I

Daughter Cells: Late Interphase

a.

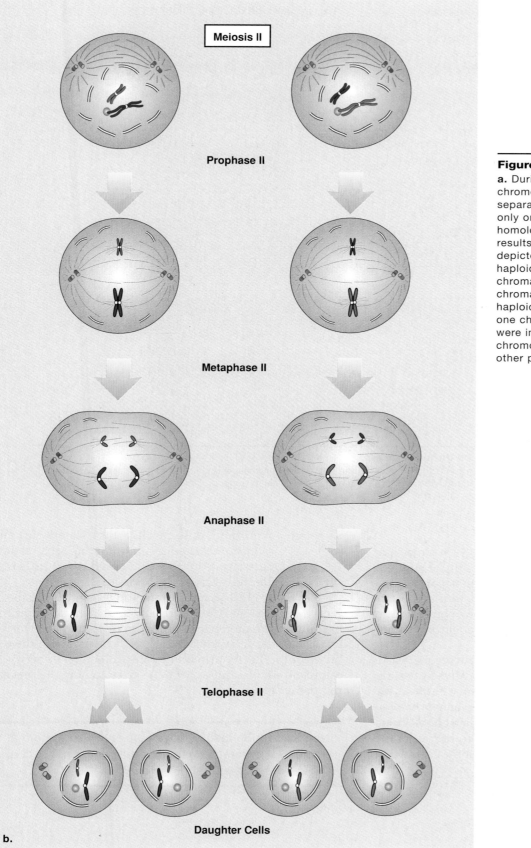

Figure 18.12   Meiosis I and meiosis II.
a. During meiosis I, homologous chromosomes undergo synapsis and then separate so that each daughter cell has only one chromosome from each original homologous pair. For simplicity's sake, the results of crossing-over have not been depicted. Notice that each daughter cell is haploid and each chromosome still has two chromatids. **b.** During meiosis II, sister chromatids separate. Each daughter cell is haploid, and each chromosome consists of one chromatid. (The blue chromosomes were inherited from one parent, and the red chromosomes were inherited from the other parent.)

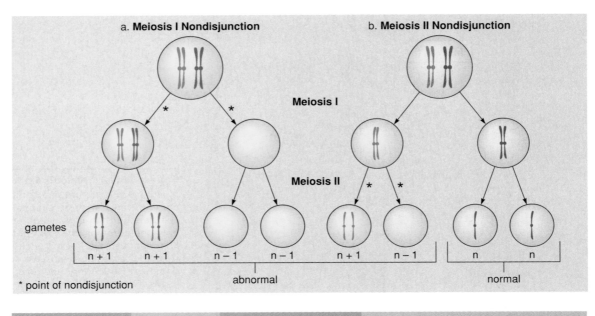

c.

Syndrome	Sex	Chromosomes	Frequency	
			Abortuses	*Births*
Down	M or F	Trisomy 21	1/40	1/800
Turner	F	XO	1/18	1/6,000
Klinefelter	M	XXY (or XXXY)	0	1/1,500
Triplo-X	F	XXX (or XXXX)	0	1/1,500
Jacob	M	XYY	?	1/1,000

Figure 18.13   Nondisjunction of autosomes during meiosis.
a. Nondisjunction can occur during meiosis I if homologous chromosomes fail to separate and **(b)** during meiosis II if the sister chromatids fail to separate completely. In either case, certain abnormal gametes carry an extra chromosome (n + 1) or lack a chromosome (n − 1). **c.** Frequency of syndromes.

Nondisjunction

Abnormal chromosome constitutions (Fig. 18.13) can be due to nondisjunction. **Nondisjunction** is the failure of homologous chromosomes or sister chromatids to separate during the formation of gametes. Nondisjunction can occur during meiosis I if the homologous chromosomes fail to separate or during meiosis II if the daughter chromosomes fail to separate.

Figure 18.13a shows nondisjunction during meiosis I. The homologous chromosomes are duplicated in the parental cell. The daughter cells are abnormal because the daughter cell on the left received both members of the homologous pair and the daughter cell on the right received neither. Following meiosis II, all four daughter cells are abnormal. If the chromosomes are chromosome 21, the first two daughter cells could result in Down syndrome after fertilization. If the chromosomes were X chromosomes,

Klinefelter syndrome could result. If they were Y chromosomes, XYY syndrome could result following fertilization. The two daughter cells on the right could result in Turner syndrome after fertilization (YO zygotes die off).

In Figure 18.13b, nondisjunction occurs during meiosis II. Meiosis I occurred normally and each daughter cell has one of the members of the homologous pair. During meiosis II, the sister chromatids fail to separate in the cell on the left. The abnormal daughter cells (n + 1 and n − 1) can result in the same syndromes discussed in the preceding paragraph.

Nondisjunction results in gametes with an abnormal number of chromosomes and, following fertilization, the syndromes listed in Figure 18.13.

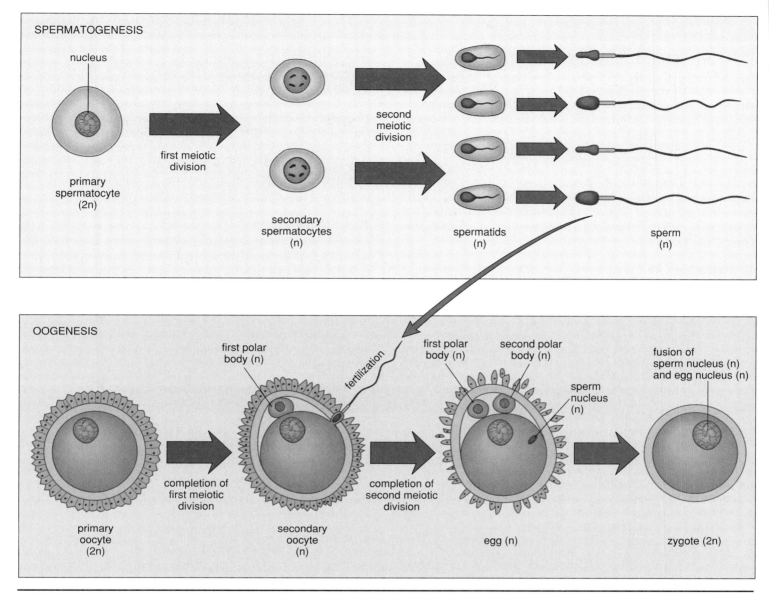

Figure 18.14 Spermatogenesis and oogenesis.
Spermatogenesis produces four viable sperm, whereas oogenesis produces one egg and two polar bodies. Notice that oogenesis does not go to completion unless the secondary oocyte is fertilized. In humans, both sperm and egg have 23 chromosomes each; therefore, following fertilization, the zygote has 46 chromosomes.

Spermatogenesis and Oogenesis

Spermatogenesis and oogenesis occur in the sex organs—the testes in males and the ovaries in females. During **spermatogenesis,** sperm are produced, and during **oogenesis,** eggs are produced. The gametes appear differently in the two sexes (Fig. 18.14), and meiosis is different, too. The process of meiosis in males always results in four cells that become sperm. Meiosis in females produces only one egg. Meiosis I results in one large cell called a secondary oocyte and one polar body. After meiosis II, there is one egg and two polar bodies. The **polar bodies** are a way to discard unnecessary chromosomes while retaining much of the cytoplasm in the egg. The cytoplasm serves as a source of nutrients for the developing embryo.

Spermatogenesis, once started, continues to completion, and mature sperm result. In contrast, oogenesis does not necessarily go to completion. Only if a sperm fertilizes the secondary oocyte does it undergo meiosis II and become an egg. Regardless of this complication, however, both the sperm and the egg contribute the haploid number of chromosomes to the zygote (fertilized egg). In humans, each contributes 23 chromosomes.

Spermatogenesis, which occurs in the testes of males, produces sperm. Oogenesis, which occurs in the ovaries of females, produces eggs. Meiosis is a part of spermatogenesis and oogenesis; therefore, both sperm and egg are haploid.

SUMMARY

18.1 Chromosome Inheritance

A human karyotype ordinarily shows 22 pairs of autosomes and one pair of sex chromosomes. The sex pair is an X and a Y chromosome in males and two X chromosomes in females. Abnormalities do occur. The major inherited autosomal abnormality is Down syndrome, in which the individual inherits three copies of chromosome 21. Cri du chat occurs when part of chromosome 5 is deleted. Examples of abnormal sex chromosome inheritance are a fragile X chromosome and abnormal chromosome numbers: Turner syndrome (XO), Klinefelter syndrome (XXY), triplo-X syndrome (XXX) , and Jacob syndrome (XYY).

18.2 Human Life Cycle

The life cycle of higher organisms requires two types of cell divisions: mitosis and meiosis.

18.3 Mitosis

Mitosis assures that all cells in the body have the diploid number and the same kinds of chromosomes. It is made up of four stages: prophase, metaphase, anaphase, and telophase. The cell cycle includes an additional stage termed interphase. During interphase, DNA replication causes each chromosome to have sister chromatids. When mitosis occurs, the chromatids separate, and each newly forming cell receives the same number and kinds of chromosomes as the original cell. The cytoplasm is partitioned by furrowing in human cells.

18.4 Meiosis

Meiosis involves two cell divisions. During meiosis I, the homologous chromosomes (following crossing-over between nonsister chromatids) separate, and during meiosis II, the sister chromatids separate. The result is four cells with the haploid number of chromosomes in a single copy. Meiosis is a part of gamete formation in humans.

Spermatogenesis in males usually produces four viable sperm, while oogenesis in females produces one egg and two polar bodies.

STUDYING THE CONCEPTS

1. Describe the normal karyotype of a human being. What is the difference between a male and a female karyotype?   380

2. Describe Down syndrome, the most common autosome abnormality in humans. What is cri du chat syndrome?   381–82

3. Name a chromosome abnormality other than an abnormal number of sex chromosomes. List four sex chromosome abnormalities that have to do with an abnormal number of sex chromosomes.   382–83

4. Draw a diagram describing the human life cycle. What is mitosis? What is meiosis?   384

5. Describe the stages of mitosis, including in your description the terms centrioles, nucleolus, spindle, and furrowing.   386–87

6. How do the terms diploid (2n) and haploid (n) pertain to meiosis?   388

7. Describe the stages of meiosis I, including the terms tetrad and dyad in your description.   390

8. Compare the stages of meiosis II to a mitotic division.   390

9. What is the importance of mitosis and meiosis in the life cycle of humans?   384, 89

10. What is nondisjunction, and how does it occur?   392

11. How does spermatogenesis in males compare to oogenesis in females?   393

APPLYING TECHNOLOGY

Your study of chromosome inheritance is supported by these available technologies:

Exploring the Internet

The Mader Home Page provides further resources for studying this chapter.

`http://www.mhhe.com/sciencemath/biology/mader/`

(Click on *Human Biology.*)

Explorations in Cell Biology & Genetics CD-ROM

Exploring Meiosis: Down Syndrome (#10)   The projected incidence of Down syndrome varies as students change the age of the mother and the age of the father separately. An animation shows the movement of chromosomes during normal meiosis I and during nondisjunction.   (1)*

*Level of difficulty

APPLYING YOUR KNOWLEDGE

Concepts

1. What is the reason that Turner, Klinefelter, and triplo-X syndromes can all occur as a result of nondisjunction in either the sperm or egg, but XYY can occur only as a result of nondisjunction in the sperm?

2. What are four changes that occur during prophase that are reversed during telophase?

3. Turner syndrome and Klinefelter syndrome are referred to as 2n − 1 and 2n + 1, respectively. What is the meaning of this terminology or symbol?

4. Normally, the two cells that result from the first mitotic division of the zygote stay together. If, instead, the two cells separate and each develops on its own, what is the result?

Bioethical Issue

For years, mothers around the age of 40 have been warned that they hold an increased risk of delivering a baby with Down syndrome. One option for these mothers-to-be is a test to determine whether their fetus carries Down syndrome.

Sometimes couples decide they lack the emotional energy or financial strength to raise a Down syndrome child. In these situations, parents may use such tests to determine whether to keep an unborn fetus. The decision is never easy or painless.

Still, some human rights groups argue that prenatal tests to uncover genetic diseases are unethical. They argue that these babies deserve to live, regardless of their mental disabilities. Such people may say parents have no right to "change their mind" about a child upon discovering that the child is Down syndrome.

Are tests for Down syndrome and other genetic diseases ethical? Do parents have the right to use such tests to decide whether to continue a pregnancy? Why or why not?

TESTING YOUR KNOWLEDGE

1. The arrangement of an individual's chromosomes according to homologous pairs is called a _____.

2. The karyotype of males includes the sex chromosomes _____, and the karyotype of females includes the sex chromosomes _____.

3. The shorthand way to designate the diploid number of chromosomes is _____.

4. If the parent cell has 24 chromosomes, the daughter cells following mitosis will have _____ chromosomes.

5. As the organelles called _____ separate and move to the poles, the spindle fibers appear.

6. During meiosis I, the _____ separate, and during meiosis II the _____ separate.

7. Meiosis in males is a part of _____, and meiosis in females is a part of _____.

8. There is a _____ chance of a newborn being a male or a female.

To answer questions 9–15, use this key:
   a. Down syndrome
   b. Turner syndrome
   c. Klinefelter syndrome
   d. cri du chat syndrome
   e. triplo-X syndrome

9. XYY _____

10. extra chromosome 21 _____

11. XXX _____

12. deletion in chromosome 5 _____

13. XO _____

14. due to an autosomal nondisjunction _____

15. due to a chromosomal mutation _____

SELECTED KEY TERMS

anaphase Stage in mitosis during which sister chromatids separate, forming daughter chromosomes.   387

aster An array of short microtubules that extend outward from a spindle pole in animal cells during cell division.   386

autosome Chromosome other than a sex chromosome.   380

cell cycle Repeating sequence of events in eukaryotic cells consisting of interphase, when growth and DNA synthesis occurs, and mitosis, when cell division occurs.   385

centromere Constricted region of a chromosome where sister chromatids are attached to one another and where the chromosome attaches to a spindle fiber.   385

chromatid (KROH-muh-tid) One of the two identical parts of a chromosome following replication of DNA.   385

cleavage furrow Indentation that deepens to divide the cytoplasm during cell division.   387

crossing-over Exchange of corresponding segments of genetic material between nonsister chromatids of homologous chromosomes during synapsis of meiosis I.   388

diploid (2n) Number of chromosomes in the body cells; twice the number of chromosomes found in gametes.   384

gamete One of two types of reproductive cells that join in fertilization to form a zygote; most often an egg or a sperm.   384

haploid (n) Half the diploid number; the number of chromosomes in the gametes.   384

homologous chromosome Similarly constructed; with chromosomes, the same appearance and containing genes for the same traits.   388

interphase Interval between successive cell divisions; during this time, the chromosomes are extended and DNA replication and growth are occurring.   385

karyotype (KAR-ee-uh-typ) Arrangement of all the chromosomes within a cell by pairs in a fixed order.   380

meiosis (my-OH-sis) Type of cell division occurring during the production of gametes; results in four daughter cells with the haploid number of chromosomes.   388

meiosis I That portion of meiosis during which homologous chromosomes come together and then later separate.   388

meiosis II That portion of meiosis during which sister chromatids separate.   388

metaphase Stage in mitosis during which chromosomes are at the equator of the mitotic spindle.   387

mitosis Type of cell division in which daughter cells receive the exact chromosome and genetic makeup of the parent cell; occurs during growth and repair.   385

nondisjunction Failure of homologous chromosomes to separate during meiosis I or sister chromatids to separate during meiosis II when gametogenesis is occurring.   392

oogenesis (oh-oh-JEN-eh-sis) Production of an egg in females by the processes of meiosis and maturation.   393

polar body Nonfunctioning daughter cell that has little cytoplasm and is formed during oogenesis.   393

prophase Early stage in mitosis during which chromatin condenses so that chromosomes appear.   386

sex chromosome Chromosome responsible for the development of characteristics associated with gender; an X or Y chromosome.   380

somatic cell In animals, a body cell excluding those that undergo meiosis and become a sperm or egg.   384

spermatogenesis (SPUR-mah-toh-JEN-eh-sis) Production of sperm in males by the process of meiosis and maturation.   393

spindle Structure consisting of fibers, poles, and asters (if animal cell) that brings about the movement of chromosomes during cell division.   386

spindle fiber Microtubule bundle in eukaryotic cells that is involved in the movement of chromosomes during mitosis and meiosis.   386

synapsis Attracting and pairing of homologous chromosomes during prophase I of meiosis.   390

telophase Stage of mitosis during which diploid number of daughter chromosomes are located at each pole.   387

Chapter 19

Genes and Medical Genetics

Chapter Outline

Figure 19.1   Inheritance.
We know that physical traits pass from parent to child. What about behavioral traits? Does the little girl like arts and crafts because of her mother's influence or because of the genes she inherited?

ntelligent. Homosexual. Loving. Aggressive. Over-weight. Do any of these characteristics describe you? If so, you can probably place some of the responsibility on your genes.

Biologists have long linked physical **traits,** such as facial characteristics, to the genes. Now, after deciphering the human body's vast network of genes, they have extended the influence of genes to our most intimate traits. It's possible the findings could lead to new treatments for various ills. For example, some biologists are studying newfound genes involved in obesity in order to create drugs that fight fat. Others are engineering lab mice without specific genes for aggression, watching the animals to see how they act.

These studies have irked some social scientists who argue that behavioral traits may be controlled to a degree by genes but childrearing, peer groups, and other social conditions also shape the personality. Even though it might be a 50/50 situation, the classic "nature" versus "nurture" debate continues (Fig. 19.1).

Although we may never know just how environment and genetics combine to make us who we are, the hunt goes on. One thing is certain: dozens of new behavior-related genes are sure to surface in the future. Even so, very few believe that behavior is predetermined by our genes.

Certainly physical features of human beings and perhaps behavioral features are controlled to a degree by the genes. But what are the genes? This chapter shows that it is sometimes helpful to think of genes as being units of

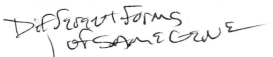

chromosomes, particularly when we want to calculate the chances of passing on a genetic disorder. But we have already mentioned that genes are actually DNA molecules that specify protein synthesis. We are now beginning to decipher how this function can directly affect the structure of the cell and the organism.

Alternate forms of a gene having the same position (locus) on a pair of chromosomes and affecting the same trait are called **alleles**. It is customary to designate an allele by a letter, which represents the specific characteristic it controls; a **dominant allele** is assigned an uppercase (capital) letter, while a **recessive allele** is given the same letter, lowercase. In humans, for example, unattached (free) earlobes are dominant over attached earlobes, so a suitable key would be *E* for unattached earlobes and *e* for attached earlobes.

Since autosomal alleles occur in pairs, the individual normally has two alleles for a trait. Just as one of each pair of chromosomes is inherited from each parent, so too is one of each pair of alleles inherited from each parent.

19.1 Genotype and Phenotype

Figure 19.2 shows three possible fertilizations and the resulting genetic makeup of the zygote and, therefore, the individual. In the first instance, the chromosome of both the sperm and egg carries an *E*. Consequently, the zygote and subsequent individual have the alleles *EE*, which may be called a *homozygous* (pure) *dominant* genotype. The word **genotype** refers to the genes of the individual. A person with genotype *EE* obviously has unattached earlobes. The physical appearance of the individual, in this case unattached earlobes, is called the **phenotype.**

In the second fertilization, the zygote received two recessive alleles (*ee*), and the genotype is called *homozygous* (pure) *recessive*. An individual with this genotype has attached earlobes. In the third fertilization, the resulting individual has the alleles *Ee,* which is called a *heterozygous* genotype. A heterozygote shows the dominant characteristic; therefore, the phenotype of this individual is unattached earlobes.

These examples show that a dominant allele contributed from only one parent can bring about a particular phenotype. A recessive allele must be received from both parents to bring about the recessive phenotype.

The genotype, whether homozygous dominant (**EE**), homozygous recessive (**ee**), or heterozygous (**Ee**), tells what genes a person carries. The phenotype—for example, attached or free earlobes—tells what the person looks like.

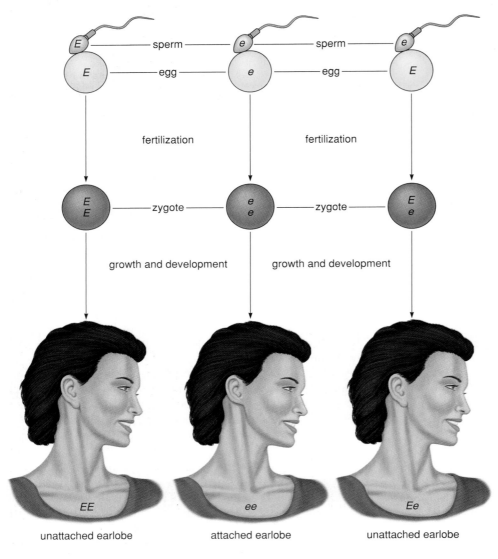

sperm — sperm

egg — egg

fertilization

fertilization

zygote — zygote

growth and development

growth and development

EE
unattached earlobe

ee
attached earlobe

Ee
unattached earlobe

Figure 19.2 Genetic inheritance.
Individuals inherit a minimum of two alleles for every characteristic of their anatomy and physiology. The inheritance of a single dominant allele (*E*) causes an individual to have unattached earlobes; two recessive alleles (*ee*) cause an individual to have attached earlobes. Notice that each individual receives one allele from the father (by way of a sperm) and one allele from the mother (by way of an egg).

CHAPTER NINETEEN *Genes and Medical Genetics*

CLASSIFIES OF GENES (handwritten)

399

19.2 Dominant/Recessive Traits

The alleles designated by *E* and *e* are on the chromosomes. An individual has two alleles for each trait because a chromosome pair carries alleles for the same traits. How many alleles for each trait will be in the gametes?

Forming the Gametes

During gametogenesis, the chromosome number is reduced. Whereas the individual has 46 chromosomes, a gamete has only 23 chromosomes. (If this did not happen, each new generation of individuals would have twice the number of chromosomes as their parents.) Reduction of the chromosome number occurs when the pairs of chromosomes separate as meiosis occurs. Since the alleles are on the chromosomes, they also separate during meiosis, and therefore, the gametes carry only one allele for each trait. If an individual carried the alleles *EE*, all the gametes would carry an *E* since that is the only choice. Similarly, if an individual carried the alleles *ee*, all the gametes would carry an *e*. What if an individual were *Ee*? Figure 19.3 shows that half of the gametes would carry an *E* and half would carry an *e*. Figure 19.4 gives the genotypes for certain other traits in humans, and you can practice deciding what alleles the gametes would carry for these genotypes.

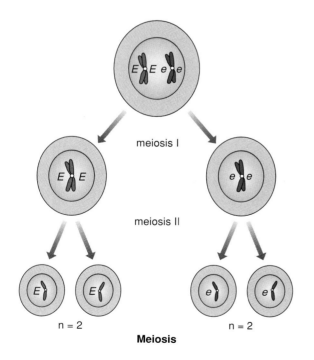

Figure 19.3 Gametogenesis.
Because the pairs of chromosomes separate during meiosis, which occurs during gametogenesis, the gametes have only one allele for each trait.

a.   Widow's peak: *WW* or *Ww*

b.   Continuous hairline: *ww*

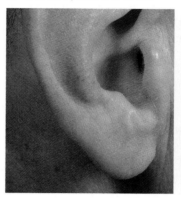
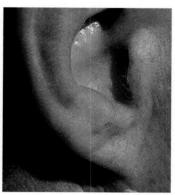

c.   Unattached earlobes: *EE* or *Ee*

d.   Attached earlobes: *ee*

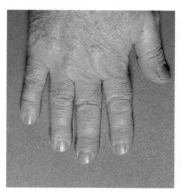
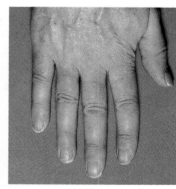

e.   Short fingers: *SS* or *Ss*

f.   Long fingers: *ss*

g.   Freckles: *FF* or *Ff*

h.   No freckles: *ff*

Figure 19.4 Common inherited characteristics in human beings.
The notations indicate which characteristics are dominant and which are recessive.

Figuring the Odds

Many times parents would like to know the chances of a child having a certain genotype and, therefore, a certain phenotype. If one of the parents is homozygous dominant (*EE*), the chances of their having a child with unattached earlobes is 100%, because this parent has only a dominant allele (*E*) to pass on to the offspring. On the other hand, if both parents are homozygous recessive (*ee*), there is a 100% chance that each of their children will have attached earlobes. However, if both parents are heterozygous, then what are the chances that their child will have unattached or attached earlobes? To solve a problem of this type, it is customary first to indicate the genotype of the parents and their possible gametes.

Genotypes:	*Ee*	*Ee*
Gametes:	*E* and *e*	*E* and *e*

Second, a **Punnett square** is used to determine the phenotypic ratio among the offspring when all possible sperm are given an equal chance to fertilize all possible eggs (Fig. 19.5). The possible sperm are lined up along one side of the square, and the possible eggs are lined up along the other side of the square (or vice versa). The ratio among the offspring in this case is 3:1 (three children with unattached earlobes to one with attached earlobes). This means that there is a 3/4 chance (75%) for each child to have unattached earlobes and a 1/4 chance (25%) for each child to have attached earlobes.

Another cross of particular interest is that between a heterozygous individual (*Ee*) and a pure recessive (*ee*). In this case, the Punnett square shows that the ratio among the offspring is 1:1, and the chance of the dominant or recessive phenotype is 1/2, or 50% (Fig. 19.6).

Comparing the two crosses (Fig. 19.5 and Fig. 19.6), each child has a 75% chance of having the dominant phenotype if the two parents are heterozygous and a 50% chance if one parent is heterozygous and the other is recessive. Each child has a 25% chance of having the recessive phenotype if the parents are heterozygous and a 50% chance if one parent is heterozygous and the other is homozygous recessive.

> If both parents are heterozygous, each child has a 25% chance of exhibiting the recessive phenotype. If one parent is heterozygous and the other is homozygous recessive, each child has a 50% chance of exhibiting the recessive phenotype.

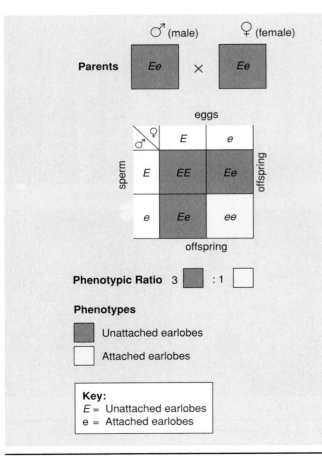

Figure 19.5   Heterozygous-by-heterozygous cross.
When the parents are heterozygous, each child has a 75% chance of having the dominant phenotype and a 25% chance of having the recessive phenotype.

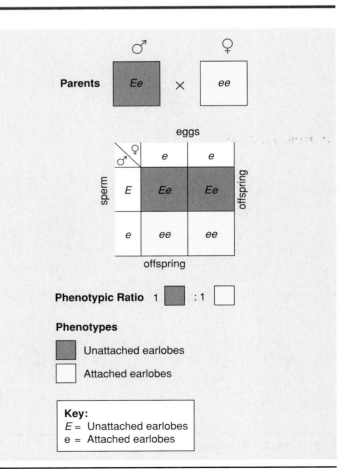

Figure 19.6   Heterozygous-by-homozygous recessive cross.
When one parent is heterozygous and the other recessive, each child has a 50% chance of having the dominant phenotype and a 50% chance of having the recessive phenotype.

Some Disorders Are Dominant

Of the many autosomal dominant disorders, we will discuss only two.

Neurofibromatosis

Neurofibromatosis, sometimes called von Recklinghausen[1] disease, is one of the most common genetic disorders. It affects roughly one in 3,000 people, including an estimated 100,000 in the United States. It is seen equally in every racial and ethnic group throughout the world.

At birth or later, the affected individual may have six or more large, tan spots (known as cafe-au-lait) on the skin. Such spots may increase in size and number and may get darker. Small benign tumors (lumps) called neurofibromas may occur under the skin or in various organs. Neurofibromas are made up of nerve cells and other cell types.

The expression of neurofibromatosis varies. In most cases, symptoms are mild, and patients live a normal life. In some cases, however, the effects are severe. Skeletal deformities, including a large head, are seen, and eye and ear tumors can lead to blindness and hearing loss. Many children with neurofibromatosis have learning disabilities and are hyperactive.

In 1990, researchers isolated the gene for neurofibromatosis, which was known to be on chromosome 17. By analyzing the DNA (deoxyribonucleic acid), they determined that the gene was huge and actually included three even smaller genes. This was only the second time that nested genes have been found in humans. The gene for neurofibromatosis is a tumor-suppressor gene active in controlling cell division. When it mutates, a benign tumor develops.

Huntington Disease

One in 20,000 persons in the United States has *Huntington disease,* a neurological disorder that leads to progressive degeneration of brain cells, which in turn causes severe muscle spasms and personality disorders (Fig. 19.7). Most people appear normal until they are of middle age and have already had children who might also be stricken. Occasionally, the first signs of the disease are seen in these children when they are teenagers or even younger. There is no effective treatment, and death comes ten to fifteen years after the onset of symptoms.

Several years ago, researchers found that the gene for Huntington disease was located on chromosome 4. A test was developed for the presence of the gene, but few want to know if they have inherited the gene because as yet there is no treatment for Huntington disease. After the gene was isolated in 1993, an analysis revealed that it contains many repeats of the base triplet CAG (cytosine, adenine, and

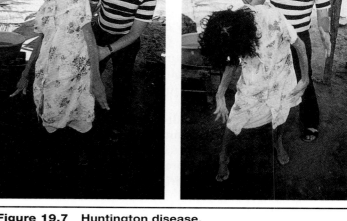

Figure 19.7 Huntington disease.
Persons with this condition gradually lose psychomotor control of the body. At first there are only minor disturbances, but the symptoms become worse over time.

guanine). Normal persons have 11 to 34 copies of the triplet, but affected persons tend to have 42 to more than 120 copies. The more repeats present, the earlier the onset of Huntington disease and the more severe the symptoms. It also appears that persons most at risk have inherited the disorder from their fathers. The latter observation is consistent with a new hypothesis called genomic imprinting. The genes are imprinted differently during formation of sperm and egg, and therefore, the sex of the parent passing on the disorder becomes important.

It is now known that there are a number of other genetic diseases whose severity and time of onset vary according to the number of triplet repeats present within the gene. Moreover, the genes for these disorders are also subject to genomic imprinting.

There are many autosomal dominant disorders in humans. Among these are neurofibromatosis and Huntington disease.

---

1 Although neurofibromatosis is commonly associated with Joseph Merrick, the severely deformed nineteenth-century Londoner depicted in *The Elephant Man*, researchers today believe Merrick actually suffered from a much rarer disorder called Proteus syndrome.

Some Disorders Are Recessive

Of the many autosomal recessive disorders, we will discuss only three.

Tay-Sachs Disease

Tay-Sachs disease is a well-known genetic disease that usually occurs among Jewish people in the United States, most of whom are of central and eastern European descent. At first, it is not apparent that a baby has Tay-Sachs disease. However, development begins to slow down between four months and eight months of age, and neurological impairment and psychomotor difficulties then become apparent. The child gradually becomes blind and helpless, develops uncontrollable seizures, and eventually becomes paralyzed. There is no treatment or cure for Tay-Sachs disease, and most affected individuals die by the age of three or four.

Tay-Sachs disease results from a lack of the enzyme hexosaminidase A (Hex A) and the subsequent storage of its substrate, a fatty substance known as glycosphingolipid, in lysosomes. Although more and more lysosomes build up in many body cells, the primary sites of storage are the cells of the brain, which account for the onset and the progressive deterioration of psychomotor functions.

Persons heterozygous for Tay-Sachs have about half the level of Hex A activity found in normal individuals. Prenatal diagnosis of the disease also is possible following either amniocentesis or chorionic villi sampling.

Cystic Fibrosis

Cystic fibrosis is the most common lethal genetic disease among Caucasians in the United States. About one in 20 Caucasians is a carrier, and about one in 2,500 births has the disorder. In these children, the mucus in the bronchial tubes and pancreatic ducts is particularly thick and viscous, interfering with the function of the lungs and pancreas. To ease breathing, the thick mucus in the lungs has to be manually loosened periodically (Fig. 19.8), but still the lungs become infected frequently. The clogged pancreatic ducts prevent digestive enzymes from reaching the small intestine, and to improve digestion, patients take digestive enzymes mixed with applesauce before every meal.

In the past few years, much progress has been made in our understanding of cystic fibrosis, and new treatments have raised the average life expectancy to twenty-eight years of age. Research has demonstrated that chloride ions (Cl⁻) fail to pass through plasma membrane channel proteins in these patients. Ordinarily, after chloride ions have passed through the membrane, water follows. It is believed that lack of water is the cause of abnormally thick mucus in bronchial tubes and pancreatic ducts. The cystic fibrosis gene, which is located on chromosome 7, has been isolated, and attempts have been made to insert it into nasal epithelium, so far with little success. Genetic testing for the gene in adult carriers and in fetuses is possible; if present, couples have to decide whether to risk having a child with the condition or whether abortion is an option.

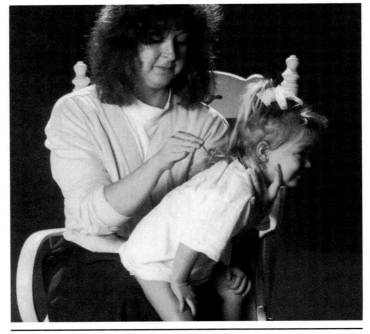

Figure 19.8   Cystic fibrosis.
The mucus in the lungs of a child with cystic fibrosis should be periodically loosened by clapping the back. A new treatment destroys the cells that tend to build up in the lungs. The white blood cells are destroyed by one drug, and another drug does away with the DNA the cells leave behind. Judicious use of antibiotics controls pulmonary infection, and aggressive use of other drugs thins mucous secretions.

Phenylketonuria (PKU)

Phenylketonuria (PKU) occurs once in 5,000 births, so it is not as frequent as the disorders previously discussed. However, it is the most commonly inherited metabolic disorder to affect nervous system development. Close relatives are more apt to have a PKU child.

Affected individuals lack an enzyme that is needed for the normal metabolism of the amino acid phenylalanine, and an abnormal breakdown product, phenylketone, accumulates in the urine. The PKU gene is located on chromosome 12, and there is a prenatal DNA test for the presence of this allele. Years ago, the urine of newborns was tested at home for phenylketone in order to detect PKU. Presently, newborns are routinely tested in the hospital for elevated levels of phenylalanine in the blood. If elevated levels are detected, newborns are placed on a diet low in phenylalanine, which must be continued until the brain is fully developed, around age seven, or else severe mental retardation develops.

There are many autosomal recessive disorders in humans. Among these are Tay-Sachs disease, cystic fibrosis, and phenylketonuria (PKU).

Pedigree Charts

When a genetic disorder is autosomal dominant, an individual with the alleles *AA* or *Aa* will have the disorder. When a genetic disorder is recessive, only individuals with the alleles *aa* will have the disorder. Genetic counselors often construct pedigree charts to determine whether a condition is dominant or recessive. A pedigree chart shows the pattern of inheritance for a particular condition. Consider these two possible patterns of inheritance:

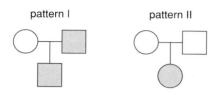

In both patterns, males are designated by squares and females by circles. Shaded circles and squares are affected individuals. A line between a square and a circle represents a union. A vertical line going downward leads, in these patterns, to a single child. (If there are more children, they are placed off a horizontal line.) Which pattern of inheritance do you suppose represents an autosomal dominant characteristic, and which represents an autosomal recessive characteristic?

In pattern I, the child is affected, as is one of the parents. When a disorder is dominant, an affected child usually has at least one affected parent. Of the two patterns, this one shows a dominant pattern of inheritance. Figure 19.9 shows a typical pedigree chart for a dominant disorder. Other ways to recognize an autosomal dominant pattern of inheritance are also given.

In pattern II, the child is affected, but neither parent is; this can happen if the condition is recessive and the parents are *Aa*. Notice that the parents are carriers because they appear to be normal but are capable of having a child with a genetic disorder. Figure 19.10 shows a typical pedigree chart for a recessive genetic disorder. Other ways to recognize an autosomal recessive pattern of inheritance are also given in the figure.

It is important to realize that "chance has no memory," therefore, each child born to heterozygous parents has a 25% chance of having the disorder. In other words, it is possible that if a heterozygous couple has four children, each child might have the condition.

Dominant and recessive alleles have different patterns of inheritance.

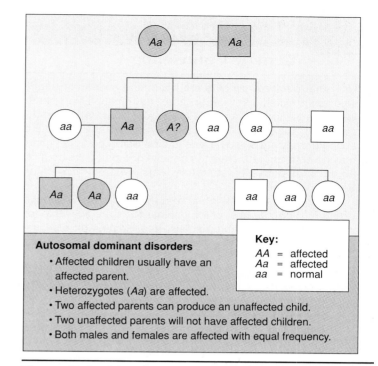

Autosomal dominant disorders

- Affected children usually have an affected parent.
- Heterozygotes (*Aa*) are affected.
- Two affected parents can produce an unaffected child.
- Two unaffected parents will not have affected children.
- Both males and females are affected with equal frequency.

Key:
AA = affected
Aa = affected
aa = normal

Figure 19.9 Autosomal dominant pedigree chart.
The list gives ways to recognize an autosomal dominant disorder.

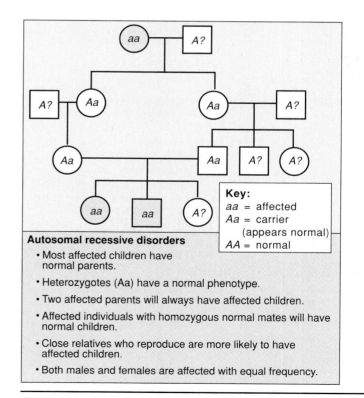

Autosomal recessive disorders

- Most affected children have normal parents.
- Heterozygotes (Aa) have a normal phenotype.
- Two affected parents will always have affected children.
- Affected individuals with homozygous normal mates will have normal children.
- Close relatives who reproduce are more likely to have affected children.
- Both males and females are affected with equal frequency.

Key:
aa = affected
Aa = carrier
(appears normal)
AA = normal

Figure 19.10 Autosomal recessive pedigree chart.
Only those affected with the recessive genetic disorder are shaded. The list gives ways to recognize an autosomal recessive disorder.

Health Focus

Genetic Counseling

Now that potential parents are becoming aware that many illnesses are caused by faulty genes, more couples are seeking genetic counseling. The counselor studies the backgrounds of the couple and tries to determine if any immediate ancestor may have had a genetic disorder. A pedigree chart may be constructed. Then the counselor studies the couple. As much as possible, laboratory tests are performed on all persons involved.

Tests are now available for a large number of genetic diseases. For example, chromosomal tests are available for cystic fibrosis, neurofibromatosis, and Huntington disease. Blood tests can identify carriers of thalassemia and sickle-cell

disease. By measuring enzyme levels in blood, tears, or skin cells, carriers of enzyme defects can also be identified for certain inborn metabolic errors, such as Tay-Sachs disease. From this information, the counselor can sometimes predict the chances of a child having the disorder.

Whenever the woman is pregnant, chorionic villi sampling can be done early, and amniocentesis can be done later in the pregnancy. These procedures, which were illustrated in chapter 17, allow the testing of embryonic and fetal cells, respectively, to determine if the unborn child has a genetic disorder. If so, treatment may be available even before birth, or parents may decide whether or not to end the pregnancy (Table 19A).

TABLE 19A

Test and Treatment for Some Human Genetic Disorders

Name	Description	Chromosome	Incidence Among Newborns	Status
Autosomal Recessive Disorders				
Cystic fibrosis	Mucus in the lungs and digestive tract is thick and viscous, making breathing and digestion difficult	7	One in 2,500 Caucasians	Allele located; chromosome test now available;* treatment being investigated
Tay-Sachs disease	Neurological impairment and psychomotor difficulties develop early, followed by blindness and uncontrollable seizures before death occurs, usually before age 5	15	One in 3,600 eastern European Jews	Biochemical test now available*
Phenylketonuria	Inability to metabolize phenylalanine, and if a special diet is not begun, mental retardation develops	12	One in 5,000 Caucasians	Biochemical test now available; treatment available
Autosomal Dominant Disorders				
Neurofibromatosis	Benign tumors occur under the skin or deeper	17	One in 3,000	Allele located; chromosome test now available*
Huntington disease	Minor disturbances in balance and coordination develop in middle age and progress towards severe neurological disturbances leading to death	4	One in 20,000	Allele located; chromosome test now available*
Incomplete Dominance				
Sickle-cell disease	Poor circulation, anemia, internal hemorrhaging, due to sickle-shaped red blood cells	11	One in 500 African Americans	Chromosome test now available*
X-Linked Recessive				
Hemophilia A	Propensity for bleeding, often internally, due to the lack of a blood clotting factor	X	One in 15,000 male births	Treatment available
Duchenne muscular dystrophy	Muscle weakness develops early and progressively intensifies until death occurs, usually before age 20	X	One in 5,000 male births	Allele located; biochemical tests of muscle tissue available; treatment being investigated

*Prenatal testing is done.

19.3 Polygenic Traits

Polygenic inheritance occurs when one trait is governed by two or more sets of alleles, and the individual has a copy of all allelic pairs. Each dominant allele has a quantitative effect on the phenotype, and these effects are additive. The result is a continuous variation of phenotypes, resulting in a distribution of these phenotypes that resembles a bell-shaped curve. The more genes involved, the more continuous the variation and the distribution of the phenotypes. Also, environmental effects cause many intervening phenotypes; in the case of height, differences in nutrition assures a bell-shaped curve (Fig. 19.11).

Skin Color

Just how many pairs of alleles control skin color is not known, but a range in colors can be explained on the basis of two pairs. When a very dark person reproduces with a very light person, the children have medium-brown skin; when two people with medium-brown skin reproduce with one another, the children range in skin color from very dark to very light. This can be explained by assuming that skin color is controlled by two pairs of alleles and that each capital letter contributes to the color of the skin:

Phenotype	Genotypes
Very dark	*AABB*
Dark	*AABb* or *AaBB*
Medium brown	*AaBb* or *AAbb* or *aaBB*
Light	*Aabb* or *aaBb*
Very light	*aabb*

Notice again that there is a range in phenotypes and that there are several possible phenotypes in between the two extremes. Therefore, the distribution of these phenotypes is expected to follow a bell-shaped curve—few people have the extreme phenotypes, and most people have the phenotype that lies in the middle between the extremes.

Polygenic Disorders

Many human disorders, such as cleft lip and/or palate, clubfoot, congenital dislocations of the hip, hypertension, diabetes, schizophrenia, and even allergies and cancers, are most likely controlled by polygenes and are subject to environmental influences. Therefore, many investigators are in the process of considering the nature versus nurture question; that is, what percentage of the trait is controlled by genes and what percentage is controlled by the environment? Thus far, it has not been possible to come to precise, generally accepted percentages for any particular trait.

In recent years, reports have surfaced that all sorts of behavioral traits such as alcoholism, homosexuality, phobias, and even suicide can be associated with particular genes. No doubt behavioral traits are to a degree controlled by genes, but again, it is impossible at this time to determine to what degree. And very few scientists would support the idea that these traits are predetermined by our genes.

Many human traits most likely controlled by polygenes are subject to environmental influences. The frequency of the phenotypes of such traits follows a bell-shaped curve.

a.

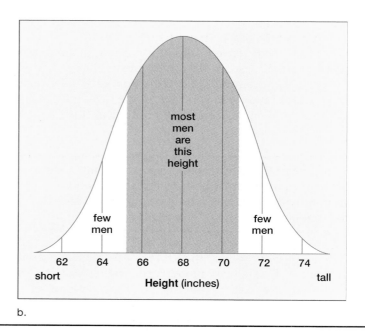

b.

Figure 19.11   Polygenic inheritance.
When you record the heights of a large group of young men **(a)** the values follow a bell-shaped curve **(b).** Such a continuous distribution is due to control of a trait by several sets of alleles. Environmental effects are also involved.

19.4 Multiple Allelic Traits

When a trait is controlled by **multiple alleles,** the gene exists in several allelic forms. But each person usually has only two of the possible alleles. A common example of multiple alleles concerns a person's blood type.

ABO Blood Types

Three alleles for the same gene control the inheritance of ABO blood types. These alleles determine the presence or absence of antigens on the red blood cells.

> A = A antigen on red blood cells
> B = B antigen on red blood cells
> O = Neither A nor B antigen on red blood cells

Alleles *A* and *B* are dominant over *O*, and are fully expressed in the presence of the other. This is called codominance. Therefore,

Phenotype	Possible Genotype
A	*AA, AO*
B	*BB, BO*
AB	*AB*
O	*OO*

An examination of possible matings between different blood types sometimes produces surprising results; for example,

> Genotypes of parents: *AO* × *BO*
> Possible genotypes of children: *AB, OO, AO, BO*

Therefore, from this particular mating, every possible phenotype (types AB, O, A, B blood) is possible.

Blood typing can sometimes aid in paternity suits. However, a blood test of a supposed father can only suggest that he might be the father, not that he definitely is the father. For example, it is possible, but not definite, that a man with type A blood (having genotype *AO*) is the father of a child with type O blood. On the other hand, a blood test sometimes can definitely prove that a man is not the father. For example, a man with type AB blood cannot possibly be the father of a child with type O blood. Therefore, blood tests can be used legally only to exclude a man from possible paternity.

Inheritance by multiple alleles occurs when a gene exists in more than two allelic forms. However, each individual usually inherits only two alleles for these genes.

As a point of interest, the Rh factor is inherited separately from A, B, AB, or O blood types. When you are Rh positive, there is a particular antigen on the red blood cells, and when you are Rh negative, it is absent. It can be assumed that the inheritance of this antigen is controlled by a single allelic pair in which simple dominance prevails: the Rh-positive allele is dominant over the Rh-negative allele.

19.5 Dominance Has Degrees

The field of human genetics also has examples of codominance and incomplete dominance. *Codominance* occurs when alleles are equally expressed in a heterozygote. We have already mentioned that the multiple alleles controlling blood type are codominant. An individual with the genotype *AB* has type AB blood. Recall that skin color is controlled by polygenes in which all dominants (capital letters) add equally to the phenotype. For this reason, it is possible to observe a range of skin colors from very dark to very light.

Incomplete dominance is exhibited when the heterozygote has an intermediate phenotype between that of either homozygote. For example, incomplete dominance pertains to the inheritance of curly versus straight hair in Caucasians. When a curly-haired Caucasian reproduces with a straight-haired Caucasian, their children will have wavy hair. When two wavy-haired persons reproduce, the expected phenotypic ratio among the offspring is 1:2:1; that is, one curly-haired child to two with wavy hair to one with straight hair (Fig. 19.12).

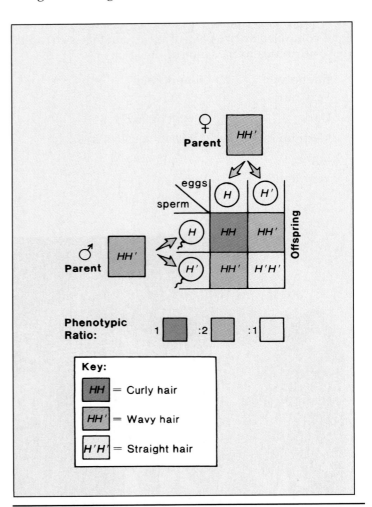

Figure 19.12   Incomplete dominance.
Among Caucasians, neither straight nor curly hair is dominant. When two wavy-haired individuals reproduce, each offspring has a 25% chance of having either straight or curly hair and a 50% chance of having wavy hair, the intermediate phenotype.

Sickle-Cell Disease

Sickle-cell disease is an example of a human disorder that is controlled by incompletely dominant alleles. Individuals with the Hb^AHb^A genotype are normal, those with the Hb^SHb^S genotype have sickle-cell disease, and those with the Hb^AHb^S genotype have the sickle-cell trait. Two individuals with sickle-cell trait can produce children with all three phenotypes, as indicated in Figure 19.13.

In persons with sickle-cell disease, the red blood cells aren't biconcave disks like normal red blood cells; they are irregular. In fact, many are sickle shaped. The defect is caused by an abnormal hemoglobin that accumulates inside the cells. Because the sickle-shaped cells can't pass along narrow capillary passageways like disk-shaped cells, they clog the vessels and break down. This is why persons with sickle-cell disease suffer from poor circulation, anemia, and poor resistance to infection. Internal hemorrhaging leads to further complications, such as jaundice, episodic pain of the abdomen and joints, and damage to internal organs.

Persons with sickle-cell trait do not usually have any sickle-shaped cells unless they experience dehydration or mild oxygen deprivation. Although a recent study found that army recruits with sickle-cell trait are more likely to die when subjected to extreme exercise, previous studies of athletes do not substantiate these findings. At present, most investigators believe that no restrictions on physical activity are needed for persons with the sickle-cell trait.

Among regions of malaria-infested Africa, infants with sickle-cell disease die, but infants with sickle-cell trait have a better chance of survival than the normal homozygote. Their sickled cells provide protection against the malaria-causing parasite, which uses red blood cells during its life cycle. The parasite dies when potassium leaks out of the red blood cells as the cells become sickle shaped. The protection afforded by the sickle-cell trait keeps the allele prevalent in populations exposed to malaria. As many as 60% of the population in malaria-infested regions of Africa have the allele. In the United States, about 10% of the African-American population carries the allele.

In a recent study, 22 children around the world were given bone marrow transplants from healthy siblings, and 16 were completely cured of sickle-cell disease. Optimism over these results is dampened by the knowledge that only 6% of patients met the medical criteria for receiving treatment and the drugs used in the procedure result in infertility. Also, if unsuccessful, health could worsen instead of improve, and there is a 10% risk of death from the treatment.

Innovative therapies are still being explored. For example, persons with sickle-cell disease produce normal fetal hemoglobin during development, and drugs that turn on the genes for fetal hemoglobin in adults are being developed. Mice have been genetically engineered to produce sickled red blood cells in order to test new antisickling drugs and various genetic therapies.

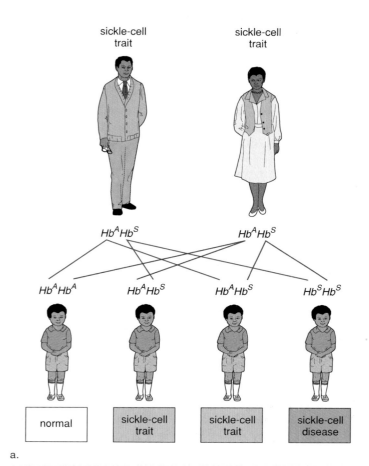

a.

b.

Figure 19.13 Inheritance of sickle-cell disease.
a. In this example, both parents have the sickle-cell trait. Therefore, each child has a 25% chance of having sickle-cell disease or of being perfectly normal and a 50% chance of having the sickle-cell trait. **b.** Sickled cells. Individuals with sickle-cell disease have sickled red blood cells that tend to clump, as illustrated here.

Sickle-cell disease is an inherited, lifelong disorder that is being investigated on many fronts.

19.6 Sex-Linked Traits

The sex chromosomes contain genes just as the autosomal chromosomes do. Some of these genes determine whether the individual is a male or a female. Investigators have found that the Y chromosome has an *SRY* gene (sex-determining region Y gene). When this gene is lacking from the Y chromosome, the individual is a female even though the chromosomal inheritance is XY.

Traits controlled by alleles on the **sex chromosomes** are said to be **sex-linked;** an allele that is only on the X chromosome is **X-linked,** and an allele that is only on the Y chromosome is Y-linked. Most sex-linked alleles are on the X chromosome, and the Y chromosome is blank for these. Very few alleles have been found on the Y chromosome, as you might predict, since it is much smaller than the X chromosome.

The X chromosomes carry many genes unrelated to the sex of the individual, and we will look at a few of these in depth. It would be logical to suppose that a sex-linked trait is passed from father to son or from mother to daughter, but this is not the case. A male always receives a sex-linked condition from his mother, from whom he inherited an X chromosome. The Y chromosome from the father does not carry an allele for the trait. Usually the trait is recessive; therefore, a female must receive two alleles, one from each parent, before she has the condition.

X-Linked Alleles

When considering X-linked traits, the allele on the X chromosome is shown as a letter attached to the X chromosome. For example, the key for red-green color blindness is as follows:

$$X^B = \text{normal vision}$$
$$X^b = \text{color blindness}$$

The possible genotypes in both males and females are as follows:

$X^B X^B$ = female who has normal color vision
$X^B X^b$ = carrier female who has normal color vision
$X^b X^b$ = female who is color blind
$X^B Y$ = male who has normal vision
$X^b Y$ = male who is color blind

Carriers are individuals that appear normal but can pass on an allele for a genetic disorder. Note that the second genotype is a carrier female because although a female with this genotype appears normal, she is capable of passing on an allele for color blindness. Color-blind females are rare because they must receive the allele from both parents; color-blind males are more common since they need only one recessive allele to be color blind. The allele for color blindness has to be inherited from their mother because it is on the X chromosome; males only inherit the Y chromosome from their father.

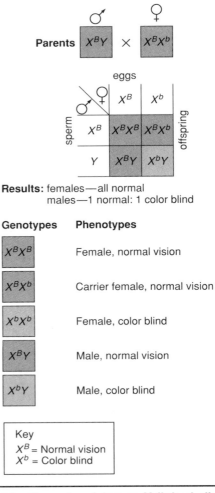

Results: females—all normal
males—1 normal: 1 color blind

Genotypes	Phenotypes
$X^B X^B$	Female, normal vision
$X^B X^b$	Carrier female, normal vision
$X^b X^b$	Female, color blind
$X^B Y$	Male, normal vision
$X^b Y$	Male, color blind

Key
X^B = Normal vision
X^b = Color blind

Figure 19.14 Cross involving an X-linked allele.
The male parent is normal, but the female parent is a carrier; an allele for color blindness is located on one of her X chromosomes. Therefore, each son stands a 50% chance of being color blind.

Now, let us consider a mating between a man with normal vision and a heterozygous woman (Fig. 19.14). What are the chances of this couple having a color-blind daughter? A color-blind son? All daughters will have normal color vision because they all receive an X^B from their father. The sons, however, have a 50% chance of being color blind, depending on whether they receive an X^B or an X^b from their mother. The inheritance of a Y chromosome from their father cannot offset the inheritance of an X^b from their mother. Notice in Figure 19.14, the phenotypic results for sex-linked problems are given separately for males and females.

The X chromosome carries alleles that are not on the Y chromosome. Therefore, a recessive allele on the X chromosome is expressed in males.

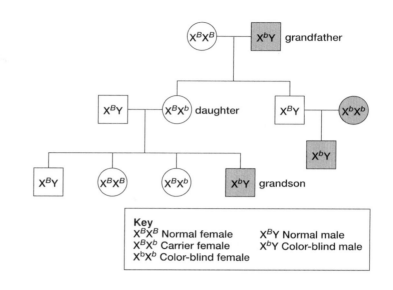

X-linked Recessive Genetic Disorders

- More males than females are affected.

- An affected son can have parents who have the normal phenotype.

- For a female to have the characteristic, her father must also have it. Her mother must have it or be a carrier.

- The characteristic often skips a generation from the grandfather to the grandson.

- If a woman has the characteristic, all of her sons will have it.

Key
X^BX^B Normal female X^BY Normal male
X^BX^b Carrier female X^bY Color-blind male
X^bX^b Color-blind female

Figure 19.15   X-linked recessive pedigree chart.
The list gives ways of recognizing an X-linked recessive disorder.

Some Disorders Are X-Linked

Figure 19.15 gives a pedigree chart for an X-linked recessive condition. It also lists ways to recognize this pattern of inheritance. X-linked conditions can be dominant or recessive, but most known are recessive. More males than females have the trait because recessive alleles on the X chromosome are always expressed in males since the Y chromosome does not have a corresponding allele. If a male has an X-linked recessive condition, his daughters are often carriers; therefore, the condition passes from grandfather to grandson. Females who have the condition inherited the allele from both their mother and their father; and all the sons of such a female will have the condition. Three well-known X-linked recessive disorders are color blindness, muscular dystrophy, and hemophilia.

Color Blindness

In humans, there are three different classes of cone cells, the receptors for color vision in the retina of the eyes. Only one pigment protein is present in each type of cone cell; there are blue-sensitive, red-sensitive, and green-sensitive cone cells. The gene for the blue-sensitive protein is autosomal, but the genes for the red- and green-sensitive proteins are on the X chromosome. About 8% of Caucasian men have red-green color blindness. Most of these see brighter greens as tans, olive greens as browns, and reds as reddish-browns. A few cannot tell reds from greens at all. They see only yellows, blues, blacks, whites, and grays. Opticians have special charts by which they detect those who are color blind.

Muscular Dystrophy

Muscular dystrophy, as the name implies, is characterized by a wasting away of the muscles. The most common form, *Duchenne muscular dystrophy,* is X-linked and occurs in about one out of every 3,600 male births. Symptoms, such as waddling gait, toe walking, frequent falls, and difficulty in rising, may appear as soon as the child starts to walk. Muscle weakness intensifies until the individual is confined to a wheelchair. Death usually occurs by age 20; therefore, affected males are rarely fathers. The recessive allele remains in the population by passage from carrier mother to carrier daughter.

Recently, the gene for muscular dystrophy was isolated, and it was discovered that the absence of a protein now called dystrophin is the cause of the disorder. Much investigative work determined that dystrophin is involved in the release of calcium from the calcium-storage sacs in muscle fibers. The lack of dystrophin causes calcium to leak into the cell, which promotes the action of an enzyme that dissolves muscle fibers. When the body attempts to repair the tissue, fibrous tissue forms, and this cuts off the blood supply so that more and more cells die.

A test is now available to detect carriers for Duchenne muscular dystrophy. Also, various treatments are being attempted. Immature muscle cells can be injected into muscles, and for every 100,000 cells injected, dystrophin production occurs in 30–40% of muscle fibers. The gene for dystrophin has been inserted into the thigh muscle cells of mice, and about 1% of these cells then produced dystrophin.

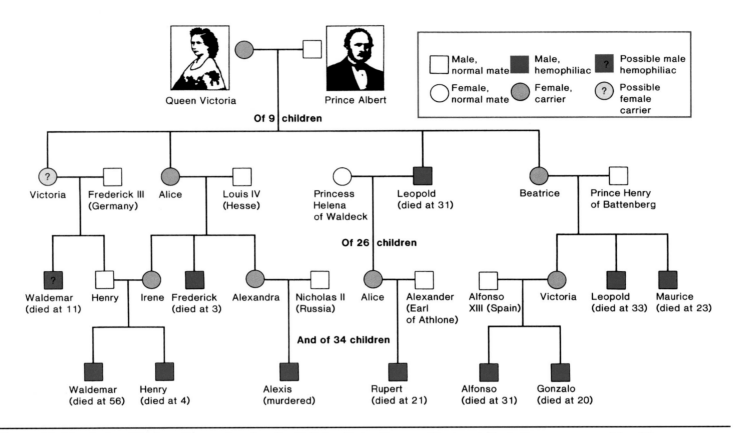

Figure 19.16   A simplified pedigree showing the X-linked inheritance of hemophilia in European royal families.
Because Queen Victoria was a carrier, each of her sons had a 50% chance of having the disease, and each of her daughters had a 50%
chance of being a carrier. This pedigree shows only the affected individuals. Many others are unaffected, such as the members of the present
British royal family.

Hemophilia

About one in 10,000 males is a hemophiliac. There are two
common types of hemophilia: hemophilia A is due to the
absence or minimal presence of a clotting factor known as
factor IX, and hemophilia B is due to the absence of clotting factor VIII.

Hemophilia is called the bleeder's disease because the
affected person's blood does not clot. Although hemophiliacs bleed externally after an injury, they also suffer from internal bleeding, particularly around joints. Hemorrhages can
be checked with transfusions of fresh blood (or plasma) or
concentrates of the clotting protein. Unfortunately, some hemophiliacs have contracted AIDS after receiving blood or using
a blood concentrate, but donors are now screened more closely,
and donated blood is now tested for HIV. Also, factor VIII is
now available as a biotechnology product.

At the turn of the century, hemophilia was prevalent
among the royal families of Europe, and all of the affected
males could trace their ancestry to Queen Victoria of England. Figure 19.16 shows that of Queen Victoria's 26 grand-
children, five grandsons had hemophilia and four grand-
daughters were carriers. Because none of Queen Victoria's
forebears or relatives were affected, it seems that the faulty
allele she carried arose by mutation either in Victoria or in
one of her parents. Her carrier daughters, Alice and Beatrice,
introduced the gene into the ruling houses of Russia and
Spain, respectively. Alexis, the last heir to the Russian throne
before the Russian Revolution, was a hemophiliac. There
are no hemophiliacs in the present British royal family
because Victoria's eldest son, King Edward VII, did not
receive the gene and therefore could not pass it on to any
of his descendants.

Certain traits that have nothing to do with the
gender of the individual are controlled by genes
on the X chromosomes. Males have only one X
chromosome, and therefore, X-linked recessive
alleles are expressed.

Some Traits Are Sex-Influenced

Not all traits we associate with the gender of the individual are sex-linked traits. Some are simply **sex-influenced traits;** that is, the phenotype is determined by autosomal genes that are expressed differently in males and females.

It is possible that the sex hormones determine whether these genes are expressed or not. Pattern baldness (Fig. 19.17) is believed to be influenced by the male sex hormone testosterone because males who take the hormone to increase masculinity begin to lose their hair. A more detailed explanation has been suggested by some investigators. It has been reasoned that due to the effect of hormones, men require only one allele for baldness in order for the condition to appear, whereas women require two alleles. In other words, the allele for baldness acts as a dominant in men but as a recessive in women. This means that men who have a bald father and a mother with hair have a 50% chance at best and a 100% chance at worst of going bald. Women who have a bald father and a mother with hair have no chance at best and a 50% chance at worst of going bald.

Another sex-influenced trait of interest is the length of the index finger. In females, an index finger longer than the fourth finger (ring finger) seems to be dominant. In males, an index finger longer than the fourth finger seems to be recessive.

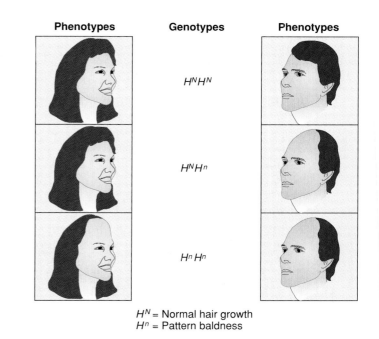

Phenotypes   Genotypes   Phenotypes

$H^N H^N$

$H^N H^n$

$H^n H^n$

H^N = Normal hair growth
H^n = Pattern baldness

Figure 19.17   Baldness, a sex-influenced characteristic.
Due to hormonal influences, the presence of only one gene for baldness causes the condition in men, whereas the condition does not occur in women unless they possess both genes for baldness.

SUMMARY

19.1 Genotype and Phenotype

It is customary to use letters to represent the genotype of individuals. Homozygous dominant (two capital letters) and heterozygous (a capital letter and a lowercase letter) exhibit the dominant phenotype. Homozygous recessive (two lowercase letters) exhibit the recessive phenotype.

19.2 Dominant/Recessive Traits

Whereas the individual has two alleles (copies of a gene) for each trait, the gametes have only one allele for each trait. A Punnett square can help determine the phenotype ratio among offspring because all possible sperm types of a particular male are given an equal chance to fertilize all possible egg types of a particular female. When a heterozygous individual reproduces with another heterozygote, there is a 75% chance the offspring will have the dominant phenotype and a 25% chance the child will have the recessive phenotype. When a heterozygous individual reproduces with a pure recessive, the offspring has a 50% chance of either phenotype.

19.3 Polygenic Traits

Polygenic traits are controlled by more than one pair of alleles, and the individual inherits each allelic pair. Skin color illustrates polygenic inheritance.

19.4 Multiple Allelic Traits

When a trait is controlled by multiple alleles, each person has one pair of all possible alleles. ABO blood types illustrate inheritance by multiple alleles.

19.5 Dominance Has Degrees

Determination of inheritance is sometimes complicated by codominance (e.g., blood type) and by incomplete dominance (e.g., hair curl among Caucasians).

19.6 Sex-linked Traits

Sex-linked genes are located on the sex chromosomes, and most of these alleles are X-linked because they are on the X chromosome. Since the Y chromosome is blank for X-linked alleles, males are more apt to exhibit the recessive phenotype, which they inherited from their mother. If a female does have the phenotype, then her father must also have it, and her mother must be a carrier or have the trait. Usually, X-linked traits skip a generation and go from maternal grandfather to grandson by way of a carrier daughter.

Sex-influenced traits are controlled by an allele that acts as if it were dominant in one sex but recessive in the other. Most likely the activity of such genes is influenced by the sex hormones.

STUDYING THE CONCEPTS

1. What is the difference between the genotype and the phenotype of an individual? For which phenotype are there two possible genotypes?   398

2. Why do the gametes have only one allele for each trait?   399

3. What is the chance of a child with the dominant phenotype from these crosses?   400

> $AA \times AA$
> $Aa \times AA$
> $Aa \times Aa$
> $aa \times aa$

4. From which of the crosses in question three can there be an offspring with the recessive phenotype? Explain.   400

5. Give examples of genetic disorders caused by inheritance of a single dominant allele and disorders caused by inheritance of two recessive alleles.   401–2

6. List ways to recognize an autosomal dominant genetic disorder and an autosomal recessive genetic disorder when examining a pedigree chart.   403

7. Give examples of these patterns of inheritance: polygenic inheritance, multiple alleles, and incomplete dominance.   405–6

8. Give examples of genetic disorders caused by the patterns of inheritance in question seven.   405–7

9. If a trait is on an autosome, how would you designate a homozygous dominant female? If a trait is on an X chromosome, how would you designate a homozygous dominant female?   408

10. List ways to recognize an X-linked recessive disorder when examining a pedigree chart. Why do males exhibit such disorders more often than females?   409

11. Give examples of genetic disorders caused by X-linked recessive alleles.   409–10

APPLYING YOUR KNOWLEDGE

Concepts

1. You sometimes hear the expression "The child gets his eyes from his grandmother." Is there any scientific basis for this statement? Explain.

2. Some genetic disorders are more prevalent in people of a particular ethnic group. Present a possible explanation for this.

3. It is said that there are no carriers for Huntington disease. What is the justification for this statement?

4. A woman with type AB blood gives birth to a type B child. She identifies a man with type O blood as the father. He claims on the basis of his blood type that he could not be the father. As a well-informed judge, how would you rule in this case? Explain.

Bioethical Issue

When researchers first discovered BRCA1—a gene linked to hereditary breast cancer—they were thrilled. Scientists envisioned blood tests that could determine whether a woman carried the risk of breast cancer. Armed with that knowledge, high-risk women could undergo regular mammograms. Doctors could catch breast cancer early.

But amid the excitement, questions about genetic screening have emerged. Researchers worry that companies might abuse genetic information. Insurance companies, for example, might refuse to accept clients carrying a gene for cancer. Employers might hesitate to hire someone likely to get ill.

Should the results from a genetic test be strictly confidential? Does anyone have the right to know that you carry a gene predisposing you toward a certain disease? Your boss, for example? If you get married, do you have the right to keep genetic information from your spouse? Is there any situation in which a person should be forced to divulge such information? Defend your answers.

TESTING YOUR KNOWLEDGE

1. Whereas an individual has two alleles for every trait, the gametes have _____ allele for every trait.

2. The recessive allele for the dominant gene *W* is _____.

3. Mary has a widow's peak, and John has a continuous hairline. This would be a description of their _____.

4. *W* = widow's peak, and *w* = continuous hairline; therefore, only the phenotype _____ could be heterozygous.

5. Two heterozygotes, each having a widow's peak, already have a child with a continuous hairline. The next child has what chance of having a continuous hairline? _____

6. Most sex-linked genes are attached to the _____ chromosome.

7. If a male is color blind, he inherited the allele for color blindness from his _____.

8. What is the genotype of a female who has a color-blind father but a homozygous normal mother? _____

9. In a pedigree chart, it is observed that although the children have a characteristic, neither parent has it. The characteristic must be inherited as a _____ gene.

10. Determine if the characteristic possessed by the shaded males and females in the pedigree chart below is an autosomal dominant, autosomal recessive, or X-linked condition:

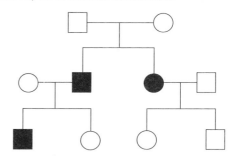

11. Identify each of these genetic disorders.

    a. Mucus in lungs and digestive tract is thick and viscous.

    b. Neurological impairment and psychomotor difficulties develop early.

    c. Benign tumors occur under the skin or deeper.

    d. Minor disturbances in balance and coordination develop in middle age and progress toward severe mental disturbances.

    e. Propensity for bleeding due to lack of blood clotting factor.

    f. Muscle weakness develops early and intensifies until death occurs.

PRACTICING GENETICS PROBLEMS

The trait is dominant for questions 1–5:

1. A woman heterozygous for polydactyly, a condition that produces six fingers and toes, reproduces with a man without this condition. What are the chances that their children will have six fingers and toes?

2. A young man's father has just been diagnosed as having Huntington disease. What are the chances that the son will inherit this condition?

3. Black hair is dominant over blond hair. A woman with black hair whose father had blond hair reproduces with a blond-haired man. What are the chances of this couple having a blond-haired child?

4. Your maternal grandmother Smith had Huntington disease. Aunt Jane, your mother's sister, also had the disease. Your mother dies at age 75 with no signs of Huntington disease. What are your chances of getting the disease assuming that your father is perfectly normal?

5. Could parents who can curl their tongues have children who cannot curl their tongues? Explain your answer.

The trait is recessive for questions 6–8:

6. Parents who do not have Tay-Sachs disease produce a child who has Tay-Sachs disease. What is the genotype of each parent? What are the chances each child will have Tay-Sachs disease?

7. One parent has lactose intolerance, the inability to digest lactose, the sugar found in milk, and the other parent is heterozygous. What are the chances that their child will have lactose intolerance?

8. A child has cystic fibrosis. His parents are normal. What is the genotype of all persons mentioned?

The trait is incompletely dominant for questions 9–12:

9. What are the chances that a person homozygous for straight hair who reproduces with a person homozygous for curly hair will have children with wavy hair?

10. One parent has sickle-cell disease, and the other is perfectly normal. What are the genotypes of their children?

11. A child has sickle-cell disease, but her parents do not. What is the genotype of each parent?

12. Both parents have the sickle-cell trait. What are their chances of having a perfectly normal child?

The trait is either controlled by multiple alleles or by more than one pair of alleles for problems 13–16:

13. The genotype of a woman with type B blood is *BO*. The genotype of her husband is *AO*. What could be the genotypes and phenotypes of the children?

14. A man has type AB blood. What is his genotype? Could this man be the father of a child with type B blood? If not, why not? If so, what blood types could the child's mother have?

15. Baby Susan has type B blood. Her mother has type O blood. What type blood could her father have?

16. Fill in the following pedigree chart to give the two probable genotypes of the twins:

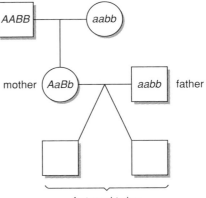

fraternal twins

APPLYING TECHNOLOGY

Your study of genes and medical genetics is supported by these available technologies:

Exploring the Internet
The Mader Home Page provides further resources for studying this chapter.

`http://www.mhhe.com/sciencemath/biology/mader/`

(Click on *Human Biology*.)

Explorations in Cell Biology & Genetics CD-ROM

Heredity in Families (#12)   Thirty pedigree charts are available for students to test their skill to determine the particular pattern of inheritance.   (4)*

---
*Level of difficulty

Gene Segregation Within Families (#13)   Students use the binomial distribution to predict the distribution of phenotypes among a certain number of offspring.   (2)

Explorations in Human Biology CD-ROM

Cystic Fibrosis (#1)   Students observe the diffusion of chloride ions and water across the membrane in relation to the buildup of mucus in persons that are homozygous normal, heterozygous, or homozygous recessive for cystic fibrosis.   (1)

SELECTED KEY TERMS

allele (uh-LEEL) Alternative form of a gene located at the same position on a pair of homologous chromosome.   398

carrier Individual that appears normal but is capable of transmitting an allele for a genetic disorder.   408

dominant allele Hereditary factor that expresses itself in the phenotype when the genotype is heterozygous.   398

genotype Genes of any individual for (a) particular trait(s).   398

multiple allele Pattern of inheritance in which there are more than two alleles for a particular trait, although each individual has only two of these alleles.   406

phenotype (FEE-noh-typ) Outward appearance of an organism caused by the genotype and environmental influences.   398

polygenic inheritance Pattern of inheritance in which more than one allelic pair controls a trait; each dominant allele has a quantitative effect on the phenotype.   405

Punnett square Gridlike device used to calculate the expected results of simple genetic crosses.   400

recessive allele Hereditary factor that expresses itself in the phenotype only when the genotype is homozygous for this allele.   398

sex chromosome Chromosome responsible for the development of characteristics associated with gender; an X or Y chromosome.   408

sex-influenced trait Autosomal trait that is expressed differently in the two sexes. Usually attributed to hormonal influence.   411

sex-linked Allele located on a sex chromosome, usually the X chromosome.   408

trait Phenotypic feature studied in heredity.   397

X-linked Allele located on the X chromosome.   408

Chapter 20

DNA and Biotechnology

Chapter Outline

20.1 DNA AND RNA STRUCTURE AND FUNCTION

- DNA is the genetic material, and therefore, its structure and function constitute the molecular basis of inheritance. 416

- When DNA replicates, two exact molecules result. RNA structure is similar to, but also different from, DNA. RNA occurs in three forms, each with a specific function. 418

20.2 DNA SPECIFIES PROTEINS

- Proteins are composed of amino acids and function, in particular, as enzymes and as structural elements of membranes and organelles in cells. 420

- DNA's genetic information codes for the sequence of amino acids in a protein during protein synthesis. 420

- There are various levels of genetic control in human cells. 425

20.3 BIOTECHNOLOGY

- Recombinant DNA technology is the basis for biotechnology, an industry that produces many products. 426

- Modern-day biotechnology is expected to revolutionize medical care and agriculture. It also permits gene therapy and the mapping of human chromosomes. 429

Figure 20.1 Drugstore DNA?
Will pharmacists one day be able to dispense gene therapy drugs? Some say that day will soon be here.

At the time, four-year-old Ashanti DeSilva could not have known that she was making history. But little Ashanti was about to change the medical world. Born with a rare immune disorder called adenosine deaminase (ADA) deficiency, Ashanti was susceptible to chronic infection. In and out of the hospital, the cheerful little girl knew what it was like to be sick—a lot. Her parents knew she would probably die young.

And so, in 1993, the DeSilvas agreed to let Ashanti become the first patient to undergo gene therapy, sponsored by the National Institutes of Health (NIH), the country's top biomedical funding agency. In a series of operations, NIH researchers took white blood cells from Ashanti, added the gene that specifies ADA, and injected the improved cells back into her. Combined with drug treatment, the therapy apparently worked. Ashanti's immune system grew stronger. The field of gene therapy was launched.

Today, over 600 patients have received some form of gene therapy in treatment for AIDS, cancer, cystic fibrosis, and other diseases. Like its first patient, gene therapy is young but growing. Hopefully, it will one day improve the lives of millions (Fig. 20.1).

The gene that Ashanti received was a specific piece of **DNA (deoxyribonucleic acid)** taken from a human chromosome. This piece of DNA is responsible for the production of ADA, the protein that Ashanti needed in her cells. In this chapter, we will learn that any particular gene is just one section of one of the chromosomes, and that a **gene** codes for a specific sequence of amino acids in a protein.

20.1 DNA and RNA Structure and Function

DNA, the genetic material, is found principally in the chromosomes, which are located in the nucleus of a cell. (Small amounts of DNA are also found in mitochondria.)

Figure 20.2 shows the levels of organization of chromosomes. While the metaphase chromosome is a very

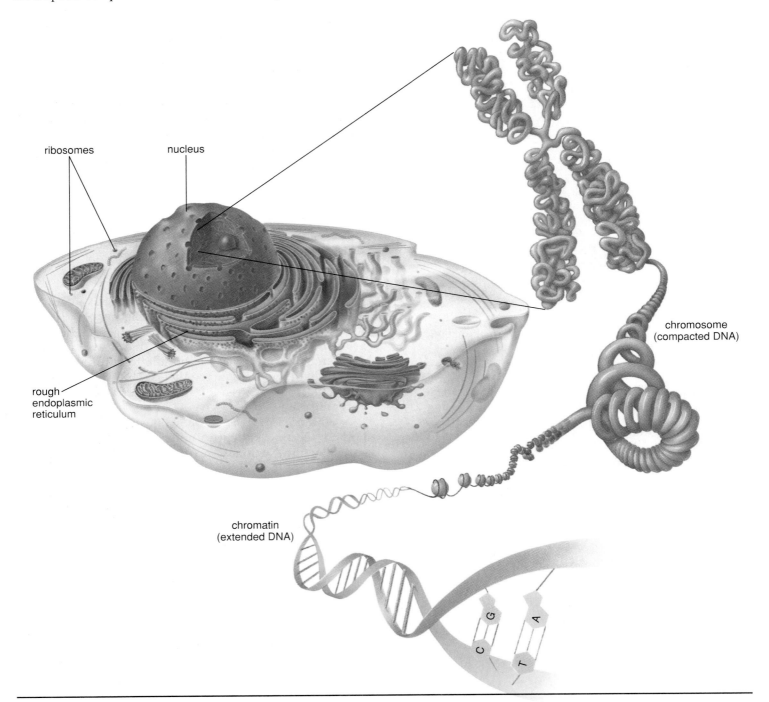

Figure 20.2 DNA location and structure.
DNA is highly compacted in chromosomes, but it is extended as chromatin during interphase. It is during interphase that DNA can be extracted from a cell and its structure and function studied.

compact structure, chromatin is fine threads in which the DNA is extended. The DNA molecule is a double-stranded helix in which the strands are held together by hydrogen bonding. The structure of DNA facilitates replication of the molecule prior to cell division.

DNA Structure and Replication

DNA is a type of nucleic acid, and like all nucleic acids, it is formed by the sequential joining of molecules called nucleotides. Nucleotides, in turn, are composed of three smaller molecules—a *phosphate,* a *sugar,* and a *base.* The sugar in DNA is deoxyribose, which accounts for its name, *deoxyribonucleic acid.* There are four different nucleotides in DNA, each having a different base: **adenine (A), thymine (T), cytosine (C),** and **guanine (G)** (Fig. 20.3).

When nucleotides join, the sugar and the phosphate molecules become the backbone of a strand, and the bases project to the side. In DNA, there are two strands of nucle-

otides; consequently, DNA is double stranded. Weak hydrogen bonds between the bases hold the strands together. Each base is bonded to another particular base, called **complementary base pairing**—adenine (A) is always paired with thymine (T), and cytosine (C) is always paired with guanine (G), and vice versa. The dotted lines in Figure 20.4*a* represent the hydrogen bonds between the bases. The structure of DNA is said to resemble a ladder; the sugar-phosphate backbone makes up the sides of a ladder, and the paired bases are the rungs. The ladder structure of DNA twists to form a spiral staircase called a *double helix* (Fig. 20.4*b*).

DNA is a double helix with a sugar-phosphate backbone on the outside and paired bases on the inside. Complementary base pairing occurs: adenine (A) pairs with thymine (T), and guanine (G) pairs with cytosine (C).

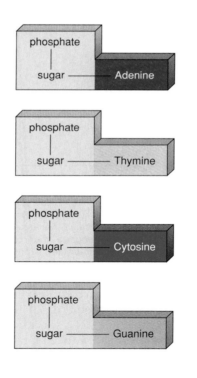

Figure 20.3   DNA nucleotides.
DNA contains four different kinds of nucleotides, molecules that in turn contain a phosphate, a sugar, and a base. The base of a DNA nucleotide can be either adenine (A), thymine (T), cytosine (C), or guanine (G).

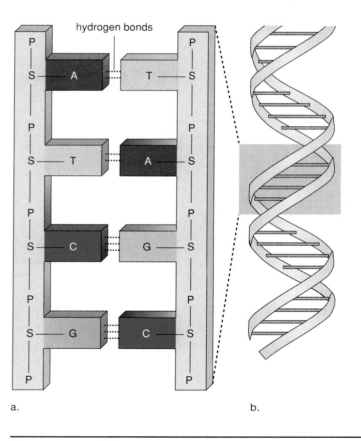

a.   b.

Figure 20.4   DNA is double stranded.
a. When the DNA nucleotides join, they form two strands so that the structure of DNA resembles a ladder. The phosphate (P) and sugar (S) molecules make up the sides of the ladder, and the bases make up the rungs of the ladder. Each base is weakly bonded (dotted lines) to its complementary base (T is bonded to A and vice versa; G is bonded to C and vice versa). **b.** The DNA ladder twists to give a double helix. Each chromatid of a duplicated chromosome contains one double helix.

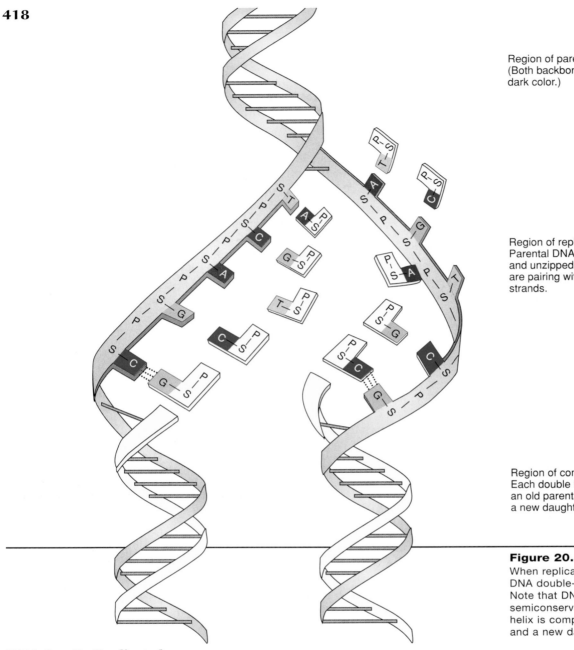

Region of parental DNA helix.
(Both backbones are shown in
dark color.)

Region of replication (simplified).
Parental DNA helix is unwound
and unzipped. New nucleotides
are pairing with those in parental
strands.

Region of completed replication.
Each double helix is composed of
an old parental strand (dark) and
a new daughter strand (light).

Figure 20.5   DNA replication.
When replication is finished, two complete
DNA double-helix molecules are present.
Note that DNA replication is
semiconservative because each new double
helix is composed of an old parental strand
and a new daughter strand.

DNA Can Be Replicated

Between cell divisions, chromosomes must duplicate before cell division can occur again. **Replication** of DNA occurs as a part of chromosome duplication. Replication has been found to require the following steps:

1. The hydrogen bonds between the two strands of DNA break as enzymes unwind and "unzip" the molecule.

2. New nucleotides, always present in the nucleus, move into place beside each old (parental) strand by the process of complementary base pairing.

3. These new nucleotides become joined by an enzyme called *DNA polymerase.*

4. When the process is finished, two complete DNA molecules are present, identical to each other and to the original molecule.

Each new double helix is composed of an old (parental) and a new (daughter) strand (Fig. 20.5). Because of this, it is said that each strand of DNA serves as a **template,** or mold, for the production of a complementary strand. Rarely, a replication error occurs—the new sequence of the bases is not exactly like that of a parental strand. Replication errors are a source of **mutations,** a permanent change in a gene causing a change in the phenotype.

DNA replication results in two double helixes. DNA unwinds and unzips, and new (daughter) strands, each complementary to an old (parental) strand, form.

Structure and Function of RNA

RNA (ribonucleic acid) is made up of nucleotides containing the sugar ribose. This sugar accounts for the scientific name of this polynucleotide. RNA contains four different nucleotides, each with a different base: adenine (A), uracil (U), cytosine (C), and guanine (G) (Fig. 20.6a). Notice that in RNA, the base uracil replaces the base thymine.

RNA, unlike DNA, is single stranded (Fig. 20.6b), but the single RNA strand sometimes doubles back on itself. Similarities and differences between these two nucleic acid molecules are given in Table 20.1.

In general, RNA is a helper to DNA, allowing protein synthesis to occur according to the blueprint that DNA provides. There are three types of RNA, each with a specific function in protein synthesis.

Ribosomal RNA

Ribosomal RNA (rRNA) forms off a DNA template in the nucleolus of a nucleus. Ribosomal RNA joins with proteins made in the cytoplasm to form the subunits of ribosomes. The subunits leave the nucleus and come together in the cytoplasm when protein synthesis is about to begin. Protein synthesis occurs at the ribosomes, which in low-power electron micrographs look like granules arranged along the endoplasmic reticulum, a system of tubules and saccules within the cytoplasm. Some ribosomes appear free in the cytoplasm or in clusters called polyribosomes.

Messenger RNA

Messenger RNA (mRNA) forms off a DNA template in the nucleus and then carries its genetic information from the chromosome to the ribosomes in the cytoplasm where protein synthesis occurs. Messenger RNA occurs as a linear molecule.

Transfer RNA

Transfer RNA (tRNA) also forms off a DNA template in the nucleus. Appropriate to its name, tRNA transfers amino acids to the ribosomes, where the amino acids are joined, forming a protein. There are 20 different types of amino acids in proteins; therefore, there has to be at least this number of tRNAs functioning in the cell. Each type of tRNA is designed to carry only one type of amino acid.

RNA is a polynucleotide that functions during protein synthesis in various ways: there are three types of RNA, each with a specific role.

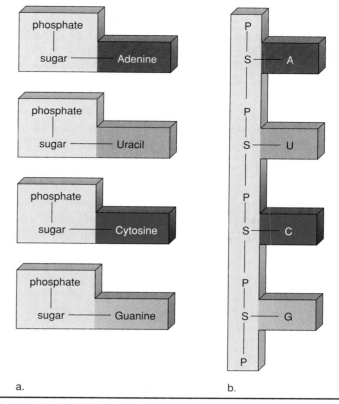

a.   b.

Figure 20.6   RNA structure.
a. The four nucleotides in RNA each have a phosphate (P) molecule; the sugar (S) is ribose; and the base may be either adenine (A), uracil (U), cytosine (C), or guanine (G). **b.** RNA is single stranded. The sugar and phosphate molecules join to form a single backbone, and the bases project to the side.

TABLE 20.1	
DNA-RNA Similarities and Differences	
DNA-RNA Similarities	
Both are nucleic acids	
Both are composed of nucleotides	
Both have a sugar-phosphate backbone	
Both have four different type bases	
DNA-RNA Differences	
DNA	*RNA*
Found in nucleus	Found in nucleus and cytoplasm
The genetic material	Helper to DNA
Sugar is deoxyribose	Sugar is ribose
Bases are A, T, C, G	Bases are A, U, C, G
Double stranded	Single stranded
Is transcribed	Is translated
(to give mRNA)	(to give proteins)

20.2 DNA Specifies Proteins

DNA provides the cell with a blueprint for the synthesis of proteins in a cell. Before discussing mechanics of protein synthesis, let's review the structure of proteins.

Structure and Function of Proteins

Proteins are composed of individual units called *amino acids*. Twenty different amino acids are commonly found in proteins, which are synthesized at the ribosomes in the cytoplasm of cells. Proteins differ because the number and order of their amino acids differ. Notice in Figure 20.7 that the sequence of amino acids in a portion of one protein can differ completely from the sequence of amino acids in a portion of another protein. The unique sequence of amino acids in a protein leads to it having a particular shape, and shape of a protein helps determine its function.

Proteins are found in all parts of the body; some are structural proteins and some are enzymes. The protein hemoglobin is responsible for the red color of red blood cells. The antibodies, as well as albumin, are well-known plasma proteins. Muscle cells contain the proteins actin and myosin, which give them substance and their ability to contract.

Enzymes are organic catalysts that speed reactions in cells. The reactions in cells form metabolic or chemical pathways. A pathway can be represented as follows:

$$A \xrightarrow{E_A} B \xrightarrow{E_B} C \xrightarrow{E_C} D \xrightarrow{E_D} E$$

In this pathway, the letters are molecules, and the notations over the arrows are enzymes: molecule A becomes molecule B, and enzyme E_A speeds the reaction; molecule B becomes molecule C, and enzyme E_B speeds the reaction, and so forth. Notice that each reaction in the pathway has its own enzyme: enzyme E_A can only convert A to B, enzyme E_B can only convert B to C, and so forth. For this reason, enzymes are said to be *specific*.

DNA Code

The occurrence of inherited metabolic disorders first suggested that genes are responsible for the metabolic workings of a cell. In phenylketonuria (PKU), mental retardation is caused by the inability to convert phenylalanine to tyrosine. In albinism, there is no natural pigment in the skin because tyrosine cannot be converted to melanin. Each condition is caused by the deficiency of a particular enzyme. Even in the early 1900s, these conditions were called inborn errors of metabolism.

After many laboratory experiments, it became clear that a gene is a segment of DNA that codes for a protein. If a person inherits a faulty gene, the code is abnormal, and a particular enzyme does not function as it should. We now know that DNA contains a triplet code—every three bases (*triplet*) represents (*codes*) one amino acid (Table 20.2).

The genetic code is essentially universal. The same codons stand for the same amino acids in most organisms from bacteria to humans. This illustrates the remarkable biochemical unity of living things and suggests that all living things have a common evolutionary ancestor.

Transcription

Protein synthesis requires two steps, called transcription and translation. It is helpful to recall that transcription is often used to signify making a copy of certain information, while translation means to put this information into another language.

In the same manner that DNA serves as a template for the production of itself, it also serves as a template for the production of RNA. **Transcription** is the synthesis of an RNA molecule off a DNA template. Although all three types of RNA are transcribed off a DNA template, just now we are interested in the transcription of *messenger RNA* (*mRNA*). During transcription, a segment of DNA helix unwinds and unzips, and complementary

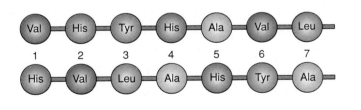

Figure 20.7 Structure of proteins.
Proteins differ by the sequence of their amino acids; the top row shows one possible sequence, and the bottom row shows another possible sequence.

TABLE 20.2		
Some DNA Codes and RNA Codons		

DNA Code	mRNA Codon	tRNA Anticodon	Amino Acid
TTT	AAA	UUU	Lysine
TGG	ACC	UGG	Threonine
CCG	GGC	CCG	Glycine
CAT	GUA	CAU	Valine

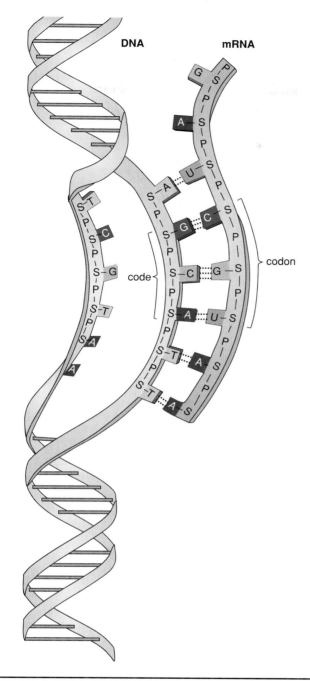

DNA   **mRNA**

code

codon

Figure 20.8   Transcription.
During transcription, mRNA is formed when nucleotides
complementary to the sequence of bases in a portion of DNA (i.e.,
a gene) join. Note that DNA contains a code (sequences of three
bases) and that mRNA contains codons (sequences of three
complementary bases). *Top*—mRNA transcript is ready to move
into the cytoplasm. *Middle*—transcription has occurred, and mRNA
nucleotides have joined together. *Bottom*—rest of DNA molecule.

RNA nucleotides pair with the DNA nucleotides of one
strand: G pairs with C, U pairs with A, and A pairs with
T in DNA. An enzyme called *RNA polymerase* joins the
nucleotides together, and the mRNA that results has a
sequence of bases complementary to those of a gene.
While DNA contains a **triplet code** in which every three
bases stand for one amino acid, mRNA contains **codons,**
each of which is made up of three bases that also stand
for the same amino acid (Table 20.2).

In Figure 20.8, mRNA has formed and is ready to move
away from DNA and make its way to the cytoplasm. Referring again to the structure of the cell, recall that DNA is
found in the nucleus while protein synthesis occurs at the
ribosomes, which are located free in the cytoplasm or on
the endoplasmic reticulum (ER)(see Fig. 20.2). (Proteins produced by ribosomes free in the cytoplasm stay in the cell.
Those produced by ribosomes attached to the ER are for
transport outside the cell.)

Processing mRNA

Most genes in humans are interrupted by segments of DNA
that are not part of the gene. These portions are called *introns* because they are intragene segments. The other portions of the gene are called *exons* because they are ultimately
*ex*pressed. They result in a protein product.

When DNA is transcribed, the mRNA contains bases
that are complementary to both exons and introns, but
before the mRNA exits the nucleus, it is *processed*. During processing, the nucleotides complementary to the
introns are enzymatically removed. There has been much
speculation about the role of introns. It is possible that
they allow crossing-over within a gene during meiosis. It
is also possible that introns divide a gene into domains
that can be joined in different combinations to give different protein products in different cells. The enzymes that
remove introns are called **ribozymes** because they are RNA
molecules. The discovery of ribozymes tells us that not
all enzymes are proteins.

In human cells, processing occurs in the nucleus. After
the mRNA strand is processed, it passes from the cell
nucleus into the cytoplasm. There it becomes associated
with the ribosomes.

Following transcription, mRNA has a sequence of
bases complementary to one of the DNA strands. It
contains codons and moves into the cytoplasm
where it becomes associated with the ribosomes.

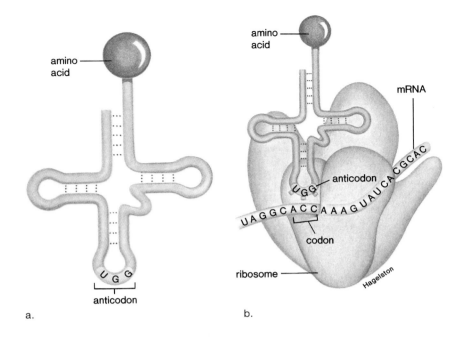

Figure 20.9   Anticodon-codon base pairing.
a. tRNA molecules have an amino acid attached to one end and an anticodon at the other end. **b.** The anticodon of a tRNA molecule is complementary to a codon. The pairing between codon and anticodon at a ribosome ensures that the sequence of amino acids in a polypeptide is that directed originally by DNA. If the codon is ACC, the anticodon is UGG, and the amino acid is threonine.

Translation

Translation is the synthesis of a polypeptide under the direction of an mRNA molecule. During translation, *transfer RNA (tRNA)* molecules bring amino acids to the ribosomes. Usually, there is more than one tRNA molecule for each of the 20 amino acids found in proteins. The amino acid binds to one end of the molecule (Fig. 20.9*a*). It takes energy to bind an amino acid to a tRNA; therefore, the entire complex is designated as tRNA–amino acid.

At the other end of each tRNA, there is a specific **anticodon,** a group of three bases that is complementary to an mRNA codon (Fig. 20.9*b*). The tRNA molecules come to the ribosome where each anticodon pairs with a codon in the order directed by the sequence of the mRNA codons. In this way, the order of codons in mRNA brings about a particular order of amino acids in a protein.

If the codon is ACC, what is the anticodon, and what amino acid will be attached to the tRNA molecule? Inspection of Table 20.2 (p. 420) allows us to determine this:

Codon	Anticodon	Amino Acid
ACC	UGG	Threonine

Translation Requires Three Steps

Polypeptide synthesis requires three steps: initiation, elongation, and termination.

1. During *initiation*, mRNA binds to the smaller of the two ribosomal subunits; then the larger subunit joins the smaller one.
2. During *elongation*, the polypeptide lengthens one amino acid at a time (Fig. 20.10). A ribosome is large enough to accommodate two tRNA molecules: the incoming tRNA molecule and the outgoing tRNA molecule. The incoming tRNA–amino acid complex receives the peptide from the outgoing tRNA. The ribosome then moves laterally so that the next mRNA codon is available to receive an incoming tRNA–amino acid complex. In this manner, the peptide grows, and the linear structure of a polypeptide comes about. (The particular shape of a polypeptide comes about after termination, as the amino acids interact with one another.)
3. Then *termination* of synthesis occurs at a codon that means stop and does not code for an amino acid. The ribosome dissociates into its two subunits and falls off the mRNA molecule.

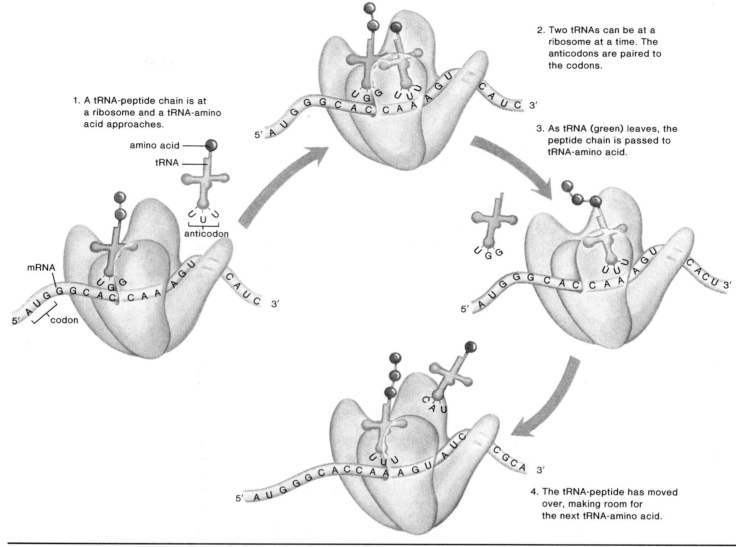

1. A tRNA-peptide chain is at a ribosome and a tRNA-amino acid approaches.

amino acid

tRNA

anticodon

mRNA

codon

2. Two tRNAs can be at a ribosome at a time. The anticodons are paired to the codons.

3. As tRNA (green) leaves, the peptide chain is passed to tRNA-amino acid.

4. The tRNA-peptide has moved over, making room for the next tRNA-amino acid.

Figure 20.10   Translation.
Transfer RNA (tRNA) amino acid molecules arrive at the ribosome, and the sequence of messenger RNA (mRNA) codons dictates the order in which amino acids become incorporated into a polypeptide.

As soon as the initial portion of mRNA has been translated by one ribosome, and the ribosome has begun to move down the mRNA, another ribosome attaches to the mRNA. Therefore, several ribosomes, collectively called a **polyribosome,** can move along one mRNA at a time. And several polypeptides of the same type can be synthesized using one mRNA molecule (Fig. 20.11).

Figure 20.11   Polyribosome structure.
Several ribosomes, collectively called a polyribosome, move along a messenger RNA (mRNA) molecule at one time. They function independently of one another; therefore, several polypeptides can be made at the same time.

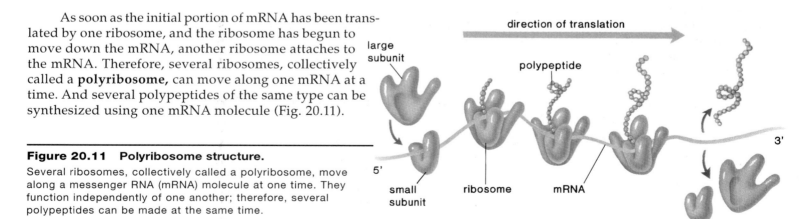

direction of translation

large subunit

polypeptide

small subunit

ribosome

mRNA

5'

3'

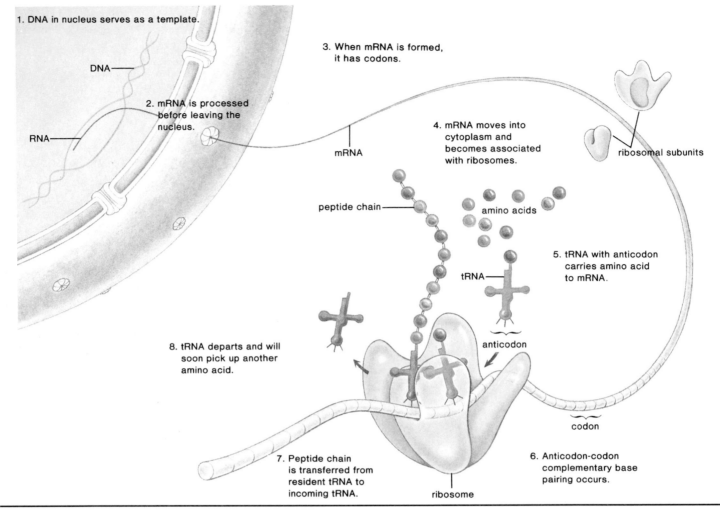

1. DNA in nucleus serves as a template.

DNA

RNA

2. mRNA is processed before leaving the nucleus.

3. When mRNA is formed, it has codons.

mRNA

4. mRNA moves into cytoplasm and becomes associated with ribosomes.

ribosomal subunits

peptide chain

amino acids

tRNA

5. tRNA with anticodon carries amino acid to mRNA.

anticodon

8. tRNA departs and will soon pick up another amino acid.

codon

7. Peptide chain is transferred from resident tRNA to incoming tRNA.

ribosome

6. Anticodon-codon complementary base pairing occurs.

Figure 20.12   Summary of protein synthesis.

Let's Review Protein Synthesis

The following list, along with Table 20.3 and Figure 20.12, provides a brief summary of the events involved in gene expression that results in a protein product.

1. DNA in the nucleus contains a *triplet code.* Each group of three bases stands for a specific amino acid.

2. During transcription, a segment of a DNA strand serves as a template for the formation of mRNA. The bases in mRNA are complementary to those in DNA; every three bases is a *codon* for a certain amino acid.

3. Messenger RNA (mRNA) is processed before it leaves the nucleus, during which time the introns are removed.

4. Messenger RNA (mRNA) carries a sequence of codons to the *ribosomes,* which are composed of rRNA and proteins.

5. Transfer RNA (tRNA) molecules, each of which is bonded to a particular amino acid, have anticodons that pair complementarily to the codons in mRNA.

TABLE 20.3		
Steps in Protein Synthesis		
Molecule	**Special Significance**	**Definition**
DNA	Code	Sequence of bases in threes
mRNA	Codon	Complementary sequence of bases in threes
tRNA	Anticodon	Sequence of three bases complementary to codon
Amino acids	Building blocks	Transported to ribosomes by tRNAs
Protein	Enzymes and structural proteins	Amino acids joined in a predetermined order

6. During translation, tRNA molecules and their attached amino acids arrive at the ribosomes, and the linear sequence of codons of the mRNA determines the order in which the amino acids become incorporated into a protein.

Control of Gene Expression

When a gene is expressed, its protein is being produced in the cytoplasm. Not all genes are expressed at the same time. During the development of humans or any organism, different genes are turned on and then off as organs take shape and become functional. Although all cells receive a copy of all the genes, differentiation involves the turning on of certain genes and the turning off of other genes. As an example, some genes are active in muscle cells, and other genes are active in nerve cells. Even when a gene is ordinarily active in a specialized cell, it needs to be turned off when there is already enough of a particular protein (e.g., enzyme) in the cell. Human cells have four primary means by which to control the expression of their genes.

1. *Transcriptional control:* In the nucleus, a number of mechanisms, two of which are discussed below, serve to control which genes are transcribed and/or the rate at which transcription of the genes occurs.

2. *Posttranscriptional control:* Posttranscriptional control occurs in the nucleus after DNA is transcribed and mRNA is formed. Differential processing of mRNA before it leaves the nucleus and also the speed with which mature mRNA leaves the nucleus can affect the amount of gene expression.

3. *Translational control:* Translational control occurs in the cytoplasm after mRNA leaves the nucleus and before there is a protein product. The life expectancy of mRNA molecules (how long they exist in the cytoplasm) can vary, as can their ability to bind ribosomes. It is also possible that some mRNAs may need additional changes before they are translated at all.

4. *Posttranslational control:* Posttranslational control, which also occurs in the cytoplasm, occurs after protein synthesis. The polypeptide product may have to undergo additional changes before it is biologically functional. Also, a functional enzyme is subject to feedback control—the binding of an enzyme's product can change its shape so that it is no longer able to carry out its reaction.

Activated Chromatin

For a gene to be transcribed in human cells, the chromosome in that region must first decondense. The chromosomes within the developing egg cells of many vertebrates are called lampbrush chromosomes because they have many loops that appear to be bristles (Fig. 20.13). Here mRNA is being synthesized in great quantity; then protein synthesis can be carried out, despite rapid cell division during development.

Transcription Factors

In human cells, *transcription factors* are DNA-binding proteins. Every cell contains many different types of transcription factors, and a specific combination is believed to regulate the activity of any particular gene. After the right combination of transcription factors binds to DNA, an RNA polymerase attaches to DNA and begins the process of transcription.

As cells mature, they become specialized. Specialization is determined by which genes are active, and therefore, perhaps, by which transcription factors are present in that cell. Signals received from inside and outside the cell could turn on or off genes that code for certain transcription factors. For example, the gene for fetal hemoglobin ordinarily gets turned off as a newborn matures—one possible treatment for sickle-cell disease is to turn this gene on again.

Regulatory proteins in human cells consist of transcription factors that bind to DNA.

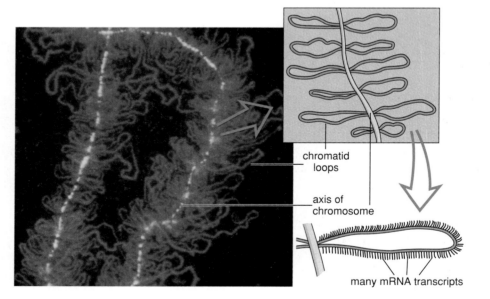

chromatid loops

axis of chromosome

many mRNA transcripts

Control of gene expression occurs at four levels in human cells. In the nucleus, there is transcriptional and posttranscriptional control; in the cytoplasm, there is translational and posttranslational control.

Figure 20.13 Lampbrush chromosomes.
These chromosomes, seen in amphibian egg cells, give evidence that when mRNA is being synthesized, chromosomes decondense. Many mRNA transcripts are being made off of these DNA loops (red).

20.3 Biotechnology

Genetic engineering is the use of technology to alter the genome of viruses, bacteria, and other cells for medical or industrial purposes. **Biotechnology** includes genetic engineering and other techniques that make use of natural biological systems to produce a product or to achieve an end desired by human beings. Since the dawn of civilization, humans have bred plants and animals to produce a particular phenotype. The biochemical capabilities of microorganisms have also been exploited for a very long time. Today, through genetic engineering, bacteria now produce drugs that promote human health, proteins that are useful as vaccines, and nucleic acids for laboratory research. Biotechnology also extends beyond unicellular organisms; ways have been found to alter the genotype and, subsequently, the phenotype of plants and animals, including ourselves.

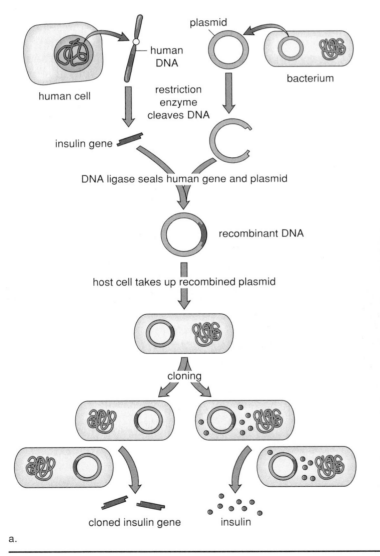

Cloning of a Gene

Genetic engineering allows the insertion of a foreign gene into cells, and this makes them capable of producing a new and different protein.

Choosing a Vector

Recombinant DNA (rDNA) contains DNA (deoxyribonucleic acid) from two or more different sources. To make rDNA, a technician often begins by selecting a **vector,** the means by which rDNA is introduced into a host cell. One common type of vector is a plasmid. **Plasmid**s are small accessory rings of DNA in which the double helix is joined into a circle. The ring is not part of the bacterial chromosome and can replicate independently. Plasmids were discovered by investigators studying the sex life of the intestinal bacterium *Escherichia coli* (*E. coli*).

Plasmids used as vectors have been removed from bacteria and have had a foreign gene inserted into them (Fig. 20.14). Bacteria can take up these engineered plasmids, and when the bacteria divide, the plasmids are replicated. Eventually, there are many copies of the plasmid and therefore many copies of the foreign gene. The gene is now said to have been **cloned.**

Viruses are often the vector of choice for animal cells. After a virus has entered a cell, the DNA is released from the virus and enters the cell proper. Here it may direct the reproduction of many more viruses. Each virus derived from a viral vector contains a copy of the foreign gene. Therefore, viral vectors also allow cloning of a particular gene.

Figure 20.14 Cloning of a human gene.
a. Human DNA and plasmid DNA are cleaved by a specific type of restriction enzyme and spliced together by the enzyme DNA ligase. Gene cloning is achieved when a host cell takes up the recombinant plasmid, and as the cell reproduces, the plasmid is replicated. Multiple copies of the gene are now available to an investigator. If the insulin gene functions normally as expected, the product (insulin) may also be retrieved and **(b)** packaged for sale to the public.

Making Recombinant DNA

The introduction of foreign DNA into vector DNA to produce rDNA requires two enzymes (Fig. 20.14). The first enzyme, called a **restriction enzyme,** cleaves plasmid DNA, and the second, called **DNA ligase,** seals foreign DNA into the opening created by the restriction enzyme.

Hundreds of restriction enzymes occur naturally in bacteria, where they stop viral reproduction by cutting up viral DNA. They are called restriction enzymes because they *restrict* the growth of viruses, but they are also called *molecular scissors* because each one cuts DNA at a specific cleavage site. For example, the restriction enzyme called *EcoRI* always cuts double-stranded DNA in this manner when it has this sequence of bases:

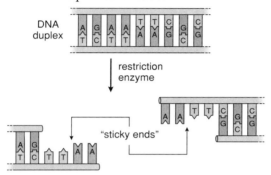

Notice there is now a gap into which a piece of foreign DNA can be placed if it ends in bases complementary to those exposed by the restriction enzyme. To assure this, it is only necessary to cleave the foreign DNA with the same type of restriction enzyme. The single-stranded, but complementary, ends of the two DNA molecules are called "sticky ends" because they can bind a piece of foreign DNA by complementary base pairing. Sticky ends facilitate the insertion of foreign DNA into vector DNA.

The second enzyme needed for preparation of rDNA, DNA ligase, is a cellular enzyme that seals any breaks in a DNA molecule. Genetic engineers use this enzyme to seal the foreign piece of DNA into the vector. DNA splicing is now complete; an rDNA molecule has been prepared.

Getting the Product

Bacterial cells take up recombinant plasmids, especially if they are treated to make them more permeable. Thereafter, if the inserted foreign gene is replicated and actively expressed, the investigator can recover either the cloned gene or a protein product (Fig. 20.14).

In order for gene expression to occur in a bacterium, the gene has to be accompanied by the proper regulatory regions. Also, the gene should not contain introns because bacterial cells do not have the necessary enzymes to process primary messenger RNA (mRNA). However, it's possible to make a mammalian gene that lacks introns. The enzyme called reverse transcriptase can be used to make a DNA copy of mature mRNA. This DNA molecule, called *complementary DNA* (*cDNA*), does not contain introns. Alternatively, it is possible to manufacture small genes in the laboratory. A machine called a DNA synthesizer joins together the correct sequence of nucleotides, and this resulting gene also lacks introns.

Many biotechnology products are now available. These products include hormones and similar types of proteins (Table 20.4) or vaccines (Table 20.5).

Biotechnology products include hormones and similar types of proteins and vaccines. These products are of enormous importance to the fields of medicine and animal husbandry.

TABLE 20.4

Biotechnology Products:
Hormones and Similar Types of Proteins

Treatment of Humans	For
Insulin	Diabetes
Growth hormone	Pituitary dwarfism
tPA (tissue plasminogen activator)	Heart attack
Interferons	Cancer
Erythropoietin	Anemia
Interleukin-2	Cancer
Clotting factor VIII	Hemophilia
Human lung surfactant	Respiratory distress syndrome
Atrial natriuretic factor	High blood pressure
Tumor necrosis factor	Cancer
Ceredase	Gaucher disease

TABLE 20.5

Biotechnology Products:
Potential Vaccines

Prevention in Humans of
Hepatitis B
Lyme disease
Whooping cough
Chlamydia*
Herpes (oral and genital)*
AIDS*

*Planned in the future

Replicating Small DNA Segments

The **polymerase chain reaction (PCR)** can create millions of copies of a single gene, or any specific piece of DNA, in a test tube. PCR is very specific—the targeted DNA sequence can be less than one part in a million of the total DNA sample! This means that a single gene among all the human genes can be amplified (copied) using PCR.

Before carrying out PCR, double-stranded DNA is denatured into single strands by heating. *Primers,* sequences of about 20 bases that are complementary to the bases on one side of the "target DNA," are then added. The primers are needed because DNA polymerase does not start the replication process; it only continues or extends the process. After the primers bind by complementary base pairing to the single-stranded DNA, DNA polymerase (the enzyme that carries out DNA replication) copies the target DNA (Fig. 20.15*a* and *b*).

Analyzing DNA Segments

Following PCR, the nucleotide sequence of the sample can be determined. There are automated DNA sequencers that make use of computers to sequence genes at a fairly rapid rate. Therefore, PCR is carried out in order to

- determine the nucleotide sequence of human genes. Many laboratories around the world are seeking to determine (1) the order of the genes on the human chromosomes and (2) the sequence of the nitrogen bases adenine, thymine, guanine, and cytosine.

- conduct molecular paleontology. Amplified mitochondrial DNA sequences in modern living populations are used to determine the evolutionary history of human populations.

Following PCR, the sample can be used to detect viral infections; diagnose genetic disorders; and diagnose cancer. When the amplified DNA matches that of a virus, mutated gene, or oncogene, then we know that a viral infection, genetic disorder, or cancer is present.

Amplified DNA can be subjected to **DNA profiling** (fingerprinting). When the amplified DNA of a person is treated with restriction enzymes, the result is a unique collection of different-sized fragments. During a process called gel electrophoresis, the fragments can be separated according to their lengths, and the result is a pattern of distinctive bands. If two DNA patterns match, then we know the DNA came from the same person. DNA profiling was used to successfully identify a teenage murder victim from

eight-year-old remains. Skeletal DNA was compared to that obtained from blood samples donated by the victim's parents. A DNA fingerprint of an individual resembles that of the parents because it is inherited. DNA from a single sperm is enough to identify a suspected rapist when PCR amplification precedes DNA profiling.

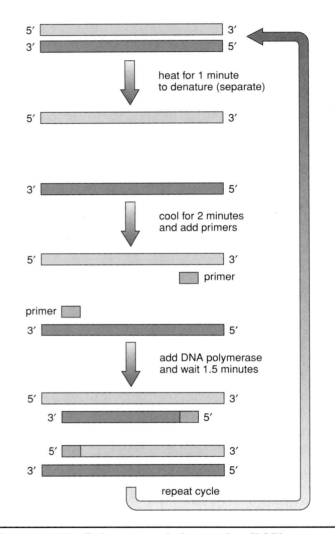

Figure 20.15   Polymerase chain reaction (PCR).
PCR is performed in laboratory glassware. Primers (pink), which are DNA sequences complementary to the bases at one end of target DNA, are necessary for DNA polymerase to make a copy of each DNA strand.

Making Transgenic Organisms

Free-living organisms in the environment that have had a foreign gene inserted into them are called **transgenic organisms.**

Transgenic Bacteria Perform Services

As you know, bacteria are used to clone a gene or to mass-produce a product. They are also genetically engineered to perform other services.

Protection and Enhancement of Plants Genetically engineered bacteria can be used to promote the health of plants. For example, bacteria that normally live on plants and encourage the formation of ice crystals have been changed from frost-plus to frost-minus bacteria. Field tests showed that these genetically engineered bacteria protect the vegetative parts of plants from frost damage. Also, a bacterium that normally colonizes the roots of corn plants has now been endowed with genes (from another bacterium) that code for an insect toxin. The toxin is expected to protect the roots from insects.

Many other rDNA applications in agriculture are thought to be possible. For example, *Rhizobium* is a bacterium that lives in nodules on the roots of leguminous plants, such as bean plants. Here, the bacteria fix atmospheric nitrogen into a form that can be used by the plant. It might be possible to genetically engineer bacteria and nonleguminous plants, such as corn, rice, and wheat, so that bacteria can infect them and produce nodules. This would reduce the amount of fertilizer needed on agricultural fields.

Bioremediation Bacteria can be selected for their ability to degrade a particular substance, and then this ability can be enhanced by genetic engineering. For instance, naturally occurring bacteria that eat oil can be genetically engineered to do an even better job of cleaning up beaches after oil spills (Fig. 20.16). Industry has found that bacteria can be used as biofilters to prevent airborne chemical pollutants from being vented into the air. They can also remove sulfur from coal before it is burned and help clean up toxic waste dumps. One such strain was given genes that allowed it to clean up levels of toxins that would have killed other strains. Further, these bacteria were given "suicide" genes that caused them to self-destruct when the job had been accomplished.

Chemical Production Organic chemicals are often synthesized by having catalysts act on precursor molecules or by using bacteria to carry out the synthesis. Today, it is possible to go one step further and to manipulate the genes that code for these enzymes. For instance, biochemists

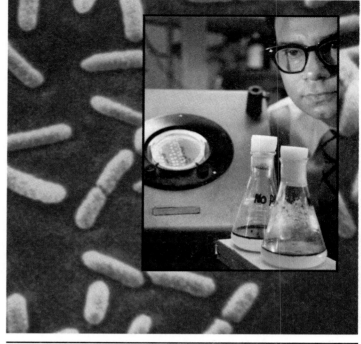

Figure 20.16 Bioremediation.
Bacteria capable of decomposing oil have been engineered and patented by the investigator, Dr. Chakrabarty. In the inset, the flask toward the rear contains oil and no bacteria; the flask toward the front contains the bacteria and is almost clear of oil. Now that engineered organisms (e.g., bacteria and plants) can be patented, there is an even greater impetus to create them.

discovered a strain of bacteria that is especially good at producing phenylalanine, an organic chemical needed to make aspartame, the dipeptide sweetener better known as NutraSweet™. They isolated, altered, and formed a vector for the appropriate genes so that various bacteria could be genetically engineered to produce phenylalanine.

Mineral Processing Many major mining companies already use bacteria to obtain various metals. Genetic engineering may enhance the ability of bacteria to extract copper, uranium, and gold from low-grade sources. At least two mining companies are testing genetically engineered organisms having improved bioleaching capabilities.

Bacteria are being genetically modified to perform all sorts of tasks, not only in the factory, but also in the environment. Safety concerns are discussed in the Ecology reading on page 431.

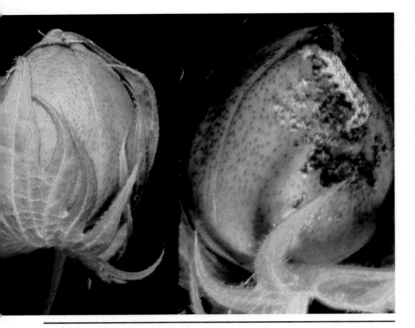

Figure 20.17   Genetically engineered plants.
The cotton boll on the *right* is from a plant that has not been genetically engineered; the boll on the *left* is from a plant that was genetically engineered to resist cotton bollworm larvae and will go on to produce a normal yield of cotton.

Figure 20.18   Genetically engineered animals.
This goat is genetically engineered to produce antithrombin III, which is secreted in her milk. This researcher and many others are involved in the project

Transgenic Plants Are Here

The only possible plasmid for genetically engineering plant cells belongs to the bacterium *Agrobacterium,* which will infect many, but not all, plants. Therefore, other techniques have been developed to introduce foreign DNA into plant cells that have had the cell wall removed and are called protoplasts. It is possible to treat protoplasts with an electric current while they are suspended in a liquid containing foreign DNA. The electric current makes tiny, self-sealing holes in the plasma membrane through which genetic material can enter. Then a protoplast will develop into a complete plant.

Presently, about 50 types of genetically engineered plants that resist insects, viruses, or herbicides have entered small-scale field trials (Fig. 20.17). The major crops that have been improved in this way are soybean, cotton, alfalfa, rice, and potato; however, even genetically engineered corn may reach the marketplace by the year 2000.

It is hoped that one day genetically engineered plants will have one or more of these attributes: (1) they will be heat-, cold-, drought-, or salt-tolerant; (2) they will be more nutritious; (3) they can be stored and transported without fear of damage; (4) they will require less fertilizer; or (5) they will produce chemicals and drugs that are of interest to humans. Plants have been engineered to produce human proteins, such as hormones, in their seeds. A weed called mouse-eared cress has been engineered to produce a biodegradable plastic (polyhydroxybutyrate, or PHB) in cell granules.

Transgenic Animals Are Here

Animals, too, are being genetically engineered. Because animal cells will not take up bacterial plasmids, other methods are used to insert genes into their eggs. It is possible to microinject foreign genes into eggs by hand, but another method uses vortex mixing. The eggs are placed in an agitator with DNA and silicon-carbide needles, and the needles make tiny holes through which the DNA can enter. Using this technique, many types of animal eggs have taken up the gene for bovine growth hormone (bGH). The procedure has been used to produce larger fishes, cows, pigs, rabbits, and sheep. Genetically engineered fishes are now being kept in ponds that offer no escape to the wild because there is much concern that they will upset or destroy natural ecosystems.

Gene pharming, the use of transgenic farm animals to produce pharmaceuticals, is being pursued by a number of firms. It is advantageous to use animals because the product is obtainable from the milk of females. Genes that code for therapeutic and diagnostic proteins are incorporated into the animal's DNA, and the proteins appear in the animal's milk. In one instance, a bull was genetically engineered to carry a gene for human lactoferrin, a drug for gastrointestinal tract infections, and he passed the gene to many offspring, among them several females. There are also plans to produce drugs for the treatment of cystic fibrosis, cancer, blood diseases, and other disorders. Antithrombin III, for preventing blood clots during surgery, is currently being produced by a herd of goats, and clinical trials are to begin soon (Fig. 20.18).

Ecology Focus

Biotechnology: Friend or Foe?

Popular magazines often report biotechnology news such as this example:

> In a feat that could boost wheat production worldwide, plant biologists have for the first time permanently transferred a foreign gene into wheat. The gene makes wheat resistant to the herbicide phosphinothricin, which normally kills any plant it touches, weed or crop. Plant breeders say the herbicide-resistant wheat should enable farmers to spray their fields with the powerful herbicide to eradicate weeds without harming their harvest.*

In a hungry world you would expect such news to be greeted with enthusiasm—this new strain of wheat offers the possibility of a more bountiful harvest and the feeding of many more people. There are some, however, who see dangers lurking in the use of biotechnology to develop new and different strains of plants and animals. And these doomsayers are not just anybody—they are ecologists.

First, we have to consider that herbicide-resistant wheat will allow farmers to use more herbicide than usual in order to kill off weeds. Then, too, suppose, this new form of wheat is better able to compete in the wild. Certainly we know of plants that have become pests when transported to a new environment: prickly pear cactus took over many acres of Australia; an ornamental tree, the melaleuca, has invaded and is drying up many of the

Figure 20A   Transgenic squash.
This squash may look like any other plump yellow squash, but it has been genetically engineered to resist viruses and therefore outproduce its cousins which grow in the wild. Could hybridization with one of its cousins create a "superweed" that would overgrow in the wild, choke out native plants, and become an environmental menace?

swamps in Florida. Such plants spread because they are able to overrun the native plants of an area. Perhaps genetically engineered plants will also spread beyond their intended areas and be out of control. Or worse, suppose herbicide-resistant wheat were to hybridize with a weed, making the weed also resistant and able to take over other agricultural fields. As more and different herbicides are used to kill off the weed, the environment would be degraded. And similar concerns pertain to any transgenic organism, whether a bacterium, animal, or plant (Fig. 20A).

In the past, humans have been quick to believe that a new advance was the answer to a particular problem. The new pesticide DDT was going to kill off mosquitoes, making malaria a disease of the past. Instead, mosquitoes become resistant, and DDT accumulates in the tissues of humans, possibly contributing to all manner of health problems, from reduced immunity to reproductive infertility. When antibiotics were first introduced, it was hoped a disease like tuberculosis would be licked forever. Resistant strains of tuberculosis have now evolved to threaten us all.

More and more transgenic varieties have been developed and are being tested in agricultural fields. A few of these, like Freedom2 Squash, have a weedy relative in the wild with which they could hybridize. Agricultural officials point out, however, that Freedom2 Squash has been growing in fields since 1994, and although hybridization with its weedy relative, a Texas gourd, may have occurred, a "superweed" has not taken over Texas yet. Still, say botanists, it could happen in the future. Laboratory studies in Denmark showed that a hybrid of transgenic oilseed rape and field mustard did resist herbicides and was able to produce highly fertile pollen. Ecologists maintain it may be only a matter of time before a "superweed" does appear in the wild.

*Research Notes, "Biology," Science News 141(1992):379.

Ecological Concerns

There are those who are very concerned about the deliberate release of genetically modified microorganisms (GMMs) into the environment. If these bacteria displaced those that normally reside in an ecosystem, the effects could be deleterious. However, tools are now available to detect, measure, and even stop cell activity in the natural environment. It is hoped that these tools will eventually pave the way for GMMs to play a significant role in agriculture and in environmental protection.

The same ecological concerns regarding genetically engineered bacteria also pertain to genetically engineered plants and animals. The Ecology reading for this chapter uses herbicide-resistant wheat as an example, but the very same arguments can be waged against any genetically engineered organism.

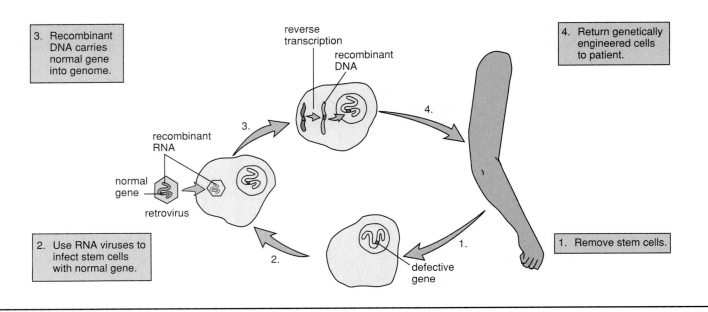

3. Recombinant DNA carries normal gene into genome.

reverse transcription

recombinant DNA

4. Return genetically engineered cells to patient.

recombinant RNA

normal gene

retrovirus

3.

4.

2. Use RNA viruses to infect stem cells with normal gene.

2.

1.

defective gene

1. Remove stem cells.

Figure 20.19   Ex vivo gene therapy in humans.
Stem cells are withdrawn from the body, a virus is used to insert a normal gene into them, and they are returned to the body.

Gene Therapy Is a Reality

Gene therapy gives a patient healthy genes to make up for faulty genes. Gene therapy also includes the use of genes to treat genetic disorders and various other human illnesses. There are ex vivo (outside the living organism) and in vivo (inside the living organism) methods of gene therapy.

Some Methods Are Ex Vivo

Ashanti DeSilva, discussed previously (p. 415), and another young girl with severe combined immunodeficiency syndrome (SCID) underwent ex vivo gene therapy several years ago. These girls lacked an enzyme that is involved in the maturation of T and B cells, and therefore, they were subject to life-threatening infections. Bone marrow stem cells were removed from their blood and infected with a retrovirus that carried a normal gene for the enzyme (Fig. 20.19). Then the cells were returned to the girls. Genetically engineered stem cells are preferred because they produce other cells with the same genes.

Among the 100-plus gene therapy trials, gene therapy is being used for treatment of familial hypercholesterolemia, a condition that develops when liver cells lack a receptor for removing cholesterol from the blood. The high levels of blood cholesterol make the patient subject to fatal heart attacks at a young age. In a newly developed procedure, a small portion of the liver is surgically excised and infected with a retrovirus containing a normal gene for the receptor. Chemotherapy in cancer patients often kills off healthy cells as well as cancer cells. In clinical trials, researchers have given genes to cancer patients that either make healthy cells more tolerant of chemotherapy or make tumors more vulnerable to it. In one trial, bone marrow stem cells from about 30 women with late-stage ovarian cancer were infected with a virus carrying a gene for multiple-drug resistance.

Some Methods Are In Vivo

Other gene therapy procedures use viruses, laboratory-grown cells, or even synthetic carriers to introduce genes directly into patients. If in vivo therapy is used, no cells are removed from the patient. For example, liposomes—microscopic vesicles that spontaneously form when lipoproteins are put into a solution—have been coated with the gene that is defective in cystic fibrosis patients and have then been sprayed into patients' nostrils. Retroviruses can be used to carry genes for cytokines, soluble hormones of the immune system, directly into the tumors of patients. It has been observed that the presence of cytokines stimulates the immune system to rid the body of cancer cells.

Perhaps it will be possible also to use in vivo therapy to cure hemophilia, diabetes, Parkinson disease, or AIDS. To treat hemophilia, patients could get regular doses of cells that contain normal clotting-factor genes. Or such cells could be placed in *organoids,* artificial organs that can be implanted in the abdominal cavity. To cure Parkinson disease, dopamine-producing cells could be grafted directly into the brain. These procedures will use laboratory-grown cells that have been stripped of antigens (to decrease the possibility of an immune system attack).

Gene therapy is now being actively investigated, and researchers are envisioning all sorts of applications aimed at curing human genetic disorders as well as many other types of illnesses.

Mapping Human Chromosomes

Investigators for many years have been trying to find the precise location of the estimated 100,000 human genes on the chromosomes (Fig. 20.20). This knowledge would facilitate laboratory research and medical diagnosis and treatment. Several methods have been used to map the human chromosomes, and one of the latest involves using a **DNA probe,** a sequence of bases that will bind to a complementary sequence. It is possible for investigators to work backwards from protein to mRNA to forming a DNA fluorescent probe that will bind to a chromosome right on a microscope slide.

Those involved in the Human Genome Project have used laboratory equipment to find STSs (sequence-tagged sites) that immediately identify a particular portion of the chromosomes. Investigators can determine if the gene they are working with has one of these unique base sequences; if so, they know where that gene belongs in the human genome. These and other scientists are in the process of sequencing the 3 billion bases in the human genome.

> The Human Genome Project seeks to establish the sequence of all the genes on all the chromosomes and the sequence of all the DNA base pairs on all the chromosomes.

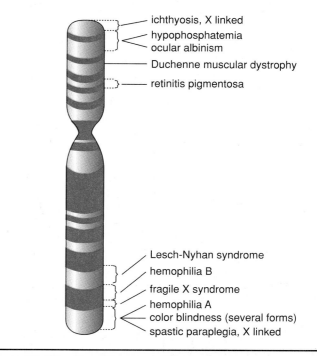

ichthyosis, X linked
hypophosphatemia
ocular albinism
Duchenne muscular dystrophy
retinitis pigmentosa

Lesch-Nyhan syndrome
hemophilia B
fragile X syndrome
hemophilia A
color blindness (several forms)
spastic paraplegia, X linked

Figure 20.20   Genetic map of X chromosome.
The human X chromosome has been partially mapped, and this is the order of some of the genes now known to be on this chromosome.

SUMMARY

20.1 DNA and RNA Structure and Function

DNA is a double helix composed of two nucleic acid strands that are held together by weak hydrogen bonds between the bases: A is bonded to T, and C is bonded to G. During replication, the DNA strands unzip, and then a new complementary strand forms opposite to each old strand. In the end there are two identical DNA molecules.

20.2 DNA Specifies Proteins

Proteins differ from one another by the sequence of their amino acids. DNA has a code that specifies this sequence. Protein synthesis requires transcription and translation. During transcription, the DNA code (triplet of three bases) is passed to an mRNA that then contains codons. Introns are removed from mRNA during mRNA processing. During translation, tRNA molecules bind to their amino acids, and then their anticodons pair with the mRNA codons. In the end, each protein has a sequence of amino acids according to the blueprint provided by the sequence of nucleotides in DNA.

Control of gene expression can occur at four levels in a human cell: at the time of transcription; after transcription and during mRNA processing; during translation; and after translation, before, at, or after protein synthesis. It is known that chromatin has to be extended for transcription to occur and that there are transcription factors that control the activity of genes in human cells.

20.3 Biotechnology

Recombinant DNA contains DNA from two different sources. Human DNA can be inserted into a plasmid, which is taken up by bacteria. When the plasmid replicates, the gene is cloned and its protein is produced. This is the basis for the production of biotechnology products, such as hormones and vaccines.

The polymerase chain reaction (PCR) uses the enzyme DNA polymerase to carry out multiple replications of target DNA. Then it is possible to determine if the DNA of an infectious organism or any particular sequence of DNA is present. The DNA copies can be subjected to DNA profiling to see if it matches DNA from another source.

Transgenic organisms also have been made. More nutritious crops resistant to pests and herbicides are commercially available. Transgenic animals have been supplied with various genes, in particular the one for bovine growth hormone (bGH). Animals are also being used to produce protein products of interest.

Human gene therapy is undergoing clinical trials. Ex vivo therapy involves withdrawing cells from the patient, inserting a functioning gene, usually via a retrovirus, and then returning the treated cells to the patient. Many investigators are trying to develop in vivo therapy in which viruses, laboratory-grown cells, or synthetic carriers will be used to carry healthy genes directly into the patient.

STUDYING THE CONCEPTS

1. Describe the structure of DNA and how this structure contributes to the ease of DNA replication.   417–18

2. Describe the structure of RNA and compare it to the structure of DNA.   419

3. Name and discuss the role of three different types of RNA.   419

4. Describe the structure and function of a protein and the manner in which DNA codes for a particular protein.   420

5. Describe the process of transcription and the three steps of translation. If the code is TTT; CAT; TGG; CCG, what are the codons, and what is the sequence of amino acids?   420–23

6. What are the four levels of genetic control in human cells? Describe two means by which transcription is regulated.   425

7. You are a scientist who has decided to "clone a gene." Tell precisely how you would proceed.   426–27

8. Name two categories of biotechnological products that are now available, and discuss their advantages.   427

9. Naturally occurring bacteria have been genetically engineered to perform what services?   429

10. What types of genetically engineered plants are now available and/or are expected in the near future? What types of animals have been genetically engineered and for what purposes?   430

11. How is gene therapy in humans currently being done?   432

12. What is the Human Genome Project, and in what ways might the project be useful?   433

APPLYING YOUR KNOWLEDGE

Concepts

1. The backbone of DNA strands contains atoms joined by covalent bonds, whereas the bases between the two strands are hydrogen bonded. Based upon your knowledge of bonds, what is the significance of the type of bonding existing in a DNA molecule?

2. Recall that ribosomes can be free in the cytoplasm or attached to the endoplasmic reticulum. Why would you expect a pancreatic cell to have many attached ribosomes?

3. A change in a sequence of DNA occurs so that the mRNA codon reads AUC rather than AUU. Both of these code for the amino acid isoleucine. Argue that this is not a mutation.

4. In attempting to produce a transgenic organism, what is the reason that the same restriction enzyme is utilized to cut both the plasmid and the foreign genetic material?

Bioethical Issue

Genetic experiments can make plants and animals healthier or more productive. Cows can be engineered to produce more milk, and plants can be engineered to resist pests. In some cases, scientists have completely transformed a plant's usual role and given traditional plants genes that will allow them to produce chemicals used in soap, detergent, plastic, and other goods. The researchers hope these green machines can churn out chemicals more cheaply than current methods.

Transgenic plants (and animals) scare some environmental watchdogs who worry that these organisms might dangerously mutate or spread their engineered genes to other organisms. Some people argue that exotic plants and animals can overrun an ecosystem, and worse, could endanger the health of humans. Therefore, biotechnology should not be used.

Should researchers have the right to change the genetic makeup of organisms, whether plant or animal? Is it wrong to release transgenic plants and animals into the wild? Who has the responsibility to keep transgenic plants and animals confined? Defend your answers.

TESTING YOUR KNOWLEDGE

1. In DNA, the base G is always paired with the base _____ , and the base A is always paired with the base _____ .

2. Replication of DNA is semiconservative, meaning that each new double helix is composed of an _____ strand and a _____ strand.

3. The DNA code is a _____ code, meaning that every three bases stands for an _____ .

4. The three types of RNA that are necessary for protein synthesis are _____ , _____ , and _____ .

5. Which of the types of RNA from question four carries amino acids to the ribosomes? _____

6. The sequence of mRNA codons dictates the sequence of amino acids in a protein. This step in protein synthesis is called _____ .

7. The two types of enzymes needed to make recombinant DNA are _____ and _____ .

8. Bacteria, plants, and animals that have been genetically engineered are called _____ organisms.

9. The current ex vivo gene therapy clinical trials use a(n) _____ as a vector to insert healthy genes in the patient's cells.

10. This is a segment of a DNA molecule. (Remember that only the transcribed strand serves as the template.) What are (a) the RNA codons and (b) the tRNA anticodons.

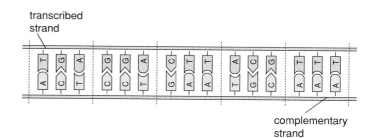

transcribed strand

complementary strand

APPLYING TECHNOLOGY

Your study of DNA and Biotechnology is supported by these available technologies:

Exploring the Internet

The Mader Home Page provides further resources for studying this chapter.

`http://www.mhhe.com/sciencemath/biology/mader/`

(Click on *Human Biology.*)

Explorations in Cell Biology & Genetics CD-ROM

DNA Fingerprinting: You Be the Judge (#14) Students use probes to create DNA fingerprinting patterns by which to decide whether an individual is most likely innocent or guilty of a crime. (4)*

Gene Regulation (#16) Students design regulatory mechanisms and observe the cellular effect. (5)

Making a Restriction Map (#17) Students construct a chromosome map by using DNA restriction fragment length data. (4)

*Level of difficulty

Life Science Animations Video

Video #2: Cell Division, Heredity, Genetics, Reproduction and Development

DNA Replication (#15) DNA structure is reviewed before complementary nucleotides move into place and join, resulting in replication of the original DNA molecule. (1)

Transcription of a Gene (#16) RNA structure is reviewed before complementary nucleotides move into place and RNA polymerase joins them, forming an mRNA molecule. (1)

Protein Synthesis (#17) Initiation, elongation, and termination occur as a part of polypeptide synthesis. (2)

SELECTED KEY TERMS

adenine (A) One of four organic bases in the nucleotides composing the structure of DNA and RNA. 417

anticodon "Triplet" of three bases in transfer RNA that pairs with a complementary triplet (codon) in messenger RNA. 422

biotechnology Use of a natural biological system to produce a commercial product or achieve an end desired by humans. 426

cloned Production of identical copies; in genetic engineering, the production of many identical copies of a gene. 426

codon "Triplet" of bases in messenger RNA that directs the placement of a particular amino acid into a protein. 421

complementary base pairing Pairing of bases between nucleic acid strands; adenine (A) pairs with either thymine (T) if DNA, or uracil (U) if RNA, and cytosine (C) pairs with guanine (G). 417

cytosine (C) One of four organic bases in the nucleotides composing the structure of DNA and RNA. 417

DNA (deoxyribonucleic acid) Nucleic acid found in the cells; the genetic material that directs protein synthesis in cells. 416

DNA ligase (LY-gays) Enzyme that links DNA from two sources; used in genetic engineering to put a gene into plasmid DNA. 427

DNA probe Single strand of DNA that can be used to locate a particular stretch of DNA. 433

DNA profiling Using fragment lengths resulting from restriction enzyme cleavage to identify particular individuals; also called DNA fingerprinting. 428

gene Unit of heredity that codes for a polypeptide and is passed on to offspring. 416

gene therapy Use of genetically engineered cells or other biotechnology techniques to treat genetic or other disorders. 432

genetic engineering Use of technology to alter the genome of a living cell for medical or industrial use. 426

guanine (G) One of four organic bases in the nucleotides composing the structure of DNA and RNA. 417

messenger RNA (mRNA) Ribonucleic acid whose sequence of codons specifies the sequence of amino acids during protein synthesis. 419

mutation Change in the genetic material. 418

plasmid Circular DNA segment that is present in bacterial cells that replicates independent from the bacterial chromosome. 426

polymerase chain reaction (PCR) Technique that uses the enzyme DNA polymerase to produce millions of copies of a particular piece of DNA. 428

polyribosome Cluster of ribosomes attached to the same mRNA molecule; each ribosome is producing a copy of the same polypeptide. 423

recombinant DNA DNA that contains genes from more than one source. 426

replication Making an exact copy, as in the duplication of DNA. 418

restriction enzyme Enzyme that stops viral reproduction by cutting viral DNA; used in genetic engineering to cut DNA at specific points. 427

ribosomal RNA (rRNA) RNA occurring in ribosomes, structures involved in protein synthesis. 419

ribozyme Enzyme that carries out mRNA processing. 421

RNA (ribonucleic acid) Nucleic acid found in cells that assists DNA in controlling protein synthesis. 419

template Pattern that serves as a mold for the production of an oppositely shaped structure; one strand of DNA is a template for a complementary strand. 418

thymine (T) One of four organic bases in the nucleotides composing the structure of DNA and RNA. 417

transcription Process resulting in the production of a strand of RNA that is complementary to a segment of DNA. 420

transfer RNA (tRNA) Molecule of RNA that carries an amino acid to a ribosome engaged in the process of protein synthesis. 419

transgenic organism Free-living organism in the environment that has had a foreign gene inserted into it. 429

translation Process by which the sequence of codons in mRNA directs the sequence of amino acids in a protein. 422

triplet code Genetic code in which sets of three bases stand for specific amino acids. 421

vector Carrier, such as a plasmid or a virus, for recombinant DNA that introduces a foreign gene into a host cell. 426

Chapter 21

Cancer

Chapter Outline

Figure 21.1 Sunbathing.
Sunbathing can lead to skin cancer. Melanoma is a particularly malignant skin cancer.

If there's one thing 25-year-old Emily loves, it's lying in the sun. "Baking on the beach" is what she calls it (Fig. 21.1). A little baby oil, a radio, some sunglasses. She does it for hours. Unfortunately, while Emily rubs in the oil, she doesn't notice the small mole on the back of her right arm. Dark and curvy, the little mole looks like a splotch of spilled ink. Doctors call it melanoma.

If Emily is lucky, she or her physician will see the mole before the cancer cells underneath break away and spread. If no one notices the mole, the cells will likely travel through Emily's circulatory system, making their way to her lungs, liver, brain, and ovaries. When that happens, it will be too late.

About 32,000 cases of melanoma are diagnosed each year. One in five persons diagnosed dies within five years. For most people, it's easy to avoid the disease. Wear sunscreen and protective clothing. Don't visit tanning machines. Most important, don't "bake on the beach."

The melanoma that developed on Emily's right arm went through certain phases. During *initiation*, a single cell developed a mutation that caused it to begin to divide repeatedly. Then, *promotion* occurred as other factors encouraged a tumor to develop, and the tumor cells continued to divide. As they divided they developed other mutations. Finally, *progression* took place as mutations gave one cell a selective advantage over the other cells. This process can be repeated several times, until there is a cell that has the ability to invade surrounding tissues. This is called *metastasis*.

21.1 Causes and Prevention of Cancer

A **mutagen** is an agent that increases the chances of a mutation, while a **carcinogen** is an environmental agent that can contribute to the development of cancer. Carcinogens are often mutagenic.

Carcinogens, heredity, and immunodeficiencies all contribute to the development of **cancer.**

Carcinogens

Among the best-known mutagenic carcinogens are (1) radiation, (2) organic chemicals (e.g., tobacco smoke, foods, pollutants), and (3) viruses.

Radiation

Some forms of radiation are carcinogenic. Ultraviolet radiation in sunlight and tanning lamps is most likely responsible for the dramatic increases seen in skin cancer the past several years. Today there are at least six cases of skin cancer for every one case of lung cancer. Nonmelanoma skin cancers are usually curable through surgery, but melanoma skin cancer tends to spread and is responsible for 1–2% of total cancer deaths in the United States.

Another natural source of radiation is radon gas. In the very rare house with an extremely high level of exposure, the risk of developing lung cancer is thought to be equivalent to smoking a pack of cigarettes a day. Therefore, the combination of radon gas and smoking cigarettes can be particularly dangerous. Radon levels can be lowered by improving the ventilation of a building.

Most of us have heard about the damaging effects of the nuclear bomb explosions or accidental emissions from nuclear power plants. For example, more cancer deaths are expected in the vicinity of the Chernobyl Power Station (in the former U.S.S.R.), which suffered a terrible accident in 1986. Usually, however, diagnostic X rays account for most of our exposure to artificial sources of radiation. The benefits of these procedures can far outweigh the possible risk, but it is still wise to avoid any X-ray procedures that are not medically warranted.

Despite much publicity, scientists have not been able to show a clear relationship between cancer and radiation from electric power lines, household appliances, and cellular telephones.

Organic Chemicals

Tobacco smoke, foods, and pollutants all contain organic chemicals that can be carcinogenic.

Tobacco Smoke Tobacco smoke contains a number of organic chemicals that are known carcinogens, and it is estimated that one-third of all cancer deaths can be attributed to smoking (Table 21.1). Lung cancer is the most frequent lethal cancer in the United States, and smoking is also implicated in the development of cancers of the mouth, larynx, bladder, kidney, and pancreas. The greater the number of cigarettes smoked per day, the earlier the habit starts,

Figure 21.2   Tobacco smoke and alcohol.
Smoking cigarettes, especially when combined with drinking of alcohol, is associated with cancer of the lungs, mouth, larynx, kidney, bladder, pancreas, and many other cancers.

TABLE 21.1	
Carcinogens in Cigarette Smoke	
Aminostilbene	N-Dibutylnitrosamine
Arsenic	2, 3-Dimethylchrysene
Benz (a) anthracene	Indenol (1, 2, 3-cd) pyrene
Benz (a) pyrene	5-Methylchrysene
Benzene	Methylfluoranthene
Benzo (b) fluoranthene	B-Napthylamine
Benzo (c) phenanthrene	Nickel compounds
Cadmium	N-Nitrosodiethylamine
Chrysene	N-Nitrosodimethylamine
Dibenz (a,c) anthracene	N-Nitrosomethylethylamine
Dibenz (a,e) fluoranthene	N-Nitrosonanabasine
Dibenz (a,h) acridine	Nitrosonornicotine
Dibenz (a,j) acridine	N-Nitrosopiperidine
Dibenz (c,g) carbazone	N-Nitrosopyrrolidine
	Polonium-210

From Steven B. Oppenheimer "Advances in Cancer Biology," in American Biology Teacher, *49(1):13, January 1987. Copyright © 1987 National Association of Biology Teachers, Reston, VA. Reprinted by permission.*

and a high tar content all contribute to the possibility of cancer. When smoking is combined with drinking alcohol, the risk of these cancers increases even more (Fig. 21.2).

Passive smoking, or inhalation of someone else's tobacco smoke, is also dangerous and probably causes a few thousand deaths each year.

Foods Statistically, studies suggest that a diet rich in saturated fats and low in fiber rivals tobacco in causing cancer, especially colon, rectal, and prostate cancer. Obesity seems to increase the risk of colon, kidney, and gallbladder cancers.

Health Focus

Prevention of Cancer

There is clear evidence that the risk of certain types of cancer can be reduced by adopting protective behaviors and the right diet.

PROTECTIVE BEHAVIORS

These behaviors help prevent cancer:

Don't Smoke Cigarette smoking accounts for about 30% of all cancer deaths. Smoking is responsible for 90% of lung cancer cases among men and 79% among women—about 87% altogether. Those who smoke two or more packs of cigarettes a day have lung cancer mortality rates 15 to 25 times greater than nonsmokers. Smokeless tobacco (chewing tobacco or snuff) increases the risk of cancers of the mouth, larynx, throat, and esophagus.

Don't Sunbathe Almost all cases of basal and squamous-cell skin cancers are considered to be sun related. Further, sun exposure is a major factor in the development of melanoma, and the incidence of this cancer increases for those living near the equator.

Avoid Alcohol Cancers of the mouth, throat, esophagus, larynx, and liver occur more frequently among heavy drinkers, especially when accompanied by tobacco use (cigarettes or chewing tobacco).

Avoid Radiation Excessive exposure to ionizing radiation can increase cancer risk. Even though most medical and dental X rays are adjusted to deliver the lowest dose possible, unnecessary X rays should be avoided. Excessive radon exposure in homes increases the risk of lung cancer, especially in cigarette smokers. It is best to test your home and take the proper remedial actions.

Be Tested for Cancer Do the shower check for breast cancer or testicular cancer. Have other exams done regularly by a physician.

Be Aware of Occupational Hazards Exposure to several different industrial agents (nickel, chromate, asbestos, vinyl chloride, etc.) and/or radiation increases the risk of various cancers. Risk from asbestos is greatly increased when combined with cigarette smoking.

Be Aware of Hormone Therapy Estrogen therapy to control menopausal symptoms increases the risk of endometrial cancer. However, including progesterone in estrogen replacement therapy helps to minimize this risk.

THE RIGHT DIET

Statistical studies have suggested that persons who follow certain dietary guidelines are less likely to have cancer. The following dietary guidelines greatly reduce your risk of developing cancer:

1. Avoid obesity. The risk of cancer (especially colon, breast, and uterine cancers) is 55% greater among obese women and 33% greater among obese men, compared to people of normal weight.

2. Lower total fat intake. A high-fat intake has been linked to development of colon, prostate, and possibly breast cancers.

3. Eat plenty of high-fiber foods. These include whole-grain cereals, fruits, and vegetables. Studies have indicated that a high-fiber diet protects against colon cancer, a frequent cause of cancer deaths. It is worth noting that foods high in fiber also tend to be low in fat!

4. Increase consumption of foods that are rich in vitamins A and C. Beta-carotene, a precursor of vitamin A, is found in dark green, leafy vegetables; carrots; and various fruits. Vitamin C is present in citrus fruits. These vitamins are called antioxidants because in cells they prevent the formation of free radicals (organic ions that have an unpaired electron) that can possibly damage DNA. Vitamin C also prevents the conversion of nitrates and nitrites into carcinogenic nitrosamines in the digestive tract.

5. Cut down on consumption of salt-cured, smoked, or nitrite-cured foods. Salt-cured or pickled foods may increase the risk of stomach and esophageal cancers. Smoked foods, like ham and sausage, contain chemical carcinogens similar to those in tobacco smoke. Nitrites are sometimes added to processed meats (e.g., hot dogs and cold cuts) and other foods to protect them from spoilage; as mentioned previously, nitrites are converted to nitrosamines in the digestive tract.

6. Include vegetables from the cabbage family in the diet. The cabbage family includes cabbage, broccoli, brussels sprouts, kohlrabi, and cauliflower. These vegetables may reduce the risk of gastrointestinal and respiratory tract cancers.

7. Be moderate in the consumption of alcohol. People who drink and smoke are at an unusually high risk for cancers of the mouth, larynx, and esophagus.

Animal testing of certain food additives, such as red dye #2 and the synthetic sweetener saccharin, has shown that these substances are cancer-causing in very high doses. In humans, only salty foods appear to make a significant contribution to cancer.

The good news is that eating vegetables and fruits is protective against the development of cancer. Antioxidants in these foods neutralize free radicals that are released during metabolism and can damage a cell's DNA.

Pollutants   Industrial chemicals, such as benzene and carbon tetrachloride, and industrial materials, such as vinyl chloride and asbestos fibers, are also associated with the development of cancer. Pesticides and herbicides are dangerous not only to pets and plants but also to our own health because they contain organic chemicals that can cause mutations. A panel of experts assembled by the National Academy of Sciences' Institute of Medicine has found conclusive evidence that exposure to dioxin, a contaminant of the herbicide Agent Orange used during the Vietnam War, can be linked to cancers of lymphoid tissues and those of muscles and certain connective tissues. Similarly, another study has found that these and other cancers were found more often in those exposed to dioxin accidently released after an explosion in a chemical plant over Seveso, Italy, on July 10, 1976, than in the rest of the population.

Viruses   At least three DNA viruses have been linked to human cancers: hepatitis B virus to liver cancer, human papillomavirus to cancer of the cervix, and Epstein-Barr virus to Burkitt's lymphoma (a cancer of the lymphoid tissues) and nasopharyngeal cancer.

In China, almost all persons have been infected with the hepatitis B virus, and this correlates with the high incidence of liver cancer in that country. For a long time, circumstances suggested that cervical cancer was a sexually transmitted disease, and now human papillomaviruses are routinely isolated from cervical cancers. Burkitt's lymphoma occurs frequently in Africa, where virtually all children are infected with the Epstein-Barr virus. In China, the Epstein-Barr virus is isolated in nearly all nasopharyngeal cancer specimens. It's believed that environmental factors must be involved in determining the final effect of an Epstein-Barr viral infection because the virus is associated with different diseases in different countries (in the United States, this virus causes mononucleosis).

RNA-containing retroviruses, in particular, are known to cause cancers in animals. In humans, the retrovirus HTLV (human T-cell lymphotropic virus, type 1) has been shown to cause adult T-cell leukemia. This disease occurs frequently in parts of Japan, the Caribbean, and Africa, particularly in those regions where people are known to be infected with the virus.

Heredity

Particular types of cancer seem to run in families. For instance, the risk of developing breast, lung, and colon cancers increases two- to threefold when first-degree relatives have had these cancers. Investigators have pinpointed the location of a gene called BRCA1 (Breast Cancer gene #1), which was first identified in a large family whose female members are prone to breast cancer.

Certain childhood cancers seem to be due to the inheritance of a dominant gene. Retinoblastoma is an eye tumor that usually develops by age three; Wilm's tumor is characterized by numerous tumors in both kidneys. In adults, several family syndromes (e.g., Li-Fraumeni cancer family syndrome, Lynch cancer family syndrome, and Warthin cancer family syndrome) are known. Those who inherit a dominant allele develop tumors in various parts of the body.

Immunodeficiencies

Cancer is apt to develop in individuals who exhibit an immunodeficiency. For example, cervical cancer develops in women with AIDS; and Kaposi's sarcoma, a cancer of the blood vessels, develops in many persons with AIDS. Transplant patients who are on immunosuppressive drugs are more apt to develop lymphomas and Kaposi's sarcoma. Persons who inherit an immunodeficiency are also more apt to develop these cancers.

It appears, then, that an active immune system can help protect us from cancer. Mutated cells may display antigens that normally subject them to attack by T lymphocytes and possibly also antibodies. Cancer is seen more often among the elderly perhaps because the immune system weakens as we age.

Specific mutations cause cancer. Carcinogens (e.g., tobacco smoke and fats), heredity, and immunodeficiency all play a role in the development of cancer.

Figure 21.3 shows that **carcinogenesis,** the development of cancer, is a multistep process with some factors causing *initiation* (an initial mutation) and others causing *promotion* (promoting the process), until *progression* leads to the ability of cells to invade the blood vessels and begin a new tumor at another location.

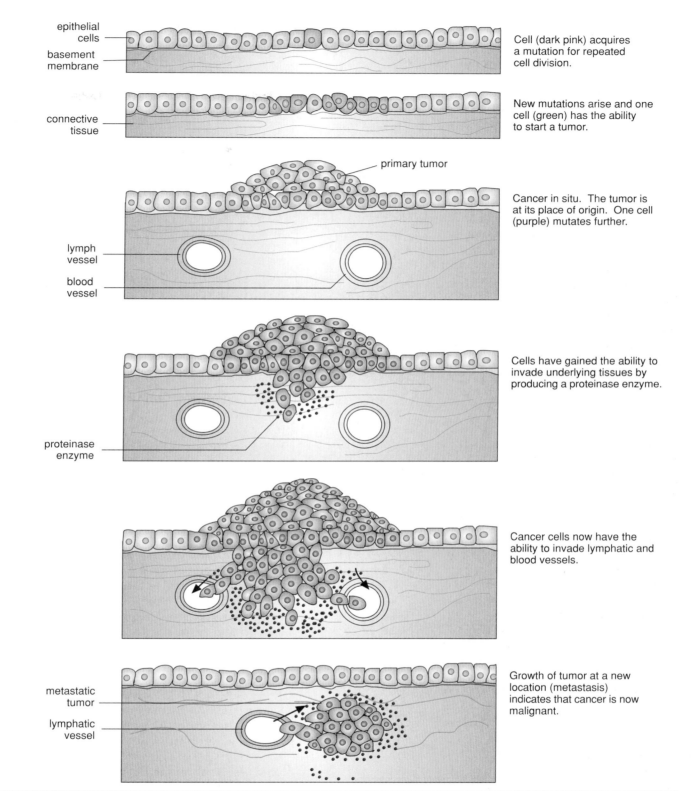

epithelial cells

basement membrane

Cell (dark pink) acquires a mutation for repeated cell division.

connective tissue

New mutations arise and one cell (green) has the ability to start a tumor.

primary tumor

Cancer in situ. The tumor is at its place of origin. One cell (purple) mutates further.

lymph vessel

blood vessel

Cells have gained the ability to invade underlying tissues by producing a proteinase enzyme.

proteinase enzyme

Cancer cells now have the ability to invade lymphatic and blood vessels.

metastatic tumor

lymphatic vessel

Growth of tumor at a new location (metastasis) indicates that cancer is now malignant.

Figure 21.3 Carcinogenesis.
The development of cancer requires a series of mutations leading first to a localized tumor and then metastatic tumors.

21.2 Cancer Cells

When cells become cancerous, a tumor often results. Tumors are classified according to the affected tissue.

Characteristics of Cancer Cells

Cancer cells exhibit characteristics that distinguish them from normal cells (Table 21.2). In general, cancer cells undergo uncontrolled growth and are not regulated by the signals and constraints that usually control the growth of cells.

Cancer Cells Lack Differentiation

Most cells are specialized; they have a specific form and function that suits them to the role they play in the body. Cancer cells are nonspecialized and do not contribute to the functioning of a body part. A cancer cell does not look like a differentiated epithelial, muscular, nervous, or connective tissue cell and instead has a shape and form that is distinctly abnormal (Fig. 21.4). Normal cells can undergo the cell cycle for about 50 times and then they die. Cancer cells can enter the cell cycle repeatedly, and in this way, they are potentially immortal. In cell culture, they die only because they run out of nutrients or are killed by their own toxic waste products.

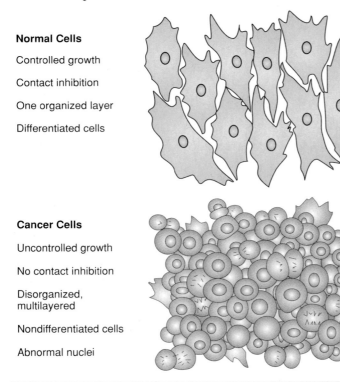

Normal Cells

Controlled growth

Contact inhibition

One organized layer

Differentiated cells

Cancer Cells

Uncontrolled growth

No contact inhibition

Disorganized, multilayered

Nondifferentiated cells

Abnormal nuclei

Figure 21.4   Cancer cells compared to normal cells.
Cancer cells differ from normal cells in the ways noted.

	TABLE 21.2	
Characteristics of Normal Cells Versus Cancer Cells		
Characteristic	**Normal Cells**	**Cancer Cells**
Differentiation	Yes	No
Nuclei	Normal	Abnormal
Growth	Controlled	Uncontrolled
Contact inhibition	Yes	No
Growth factors	Required	Not required
Angiogenesis	No	Yes
Metastasis	No	Yes

Cancer Cells Have Abnormal Nuclei

The nuclei of cancer cells are enlarged, and there may be an abnormal number of chromosomes. The chromosomes have mutated; some parts may be duplicated and some may be deleted, for example. In addition, *gene amplification* (extra copies of specific genes) is seen much more frequently than in normal cells.

Cancer Cells Form Tumors

Normal cells anchor themselves to a substratum and/or adhere to their neighbors. They exhibit *contact inhibition*—when they come in contact with a neighbor they stop dividing. In culture, normal cells form a single layer that covers the bottom of a petri dish. Cancer cells have lost all restraint; they pile up on top of each other to grow in multiple layers. Normal cells do not grow and divide unless they are stimulated to do so by a growth factor. Cancer cells have a reduced need for a growth factor, such as epidermal growth factor, in order to grow and divide.

In the body, a cancer cell divides to form an abnormal mass of cells called a **tumor,** which invades and destroys neighboring tissue (see Fig. 21.3). This new growth, termed *neoplasia,* is made up of cells that are disorganized, a condition termed *anaplasia.* A *benign tumor* is a disorganized, usually encapsulated, mass that does not invade adjacent tissue.

Cancer Cells Undergo Angiogenesis and Metastasis

Angiogenesis, the formation of new blood vessels, is required to bring nutrients and oxygen to a cancerous tumor whose growth is not contained within a capsule. Cancer cells release a growth factor that causes neighboring blood vessels to branch into the cancerous tissue. Some modes of cancer treatment are aimed at preventing angiogenesis from occurring.

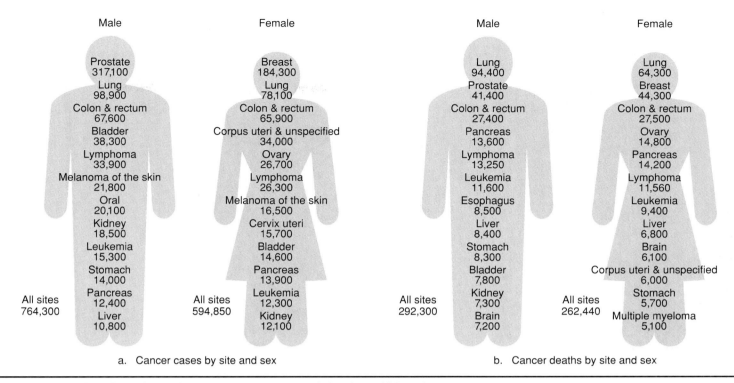

Male	Female	Male	Female
Prostate 317,100	Breast 184,300	Lung 94,400	Lung 64,300
Lung 98,900	Lung 78,100	Prostate 41,400	Breast 44,300
Colon & rectum 67,600	Colon & rectum 65,900	Colon & rectum 27,400	Colon & rectum 27,500
Bladder 38,300	Corpus uteri & unspecified 34,000	Pancreas 13,600	Ovary 14,800
Lymphoma 33,900	Ovary 26,700	Lymphoma 13,250	Pancreas 14,200
Melanoma of the skin 21,800	Lymphoma 26,300	Leukemia 11,600	Lymphoma 11,560
Oral 20,100	Melanoma of the skin 16,500	Esophagus 8,500	Leukemia 9,400
Kidney 18,500	Cervix uteri 15,700	Liver 8,400	Liver 6,800
Leukemia 15,300	Bladder 14,600	Stomach 8,300	Brain 6,100
Stomach 14,000	Pancreas 13,900	Bladder 7,800	Corpus uteri & unspecified 6,000
All sites 764,300 — Pancreas 12,400	All sites 594,850 — Leukemia 12,300	All sites 292,300 — Kidney 7,300	All sites 262,440 — Stomach 5,700
Liver 10,800	Kidney 12,100	Brain 7,200	Multiple myeloma 5,100

a. Cancer cases by site and sex

b. Cancer deaths by site and sex

Figure 21.5   Leading sites of new cancer cases and deaths, 1996 estimates.

Cancer in situ is a tumor located in its place of origin, before there has been any invasion of normal tissue. Malignancy is present when **metastasis** establishes new tumors distant from the primary tumor. To accomplish metastasis, cancer cells must make their way across a basement membrane and into a blood vessel or lymphatic vessel. It has been discovered that cancer cells have receptors that allow them to adhere to basement membranes; they also produce proteinase enzymes that degrade the membrane and allow them to invade underlying tissues. Cancer cells tend to be motile. They have a disorganized internal cytoskeleton and lack intact actin filament bundles. After traveling through the blood or lymph, cancer cells may start tumors elsewhere in the body (Fig. 21.5).

The patient's prognosis (probable outcome) is dependent on the degree to which the cancer has progressed: (1) whether the tumor has invaded surrounding tissues; (2) if so, whether there is any lymph node involvement; and (3) whether there are metastatic tumors in distant parts of the body. With each progressive step, the prognosis becomes less favorable.

Cancer cells are nonspecialized, have abnormal chromosomes, and divide uncontrollably. Because they are not constrained by their neighbors, they form a tumor. Then they metastasize, forming new tumors wherever they relocate.

Classification of Cancers

Cancers are classified according to the type of tissue from which they arise. **Carcinomas** are cancers of the epithelial tissues, and adenocarcinomas are cancers of glandular epithelial cells. Carcinomas include cancer of the skin, breast, liver, pancreas, intestines, lung, prostate, and thyroid. **Sarcomas** are cancers that arise in muscles and connective tissue, such as bone and fibrous connective tissue. **Leukemias** are cancers of the blood, and **lymphomas** are tumors of lymphoid tissue.

The size of a tumor and the degree of metastasis is of critical importance when deciding on a plan of treatment. Currently, different ways to describe staging—the extent of the disease—are frequently used for different cancers. However, the TNM system is applicable to all cancers. In this system, the extent of disease is described in terms of three parameters: T, the condition of the primary tumor, that is, the extent to which the tumor has invaded proximal tissues; N, the extent of lymph node involvement; and M, the extent of distant metastases.

Cancers are classified as carcinomas (epithelial tissue cancers), sarcomas (muscles and connective tissue), leukemias (blood), and lymphomas (lymphoid tissue). Appendix E gives pertinent information about cancers at specific sites.

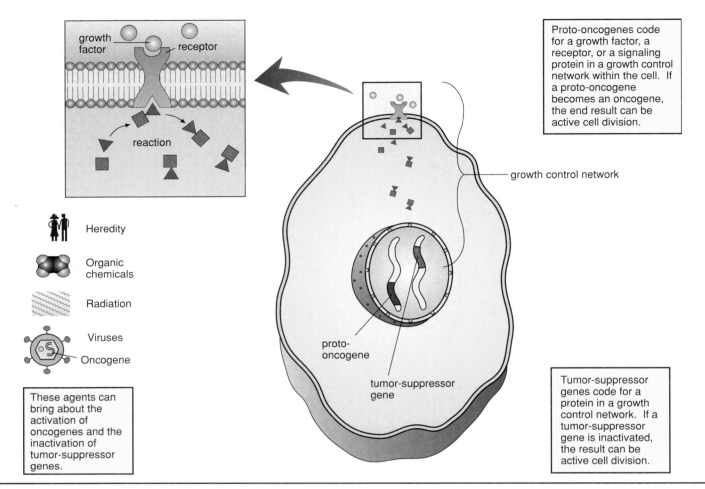

Proto-oncogenes code for a growth factor, a receptor, or a signaling protein in a growth control network within the cell. If a proto-oncogene becomes an oncogene, the end result can be active cell division.

growth factor

receptor

reaction

growth control network

Heredity

Organic chemicals

Radiation

Viruses

Oncogene

These agents can bring about the activation of oncogenes and the inactivation of tumor-suppressor genes.

proto-oncogene

tumor-suppressor gene

Tumor-suppressor genes code for a protein in a growth control network. If a tumor-suppressor gene is inactivated, the result can be active cell division.

Figure 21.6 Growth control network.
The growth control network includes growth factors, their plasma membrane receptors, intracellular reactions, and the genes, notably proto-oncogenes and tumor-suppressor genes. Changes in the growth control network can lead to uncontrolled growth and a tumor.

21.3 Oncogenes and Tumor–Suppressor Genes

A *growth control network* controls cell division in cells. The growth control network influences whether a cell enters and/or completes the cell cycle, which includes the stages of cell division. As Figure 21.6 shows, the growth control network begins with a growth factor and its receptor in the plasma membrane. A growth factor is a signaling protein that causes cells to divide. Cells can secrete growth factors that stimulate themselves and/or other cells.

Reception of a growth factor causes the receptor to set in motion a whole series of responses in which one metabolic reaction leads to another, until finally a gene is turned on or off. Any change occurring anywhere along the pathway of the growth control network can possibly contribute to the occurrence of abnormal cellular growth.

Proto-oncogenes and tumor-suppressor genes, which are discussed next, code for various elements that are active in the growth control network.

Proto-oncogenes and Oncogenes

Proto-oncogenes are normal genes that code for proteins in the growth control network. Some proto-oncogenes code for growth factors; others code for growth factor receptors, or stimulatory proteins in the network that are located in the cytoplasm or even the nucleus.

When a proto-oncogene mutates due to one of the environmental factors outlined in Figure 21.6, an oncogene can result. **Oncogenes** are cancer-causing genes because their products bring about unbridled cell division. For example, an oncogene might cause the cell to produce too much growth factor, or it might cause the growth control network to always lead to cell division. For example, the receptor might promote reactions in the cytoplasm, or certain proteins in the cytoplasm might bring about stimulatory reactions or turn on genes in the nucleus even though no growth factor has been received.

The *ras* family of oncogenes has been implicated in several types of cancers. An alteration of only a single nucleotide pair is sufficient to convert a normally functioning *ras* proto-oncogene to an oncogene. The *ras*K

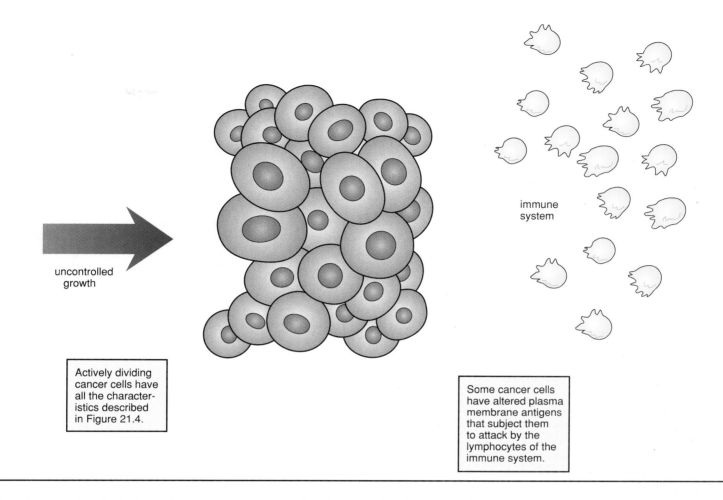

uncontrolled growth

Actively dividing cancer cells have all the character- istics described in Figure 21.4.

immune system

Some cancer cells have altered plasma membrane antigens that subject them to attack by the lymphocytes of the immune system.

Cancer still does not develop unless the immune system fails to respond and kill these abnormal cells.

oncogene is found in about 25% of lung cancers, 50% of colon cancers, and 90% of pancreatic cancers. The *ras*N oncogene is associated with leukemias (cancer of blood-forming cells) and lymphomas (cancers of lymphoid tissue), and both *ras* oncogenes are frequently found in thyroid cancers. The proteins encoded by mutant *ras* genes continuously stimulate the reactions in the growth control network, even when no growth factor has been received.

Tumor-Suppressor Genes

Another class of genes play a major role in triggering cancer. **Tumor-suppressor gene**s code for regulatory proteins in the growth control network that can stop stimulatory reactions leading to cell division. Tumor-suppressor genes can turn off the effects of oncogenes. Mutated tumor-suppressor genes are no longer able to stop the stimulatory effects of oncogenes, and cancer results.

Researchers have identified about a half-dozen or so tumor-suppressor genes that, when they malfunction, can result in increased growth. The *RB* tumor-suppressor gene was discovered by studying the inherited condition retino-

blastoma. When a child receives only one normal *RB* gene and that gene mutates, eye tumors develop in the retina by age three. *RB* has now been found to malfunction in other cancers such as breast, prostate, and bladder cancers.

We now know that there are signaling proteins that act like negative growth factors. When the substance TGFβ attaches to a receptor, the *RB* protein is activated. An active *RB* protein turns off the expression of the proto-oncogene *c-myc*. When the *RB* protein is not present, the protein product of the *c-myc* gene is thought to cause the expression of other genes whose products lead to abnormal cell division and cancer.

Each cell contains a growth control network involving proto-oncogenes and tumor-suppressor genes that code for growth factor receptors and signaling proteins active in intracellular reactions. When proto-oncogenes mutate, becoming oncogenes, and tumor-suppressor genes mutate, the growth control network no longer functions as it should and uncontrolled growth results.

21.4   Diagnosis and Treatment

It is estimated that about one out of every three people in the United States will develop cancer and one out of every four will die from it. Appendix E gives the warning signs, risk factors, methods for diagnosis, and the suggested treatment for various types of cancer.

Diagnosis of Cancer

Diagnosis of cancer before metastasis is difficult although treatment at this stage is usually more successful. The American Cancer Society, Inc., publicizes seven warning signals that spell out the word CAUTION and that everyone should be aware of:

C hange in bowel or bladder habits;
A sore that does not heal;
U nusual bleeding or discharge;
T hickening or lump in breast or elsewhere;
I ndigestion or difficulty in swallowing;
O bvious change in wart or mole;
N agging cough or hoarseness.

Unfortunately, some of these symptoms are not obvious until cancer has progressed to one of its later stages.

Routine Screening Tests

The aim of medicine is to develop tests for cancer that are relatively easy to do, cost little, and are fairly accurate. So far, only the Pap smear for cervical cancer fulfills these three requirements. A physician merely takes a sample of cells from the cervix, which are examined microscopically for signs of abnormality. Regular Pap smears are credited with preventing over 90% of deaths from cervical cancer.

Breast cancer is not as easily detected, but three procedures are recommended. Every woman should do a monthly breast self-examination (see the Health reading on page 447). During an annual physical examination, recommended especially for women above age 40, a physician does this same procedure. While helpful, an examination may not detect lumps before metastasis has already taken place. The third recommended procedure, *mammography,* which is an X-ray study of the breast, is expected to do this (Fig. 21.7). However, mammograms do not show all cancers, and new tumors may develop during the interval between mammograms. The objective is that a mammogram will reveal a lump that is too small to be felt and at a time when the cancer is still highly curable.

Screening for colon cancer is also dependent upon three types of testing. A digital rectal examination performed by a physician is actually of limited value because only a portion of the rectum can be reached by finger. With flexible sigmoidoscopy, the second procedure, a much larger portion of the colon can be examined by using a thin, pliable, lighted tube. Finally, a stool blood test (fecal occult blood test) consists of examining

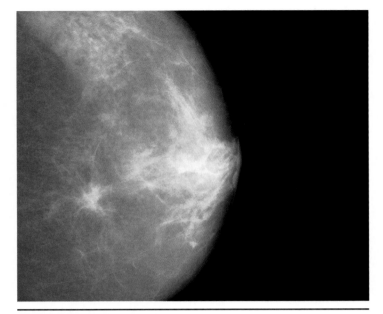

Figure 21.7   Mammogram.
X-ray image of the breast can find tumors too small to be palpable.

a stool sample to detect any hidden blood. The sample is smeared on a slide, and a chemical is added that changes color in the presence of hemoglobin. This procedure is based on the supposition that a cancerous polyp bleeds, although some polyps do not bleed, and bleeding is not always due to a polyp. Therefore, the percentage of false negatives and false positives is high. All positive tests are followed up by a colonoscopy, an examination of the entire colon, or by X ray after a barium enema. The colonoscope can detect polyps that are destroyed by laser as soon as they are detected.

Other tests in routine use are blood tests to detect leukemia and urinalysis for the diagnosis of bladder cancer. Newer tests under consideration are tumor marker tests and tests for oncogenes.

Tumor Marker Tests

Blood tests for tumor antigens/antibodies are called *tumor marker tests.* They are possible because tumors release substances that provoke an antibody response in the body. For example, if an individual has already had colon cancer, it is possible to use the presence of an antigen, called CEA (for carcinoembryonic antigen), to detect any relapses. When the CEA level rises, additional tumor growth has occurred.

There are also tumor marker tests that can be used as an adjunct procedure to detect cancer in the first place. They are not reliable enough to count on solely, but in conjunction with physical examination and ultrasound (see following), they are considered useful. There is a prostate-specific antigen (PSA) test for prostate cancer, a CA 125 test for ovarian cancer, and an alpha-fetoprotein (AFP) test for liver tumors, for example.

Health Focus

Shower Check for Cancer

The American Cancer Society urges women to do a breast self-exam and men to do a testicle self-exam every month. Breast cancer and testicular cancer are far more curable if found early, and we must all take on the responsibility of checking for one or the other.

BREAST SELF-EXAM FOR WOMEN

1. Check your breasts for any lumps, knots, or changes about one week after your period.
2. Place your right hand behind your head. Press firmly with the pads of your fingers (Fig. 21A). Move your *left* hand over your *right* breast in a circle. Also check the armpit.
3. Now place your left hand behind your head and check your *left* breast with your *right* hand in the same manner as before. Also check the armpit.
4. Check your breasts while standing in front of a mirror right after you do your shower check. First, put your hands on your hips and then raise your arms above your head (Fig. 21B). Look for any changes in the way your breasts look; dimpling of the skin, changes in the nipple, or redness or swelling.

5. If you find any changes during your shower or mirror check, see your doctor right away.

You should know that the best check for breast cancer is a mammogram. When your doctor checks your breasts, ask about this. See Appendix E, which gives the warning signals for breast cancer.

TESTICLE SELF-EXAM FOR MEN

1. Check your testicles once a month.
2. Roll each testicle between your thumb and finger as shown in Figure 21C. Feel for hard lumps or bumps.
3. If you notice a change or have aches or lumps, tell your doctor right away so he or she can recommend proper treatment.

Cancer of the testicles can be cured if you find it early. You should also know that prostate cancer is the most common cancer in men. Men over age 50 should have an annual health checkup that includes a prostate examination. See Appendix E, which gives the warning signs for prostate cancer.

Information provided by the American Cancer Society. Used by permission.

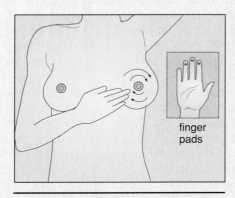

finger pads

Figure 21A Shower check for breast cancer.

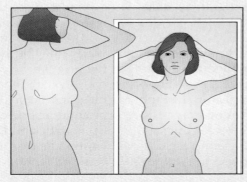

Figure 21B Mirror check for breast cancer.

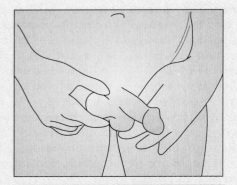

Figure 21C Shower check for testicular cancer.

Tests for Cancer Genes

The Pap smear has proven to be a reliable and helpful test for cervical cancer, but other tests are needed to detect other cancers at an early stage. Some believe that testing for cancer genes will save many lives. Tests are available for the presence of mutations that may raise the risk of colon, breast, and thyroid cancers and melanoma.

Researchers believe that the presence of the *ras* gene can be used to indicate the possibility of colon cancer. They have found that when a person has a tumor that contains a *ras* oncogene, this same oncogene can be detected in stool samples. Genetic testing for inherited breast cancer may also be possible. Researchers have been able to develop a genetic marker test for the oncogene BRCA1 (Breast Cancer gene #1) mentioned earlier. Those who test positive for inheritance of the gene can choose either to have prophylactic surgery or to be frequently examined for signs of breast cancer. Mutations in the RET gene seem to signify thyroid cancer, and mutations in the p16 gene appear to be linked to melanoma.

Perhaps genetic analysis for the likelihood of all sorts of cancer in the general populace may one day be routine.

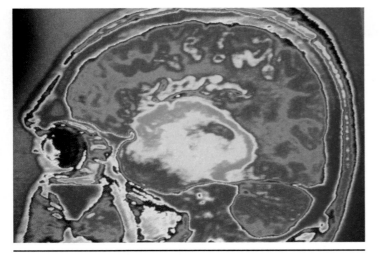

Figure 21.8 MRI image of brain.
The yellow region is a tumor.

Confirming the Diagnosis

There are ways to confirm a diagnosis of cancer without major surgery. Needle biopsies allow removal of a few cells for examination, and sophisticated imaging techniques such as laparoscopy permit a viewing of body parts. Computerized axial tomography (or CAT scan) uses computer analysis of scanning X-ray images to create cross-sectional pictures that portray a tumor's size and location. Magnetic resonance imaging (MRI) is another type of imaging technique that depends on computer analysis. MRI is particularly useful for analysis of tumors in tissues surrounded by bone, such as tumors of the brain or spinal cord (Fig. 21.8). A radioactive scan obtained after a radioactive isotope is administered can reveal any abnormal isotope accumulation due to a tumor. During ultrasound, echoes of high-frequency sound waves directed at a part of the body are used to reveal the size, shape, and location of tissue masses. Ultrasound can confirm tumors of the stomach, prostate, pancreas, kidney, uterus, and ovary.

There are standard procedures to detect specific cancers; for example, the Pap smear for cervical cancer, mammograms for breast cancer, and the stool blood test for colon cancer. Tumor marker tests and oncogene tests are new ways to test for cancer. Biopsy and imaging are used to confirm the diagnosis of cancer.

Treatment of Cancer

Surgery, radiation, and chemotherapy are the standard methods of treatment (Fig. 21.9). Surgery alone is sufficient for cancer in situ. But because there is always the danger that some cancer cells were left behind, surgery is often preceded by and/or followed by radiation therapy.

Radiation is mutagenic, and dividing cells such as cancer cells are more susceptible to its effects. The theory is that cancer cells will mutate to the point of self-destruction. Powerful X rays or gamma rays can be administered by using an externally applied beam or, in some instances, by implanting tiny radioactive sources into the patient's body. Cancer of the cervix and larynx, early stages of prostate cancer, and Hodgkin's disease are often treated with radiation therapy only. Proton beams are a fairly new form of radiation therapy that can be aimed at the tumor like a rifle bullet hitting the bull's eye of a target.

Chemotherapy is a way to catch cancer cells that have spread throughout the body. Most chemotherapeutic drugs kill cells by damaging their DNA or interfering with DNA synthesis. The hope is that all cancer cells will be killed while leaving untouched enough normal cells to allow the body to keep functioning. Whenever possible, chemotherapy is specifically designed for the particular cancer. For example, in Allen's lymphoma/leukemia, it is known that a small portion of a chromosome 9 is missing, and, therefore, DNA metabolism differs in the cancerous cells compared to normal cells. Specific chemotherapy for this cancer provides the patient with a drug designed to exploit this metabolic difference and destroy the cancerous cells.

A few years ago, a drug called taxol, extracted from bark of the Pacific yew tree, was found to be particularly effective against advanced ovarian cancers as well as breast, head, and neck tumors. Taxol interferes with microtubules needed for cell division. Now chemists have synthesized a family of related drugs, called taxoids, which may be more powerful with less side effects than the original.

Certain types of cancer, such as leukemias, lymphomas, and testicular cancer, are now successfully treated by combination chemotherapy alone. Almost 75% of children with childhood leukemia are completely cured. Hodgkin's disease, a lymphoma, once killed two out of three patients. Now, combination therapy of four different drugs can wipe out the disease in a matter of months in three out of four patients, even when the cancer is not diagnosed immediately. In other cancers—most notably breast and colon cancer—chemotherapy can reduce the chance of recurrence after surgery has removed all detectable traces of the disease.

Chemotherapy sometimes fails because cancer cells become resistant to one or several chemotherapeutic drugs. When cancer cells become resistant to combinations of drugs, it is called multidrug resistance. This occurs because all the drugs are capable of interacting with a plasma membrane carrier that pumps them out of the cell. Researchers are testing drugs known to poison the pump in an effort to restore efficacy of the drugs. In the meantime, combinations of drugs with nonoverlapping patterns of toxicity are still helpful because cancer cells can't become resistant to many different types at once. Also, when smaller doses of each type drug are used, more normal cells survive.

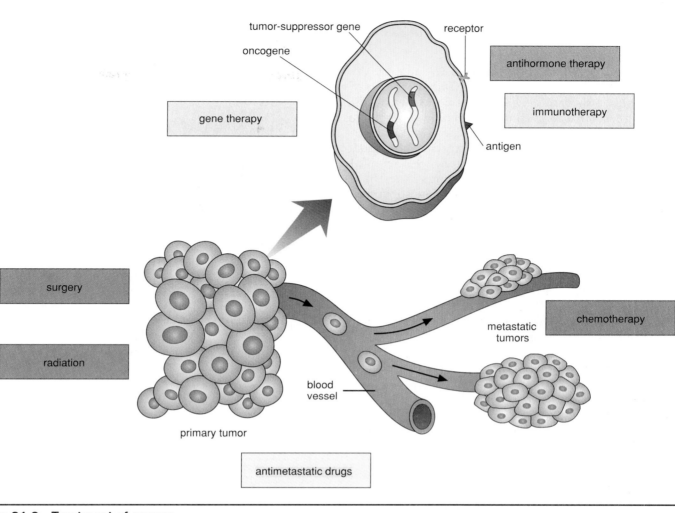

Figure 21.9 Treatment of cancer.
In standard use (dark blue), surgery and radiation are used to rid the body of a localized tumor. Chemotherapy is more effective and used when the cancer has metastasized. Also used (medium blue), antihormone therapy inactivates the plasma membrane receptors that are active when cancer develops. Still experimental (light blue), are antimetastatic drugs, gene therapy, and immunotherapy. Immunotherapy zeros in on plasma membrane receptors and antigens as a way to identify cancer cells.

Other Current Forms of Therapy

Aside from surgery, radiation, and chemotherapy, antihormone therapy and bone marrow transplants are currently in use for the treatment of cancer.

Antihormone Therapy   As mentioned previously, hormones are promoters for certain types of cancers. Breast and uterine cancer cells have receptors for estrogen, which can be blocked by the administration of a drug called tamoxifen. Hormone neutralization therapy is also standard treatment for metastatic prostate cancer because this type of tumor cell is stimulated by testosterone. GnRH (gonadotropin-releasing hormone) analogs are used to prevent the production of testosterone in the first place, or antiandrogens are used to block the reception of testosterone by tumor cells. Either way, testosterone is neutralized to prevent tumor growth.

In promyelocytic leukemia, cells have too many retinoic acid (a cousin of vitamin A) receptors. Strangely enough, the treatment is the administration of retinoic acid because this leads to differentiation and the concomitant cessation of cell growth.

Bone Marrow Transplants   The red bone marrow contains large populations of dividing cells; therefore, red bone marrow is particularly prone to destruction by chemotherapeutic drugs. In bone marrow autotransplantation, a patient's own bone marrow is harvested, treated to remove cancer cells, and stored before chemotherapy begins. Quite high doses of radiation or chemotherapeutic drugs are then given within a relatively short period of time. This prevents multidrug resistance from occurring, and the treatment is more likely to catch each and every cancer cell. Then, the stored bone marrow, which is needed to produce blood cells, is returned to the patient.

Future Forms of Therapy

Immunotherapy, turning off angiogenesis, antimetastatic drugs, and gene therapy may one day prove to be useful cancer therapies.

Immunotherapy Immunotherapy is the use of the body's immune system to promote the health of the body. Figure 21.9 shows that antigens are present on cancer cells, and this suggests that immunotherapy could possibly be successful in the treatment of cancer.

Various approaches to immunotherapy have been tried, but none have been highly successful. Lymphocytes removed from tumor cells (called tumor-infiltrating lymphocytes) have been genetically engineered to express a gene for tumor necrosis factor, a lymphokine, before being returned to the patient. Such treatments have been about 20% effective in patients with renal cancers and melanoma.

Monoclonal antibodies are antibodies of the same type because they are produced by the same plasma cell. Monoclonal antibodies can be designed to zero in on plasma membrane receptors of cancer cells, but alone they are not effective at killing cancer cells. One idea is to combine them with a radiation source called yttrium 90, which travels only a very short distance. Another is to produce monoclonal antibodies linked with a chemotherapeutic drug such as doxorubicin.

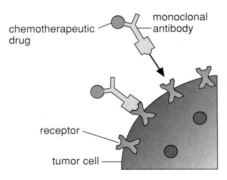

Turning off angiogenesis Tumor cells have the ability to promote angiogenesis, a proliferation of capillaries, to bring them the nutrients and growth factors they need to keep on expanding. Blood appearing between menstrual periods or in the urine, stool, or sputum is a danger signal that angiogenesis may have occurred in the cervix, bladder, colon, or lung, respectively. Antiangiogenic drugs confine and reduce tumors by breaking up the network of new capillaries in the vicinity of a tumor. A number of antiangiogenic compounds are currently being tested in clinical trials.

Antimetastatic Drugs Usually primary tumors can be successfully treated by surgery and radiation, but more accurate methods are needed for metastatic tumors. It has come as a surprise to learn that cancer cells themselves produce inhibitors of proteinase enzymes that allow invasion of nearby tissues. Proteinase action occurs only if the number of enzyme molecules is greater than the number of TIMP (tissue inhibitor of metalloproteinase) molecules. This means that TIMPs or drugs that act like them may offer an approach to prevent metastasis.

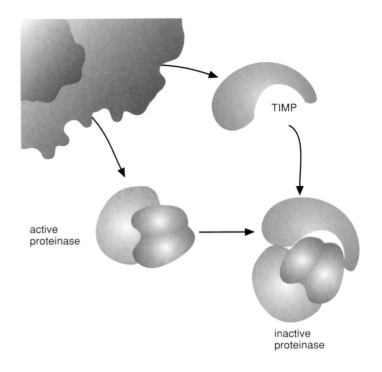

In other studies, investigators found that cells having a high level of a protein called nm23 (nonmetastatic 23) do not metastasize. Someday this protein might be used as a drug. In the meantime, human trials of the drug CAI (carboxyamide aminoimidazoles) are underway because this drug prevents metastasis in some, as yet, unknown way.

Gene Therapy With greater understanding of genes and carcinogenesis, it is not unreasonable to anticipate that gene therapy may eventually be applied in curing human cancer. For example, researchers think one therapy might be to inject tumors with genetically engineered viruses carrying an antisense polynucleotide to combine with and shut down the action of *ras*K and a second gene to replace a defective p53 gene in tumors. These genes will become effective once the viruses have entered cancer cells.

Surgery followed by radiation and/or chemotherapy is the standard method of treating cancer. Antihormone and bone marrow transplant therapies are also used. Other methods are being investigated.

SUMMARY

21.1 Causes and Prevention of Cancer

Cancer-causing mutations accumulate during the development of cancer, which is a multistage process involving initiation, promotion, and progression. A cancer-causing mutation can arise due to exposure to radiation, such as sunlight; certain organic chemicals, such as tobacco smoke and dietary fat; and a few types of viruses. Also, a mutation leading to cancer can be inherited. Even so, cancer is most apt to occur in the immunodeficient individual.

21.2 Cancer Cells

Cancer cells have characteristics that distinguish them from normal cells. They are nondifferentiated, have abnormal nuclei, do not require growth factors, and divide repeatedly. Because they are not constrained by their neighbors, they form a tumor. Finally, they metastasize and start new tumors elsewhere in the body.

Cancers are classified as carcinomas (epithelial tissue cancers), sarcomas (muscles and connective tissue), leukemias (blood), and lymphomas (lymphoid tissue). Appendix E gives pertinent information about cancers at specific sites.

21.3 Oncogenes and Tumor-Suppressor Genes

Each cell contains a growth control network involving proto-oncogenes and tumor-suppressor genes, which code for a growth factor receptor, and cellular proteins active in intracellular reactions that lead to cell division and growth. When proto-oncogenes mutate, becoming oncogenes, and tumor-suppressor genes mutate, the growth control network no longer functions as it should, and uncontrolled growth results.

21.4 Diagnosis and Treatment

There are procedures to detect specific cancers; for example, the Pap smear for cervical cancer, mammograms for breast cancer, and the stool blood test for colon cancer, to name a few. But new ways, such as tumor marker tests and oncogene tests, are being developed. Biopsy and imaging are used to confirm the diagnosis of cancer.

Surgery followed by radiation and/or chemotherapy is the standard method of treating cancer. Chemotherapy followed by bone marrow transplants and antihormone therapy is also used. New therapies under investigation are immunotherapy, turning off angiogenesis, antimetastatic drugs, and gene therapy.

STUDYING THE CONCEPTS

1. Name three stages of cancer development, and describe each phase. 437

2. Name three types of carcinogens, and give examples of each type. What role does heredity and immunodeficiency play in the development of cancer? 438–40

3. List and discuss four characteristics of cancer cells that distinguish them from normal cells. 442

4. Tell, in general, how cancers are classified, and give a system of clinical staging that can apply to all cancers. 443

5. What are oncogenes and tumor-suppressor genes? What role do they play in a growth control network that controls cell division and involves plasma membrane receptors and metabolic reactions? 444–45

6. What are the standard ways to detect cervical cancer, breast cancer, and colon cancer? 446

7. Describe and give examples of tumor marker tests and oncogene tests. 446–47

8. What are the standard methods of treatment for cancer? 448

9. Explain the rationale for antihormone therapy and bone marrow transplants. 449

10. List and describe four forms of therapy now under investigation. How will each type be used to treat cancer? 450

APPLYING YOUR KNOWLEDGE

Concepts

1. What is the reason that a person with AIDS is more prone to develop certain types of cancer than the uninfected person?

2. What is the reason that a person undergoing chemotherapy often loses her/his hair?

3. One is advised to eat plenty of high-fiber foods as a prevention against colon cancer. What is the reason that high-fiber foods have this effect?

4. Estrogen therapy to control menopausal symptoms increases the risk of endometrial cancer. What is the explanation for this?

Bioethical Issue

Turn on the evening news or read a paper, and you're bound to hear about the latest research on breast cancer. Long overlooked as a disease, breast cancer is now winning national attention as a major factor in women's health. That, undoubtedly, is good. But attention tends to focus on one thing, at the expense of others. Some people suggest breast cancer has stolen the spotlight, leaving prostate and other forms of cancer in the dark. For example, we rarely hear about liver or stomach cancer, which also are deadly.

How should the nation prioritize cancer research? Is it fair that one form of the disease tends to dominate the news? Should attention focus on diseases that affect the highest number of people—or diseases that are most deadly? And how should research funding be allocated among cancers that affect only women or only men?

TESTING YOUR KNOWLEDGE

1. To prevent cancer, you should avoid _____, which are environmental agents associated with the development of cancer.

2. The _____ virus is a carcinogen for cancer of the cervix.

3. _____ contains many organic chemical carcinogens and is associated with one-third of all cancers.

4. Cancer cells _____; they travel to distant body parts and start new tumors.

5. The mutation of _____ and _____ genes leads to uncontrolled growth and cancer.

6. The *RB* gene is a _____; it normally keeps *c-myc* from being an oncogene.

7. Cancer cells are sensitive to radiation therapy and chemotherapy because they are constantly _____.

8. Autotransplants of bone marrow permit a much higher dosage of _____ than otherwise.

9. Immunotherapy includes genetic engineering of tumor-infiltrating cells to carry tumor necrosis factor, a _____.

10. (a–e) Identify these portions of a cancer cell.

    (1–7) Tell what type of therapy is appropriate for each abnormality in anatomy and physiology.

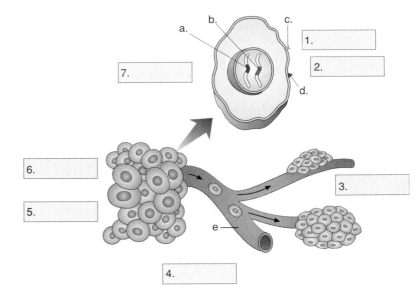

APPLYING TECHNOLOGY

Your study of cancer is supported by these available technologies:

Exploring the Internet

The Mader Home Page provides further resources for studying this chapter.

`http://www.mhhe.com/sciencemath/biology/mader/`

(Click on *Human Biology.*)

Explorations in Cell Biology & Genetics CD-ROM

Mitosis: Regulating the Cell Cycle (#5) Students choose a particular tissue and then alter the age, physical exercise, tissue injury/loss, adequacy of nutrition, and exposure to mutagens, while noting the effect on the frequency of the cell cycle. (2)*

Explorations in Human Biology CD-ROM

Life Span and Lifestyle (#3) Life expectancy varies as students change the amount of smoking, drinking, fat in diet, and exercise. (2)

Smoking and Cancer (#6) The numbers of years it takes for cancer to develop decreases as students increase the years of smoking and the number of cigarettes smoked a day. (2)

*Level of difficulty

SELECTED KEY TERMS

cancer Malignant tumor that metastasizes. 438

carcinogen (kar-SIN-uh-jen) Environmental agent that contributes to the development of cancer. 438

carcinogenesis Development of cancer. 440

carcinoma Cancer arising in epithelial tissue. 443

chemotherapy Use of a drug to selectively kill off dividing cancer cells as opposed to nondividing cells. 448

immunotherapy Use of any immune system component such as antibodies, cytotoxic T cells, or lymphokines to promote the health of the body, such as curing cancer. 450

leukemia (loo-KEE-mee-ah) Cancer of the blood-forming tissues leading to the overproduction of abnormal white blood cells. 443

lymphoma (lim-FOH-mah) Cancer of the lymphoid organs such as lymph nodes, spleen, and thymus gland. 443

metastasis (meh-TAS-tuh-sus) Spread of cancer from the place of origin throughout the body caused by the ability of cancer cells to migrate and invade tissues. 443

mutagen (MYOOT-uh-jen) Environmental agent that induces mutations. 438

oncogene (ONG-koh-jeen) Gene that contributes to the transformation of a normal cell into a cancer cell. 444

proto-oncogene (PROH-toh-ONG-koh-jeen) Normal gene involved in cell growth and differentiation that becomes an oncogene through mutation. 444

sarcoma (sar-KOH-mah) Cancer that arises in connective tissue, such as muscle, bone, and fibrous connective tissue. 443

tumor Growth that contains cells derived from a single mutated cell that has repeatedly undergone cell division; benign tumors remain at the site of origin, and malignant tumors metastasize. 442

tumor-suppressor gene Genes that, when expressed, prevent abnormal cell division and cancer. 445

FURTHER READINGS FOR PART SIX

Anderson, W. F. September 1995. Gene therapy. *Scientific American* 273(3):124. Presents gene therapy now and discusses prospects for the future.

Beardsley, T. March 1996. Vital data. *Scientific American* 274(3):100. DNA tests for a wide array of conditions are becoming available.

Black, P. H. November/December 1995. Psychoneuroimmunology: Brain and immunity. *Scientific American Science & Medicine* 2(6):16. Article discusses the role of stress in susceptibility to infections, and cancer and HIV progression.

Blaser, M. J. February 1996. The bacteria behind ulcers. *Scientific American* 274(2):104. Acid-loving pathogens are linked to stomach ulcers and stomach cancer.

Boon, T. March 1993. Teaching the immune system to fight cancer. *Scientific American* 268(3):82. The search for ways to direct the immune system against cancer cells is promising.

Capecchi, M. March 1994. Targeted gene replacement. *Scientific American* 270(3):52. Researchers are deciphering DNA segments that control development and immunity.

Cavanee, W. K., and White, R. L. March 1995. The genetic basis of cancer. *Scientific American* 272(3):72. Medical intervention may be possible during the mutation process of cancerous cells.

Cohen, J. S., and Hogan, M. E. December 1994. The new genetic medicines. *Scientific American* 271(6):76. Artificial strings of nucleic acids can silence genes responsible for many illnesses.

Cooper, G. 1993. *The cancer book*. Boston: Jones and Bartlett Publishers. This is a nontechnical presentation of the basic nature and causes of cancer and current strategies for its prevention and treatment.

Cummings, M. R. 1994. *Human heredity.* 3d ed. St. Paul, Minn.: West Publishing. This introductory text is written in a straightforward, interesting manner and contains current information in the changing field of genetics.

Davis, D. L., and Bradlow, H. L. April 1995. Can environmental estrogens cause breast cancer? *Scientific American* 273(4):166. Estrogenlike compounds found in the environment may contribute to breast cancer.

Dawkins, R. November 1995. God's utility function. *Scientific American* 273(5):80. The role of genetics in evolution and natural selection is discussed.

Frederick, R. J., and Egan, M. September 1994. Environmentally compatible applications of biotechnology. *BioScience* 44(8):529. Explains the use of living organisms to restore and safeguard the environment.

Garnick, M. April 1994. The dilemmas of prostate cancer. *Scientific American* 270(4):72. Prostate cancer has been detected with increasing frequency in recent years.

Gasser, C. S., and Fraley, R. T. June 1992. Transgenic crops. *Scientific American* 266(6):62. Genetic engineering helps the world produce better crops.

Gibbs, W. W. August 1996. Gaining on fat. *Scientific American* 275(2):88. Some weight problems are genetic or physiological in origin. New treatments might help.

Greenspan, R. J. April 1995. Understanding the genetic construction of behavior. *Scientific American* 272(4):72. Research indicates that even in simple organisms, behavior is influenced by a number of genes.

Greider, C. W., and Blackburn, E. H. February 1996. Telomeres, telomerase, and cancer. *Scientific American* 274(2):92. The enzyme telomerase rebuilds the chromosomes of tumor cells; this enzyme is being researched as a target for anticancer treatments.

Hartl, D. 1994. *Genetics.* 3d ed. Boston: Jones and Bartlett Publishers. Provides a clear, comprehensive, and straightforward introduction to the principles of genetics at the college level.

Johnson, G. B. 1996. *How scientists think.* Dubuque, Iowa: Wm. C. Brown Publishers. Presents the rationale behind 21 important experiments in genetics and molecular biology that became the foundation for today's research.

Langer, R., and Vacanti, J. P. September 1995. Artificial organs. *Scientific American* 273(3):130. Discusses engineering artificial tissue using the body's own cells.

Lanza, R. P., and Chick, W. L. 1995. Encapsulated cell therapy. *Scientific American Science & Medicine* 2(4):16. Implants containing living cells (tissue engineering) within a selectively permeable membrane could provide drug or hormone doses as required for therapy.

Lasic, D. D. May/June 1996. Liposomes. *Science & Medicine* 3(3):34. Liposomes can be used to deliver drugs or genes for gene therapy.

Leffell, D. J., and Brash, D. E. July 1996. Sunlight and skin cancer. *Scientific American* 275(1):52. Discusses the sequence of changes that may occur in skin cells after exposure to UV rays.

Liotta, L. A. February 1992. Cancer cell invasion and metastasis. *Scientific American* 266(2):54. Regulatory genes and proteins that control metastasis have produced a promising class of synthetic drugs.

Lyon, J., and Gorner, P. 1995. *Altered fates: Gene therapy and the retooling of human life.* New York: W. W. Norton and Company. This nonfiction work details the saga of gene therapy, from the scientists to the patients and their families.

Mange, E., and Mange, A. 1994. *Basic human genetics.* Sunderland, Mass.: Sinauer Associates. Presents the general principles of genetics and discusses the implications for individuals and society.

McConkey, E. H. 1993. *Human genetics: The molecular revolution.* Boston: Jones and Bartlett Publishers. This text for professionals and advanced students emphasizes the impact of molecular information while surveying the current status of understanding and treatment of genetic disease.

McGinnis, W., and Kuziora, M. February 1994. The molecular architects of body design. *Scientific American* 270(2):58. The ability to transfer genes between species provides a way to study how genes control development.

Moses, V., and Moses, S. 1995. *Exploiting biotechnology.* Chur, Switzerland: Harwood Academic Publishers. Provides a general understanding of biotechnology and presents its commercial and industrial applications.

Nicolaou, K. C., et al. June 1996. Taxoids: New weapons against cancer. *Scientific American* 274(6):94. Chemists are synthesizing a family of drugs related to taxol for the treatment of cancer.

Noble, E. P. March/April 1996. The gene that rewards alcoholism. *Scientific American Science & Medicine* 3(2):52. A dopamine receptor gene is linked to severe alcoholism.

Perera, F. P. May 1996. Uncovering new clues to cancer risk. *Scientific American* 274(5):54. Molecular epidemiology finds biological markers that explain what makes people susceptible to cancer.

Peters, P. 1994. *Biotechnology: A guide to genetic engineering.* Dubuque, Iowa: Wm. C. Brown Publishers. DNA structure and function are reviewed before biotechnology, including genetic engineering, is explained.

Rennie, J. July 1994. Immortal's enzyme. *Scientific American* 271(1):14. An enzyme that maintains telomeres may make tumor cells immortal.

Rennie, J. June 1994. Grading the gene tests. *Scientific American* 270(6):88. Raises ethical questions about screening embryos for genetic disease before implantation.

Rodgers, G. P., et al. October 1994. Sickle cell anemia. *Scientific American Science & Medicine* 1(4):48. Promising therapies may be forthcoming, despite obstacles to research.

Rothwell, N. 1993. *Understanding genetics: A molecular approach.* New York: John Wiley & Sons. Focuses on the human genome and provides an insight into approaches that have led to the isolation of genetic regions associated with human afflictions.

Rusting, R. L. December 1992. Why do we age? *Scientific American* 267(6):130. Genetic mechanisms that contribute to deterioration and death are being researched.

Scientific American Special Issue. What you need to know about cancer. September 1996. 275(3). The entire issue is devoted to the causes, prevention, and early detection of cancer, and cancer therapies—conventional and future.

Shcherbak, Y. M. April 1996. Ten years of the Chernobyl Era. *Scientific American* 274(4):44. Article discusses the medical aftermath of the accident.

Spicer, D. V., and Pike, M. C. July/August 1995. Hormonal manipulation to prevent breast cancer. *Scientific American Science & Medicine* 2(4):58. This contraceptive regimen could reduce the risk of breast cancer.

Welsh, M. J., and Smith, A. E. December 1995. Cystic fibrosis. *Scientific American* 273(6):52. The genetic defects that cause CF hamper or prevent ion transport by affected lung cells.

Wolffe, A. P. November/December 1995. Genetic effects of DNA packaging. *Science & Medicine* 2(6):68. The regulation of DNA coiling in the chromosomes adds to the properties of the genes involved in several genetic diseases.

Part

7

Human Evolution and Ecology

Evidence for the theory of evolution is drawn from many areas of biology. Charles Darwin was the first to present extensive evidence, and he suggested that those organisms best suited to a particular environment are the ones that survive and reproduce most successfully. Evolution causes life to have a history, and it is possible to trace the ancestry of humans even from the first cell or cells.

The world's diverse forms of life live within ecosystems, where energy flows and chemicals cycle. Humans have greatly modified the ecosystems, and worldwide, the human population keeps increasing in size so that an ever-greater amount of energy and raw materials are needed each year. Since 1850, the human population has expanded so rapidly that some doubt there will be sufficient energy and food to permit the same degree of growth in the future. The human-impacted ecosystem depends on natural ecosystems not only because they absorb pollutants but also because natural ecosystems are inherently stable. Every possible step should be taken to protect natural ecosystems to help ensure the continuance of the human species.

Chapter 22

Evolution

Chapter Outline

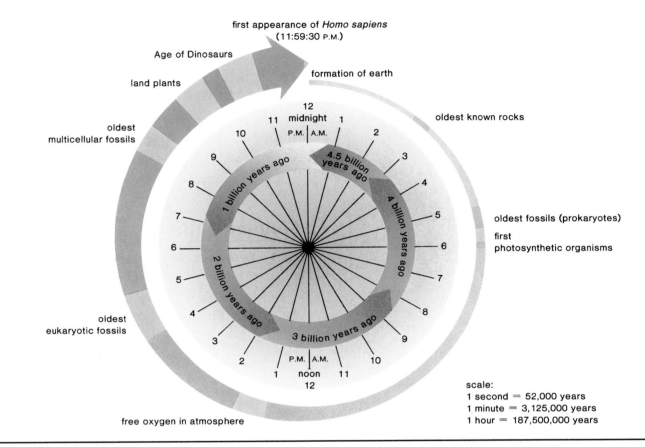

Figure 22.1   History of earth.
History of earth measured on a 24-hour time scale (yellow) compared to actual years (red). A very large portion of life's history is devoted to the evolution of unicellular organisms. (Prokaryotes do not have a nucleus; eukaryotes do have a true nucleus.) The first multicellular organisms do not appear until just after 8 P.M., and humans are not on the scene until less than a minute before midnight.

O n Christmas Day, 1994, archaeologists stumbled upon the discovery of a lifetime: an underground cave in southeast France that had lain forgotten for 20,000 years. Making their way through the dark, the researchers were amazed to find nearly 300 intricate drawings and engravings of animals. The researchers also noticed preserved footprints, bone pieces, and what appeared to be bits of torches left on the cave floor. These finds should yield clues to the way regional inhabitants lived—from the food they ate to the clothes they wore or how they spent their daily lives.

The newfound French cave, called "Grotte Chauvet," is important because scientists are still learning about human evolution—both physical and cultural. How did we look? How did we live when we first evolved? Like an unfinished puzzle, the story of evolution continues to intrigue us. This chapter gives evidence for the evolution of all living things and discusses the process of evolution before considering the history of evolution from the first cell to human beings.

22.1   Evidence for Evolution

The fossil record, comparative biochemistry and anatomy, and biogeography all support the theory of evolution.

Fossils Show the History of Life

Our knowledge of the history of life is based primarily on the fossil record. **Fossils** are the remains or evidence of some organism that lived long ago. Nearly all fossils are of organisms that are now extinct (there are no living forms), and some fossils are older than others. Today, it is possible to date fossils by using isotopes. The oldest fossils found are of bacteria dated some 3.5 billion years ago. Thereafter, the fossils get more and more complex. For example, bacteria were followed by nucleated cells, and among animals, invertebrates (no backbone) were followed by vertebrates (having a backbone). Humans began evolving about 5 million years ago, but modern humans (*Homo sapiens*) do not appear in the fossil record until about 100,000 years ago (Fig. 22.1).

In some instances, it is even possible to trace a line of descent over vast amounts of time. Researchers have now traced the modern-day horse to an animal that was about the size of a dog with four toes on each front foot and three toes on each hind foot. When grasslands replaced the forest home of this animal, it was millions of years before a larger size provided the strength needed for combat; a larger skull made room for a larger brain; elongated legs ending in hooves provided greater speed to escape enemies; and durable, grinding teeth enabled the animal to feed efficiently on grasses.

The fossil record broadly traces the history of life and, more specifically, allows us to study the history of particular groups.

Comparative Biochemical Evidence

Almost all living organisms use the same basic biochemical molecules including DNA, ATP, and many identical or nearly identical enzymes. Further, almost all organisms utilize the same nuclear DNA triplet code and the same 20 amino acids in their proteins. Organisms even share the same introns and hypervariable regions. There is obviously no functional reason why these elements need be so similar. But their similarity can be explained by descent from a common ancestor.

Further, the degree of similarity in amino acid sequences of proteins and the degree of similarity in DNA base sequences is consistent with data regarding the anatomical similarities and, therefore, the relatedness of organisms.

All organisms have certain biochemical molecules in common. The degree of similarity between DNA base sequences and amino acid sequences indicates degree of relatedness between organisms.

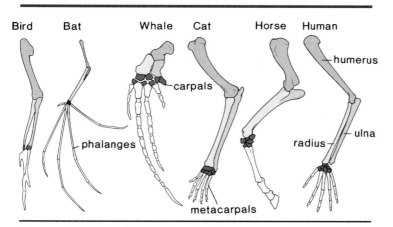

Figure 22.2 Vertebrate forelimbs.
The same bones are present (they are color-coded) in all vertebrates (animals with a backbone) but designed for different functions. The unity of plan is evidence of a common ancestor.

Comparative Anatomical Evidence

Diverse organisms sometimes share anatomical similarities. Vertebrate forelimbs are used for flight (birds and bats), orientation during swimming (whales and seals), running (horses), climbing (arboreal lizards), or swinging from tree branches (monkeys). Yet all vertebrate forelimbs contain the same sets of bones organized in similar ways, despite their dissimilar functions (Fig. 22.2). The most plausible explanation for this unity is that the basic forelimb plan originated with a **common ancestor,** and then the plan was modified in the succeeding groups as each continued along its own evolutionary pathway. Structures that are similar because they were inherited from a common ancestor are called **homologous structures.** The wing of a bird and insect are analogous structures—they are all adaptations for flying but are structurally unrelated.

The unity of plan shared by vertebrates extends to their embryological development (Fig. 22.3). At some time during development, all vertebrates have a supporting dorsal rod, called a notochord, and exhibit paired pharyngeal pouches. In fishes and amphibian larvae, these pouches develop into functioning gills. In humans, the first pair of pouches becomes the cavity of the middle ear and the auditory tube. The second pair becomes the tonsils, while the third and fourth pairs become the thymus and parathyroid glands. Why do pharyngeal pouches appear at all during vertebrate development, since they later undergo modification? The most likely explanation is that fishes are ancestral to other vertebrate groups.

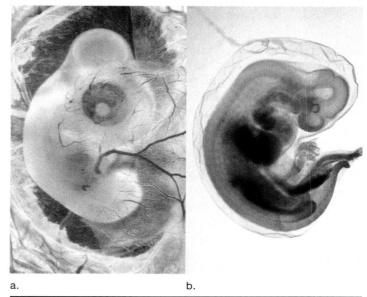

a.   b.

Figure 22.3 Early developmental stage of chick and pig.
a. Chick embryo. **b.** Pig embryo. At these comparable early developmental stages, the two have many features in common, although eventually they are completely different animals. This is evidence they evolved from a common ancestor.

Vestigial structures are anatomical features that are fully developed in one group of organisms but are reduced and may have no function in similar groups. Most birds, for example, have well-developed wings used for flight. Some bird species (e.g., ostrich), however, have wings that are greatly reduced, and they do not fly. Similarly, snakes do not use limbs for locomotion and yet some have remnants of a pelvic girdle and legs. Humans have a tail bone but no tail. The presence of vestigial structures can be explained by the common descent hypothesis. Vestigial structures occur because organisms inherit their anatomy from their ancestors; they are traces of an organism's evolutionary history.

Organisms share a unity of plan when they are closely related because of common descent. This is substantiated by comparative anatomy and embryological development.

Biogeographical Evidence

Biogeography is the study of the distribution of plants and animals throughout the world. Such distributions are consistent with the hypothesis that related forms evolved in one locale and then spread out into other accessible regions. Why are there no rabbits in South America even though the environment is quite suitable to them? Most likely because rabbits originated someplace else, and they had no means to reach South America.

The islands of the world have many unique animals and plants found no place else, even when the soil and climate are the same as other places. Why are there so many species of finches on the Galápagos Islands located off the coast of Ecuador? Since they are not on the mainland, the reasonable explanation is that all these types of finches are descended from a common ancestor that came to the islands by chance.

Physical factors, such as the location of continents, often determine where a population can spread. Both cacti and euphorbia are plants adapted similarly to a hot, dry environment— they both are succulent, spiny, flowering plants. Why do cacti grow in North American deserts and euphorbia grow in African deserts when each would do well on the other continent?

It seems obvious that they just happened to evolve on their respective continents.

At one time in the history of the earth, South America, Antarctica, and Australia were all connected. Marsupials (pouched mammals) arose at this time and today are found in both South America and Australia. When Australia separated and drifted away, the marsupials diversified into many different forms suited to various environments (Fig. 22.4). They were free to do so because there were no placental mammals in Australia. In other regions, such as South America where there are placental mammals, marsupials are not as diverse.

The distribution of organisms on the earth is explainable by assuming that related forms evolved in one locale, where they then diversified and/or spread out into other accessible areas.

Evolution is one of the great unifying theories of biology. In science, the word theory is reserved for those conceptual schemes that are supported by a large number of observations and have not yet been found lacking. The theory of evolution has the same status in biology that the germ theory of disease has in medicine.

Kangaroo, *Macropus*, a herbivore of plains and forests

Tasmanian wolf, *Thylacinus*, a nocturnal carnivore of deserts and plains

Australian native cat, *Kasyurus*, a carnivore of forests

Sugar glider, *Petaurista*, a tree dweller

Coarse-haired wombat, *Vombatus*, nocturnal and living in burrows

Figure 22.4   Marsupials from Australia.
Each type of marsupial is adapted to a different way of life. All of the marsupials of Australia presumably evolved from a common ancestor that entered Australia some 60 million years ago.

22.2 The Evolutionary Process

Based on much evidence he collected, Charles Darwin formulated a theory of **natural selection** around 1860 to explain the evolutionary process. The following are critical to understanding natural selection.

Variations Come First

Individual members of a population vary in physical characteristics. Such variations can be passed on from generation to generation. Darwin was never able to determine the cause of variations or how they are passed on. Today, we realize that genes determine the appearance of an organism and that mutations can cause new variations to arise.

Struggle for Existence

Darwin knew that a socioeconomist, Thomas Malthus, had stressed the reproductive potential of human beings. He proposed that death and famine were inevitable because the human population tended to increase faster than the supply of food. Darwin applied this concept to all organisms and saw that members of plant or animal populations must compete with one another for available resources. Darwin calculated the reproductive potential of elephants. Assuming a life span of about 100 years and a breeding span of from 30–90 years, a single female will probably bear no fewer than six young. If all these young survived and continued to reproduce at the same rate, after only 750 years, the descendants of a single pair of elephants would number about 19 million! Such reproductive potential necessitates a *struggle for existence;* only certain members of a population survive and reproduce those characteristics that give them a competitive advantage.

Survival of the Fittest

Darwin noted that in *artificial selection* humans choose which plants or animals will reproduce. This selection process brings out certain traits. For instance, there are many varieties of dogs, each of which was derived from the wild wolf. In a similar way, several varieties of vegetables can be traced to a single type.

In contrast to artificial selection, *natural selection* occurs because certain members of a population happen to have a variation that makes them more suited to the environment. For example, any variation that increases the speed of a hoofed animal will help it escape predators and live longer; a variation that reduces water loss will help a desert plant survive; and one that increases the sense of smell will help a coyote find its prey. Therefore, we would expect organisms with these traits to live longer and, consequently, reproduce to a greater extent.

Adaptation to the Environment

An **adaptation** is a trait that helps an organism be more suited to its environment. We can especially recognize an adaptation when unrelated organisms living in a particular environment display similar characteristics (Fig. 22.5).

Natural selection results in the adaptation of populations to their specific environments. Because of differential reproduction generation after generation, adaptive traits are more and more common in each succeeding generation.

The following listing summarizes the theory of evolution as developed by Darwin.

1. There are inheritable variations among the members of a population.

2. Many more individuals are produced each generation than can survive and reproduce.

3. Individuals with adaptive characteristics are more likely to be selected to reproduce by the environment.

4. Gradually, over long periods of time, a population can become well adapted to a particular environment.

5. The end result of organic evolution is many different species, each adapted to specific environments.

By Gills   **By Gills with Arches**

Figure 22.5 Adaptation to aquatic environment.
Distantly related animals living in the same environment have similar adaptations. Both crayfishes (arthropods) and fishes (chordates) breathe by gills that are finely divided and vascularized outgrowths of the body.

22.3 Organic Evolution

Organic evolution began with the evolution of the first form of life, that is, the first cell or cells (Fig. 22.6).

Origin of Life

The sun and the planets probably formed from aggregates of dust particles and debris about 4.6 billion years ago. Intense heat produced by gravitational energy and radioactivity caused the earth to become stratified into several layers. Heavier atoms of iron and nickel became the molten liquid core, and dense silicate minerals became the semiliquid mantle. Upwellings of volcanic lava produced the first crust. The size of the earth is such that its gravitational field is strong enough to have an atmosphere. The earth's primitive atmosphere was not the same as today's atmosphere. It is thought that the primitive atmosphere was produced by outgassing from the interior, particularly by volcanic action. In that case, the primitive atmosphere would have consisted mostly of water vapor (H_2O), nitrogen (N_2), and carbon dioxide (CO_2), with only small amounts of hydrogen (H_2) and carbon monoxide (CO). The primitive atmosphere had little, if any, free oxygen.

Small Organic Molecules Evolve

At first the earth was so hot that water was present only as a vapor that formed dense, thick clouds. Then as the earth cooled, water vapor condensed to liquid water, and rain began to fall. It rained in such enormous quantity over hundreds of millions of years that the oceans of the world were produced. The atmospheric gases, dissolved in rain, were carried down into newly forming oceans. The energy sources on the primitive earth included heat from volcanoes and meteorites, radioactivity from isotopes in the earth's crust, powerful electric discharges in lightning, and solar radiation, especially ultraviolet radiation. In the presence of such abundant energy, the primitive gases reacted with one another and produced small organic compounds. With the accumulation of these small organic compounds, the oceans became a warm, organic soup containing a variety of organic molecules.

Macromolecules Evolve and Interact

The newly formed organic molecules likely joined to produce still larger molecules and then macromolecules. There are two hypotheses of interest concerning this stage in the origin of life. One is the RNA-first hypothesis, which suggests that only the macromolecule RNA (ribonucleic acid) was needed at this time to progress toward formation of the first cell or cells. This hypothesis was formulated after the discovery that RNA can sometimes be both a substrate and an enzyme. Such RNA molecules are called ribozymes. It would seem, then, that RNA could have carried out the processes of life commonly associated today with DNA (deoxyribonucleic acid, the genetic material)

and proteins (enzymes). Some viruses today have RNA genes; therefore, the first genes could have been RNA. And the first enzymes also could have been RNA molecules, since we now know that ribozymes exist. Those who support this hypothesis say that it was an "RNA world" some 4 billion years ago.

Another hypothesis is termed the protein-first hypothesis. Sidney Fox has shown that amino acids polymerize abiotically when exposed to dry heat. He suggests that amino acids collected in shallow puddles along the rocky shore and the heat of the sun caused them to form proteinoids, small polypeptides that have some catalytic properties. When proteinoids are returned to water, they form microspheres, structures composed only of protein that have many properties of a cell. It is possible that the first polypeptides had enzymatic properties, and some proved to be more capable than others. Those that led to the first cell or cells had a selective advantage. This hypothesis assumes that DNA genes came after protein enzymes arose. After all, it is protein enzymes that are needed for DNA replication.

A Protocell Evolves

Before the first true cell arose, there would have been a protocell, a structure that had a lipid-protein membrane and carried on energy metabolism (Fig. 22.6). Fox has shown that if lipids are made available to microspheres, lipids tend to become associated with microspheres, producing a lipid-protein membrane.

Eventually, a semipermeable-type boundary may have formed about the droplet. In a liquid environment, phospholipid molecules automatically form droplets called liposomes. Perhaps the first membrane formed in this manner. In that case, the protocell could have contained only RNA, which functioned as both genetic material and enzymes.

The protocell would have had to carry on nutrition so that it could grow. Nutrition was no problem because the protocell existed in the ocean, which at that time contained small organic molecules that could have served as food. Therefore, the protocell likely was a heterotroph, an organism that takes in preformed food. Notice that this suggests that heterotrophs are believed to have preceded autotrophs, organisms that make their own food.

A Self-Replication System Evolves

A true cell is a membrane-bounded structure that can carry on protein synthesis needed to produce the enzymes that allow DNA to replicate. The central dogma of genetics states that DNA directs protein synthesis and that there is a flow of information from DNA → RNA → protein. It is possible that this sequence developed in stages.

According to the RNA-first hypothesis, RNA would have been the first to evolve, and the first true cell would have had RNA genes. These genes would have directed and enzymatically carried out protein synthesis. Viruses

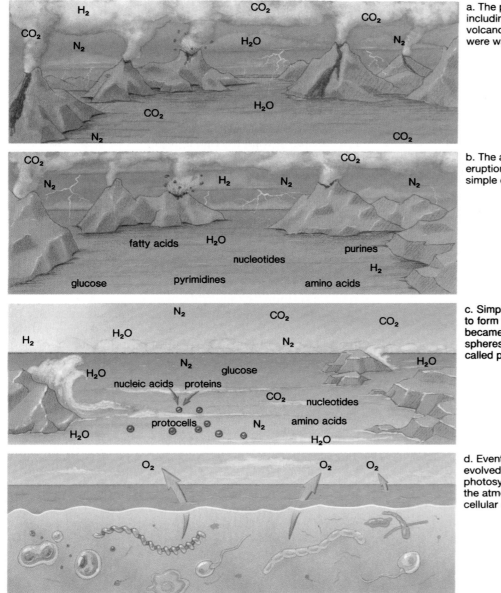

a. The primitive atmosphere contained gases, including water vapor, that escaped from volcanoes; as the water vapor cooled, some gases were washed into the oceans by rain.

b. The availability of energy from volcanic eruption and lightning allowed gases to form simple organic molecules.

c. Simple organic molecules could have joined to form proteins and nucleic acids, which became incorporated into membrane-bounded spheres. The spheres became the first cells, called protocells.

d. Eventually, various types of organisms evolved. Some of these were oxygen-producing photosynthesizers. The presence of oxygen in the atmosphere was necessary for aerobic cellular respiration to evolve.

Figure 22.6   A model for the origin of life.

with RNA genes have a protein enzyme called reverse transcriptase that uses RNA as a template to form DNA. Perhaps with time, reverse transcription occurred within the protocell, and this is how DNA genes arose. Once there were DNA genes, then protein synthesis would have been carried out in the manner dictated by the central dogma of genetics.

According to the protein-first hypothesis, proteins, or at least polypeptides, were the first of the three (i.e., DNA, RNA, and protein) to arise. Only after the protocell developed sophisticated enzymes did it have the ability to synthesize DNA and RNA from small molecules provided by the ocean. Researchers point out that because a nucleic acid is a very complicated molecule, the likelihood that RNA arose de novo (on its own) is minimal. It seems more likely that enzymes were needed to guide the synthesis of nucleotides and then nucleic acids.

Once the protocells acquired genes that could replicate, they became cells capable of reproducing, and evolution began.

Evolution and Classification of Living Things

As mentioned, the fossil record and molecular data allow us to trace the evolution of life, including particular groups of organisms, since life began. These same data are used to classify organisms. **Taxonomy** is that part of biology dedicated to naming, describing, classifying, and relating species. A species is a group of similarly constructed organisms capable of interbreeding and producing fertile offspring. Table 22.1 shows you how taxonomists classify human beings into different categories. As we move from genus to kingdom, more and more different types of species are included in each successive category. Only human beings

are in the genus *Homo,* but many different types of animals are in the animal kingdom. Notice that in the example given, species within the same genus share very specific characteristics, but those that are in the same kingdom have only general characteristics in common.

Taxonomists give each species a scientific name in Latin. The scientific name is a binomial (*bi* means *two, nomen* means *name*). For example, the name for humans is *Homo sapiens.* The first word is the genus, and the second word is a specific epithet for that species. (Note that both words are in italic but only the genus is capitalized.) Scientific names are universally used by biologists so as to avoid confusion. Common names tend to overlap and often are in the language of a particular country.

Taxonomy is an attempt to make sense out of the bewildering variety of life on earth. Species are classified according to their presumed evolutionary relationship; those placed in the same genus are the most closely related, and those placed in separate kingdoms are the most distantly related. As more is known about evolutionary relationships between species, taxonomy changes. Presently, many biologists recognize five kingdoms, which they believe to be related as shown in Figure 22.7. Such a diagram, called an evolutionary tree, portrays the evolutionary history of a group of organisms; in this case, all living things.

Taxonomists classify living things into categories according to their evolutionary relationships and use evolutionary trees to depict these relationships.

TABLE 22.1

Classification Categories

Categories	For Humans	Description
Kingdom	Animalia	Multicellular, moves, ingests food
Phylum	Chordata	Dorsal supporting rod and nerve cord
Class	Mammalia	Hair, mammary glands
Order	Primates	Adapted to climb trees
Family	Hominidae	Adapted to walk erect
Genus	*Homo*	Large brain, tool use
Species	*H. sapiens*	

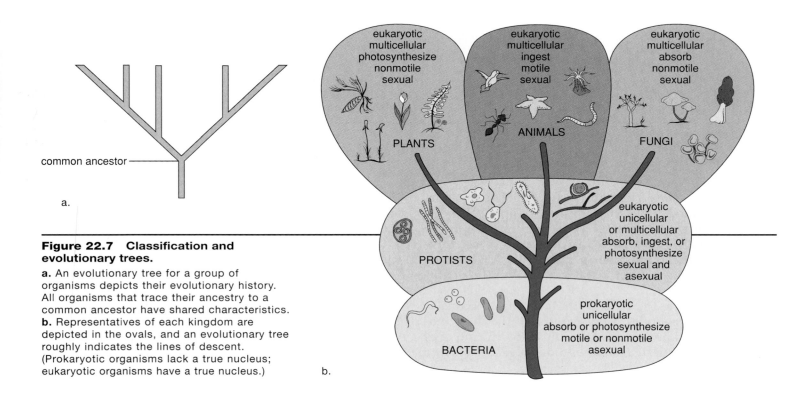

Figure 22.7   Classification and evolutionary trees.
a. An evolutionary tree for a group of organisms depicts their evolutionary history. All organisms that trace their ancestry to a common ancestor have shared characteristics. **b.** Representatives of each kingdom are depicted in the ovals, and an evolutionary tree roughly indicates the lines of descent. (Prokaryotic organisms lack a true nucleus; eukaryotic organisms have a true nucleus.)

22.4 Modern Humans Evolve

Humans are **primates** (order primates), which are placental mammals adapted to living in trees.

Primates Live in Trees

The limbs of primates are mobile, as are the hands, because the thumb (and in nonhuman primates, the big toe as well) is opposable; that is, the thumb can touch each of the other fingers. In primates, the snout is shortened considerably, allowing the eyes to move to the front of the head. The stereoscopic vision (or depth perception) that results permits primates to make accurate judgments about the distance and position of adjoining tree limbs. Gestation is lengthy in a primate, allowing time for good forebrain development. One birth at a time is the norm in primates; it is difficult to care for several offspring while moving from limb to limb. The juvenile period of dependency is extended, and there is an emphasis on learned behavior and complex social interactions.

Among primates, humans are most closely related to the apes. There are four types of modern apes: gibbons, orangutans, gorillas, and chimpanzees. Gibbons, the smallest of the apes, have extremely long arms, which are specialized for swinging between tree limbs. The orangutan is a large ape but nevertheless spends a great deal of time in trees. In contrast, the gorilla, the largest of the apes, spends most of its time on the ground. Chimpanzees, which are at home both in the trees and on the ground, are the most humanlike of the apes in appearance.

Humans can be distinguished from modern apes by locomotion and posture, dental features, and other characteristics (Fig. 22.8).

These characteristics especially distinguish primates from other mammals: opposable thumb (and in some cases, big toe); expanded forebrain; emphasis on learned behaviors; single birth; extended period of parental care.

Hominids Walk Erect

Modern humans are **hominids** (family Hominidae) as are their immediate ancestors. The hominid line of descent begins with the **australopithecines,** which evolved and diversified in eastern Africa from about 4.4 MYA (millions of years ago) to almost 1 MYA.

The most famous of these fossils, dated about 3.4 MYA, is scientifically called *Australopithecus afarensis* but is better known by its field name, Lucy. (The name derives from the Beatles' song "Lucy in the Sky with Diamonds.") Although her brain was quite small, the shapes and relative proportions of her limbs indicate that Lucy walked upright. There are even footprints that show how Lucy walked. *A. afarensis* males, which were discovered later, are much larger than the females. Taking into account their smaller body size, the relative brain size is about half of ours, and the jaw is heavy. The cheek teeth are enormous, and in males, large canine teeth project forward. *A. afarensis* had descendants, and there may have been as many as ten species of hominids about 2 MYA in Africa, some of which were quite robust. One of these descendents is classified in the genus *Homo*.

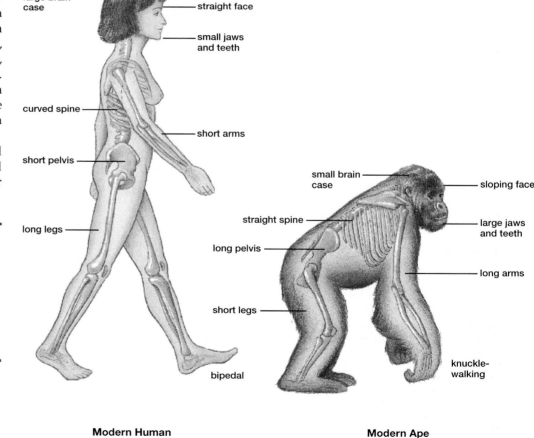

Modern Human   **Modern Ape**

Figure 22.8   Comparison of human skeletal features with those of a gorilla.

Homo habilis Made Tools

Homo habilis is dated as early as 2 MYA (Fig. 22.9). The name means handy man, and this hominid is often accompanied by stone tools. Why is this fossil classified within our own genus? *H. habilis* was small—about the size of Lucy—but the brain at 700 cc is about 45% larger.

The stone tools made by *H. habilis* are called Oldowan tools because they were first identified as tools at a place called Olduvai Gorge. Oldowan tools are simple and look rather clumsy (Fig. 22.10), but perhaps stone flakes were also used. The flakes would have been sharp and able to scrape away hide and cut tendons to easily remove meat from a carcass. Perhaps a division of labor arose with certain individuals serving as hunters and others as gatherers. Speech would have facilitated their cooperative efforts, and later they most likely shared their food and ate together. In this way, society and culture could have begun. Prior to the development of culture, adaptation to the environment necessitated a biological change. The acquisition of culture provided an additional way by which adaptation was possible. And the possession of culture by *H. habilis* may have hastened the extinction of the australopithecines.

> **H. habilis** warrants classification as a *Homo* because of brain size, posture, and dentition. Circumstantial evidence suggests the use of tools and also the development of culture.

Homo erectus Traveled

Homo erectus is the name assigned to hominid fossils found in Africa, Asia, and Europe and dated between 1.9 and 0.5 MYA. Compared to *H. habilis,* *H. erectus* had a larger brain (about 1,000 cc), more pronounced brow ridges, a flatter face, and a nose that projects like ours. These hominids not only stood erect, they most likely had a striding gait like ours.

It is believed that *H. erectus* first appeared in Africa and then migrated into Asia and Europe. Such an extensive population movement is a first in the history of humankind and a tribute to the intellectual and physical skills of the species. *H. erectus* was also the first hominid to use fire and fashion more advanced tools, called Acheulean tools after a site in France (Fig. 22.10). There are heavy teardrop-shaped axes and cleavers as well as flakes, which were probably used for cutting and scraping. Some believe that *H. erectus* was a systematic hunter

and brought kills to the same site over and over again. In one location, there are over 40,000 bones and 2,647 stones. These sites could have been "home bases" where social interaction occurred and a prolonged childhood allowed time for much learning. Perhaps a language evolved and a culture more like our own developed.

> **H. erectus,** which evolved from **H. habilis,** had a striding gait, made well-fashioned tools (perhaps for hunting), and could control fire. This hominid migrated into Europe and Asia from Africa about 1 MYA.

Modern Humans Originate

The out-of-Africa hypothesis proposes that *H. sapiens* became fully modern only in Africa, and thereafter migrated to Europe and Asia about 100,000 years before present. Modern humans may have interbred to a degree with populations of *H. erectus* in Europe and Asia, but in effect, they supplanted them.

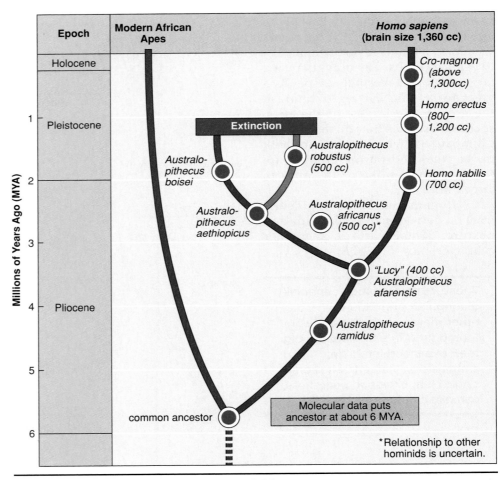

Figure 22.9 Evolutionary tree for hominids.
This tree suggests that African apes and hominids had a common ancestor about 6 million years ago (MYA). Each group has been on their own evolutionary pathway since that time. *Australopithecus afarensis* (Lucy) is a common ancestor for all the other australopithecines and for the human line of descent.

Two *Homo sapiens* of Interest A brain capacity larger than 1,000 cc allows a species to be classified as ***Homo sapiens.*** The **Neanderthals** (*H. sapiens neanderthalensis*) take their name from Germany's Neander Valley, where one of the first Neanderthal skeletons, dated some 200,000 years ago, was discovered. The Neanderthals had massive brow ridges, and the nose, the jaws, and the teeth protruded far forward. The forehead was low and sloping, and the lower jaw sloped back without a chin.

The Neanderthals are thought to be an archaic *H. sapiens* and most likely not in the main line of Homo descent. Surprisingly, however, the Neanderthal brain was, on the average, slightly larger than that of modern humans (1,400 cc compared to 1,360 cc in most modern humans). The Neanderthals were heavily muscled, especially in the shoulders and the neck. The bones of the limbs were shorter and thicker than those of modern humans. It is hypothesized that a larger brain than that of modern humans was required to control the extra musculature. They lived in Eurasia during the last Ice Age, and their sturdy build could have helped conserve heat.

The Neanderthals give evidence of being culturally advanced. They most likely successfully hunted bears, woolly mammoths, rhinoceroses, reindeer, and other contemporary animals. They used and could control fire, and they even buried their dead with flowers and tools, indicating that they may have had a religion.

In keeping with the out-of-Africa hypothesis, it is increasingly believed that modern humans, by custom called **Cro-Magnon** after a fossil location in France, entered Eurasia 100,000 years ago or even earlier. Cro-Magnons (*H. sapiens sapiens*) had a thoroughly modern appearance. They made advanced stone tools, called Aurignacian tools (Fig. 22.10), and were such accomplished hunters they may have caused the extinction of many larger mammals, such as the giant sloth, the mammoth, the saber-toothed tiger, and the giant ox.

Cro-Magnons hunted cooperatively and most likely lived in small groups, with the men hunting by day while the women remained at home with the children. The Cro-Magnon culture included art. They sculpted small figurines out of reindeer bones and antlers. They also painted beautiful drawings of animals on cave walls in Spain and France.

> The main line of hominid descent is now believed to include *A. afarensis*, *H. habilis*, *H. erectus*, and Cro-Magnon.

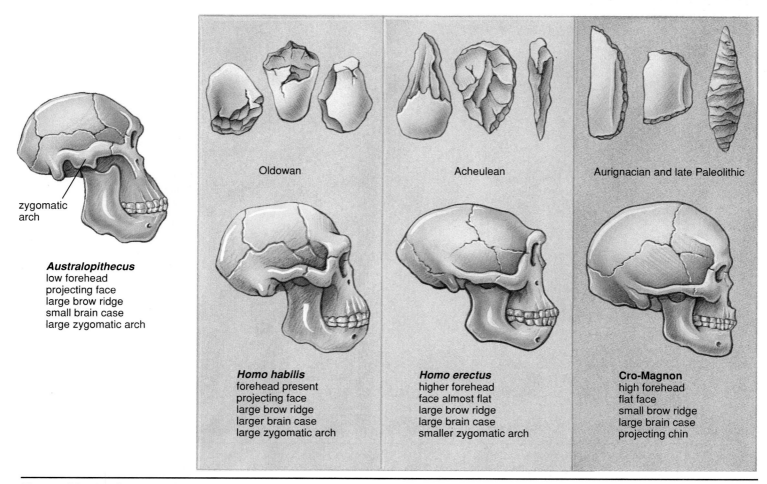

zygomatic arch

Australopithecus
low forehead
projecting face
large brow ridge
small brain case
large zygomatic arch

Oldowan

Acheulean

Aurignacian and late Paleolithic

Homo habilis
forehead present
projecting face
large brow ridge
larger brain case
large zygomatic arch

Homo erectus
higher forehead
face almost flat
large brow ridge
large brain case
smaller zygomatic arch

Cro-Magnon
high forehead
flat face
small brow ridge
large brain case
projecting chin

Figure 22.10   Comparative hominid skull anatomy and tools.
Oldowan tools are crude; Acheulean tools are better made and more varied; Aurignacian tools are well designed for their specific purposes.

We Are One Species

Human beings are diverse, but even so, we are all classified as *H. sapiens sapiens* (Fig. 22.11). This is consistent with the biological definition of species because it is possible for all types of humans to interbreed and to bear fertile offspring. While it may appear that there are various "races," molecular data show that the DNA base sequence varies as much between individuals of the same ethnicity as between individuals of different ethnicity.

It is generally accepted that the phenotype is adapted to the climate of a region. Although it might seem as if dark skin is a protection against the hot rays of the sun, it has been suggested that it is actually a protection against ultraviolet ray absorption. Dark-skinned persons living in southern regions and light-skinned persons living in northern regions absorb the same amount of radiation. (Some absorption is required for vitamin D production.) Other features that correlate with skin color, such as hair type and eye color, may simply be side effects of genes that control skin color.

Differences in body shape represent adaptations to temperature. A squat body with short limbs and a short nose retains more heat than an elongated body with long limbs and a long nose. Also, almond-shaped eyes, a flat nose and forehead, and broad cheeks are believed to be adaptations to the last Ice Age.

While it always has seemed to some that physical differences warrant assigning humans to different "races," this contention is not borne out by the molecular data mentioned previously.

Figure 22.11 Modern *Homo sapiens.*
Despite their apparent diversity, all modern-day human beings belong to one species. There is as much molecular diversity between individuals of the same ethnicity as between individuals of different ethnicities.

SUMMARY

22.1 Evidence for Evolution

The theory of evolution explains the history and diversity of life. Evidence for evolution can be taken from the fossil record, comparative biochemistry and anatomy, and biogeography.

22.2 The Evolutionary Process

Darwin not only presented evidence in support of organic evolution, he showed that evolution was guided by natural selection. Due to reproductive potential, there is a struggle for existence among members of the same species. Those members that possess variations more suited to the environment will most likely have more offspring than other members. Because of this natural selection process, there is a gradual change in species composition, which leads to adaptation to the environment.

22.3 Organic Evolution

Life came into being with the first cell(s), and thereafter diversification accounts for the evolution of all other life-forms. The classification of organisms reveals their evolutionary history.

22.4 Modern Humans Evolve

Humans are mammalian primates, as are apes. Humans differ from apes in regard to mode of locomotion, shape of jaw, and brain size.

The first hominid (humans and immediate ancestors) was *A. afarensis*, which could walk erect but had a small brain. *H. habilis* made tools, but *H. erectus* was the first fossil to have a brain size of more than 1,000 cc. *H. erectus* migrated from Africa into Europe and Asia. They used fire and may have been big-game hunters.

Most likely, *H. sapiens* evolved in Africa but then migrated to Europe and Asia, where this species supplanted the previous humans living there. One of the species was *H. sapiens neanderthalensis*. The Neanderthals did not have the physical traits of modern humans, but they did have culture.

STUDYING THE CONCEPTS

1. Show that the fossil record, comparative biochemistry and anatomy, and biogeography all give evidence of evolution.   456–58

2. What are the five aspects of Darwin's theory of evolution?   459

3. Describe the events that led to the origin of the first cell(s).   460–61

4. What are the major categories of classification? What is the goal of modern taxonomists?   462

5. Name several primate characteristics still retained by humans.   How do modern humans differ from modern apes?   463

6. Draw a hominid evolutionary tree.   464

7. How do the australopithecines, *Homo erectus,* and Cro-Magnon differ anatomically from one another and from apes?   463–65

8. Which humans were tool users? Walked erect? Used fire? Drew pictures?   463–65

9. Discuss the reason(s) we know that all humans belong to the same species and why the concept of "race" is unnecessary.   466

APPLYING YOUR KNOWLEDGE

Concepts

1. On the basis of Darwin's theory of evolution, how would you explain the fact that certain insects, flies in particular, have become resistant to DDT?

2. Organisms can be classified as anaerobic or aerobic depending on whether they do not require oxygen or do require oxygen for respiration. Yet, even aerobic organisms start the respiratory process with anaerobic reactions. On an evolutionary basis, explain this.

3. Many individuals have great difficulty with the concept that humans evolved from apes. They do not like to be that closely associated with gorillas or chimpanzees. What information could you provide that might ease their concern?

Bioethical Issue

No matter how much evidence researchers find to support evolutionary theory, some questions remain. For example, what is the rate of change necessary for one animal to evolve into another; and what are the effects of aging and population size on the ability of certain mutations to spread throughout a group?

Disbelieving evolution, many religions steadfastly believe in creationism—the idea that a god or gods created the world. To religious people, creationism may seem as plausible as evolution.

Is it ethical for science teachers to disregard creationist beliefs in favor of evolutionary theory? Or should they teach students about both systems of thought? Is it appropriate for teachers to explain that evolutionary theory is based on scientific fact, while creationism is based on belief? Does the distinction matter? Explain.

TESTING YOUR KNOWLEDGE

Match the phrases in questions 1–4 with those in this key:

    a. biogeography
    b. fossil record
    c. comparative biochemistry
    d. comparative anatomy

1. Species change over time.
2. Forms of life are variously distributed.
3. A group of related species has a unity of plan.
4. The same types of molecules are found in all living things.
5. Evolutionary success is judged by _____ success, or the number of offspring.
6. The end result of natural selection is _____ to the environment.
7. A _____ evolution is believed to have produced the first cell(s).

8. The australopithecines could probably walk _____, but they had a _____ brain.
9. The two varieties of *Homo sapiens* from the fossil record are _____ and _____.
10. Complete this table to describe the classification of humans:

Category	Name	Examples
Kingdom	_____	_____
Phylum	_____	_____
Class	_____	_____
Order	_____	_____
Family	_____	_____
Genus	_____	_____
Species	_____	_____

APPLYING TECHNOLOGY

Your study of evolution is supported by these available technologies:

Exploring the Internet

The Mader Home Page provides further resources for studying this chapter.

`http://www.mhhe.com/sciencemath/biology/mader/`

(Click on *Human Biology*.)

SELECTED KEY TERMS

adaptation Modification in structure, function, or behavior that increases likelihood of surviving and reproducing. 459

australopithecine (aw-strah-loh-PITH-uh-seen) The first generally recognized hominid. 463

common ancestor Ancestor to two or more branches of evolution. 457

Cro-Magnon Common name for the first fossils to be accepted as representative of modern humans. 465

fossil Any remains of an organism that have been preserved in the earth's crust. 456

hominid Common term for member of a family of upright, bipedal primates that includes australopithecines and humans. 463

Homo erectus Extinct earliest nondisputed species of humans, named for their erect posture, which allowed them to walk as we do. 464

Homo habilis Extinct species that may include the earliest humans; had a small brain but made quality tools. 464

Homo sapiens Present-day species of humans that has a brain capacity larger than 1,000 cc. 465

natural selection Process by which populations become adapted to their environment. 459

Neanderthal Common name for an extinct subspecies of humans whose remains are found in Europe and Asia. 465

primate Animal that belongs to the order Primates; the order of mammals that includes monkeys, apes, and humans. 463

taxonomy Science of naming, classifying, and showing relationships of organisms. 462

vestigial structure Remains of a structure that was functional in some ancestor but is no longer functional in the organism in question. 458

Chapter 23

Ecosystems

Chapter Outline

Figure 23.1   Piñon trees.
Piñon trees are a population of organisms in an ecosystem that also includes deer mice, which feed on piñon nuts.

When people first began dying in the Southwest of an unknown cause in 1993, no one knew to blame the warm, wet weather. Or the fat little deer mice scampering about. And absolutely no one would have pointed a finger at a new strain of a Korean hemorrhagic fever. But these unlikely factors combined to launch one of the decade's most deadly emerging diseases: hantavirus.

First found in Korea some 40 years ago, hantavirus—which is deadly in humans—can be carried by common deer mice. Normally, that's not a problem. But in 1993, deer mice in New Mexico and nearby states experienced a population boom. The boom stemmed from unusually heavy spring rains, which nourished trees carrying piñon nuts—a favorite food of the deer mouse (Fig. 23.1). This slight change in the weather was enough to spur a tenfold increase in the deer mouse population. Suddenly, the mice were everywhere—inside garages, in the backyard, on Indian reservations in the region. With the mice came hantavirus, carried in rodent feces and urine.

This example shows that human beings are indeed a part of natural **ecosystems,** which contain populations of organisms such as deer mice, piñon trees, and yes, humans. In ecosystems, populations interact among themselves and the physical environment. The saying goes that in an ecosystem everything is connected to everything else: warm, wet weather led to plentiful piñon nuts, which led to a deer mouse explosion and finally, illness in humans. This chapter examines the interworkings of ecosystems and how they have been impacted by human beings. While an example like this might make it seem as if we should further restrict natural ecosystems, this is not the case. The case will be made that they need to be preserved, albeit managed, if the human species is to continue.

Thankfully, the illness outbreak caused by the hantavirus ended. Rodent control efforts, disease surveillance programs, and research on a hantavirus vaccine should prevent similar outbreaks in the future.

23.1 The Nature of Ecosystems

When the earth was formed, the outer crust was covered by ocean and barren land. Over time, aquatic organisms filled the seas and terrestrial organisms colonized the land so that eventually there were many complex communities of living things. A **community** is made up of all the **populations** in a particular area, such as a forest or pond. When we study a community, we are considering only the populations of organisms that make up that community, but when we study an *ecosystem*, we are concerned with the community and its physical environment (Table 23.1).

Succession: Change Over Time

To the human eye, it seems as if the communities stay pretty much the same from year to year. Actually, one of the most outstanding characteristics of natural communities is their changing nature. The dynamic nature of communities is dramatically illustrated by succession. **Succession** is a sequential change in the relative dominance of species within a community. Most likely you have observed that an abandoned field slowly changes to be more like the surrounding natural area. Although the final result will be determined by the species that migrate there, it is possible to suggest the interim stages of succession.

Primary succession on land begins on bare rock. At first the rock is subjected to weathering by wind and rain, followed by the invasion of lichens and mosses (Fig. 23.2). They cause the buildup of soil, permitting low-lying grasses and then larger plants to take over. Depending on the area, pine trees and then broad-leaf trees will eventu-

Table 23.1

Ecological Terms

Term	Definition
Ecology	Study of the interactions of organisms with each other and with the physical environment
Population	All the members of the same species that inhabit a particular area
Community	All the populations that are found in a particular area
Ecosystem	A community and its physical environment; nonliving (abiotic) and living (biotic) components
Biosphere	All the communities on earth whose members exist in air, water, and on land

ally take root. **Secondary succession,** which begins with an abandoned field, will also go through a herbaceous stage before shrubs and then trees possibly appear. When ecologists first observed succession, they called the final stage of succession the *climax community*. They felt that each particular area had its own climax community. For example, in the United States, a deciduous forest is typical of the Northeast, a prairie is natural to the Midwest, and a semidesert covers the Southwest (Fig. 23.3). Today, we realize that the climax community is simply dependent on what species happen to inhabit the surrounding area. In other words, the composition and diversity of a climax community is in part simply due to chance.

Researchers find that the early stages of succession show the most growth and are therefore the most productive. But these stages are unable to make full use of inputs such as energy and nutrients, and therefore, the outputs (heat and loss of nutrients) are high. As succession proceeds, species variety increases, nutrients recycle more, and there is more efficient energy use. Humans often replace climax communities with simpler, often monolithic, communities such as agricultural fields and managed forests. The price paid is a certain amount of energy and nutrient waste. We should keep in mind that climax communities

a.   b.   c.   d.

Figure 23.2   Example of primary succession.
a. Lichens can grow on bare rock. **b.** Grasses take hold. **c.** Larger, nonwoody plants spread throughout the area. **d.** Trees become established.

aid in many ecological services, such as purification of the air and water, that are necessary to our well-being. For this reason, it would be wise to make sure that climax communities remain undisturbed.

Ecologists find that when climax communities are disturbed, they actually have difficulty in recovering. At one time it was believed that the variety of species and the interactions within a tropical rain forest would help it resist disturbance by human beings. But this has proved to be wrong. The more diverse the community, the more fragile and sensitive it becomes. Removal of one link can result in the whole community collapsing.

> The dynamic nature of communities is witnessed by their changing nature when succession occurs. Climax communities are threatened by disturbances.

Figure 23.3 Biological communities of the United States.
a. A desert gets limited rainfall a year. Because daytime temperatures are high, most animals are active at night when it is cooler. **b.** The California chaparral recovers quickly after a fire. **c.** A coastal rain forest of the Pacific Northwest receives a plentiful amount of rain. **d.** Midwest prairies contain grasses and rich soil built up by the remains of grasses.

Components of an Ecosystem

An ecosystem possesses both living (biotic) and nonliving (abiotic) components. The abiotic components include resources, such as sunlight and inorganic nutrients, and conditions, such as type of soil, water availability, prevailing temperature, and amount of wind (Fig. 23.4).

Biotic Components of an Ecosystem

Each population in an ecosystem has a habitat and a niche. The habitat of an organism is its place of residence; that is, where it can be found, such as under a log or at the bottom of the pond. The **niche** of an organism is its profession or total role in the community. A description of an organism's niche includes its interactions with the physical environment and with the other organisms in the community. For example, the trees in Figure 23.4*a* take solar energy, carbon dioxide, and water from the abiotic environment but have many interactions with other organisms, such as woodpeckers (Fig. 23.4*b*). Woodpeckers feed on parasitic grubs from a tree, which also provides a habitat for the woodpecker's young. All of these behaviors are aspects of the woodpecker's niche, as is apparent from reviewing the listing given in the figure.

The biotic components are often categorized in this way:

Producers are autotrophic organisms with the capability of carrying on photosynthesis and making food for themselves (and indirectly for the other populations as well). In terrestrial ecosystems, the producers are predominantly green plants, while in freshwater and marine ecosystems, the dominant producers are various species of algae.

Consumers are heterotrophic organisms that use preformed food. It is possible to distinguish four types of consumers, depending on their food source. **Herbivores** feed directly on green plants; they are termed primary consumers. **Carnivores** feed only on other animals and are thus secondary, or tertiary, consumers. **Omnivores** feed on both plants and animals. Therefore, a caterpillar feeding on a leaf is a herbivore; a green heron feeding on a fish is a carnivore; a human being eating both leafy green vegetables and beef is an omnivore. **Decomposers,** a fourth type of consumer, feed on detritus. **Detritus** is the remains of plants and animals following their death and fragmentation by soil organisms. The bacteria and fungi of decay are important detritus feeders, but so are other soil organisms, such as earthworms and various small arthropods. The importance of the latter can be demonstrated by placing leaf litter in bags with mesh too fine to allow soil animals to enter; the leaf litter does not decompose well, even though bacteria and fungi are present. Small soil organisms precondition the detritus so that bacteria and fungi can break it down to inorganic matter that producers can use again.

When we diagram all the biotic components of an ecosystem, as in Figure 23.5, it is possible to illustrate that ecosystems are characterized by two fundamental phenomena: energy flow and chemical cycling. *Energy flow* begins when producers absorb solar energy, and chemical cycling begins when producers take in inorganic nutrients from the physical environment. Thereafter, producers make food for themselves and indirectly for the other populations of the ecosystem. Energy flow occurs because all the energy content of organic food is eventually converted to heat, which dissipates in the environment. Therefore, most ecosystems cannot exist without a continual supply of solar energy. *Chemicals cycle*, that is, the original inorganic elements are returned to the producers within and between ecosystems.

Within ecosystems, energy flows and chemicals cycle.

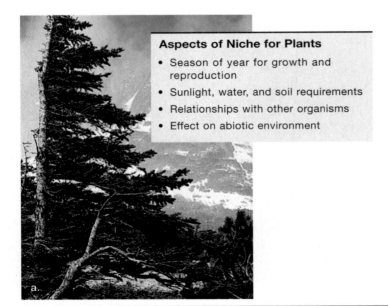

Aspects of Niche for Plants
- Season of year for growth and reproduction
- Sunlight, water, and soil requirements
- Relationships with other organisms
- Effect on abiotic environment

Aspects of Niche for Animals
- Time of day for feeding and season of year for reproduction
- Habitat and food requirements
- Relationships with other organisms
- Effect on abiotic environment

Figure 23.4 Biotic components of an ecosystem.
The biotic components of an ecosystem are influenced by the abiotic components as in **(a)**, where the force of the wind has affected tree growth. Each biotic component has a niche whose aspects are listed for plants **(a)** and animals **(b)**.

If all goes well, an ecosystem perpetuates itself. Approximately the same amount of energy enters, and the same amount of matter cycles, so that the populations continue to interact as usual. Ecosystems are resilient in that they can resist disturbances. However, human activities often overcome the resilience of an ecosystem to the point that a total collapse is possible. The following human activities are some of the ways that humans disturb ecosystems.

1. Humans do away with natural habitats by replacing them with agricultural fields, housing projects, and cities; channelizing rivers; building highways; and changing natural areas in innumerable ways to the point that they no longer exist.

2. Humans introduce pollutants by spraying for pests, such as mosquitoes, and by deliberately discharging wastes into the air and water. They have established dumps from which dangerous pollutants seep into the ground and then into underground streams called aquifers. Shotgun pellets stored as stones in the gizzards of water fowl cause them to die of lead poisoning, and plastic from a six-pack of beer strangles them. Pollutants that don't kill indigenous organisms outright often affect their ability to reproduce.

3. Humans hunt and harvest species until some of them even become extinct. Several types of whales (blue whales, gray whales, sperm whales, and others) are threatened because humans harvest them for commercial gain, and many types of sharks are on the brink of extinction simply because humans just enjoy killing them. Highly efficient but destructive modern fishing techniques play a significant role in the decline of many marine populations. Shrimp trawlers catch several tons of "trash fish" for every ton of shrimp they keep. The coats of big cats, such as leopards and cheetahs, are used to make fashionable clothing, and the horns of several species of rhinoceros are used to make medicines desired in the Far East.

4. One extreme problem today is also the introduction of exotic species. The American chestnuts and elms are in limited number because of diseases introduced from China and Europe, respectively. The starling and house sparrow introduced into the United States from Europe have displaced many of our native songbirds. The Melaleuca tree was intentionally introduced in southern Florida and has overgrown to threaten the existence of many native plants. So many exotic species have been introduced into Hawaii it has become almost impossible to counter their effects.

Various activities of humans threaten ecosystems by destroying them outright or by decimating selected populations of organisms.

Figure 23.5 Ecosystem composition.
Chemicals cycle within and between ecosystems, but energy flows through ecosystems because all the energy derived from the sun eventually dissipates as heat.

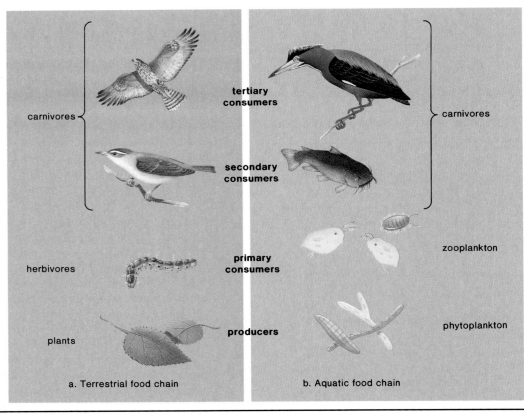

a. Terrestrial food chain

b. Aquatic food chain

Figure 23.6   Examples of food chains.
a. Terrestrial. **b.** Aquatic.

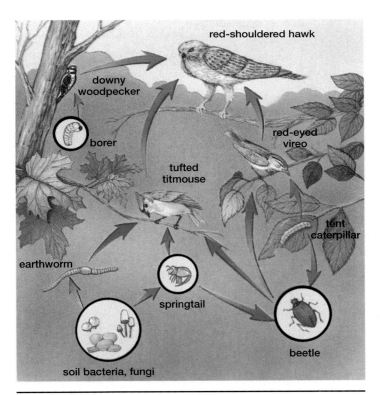

Figure 23.7   Deciduous forest ecosystem.
The arrows indicate the flow of energy in a food web.

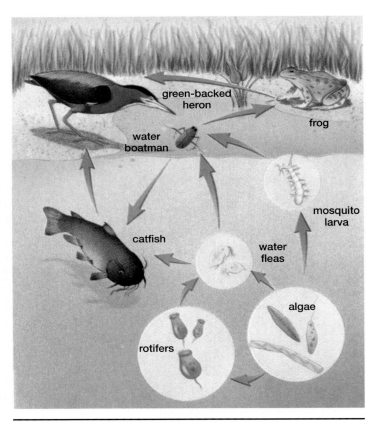

Figure 23.8   Freshwater pond ecosystem.
The arrows indicate the flow of energy in a food web.

23.2 Energy Flow in an Ecosystem

Energy flow in an ecosystem is a consequence of two fundamental laws of thermodynamics. The first law states that energy cannot be created or destroyed, and the second law states that when energy is transformed from one form to another, there is always a loss of some usable energy as heat. This means that as primary consumers feed on producers, and secondary consumers feed on primary consumers, as shown in Figure 23.6, some energy is lost to the environment as heat. Eventually, all the energy that entered the system is dissipated.

Since energy cannot be created or completely transformed from one form to another, ecosystems are unable to function unless there is a constant energy input. This input usually comes from the sun, the ultimate source of energy for our planet. It can't be said that the sun always supplies energy for an ecosystem because there are well-known exceptions. A unique community of organisms existing around deep-sea hypothermal vents in the ocean rely on chemosynthetic bacteria rather than photosynthetic algae and plants to produce food. Scalding-hot jets of water spew out of cracks in the earth's crust and supply chemosynthetic bacteria with hydrogen sulfide, a molecule they can oxidize to supply energy for synthesis of organic molecules. These bacteria that exist freely or mutualistically within the tissues of organisms are the start of food chains within a rich community of organisms.

Food Chains Become Food Webs

Energy flows through an ecosystem as the members of one population feed on those of another population. A **food chain** indicates who eats whom in an ecosystem. Figure 23.6a depicts a possible terrestrial food chain, and Figure 23.6b depicts a possible aquatic food chain. It is important to realize that each represents just one path of energy flow through an ecosystem. Natural ecosystems have numerous food chains, each linked to others to form a complex food web. For example, Figure 23.7 shows a deciduous forest **food web** in which plants are eaten by a variety of insects, and these, in turn, are eaten by several different birds, which may then be eaten by a larger bird, such as a hawk. Therefore, energy flow is better described in terms of **trophic levels** (feeding levels), each one further removed from the producer population, the first (photosynthetic) level. All animals acting as primary consumers are part of a second trophic level; all animals acting as secondary consumers are part of the third trophic level; and so on. The populations in an ecosystem form food chains in which the producers produce food for the other populations, which are consumers. While it is convenient to study food chains, the populations in an ecosystem actually form a food web in which food chains join and overlap with one another.

The food chain depicted in Figure 23.6a is part of the forest food web shown in Figure 23.7, and the food chain depicted in Figure 23.6b is part of the aquatic food web shown in Figure 23.8. Both these food chains are called *grazing food chains* because the primary consumer feeds on a photosynthesizer. In some ecosystems (forests, rivers, and marshes), there are food chains in which the primary consumer feeds mostly on detritus, partially decomposed organic matter. *Detritus food chains* can account for more energy flow than grazing food chains, because most organisms die without having been eaten. In the forest, an example of a detritus food chain is

detritus ⟶ soil bacteria ⟶ earthworms

A detritus food chain is often connected to a grazing food chain, as when earthworms are eaten by a tufted titmouse. When dead organisms have decomposed, all the solar energy that was taken up by the producer populations dissipates as heat. Therefore, energy does not cycle.

> In an ecosystem, food chains containing producers and consumers join and overlap in a food web. Decomposers break down the remains of these organisms when they die.

Ecosystems Respond to Disturbances

In an undisturbed ecosystem, there is a dynamic balance of populations caused by a limited inorganic nutrient supply and by interactions such as competition, predation, and parasitism. In Figure 23.7, tufted titmice and red-eyed vireos both prey on beetles. What would happen if a selective pesticide was used to kill off beetles? More intense competition between tufted titmouse and red-eyed vireo populations might cause their populations to decrease. How might the earthworm population size be affected? Tufted titmice might increase their predation on earthworms, whose population might now decline. How might the red-shouldered hawk population be affected? A reduction in the number of prey might eventually cause a reduction in the number of hawks. Would the size of the decomposers change if many more beetles die than usual?

Our discussion supports the ecological truism "everything is connected to everything else." A disturbance in the food chains within a food web can have diverse consequences that are hard to predict exactly.

> Because food chains join and overlap, a disturbance in one chain can affect the other chains.

Populations Form a Pyramid

The trophic structure of an ecosystem can be summarized in the form of an **ecological pyramid.** The base of the pyramid represents the producer trophic level, and the apex is the highest level of consumer, called the top predator. The other consumer trophic levels are in between the producer and the top-predator level.

There are three possible kinds of pyramids. One is the *pyramid of numbers,* based on the number of organisms at each trophic level. A second is the *pyramid of biomass.* Biomass is the weight of living material at some particular time. To calculate the biomass for each trophic level, first an average weight for the organisms at each level is determined, and then the number of organisms at each level is estimated. Multiplying the average weight by the estimated number gives the approximate biomass for each trophic level. A third pyramid, the *pyramid of energy,* shows that there is a decreasing amount of energy available at each successive trophic level (Fig. 23.9). Less energy is found in each succeeding trophic level for the following reasons:

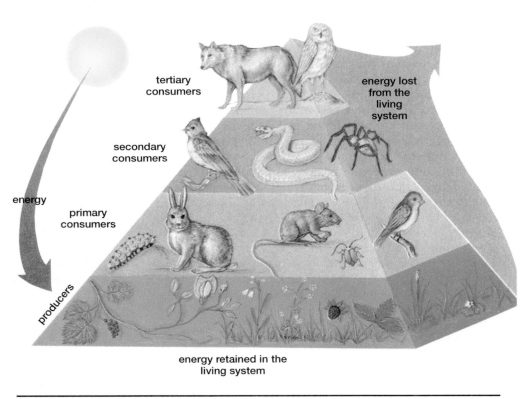

Figure 23.9.   Pyramid of energy.
At each succeeding trophic level in the pyramid, an appreciable portion of energy originally trapped by the producer is dissipated as heat. Accordingly, organisms in each trophic level pass on less energy than they received.

1. Only a certain amount is captured and eaten by the next trophic level.

2. Some of the food that is eaten cannot be digested and exits the digestive tract as waste.

3. Only a portion of the food that is digested becomes part of the organism's body. The rest is used as a source of energy.

In regard to the last point, we have to realize that a significant portion of food molecules is used as an energy source for ATP (adenosine triphosphate) buildup in mitochondria. This ATP is needed to build the proteins, carbohydrates, and lipids that compose the body. ATP is also needed for such activities as muscle contraction, nerve conduction, and active transport.

The unavoidable loss of usable energy between feeding levels explains why food chains are relatively short—at most, four or five links—and why mice (herbivores) are more common than weasels, foxes, or hawks (carnivores). The producer populations at the base of the pyramid have the largest amount of biomass, and the top predators at the top of the pyramid have the smallest amount of biomass. The amount of energy available at each trophic level is appropriate to the biomass until, finally, there is an insufficient amount of energy to support another level. As with all trophic levels, the amount of biomass is controlled by the amount of food energy available to the populations at that trophic level.

The biomass and energy considerations associated with ecological pyramids have implications for the human population. In general, only about 10% of the biomass and energy available at a particular trophic level is incorporated into the tissues of the next level. This being the case, it can be estimated that if 100 kg of grain were eaten by humans, 10% of its energy would be realized, but if this grain were fed to cattle and humans ate the cattle, then only 1% of its energy would be realized. Therefore, a larger human population can be sustained by eating grain than by eating grain-consuming animals.

> In a food web, each successive trophic level has less total biomass and energy content, because when energy is transferred from one level to the next, some energy is lost to the environment as heat.

23.3 Chemical Cycling in an Ecosystem

In contrast to energy, inorganic nutrients do cycle through large natural ecosystems. Because there is minimal input from the outside, the various chemical elements essential to life are used over and over. Since the pathways by which chemicals circulate through ecosystems involve both living (biosphere) and nonliving (geological) areas, they are known as **biogeochemical cycles.** For example, water cycles through the biosphere in the following manner. Fresh water is distilled from salt water. The sun's rays cause fresh water to evaporate from seawater, and the salts are left behind. Vaporized fresh water rises into the atmosphere, cools, and falls as precipitation (rain, snow, etc.) over the oceans and the land. A lesser amount of water also evaporates from bodies of fresh water and the bodies of plants and animals. Plants act as ground cover and retard the evaporation of water from the soil. In some ecosystems, such as deserts, succulent plants hold water and make it available to organisms that feed on them.

Since land lies above sea level, gravity eventually returns all fresh water to the sea, but in the meantime, it is contained within standing waters (lakes and ponds), flowing water (streams and rivers), and groundwater.

When rain falls, some of the water sinks or percolates into the ground and saturates the earth to a certain level. The top of the saturation zone is called the groundwater table, or simply, the water table. Sometimes groundwater is also located in a porous layer, called an **aquifer,** that lies between two sloping layers of impervious rock. Wells can be used to remove some of this water.

Fresh water, which makes up only about 3% of the world's supply of water, is called a renewable resource because a new supply is always being produced. Even so, we can run out of fresh water when the available supply is not adequate and/or is polluted so that it is not usable. Groundwater mining removes water from aquifers for the purpose of irrigation or for household use.

In the water (hydrologic) cycle, fresh water evaporates from the bodies of water and falls to the earth. Water that falls on land enters the ground- or surface waters or aquifers. Water remains in aquifers for some time; otherwise, it soon returns to the ocean.

In the phosphorus, nitrogen, and carbon cycles, the cycling process involves (1) a reservoir—that portion of the earth that acts as a storehouse for the element; (2) an exchange pool—that portion of the environment from which the producers take their nutrients; and (3) the biotic community—through which chemicals move along food chains to and from the exchange pool (Fig. 23.10). In the phosphorus cycle, phosphate is ordinarily trapped in sediments. It becomes available to terrestrial plants following an upheaval that deposits sedimentary rocks on land. Then it enters the exchange pool and is passed from organism to organism in a cycling manner. In the nitrogen cycle, the atmosphere acts as the storage area, and nitrogen becomes available to plants only after it is fixed by bacteria. Thereafter, it cycles within the biotic community. In the carbon cycle, living and dead organisms, including their shells, serve as important reservoirs for carbon. The atmosphere is the exchange pool into which organisms deposit and withdraw carbon, which is also passed from one organism to another along food chains.

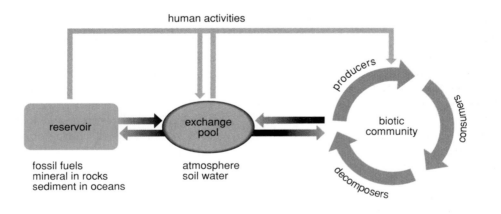

Figure 23.10 Components of a chemical cycle.
The reservoir stores the chemical, and the exchange pool makes it available to producers. The chemical then cycles through food chains. Decomposition returns the chemical to the exchange pool once again (if it has not already returned by another process).

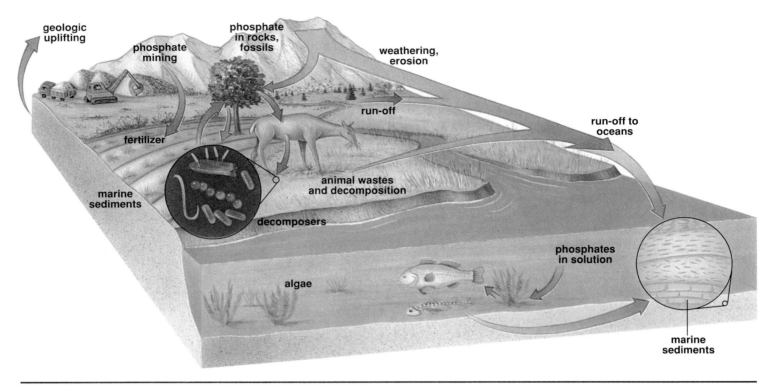

Figure 23.11   Phosphorus cycle.
Weathering of rocks releases phosphate, which enters producers and then cycles through organisms. Runoff from the land takes phosphates into the oceans, where it is incorporated into sediments that are sometimes uplifted by geological upheavals. When humans mine phosphate for inclusion in fertilizers, they increase the amount of available phosphate ions in ecosystems.

Phosphorus Cycle

On land, the weathering of rocks makes phosphate ions ($PO_4^{\equiv}$ and $HPO_4^{=}$) available to plants, which take it up from the soil (Fig. 23.11). Some of this phosphate runs off into aquatic ecosystems, where algae acquire phosphate from the water before it becomes trapped in sediments. Phosphate in sediments becomes available only when a geological upheaval exposes sedimentary rocks to weathering once more. Phosphorus does not enter the atmosphere and, therefore, the phosphorus cycle is called a sedimentary cycle.

Producers use phosphate in a variety of molecules, including phospholipids, ATP, and the nucleotides that become a part of DNA and RNA. Animals eat producers and incorporate some of the phosphate into teeth, bones, and shells that take many years to decompose. Death and decay of all organisms and also decomposition of animal wastes do, however, make phosphate ions available to producers once again. Phosphate is usually a limiting nutrient in most ecosystems because the available amount has already been taken up by organisms.

Humans Influence the Phosphorus Cycle

Human beings boost the supply of phosphate by mining phosphate ores for fertilizer production, animal feed supplements, and detergents. One well-known region where phosphorus is strip-mined (so-called because earth is removed in strips) lies east of Tampa, Florida. Here, fossilized remains of marine animals laid down some 10 to 15 million years ago provide an accessible source of phosphate. Phosphate ore is slightly radioactive, and therefore, the mining of phosphate poses a health threat to all organisms, even those that do the mining! Also, only a portion of the disturbed land has been properly reclaimed, and the rest is subject to severe soil erosion.

Animal wastes from livestock feedlots, fertilizers from cropland, and the discharge of untreated and treated municipal sewage all add excess phosphate to nearby waters. Then, when algae grow in excess, a condition called an algal bloom occurs.

In the phosphorus cycle, weathering makes phosphate available to producers, followed by consumers. Death and decay of all organisms make phosphate available to producers once again. Phosphate trapped in sedimentary rock becomes available only following a geological upheaval.

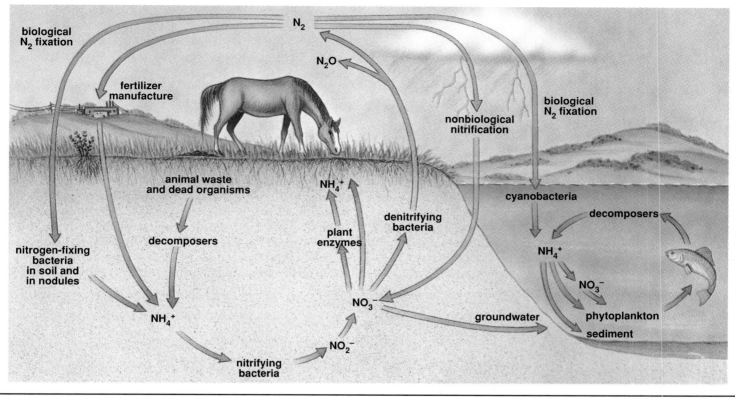

Figure 23.12   Nitrogen cycle.
Several types of bacteria are at work: nitrogen-fixing bacteria reduce nitrogen gas (N_2); nitrifying bacteria, which include both nitrite-producing and nitrate-producing bacteria, convert ammonium (NH_4^+) to nitrate (NO_3^-) ; and the denitrifying bacteria convert nitrate back to nitrogen gas. Humans contribute to the cycle by using nitrogen gas to produce nitrate for fertilizers.

Nitrogen Cycle

Nitrogen is an abundant element in the atmosphere. Nitrogen gas (N_2) makes up about 78% of the atmosphere by volume, yet nitrogen deficiency commonly limits plant growth. Plants cannot make use of nitrogen gas, and therefore, they depend on various types of bacteria that are able to take up nitrogen gas from the atmosphere (Fig. 23.12).

Nitrogen Gas Becomes Fixed

Nitrogen fixation occurs when nitrogen gas (N_2) is reduced and added to organic compounds, which can be utilized by plants. Some cyanobacteria in aquatic ecosystems and some free-living bacteria in soil are able to reduce nitrogen gas to ammonium (NH_4^+). Other nitrogen-fixing bacteria live in nodules on the roots of legumes (Fig. 23.12). They make reduced nitrogen and organic compounds available to the host plant.

Plants take up both NH_4^+ and nitrate (NO_3^-) from the soil. After plants take up NO_3^- from the soil, it is enzymatically reduced to NH_4^+ and is used to produce amino acids and nucleic acids.

Nitrogen Gas Becomes Nitrates

Nitrification is the production of nitrates. Nitrogen gas (N_2) is converted to nitrate (NO_3^-) in the atmosphere when cosmic radiation, meteor trails, and lightning provide the high energy needed for nitrogen to react with oxygen. Also, hu-

mans make a most significant contribution to the nitrogen cycle when they convert nitrogen gas to ammonium and urea for use in fertilizers.

Ammonium (NH_4^+) in the soil is converted to nitrate by certain soil bacteria in a two-step process. First, nitrite-producing bacteria convert ammonium to nitrite (NO_2^-), and then nitrate-producing bacteria convert nitrite to nitrate. These two groups of bacteria are called the nitrifying bacteria. Notice the subcycle in the nitrogen cycle that involves only ammonium, nitrites, and nitrates. This subcycle does not depend on the presence of nitrogen gas at all (Fig. 23.12).

Denitrification is the conversion of nitrate to nitrous oxide and nitrogen gas, which enters the atmosphere. There are denitrifying bacteria that act under anaerobic conditions in both aquatic and terrestrial ecosystems. In the nitrogen cycle, denitrification counterbalanced nitrogen fixation until humans started making fertilizer, which requires nitrogen fixation. Now, excess nitrates run off into bodies of fresh water and cause water pollution.

In the nitrogen cycle, nitrogen-fixing bacteria (in nodules and in the soil) reduce nitrogen gas; nitrifying bacteria convert ammonium to nitrate; denitrifying bacteria convert nitrate back to nitrogen gas.

Carbon Cycle

The relationship between photosynthesis and aerobic cellular respiration should be kept in mind when discussing the carbon cycle. Recall that this equation in the forward direction represents aerobic cellular respiration, and in the reverse direction, for simplicity's sake, is used to represent photosynthesis.

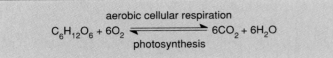

$$C_6H_{12}O_6 + 6O_2 \underset{\text{photosynthesis}}{\overset{\text{aerobic cellular respiration}}{\rightleftharpoons}} 6CO_2 + 6H_2O$$

The equation tells us that aerobic cellular respiration releases carbon dioxide, the molecule needed for photosynthesis, and photosynthesis releases oxygen, the molecule needed for aerobic respiration. Animals are dependent on green organisms, not only to produce organic food and energy, but also for a supply of oxygen. However, since producers both photosynthesize and respire, they can function independently of the animal world.

In the carbon cycle, organisms in both terrestrial and aquatic ecosystems exchange carbon dioxide with the atmosphere (Fig. 23.13). On land, plants take up carbon dioxide from the air, and through photosynthesis, they incorporate carbon into food that is used by autotrophs and heterotrophs alike. When organisms respire, a portion of this carbon is returned to the atmosphere as carbon dioxide.

In aquatic ecosystems, the exchange of carbon dioxide with the atmosphere is indirect. Carbon dioxide from the air combines with water to produce bicarbonate (HCO_3^-), a source of carbon for algae that produce food for themselves and for heterotrophs. Similarly, when aquatic organisms respire, the carbon dioxide they give off becomes bicarbonate. The amount of bicarbonate in the water is in equilibrium with the amount of carbon dioxide in the air.

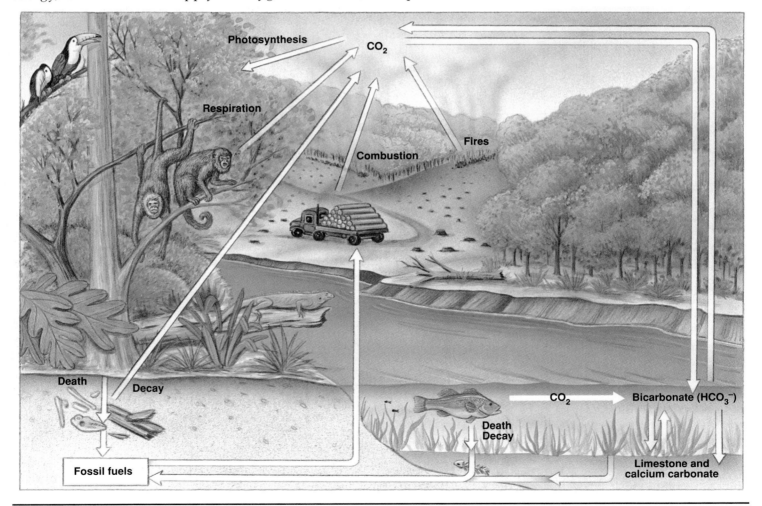

Figure 23.13 Carbon cycle.
Photosynthesizers take up carbon dioxide (CO_2) from the air or bicarbonate ions (HCO_3^-) from the water. They and all other organisms return carbon dioxide to the environment. The carbon dioxide level is also increased when volcanoes erupt and fossil fuels are burned. Presently, the oceans are a primary reservoir for carbon in the form of limestone and calcium carbonate shells.

Reservoirs Hold Carbon

Living and dead organisms contain organic carbon and serve as one of the reservoirs for the carbon cycle. The world's biota, particularly trees, contains 800 billion tons of organic carbon, and an additional 1,000–3,000 billion metric tons are estimated to be held in the remains of plants and animals in the soil. Before decomposition can occur, some of these remains are subjected to physical processes that transform them into coal, oil, and natural gas. We call these materials the **fossil fuels.** Most of the fossil fuels were formed during the Carboniferous period, 286–360 million years ago, when an exceptionally large amount of organic matter was buried before decomposing. Another reservoir is the inorganic carbonate that accumulates in limestone and in calcium carbonate shells. Many marine organisms have calcium carbonate shells that remain in bottom sediments long after the organisms have died. Geological forces change these sediments into limestone.

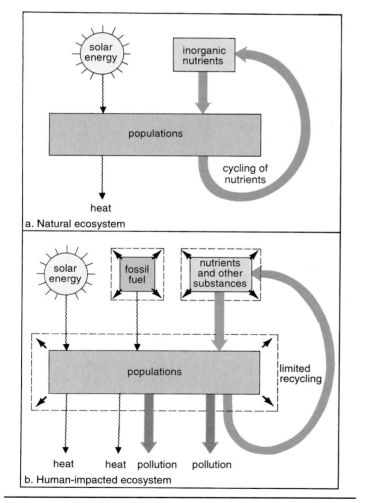

Figure 23.14 The natural ecosystem compared to the human-impacted ecosystem.
a. In natural ecosystems, the inputs and outputs are minimal.
b. In the human-impacted ecosystem, the ever-increasing size of the human population and the concomitant use of more fossil fuels and other supplements cause an ever-greater amount of wastes, including pollutants.

Humans Alter the Balances

The activities of human beings have increased the amount of carbon dioxide and other gases in the atmosphere. Data from monitoring stations record an increase of 20 ppm (parts per million) in carbon dioxide in only 22 years. (This is equivalent to 42 billion metric tons of carbon.) This buildup is primarily attributed to the burning of fossil fuels and wood. The oil-well fires that were set during the Persian Gulf War of 1991 added carbon dioxide to the atmosphere, but the ongoing burning of tropical rain forests is of even more concern. When we do away with forests, we reduce a reservoir that takes up excess carbon dioxide. At this time, the oceans are believed to be taking up most of the excess carbon dioxide; the burning of fossil fuels in the last 22 years has probably released 78 billion metric tons of carbon, yet the atmosphere registers an increase of "only" 42 billion metric tons.

There is much concern that an increased amount of carbon dioxide (and other gases) in the atmosphere is causing a global warming. These gases allow the sun's rays to pass through, but they absorb and reradiate heat back to the earth, a phenomenon called the greenhouse effect.

In the carbon cycle, photosynthesis removes, but respiration returns, carbon dioxide to the atmosphere. Forests and dead organisms are carbon reservoirs, as is the ocean.

23.4 Human–Impacted Ecosystems

Mature natural ecosystems are stable in the sense that they perpetuate themselves and require little, if any, additional materials each year. The many and varied populations are held in dynamic balance by the interactions between species, such as competition and predation. The amount of energy that enters and the amount of matter that cycles is appropriate to support these populations. **Pollution,** defined as any undesirable change in the environment that can be harmful to humans and other life, does not normally occur. Human-impacted ecosystems, however, are quite different as is shown in Figure 23.14.

Human-impacted ecosystems essentially have two added parts: the *country,* where agriculture and animal husbandry are found, and the *city,* where most people live and where industry is carried on. This representation of human-impacted ecosystems, although simplified, allows us to see that these systems require two major inputs: *fuel energy* and *raw materials* (e.g., metals, wood, synthetic materials). The use of these necessarily results in *waste* and *pollution* as outputs.

Figure 23.15 The country.
Crops are tended with heavy farming equipment that operates on fossil fuel. High yields are dependent upon a generous supply of fertilizers, pesticides, herbicides, and water.

The Country

Modern U.S. agriculture produces exceptionally high yields per acre, but this bounty is dependent on a combination of the five variables given here.

1. **Planting of a few genetic varieties.** The majority of farmers specialize in growing only one variety of a crop. Wheat farmers plant the same type of wheat, and corn farmers plant the same type of corn (Fig. 23.15). This so-called monoculture agriculture is subject to attack by a single type of parasite. For example, a single parasitic mold reduced the 1970 corn crop by 15%, and the results could have been much worse because 80% of the nation's corn acreage was susceptible.

2. **Heavy use of fertilizers, pesticides, and herbicides.** Fertilizer production requires a large energy input, and fertilizer runoff contributes to water pollution. Pesticides reduce soil fertility because they kill off beneficial soil organisms as well as pests, and some pesticides, like alar, have been accused of increasing the long-term risk of cancer, particularly in children. Herbicides,

especially those containing the contaminant dioxin, have been charged with causing adverse reproductive effects and cancer.

3. **Generous irrigation.** River waters sometimes are redirected for the purpose of irrigation, in which case "used water" returns to the river carrying a heavy concentration of salt. The salt content of the Rio Grande River in the Southwest is so high that the government has built a treatment plant to remove the salt. Water also is sometimes taken from aquifers (underground rivers), whose water content can be so reduced that it becomes too expensive to pump out more water. Farmers in Texas already are facing this situation.

4. **Excessive fuel consumption.** Energy is consumed on the farm for many purposes. Irrigation pumps already have been mentioned, but large farming machines also are used to spread fertilizers, pesticides, and herbicides, and to sow and harvest the crops. It is not incorrect to suggest that modern farming methods transform fossil fuel energy into food energy. Supplemental fossil fuel energy also

Figure 23.16 The city.
The city is dependent upon the country to supply it with food and other resources. The larger the city, the more resources required to support it.

contributes to animal husbandry yields. At least 50% of all cattle are kept in feedlots, where they are fed grain. Chickens are raised in a completely artificial environment, where the climate is controlled and each bird has its own cage to which food is delivered on a conveyor belt. Animals raised under these conditions often have antibiotics and hormones added to their feed to increase yield.

5. **Loss of land quality.** Evaporation of excess water on irrigated lands can result in a residue of salt. This process, termed salinization, makes the land unsuitable for the growth of crops. Between 25 and 35% of the irrigated western croplands are thought to have excessive salinity. Soil erosion is also a serious problem. It is said that we are mining the soil because farmers are not taking measures to prevent the loss of topsoil. The Department of Agriculture estimates that erosion is causing a steady drop in the productivity of land equivalent to the loss of 1.25 million acres per year. Even more fertilizers, pesticides, and energy supplements will be required to maintain present-day yield.

The City

The city (Fig. 23.16) is dependent on the country to meet its needs. For example, each person in the city requires several acres of land for food production. Overcrowding in cities does not mean that less land is needed; each person still requires a certain amount of land to ensure survival. Unfortunately, however, as the population increases, the suburbs and the cities tend to encroach on agricultural areas and rangeland.

The city houses workers for both commercial businesses and industrial plants. Solar and other renewable types of energy rarely are used; cities currently rely mainly on fossil fuel in the form of coal and oil. The city does not conserve resources. An office building with continuously burning lights and windows that cannot be opened shows how energy is wasted. Another example is people who drive cars long distances instead of carpooling or taking public transportation and who drive short distances instead of walking or bicycling. Also, materials are not recycled, and products are designed for rapid replacement.

The burning of fossil fuels for transportation, commercial needs, and industrial processes causes air and water pollution. This pollution is compounded by the chemical and solid waste pollution that results from the

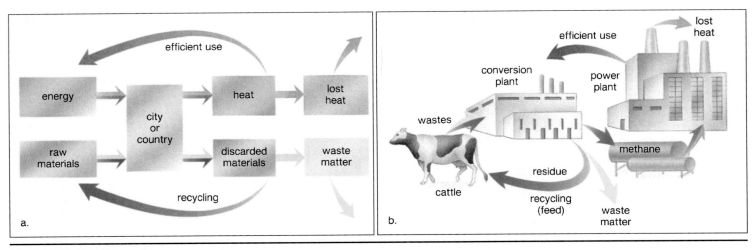

Figure 23.17   Recycling.
a. In order to cut down on the amount of lost heat and waste matter, heat could be used more efficiently, and discarded materials could be recycled. **b.** For example, instead of allowing cattle waste to enter a water supply, it could be sent to a conversion plant that produces methane gas. (The remaining residue could be converted into feed for cattle.) Excess heat, which arises from the burning of methane gas to produce electricity, could be cycled back to the conversion plant.

manufacture of many products. Consider that any product used by the average consumer (house, car, washing machine) causes pollution and waste, both during its production and when it is discarded. Humans themselves produce much sewage that is discharged into bodies of water, often after only minimal treatment.

The Solution

In human-impacted ecosystems, fuel combustion by-products, sewage, fertilizers, pesticides, and solid wastes all are added to the environment in the hope that natural cycles will cleanse the biosphere of these pollutants. But we have exploited natural ecosystems to the extent that the environment is overloaded.

More and more natural ecosystems are impacted because an ever-increasing number of people want to maintain a standard of living that requires many goods and services. But we can call a halt to this spiraling process if we achieve zero population growth and if we conserve energy and raw materials. Conservation can be achieved in three ways: (1) wise use of only what is actually needed; (2) recycling of nonfuel minerals, such as iron, copper, lead, and aluminum; and (3) use of renewable energy resources and development of more efficient ways to utilize all forms of energy (Fig. 23.17*a*).

Some farmers have adopted organic farming methods and do not apply fertilizers, pesticides, or herbicides to their crops. They use cultivation of row crops to control weeds, crop rotation to combat major pests, and the growth of legumes to supply nitrogen fertility to the soil. Some farmers use natural predators and parasites instead of pesticides to

control insects. Most of these farmers switched farming methods because they were concerned about the health of their families and livestock and had found that the chemicals were sometimes ineffective. A study of about 40 farms showed that organic farming, for the most part, was just as profitable as conventional farming. Crop yields were lower, but so were operating costs. Organic farms required about two-fifths as much fossil energy to produce one dollar's worth of crop. The method of plowing and the utilization of crop rotation resulted in one-third less soil erosion. The researchers concluded it would be well to determine how far farmers can move in the direction of reduced agricultural chemical use and still maintain the quality of the product. They noted that a modest application of fertilizer would have improved the protein content of the crop.

As another example, consider a plant that was built in Lamar, Colorado, which produces methane from feedlot animals' wastes (Fig. 23.17*b*). The methane is burned in the city's electrical power plant, and the heat given off is used to incubate the anaerobic digestion process that produces the methane. In addition, a protein feed supplement is produced from the residue of the digestion process. This system represents a cyclical use of material and an efficient use of energy similar to that found in nature. Many other such processes for achieving this end have been, and could be, devised.

It is important to consider that ecosystems should be developed and managed with care so that their integral nature is preserved. This is often called "working with nature" rather than against nature. This principle is explored in the Ecology reading on the next page.

Ecology Focus

Preservation of the Everglades

Originally, the Everglades encompassed the whole of southern Florida from Lake Okeechobee down to Florida Bay (Fig. 23A). Now, largely in the Everglades National Park alone do we find the vast sawgrass prairie, interrupted occasionally by a cypress dome or hardwood tree island. Within these islands, both temperate and tropical evergreen trees grow amongst the dense and tangled vegetation. Mangrove, or salt-tolerant trees, are found along sloughs (creeks) and at the shoreline. Only the roots of the red mangrove can tolerate the sea constantly. The prop roots of this tree protect over 40 different types of juvenile fishes as they grow to maturity. During the wet season, from May to November, animals are dispersed throughout the region, but in the dry season, from December to April, they congregate wherever pools of water are found. Alligators are famous for making "gator holes" where water collects, and fish, shrimps, crabs, birds, and a host of living things survive until the rains come again. Almost everyone is captivated by the birds that find the ready supply of fish they need for daily existence at these holes. The large and beautiful herons, egrets, roseate spoonbill, and anhinga are awesome. These birds once numbered in the millions; now they number only in the thousands. Why is this?

At the turn of the century, settlers began to drain the land just south of Lake Okeechobee to grow crops on the soil enriched by partially decomposed saw grass. The large dike that now rings the lake prevents the water from taking its usual course: over the banks of Lake Okeechobee and slowly southward. In times of flooding, water can be shunted through the St. Lucie Canal to the Atlantic Ocean or through the canalized Caloosahatchee River to the Gulf of Mexico. In times of drought, water is contained not only in the lake but also in three so-called conservation areas established to the south of the lake. Water must be conserved for the irrigation of the farmland and to recharge the Biscayne aquifer (underground river), which supplies drinking water for the cities on the east coast of Florida. Containing and moving the water from place to place has required the construction of over 2,250 kilometers of canals, 125 water control stations, and 18 large pumping stations. Now the Everglades National Park receives water only when it is discharged artificially from a conservation area. This disruption of the natural flow of water has affected the reproduction pattern of the birds, which is attuned to the natural wet-dry season turnover.

It took considerable human effort and a huge financial investment to control nature and to establish the Everglades Agricultural Area. Has this attempt to bend nature to human will been worthwhile? The area does, in fact, produce more sugar than Hawaii and a large proportion of the vegetables consumed in the United States each winter. But this has not been without a price. The rich soil, built up over thousands of years, is disappearing and most likely will be unable to sustain conventional agriculture after the year 2000. It has been suggested that at that time we might use the Everglades Agricultural Area for the growth of aquatic plants. Perhaps it should have been decided in the beginning to work with nature by growing aquatic plants instead of conventional plants. Then all the canals and pumping stations would have been unnecessary, the water would still flow from Lake Okeechobee to the Everglades as it had for eons, and the birds today would still number in the millions.

Figure 23A   The Everglades.
The Everglades once extended from Lake Okeechobee south to Florida Bay. Now it only encompasses three conservation areas and the Everglades National Park.

SUMMARY

23.1 The Nature of Ecosystems

The process of succession from either bare rock or disturbed land results in a climax community. An ecosystem is a community of organisms plus the physical environment. Each population in an ecosystem has a habitat and niche. Some populations are producers and some are consumers. Consumers may be herbivores, carnivores, omnivores, or decomposers.

Energy flow and chemical cycling are two phenomena observed in ecosystems.

23.2 Energy Flow in an Ecosystem

Food chains are paths of energy flow through an ecosystem. Grazing food chains always begin with a producer population, which is capable of producing organic food, followed by a series of consumer populations. When members of food chains die and decompose, the very same chemicals are then made available to the producer population again, but the energy has been dissipated as heat. Therefore, energy does not cycle through an ecosystem.

If we consider the various food chains together, we realize that they form an intricate food web in which there are various trophic (feeding) levels. All producers are on the first level, all primary consumers are on the second level, and so forth. To illustrate that energy does not cycle, it is customary to arrange the various trophic levels to form an energy pyramid.

23.3 Chemical Cycling in an Ecosystem

Chemical cycling through an ecosystem involves not only the biotic components of an ecosystem but also the physical environment. Most cycles involve a reservoir, where

the element is stored; an exchange pool, from which the populations take and return nutrients; and the populations themselves, where the chemical moves from producers to consumers before returning to producers once again.

In the phosphorus cycle, the reservoir is sediments at the bottom of the ocean. Only upheavals make sedimentary rock available on land to plants.

In the nitrogen cycle, the reservoir is the atmosphere, and nitrogen (N_2) gas must be converted to nitrate (NO_3^-) for use by producers. Nitrogen-fixing bacteria, particularly in root nodules, make organic nitrogen available to plants. Other bacteria active in the nitrogen cycle are the nitrifying bacteria, which convert ammonium to nitrate, and the denitrifying bacteria, which reconvert nitrate to N_2.

In the carbon cycle, the reservoir is organic matter, calcium carbonate shells, and limestone. The exchange pool is the atmosphere: photosynthesis removes carbon dioxide, and respiration and combustion add carbon dioxide.

23.4 Human-Impacted Ecosystems

If undisturbed, mature, natural ecosystems remain in dynamic balance and need the same amount of energy each year. Additional nutrient inputs are minimal because chemicals cycle within and between ecosystems. In human-impacted ecosystems, the human population size constantly increases. In the country, farmers plant only certain high-yield varieties of plants, which require such supplements as fertilizers, pesticides, and water. In the city, the populace is wasteful of energy and materials. Therefore, there is much pollution. We and future generations must find ways to conserve energy, use excess heat, recycle materials, and reduce pollution.

STUDYING THE CONCEPTS

1. What is succession, and how does it result in a climax community? 470

2. Distinguish between abiotic and biotic components of an ecosystem. What are the aspects of niche for a plant? An animal? 472

3. Name four different types of consumers found in natural ecosystems. 472

4. Give an example of a grazing and a detritus food chain for a terrestrial and for an aquatic ecosystem. 475

5. Why is the term trophic level more appropriately applied to a food web rather than a food chain? 475

6. How would other population sizes be affected if the tufted titmouse population were to be decimated in Figure 23.7? If a pesticide were used to kill the mosquito larvae in Figure 23.8? 474–75

7. What are the three possible kinds of pyramids? Draw a pyramid of energy, and explain why such a pyramid can be used to verify that energy does not cycle. 476

8. Explain what is meant by a reservoir and exchange pool for a biogeochemical cycle. 477

9. Draw a diagram to illustrate the phosphorus, nitrogen, and carbon biogeochemical cycles, and include in your diagram the manner in which humans disturb the equilibrium of these cycles. 478–81

10. Compare and contrast the basic characteristics of a natural ecosystem to the characteristics of the human-impacted ecosystem. 481

11. Suggest ways by which it would be possible to make the human-impacted ecosystem more cyclical so pollution might be better controlled. 484

APPLYING YOUR KNOWLEDGE

Concepts

1. A group of concerned citizens found out that a plant growing wild in a rather large natural area was being used as a stimulant by some of the high-school students. They proposed a systematic elimination of all these plants. The local Audubon group opposed this action, saying that it could affect the entire ecosystem. Explain the reasoning behind the Audubon's statement.

2. Pigs are fed corn. Use this information to suggest why pork costs much more per pound than corn.

3. Indicate the reasons that replacing a forest or grassland area with a factory and parking lot increases the amount of carbon dioxide in the atmosphere.

4. What is the reason(s) that introduction of an exotic species into an ecosystem often results in the new species overrunning or displacing the native species?

Bioethical Issue

Doctors and researchers spend countless hours trying to save lives and spare people from disease. But while we focus our attention on human life, some of our animal counterparts quietly fade away.

Many endangered species become extinct. That's why some animal-lovers devote themselves to saving endangered species. Through captive breeding programs and rescue efforts, researchers have managed to save gorillas and other animals.

How much effort should we make to save endangered species? Is it ethical to contribute to their decline—say, by killing the animals for their fur? Is experimentation with animals justified in an effort to improve the medical treatment of humans? Should governments allocate money to find and preserve endangered species? Whose responsibility is it to seek out and save these animals?

TESTING YOUR KNOWLEDGE

1. Label this diagram of an ecosystem.

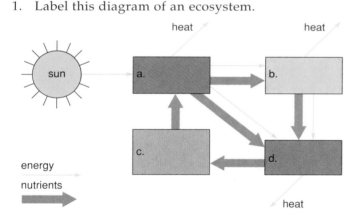

2. Chemicals cycle through the populations of an ecosystem, but energy is said to _____ because all of it eventually dissipates as heat.

3. When organisms die and decay, chemical elements are made available to _____ populations once again.

4. Organisms that feed on plants are called _____.

5. A pyramid of energy illustrates that there is a loss of energy from one _____ level to the next.

6. Add these trophic levels to the diagram, representing a pyramid of energy: algae, large fishes, humans, small fishes, zooplankton.

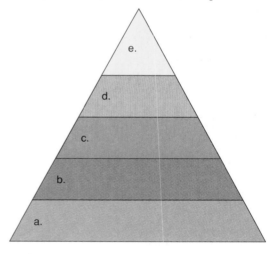

7. In the carbon cycle, when living organisms _____, carbon dioxide (CO_2) is returned to the exchange pool.

8. Humans make a significant contribution to the nitrogen cycle when they convert nitrogen (N_2) to _____ for use in fertilizers.

9. During the process of denitrification, nitrate is converted to _____.

10. Human-impacted ecosystems are characterized by an ever-increasing use of _____ and _____.

APPLYING TECHNOLOGY

Your study of ecosystems is supported by these available technologies:

Exploring the Internet

The Mader Home Page provides further resources for studying this chapter.

`http://www.mhhe.com/sciencemath/biology/mader/`

(Click on *Human Biology*.)

Life Science Animations Video

Video #5:   Plant Biology, Evolution, and Ecology

Carbon and Nitrogen Cycles (#51)   Carbon and nitrogen compounds cycle from the atmosphere and through a biological community.   (2)*

Energy Flow Through an Ecosystem (#52)   Energy loss occurs as one trophic level feeds on another.   (1)

*Level of difficulty

SELECTED KEY TERMS

carnivore Animal that feeds only on animals.   472

community Group of many different populations that interact with one another within the same environment.   470

consumer Member of a population that feeds on members of other populations in an ecosystem.   472

decomposer Organism of decay (fungus and bacterium) in an ecosystem.   472

detritus (dih-TRYT-us) Nonliving organic matter.   472

ecological pyramid Pictorial graph representing the biomass, organism number, or energy content of each trophic level in a food web, from the producer to the final consumer populations.   476

ecology Study of the interactions of organisms with each other and with the physical environment.   470

ecosystem Setting in which populations interact with each other and with the physical environment.   469

food chain Pathway by which nutrients and energy are transferred from one organism to another at successive trophic levels.   475

food web Complete set of food links between populations in a community.   475

herbivore Animal that feeds directly on plants.   472

niche (nich) Functional role of an organism, including how it interacts with both the biotic and the abiotic components of an ecosystem.   472

nitrogen fixation Process whereby nitrogen is reduced to ammonia, which is then converted to organic compounds.   479

omnivore Animal that feeds on both plants and animals.   472

pollution Detrimental alteration of the normal constituents of air, land, and water due to human activities.   481

population All the organisms of the same species in one place.   470

producer Organism at the start of a food chain that makes its own food (e.g., green plants on land and algae in water).   472

succession Series of ecological stages by which the community in a particular area gradually changes until there is a climax community that can maintain itself.   470

trophic level (TRO-fik) Feeding level of one or more populations in a food web.   475

Chapter 24

Population Concerns

Chapter Outline

Figure 24.1   The American family.
Should young couples think about how their children will contribute to population statistics and environmental stress before having a family?

L inda L. is worried. At age 30, she expected to have made it. She expected to be successful. Comfortable. Peaceful. Instead, Linda—like millions of other Americans—worries about two major things: her kids, who will soon be entering high school, and her parents, who may soon need live-in help.

Linda is caught in the aftermath of the baby boom, a period of time in the 1950s when the American birth-rate skyrocketed (Fig. 24.1). When the so-called baby boomers were in their 20s, they were young and vibrant, and they added to American productivity. Today, however, the boomers are aging. Their children worry that elderly boomers will drain resources—not to mention national attention—from younger adults.

Suddenly, Linda fears that nursing homes, Alzheimer disease, and other topics involving the aged will swamp political agendas and public issues in the near future. Meanwhile, adults her age will find themselves taking care of both their children and parents at the same time.

By international standards, the U.S. population age shift is not devastating. Far worse population concerns have long haunted some countries. In overpopulated China, for example, couples are punished for conceiving more than one child. The countries of the world today are divided into two groups. The more-developed countries (MDCs), typified by countries in North America and Europe, are those in which population growth is under control and the people enjoy a good standard of living. The less-developed countries (LDCs), typified by countries in Latin America, Africa, and Asia, are those in which population growth is out of control and the majority of people live in poverty. (Sometimes the term *third-world countries* is used to mean the less-developed countries. This term was introduced by those who thought of the United States and Europe as the first world and the former USSR as the second world.)

Before we explore the reasons that the world is now divided into MDCs and LDCs, it is necessary to study exponential population growth in general.

24.1 Exponential Population Growth

The human population growth curve is a J-shaped growth curve (Fig. 24.2). In the beginning, growth of the human population was relatively slow, but as more reproducing individuals were added, growth increased, until the curve began to slope steeply upward. It is apparent from the position of 1995 on the growth curve in Figure 24.2 that growth is quite rapid now. The world population increases at least the equivalent of a medium-sized city (200,000) every day and the equivalent of the combined populations of the United Kingdom, Norway, Ireland, Iceland, Finland, and Denmark every year. These startling figures are a reflection of the fact that a very large world population is undergoing exponential growth.

Mathematically speaking, **exponential growth,** or geometric increase, occurs in the same manner as compound interest; that is, the percentage increase is added to the principal before the next increase is calculated. With regard to populations, the percentage increase is termed the **growth rate,** which is applied per year. Because of exponential growth, each year's increase will be greater than the previous year's increase since the population size has increased in the meantime. In fact, the world's population grows by a larger amount each year even when the growth rate decreases slightly. The increase in size is dramatically large because the world population is very large.

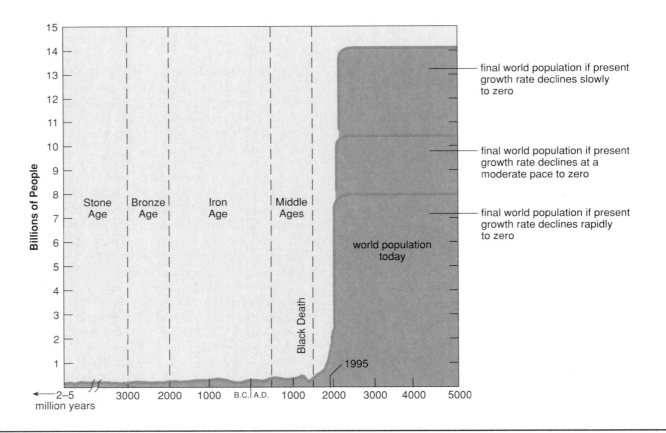

Figure 24.2 Growth curve for human population.
The human population is now undergoing rapid exponential growth. Since the growth rate is declining, it is predicted that the population size will level off at 8 billion, 10.5 billion, or 14.2 billion, depending upon the speed with which the growth rate declines.

The Growth Rate Depends on Births Versus Deaths

The growth rate of a population is determined by considering the difference between the number of persons born per year (birthrate, or natality) and the number of persons who die per year (death rate, or mortality). It is customary to record these rates per 1,000 persons. For example, Canada at the present time has a birthrate of 13 per 1,000 per year, but it has a death rate of 6 per 1,000 per year. This means that Canada's population growth, or simply its growth rate, is

$$\frac{13-7}{1,000} = \frac{6}{1,000} = 0.006 \times 100 = 0.6\%$$

Notice that while birthrate and death rate are expressed in terms of 1,000 persons, the growth rate is expressed per 100 persons, or as a percentage.

After 1750, the world population growth rate steadily increased, until it peaked at 2% in 1965. It has fallen since then to 1.5%. Yet, there is an ever-greater increase in the world population each year because of exponential growth. The explosive potential of the present world population can be appreciated by considering the doubling time.

Doubling Time Depends on Growth Rate

The **doubling time** (*d*)—the length of time it takes for the population size to double—can be calculated by dividing 70 (the demographic constant) by the growth rate (*gr*):

$$d = 70/gr$$

  d = Doubling time
  gr = Growth rate
  70 = Demographic constant

If the present world growth rate of 1.5% continues, the world population will double in 47 years.

  d = 70/1.5 = 47 years

This means that in 47 years, the world will need double the amount of food, jobs, water, energy, and so on to maintain the same standard of living.

It is of grave concern to many that the amount of time needed to add each additional billion persons to the world population has taken less and less time. The first billion didn't occur until 1800; the second billion arrived in 1930; the third billion in 1960; and today there are 5.8 billion. Only if the growth rate continues to decline can there be zero population growth when births equal deaths and the population size remains steady. Figure 24.2 shows the population may level off at 8, 10.5, or 14.2 billion, depending on the speed with which the growth rate declines.

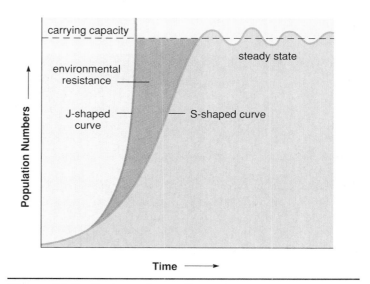

Figure 24.3   S-shaped growth curve.
In nature, populations initially grow exponentially; but instead of following a J-shaped growth curve, they follow an S-shaped curve because of environmental resistance. Population size enters a steady state when the carrying capacity is reached.

Carrying Capacity Limits Population Size

The growth curve for many nonhuman populations is S-shaped—the population tends to level off at a certain size. Figure 24.3 illustrates an example based on actual data for the growth of a fruit fly population reared in a culture bottle. In the beginning, the numbers were small, and population growth was minimal. Then the flies began to multiply rapidly. Notice that the curve began to rise dramatically, just as the human population curve does now. At this time, a population is demonstrating its biotic potential. **Biotic potential** is the maximum growth rate under ideal conditions. It usually is not demonstrated for long because of an opposing force called environmental resistance. **Environmental resistance** includes all the factors that cause early death of organisms and therefore prevent the population from producing as many offspring as it might otherwise. In the fruit fly example, we can speculate that environmental resistance included the limiting factors of food and space. Fruit fly wastes also may have limited the population size. When environmental resistance sets in, biotic potential is overcome, and the slope of the growth curve begins to decline. This is the inflection point of the curve.

The eventual size of any population represents a compromise between biotic potential and environmental resistance. This compromise occurs at the carrying capacity of the environment. The **carrying capacity** is the maximum population that the environment can support for an indefinite period. The carrying capacity of the earth for humans has not been determined. Some authorities think the earth is potentially capable of supporting 50–100 billion people. Others think we already have more humans than the earth can adequately support.

Figure 24.4   More-developed countries (MDCs) versus less-developed countries (LDCs).
a. In the MDCs, most people enjoy a high standard of living. **b.** In the LDCs, the majority of people are poor and have few amenities.

24.2 Human Population Growth

The human population has undergone three periods of exponential growth. *Toolmaking* may have been the first technological advance that enabled the human population to enter a period of exponential growth; farming may have resulted in a second phase of growth; and the *Industrial Revolution,* which began about 1850, promoted the third phase.

Presently, population growth in the *more-developed countries (MDCs)* is minimal, while population growth in the *less-developed countries (LDCs)* is increasing rapidly (Fig. 24.4).

More-Developed Versus Less-Developed Countries

The *MDCs* doubled their populations between 1850 and 1950. This was largely due to a decline in the death rate, the development of modern medicine, and improved socioeconomic conditions. The decline in the death rate was followed shortly thereafter by a decline in the birthrate, so that populations in the MDCs experienced only modest growth between 1950 and 1975. This sequence of events (i.e., decreased death rate followed by decreased birthrate) is termed a **demographic transition.**

The growth rate for the MDCs is now about 0.1%. The populations of a few of the MDCs—Italy, Denmark, Hungary, Sweden—are not growing or are actually decreasing in size. In contrast, there is no leveling off and no end in sight to U.S. population growth. Although the United States has a growth rate of 0.6%, many people immigrate to the United States each year. In addition, a baby boom between 1947 and 1964 meas that a large number of women are still of reproductive age.

Although the death rate began to decline steeply in the LDCs following World War II with the importation of modern medicine from the MDCs, the birthrate remained high. The growth rate of the LDCs peaked at 2.5% between 1960 and 1965. Since that time, a demographic transition has begun: the decline in the death rate has slowed and the birthrate has fallen. A growth rate of 1.8% is expected by the end of the century. Still, because of exponential growth, discussed previously, the population of the LDCs may explode from 4.6 billion today to 10.2 billion in 2100. Most of this growth will occur in Africa, Asia, and Latin America.

Although the less-developed countries are now undergoing demographic transition and the growth rate is declining, the populations of Africa, Asia, and Latin America are expanding dramatically because of exponential growth.

It will be almost impossible to reduce poverty and bring about sustainable development in Africa, Asia, and Latin America if more than 6 billion people are added to this area within the next century. However, ways to greatly reduce the expected increase have been suggested.

1. Establish and/or strengthen family planning programs. A decline in growth rate is seen in countries with good family planning programs supported by community leaders. Presently, 25% of women in sub-Saharan Africa say they would like to delay or stop childbearing, yet they are not practicing birth control; likewise, 15% of women in Asia and Latin America have an unmet need of birth control. There are also some users who are not satisfied with their current method of contraception. It is estimated that accessibility to family planning could lower the expected population increase of LDCs to 8.3 billion instead of 10.2 billion by 2100.

2. Use social progress to reduce the desire for large families. Many couples in the LDCs presently desire as many as four to six children. But providing available education, raising the status of women, and reducing child mortality are desirable social improvements that could cause them to think differently. It is estimated that such improvements could further reduce the expected population increase in the LDCs to 7.3 billion by the year 2100.

3. Delay the onset of childbearing. A delay in the onset of childbearing and wider spacing of births could cause a temporary decline in fertility and hence reduce the present population growth rate. If childbearing begins five years later than usual, a further reduction of 1.2 billion births might be possible, so that 6.1 rather 10.2 billion persons would result in the LDCs by the year 2100.

Comparing Age Structure

The LDCs are experiencing a population momentum because they have more women entering the reproductive years than there are older women leaving them behind. Populations have three age groups: dependency, reproductive, and postreproductive. One way of characterizing population is by these age groups. This is best visualized when the proportion of individuals in each group is plotted on a bar graph, thereby producing an age-structure diagram (Fig. 24.5).

Laypeople are sometimes under the impression that if each couple has two children, zero population growth will take place immediately. However, **replacement reproduction,** as it is called, will still cause most countries today to continue growing due to the age structure of the

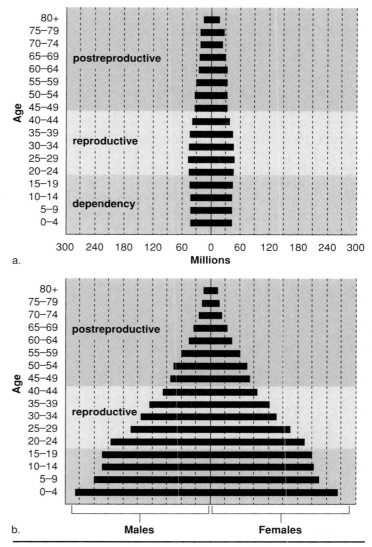

a.

b.

Figure 24.5 Age-structure diagrams for MDCs and LDCs, 1989.
The diagrams illustrate that **(a)** the populations of MDCs are approaching stabilization, whereas **(b)** the populations of LDCs will expand rapidly due to the shape of their age-structure diagram.
Source: Data from World Population Profile: 1989, WP-89.

population. If there are more young women entering the reproductive years than there are older women leaving them behind, then replacement reproduction will still give a positive growth rate.

Many MDCs have a stabilized age-structure diagram (Fig. 24.5a), but most LDCs have a youthful profile—a large proportion of the population is younger than the age of 15. Since there are so many young women entering the reproductive years, the population will still expand greatly, even after replacement reproduction is attained. The more quickly replacement reproduction is achieved, however, the sooner zero population growth will result.

24.3 Human Population and Pollution

As the human population increases in size, more and more ecosystems are impacted by humans. Ecosystems are then no longer able to process and rid the biosphere of wastes, which accumulate and are called pollutants. **Pollutants** are substances added to the environment, particularly by human activities, that lead to undesirable effects for living things, including humans. Human beings add pollutants to all parts of the biosphere—air, water, and land.

Air Pollution

Four major concerns (destruction of the ozone shield, global warming, acid deposition, and photochemical smog) are associated with the air pollutants listed in Figure 24.6. You can see that fossil fuel burning and vehicle exhaust are primary sources of air pollution gases. These two are

related because gasoline is derived from petroleum (oil), a fossil fuel. The fossil fuels (oil, coal, and natural gas) are burned in the home to provide heat and in power plants to generate electricity.

U.S. citizens have begun to practice energy conservation, and it is estimated that in so doing they have saved $150 billion a year. But it is clear that even more energy conservation is possible by making use of renewable energy sources such as solar and wind energy, falling water, and geothermal energy.

Global Warming from Greenhouse Gases

Certain air pollutants allow the sun's rays to pass through, but then they absorb and reradiate the heat back toward the earth (Fig. 24.7*a*). This is called the **greenhouse effect** because it is like the glass of a greenhouse that allows sunlight to pass through and then traps the resulting heat

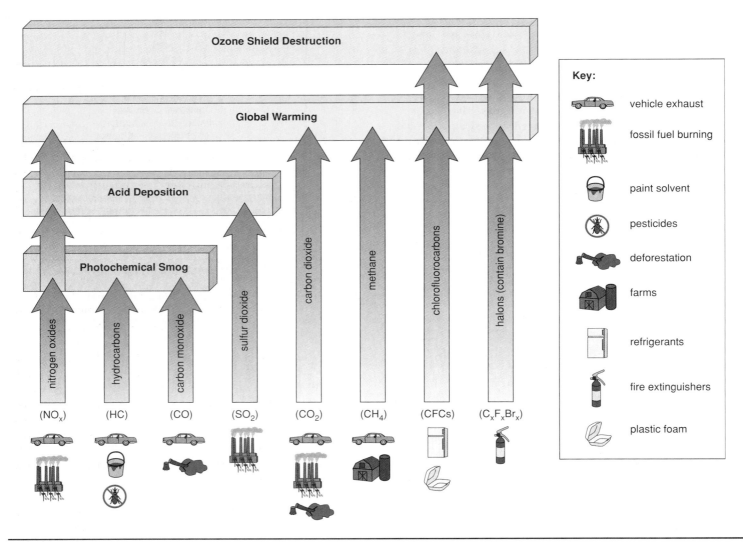

Figure 24.6 Air pollutants.
These are the gases, along with their sources, that contribute to four environmental effects of major concern: global warming, ozone shield destruction, acid deposition, and photochemical smog. An examination of the sources of these gases shows that vehicle exhaust and fossil fuel burning are the chief contributors.

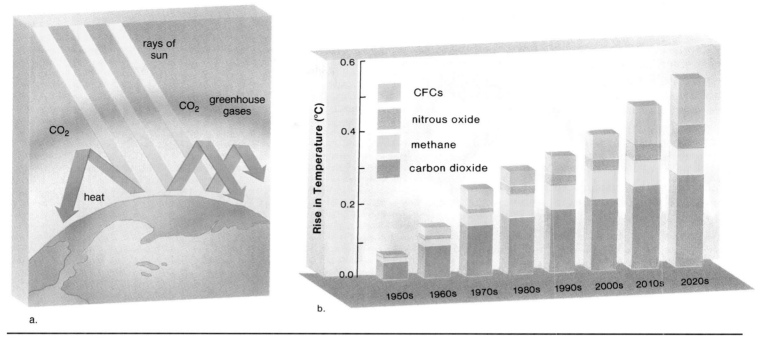

Figure 24.7 Global warming.
a. The greenhouse effect is caused by the atmospheric accumulation of certain gases that allow the rays of the sun to pass through but absorb and reradiate heat to the earth. **b.** The greenhouse gases. The increase in warming caused by these gases projected to the year 2020.

inside the structure. The air pollutants responsible for the greenhouse effect are known as the greenhouse gases. They are as follows:

Gas	From
Carbon dioxide (CO_2)	Fossil fuel and wood burning
Nitrous oxide (NO_2)	Fertilizer use and animal wastes
Methane (CH_4)	Biogas (bacterial decomposition, particularly in the guts of animals, in sediments, and in flooded rice paddies)
Chlorofluorocarbons (CFCs)	Freon, a refrigerant
Halons (halocarbons $C_xF_xBr_x$)	Fire extinguishers

If nothing is done to control the level of greenhouse gases in the atmosphere, a rise in global temperature is expected. Figure 24.7*b* predicts a rise of over 0.5°C by 2020, but some authorities predict the rise in temperature could be as much as 3.5°C during the next century.

The ecological effects of a 3.5°C rise in global temperature would be severe. The sea level would rise—melting of the polar ice caps would add more water to the sea, and, in addition, water expands when it heats up. There would be coastal flooding and the possible loss of many cities, like New York, Boston, Miami, and Galveston in the United States. Coastal ecosystems, such as marshes, swamps, and bayous, would normally move inland to higher ground as the sea level rises, but many of these ecosystems are blocked by artificial structures and may be unable to move inland. If so, the loss of fertility would be immense.

There may also be food loss because of regional changes in climate. Because of greater temperatures and also drought in the midwestern United States, the suitable climate for growing wheat and corn could shift as far north as Canada, where the soil is not as suitable. The occurrence of droughts will reduce agricultural yields and also cause trees to die off. Expansion of forests into Arctic areas will most likely not be able to offset the loss of forests in the temperate zones.

It is clear from Figure 24.7*b* that carbon dioxide accounts for at least 50% of the predicted rise in global temperature. To prevent this occurrence, a sharp decrease in the consumption of fossil fuels and more efficient ways to utilize cleaner fuels, such as natural gas, are recommended. The use of alternative energy sources, such as solar and geothermal energy and even perhaps nuclear power, is needed. Deforestation, like fossil fuel burning, is a major contributor to a rise in carbon dioxide and therefore to global warming. Burning one acre of primary forest releases 200,000 kg of carbon dioxide into the air; moreover, the trees are no longer available to act as a sink to take up carbon dioxide during photosynthesis. Therefore, tropical rain forest deforestation should be halted, and extensive reforestation should take place all over the globe.

The other greenhouse gases combined account for the other 50% predicted rise in global temperature. A complete phaseout of chlorofluorocarbon use would be most beneficial. Fortunately, in an effort to arrest ozone shield destruction, the United States and the European countries have agreed to reduce CFC production by 85% as soon as possible and to stop their production altogether by the end of the century.

Destroying the World's Ozone Shield

The earth's atmosphere is divided into layers. The troposphere envelops us as we go about our day-to-day lives. Ozone in the troposphere is a pollutant, but in the stratosphere, some 50 km above the earth, ozone (O_3) forms a layer called the **ozone shield** that absorbs the ultraviolet (UV) rays of the sun so that they do not strike the earth. UV radiation causes mutations that can lead to skin cancer and can make the lens of the eyes develop cataracts. It also is believed to adversely affect the immune system and our ability to resist infectious diseases. UV radiation impairs crop and tree growth and also kills off plankton (microscopic plant and animal life) that sustain oceanic life. Without an adequate ozone shield, our health and food sources are threatened.

Depletion of the ozone layer within the stratosphere in recent years is of serious concern. It became apparent in the 1980s that some worldwide depletion of ozone had occurred and that there was a severe depletion of some 40–50% above the Antarctic every spring. Severe depletions of the ozone layer are commonly called "ozone holes." Detection devices now tell us that there is an ozone hole above the Arctic as well, and ozone holes could also develop within northern and southern latitudes where many people live. Whether or not these holes develop depends on prevailing winds, weather conditions, and the type of particles in the atmosphere. A United Nations Environment Program report predicts a 26% rise in cataracts and nonmelanoma skin cancers for every 10% drop in the ozone level. A 26% increase translates into 1.75 million additional cases of cataracts and 300,000 more cases of skin cancer every year, worldwide.

The cause of ozone depletion can be traced to the release of chlorine atoms (Cl) into the stratosphere (Fig. 24.8). Chlorine atoms combine with ozone and strip away the oxygen atoms one by one. One atom of chlorine can destroy up to 100,000 molecules of ozone before settling to the earth's surface as chloride years later. These chlorine atoms come from the breakdown of chlorofluorocarbons (CFCs), chemicals much in use by humans. The best known CFC is Freon, a heat transfer agent found in refrigerators and air conditioners. CFCs are also used as cleaning agents and foaming agents during the production of styrofoam found in coffee cups, egg cartons, insulation, and paddings. Formerly, CFCs were used as propellants in spray cans, but this application is now banned in the United States and several European countries. Most countries of the world have agreed to stop using CFCs by the year 2000, but the United States halted production by 1995. Scientists are now searching for CFC substitutes that will not release chlorine atoms (nor bromine atoms) to harm the ozone shield.

Acid Deposition Destroys Ecosystems

The coal and oil routinely burned by power plants release sulfur dioxide into the air. Kuwait oil has a high sulfur content, and therefore, the oil-well fires started during the Persian Gulf War released much sulfur dioxide into the

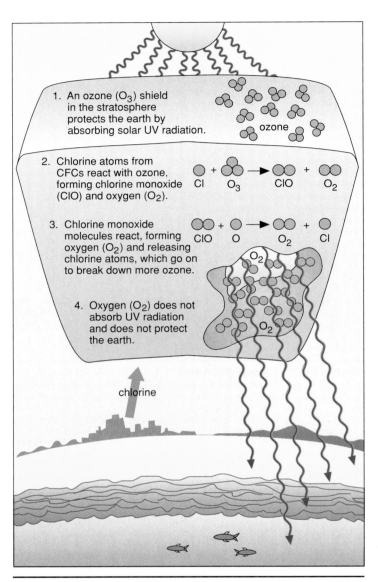

Figure 24.8   Ozone hole.
The development of an ozone hole due to the release of chlorine atoms from CFCs.

atmosphere. Automobile exhaust routinely puts nitrogen oxides in the air. Both sulfur dioxide and nitrogen oxides are converted to acids when they combine with water vapor in the atmosphere, a reaction that is promoted by ozone in smog. These acids return to earth as either wet deposition (acid rain or snow) or dry deposition (sulfate and nitrate salts).

Acid deposition is now associated with dead or dying lakes and forests, particularly in North America and Europe (Fig. 24.9). Acid deposition also corrodes marble, metal, and stonework, an effect that is noticeable in cities. It can also degrade our water supply by leaching heavy metals from the soil into drinking-water supplies. Similarly, acid water dissolves copper from pipes and lead from lead solder, which is used to join pipes.

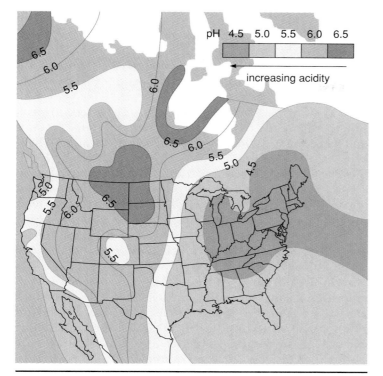

Figure 24.9   Acid deposition in the United States.
The numbers in this map are the average pH of precipitation. The northeast is the hardest hit area because winds carry the acid pollutants from other parts of the country in this direction.

a.   b.

Figure 24.10   Ozone effects.
a. This milkweed was exposed to ozone and appears unhealthy—leaf mottling is apparent. **b.** This milkweed was grown in an enclosure with filtered air and appears healthy.

Photochemical Smog Affects Health

Photochemical smog contains two air pollutants—nitrogen oxides (NO_x) and hydrocarbons (HC)—that react with one another in the presence of sunlight to produce ozone (O_3) and PAN (peroxyacetyl nitrate). Both nitrogen oxides and hydrocarbons come from fossil fuel combustion, but additional hydrocarbons come from various other sources as well, including paint solvents and pesticides.

Ozone and PAN are commonly referred to as oxidants. Breathing ozone affects the respiratory and nervous systems, resulting in respiratory distress, headache, and exhaustion. These symptoms are particularly apt to appear in young people; therefore, in Los Angeles, where ozone levels are often high, schoolchildren must remain inside the school building whenever the ozone level reaches 0.24 ppm (parts per million by weight). Ozone is especially damaging to plants, resulting in leaf mottling and reduced growth (Fig. 24.10).

Carbon monoxide (CO) is another gas that comes from the burning of fossil fuels in the industrial Northern Hemisphere. High levels of carbon monoxide promote the formation of ozone. Carbon monoxide also combines preferentially with hemoglobin and thereby prevents hemoglobin from carrying oxygen. Breathing large quantities of automobile exhaust can even result in death because of this effect. Of late, it has been discovered that the amount of carbon monoxide over the Southern Hemisphere—from the burning of tropical forests—is equal to that over the Northern Hemisphere.

Normally, warm air near the ground is able to escape into the atmosphere. Sometimes, however, air pollutants, including smog and soot, are trapped near the earth due to a long-lasting thermal inversion. During a **thermal inversion,** there is the cold air at ground level beneath a layer of warm, stagnant air above. This often occurs at sunset because turbulence usually mixes these layers during the day. Some areas surrounded by hills are particularly susceptible to the effects of a temperature inversion because the air tends to stagnate and there is little turbulent mixing.

Air pollutants are involved in causing four major environmental effects: global warming, ozone shield destruction, acid deposition, and photochemical smog. Each pollutant may be involved in more than one of these.

While each of these environmental effects is bad enough when considered separately, they actually feed on one another, making the total effect much worse than is predicted for each separately.

Water Use and Pollution

Fresh water is required not only for domestic purposes, including drinking water, but also for crop irrigation, industrial use, and energy production. Surface water from rivers, lakes, and underground rivers called **aquifers** are used to meet these needs (Fig. 24.11). The water in aquifers is a vast natural resource, but to ensure a continual supply, withdrawals cannot exceed deposits. In this country, the farmers of the Midwest withdraw water from aquifers up to 50 times faster than nature replaces it. China, with a population of one billion, is also mining its water to meet the needs of its people, despite the estimate that its aquifers can sustain only 650 million people. In the United States, the government still heavily subsidizes water so that the incentive to use water carefully and efficiently is lacking.

Pollution of surface water, groundwater, and the oceans is another reason why we are running out of fresh water.

Surface Water Pollution

All sorts of pollutants from various sources enter surface waters, as depicted in Figure 24.12. Sewage treatment plants help degrade organic wastes, which otherwise can cause oxygen depletion in lakes and rivers. As the oxygen level

Figure 24.11   Crop irrigation.
Forty percent of the water used for irrigation comes from under the ground. In the Texas high plains, central Arizona, and southern Florida, withdrawals from aquifers exceed any possibility of recharge. This is called "groundwater mining," which causes sinkholes due to the collapse of underground caverns and saltwater intrusion into freshwater supplies.

decreases, the diversity of life is greatly reduced. Also, human feces can contain pathogenic microorganisms that cause cholera, typhoid fever, and dysentery. In less-developed countries, where the population is growing and where waste treatment is practically nonexistent, many children die each year from these diseases.

Typically, sewage treatment plants use bacteria to break down organic matter to inorganic nutrients, like nitrates and phosphates, which then enter surface waters. These types of nutrients, which also can enter waters by fertilizer runoff and soil erosion, lead to *cultural eutrophication,* an acceleration of the natural process by which bodies of water fill in and disappear. First, the nutrients cause overgrowth of algae. Then, when the algae die, oxygen is used up by the decomposers, and the water's capacity to support life is reduced. Massive fish kills are sometimes the result of cultural eutrophication.

Industrial wastes include heavy metals and organochlorides, such as those in some pesticides. These materials are not degraded readily under natural conditions or in conventional sewage treatment plants. Sometimes, they accumulate in the mud of deltas and estuaries of highly polluted rivers and cause environmental problems if they are disturbed. Industrial pollution is being addressed in many industrialized countries but usually has low priority in less-developed countries.

As discussed previously, some pollutants enter bodies of water from the atmosphere. Acid deposition has caused many lakes to become sterile in the industrialized world because acid leaches aluminum and iron out of the soil. A high concentration of these ions kills fishes and other forms of aquatic life. Lime is sometimes helpful against acidification of a lake.

Groundwater Pollution

In areas of intensive farming or where there are many septic tanks, ammonium (NH_4^+) released from animal and human waste is converted by soil bacteria to soluble nitrate, which moves down through the soil (percolates) into underground water supplies. Between 5 and 10% of all wells examined in the United States have nitrate levels higher than the recommended maximum.

Industry also pollutes aquifers. Previously, industry ran wastewater into a pit, from which pollutants could seep into the ground. Wastewater and chemical wastes were also injected into deep wells, from which pollutants constantly discharged. Both of these customs have been, or are in the process of being, phased out. It is very difficult for industry to find other ways to dispose of wastes, especially since citizens do not wish to live near waste treatment plants. As mentioned previously, the emphasis today is on prevention of wastes in the first place. Industry is trying to use processes that do not create wastes and/or to recycle the wastes they do generate.

Pollution of the Oceans

Coastal regions are not only the immediate receptors for local pollutants, they are also the final receptors for pollutants carried by rivers that empty at a coast (Fig. 24.12). Waste dumping also occurs at sea, and ocean currents sometimes transport both trash and pollutants back to shore. Examples are the nonbiodegradable plastic bottles, pellets, and containers that now commonly litter beaches and the oceans' surfaces. Some of these, such as the plastic that holds a six-pack of cans, cause the death of birds, fishes, and marine mammals that mistake them for food and get entangled in them.

Offshore mining and shipping add pollutants to the oceans. Some 5 million metric tons of oil a year—or more than one gram per 100 square meters of the oceans' surfaces—end up in the oceans. Large oil spills kill plankton, fish fry, and shellfishes, as well as birds and marine mammals. The largest tanker spill in U.S. territorial waters occurred on March 24, 1989, when the tanker *Exxon Valdez* struck a reef in Alaska's Prince William Sound and leaked 44 million liters of crude oil. During the Persian Gulf War, 120 million liters were released from damaged onshore storage tanks into the Persian Gulf—an event that was called environmental terrorism. Although petroleum is biodegradable, the process takes a long time because the low-nutrient content of seawater does not support a large bacterial population. Once the oil washes up onto beaches, it takes many hours of work and millions of dollars to clean it up.

Losing Marine Biodiversity

Life-forms are distributed in the sea in various habitats, from the coast to the open seas and from the surface waters to the interface between soil and water. In the last 50 years, we have polluted the seas and exploited their resources to the point that many species are literally at the brink of extinction. Fisheries once rich and diverse, such as George's Bank off the coast of New England, are in severe decline. Haddock was once the most abundant species in this fishery, and now it

accounts for less than 2% of the total catch. Cod and blue-fin tuna have suffered a 90% reduction in population size. In warm, tropical regions, many areas of coral reefs are now overgrown with algae because the fish that normally keep the algae under control have been killed off. Corals cannot regrow, and the result is a reef devoid of the many kinds of animals that normally live there. Sharks are in decline particularly because their fins bring such a high price in Asia that fishermen cut off the two valuable fins and toss the helpless animal back in the sea to die.

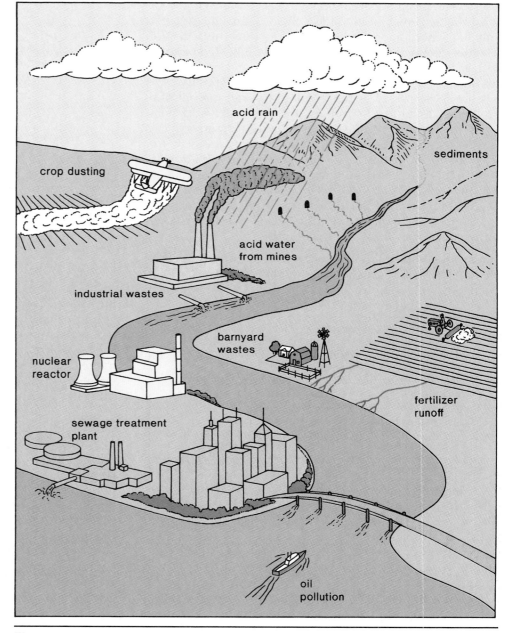

Figure 24.12 Sources of water pollution.
Many bodies of water are dying due to the introduction of sediments and surplus nutrients.

Land Pollution

Whereas 20% of the world's population lived in cities in 1950, it is predicted that 60% will live in cities by the year 2000. Near cities, the development of new housing areas has led to urban sprawl. Such areas tend to take over agricultural land or simply further degrade the land in the area. The land has been degraded in many ways. Here, we will discuss some of the greatest concerns.

Waste Disposal and Dangerous Trash

Every year, the U.S. population discards billions of tons of solid wastes, much of it on land. Solid wastes include not only household trash, but also sewage sludge, agricultural residues, mining refuse, and industrial wastes. Open dumping, sanitary landfills, or incineration have been the most common practices of disposing trash. These disposal methods are increasingly expensive and also cause pollution problems. It would be far more satisfactory to recycle materials as much as possible and/or to use organic substances as a fuel to generate electricity. One study showed that it was possible to achieve 70–90% public participation in recycling by spending only thirty cents per household. A city the size of Washington, D.C., where 500,000 tons of waste are generated per year, could have an increase of 1,300 jobs if the community utilized solid wastes as a resource instead of throwing them away.

Hazardous Wastes   Solid wastes include various hazardous wastes, which can endanger our health. Some hazardous wastes include the following:

Heavy metals, such as lead, mercury, cadmium, nickel, and beryllium, contaminate many wastes. These metals can accumulate in various organs, interfering with normal enzymatic actions and causing illness.

Chlorinated hydrocarbons, also called organochlorides, include various pesticides and PCBs (polychlorinated biphenyls), which are often cancer-producing in laboratory animals.

Nuclear waste includes radioactive elements that will be dangerous for thousands of years. ^{239}Plutonium takes 2,000,000 years to lose its radioactivity.

These wastes enter bodies of water and are subject to **biological magnification** (Fig. 24.13). Decomposers are unable to break down these wastes. They enter and remain in the body because they are not excreted. Therefore, they become more concentrated as they pass along a food chain. Notice in Figure 24.13 that the dots representing DDT become more concentrated as they pass from producer to tertiary consumer. Biological magnification is most apt to occur in aquatic food chains, since there are more links in aquatic food chains than there are in terrestrial food chains. Humans are the final consumers in both types of food chains, and in some areas, human milk contains detectable amounts of DDT and PCBs, which are organochlorides.

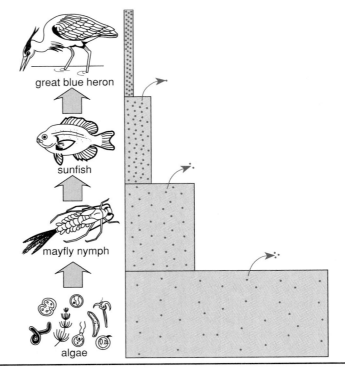

Figure 24.13   Biological magnification.
A poison (dots), such as DDT, which is minimally excreted (arrows), becomes maximally concentrated as it passes along a food chain due to the ever smaller biomass of each higher trophic level.

The United States spends $9 billion a year on cleanup but only $200 million yearly to prevent contamination. It would be best to place the emphasis on preventing contaminants from entering the environment rather than cleaning up pollution. Likewise, recycling can save industry money, as witnessed by the 3M Corporation, which reported a savings of $1.2 billion by recycling waste and preventing pollution.

Soil Erodes and Deserts Grow

In agricultural areas, wind and rain carry away about 25 billion tons of topsoil yearly, worldwide. If this rate of loss continues, we will lose practically all our topsoil by the middle of the next century. Soil erosion causes a loss of productivity that is compensated by increased use of fertilizers, pesticides, and fossil fuel energy. One answer to the problem of erosion is to adopt soil conservation measures. For example, farmers could use strip-cropping and contour farming (Fig. 24.14). Thanks to a U.S. program offering incentives to farmers to use these techniques, erosion is being significantly reduced.

Desertification is the transformation of marginal lands to desert conditions because of overgrazing and overfarming. Desertification has been particularly evident along the southern edge of the Sahara Desert in Africa, where it is estimated that 240,000 square miles of once-productive grazing land has become desert in the last 50 years. However, desertification also occurs in this country. The U.S. Bureau of Land Management, which opens up federal lands for grazing, reports that much of the rangeland it manages is in poor or bad condition, with much of its topsoil gone and with greatly reduced ability to support forage plants.

The dumping of solid wastes contributes to water and land pollution. Agricultural land quality is threatened by soil erosion, which can lead to desertification.

Losing Terrestrial Biodiversity

Loss of habitat due to human encroachment and destruction of natural ecosystems is the chief reason terrestrial biodiversity is declining.

Wars Cause Land Spoilage   Many plants and animals are killed because of wars. For example, during a civil war in Angola, rhinos and elephants were shot, and their tusks and horns were sold to buy uniforms and weapons. During a war between Uganda and Tanzania, hippos were used as target practice, and other animals were killed for food and for their ivory tusks.

However, habitat destruction caused by wars probably leads to more loss of life than direct killing. Scientists estimate that during the Vietnam War, 5.45 million acres of tropical forests were destroyed due to bombing, bulldozing, and the spraying of defoliants such as Agent Orange. Also, about half of South Vietnam's wetlands are now devoid of their mangrove trees due to the effects of war. The Persian Gulf War was similarly destructive. Although the sands of Kuwait may seem lifeless, they are actually home to a variety of spiders, snakes, and scorpions, as well as sheep, gazelles, and camels.

Deforestation of Coniferous and Tropical Forests   In Canada, vast stands of trees are scheduled to be felled and turned into paper and wood products such as posts and particleboard. Thousands of miles of new logging roads are to be built during the next five to ten years, and this no doubt will bring many visitors to hunt and fish where only indigenous people did so previously. The animals that live there—moose, porcupine, lynx, and snowshoe hare—will also be displaced. Songbirds who migrate there during the summer will no longer find shelter. Although the logging companies are to replant, conservationists wonder if companies have the necessary expertise or if wildlife can sustain themselves in the meantime.

Figure 24.14   Contour farming.
Crops are planted according to the lay of the land to reduce soil erosion. This farmer has planted alfalfa in between the strips of corn. The roots of alfalfa, a legume, hold the soil, and its root nodules contain nitrogen-fixing bacteria, which help to replenish the nitrogen content of the soil.

Tropical rain forests are much more biologically diverse than temperate forests. For example, temperate forests across the entire United States contain about 400 tree species. In the rain forest, a typical ten-hectare area holds as many as 750 types of trees. The fresh waters of South America are inhabited by an estimated 5,000 fish species; on the eastern slopes of the Andes, there are 80 or more species of frogs and toads; and in Ecuador, there are more than 1,200 species of birds—roughly twice as many as those inhabiting all of the United States and Canada. Therefore, a very serious side effect of deforestation in tropical countries is the loss of biological diversity.

A National Academy of Sciences study estimated that a million species of plants and animals are in danger of disappearing within 20 years as a result of deforestation in tropical countries. Many of these life-forms have never been studied, and yet they may be useful sources of food or medicines.

Logging of tropical forests occurs because industrialized nations prefer furniture made from costly tropical woods and because people want to farm the land (Fig. 24.15). In Brazil, the government allows citizens to own any land they clear in the Amazon forest (along the Amazon River). When they arrive, the people practice *slash-and-burn agriculture,* in which trees are cut down and burned to provide nutrients and space to raise crops. Unfortunately, the fertility of the land is sufficient to sustain agriculture for only a few years. Once the cleared land is incapable of sustaining crops, the farmer moves on to another part of the rain forest to slash and burn again. In the meantime, cattle ranchers move in. Cattle ranchers are the greatest beneficiaries of deforestation, and increased ranching is therefore another reason for tropical rain forest destruction. A newly begun pig-iron industry in Brazil also indirectly results in further exploitation of the rain forest. The pig iron must be processed before it is exported, and smelting the pig iron requires the use of charcoal (burnt wood).

There is much concern worldwide about the loss of biological diversity due to the destruction of tropical rain forests. The myriad plants and animals that live there could possibly benefit human beings.

Forest Destruction	
Loss of a CO_2 sink:	forests take up carbon dioxide from the atmosphere
Loss of biodiversity:	forests are home for plants and animals
Loss of possible medicinal plants:	tropical rain forests contain unique plants
Soil erosion:	trees hold the soil
Water pollution:	saw mills and paper mills pollute
Ecosystems destruction:	first step toward converting land to industrialized or urbanized areas

Figure 24.15   Forest destruction.
Forest destruction leads to the detrimental effects mentioned.

Ecology Focus

Ten Ways to Reduce Your Impact on the Environment

Norman Dean walks down an aisle of a busy Washington, D.C., supermarket and stops next to some shelves filled with household cleaners. "This one claims to be 'environmentally safe,'" he says, squinting at the label of one product. He points to another plastic bottle: "That one says it's 'biodegradable,' and look, here's an aerosol cleaner that claims to be 'ozone friendly.' What exactly does that mean?"

All across the country, millions of American consumers are asking that same question these days as more and more companies jump on the "green" bandwagon. "There are no uniform accepted definitions for many of these claims," says Dean, "so you can't take them at face value."

Dean is president of Green Seal, a nonprofit organization based in the nation's capital that is helping shoppers separate the honest claims from the false ones by certifying legitimate environmentally friendly products. The Green Seal logo is increasingly cropping up on products that pass muster. But Dean points out that even when you don't know which brand name to buy, you can still reduce your impact on the environment. Here are ten ways to do so.

1. Avoid excessive packaging, especially individually wrapped servings. Choose packaging that is manufactured from recycled materials. Or look for products sold in packaging that can be recycled, like aluminum, glass, or cardboard. Or better yet, buy products that have eliminated extra packaging altogether.

2. If you buy products packaged in plastic, look for those that you can recycle. Almost every plastic container manufactured these days has a code number on the bottom that indicates the type of plastic. Some types can be recycled, others cannot. If the container is not numbered, it may contain ingredients that prevent it from being recycled. Ask your local government's recycling department or community recycling center which types of plastic are recycled locally.

3. Use compact fluorescent light bulbs (CFLs). An 18-watt CFL can provide the same amount of light as a standard 60-watt bulb. Though CFLs cost more initially, they last nearly 10 times longer than incandescent bulbs, saving you at least $15 (after replacement costs) over each bulb's lifetime. What's more, because of its energy efficiency, during its lifetime each CFL bulb will spare the Earth as many as 1,500 pounds of carbon dioxide and nearly 20 pounds of sulfur dioxide from a coal-fired power plant smokestack. If you can't use CFLs throughout your house, try to use them in those light fixtures that you leave on for long periods.

4. Buy unbleached or chlorine-free paper products such as coffee filters and bath tissue. The industrial chlorine-bleaching process used to whiten paper products can create dioxin, a dangerous toxic substance that gets into our lakes, rivers, and streams.

5. To help reduce solid waste, look for labels on recycled paper products indicating a percentage of post-consumer waste, which should be no less than 20% for most products (10% for facial tissue). Some companies recycle only scrap paper from their manufacturing process and claim that their products are "recycled." Postconsumer waste refers to paper that someone has actually used and returned.

6. When you must buy products containing hazardous substances, buy only the amount you need. Try less-hazardous alternatives to common household products, such as washing windows with equal parts of water and vinegar instead of glass cleaner, and spraying equal parts of vinegar and lemon juice instead of commercial air freshener.

7. To help reduce toxic waste, use rechargeable alkaline batteries. Most of the nearly 3 billion household batteries purchased and disposed of annually in the United States contain toxic metals that can be dangerous to human health if the metals leach out of landfills. Though initially more expensive, rechargeable alkalines can be reused as many as 25 times.

8. Experiment with using less laundry soap than manufacturers recommend—even as little as half—to cut down on packaging waste. "Your wash most likely will be just as clean," says Dean.

9. To save energy, wash your clothes in cold water. "Most modern detergents are designed to work effectively in cold water," adds Dean.

10. Buy garden hoses labeled as being made from recycled tire rubber. This helps reduce landfill overflow and tire burning, which emits hazardous pollutants into the air. And when possible, use soaker hoses. By "weeping" water through thousands of tiny pores, such hoses use 50–75% less water than most conventional watering systems.

24.4   A Sustainable World

While it may seem like an either-or situation—either preservation of ecosystems or human survival—there are some who believe that this is not the case. Natives who harvest rubber from the trees of a tropical rain forest can have a sustainable income from the same trees year after year. A recent study calculated the market value of rubber and exotic produce, like the aguaje palm fruit, that can be harvested continually from the Amazon forest. It concluded that selling these products would yield more than twice the income from either lumbering or cattle ranching, and it would help achieve a sustainable income (Fig. 24.16).

While we are sometimes quick to realize that the growing populations of the LDCs are putting a strain on the environment, we should realize that the excessive resource consumption of the MDCs also stresses the environment. Environmental impact is measured not only in terms of population size, it is also measured in terms of the pollution caused by each person in the population. An average American family, in terms of per capita resource consumption and waste production, is the equivalent of 30 people in a less-developed country. However, it is possible for each of us to reduce our impact on the environment by following the guidelines given in the Ecology reading on page 503.

Before the establishment of industrialized societies, people felt connected to the plants and animals on which they depended, and they were then better able to live in a sustainable way. After the industrial revolution, we began to think of ourselves as separate from nature and endowed with the right to exploit nature as much as possible. But our industrial society lives on *borrowed carrying capacity*—our cities not only borrow resources from the country, our entire population borrows from the past and future. The forests of the Carboniferous have become the fossil fuels that sustain our way of life today, and the environmental degradation we cause is going to be paid for by our children.

Ecologists have two favorite sayings: (1) everything is connected to everything else, and (2) there is no free lunch. We have seen that if you affect one part of the carbon cycle, you affect the entire balance of carbon in the entire world. Ecological effects know no boundaries. Coal that is burned in the Midwest releases acids into the atmosphere that affect lakes in the Northeast. And plants and animals aren't the only organisms affected. Humans are dependent on natural cycles just as much as any organism in the biosphere. What we do to natural ecosystems will eventually be felt by us also. The second saying means that we have to pay for what we do. If we build a home on a flood plain, we can expect that it will be flooded once in a while. When we burn fossil fuels, we can expect acid rain and global warming as a consequence. Many times it is difficult to predict the particular consequences, but we can be assured that eventually they will become apparent.

Overpopulation and overconsumption account for increased pollution and also for the mass extinction of wildlife

Figure 24.16   Sustainable life in the tropics.
In its natural state, a tropical rain forest is immensely rich in wildlife. If the rubber trees are not cut down, rubber tappers can earn a sustainable living by tapping the same trees year after year.

that is going on. We are expected to lose one-third to two-thirds of the earth's species, any one of which could possibly have made a significant contribution to agriculture or medicine. It should never be said, "What use is this organism?" Aside from its contribution to the ecosystem in which it lives, one never knows how a particular organism might someday be useful to humans. Adult sea urchin skeletons are now used as molds for the production of small artificial blood vessels, and armadillos are used in leprosy research.

It is clearly time for a new philosophy. In a **sustainable world,** development will meet economic needs of all peoples while protecting the environment for future generations. Various organizations have singled out communities to serve as models of how to balance ecological and economic goals. For example, in Clinch Valley of southwest Virginia, the Nature Conservancy is helping to revive the traditional method of logging with draft horses. This technique, which allows the selective cutting of trees, preserves the forest and prevents soil erosion, which is so damaging to the environment. The United Nations has an established bioreserve system, a global network of sites that combine preservation with research on sustainable management for human welfare. More than 100 countries are now participants in the program.

All peoples can benefit from a sustainable world, where economic development and environmental preservation are considered complementary rather than opposing processes.

SUMMARY

24.1 Exponential Population Growth

The human population is expanding exponentially, and it is unknown when growth will level off. Presently, each year exhibits a large increase, and the doubling time is now about 47 years.

Populations have a biotic potential for increase in size. Biotic potential is normally held in check by environmental resistance, thereby producing an S-shaped growth curve, leveling off at the carrying capacity of the environment.

24.2 Human Population Growth

The MDCs underwent a demographic transition between 1950 and 1975, but the LDCs are just now undergoing demographic transition. Due particularly to increases in Africa, Asia, and Latin America, an explosion in LDC's populations from 4.3 billion to 10.2 billion is expected. Support for family planning, human development, and delayed childbearing could help prevent such a large increase.

24.3 Human Population and Pollution

An increasing human population is causing land, water, and air pollution. Various substances are associated with air pollution, such as sulfur dioxide, hydrocarbons, nitrogen oxides, methane, chlorofluorocarbons (CFCs), carbon dioxide, and carbon monoxide. These have various sources, but most come from vehicle exhaust and fossil fuel burning. Some of these, particularly CO_2, are greenhouse gases, which trap heat and lead to global warming. Some, particularly CFCs, cause destruction of the ozone shield. Some, particularly sulfur dioxide and nitrogen oxides, react with water vapor to form acids that contribute to acid deposition. Some, particularly hydrocarbons and nitrogen oxides, react to form photochemical smog.

The dumping of solid wastes, some of which are hazardous, is a cause of water pollution. Hazardous wastes, including metals, organochlorides, and nuclear wastes, may contaminate water supplies and are subject to biological magnification. The oceans are the final recipients of wastes deposited in rivers and along the coasts. Marine biodiversity is at risk because of pollution and overfishing.

Soil erosion reduces the quality of land and leads to desertification. Loss of terrestrial biodiversity is in part caused by wars and deforestation. The Canadian forests and the tropical rain forests in Southeast Asia and Oceania, Central and South America, and Africa are being cut to provide wood for export. Slash-and-burn agriculture also reduces tropical rain forests.

24.4 A Sustainable World

In a sustainable world, economic development will be tied to environmental preservation, and we will no longer borrow carrying capacity from past ages and future generations.

STUDYING THE CONCEPTS

1. Draw a growth curve to represent exponential growth, and explain why a curve representing population growth usually levels off.   490

2. Calculate the growth rate and the doubling time for a population in which the birthrate is 20 per 1,000 and the death rate is 2 per 1,000.   491

3. Distinguish between the MDCs and the LDCs. Include a reference to age-structure diagrams.   492–93

4. Explain why the population of LDCs is expected to increase tremendously. What steps could be taken to prevent this from occurring?   492–93

5. What substances contribute to air pollution? What are their sources? Which ones are associated with photochemical smog, acid deposition, global warming, and destruction of the ozone shield?   494

6. How do the greenhouse gases bring about global warming? How do CFCs break down the ozone shield? How do acids develop in the atmosphere? How does photochemical smog develop?   494–97

7. What are several ways in which underground water supplies can be polluted?   498

8. Describe and discuss two reasons for the loss of marine biodiversity.   499

9. What are three types of hazardous wastes that contribute to pollution on land? What is biological magnification?   500

10. Describe three ways the quality of the land is being degraded today. What is desertification?   500–1

11. Describe and discuss two reasons for the loss of terrestrial biodiversity.   501

12. How do less-developed countries (LDCs) contribute to environmental degradation? More-developed countries (MDCs)?   494–504

13. What is meant by the expression, "sustainable world"?   504

APPLYING YOUR KNOWLEDGE

Concepts

1. What are the two factors, either of which must change, in order to decrease the growth rate? Explain.

2. How long would it take to reduce the world's population if the growth rate is reduced to zero? Explain.

3. Humans, as well as other animals, have been dumping their wastes into the environment for thousands of years. What is the reason(s) that this appears to be such a problem today?

4. Some individuals believe that the carrying capacity of the earth is between 50–100 billion people; others believe that the present population of 5.5–6 billion people already exceeds the earth's carrying capacity. How is it possible for so-called "experts" to arrive at such different numbers?

Bioethical Issue

The "green clean," as some people call the environmental movement, hasn't always been easy for industry. Most companies, for example, have had to cut way back on waste production or find ways to recycle them. Entire countries have had to prioritize pollution control budgets—cutting the grime from transportation and tracking global warming are examples.

Along the way, some business people and politicians have argued that pollution control is too costly, even with its presumed health benefits. Company attorneys have claimed that costly pollution controls will choke the profit from industry. Public-health experts, on the other hand, have said that you can't put a price on cutting pollution.

What do you think? Is it appropriate for companies—and countries—to analyze the risks and benefits of pollution control before committing to it? Is pollution rightly a budget issue—or is it more important than that? Who has the responsibility to see that pollution stays in check in your community?

TESTING YOUR KNOWLEDGE

1. When a population is undergoing exponential growth, the increase in number of people each year is _____ (higher, lower) than the year before.

2. Label this S-shaped growth curve.

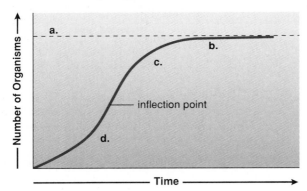

3. After a country has undergone the demographic transition, the death rate and the birthrate are both _____ (high, low).

4. If a country has a pyramid-shaped age-structure diagram, most individuals are _____ (prereproductive, reproductive, or postreproductive).

5. The designation LDC suggests that these countries are not as _____ as MDCs.

6. The gas best associated with global warming is _____.

7. The chemicals best associated with ozone depletion are _____.

8. Sewage is biodegradable, but the nutrients released in the process can lead to _____ of surface waters.

9. Pesticides and radioactive wastes are both subject to biological _____.

10. Match the boxed terms shown in the following illustration with one of the environmental problems in the key.

   Key:

   sulfur dioxide   carbon dioxide

   hydrocarbons   CFCs

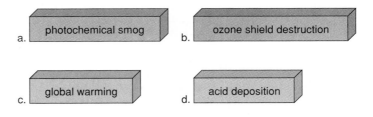

11. In a sustainable world, it is necessary to _____ the environment.

APPLYING TECHNOLOGY

Your study of population is supported by these available technologies:

Exploring the Internet

The Mader Home Page provides further resources for studying this chapter.

`http://www.mhhe.com/sciencemath/biology/mader/`

(Click on *Human Biology*.)

Explorations in Human Biology CD-ROM

Pollution of a Freshwater Lake (#16) Students vary the number of sewage treatment plants on Lake Washington and observe the effect on amount of sewage discharged, algal growth, dissolved oxygen, and phosphorus levels. **(2)***

*Level of difficulty

SELECTED KEY TERMS

acid deposition Acid rain or snow due to the presence of sulfate and/or nitrate produced by commercial and industrial activities.   496

aquifer (AHK-wuh-fur) Water-bearing stratum of permeable rock that constitutes an underground reservoir.   498

biological magnification Process by which nonexcreted substances, such as DDT, become more concentrated in organisms in the higher trophic levels of the food chain.   500

biotic potential Maximum population growth rate under ideal conditions.   491

carrying capacity Largest number of organisms of a particular species that can be maintained indefinitely in an ecosystem.   491

demographic transition Decline in the birthrate following a reduction in the death rate so that the population growth rate is lowered.   492

desertification (di-zurt-uh-fuh-KAY-shun) Transformation of marginal lands to desert conditions.   500

doubling time Number of years it takes for a population to double in size.   491

environmental resistance Sum total of factors in the environment that limit the numerical increase of a population in a particular region.   491

exponential growth Growth, particularly of a population, in which the increase occurs in the same manner as compound interest.   490

greenhouse effect Reradiation of solar heat toward the earth, caused by an accumulation of gases such as carbon dioxide in the atmosphere.   494

growth rate The yearly percentage of increase or decrease in the size of a population.   490

ozone shield Formed from oxygen in the upper atmosphere, ozone (O_3) protects the earth from ultraviolet radiation.   496

photochemical smog Air pollution that contains nitrogen oxides and hydrocarbons, which react to produce ozone and PAN (peroxyacetyl nitrate).   497

pollutant Substance that is added to the environment and leads to undesirable effects for living organisms.   494

replacement reproduction Population in which each person is replaced by only one child.   493

sustainable world Global way of life that can continue indefinitely because the economic needs of all peoples are met while still protecting the environment.   504

thermal inversion Temperature inversion that traps cold air and its pollutants near the earth, with the warm air above it.   497

FURTHER READINGS FOR PART SEVEN

Allen, J. L., editor. *Environment 96/97.* 1996. Guilford, Conn.: Dushkin Publishing Group/Brown & Benchmark Publishers. This volume is a collection of articles pertaining to the problems and issues of the environment and environmental quality.

Anderson, D. M. August 1994. Red tides. *Scientific American* 271(2):62. The frequency of toxic red tides has been increasing because pollution provides rich nutrients, which encourages algal bloom.

Balick, M. J., and Cox, P. A. 1996. *Plants, people, and culture: The science of ethnobotany.* New York: *Scientific American* Library. This interesting, well-illustrated book discusses the medicinal and cultural uses of plants, and the importance of rain forest conservation.

Begon, M., et al. 1996. Ecology: Individuals, populations, and communities. London: Blackwell Science Ltd. This text discusses the distribution and abundance of organisms and their physical and chemical interactions with their ecosystems.

Blaustein, A. R., and Wake, D. B. April 1995. The puzzle of declining amphibian populations. *Scientific American* 272(4):52. Many species are disappearing; this may be a result of changes in the environment, including pollution, disease, and diet.

Broecker, W. S. November 1995. Chaotic climate. *Scientific American* 273(5):62. Researchers are beginning to understand past climate patterns; this knowledge may be used to predict future patterns.

Charlson, R., and Wigley, T. L. February 1994. Sulfate aerosol and climatic change. *Scientific American* 270(2):48. Sulfur aerosols scatter sunlight back into space before it can contribute to global warming.

Cowen, R. 1995. *The history of life*. 2d ed. Boston: Blackwell Scientific Publications. From the origin of life through human evolution, this is an introduction to paleontology and paleobiology.

Cox, G. 1997. *Conservation Ecology*. 2d. ed. Dubuque, Iowa: Wm. C. Brown Publishers. Discusses the nature of the biosphere, the threats to its integrity, and ecologically sound responses.

Cox, P. A., and Balick, M. J. June 1994. The ethnobotanical approach to drug discovery. *Scientific American* 270(6):82. Many rain forest plants are used by indigenous cultures for medicinal purposes; these flora should be screened for pharmaceutical compounds.

Cranbrook, E., and Edwards, D. S. 1994. *Belalong: A tropical rain forest*. London: The Royal Geographic Society, and Singapore: Sun Tree Publishing. This is a readable, well-illustrated account of biodiversity in a Brunei rain forest.

Cunningham, W. P., and Saigo, B. W. 1997. *Environmental science: A global concern*. Dubuque, Iowa: Wm. C. Brown Publishers. Provides scientific principles plus insights into the social, political, and economic systems impacting the environment.

Dawkins, R. November 1995. God's utility function. *Scientific American* 273(5):80. The role of genetics in evolution and natural selection is discussed.

Dobson, A. P. 1996. *Conservation and biodiversity*. New York: *Scientific American* Library. Discusses the extent and the value of biodiversity, and describes attempts to manage endangered species.

Frosch, R. A. September 1995. The industrial ecology of the 21st century. *Scientific American* 273(3):178. Taking lessons from the natural ecological system, industries may be able to recycle materials and minimize waste.

Futuyma, D. J. 1995. *Science on trial: The case for evolution*. Sunderland, Mass.: Sinauer Associates. Presents the evidence for evolution and the operation of the evolutionary process versus that proposed by the doctrine of creation.

Getis, J. 1991. *You can make a difference: Help protect the earth*. Dubuque, Iowa: Wm. C. Brown Publishers. Presents challenges and responses with regard to environmental problems and interesting alternatives to common household chemicals and cleaners.

Gore, R. January 1996. Neanderthals. *National Geographic* 189(1):2. Archeological finds are providing much information on the lives of these early humans.

Hammond, A. L., et al., editors. 1996. *World resources 1996-97: The urban environment*. New York: Oxford University Press. This report provides accurate information on urban environment management.

Hoagland, W. September 1995. Solar energy. *Scientific American* 273(3):170. Article discusses the use of solar energy to provide fuels.

Holloway, M. April 1994. Nurturing nature. *Scientific American* 270(4):98. Florida's Everglades are serving as a testing ground and battlefield for an epic attempt to restore an environment damaged by human activity.

Holloway, M. August 1994. Diversity blues. *Scientific American* 271(2):16. Toxic tides, coastal development, and pollution are increasing; oceanic biodiversity is suffering.

Horgan, J. October 1995. The new social Darwinists. *Scientific American* 273(4):174. Psychologists and social scientists apply principles of natural selection to mind and behavioral studies.

Johanson, D. C. March 1996. Face-to-face with Lucy's family. *National Geographic* 189(3):96. New fossils from Ethiopia provide more information about human evolution.

Klinkenborg, V. December 1995. A farming revolution. *National Geographic* 188(6):60. Farmers are experiencing strong crop yields by practicing farming methods that promote sustainable agriculture.

Leakey, M. September 1995. The dawn of humans. *National Geographic* 188(3):38. An East African species that walked upright 4 million years ago has been discovered.

Mayer, E. 1991. *One long argument: Charles Darwin and the genesis of modern evolutionary thought*. Cambridge, Mass.: Harvard University Press. Provides an excellent discussion of the history of evolutionary thinking.

Miller, P. December 1995. Jane Goodall. *National Geographic* 188(6):102. This article highlights the research of primatologist, Jane Goodall.

Mitchell, J. G. February 1996. Our polluted runoff. *National Geographic* 189(2):106. Eighty percent of U.S. water pollution is due to land runoff not resulting from municipal or industrial sources.

Odum, E. 1997. *A bridge between science and society*. 3d. ed. Sunderland, Mass.: Sinauer Associates. Introduces the principles of modern ecology as they relate to threats to the biosphere.

Plucknett, D. L., and Winkelmann, D. L. September 1995. Technology for sustainable agriculture. *Scientific American* 273(3):182. The practice of sustainable agriculture, increasing productivity while protecting the environment, will be difficult to achieve in developing countries.

Primak, R. B. 1995. *A primer of conservation biology*. Sunderland, Mass.: Sinauer Associates. The relatively new discipline of conservation biology addresses the alarming loss of biological diversity throughout the world.

Prugh, T. 1995. *Natural capital and human economic survival*. Solomons, Maryland: ISEE Press. A concise treatment detailing the necessity of incorporating the dynamics of ecosystems into economic and commercial systems.

Robison, B. H. July 1995. Light in the ocean's midwaters. *Scientific American* 273(1):60. New techology allows organisms found in the very deep, dark benthic regions of the ocean to be seen.

Strickberger, M. 1995. *Evolution*. 2d ed. Boston: Jones and Bartlett Publishers. Presents the basics of evolutionary theories.

Wenke, R. 1996. *Patterns in prehistory: Humankind's first three million years*. 4th ed. New York: Oxford University Press. Provides a comprehensive review of world prehistory.

Appendix A

ANSWER KEY

This appendix contains the answers to the Applying Your Knowledge questions and the Testing Your Knowledge questions, which appear at the end of each chapter. Answers to social issue questions can vary according to the student.

Introduction
Applying Your Knowledge: Concepts

1. The type of DNA, but not its arrangement, is the same in both; therefore, if a chemical affects one, it will probably affect the other.

2. When one undergoes a medical examination, homeostasis is being checked. The medical personnel run tests on urine and blood samples plus various other tests.

3. Humans have much more control over their environment than other animals. Their population is not as subject to environmental factors; thus, it is growing rapidly with an accompanying depletion of resources and production of pollutants.

Testing Your Knowledge

1. vertebrates; 2. solar; 3. copy; 4. cultural; 5. multicellular; 6. control; 7. hypothesis; 8. Society.

Chapter 1
Applying Your Knowledge: Concepts

1. In a covalent bond, two atoms are sharing the same electrons. In an ionic bond and a hydrogen bond, there is simply an attraction between the two ions.

2. As indicated, carbon dioxide combines with water to form carbonic acid, and this makes the carbonated drink acidic; once the bottle is opened, the carbon dioxide escapes and the pH increases.

3. The liver stores glucose as glycogen; in between eating, the liver releases glucose.

4. A protein can have four levels of structure; only the first level involves the sequence of amino acids. Obviously levels two to four must be affected by heating. Heating changes the shape of the enzyme and makes it ineffective.

Testing Your Knowledge

1. protons; 2. electrons, protons; 3. share; 4. hydrogen bonding; 5. Buffers; 6. increases, decreases; 7. ionizes; 8. glucose; 9. cellulose; 10. saturated; 11. glycerol; 12. enzymes; 13. peptide; 14. primary; 15. nucleotides; 16. double; 17. a. monomers, b. condensation, c. polymer, d. hydrolysis.

Chapter 2
Applying Your Knowledge: Concepts

1. Glycoprotein. These molecules serve to identify differences in individuals, such as blood type.

2. Lysosome. The lysosome contains hydrolytic digestive enzymes.

3. A microtubule contains tubulin proteins, not actin proteins.

4. In the case of human beings, the rest of the chemical energy is tied up in the two lactate molecules still remaining following fermentation.

Testing Your Knowledge

1. c; 2. e; 3. a; 4. d; 5. b; 6. a. nucleus—DNA directs; b. nucleolus—RNA helps; c. rough ER produces; d. smooth ER transports; e. Golgi apparatus packages and secretes; 7. cytoskeleton; 8. hypotonic; 9. active site; 10. electron transport system, cristae; 11. 2, 36; 12. a. Glycolysis; b. ATP; c. Transition reaction; d. CO_2; e. Krebs cycle; f. CO_2; g. ATP; h. Electron transport system; i. ATP; j. 1/2 O_2; k. H_2O. See also Fig. 2.15, page 54, text.

Chapter 3
Applying Your Knowledge: Concepts

1. Cartilage gives more flexibility during the birth process. And it allows easier growth in the skeleton of the newborn.

2. Smooth muscle functions without conscious input from the nervous system; it is found in many of the systems that carry on "housekeeping" duties that do not require a sudden response to a stimulus. Skeletal muscle is under conscious control and generally permits rapid response to a stimulus that has been monitored by the nervous system.

3. Shaving just cuts off the shaft at the skin level, and the hair shaft continues to project above the skin as cells are produced by the hair root. Pulling out the hair actually just pulls out some of the hair shaft located in the follicle and some of the follicular cells, and the hair root is still present. It just takes a longer period of time to show than when one shaves. Electrolysis destroys the hair root; thus, the shaft will not be produced.

4. The arrector pili muscle contracts and pulls on the hair shaft in such a way that there is a small raised area of the skin, "goose bump." In many other mammals, this response would pull the hairs up from the body surface and air would be

trapped between the hair shafts, producing an insulation layer. In the case of humans, this is a vestigial response.

Testing Your Knowledge

1. tissues; **2.** cuboidal; **3.** layered (stratified), cilia, columnar (elongated); **4.** connective; **5.** striated; **6.** neurons; **7.** dermis; **8.** epithelial, loose connective; **9.** keratin; **10.** thoracic; **11.** constancy, tissue; **12. a.** columnar epithelium, lining of intestine (digestive tract), protection and absorption; **b.** cardiac muscle, wall of heart, pumps blood; **c.** compact bone, skeleton, support and protection.

Chapter 4
Applying Your Knowledge: Concepts

1. The mouth and nasal cavity are separated by the hard and soft palate in the forward part of the mouth, but the oral pharynx is connected to the nasopharynx at the rear of the mouth. The pressure of laughing forces the fluid into the nasopharynx and through the nostrils.

2. The ending *ase* indicates it is an enzyme. Dehydrogen indicates that hydrogen is removed. Succinic acid names the substrate. Therefore, it is an enzyme that removes hydrogen from succinic acid.

3. The dark color is partially due to the breakdown products of hemoglobin, which is found in significant amounts in meat, particularly those meats referred to as red meat.

4. A certain amount of fat is necessary in the diet; one of the fatty acids is necessary for the production of phospholipids, an important component of cell membranes. Just make sure to reduce the amount of fat in the diet to 30% or somewhat less.

Testing Your Knowledge

1. epiglottis; **2.** esophagus, protein; **3.** villi; **4. a.** salivary glands; **b.** esophagus; **c.** stomach; **d.** liver; **e.** gallbladder; **f.** pancreas; **g.** small intestine; **h.** large intestine; **i.** sugar and amino acids; **j.** lipids; **k.** water; **5.** duodenum; **6.** glycogen; **7.** bile, emulsifies; **8.** amylase, maltose; **9.** trypsin, pancreatic amylase, lipase; **10.** acidic, basic; **11.** essential amino acids.

Chapter 5
Applying Your Knowledge: Concepts

1. The stem cells for the production of white blood cells are located in the bone marrow. If they fail to function, the production of WBCs decreases. A bone marrow transplant will hopefully provide functioning bone marrow to produce WBCs.

2. Individuals living at the higher altitudes had higher concentrations of red blood cells than most of the other runners; therefore, running a race at a high altitude where oxygen is less concentrated gave them more RBCs to combine with the oxygen.

A low blood concentration of oxygen stimulates the production of RBCs. This can result from a low oxygen content in the air that is breathed as occurs at high altitudes.

3. Probably the mother is Rh⁻ and the father Rh⁺. Their first child may have been Rh⁻ and their second child Rh⁺. Some blood of the second child entered the mother's blood system, and she produced antibodies that acted on the red blood cells of the child. Another possibility is that their first child may have been Rh⁺, but the mother was sensitized too late in the pregnancy to have built up antibodies to affect the first child, but the antibodies were present to affect the second child.

Testing Your Knowledge

1. plasma; **2.** oxygen, fight infection; **3.** oxyhemoglobin; **4.** nucleus, 120; **5.** fibrin; **6.** neutrophil; **7.** antibodies; **8.** oxygen, amino acids, glucose; carbon dioxide, other wastes; **9.** A, B, no; **10.** Rh⁻, Rh⁺; **11. a.** arterial end; **b.** venous end; **c.** water, oxygen, nutrients; **d.** water, carbon dioxide; **e.** plasma proteins.

Chapter 6
Applying Your Knowledge: Concepts

1. The additional weight requires increased units in the circulatory system; thus, the heart must exert more pressure to deliver the blood. This puts added tension on the arteries and also additional work on the heart. Also, some of the fat collects around the heart, which restricts its pumping action.

2. A high enough level of drugs should be administered to prevent the immune system from rejecting the tissue; at the same time, if the immune system is depressed too much, the patient is unable to combat disease, producing organisms that enter the body. Transplant patients are very susceptible to the common cold and similar types of viruses.

3. The atria serve largely as collecting chambers and, as indicated by their thin walls, do not pump blood with much pressure. The expanding of the ventricles causes a negative pressure, which will result in the blood, under greater pressure in the atria, entering the ventricles without the atria pumping.

Testing Your Knowledge

1. away; **2.** aorta; **3.** high; **4.** lungs; **5.** SA; **6.** coronary; **7.** blood pressure; **8.** blood pressure, skeletal muscle contraction; **9.** valves; **10.** saturated fat, cholesterol; **11. a.** aorta; **b.** left pulmonary arteries; **c.** pulmonary trunk; **d.** left pulmonary veins; **e.** left atrium; **f.** semilunar valves; **g.** atrioventricular (mitral) valve; **h.** left ventricle; **i.** septum; **j.** inferior vena cava; **k.** right ventricle; **l.** chordae tendineae; **m.** atrioventricular (tricuspid) valve; **n.** right atrium; **o.** right pulmonary veins; **p.** right pulmonary arteries; **q.** superior vena cava. See also Fig. 6.5, page 127, text.

Chapter 7
Applying Your Knowledge: Concepts

1. The lymph node serves as a filter for foreign material, including bacteria. The reaction of the lymph node tissue with this foreign material results in the swelling and, consequently, pain from the node.

2. Immunological memory is dependent upon the presence of memory B cells. In the case of certain types of memory B cells, they decline in number so that they will not be able to respond rapidly to an antigen (disease-producing organism). A booster shot will stimulate the production of more memory B cells.

3. Immunosuppressive drugs prevent immune responses in general, and therefore, the body is more susceptible to all pathogens.

Testing Your Knowledge

1. tissue fluid, subclavian; **2.** filter; **3.** thymus; **4.** The complement system; **5.** plasma, memory; **6.** antibody; **7.** cytokines; **8.** APC; **9.** helper (produces lymphokines), suppressor (shuts down response), memory (retains ability to kill infected cells in the future), cytotoxic (kills infected cells); **10.** vaccines (or antigens); **11.** monoclonal; **12.** histamine.

Chapter 8
Applying Your Knowledge: Concepts

1. During normal tidal breathing, we don't use the abdominal muscles or the internal intercostal muscles. During heavy or deep breathing, we do use these muscles, and unless one has exercised these muscles in the past, they get sore. This is similar to what happens when one exercises a seldom-used muscle.

2. A child that was stillborn will have no air in its lungs; but a baby that has taken a minimum of one breath will have residual air in its lungs that can't be forced out.

3. The active part of the body will be releasing more CO_2 and heat; thus, the affinity of hemoglobin for oxygen will decrease, and more oxygen will be released in the tissue fluid surrounding the active cells. (Also, there will be a dilation of the blood vessels in that area.)

Testing Your Knowledge

1. larynx; **2.** alveoli; **3. a.** nasal cavity; **b.** nostril; **c.** pharynx; **d.** epiglottis; **e.** glottis; **f.** larynx; **g.** trachea; **h.** bronchus; **i.** bronchiole; See also Fig. 8.2, page 164, text; **4.** CO_2 and H^+; **5.** expanded (or lower pressure); **6.** diffusion; **7.** bicarbonate; **8.** the globin portion of hemoglobin; **9.** lungs; **10.** bronchi.

Chapter 9
Applying Your Knowledge: Concepts

1. The urea is being converted to ammonia, carbon dioxide, and water by bacteria that possess the enzyme urease.

2. Alcohol inhibits the secretion of ADH, and this results in a dilute urine being secreted. This lowers the fluid content of the body and results in a dry mouth and feeling of thirst.

3. Load your body with sugar prior to being tested, thereby overloading the reabsorptive power of the kidneys. Then, excess sugar is excreted, and urinalysis gives a positive test for diabetes.

4. All material moving through the membrane in an artificial kidney must move from a higher concentration to a lower concentration by passive transport. In the human kidney, some materials are moved against the concentration gradient by active transport.

Testing Your Knowledge

1. bile pigments, hemoglobin; **2.** urea; **3.** urethra; **4. a.** glomerulus; **b.** efferent arteriole; **c.** afferent arteriole; **d.** proximal convoluted tubule; **e.** loop of the nephron; **f.** descending limb; **g.** ascending limb; **h.** peritubular capillaries; **i.** distal convoluted tubule; **j.** vein; **k.** artery; **l.** collecting duct. **5.** microvilli, reabsorbing; **6.** Water; **7.** Urea; **8.** distal convoluted tubule; **9.** ADH; **10.** volume, pH; **11.** hemodialysis.

Chapter 10
Applying Your Knowledge: Concepts

1. Osteoclasts cause absorption of the minerals in the bone, and if there is an insufficient amount of calcium in the blood, the osteoblasts cannot repair the bone.

2. Either the spinal cord or spinal nerves were being compressed by the vertebrae or by damaged disks.

3. The knee is a synovial joint that is structured to be mobile in only one plane. Tackling from the side puts pressure on the joint in a manner to which the joint cannot respond, and consequently, the ligaments and tendons can suffer severe damage.

Testing Your Knowledge

1. c; **2.** g; **3.** a; **4.** d; **5.** f; **6.** e; **7.** e; **8.** a; **9.** b; **10.** c; **11.** f; **12.** d; **13.** osteons; **14.** red bone marrow; **15.** osteoclasts; **16.** sinuses; **17.** pelvis, rib cage; **18.** flexibility, strength; **19.** fingers, toes; **20.** synovial.

Chapter 11
Applying Your Knowledge: Concepts

1. Standing is dependent on muscle contraction, and muscle contraction requires calcium. The birthing process must have depleted the cow's supply of calcium.

2. For some reason, the neuromuscular junction is no longer working properly. Some think the junction acts like a fuse, protecting the muscle from being damaged by overstimulation.

3. The liver has no need of glucose like the muscles do. Having a ready supply of glucose means that the muscles can respond quickly to get the organism away from danger.

Testing Your Knowledge

1. muscle fibers; 2. move; 3. calcium; 4. Acetylcholine (ACh); 5. ATP; 6. tetanic; 7. does not; 8. creatine phosphate; 9. atrophy; 10. Slow; 11. ATP; 12. a. sarcolemma; b. T tubules; c. mitochondrion; d. sarcoplasmic reticulum; e. sarcomere; f. myofibril.

Chapter 12
Applying Your Knowledge: Concepts

1. Alcohol is one of the few substances that is absorbed directly through the wall of the stomach; therefore, it gets into the blood and is carried to the brain in a minute or less after being ingested.

2. The advantage lies in the rapid response to a stimulus. Many reflex actions are protective in nature. The disadvantage lies in no variation in the response to the stimulus.

3. It causes paralysis, and ultimately, atrophy of the muscles innervated by the destroyed nerves. Muscles that don't receive stimuli from nerves atrophy.

4. Anger and an excited state fall under "fight or flight." Although this complex response readies the body for action, it inhibits the digestive tract. One aspect—blood is shunted from the skin and visceral organs to the skeletal muscles.

Testing Your Knowledge

1. a. sensory neuron; b. interneuron; c. motor neuron; 2. axon; 3. sodium, inside; 4. synaptic cleft; 5. AChE; 6. muscles, glands; 7. interneuron; 8. cranial (parasympathetic), internal organs; 9. meninges; 10. cerebrum; 11. cerebellum; 12. nicotine.

Chapter 13
Applying Your Knowledge: Concepts

1. The eye can respond to stimuli other than light. (It has a low threshold to light, however.)

2. Vitamin A, the precursor of retinal, found in carrots, can help night blindness which occurs when rods are not functioning as they should. Vitamin A cannot help red-green color blindness, an inherited condition that affects the cones.

3. The eye helps in positioning the body by allowing us to line up on some visual object.

4. The ability to hear higher frequencies is lost before the ability to hear lower frequencies. Male voices are lower and result in lower frequencies that can be heard better.

Testing Your Knowledge

1. brain; 2. dermis, receptors; 3. Adaptation; 4. chemoreceptors; 5. rods, cones, retina; 6. a. ciliary body; b. lens; c. iris; d. pupil; e. cornea; f. fovea centralis; g. optic nerve; h. sclera; i. choroid; j. retina. See also Fig. 13.6, page 270; 7. color, bright (day); 8. rhodopsin; 9. rounds up (accommodation); 10. distant, concave; 11. malleus (hammer), incus (anvil), stapes (stirrup); 12. equilibrium (balance); 13. cochlear, cochlea.

Chapter 14
Applying Your Knowledge: Concepts

1. Epinephrine is a peptide hormone and has its effect on existing enzymes; whereas, testosterone is a steroid hormone and has its effect on the genetic material resulting in the production of more enzymes.

2. Although the anterior pituitary secretes hormones that control other endocrine glands, it is the hypothalamus that controls the anterior pituitary's secretion of these hormones.

3. Thyroxin increases metabolism and thus resulted in the burning of extra calories rather than them being stored as fat.

4. Melatonin is secreted by the pineal gland in larger amounts in lower light intensity, at night, and thus prepares a person for sleep. On an overseas trip, the hormone is taken about an hour before one wants to sleep.

Testing Your Knowledge

1. plasma membrane; 2. ADH, oxytocin; 3. Hypothalamic-releasing and release-inhibiting hormones; 4. negative-feedback; 5. anterior; 6. too little, thyroxin; 7. calcium; 8. cortex; 9. Cushing syndrome; 10. a. high sodium; b. inhibits; c. renin; d. angiotensin I and II; e. aldosterone. When the blood sodium level is low, the kidneys secrete renin. Renin stimulates the production of angiotensin I and II, which stimulate the adrenal cortex to release aldosterone. The kidneys absorb sodium. See also Fig. 14.15, page 298; 11. pancreas, insulin; 12. blood; 13. chemical messengers.

Chapter 15
Applying Your Knowledge: Concepts

1. The only difference in the sexual act after a vasectomy is the fact that the semen contains no sperm. The amount of semen will be almost the same and the same sensations will occur.

2. The sperm can remain viable for a period of time in the female reproductive tract and the egg can remain viable for a period of time. Therefore, to account for this, one must allow a period of time on each side of the day of ovulation.

3. Some believe that jockey shorts hold the scrotum and thus the testes so close to the body that the temperature within the scrotum is too high for the production of viable sperm.

Testing Your Knowledge

1. vas deferens; **2.** seminal vesicles; **3.** testosterone; **4.** blood; **5. a.** seminal vesicle; **b.** ejaculatory duct; **c.** prostate gland; **d.** bulbourethral gland; **e.** anus; **f.** vas deferens; **g.** epididymis; **h.** testis; **i.** scrotum; **j.** foreskin; **k.** glans penis; **l.** penis; **m.** urethra; **n.** vas deferens; **o.** urinary bladder; **6.** vagina; **7.** follicle, endometrial; **8.** estrogens, progesterone; **9.** human chorionic gonadotropin (HCG); **10.** laboratory glassware.

Chapter 16
Applying Your Knowledge: Concepts

1. STDs of bacterial origin can be treated and cured; whereas, STDs of viral origin cannot be cured but, at present, only inhibited in their multiplication.

2. Some animal viruses, including HIV, cloak themselves in molecules from the cell membrane of the host. These coatings make it more difficult for the immune system to recognize the viruses as foreign.

3. The advent of the pill makes it unnecessary to use other means of contraception, particularly the condom. Thus STDs are more readily transmitted during sexual activity.

4. Pap tests detect the presence of cervical cancer and there is a high correlation between the incidence of genital warts and cervical cancer.

Testing Your Knowledge

1. nucleic acid, protein; **2.** retrovirus; **3.** blood; **4.** cold sores, genital herpes; **5.** rod, spherical, spiral; **6.** PID; **7.** condom; **8.** gummas; **9.** gonorrhea, chlamydia, syphilis; **10.** yeast; **11. a.** gonorrhea; **b.** AIDS; **c.** syphilis.

Chapter 17
Applying Your Knowledge: Concepts

1. The fertilized egg has time to develop to a stage so that it can implant itself in the uterine wall. If fertilization occurred in the uterus, the chances of implantation would be greatly reduced.

2. In the case of a low oxygen supply in the mother's blood, the fetus will be able to "pull" the oxygen from the mother's hemoglobin. This is an insurance for the well-being of the fetus.

3. Don't smoke or drink alcohol to excess. Do get plenty of calcium in your diet and exercise regularly.

4. This condition is called polyspermy and prevents normal development of the zygote; therefore, the zygote would pass out with the menstrual flow or be absorbed.

Testing Your Knowledge

1. sperm, egg; **2.** cleavage; **3.** implant; **4.** differentiation; **5.** extraembryonic, amnion; **6.** organs; **7.** second; **8.** placenta; **9.** head; **10.** oxytocin; **11.** cross-linking; **12. a.** chorion; **b.** amnion; **c.** embryo; **d.** allantois; **e.** yolk sac; **f.** fetal portion of placenta; **g.** maternal portion of placenta; **h.** umbilical cord. See also Fig. 17.6, page 362, text.

Chapter 18
Applying Your Knowledge: Concepts

1. Only nondisjunction of the sperm can result in a cell with two Ys (YY).

2. Chromatin becomes more dense and forms chromosomes, the nucleolus disappears, spindle fibers form, and nuclear membrane fragments.

3. Turner syndrome is the result of having only one X chromosome; therefore, the cell contains 1 less than a complete 2n number. Klinefelter syndrome has 2 X plus 1 Y chromosome and thus has a complete set, 2n, plus the extra X chromosome.

4. If the two cells separate and both develop, it results in identical twins.

Testing Your Knowledge

1. karyotype; **2.** XY, XX; **3.** 2n; **4.** 24; **5.** centrioles; **6.** homologous chromosomes, sister chromatids; **7.** spermatogenesis, oogenesis; **8.** 50%; **9.** c; **10.** a; **11.** e; **12.** d; **13.** b; **14.** a; **15.** d.

Chapter 19
Applying Your Knowledge: Concepts

1. Both of the parents might be carriers for a trait which could have been expressed in their grandmothers. The grandmothers would have been homozygous recessive.

2. Gene mutation may have occurred in a specific area of the world where a particular ethnic group lived and there may have been little genetic exchange with outside groups.

3. A carrier is an individual who carries a hidden faulty gene; Huntington disease is caused by a dominant gene and thus is not hidden.

4. The blood type in this case neither proves nor disproves the man is the father. The mother could have passed the B gene to her offspring and the suspected father would have passed an O gene with the child then being genotype BO, type B.

Testing Your Knowledge

1. one; **2.** *w*; **3.** phenotypes; **4.** widow's peak (genotype *Ww*); **5.** 25%; **6.** X; **7.** mother; **8.** $X^B X^b$; **9.** autosomal recessive; **10.** autosomal recessive condition; **11. a.** cystic fibrosis; **b.** Tay-Sachs disease; **c.** neurofibromatosis (NF); **d.** Huntington disease; **e.** hemophilia; **f.** muscular dystrophy.

Practicing Genetics Problems

1. 50%.
2. 50%.
3. 50%.
4. 0%
5. Yes. Parents could be heterozygous.
6. Heterozygous; 25%.
7. 50%.
8. Child: *cc*; Parents: *Cc*, *Cc*.
9. 100%
10. $Hb^A Hb^S$
11. $Hb^A Hb^S$
12. 25%
13. A, B, AB, O
14. AB; Yes; A, B, O, AB
15. *AB*, *B*
16. *AaBb* and *aabb*

Chapter 20
Applying Your Knowledge: Concepts

1. Covalent bonds are stronger and tend to hold the backbone together. Hydrogen bonds are much weaker and lend themselves to breaking to enable replication or transcription.

2. Pancreatic cells produce many proteins for secretion.

3. If one defines a mutation as a change in genetic material resulting in a different phenotypic expression, then this is not a mutation.

4. This assures that the genetic materials will be cut in similar places and will match up when put together.

Testing Your Knowledge

1. C, T; **2.** old, new; **3.** triplet, amino acid; **4.** mRNA, tRNA, rRNA; **5.** tRNA; **6.** translation; **7.** restriction, DNA ligase; **8.** transgenic; **9.** retrovirus; **10. a.** ACU´CCU´ GAA´UGC´AAA; **b.** UGA´GGA´CUU´ACG´UUU;

Chapter 21
Applying Your Knowledge: Concepts

1. The immune system of AIDS patients is inhibited, thus not able to resist the growth of abnormal cells. It is thought that many possible cancer cells are recognized and destroyed by a healthy immune system.

2. The chemicals are particularly hard on rapidly dividing cells. The hair bulb cells divide relatively rapidly in producing the hair shaft.

3. It is thought that feces remaining in the colon for lengthy periods results in the production of toxic materials which can serve as carcinogens. High fiber materials stimulate the movement of feces from the colon.

4. Estrogen stimulates growth of endometrial tissue; it would tend to amplify a cancer developing as cancer is an uncontrolled growth of cells.

1. carcinogens; **2.** papilloma; **3.** cigarette smoke; **4.** metastasize; **5.** proto-oncogenes, tumor-suppressor genes; **6.** tumor-suppressor gene; **7.** dividing; **8.** chemotherapy; **9.** lymphokine; **10. a.** oncogene; **b.** tumor-suppressor gene; **c.** receptor; **d.** antigen; **e.** blood vessel. (1) hormone therapy; (2) immunotherapy; (3) chemotherapy; (4) antimetastatic drugs; (5) radiation; (6) surgery; (7) gene therapy. See also Fig. 21.9, page 449, text.

Chapter 22

Applying Your Knowledge: Concepts

1. The DDT killed all those flies that were not resistant to it; the genetic material of the flies that were not killed enabled them to live in the presence of DDT. These flies lived to reproduce and thus a strain of flies resistant to DDT developed.

2. It is thought that the primitive earth contained little, if any gaseous oxygen. The first cells had to carry on anaerobic respiration. As aerobic organisms evolved from the more primitive cells, they retained the enzymes for the reactions of anaerobic respiration and aerobic respiration is built on top of these reactions.

3. Both present day humans and present day great apes have evolved a great deal from their primitive ancestors; therefore, both evolved from a distant relative.

Testing Your Knowledge

1. b; **2.** a; **3.** d; **4.** c; **5.** reproductive; **6.** adaptation; **7.** chemical; **8.** erect, small; **9.** Neanderthals, Cro-Magnon; **10.**

Name	Example
Animalia	Multicellular, moves, ingests food
Chordata	Animals with dorsal supporting rod and nerve cord
Mammalia	Animals with hair, mammary glands
Primates	Apes, humans
Hominidae	hominids, *A. afarensis*
Homo	*H. habilis, H. erectus*
H. sapiens	Neanderthal, Cro-Magnon

See also the table on page 462, text.

Chapter 23

Applying Your Knowledge: Concepts

1. All plants and animals in an ecosystem are related either directly or indirectly to one another. If one member of a food web is removed, it will affect the other members of the web, for example.

2. Pigs are primary consumers and as such they trap only about 10% of the energy of corn. In addition, there is the cost of growing both corn and pigs and processing both for marketing.

3. Removing the forest or grassland results in a smaller amount of carbon dioxide being removed from the atmosphere. Building a factory and parking lot will result in more fossil fuels being consumed in the building and operation of the factory and thus more carbon dioxide being released into the atmosphere.

4. In many cases there are no natural predators to keep the new species in check and they reproduce in large numbers, taking over the food supply of other animals.

Testing Your Knowledge

1. a. producers; **b.** consumers; **c.** inorganic nutrient pool; **d.** decomposers; **2.** flow; **3.** producer; **4.** herbivores; **5.** trophic; **6. a.** algae; **b.** zooplankton; **c.** small fishes; **d.** large fishes; **e.** humans; **7.** respire; **8.** nitrate; **9.** nitrogen gas; **10.** matter, energy.

Chapter 24

Applying Your Knowledge: Concepts

1. The death rate must increase or the birthrate decrease. Either will reduce the ratio of births to deaths.

2. The population will never be reduced unless the growth rate is reduced to below zero. It will continue to grow for about one generation and then will level off at a larger population than now exists.

3. The human population is increasing rapidly, producing more wastes; and a significant amount of the wastes are non-biodegradable.

4. Some experts believe that it will be possible to solve various problems like food supply and waste disposal, particularly if we all assume a less affluent life-style. They believe that much more food can be produced by better farming methods and that the chief problem of food supply stems from inadequate distribution of food. They believe we could recycle more and develop methods so that humans would produce less waste.

Testing Your Knowledge

1. higher; **2. a.** carrying capacity; **b.** steady state; **c.** deceleration; **d.** acceleration; **3.** low; **4.** prereproductive; **5.** industrialized; **6.** carbon dioxide (CO_2); **7.** CFCs; **8.** cultural eutrophication; **9.** magnification; **10. a.** hydrocarbons; **b.** CFCs; **c.** carbon dioxide; **d.** sulfur dioxide; **11.** protect.

Appendix B
METRIC SYSTEM

Unit and Abbreviation	Metric Equivalent	Approximate English-to-Metric Conversion Factor	Units of Temperature
Length			
nanometer (nm)	$= 10^{-9}$ m		
micrometer (μm)	$= 10^{-6}$ m		
millimeter (mm)	$= 0.001\ (10^{-3})$ m		
centimeter (cm)	$= 0.01\ (10^{-2})$ m	1 inch = 2.54 cm	
		1 foot = 30.5 cm	
meter (m)	$= 100\ (10^{2})$ cm	1 foot = 0.30 m	
	$= 1{,}000$ mm	1 yard = 0.91 m	
kilometer (km)	$= 1{,}000\ (10^{3})$ m	1 mi = 1.6 km	
Weight (mass)			
nanogram (ng)	$= 10^{-9}$ g		
microgram (μg)	$= 10^{-6}$ g		
milligram (mg)	$= 10^{-3}$ g		
gram (g)	$= 1{,}000$ mg	1 ounce = 28.3 g	
		1 pound = 454 g	
kilogram (kg)	$= 1{,}000\ (10^{3})$ g	= 0.45 kg	
metric ton (t)	$= 1{,}000$ kg	1 ton = 0.91 t	
Volume			
microliter (μl)	$= 10^{-6}$ l $(10^{-3}$ ml)		
milliliter (ml)	$= 10^{-3}$ liter	1 tsp = 5 ml	
	$= 1$ cm^3 (cc)	1 fl oz = 30 ml	
	$= 1{,}000$ mm^3		
liter (l)	$= 1{,}000$ ml	1 pint = 0.47 liter	
		1 quart = 0.95 liter	
		1 gallon = 3.79 liter	
kiloliter (kl)	$= 1{,}000$ liter		

Units of Temperature

F° scale readings: 230, 220, 212°–210, 200, 190, 180, 170, 160°–160, 150, 140, 134°/131°–130, 120, 110, 105.8°, 98.6°–100, 90, 80, 68.6°–70, 60, 50, 40, 32°–30, 20, 10, 0, -10, -20, -30, -40

C° scale readings: 110, 100–100°, 90, 80, 70–71°, 60, 57°, 50, 41°/37°, 40, 30, 20–20.3°, 10, 0–0°, -10, -20, -30, -40

°C	°F	
100	212	Water boils at standard temperature and pressure
71	160	Flash pasteurization of milk
57	134	Highest recorded temperature in the United States, Death Valley, July 10, 1913
41	105.8	Average body temperature of a marathon runner in hot weather
37	98.6	Human body temperature
20.3	68.6	Lowest recorded body temperature to be survived by a human
0	32.0	Water freezes at standard temperature and pressure

To convert temperature scales:

$$°C = \frac{5(°F - 32)}{9}$$

$$°F = \frac{9°C}{5} + 32$$

Appendix C
TABLE OF CHEMICAL ELEMENTS

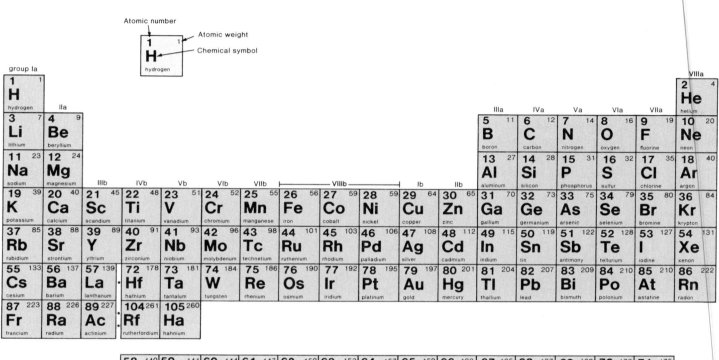

Appendix D
DRUGS OF ABUSE

	Drugs	Often Prescribed Brand Names	Medical Uses	Potential Physical Dependence	Potential Psychological Dependence	Tolerance
Narcotics	Opium	Dover's Powder, Paregoric	Analgesic, antidiarrheal	High	High	Yes
	Morphine	Morphine	Analgesic	High	High	Yes
	Codeine	Codeine	Analgesic, antitussive	Moderate	Moderate	Yes
	Heroin	None	None	High	High	Yes
	Meperidine (Pethidine)	Demerol, Pethadol	Analgesic	High	High	Yes
	Methadone	Dolophine, Methadone, Methadose	Analgesic, heroin substitute	High	High	Yes
	Other Narcotics	Dilaudid, Leritine, Numorphan, Percodan	Analgesic, antidiarrheal, antitussive	High	High	Yes
Depressants	Chloral Hydrate	Noctec, Somnos	Hypnotic	Moderate	Moderate	Probable
	Barbiturates	Amytal, Butisol, Nembutal, Phenobarbitol, Seconal, Tuinal	Anesthetic, anticonvulsant, sedation, sleep	High	High	Yes
	Glutethimide	Doriden	Sedation, sleep	High	High	Yes
	Methaqualone	Optimil, Parest, Quaalude, Somnafac, Sopor	Sedation, sleep	High	High	Yes
	Tranquilizers	Equanil, Librium, Miltown, Serax, Tranxene, Valium	Antianxiety, muscle relaxant, sedation	Moderate	Moderate	Yes
	Other Depressants	Clonopin, Dalmane, Dormate, Noludar, Placydil, Valmid	Antianxiety, sedation, sleep	Possible	Possible	Yes
Stimulants	Cocaine*	Cocaine	Local anesthetic	Possible	High	Yes
	Amphetamines	Benzedrine, Biphetamine, Desoxyn, Dexedrine	Hyperkinesis, narcolepsy, weight control	Possible	High	Yes
	Phenmetrazine	Preludin	Weight control	Possible	High	Yes
	Methylphenidate	Ritalin	Hyperkinesis	Possible	High	Yes
	Other Stimulants	Bacarate, Cylert, Didrex, Ionamin, Plegine, Pondimin, Pro-Sate, Sanorex, Voranil	Weight control	Possible	Possible	Yes
Hallucinogens	LSD	None	None	None	Degree unknown	Yes
	Mescaline	None	None	None	Degree unknown	Yes
	Psilocybin-Psilocyn	None	None	None	Degree unknown	Yes
	MDA	None	None	None	Degree unknown	Yes
	PCP†	Sernylan	Veterinary anesthetic	None	Degree unknown	Yes
	Other Hallucinogens	None	None	None	Degree unknown	Yes
Cannabis	Marijuana, Hashish, Hashish Oil	None	Glaucoma	Degree unknown	Moderate	Yes

Source: Drugs of Abuse, produced by the Affairs in Cooperation with the Office of Public Science and Technology.

*Designated a narcotic under the Controlled Substances Act.

†Designated a depressant under the Controlled Substances Act.

Duration of Effects (in hours)	Usual Methods of Administration	Possible Effects	Effects of Overdose	Withdrawal Syndrome
3–6	Oral, smoked	Euphoria, drowsiness, respiratory depression, constricted pupils, nausea	Slow and shallow breathing, clammy skin, convulsions, coma, possible death	Watery eyes, runny nose, yawning, appetite loss, irritability, tremors, panic, chills
3–6	Injected, smoked			
3–6	Oral, injected			
3–6	Injected, sniffed			
3–6	Oral, injected			
12–24	Oral, injected			
3–6	Oral, injected			
5–8	Oral			
1–16	Oral, injected	Slurred speech, disorientation, drunken behavior without odor of alcohol	Shallow respiration, cold and clammy skin, dilated pupils, weak and rapid pulse, coma, possible death	Anxiety, insomnia, tremors, delirium, convulsions, possible death
4–8	Oral			
4–8	Oral			
4–8	Oral			
4–8	Oral			
2	Injected, sniffed	Increased alertness, excitation, euphoria, dilated pupils, increased pulse rate and blood pressure, insomnia, loss of appetite	Agitation, increased body temperature, hallucinations, convulsions, possible death	Apathy, long periods of sleep, irritability, depression, disorientation
2–4	Oral, injected			
2–4	Oral			
2–4	Oral			
2–4	Oral			
Variable	Oral	Illusions and hallucinations (with the exception of MDA), poor perception of time and distance	Longer, more intense "trip" episodes, psychosis, possible death	Withdrawal syndrome not reported
Variable	Oral, injected			
Variable	Oral			
Variable	Oral, injected, sniffed			
Variable	Oral, injected, smoked			
Variable	Oral, injected, sniffed			
2–4	Oral, smoked	Euphoria, relaxed inhibitions, increased appetite, disoriented behavior	Fatigue, paranoia, possible psychosis	Insomnia, hyperactivity, and decreased appetite reported in a limited number of individuals

Appendix E

SELECTED TYPES OF CANCER

Cancer incidence and deaths by site are given in Table 21.3. Five-year survival rates for the most common cancers are given in Table E.

Breast Cancer

Signs and Symptoms: Pre-clinical radiographic signs seen on a mammogram. Breast changes, such as a lump, thickening, swelling, dimpling, skin irritation, distortion, retraction, scaliness, pain, tenderness of the nipple, or nipple discharge.

Risk Factors: The risk of breast cancer increases with age. The risk is higher in the woman who has a personal or family history of breast cancer; some forms of benign breast disease; early onset of menstruation; late menopause; lengthy exposure to cyclic estrogen; never having children or having the first live birth at a later age; and higher education and socioeconomic status. International variability in breast cancer incidence rates correlates with variations in diet, especially fat intake, although a causal role for dietary factors has not been firmly established. Additional factors that may be associated with increased breast cancer risk and that are currently under study include pesticide and other chemical exposures, alcohol consumption, induced abortion, and physical inactivity. A majority of women will have one or more risk factors for breast cancer. However, most risks are at such a low level that they only partly explain the high frequency of the disease in the population. To date, knowledge about risk factors has not translated into practical ways to prevent breast cancer. Since women may not be able to alter their personal risk factors, the best opportunity at present for reducing mortality is through early detection.

Early Detection: The American Cancer Society recommends that asymptomatic women aged 40 to 49 should have a screening mammogram every one to two years; and women aged 50 and over should have a mammogram every year. In addition, a clinical breast exam is recommended every three years for women 20 to 40, and every year for women over 40. The American Cancer Society also recommends monthly breast self-exam as a routine good health habit for women 20 years or older. Most breast lumps are not cancer, but only a physician can make a diagnosis.

Mammography is recognized as a valuable diagnostic technique for women who have findings suggestive of breast cancer. When a suspicious area is identified on a mammogram, or when a woman has a suspicious lump, mammography can help determine if there are other lesions too small to be felt in the same or opposite breast. Since a small percentage of breast cancers may not be seen on a mammogram, all suspicious lumps should be biopsied for a definitive diagnosis, even when current or recent mammography findings are described as normal.

Treatment: Taking into account the medical situation and the patient's preferences, treatment may involve lumpectomy (local removal of the tumor), mastectomy (surgical removal of the breast), radiation therapy, chemotherapy, or hormone therapy. Often, two or more methods are used in combination. Patients should discuss possible options for the best management of their breast cancer with their physicians.

In recent years, new techniques have made breast reconstruction possible after mastectomy, and the cosmetic results usually are good. Reconstruction has become an important part of treatment and rehabilitation. Bone marrow transplantation is another new type of treatment that is under study.

Lung Cancer

Signs and Symptoms: Persistent cough, sputum streaked with blood, chest pain, recurring pneumonia or bronchitis.

Risk Factors: Cigarette smoking is by far the most important risk factor in the development of lung cancer. Other factors include exposure to certain industrial substances, such as arsenic; certain organic chemicals and asbestos, particularly for persons who smoke; radiation exposure from occupational, medical, and environmental sources; air pollution; tuberculosis; radon exposure, especially in cigarette smokers; and environmental tobacco smoke in nonsmokers.

Early Detection: Because symptoms often don't appear until the disease is advanced, early detection is difficult. In smokers who stop smoking when precancerous changes are found, damaged lung tissue often returns to normal. Smokers who persist in smoking may form abnormal cell growth patterns that lead to cancer. Chest X ray, analysis of the types of cells contained in sputum, and fiberoptic examination of the bronchial passages assist diagnosis.

Treatment: Determined by the type and stage of the cancer. Options include surgery, radiation therapy, and chemotherapy. For many localized cancers, surgery is usually the treatment of choice. Because the disease has usually spread by the time it is discovered, radiation therapy and chemotherapy are often needed in combination with surgery as the treatment of choice; on this regimen, a large percentage of patients experience remission, which in some cases is long-lasting.

TABLE E

Five-Year Survival Rates by Stage at Diagnosis*

Site	All Stages %	Local %	Regional %	Distant %
Oral	52	81	42	18
Colon/rectum	61	91	63	7
Pancreas	4	12	5	2
Lung	13	47	17	2
Skin (melanoma)	87	94	6	16
Female Breast	83	96	75	20
Cervix	68	91	50	9
Uterus	83	95	65	26
Ovary	44	91	49	23
Prostate	85	98	92	30
Bladder	81	93	49	6
Kidney	58	88	59	9

*Adjusted for normal life expectancy. This chart is based on cases diagnosed in 1986–1991, followed through 1993.
Source: Cancer Statistics Branch, National Cancer Institute.

Prostate Cancer

Signs and Symptoms: Weak or interrupted urine flow; inability to urinate, or difficulty starting or stopping the urine flow; the need to urinate frequently, especially at night; blood in the urine; pain or burning on urination; continuing pain in lower back, pelvis, or upper thighs. Most of these symptoms are nonspecific and may be similar to those caused by benign conditions such as infection or prostate enlargement.

Risk Factors: The incidence of prostate cancer increases with age; over 80% of all prostate cancers are diagnosed in men over age 65. African Americans have the highest prostate cancer incidence rates in the world; the disease is common in North America and Northwestern Europe and is rare in the Near East, Africa, and South America. There may be some familial tendency, but it is unclear whether this is due to genetic or environmental factors. International studies suggest that dietary fat may also be a factor.

Early Detection: Every man aged 40 and over should have a digital rectal exam as part of his regular annual physical checkup. In addition, the American Cancer Society recommends that men aged 50 and over have an annual prostate-specific antigen blood test. If either result is suspicious, further evaluation in the form of transrectal ultrasound should be performed.

Treatment: Surgery, radiation, and/or hormones and anticancer drugs are treatment options. Hormone treatment and anticancer drugs may control prostate cancer for long periods by shrinking the size of the tumor, thus relieving pain. Careful observation without immediate active treatment may be appropriate for individuals with low-grade and/or early-stage tumors.

Colon and Rectal Cancer

Signs and Symptoms: Rectal bleeding, blood in the stool, a change in bowel habits.

Risk Factors: Personal or family history of colorectal cancer or polyps, and inflammatory bowel disease have been associated with increased colorectal cancer risk. Other possible risk factors include physical inactivity and high-fat and/or low-fiber diet. Recent studies have suggested that estrogen replacement therapy and nonsteroidal anti-inflammatory drugs such as aspirin may reduce colorectal cancer risk.

Early Detection: Digital rectal exam, stool blood test, and sigmoidoscopy are recommended by the American Cancer Society to detect colon or rectal cancer in asymptomatic patients. These tests offer the best opportunity for the diagnosis and removal of polyps, and hence the prevention of these cancers.

Digital rectal exam is performed by a physician during an office visit. The American Cancer Society recommends that this exam be performed annually after age 40.

The stool blood test is a simple method to test feces for hidden blood. The specimen is obtained by the patient at home and returned to the physician's office, a hospital, or a clinic for analysis. The Society recommends annual testing after age 50.

In sigmoidoscopy, the physician uses a hollow, lighted tube or a fiberoptic sigmoidoscope to inspect the rectum and lower colon. The American Cancer Society recommends sigmoidoscopy every three to five years after age 50.

If any of these tests reveal possible problems, more extensive studies, such as colonoscopy (exam of the entire colon) and barium enema (an X-ray procedure in which the intestines are viewed), may be needed.

Treatment: Surgery, at times combined with radiation, is the most effective method of treating colorectal cancer. The role of chemotherapy and immunologic agents may be beneficial in postoperative patients with cancerous lymph nodes.

Colostomy (creation of an abdominal opening for elimination of body wastes) is seldom needed for colon cancer and is infrequently required for rectal cancer. The American Cancer Society has a patient assistance program for those who do have permanent colostomies.

Bladder Cancer

Signs and Symptoms: Blood in the urine. Usually associated with increased frequency of urination.

Risk Factors: Smoking is the greatest risk factor in bladder cancer, with smokers experiencing twice the risk of nonsmokers. Smoking is estimated to be responsible for approximately 47% of the bladder cancer deaths among men and 37% among women. People living in urban areas and workers exposed to dye, rubber, or leather also are at higher risk. Chlorination in water has also been associated with a small increase in bladder cancer risk.

Early Detection: Bladder cancer is diagnosed by examination of the bladder wall with a cytoscope, a slender tube fitted with a lens and light that can be inserted into the tract through the urethra.

Treatment: Surgery, alone or in combination with other treatments, is used in over 90% of cases. Preoperative chemotherapy, alone or with radiation before cystectomy (bladder removal), has improved some treatment results.

Lymphoma

Signs and Symptoms: Enlarged lymph nodes, itching, fever, night sweats, anemia, and weight loss. Fever can come and go in periods of several days or weeks.

Risk Factors: Risk factors are largely unknown, but in part involve reduced immune function and exposure to certain infectious agents. Persons with organ transplants are at higher risk due to altered immune function. Human immunodeficiency virus (HIV) and human T lymphocyte leukemia/lymphoma virus-I (HTLV-I) are associated with increased risk of non-Hodgkin lymphoma. Burkitt lymphoma in Africa is partly caused by the Epstein-Barr herpes virus. Other possible risk factors include occupational exposures to herbicides and perhaps other chemicals.

Treatment: Hodgkin disease: chemotherapy and radiotherapy are useful for most patients. Non-Hodgkin lymphoma: early stage, localized lymph node disease can be treated with radiotherapy. Patients with later stage disease often benefit from the addition of chemotherapy. New programs using highly specific monoclonal antibodies directed at lymphoma cells and improved techniques in bone marrow preservation are under investigation in selected patients who relapse after standard treatment.

Cervical Cancer

Signs and Symptoms: Abnormal uterine bleeding or spotting; abnormal vaginal discharge. Pain and systemic symptoms are late manifestations of the disease.

Risk Factors: First intercourse at an early age, multiple sexual partners, or partners who have had multiple sexual partners are at increased risk of developing the disease. Other risk factors include genital warts and cigarette smoking.

Early Detection: The Pap test is a simple procedure that can be performed at appropriate intervals by health care professionals as part of a pelvic exam. A small sample of cells is swabbed from the cervix, transferred to a slide, and examined under a microscope. This test should be performed annually with a pelvic exam in women who are, or have been, sexually active or who have reached age 18 years. After three or more consecutive annual exams with normal findings, the Pap test may be performed less frequently at the discretion of the physician.

Treatment: Cervical cancers generally are treated by surgery or radiation, or by a combination of the two. In precancerous (in situ) stages, changes in the cervix may be treated by cryotherapy (the destruction of cells by extreme cold), by electrocoagulation (the destruction of tissue through intense heat by electric current), or by local surgery.

Skin Cancer

Signs and Symptoms: Any change in the skin, especially a change in the size or color of a mole or other darkly pigmented growth or spot. Scaliness, oozing, bleeding, or change in the appearance of a bump or nodule, the spread of pigmentation beyond its border, a change in sensation, itchiness, tenderness, or pain.

Risk Factors: Excessive exposure to ultraviolet radiation; fair complexion; occupational exposure to coal tar, pitch, creosote, arsenic compounds, or radium; family history.

Protective clothing and sunscreen should be worn. The sun's ultraviolet rays are strongest between 10 A.M. and 3 P.M; exposure at these times should be avoided. Sunscreen comes in various strengths, ranging from those facilitating gradual tanning to those that allow practically no tanning. Because of the possible link between severe sunburns in childhood and greatly increased risk of melanoma in later life, children, in particular, should be protected from the sun.

Early Detection: Early detection is critical. Recognition of changes in skin growths or the appearance of new growths is the best way to find early skin cancer. Adults should practice skin self-exam once a month, and suspicious lesions should be evaluated promptly by a physician. Basal and squamous cell skin cancers often take the form of a pale, waxlike, pearly nodule, or a red, scaly, sharply outlined patch. A sudden or progressive change in a mole's appearance should be checked by a physician. Melanomas often start as small, molelike growths that increase in size, change color, become ulcerated, and bleed easily from a slight injury. A simple **ABCD** rule outlines the warning signals of melanoma: **A** is for asymmetry. One-half of the mole does not match the other half. **B** is for border irregularity. The edges are ragged, notched, or blurred. **C** is for color. The pigmentation is not uniform. **D** is for diameter greater than 6 mm. Any sudden or progressive increase in size should be of special concern.

Treatment: There are five methods of treatment: surgery (used in 90% of cases), radiation therapy, electrodessication (tissue destruction by heat), cryosurgery (tissue destruction by freezing), and laser therapy for early skin cancer. For malignant melanoma, the primary growth must be adequately excised, and it may be necessary to remove nearby lymph nodes. Removal and microscopic examination of all suspicious moles is essential. Advanced cases of melanoma are treated according to the characteristics of the case.

Oral Cavity and Pharynx Cancer

Signs and Symptoms: A sore that bleeds easily and does not heal; a lump or thickening; a red or white patch that persists. Difficulty in chewing, swallowing, or moving tongue or jaws are often late symptoms.

Risk Factors: Cigarette, cigar, or pipe smoking; use of smokeless tobacco; excess use of alcohol.

Early Detection: Cancer can affect any part of the oral cavity, including the lip, tongue, mouth, and throat. Dentists and primary care physicians have the opportunity during regular checkups to see abnormal tissue changes and to detect cancer at an early, curable stage.

Treatment: Principal methods are radiation therapy and surgery. In advanced disease, chemotherapy is being studied as an adjunct to surgery.

Leukemia

Signs and Symptoms: Fatigue, paleness, weight loss, repeated infections, bruising easily, and nosebleeds or other hemorrhages. In children, these symptoms can appear suddenly. Chronic leukemia can progress slowly and with few symptoms.

Risk Factors: Leukemia strikes both sexes and all ages. Causes of most leukemias are unknown. Persons with Down syndrome and certain other genetic abnormalities have higher than usual incidence rate of leukemia. It has also been linked to excessive exposure to ionizing radiation and to certain chemicals such as benzene, a commercially used toxic liquid that is also present in lead-free gasoline. Cer-

tain forms of leukemia and lymphoma are caused by a retrovirus, HTLV-I (human T lymphocyte leukemia/lymphoma virus-I).

Early Detection: Because symptoms often resemble those of other, less serious conditions, leukemia can be difficult to diagnose early. When a physician does suspect leukemia, diagnosis can be made using blood tests and bone marrow biopsy.

Treatment: Chemotherapy is the most effective method of treating leukemia. Various anticancer drugs are used, either in combination or as single agents. Transfusions of blood components and antibiotics are used as supportive treatments. To eliminate hidden cells, therapy of the central nervous system has become standard treatment, especially in acute lymphocytic leukemia. Under appropriate conditions, bone marrow transplantation may be useful in the treatment of certain leukemias.

Pancreas Cancer

Signs and Symptoms: Cancer of the pancreas generally occurs without symptoms until it is in advanced stages and thus is considered a "silent" disease.

Risk Factors: Very little is known about what causes the disease or how to prevent it. Risk increases after age 50, with most cases occurring between ages 65 and 79. Smoking is a risk factor; incidence rates are more than twice as high for smokers as nonsmokers. Some studies have suggested associations with chronic pancreatitis, diabetes, or cirrhosis. In countries where the diet is high in fat, pancreatic cancer rates are higher.

Early Detection: At present, only surgical biopsy yields a certain diagnosis, and because of the "silent" course of the disease, the need for biopsy is likely to be obvious only after the disease has advanced. Researchers are focusing on ways to diagnose pancreatic cancer before symptoms occur. Ultrasound imaging and computerized tomography scans are under investigation.

Treatment: Surgery, radiation therapy, and anticancer drugs are treatment options, but they have had little influence on the outcome. Diagnosis is usually so late that none of these is used.

Ovarian Cancer

Signs and Symptoms: Ovarian cancer is often "silent," showing no obvious signs or symptoms until late in its development. The most common sign is enlargement of the abdomen, which is caused by the accumulation of fluid. Rarely will there be abnormal vaginal bleeding. In women over 40, vague digestive disturbances (stomach discomfort, gas, distention) that persist and cannot be explained by any other cause may indicate the need for a thorough evaluation for ovarian cancer.

Risk Factors: Risk for ovarian cancer increases with age. Women who have never had children are more likely to develop ovarian cancer than those who have. Pregnancy and the use of oral contraceptives appear to be protective against ovarian cancer. Women who have had breast cancer or have a family history of ovarian cancer are at increased risk. With the exception of Japan, industrialized countries have the highest incidence rates.

Early Detection: Periodic, thorough pelvic examinations are important. The Pap test, useful in detecting cervical cancer, only rarely uncovers ovarian cancer. Transvaginal ultrasound and a tumor marker, CA 125, may assist diagnosis. Women over the age of 40 should have a cancer-related checkup every year.

Treatment: Surgery, radiation therapy, and drug therapy are treatment options. Surgery usually includes the removal of one or both ovaries (oophorectomy), the uterus (hysterectomy), and the Fallopian tubes (salpingectomy). In some very early tumors, only the involved ovary will be removed, especially in young women. In advanced disease, an attempt is made to remove all intraabdominal disease to enhance the effect of chemotherapy.

Cancer in Children

New Cases: An estimated 8,300 new cases in 1996. As a childhood disease, cancer is rare. Common sites include the blood and bone marrow, bone, lymph nodes, brain, nervous system, kidneys, and soft tissues.

Some of the main childhood cancers are as follows:

1. Leukemia.

2. Osteogenic sarcoma is a bone cancer that may cause no pain at first; swelling in the area of the tumor is often the first sign. Ewing's sarcoma is another type of cancer that arises in bone.

3. Neuroblastoma can appear anywhere but usually in the abdomen, where a swelling occurs.

4. Rhabdomyosarcoma, the most common soft tissue sarcoma, can occur in the head and neck area, genitourinary area, trunk, and extremities.

5. Brain cancers in early stages may cause headaches, blurred or double vision, dizziness, difficulty in walking or handling objects, and nausea.

6. Lymphomas and Hodgkin disease are cancers that involve the lymph nodes but also may invade bone marrow and other organs. They may cause swelling of lymph nodes in the neck, armpit, or groin. Other symptoms may include general weakness and fever.

7. Retinoblastoma, an eye cancer, usually occurs in children under age 4. When detected early, cure is possible with appropriate treatment.

8. Wilms' tumor, a kidney cancer, may be recognized by a swelling or lump in the abdomen.

Childhood cancers can be treated by a combination of therapies. Treatment is coordinated by a team of experts including oncologic physicians, pediatric nurses, social workers, psychologists, and others who assist children and their families.

Source: From the American Cancer Society, Inc. Cancer Facts & Figures, 1996. Used with permission.

Appendix F

ACRONYMS

Acronym	Meaning
ACh	acetylcholine
AChE	acetylcholinesterase
ACTH	adrenocorticotropic hormone
ACV	acyclovir
AD	Alzheimer disease
ADH	antidiuretic hormone
ADP	adenosine diphosphate
AID	artificial insemination by donor
AIDS	acquired immunodeficiency syndrome
ANH	atrial natriuretic hormone
APC	antigen-presenting cell
ATP	adenosine triphosphate
AV node	atrioventricular node
AZT	azidothymidine
bGH	bovine growth hormone
BP	before present
CAM	crassulacean-acid metabolism
cAMP	cyclic adenosine monophosphate
CAPD	continuous ambulatory peritoneal dialysis
CCK	cholecystokinin
cDNA	complementary deoxyribonucleic acid
CFC	chlorofluorocarbon
cGMP	cyclic guanosine monophosphate
CHNOPS	carbon, hydrogen, nitrogen, oxygen, phosphorus, sulfur
CNS	central nervous system
CoA	coenzyme A
CVA	cardiovascular accident
CVD	cardiovascular disease
DES	diethylstilbestrol
DNA	deoxyribonucleic acid
DTP	diphtheria, tetanus, whooping cough
ECG (or EKG)	electrocardiogram
ELH	egg-laying hormone
ER	endoplasmic reticulum (rough ER, smooth ER)
EST	expressed sequence tag
FAD	flavin adenine dinucleotide
FAP	fixed action patterns
FAS	fetal alcohol syndrome
FSH	follicle-stimulating hormone
GABA	an inhibiting neurotransmitter
GDGF	glial-derived growth factor
GEM	genetically engineered microbe
GH	growth hormone
GIFT	gamete intrafallopian transfer
GIP	gastric inhibitory peptide
GMP	guanosine monophosphate
GnRH	gonadotropic-releasing hormone
GTP	guanosine triphosphate

Acronym	Meaning
Hb	deoxyhemoglobin
HbO_2	oxyhemoglobin
HC	hydrocarbon
HCG	human chorionic gonadotropin
HDL	high-density lipoprotein
HDN	hemolytic disease of the newborn
HepB	hepatitis B
HHb	hemoglobin
Hib	Haemophilus influenza, type b
HIV	human immunodeficiency virus
HPV	human papillomavirus
ICSH	interstitial cell-stimulating hormone
IUD	intrauterine device
IUI	intrauterine insemination
IVF	in vitro fertilization
LDC	less-developed country
LDL	low-density lipoprotein
LH	luteinizing hormone
LPN	licensed practical nurse
LSD	lysergic acid diethylamide
LVN	licensed vocational nurse
MDC	more-developed country
MHC	major histocompatibility complex
MI	myocardial infarction
MMR	measles, mumps, rubella
MPPP	1-methyl-4-phenylprionoxy-piperidine
MSH	melanocyte-stimulating hormone
MYA	millions of years ago
NAD^+	nicotinamide adenine dinucleotide
$NADP^+$	nicotinamide adenine dinucleotide phosphate
NE	norepinephrine
NGF	nerve growth factor
NGU	nongonococcal urethritis
NOX	nitrogen oxides
NPS	nail patella syndrome
OPV	oral polio vaccine
PA	physician's assistant
PAN	peroxyacetyl nitrate
PCB	polychlorinated biphenyl
P_{CO_2}	partial pressure for carbon dioxide
PCR	polymerase chain reaction
PG	prostaglandin
PGA	phosphoglycerate
PGAL	phosphoglyceraldehyde
PGAP	diphosphoglycerate
PHB	polyhydroxybutyrate
PID	pelvic inflammatory disease
PNS	peripheral nervous system
P_{O_2}	partial pressure for oxygen

Acronym	Meaning
PRL	prolactin
PS I	photosystem I
PS II	photosystem II
PTH	parathyroid hormone
rDNA	recombinant deoxyribonucleic acid
RFLP	restriction fragment length polymorphism
RN	registered nurse
RNA	ribonucleic acid
rRNA	ribosomal ribonucleic acid
RuBP	ribulose bisphosphate
SA node	sinoatrial node
SAD	seasonal affective disorder

Acronym	Meaning
SCID	severe combined immunodeficiency syndrome
SEM	scanning electron microscope
SPF	sun protection factor
STD	sexually transmitted disease
Td	tetanus
TEM	transmission electron microscope
THC	tetrahydrocannabinol
tPA	tissue plasminogen activator
TRH	thyroid-releasing hormone
TRIH	thyroid-release-inhibiting hormone
TSH	thyroid-stimulating hormone
UV	ultraviolet

Appendix G

INTERNET GUIDE

The Internet consists of computers that are electronically connected with one another for the purpose of sharing information. The World Wide Web (WWW) is a network of computers from around the world! Quite often when using the Web, you go to a particular "home page," a location from which you can get various types of information. Each home page has a particular address.

Hopefully your institution has a computer lab and you will be able to use a computer there in order to become familiar with accessing and using the Internet. The last page of this appendix lists all the hardware and software that is needed to equip your own computer to access the Internet. The software that permits you to get onto the Web is called a browser because it allows you to "browse," that is, move freely about the Web. The most frequently used browsers are Netscape and Internet Explorer, although other types are available.

This appendix will instruct you on how to use Netscape to reach the Mader Home Page by using the address:

    http://www.mhhe.com/sciencemath/biology/mader/

It will also tell you what information is available via the Mader Home Page.

Accessing the Mader Home Page

When your computer is up and running, double click on the icon for Netscape and type in the Mader Home Page address in the location box as shown in Figure G1. Then, press the return, or enter, key on your keyboard. If you get a message that indicates a problem, you probably made a mistake in typing in the address. Click on O.K. and correct your error before pressing the return key once more. If you get a busy message, simply try again by clicking on O.K. and pressing the return key again.

The Mader Home Page appears as in Figure G2. Notice that the textbook titles are highlighted. Highlighted words (called hyperlinks) allow you to move from the home page to other pages. Click on *Human Biology*, 5/e, and a screen will appear with image icons for **Instructor Information** (supplemental teaching material and additional topic information), **Student Information** (study questions and additional information of general interest), and other topics.

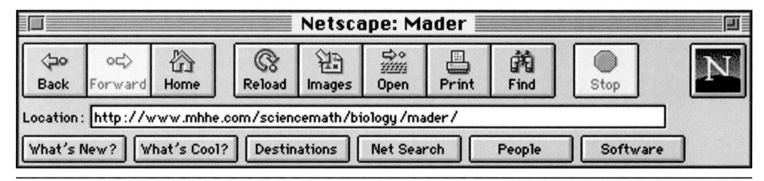

Figure G1   **Accessing the Mader Home Page with Netscape.**

Welcome to the Home Page for

textbooks

by Sylvia S. Mader

- Biology, 5/e
- Inquiry into Life, 8/e
- Human Biology, 5/e (Web site coming in 1997)
- Understanding Human Anatomy and Physiology, 3/e

Don't forget to bookmark our site!!

About the Web   Search Engines

Figure G2 The Mader Home Page.

If you click on Student, a table of contents will appear. You will have your choice of working through test questions, seeking additional information about biological topics, or reading articles specific to each chapter. The study questions are specific to each chapter of *Human Biology,* fifth edition, and will help you study the material.

For the Instructor, additional information is available about biological concepts. The lecture outline specific to the text, teaching techniques, links to science organizations, other web sites of interest, and critical thinking activities are just some of the features available for instructors.

Use the hyperlink at the bottom of the screen to return to the previous page so that you can explore each of these items in turn.

Searching the Internet

Notice that the Mader Home Page has a hyperlink entitled "About the Web." Use this resource to find out more about the World Wide Web. Also, you will probably want to do some research about a particular topic by using a search engine. A search engine is simply a service that searches for information for you. After clicking on Search Engines, type your topic in the space provided and review the responses. Click on any response that interests you. It is possible to print any one of these articles by clicking on the Netscape Print button at the top of the screen. Also, Netscape will allow you to move back and forward between pages by clicking on its forward or back arrow buttons.

Equipment and Software

This is the equipment and software that is needed to access the Internet and the World Wide Web.

PC Users:

1. A VGA or, better, a SVGA monitor.
2. Computer with a 386SX processor or better. A 486SX or a Pentium is best. Things will run pretty slow with a 386, but they will run.
3. 4 MB of RAM is sufficient, but 8 MB is better.
4. 4 MB minimum of free hard disk space.
5. A modem. A 28,800 bps modem or faster is recommended, but a 14,400 is satisfactory.
6. Windows 3.1 or Windows 95.
7. Internet access provider.
8. A Web browser.

Mac Users:

1. A 256-color monitor.
2. A Mac with a 68020, –030, or –040 processor, or a Power Mac.
3. 4 MB of RAM is sufficient, but 8 MB is better.
4. 4 MB minimum of free hard-disk space.
5. A modem. A 28,800 bps modem or faster is recommended, but a 14,400 is satisfactory.
6. System 7 or better, or MacOS.
7. Internet access provider.
8. A Web browser.

Glossary

A

accommodation Lens adjustment to see near objects. 272

acetylcholine (ACh) (uh-set-ul-KOH-leen) Neurotransmitter active in both the peripheral and central nervous systems. 246

acetylcholinesterase (AChE) (uh-set-ul-koh-luh-NES-tuh-rays) Enzyme that breaks down acetylcholine bound to postsynaptic receptors within a synapse. 247

acid Solution in which pH is less than 7; a substance that contributes or liberates hydrogen ions (protons) in a solution. 21

acid deposition Acid rain or snow due to the presence of sulfate and/or nitrate produced by commercial and industrial activities. 496

acquired immunodeficiency syndrome (AIDS) Disease caused by HIV that is characterized by failure of the immune system. 336

actin Protein that forms helical filaments, found in human cells as a cytoskeletal element, and called the thin filaments of myofibrils in muscle fibers. See myosin. 50, 227

actin filament Extremely thin fiber found within the cytoplasm that is composed of the protein actin; involved in the maintenance of cell shape and the movement of cell contents. 50

action potential Polarity changes due to the movement of ions across the plasma membrane of an active neuron. 244

active site Region on the surface of an enzyme where the substrate binds and where the reaction occurs. 52

active transport Transfer of a substance into or out of a cell from a region of lower concentration to a region of higher concentration by a process that requires a carrier and an expenditure of energy. 46

adaptation Fitness of an organism for its environment, including the process by which it becomes fit and is able to survive and reproduce; also, decrease in the excitability of a receptor in response to continuous constant-intensity stimulation. 267, 459

adenine (A) One of four organic bases in the nucleotides composing the structure of DNA and RNA. 417

adipose tissue Connective tissue in which fat is stored. 62

adrenal gland (uh-DREEN-ul) Gland that lies atop a kidney; the adrenal medulla produces the hormones epinephrine and norepinephrine, and the adrenal cortex produces the glucocorticoid and mineralocorticoid hormones. 297

adrenocorticotropic hormone (ACTH) (uh-DREEN-noh-kawrt-ih-koh-TROH-pik) Hormone secreted by the anterior lobe of the pituitary gland that stimulates activity in the adrenal cortex. 294

aerobic cellular respiration Complete breakdown of glucose to carbon dioxide and water in the presence of oxygen. 50

agglutination (uh-gloot-un-AY-shun) Clumping of cells, particularly in reference to red blood cells involved in an antigen-antibody reaction. 118

aging Progressive changes over time, leading to loss of physiological function and eventual death. 372

agranular leukocyte White blood cell that does not contain distinctive granules. 113

albumin Plasma protein of the blood having transport and osmotic functions. 116

aldosterone (al-DAHS-tuh-rohn) Hormone secreted by the adrenal cortex that regulates the sodium and potassium balance of the blood. 195, 298

allantois (ahl-un-TOE-us) Extraembryonic membrane that contributes to the formation of blood vessels. 362

allele (uh-LEEL) Alternative form of a gene located at the same locus on homologous chromosomes. 398

alveolus (pl., alveoli) (al-VEE-uh-lus) Air sac of a lung. 166

amino acid Monomer of a protein; takes its name from the fact that it contains an amino group ($-NH_2$) and an acid group ($-COOH$). 29

amniocentesis (AM-nee-oh-sen-TEE-sis) Removal of a small amount of amniotic fluid to examine the chromosomes and the enzymatic potential of fetal cells. 364

amnion (AM-nee-ahn) Extraembryonic membrane that forms a fluid-filled sac around the embryo. 362

ampulla Base of a semicircular canal in the inner ear. 280

amylase (AM-i-lays) Starch-digesting enzyme secreted by the salivary glands (salivary amylase) and the pancreas (pancreatic amylase). 90

anaphase Stage in mitosis during which sister chromatids separate, forming daughter chromosomes. 387

anemia Inefficient oxygen-carrying ability of blood due to hemoglobin or mature red blood cell shortage. 111

anorexia nervosa Eating disorder caused by the fear of becoming obese; includes loss of appetite and inability to maintain a normal minimum body weight. 103

anterior pituitary Portion of the pituitary gland that produces six types of hormones and is controlled by hypothalamic-releasing and release-inhibiting hormones. 294

antibiotic Medicine that specifically interferes with bacterial metabolism and in that way cures humans of a bacterial disease. 341

antibody Protein produced by plasma cells that binds with a specific antigen. 146

antibody-mediated immunity Specific mechanism of defense in which plasma cells derived from B cells produce antibodies that combine with antigens. 147

anticodon "Triplet" of three bases in transfer RNA that pairs with a complementary triplet (codon) in messenger RNA. 422

antidiuretic hormone (ADH) (ANT-ih-dy-yuu-RET-ik) Hormone secreted by the posterior pituitary that promotes the reabsorption of water by the kidneys. 193, 292

antigen Foreign substance, usually a protein or a polysaccharide, that stimulates the immune system to react, such as to produce antibodies.   146

anus Outlet of the digestive tract.   86

aorta Major systemic artery that receives blood from the left ventricle.   127

appendicular skeleton (ap-un-DIK-yuh-ler) Portion of the skeleton forming the pectoral girdle and arms, and the pelvic girdle and legs.   214

appendix Small, tubular projection that extends outward from the cecum of the large intestine.   86

aqueous humor Clear, watery fluid between the cornea and lens of the eye.   271

aquifer (AHK-wuh-fur) Water-bearing stratum of permeable rock that constitutes and underground reservoir.   498

arteriole Vessel that takes blood from an artery to capillaries.   124

artery Vessel that takes blood away from the heart to arterioles; characteristically possessing thick layers of connective and muscular tissue in the wall.   124

articular cartilage Hyaline cartilaginous covering over the articulating surface of the bones of synovial joints.   205

aster An array of short microtubules that extends outward from a spindle pole in animal cells during cell division.   386

astigmatism Blurred vision due to an irregular curvature of the cornea or the lens.   274

atom Smallest unit of matter that cannot be divided by chemical means.   14

ATP (adenosine triphosphate) (ah-DEN-ah-zeen try-FOS-fayt) Compound having a nitrogen base, ribose, and three phosphate groups; a carrier of energy in cells.   35

atrial natriuretic hormone (ANH) Substance secreted by the atria of the heart that promotes sodium excretion so that blood volume and blood pressure decrease.   195

atrioventricular valve Heart valve located between the atrium and the ventricle.   126

atrium Chamber; particularly an upper chamber of the heart lying above a ventricle; either the left atrium or the right atrium.   126

atrophy Wasting of a muscle due to lack of use.   235

auditory tube Extension from the middle ear to the nasopharynx for equalization of air pressure on the eardrum.   280

australopithecine (aw-strah-loh-PITH-uh-seen) The first generally recognized hominid.   463

autoimmune disease Disease that results when the immune system mistakenly attacks the body's own tissues.   157

autonomic system (awt-uh-NAHM-ik) Branch of the peripheral nervous system that has involuntary control over the internal organs; consists of the sympathetic and parasympathetic systems.   251

autosome Chromosome other than a sex chromosome.   380

AV (atrioventricular) node Small region of neuromuscular tissue that transmits impulses received from the SA node to the ventricular walls.   128

axial skeleton (AK-see-ul) Portion of the skeleton that lies in the midline: the skull, the vertebral column, and the rib cage.   210

axon Fiber of a neuron that conducts nerve impulses away from the cell body.   242

B

B lymphocyte (LIM-fuh-syt) Lymphocyte that matures in the red bone marrow and, when stimulated by the presence of a specific antigen, gives rise to antibody-producing plasma cells.   146

bacterium Unicellular organism lacking a membrane-bounded nucleus and the other membranous organelles typical of eukaryotic cells.   340

ball-and-socket joint The most freely movable type of joint (for example, the shoulder or hip joint).   216

basal body Structure in the cell cytoplasm, located at the base of a cilium or flagellum.   52

base Solution in which pH is greater than 7; a substance that tends to lower the hydrogen ion concentration of a solution; alkaline; opposite of acid. Also, a term commonly applied to one of the components of a nucleotide.   21

basophil Granular leukocyte capable of being stained with a basic dye.   113

bile Secretion of the liver that is temporarily stored in the gallbladder before being released into the small intestine where it emulsifies fat.   88

binary fission Reproduction by division into two daughter cells without the utilization of a mitotic spindle.   340

binomial system Assignment of two names to each organism, the first of which designates the genus and the second of which designates the species.   462

biodiversity Total number of species, the variability of their genes, and the ecosystems in which they live.   5

biological magnification Process by which nonexcreted substances, such as DDT, become more concentrated in organisms in the higher trophic levels of the food chain.   500

biosphere Portion of the surface of the earth where living organisms exist, including air, water, and land.   2

biotechnology Use of a natural biological system to produce a commercial product or achieve an end desired by humans.   426

biotic potential Maximum population growth rate under ideal conditions.   491

blastocyst Early stage of embryonic development that consists of a hollow ball of cells.   360

blind spot Area of the eye containing no rods or cones and where the optic nerve passes through the retina.   271

blood Type of connective tissue in which cells are separated by a liquid called plasma.   64

blood pressure Force of blood pushing against the inside wall of an artery.   130

bone Hardest and most rigid connective tissue whose matrix consists of mineral salts, notably calcium salts.   63

bradykinin (bray-dih-KY-nen) Substance found in damaged tissue that initiates nerve impulses, resulting in the sensation of pain.   145

bronchiole (BRAHNG-kee-ohl) Smaller air passages in the lungs that eventually terminate in alveoli.   166

bronchus (pl., bronchi) (BRAHNG-kus) One of two major divisions of the trachea leading to the lungs.   166

buffer Substance or compound that prevents large changes in the pH of a solution.   22

bulbourethral gland Either of two small structures located below the prostate gland in males; adds secretions to semen.   313

bulimia Eating disorder of binge eating followed by purging.   103

bursa Saclike, fluid-filled structure, lined with synovial membrane, that occurs near a joint.   216

C

calcitonin Hormone secreted by the thyroid gland that helps regulate the blood calcium level.   296

cancer Malignant tumor that metastasizes.   438

capillary Microscopic blood vessel connecting an arteriole to a venule; thin walls permit substances to either exit or enter blood.   124

carbaminohemoglobin (kar-buh-MEE-noh-HEE-muh-gloh-bun) Hemoglobin carrying carbon dioxide.   172

carbohydrate Organic compound characterized by the presence of CH_2O groups; includes monosaccharides, disaccharides, and polysaccharides.   25

carcinogen (kar-SIN-uh-jen) Environmental agent that contributes to the development of cancer.   438

carcinogenesis Development of cancer.   440

carcinoma Cancer arising in epithelial tissue.   443

cardiac muscle Striated, involuntary muscle tissue found only in the heart.   65

carnivore Animal that feeds only on animals.   472

carrier Individual that appears normal but is capable of transmitting an allele for a genetic disorder; protein that combines with and transports a molecule across the plasma membrane.   408

carrying capacity Largest number of organisms of a particular species that can be maintained indefinitely in an ecosystem.   491

cartilage Type of connective tissue having cells within lacunae separated by a flexible matrix.   63

cell Structural and functional unit of an organism; the smallest structure capable of performing all the functions necessary for life.   2, 40

cell body Portion of a nerve cell that includes a cytoplasmic mass and a nucleus, and from which the nerve fibers extend.   242

cell cycle Repeating sequence of events in eukaryotic cells consisting of interphase, when growth and DNA synthesis occurs, and mitosis, when cell division occurs.   385

cell-mediated immunity Specific mechanism of defense in which T lymphocytes destroy antigen-bearing cells.   150

cellulose Polysaccharide composed of glucose molecules; the chief constituent of a plant's cell wall.   26

central nervous system (CNS) Portion of the nervous system consisting of the brain and spinal cord.   242

centriole Short cylinder that contains microtubules in a 9 + 0 pattern; associated with the formation of basal bodies and possibly the miotic spindle.   51

centromere Constricted region of a chromosome where sister chromatids are attached to one another and where the chromosome attaches to a spindle fiber.   385

cerebellum (ser-uh-BEL-um) Part of the brain located inferior to the cerebrum that coordinates skeletal muscles to produce smooth, coordinated motions.   254

cerebral hemisphere One of the large, paired structures that together constitute the cerebrum of the brain.   255

cerebrospinal fluid (suh-ree-broh-SPYN-ul) Fluid found in the ventricles of the brain, in the central canal of the spinal cord, and in association with the meninges.   252

cerebrum (suh-REE-brum) Part of the brain that provides higher mental function and consists of two large masses, or cerebral hemispheres.   255

cervix Narrow end of the uterus, which leads into the vagina.   317

chemoreceptor Sensory receptor that is sensitive to chemical stimulation—for example, receptors for taste and smell.   266

chemotherapy Use of a drug to selectively kill off dividing cancer cells as opposed to nondividing cells.   448

chlamydia (klah-MID-ee-ah) Sexually transmitted disease caused by a tiny bacterium of the same name and particularly characterized by urethritis.   342

chorion (KOR-ee-ahn) Extraembryonic membrane that forms an outer covering around the embryo and contributes to the formation of the placenta.   362

chorionic villi Treelike extensions of the chorion of the embryo, projecting into the maternal tissues.   362

choroid (KOR-oyd) Vascular, pigmented middle layer of the eyeball.   270

chromatid (KROH-muh-tid) One of the two identical parts of a chromosome following replication of DNA.   385

chromatin Threadlike network in the nucleus that is made up of DNA and proteins.   47

chromosome Rodlike structure in the nucleus seen during cell division; contains the hereditary units, or genes.   47

ciliary body Structure associated with the choroid layer that contains ciliary muscle, which controls the shape of the lens of the eye.   270

cilium Short, hairlike extension from a cell, occurring in large numbers and used for cell mobility.   52, 60

circadian rhythm Regular physiological or behavioral event that occurs on an approximately 24-hour cycle.   303

cleavage furrow Indentation that deepens to divide the cytoplasm during cell division.   387

cloned Production of identical copies; in genetic engineering, the production of many identical copies of a gene.   426

clotting Process of blood coagulation, usually when injury occurs.   114

cochlea (KOH-klee-uh) Portion of the inner ear that resembles a snail's shell and contains the organ of Corti, the sense organ for hearing.   280

cochlear canal Canal within the cochlea that contains the organ of Corti.   280

cochlear nerve Cranial nerves that carry nerve impulses from the organ of Corti to the brain, thereby contributing to the sense of hearing; auditory nerve.   280

codon "Triplet" of bases in messenger RNA that directs the placement of a particular amino acid into a protein.   421

coelom First body cavity to appear during development of humans. It is completely lined with mesoderm, one of the first layers of cells.   69

coenzyme Nonprotein organic molecule that aids the action of the enzyme to which it is loosely bound.   53

collecting duct Tube that receives fluid from the distal convoluted tubules of several nephrons within a kidney. 189

colon Large intestine that extends from the cecum to the rectum. 86

colostrum (kul-LAHS-trum) Thin, milky fluid rich in proteins, including antibodies, that is secreted by the mammary glands a few days prior to or after delivery before true milk is secreted. 371

columnar epithelium Pillar-shaped cells usually having the nuclei near the bottom of each cell and found lining the digestive tract. 60

common ancestor Ancestor to two or more branches of evolution. 457

community Group of many different populations that interact with one another within the same environment. 470

compact bone Hard bone consisting of osteons (Haversian systems) cemented together by a matrix that contains calcium phosphate. 63, 204

complement system Group of plasma proteins that form a nonspecific defense mechanism; its cations complement the antigen-antibody reaction. 146

complementary base pairing Pairing of bases between nucleic acid strands; adenine (A) pairs with either thymine (T) if DNA, or uracil (U) if RNA, and cytosine (C) pairs with guanine (G). 417

condensation Chemical change resulting in the covalent bonding of two monomers with the accompanying loss of a water molecule. 25

cone Bright-light receptor in the retina of the eye that detects color and provides visual acuity. 271

conjunctiva Membranous covering on the anterior surface of the eye. 272

connective tissue Type of tissue characterized by cells separated by a matrix and often contains fibers. 62

consumer Member of a population that feeds on members of other populations in an ecosystem. 472

control Sample that undergoes all the steps in the experiment except the one being tested. 9

coronary artery Artery that supplies blood to the wall of the heart. 133

corpus luteum (KOR-pus LOOT-ee-um) Yellow body that forms in the ovary from a follicle that has discharged its egg; it secretes progesterone and estrogen. 319

cortisol Glucocorticoid secreted by the adrenal cortex that affects metabolism and has an anti-inflammatory and immunosupressive effect. 298

covalent bond (coh-VAY-lent) Chemical bond between atoms that results from the sharing of a pair of electrons. 18

cranial nerve Nerve that arises from the brain. 248

creatine phosphate Compound unique to muscles that contains a high-energy phosphate bond. 233

cretinism Condition resulting from improper development of the thyroid in an infant. 295

Cro-Magnon Common name for the first fossils to be accepted as representative of modern humans. 465

crossing-over Exchange of corresponding segments of genetic material between nonsister chromatids of homologous chromosomes during synapsis of meiosis I. 388

cuboidal epithelium Cube-shaped cells found lining the kidney tubules. 60

cytoplasm Semifluid medium between the nucleus and the plasma membrane that contains the organelles. 42

cytosine (C) One of four organic bases in the nucleotides composing the structure of DNA and RNA. 417

cytoskeleton Internal framework of cell that helps maintain the shape of the cell, anchor the organelles, and allow the cell and its organelles to move. 50

cytotoxic T cell T lymphocyte that attacks and kills antigen-bearing cells. 150

D

data Facts that are derived from observations and experiments pertinent to the matter under study. 8

decomposer Organism of decay (fungus and bacterium) in an ecosystem. 472

defecation Discharge of feces from the rectum through the anus. 86

demographic transition Decline in the birthrate following a reduction in the death rate so that the population growth rate is lowered. 492

denaturation Loss of normal shape by an enzyme so that it no longer functions; caused by a less than optimal pH and temperature. 32

dendrite Part of a neuron, typically branched, that conducts signals toward the cell body. 242

denitrification Process of converting nitrate to nitrogen; part of the nitrogen cycle. 479

dermis Thick layer of the skin beneath the epidermis. 66

desertification (di-zurt-uh-fuh-KAY-shun) Transformation of marginal lands to desert conditions. 500

detritus (dih-TRYT-us) Nonliving organic matter. 472

diabetes mellitus Condition characterized by a high blood glucose level and the appearance of glucose in the urine due to a deficiency of insulin secretion or decreased sensitivity of target cells to insulin. 300

dialysis Diffusion of dissolved molecules through a semi-permeable membrane. 198

diaphragm Dome-shaped horizontal sheet of muscle and connective tissue that separates the thoracic cavity from the abdominal cavity. Its contraction helps expand the thoracic cavity during inspiration. 170

diastole Relaxation of a heart chamber. 128

differentiation Process and the developmental stages by which a cell becomes specialized for a particular function. 360

diffusion Movement of molecules from a region of higher concentration to a region of lower concentration. 45

diploid (2n) Number of chromosomes in the body cells; twice the number of chromosomes found in gametes. 384

disaccharide Sugar that contains two units of a monosaccharide; e.g., maltose. 25

distal convoluted tubule Highly coiled region of a nephron that is distant from the glomerular capsule, where tubular secretion takes place. 189

DNA (deoxyribonucleic acid) Nucleic acid found in cells; the genetic material that specifies protein synthesis in cells. 34, 416

DNA ligase (LY-gays) Enzyme that links DNA from two sources; used in genetic engineering to put a gene into plasmid DNA. 427

DNA probe Radioactive single strand of DNA that can be used to locate a particular stretch of DNA. 433

DNA profiling Using fragment lengths resulting from restriction enzyme cleavage to identify particular individuals; also called DNA fingerprinting.   428

dominant allele Hereditary factor that expresses itself in the phenotype when the genotype is heterozygous.   398

doubling time Number of years it takes for a population to double in size.   491

duodenum (doo-uh-DEE-num) First portion of the small intestine into which secretions from the liver and pancreas enter.   85

E

ecological pyramid Pictorial graph representing the biomass, organism number, or energy content of each trophic level in a food web, from the producer to the final consumer populations.   476

ecology Study of the interactions of organisms with each other and with the physical environment.   470

ecosystem Region in which populations interact with each other and with the physical environment.   4, 469

ectoderm Outer germ layer of the embryonic gastrula; it gives rise to the nervous system and skin.   361

edema Swelling due to tissue fluid accumulation in the intercellular spaces.   142

effector Structure such as a muscle or a gland that allows an organism to respond to internal and external stimuli.   249

ejaculation Ejection of seminal fluid.   313

elastic cartilage Cartilage composed of elastic fibers, allowing greater flexibility.   63

electrocardiogram (ECG or EKG) Recording of the electrical activity associated with the heartbeat.   128

electron Subatomic particle that has almost no weight and carries a negative charge; orbits in a shell about the nucleus of an atom.   15

electron transport system Chain of electron carriers in the cristae of mitochondria.   The electrons release energy as they pass down the chain, and this is used to produce ATP.   54

embryo Organism in its early stages of development; first week to two months.   359

endoderm Inner germ layer that lines the archenteron/gut of the gastrula; it becomes the lining of the digestive and respiratory tracts and associated organs.   361

endometrium (en-doh-MEE-tree-um) Lining of the uterus, which becomes thickened and vascular during the uterine cycle.   317

endoplasmic reticulum (ER) (en-doh-PLAZ-mik reh-TIK-yoo-lum) Membranous system of tubules, vesicles, and sacs in cells, sometimes having attached ribosomes. Rough ER has ribosomes; smooth ER does not.   48

endorphin Neuropeptide synthesized in the pituitary gland that suppresses pain.   259

endospore Resistant body formed by bacteria when environmental conditions worsen.   340

environmental resistance Sum total of factors in the environment that limit the numerical increase of a population in a particular region.   491

enzyme Organic catalyst, usually a protein, that speeds up a reaction in cells due to its particular shape.   29

eosinophil Granular leukocyte capable of being stained with the dye eosin.   113

epidermis Outer layer of skin, composed of stratified squamous epithelium.   66

epididymis Coiled tubule next to the testes where sperm mature and may be stored for a short time.   313

epiglottis (ep-uh-GLAHT-us) Structure that covers the glottis and closes off the air tract during the process of swallowing.   82, 166

epinephrine (eh-puh-NEH-frun) Hormone secreted by the adrenal medulla in times of stress; also called adrenaline.   297

epiphyseal plate Cartilaginous layer within the epiphysis of a long bone that functions as a growing region.   206

episiotomy (eh-peez-ee-AHT-oh-mee) Surgical procedure performed during childbirth in which the opening of the vagina is enlarged to avoid tearing.   371

epithelial tissue Tissue type that covers the external surface of the body and lines its cavities.   60

erection Condition of the penis when it is turgid and erect, instead of being flaccid or lacking turgidity.   313

esophagus (i-SAHF-uh-gus) Tube that transports food from the pharynx to the stomach.   82

essential amino acids Amino acids required in the human diet because the body cannot make them.   95

estrogen Female sex hormone, which, along with progesterone, maintains the primary sex organs and stimulates development of the female secondary sexual characteristics.   318

evolution Descent of organisms from common ancestors with the development of genetic and phenotypic changes over time that make them more suited to the local environment.   2, 455

excretion Removal of metabolic wastes from the body.   184

exophthalmic goiter (ek-sahf-THAL-mik) Enlargement of the thyroid gland accompanied by secretion of too much thyroxin.   296

expiration Act of expelling air from the lungs; exhalation.   164

exponential growth Growth, particularly of a population, in which the increase occurs in the same manner as compound interest.   490

extraembryonic membranes Membranes that are not a part of the embryo but are necessary to the continued existence and health of the embryo.   362

F

facilitated transport Passive transfer of a substance into or out of a cell along a concentration gradient by a process that requires a carrier.   46

fermentation Anaerobic breakdown of carbohydrates that results in two ATP and products such as alcohol and lactic acid.   55

fertilization Union of a sperm nucleus and an egg nucleus, which creates a zygote with the diploid number of chromosomes.   358

fiber Plant material that is nondigestible; cellulose and lignin.   94

fibrinogen Plasma protein that is converted into fibrin threads during blood clotting.   114

fibrocartilage Cartilage with a matrix of strong collagenous fibers.   63

fibrous connective tissue Tissue composed mainly of closely packed collagenous fibers and found in tendons and ligaments.   62

fimbria (FIM-bree-uh) Fingerlike extension from the oviduct near the ovary. 317

flagellum Slender, long extension that propels a cell through a fluid medium, occuring as one or two on a cell. 52

focus Manner by which light rays are bent by the cornea and lens, creating an image on the retina. 272

follicle Structure in the ovary that produces the egg and, in particular, the female sex hormones, estrogen and progesterone. 318

follicle-stimulating hormone (FSH) Hormone secreted by the anterior pituitary gland that stimulates the development of an ovarian follicle in a female or the production of sperm in a male. 318

fontanel Membranous region located between certain cranial bones in the skull of a fetus or infant. 210

food chain Pathway by which nutrients and energy are transferred from one organism to another at successive trophic levels. 474

food web Complete set of food links between populations in a community. 474

formed element Constituent of blood that is either cellular (red blood cells and white blood cells) or cellular in origin (platelets). 109

fossil Any remains of an organism that have been preserved in the earth's crust. 456

fovea centralis (FOH-vee-uh sen-TRA-lus) Region of the retina consisting of densely packed cones that is responsible for the greatest visual acuity. 271

G

gallbladder Saclike organ associated with the liver that stores and concentrates bile. 89

gamete One of two types of reproductive cells that join in fertilization to form a zygote; most often an egg or a sperm. 384

ganglion (GANG-glee-un) Collection of neuron cell bodies within the peripheral nervous system. 248

gastric gland Gland within the stomach wall that secretes gastric juice. 84

gene Unit of heredity that codes for a polypeptide and is passed on to offspring. 416

gene therapy Use of genetically engineered cells or other biotechnology techniques to treat genetic or other disorders. 432

genetic engineering Use of technology to alter the genome of a living cell for medical or industrial use. 426

genital herpes Sexually transmitted disease characterized by open sores on the external genitalia. 338

genital warts Sexually transmitted disease caused by papillomavirus and characterized by warts on the genitals. 337

genotype Genes of any individual for (a) particular trait(s). 398

gerontology Study of aging. 372

gland Cell or group of cells that produces a secretion. 60

glomerular capsule (glu-MER-uh-lur) Double-walled cup that surrounds the glomerulus at the beginning of the nephron. 189

glomerular filtrate (glu-MER-uh-lur) Filtered portion of blood contained within the glomerular capsule. 191

glomerular filtration Movement of small molecules from the glomerulus into the glomerular capsule due to the action of blood pressure. 190

glomerulus (glu-MER-uh-lus) Cluster; for example, the knot of capillaries surrounded by the glomerular capsule in a nephron, where glomerular filtration takes place. 188

glottis (GLAHT-us) Opening for airflow into the larynx. 82, 166

glucagon Hormone secreted by the pancreas that causes the liver to break down glycogen and raises the blood glucose level. 300

glucose Six-carbon sugar that organisms degrade as a source of energy during cellular respiration. 25

glycogen Storage polysaccharide found in animals that is composed of glucose molecules joined in a linear fashion but having numerous branches. 25

glycolysis (gly-KOL-uh-sis) Metabolic pathway found in the cytoplasm that participates in aerobic cellular respiration and fermentation; it converts glucose to two molecules of pyruvate. 54

Golgi apparatus Organelle consisting of concentrically folded saccules that functions in the packaging, storage, and distribution of cellular products. 48

gonadotropic hormone (goh-nad-oh-TROHP-ik) Substance secreted by the anterior pituitary that regulates the activity of the ovaries and testes; principally, follicle-stimulating hormone (FSH) and luteinizing hormone (LH). 294

gonorrhea (gahn-ah-REE-ah) Contagious sexually transmitted disease caused by bacteria and leading to inflammation of the urogenital tract. 343

granular leukocyte White blood cell that contains distinctive granules. 113

greenhouse effect Reradiation of solar heat toward the earth, caused by an accumulation of gases such as carbon dioxide in the atmosphere. 494

growth hormone (GH) Hormone secreted by the anterior pituitary that promotes cell division, protein synthesis, and bone growth. 293

growth rate Yearly percentage of increase or decrease in the size of a population. 490

guanine (G) One of four organic bases in the nucleotides composing the structure of DNA and RNA. 417

gumma (GUM-ah) Large unpleasant sore that may occur during the tertiary stage of syphilis. 344

H

haploid (n) Half the diploid number; the number of chromosomes in the gametes. 384

hard palate Bony, anterior portion of the roof of the mouth. 80

heart Muscular organ located in the thoracic cavity responsible for maintenance of blood circulation. 126

helper T cell T lymphocyte that releases lymphokines and stimulates certain other immune cells to perform their respective functions. 150

hemoglobin Iron-containing protein in red blood cells that combines with and transports oxygen. 109

hemolysis Rupture of red blood cells. 111

hepatitis Inflammation of the liver. 339

herbivore Animal that feeds directly on plants. 472

herpes simplex virus Cold sore virus (type 1) and genital herpes virus (type 2). 338

hexose Six-carbon sugar. 25

hinge joint Type of joint characterized by a convex surface of one bone fitting into a concave surface of another bone so that movement is confined to one place, such as in the knee or interphalangeal joint.   216

histamine Substance produced by basophils or mast cells that promotes some of the reactions associated with inflammation and allergies.   145

HLA (human lymphocyte associated) Protein in a plasma membrane that identifies the cell as belonging to a particular individual and acts as a self-antigen.   151

homeostasis Maintenance of a dynamic equilibrium, such that temperature, blood pressure, and other body conditions remain within narrow limits.   2, 72

hominid Common term for member of a family of upright, bipedal primates that includes australopithecines and humans.   463

Homo erectus Extinct earliest nondisputed species of humans, named for their erect posture, which allowed them to walk as we do.   464

Homo habilis Extinct species that may include the earliest humans; had a small brain but made quality tools.   464

homologous chromosome Similarly constructed; with chromosomes, the same appearance and containing genes for the same traits.   388

Homo sapiens Present day species of humans that has a brain capacity larger than 1,000 cc.   465

hormone Chemical messenger produced in low concentrations that has physiological and/or developmental effects, usually in another part of the organism.   288

host Organism a parasite lives on or in, to obtain nutrients.   334

human immunodeficiency virus (HIV) Virus responsible for AIDS.   336

human papillomavirus (HPV) (pap-ih-LOH-mah-vy-rus) Human genital wart virus.   337

hydrogen bond Weak attraction between a hydrogen atom with a partial positive charge and an atom of another molecule with a partial negative charge.   19

hydrolysis (hy-DRAH-lih-sis) Splitting of a covalent bond by the addition of water.   25

hydrolytic enzyme Enzyme that catalyzes a reaction in which the substrate is broken down with the addition of water.   90

hypertonic Containing a higher concentration of solute and a lower concentration of water than the cell.   45

hypertrophy Increase in muscle size following long-term exercise.   235

hypothalamus (hy-poh-THAL-uh-mus) Part of the brain located below the thalamus that helps regulate the internal environment of the body and produces releasing factors that control the anterior pituitary.   254

hypothesis Statement that is capable of explaining present data and is to be tested by future observations and experimentation.   8

hypotonic Containing a lower concentration of solute and a higher concentration of water than the cell.   45

I

immunity Ability of the body to protect itself from specific foreign substances and cells, including infectious pathogens.   144

immunotherapy Use of any immune system component such as antibodies, cytotoxic T cells, or lymphokines to promote the health of the body, such as curing cancer.   450

implantation Attachment and penetration of the embryo into the lining of the uterus (endometrium).   325, 359

incus Middle of three ossicles of the ear that serves to conduct vibrations from the tympanic membrane to the oval window of the inner ear.   277

induction Ability of a chemical or a tissue to influence the development of another tissue.   361

inflammatory reaction Tissue response to injury that is characterized by redness, swelling, pain, and heat.   144

inner ear Portion of the ear consisting of a vestibule, semicircular canals, and the cochlea; where balance is maintained and sound is transmitted.   280

insertion End of a muscle that is attached to a movable bone.   222

inspiration Act of taking air into the lungs; inhalation.   164

insulin Hormone secreted by the pancreas that lowers the blood glucose level by promoting the uptake of glucose by cells and the conversion of glucose to glycogen by the liver and skeletal muscles.   300

interferon (in-tur-FIR-ahn) Protein formed by a cell infected with a virus that can increase the resistance of other cells to the virus.   146

interneuron Neuron found within the central nervous sytem that takes nerve impulses from one portion of the system to another.   242

interphase Interval between successive cell divisions; during this time, the chromosomes are extended, and DNA replication and growth are occurring.   385

ion Atom or group of atoms carrying a positive or negative charge.   16

ionic bond Bond created by an attraction between oppositely charged ions.   17

iris Muscular ring that surrounds the pupil and regulates the passage of light through this opening.   270

isometric contraction Muscular contraction in which the muscle fails to shorten.   231

isotonic Containing the same concentration of solute and water as the cell.   45

isotonic contraction Muscular contraction in which the muscle shortens.   231

isotope One of two or more atoms with the same atomic number that differ in the number of neutrons and therefore in weight.   15

J

joint Union of two or more bones; an articulation.   204

K

karyotype (KAR-ee-uh-typ) Arrangement of all the chromosomes within a cell by pairs in a fixed order.   380

kidney Organ in the urinary system that produces and excretes urine.   184

kingdom Classification category into which organisms are placed: Monera, Protists, Fungi, Plants, and Animals.   2

Krebs cycle Cyclical metabolic pathway found in the matrix of mitochondria that participates in aerobic cellular respiration; breaks down acetyl groups to carbon dioxide and hydrogen.   54

L

labia majora (LAY-bee-uh) Outer folds of the vulva.   317

labia minora Inner folds of the vulva.   317

lacteal (LAK-tee-ul) Lymphatic vessel in a villus of the intestinal wall.   85

lacuna Small pit or hollow cavity, as in bone or cartilage, where a cell or cells are located.   63

lanugo (lah-NOO-goh) Short, fine hair that is present during the later portion of fetal development.   369

large intestine Last major portion of the digestive tract, extending from the small intestine to the anus and consisting of the cecum, the colon, the rectum, and the anal canal.   86

larynx (LAR-ingks) Cartilaginous organ located between pharynx and trachea that contains the vocal cords; voice box.   166

lens Clear membranelike structure found in the eye behind the iris; brings light rays to focus on the retina.   270

leukemia (loo-KEE-MEE-AH) Cancer of the blood-forming tissues leading to the overproduction of abnormal white blood cells.   443

ligament Fibrous connective tissue that joins bone to bone at a joint.   62, 216

limbic system Portion of the brain concerned with memory and emotions.   255

lipase Fat-digesting enzyme secreted by the pancreas.   90

lipid (LIP-id) Organic compound that is insoluble in water; notably fats, oils, and steroids.   27

liver Large organ in the abdominal cavity that has many functions, such as production of proteins and bile and detoxification of harmful substances.   88

loop of the nephron (NEF-rahn) Portion of the nephron lying between the proximal convoluted tubule and the distal convoluted tubule that functions in water reabsorption.   189

loose connective tissue Tissue that is composed mainly of fibroblasts separated by collagenous and elastin fibers and that is found beneath epithelium.   62

lumen Cavity inside any tubular structure, such as the lumen of the digestive tract.   83

luteinizing hormone (LH) (LOOT-ee-nyzing) Hormone produced by the anterior pituitary gland that stimulates the development of the corpus luteum in the ovary in females and the production of testosterone in males.   318

lymph Fluid derived from tissue fluid that is carried in lymphatic vessels.   117, 142

lymph nodes Mass of lymphoid tissue located along the course of a lymphatic vessel.   143

lymphatic system Mammalian organ system consisting of lymphatic vessels and lymphoid organs.   142

lymphocyte Specialized leukocyte that functions in specific defense; occurs in two forms—T lymphocyte and B lymphocyte.   113

lymphokine (LIM-fuh-kyn) Molecule secreted by lymphocytes that has the ability to affect the activity of all types of immune cells.   150

lymphoma (lim-FOH-mah) Cancer of the lymphoid organs such as lymph nodes, spleen, and thymus gland.   443

lysosome Membrane-bounded organelle containing digestive enzymes.   49

M

macrophage Large phagocytic cell derived from a monocyte that ingests pathogens and debris.   145

malleus (MA-lee-us) First of three ossicles of the ear that serves to conduct vibrations from the tympanic membrane to the oval window of the inner ear.   277

maltase Enzyme produced in small intestine that breaks down maltose to two glucose molecules.   90

matrix Intercellular substance of connective tissue.   62

mechanoreceptor Sensory receptor that is sensitive to mechanical stimulation, such as that from pressure, sound waves, and gravity.   266

medulla oblongata Portion of the brain located between the pons and the spinal cord that is concerned with the control of internal organs.   253

medullary cavity Cavity within the diaphysis of a long bone occupied by yellow bone marrow.   205

meiosis (my-OH-sis) Type of cell division occurring during the production of gametes; results in four daughter cells with the haploid number of chromosomes.   388

meiosis I That portion of meiosis during which homologous chromosomes come together and then later separate.   388

meiosis II That portion of meiosis during which sister chromatids separate.   388

memory B cell Persistent population of B cells ready to produce antibodies specific to a particular antigen; accounts for the development of active immunity.   147

memory T cell Persistent population of T cells ready to recognize an antigen that previously invaded the body.   151

meninges (sing., meninx) (muh-NIN-jeez) Protective membranous coverings about the central nervous system.   252

menisci (sing., meniscus) (mun-NIS-ky) Cartilaginous wedges that separate the surfaces of bones in synovial joints.   216

menopause Termination of the ovarian and uterine cycles in older women.   322

menstruation Loss of blood and tissue from the uterus at the end of a uterine cycle.   321

mesoderm Middle germ layer of the embryonic gastrula; gives rise to the muscles, the connective tissue, and the circulatory system.   361

messenger RNA (mRNA) Ribonucleic acid whose sequence of codons specifies the sequence of amino acids during protein synthesis.   419

metabolism All of the chemical changes within a cell (or an organism) including breakdown reactions and synthetic reactions.   52

metaphase Stage in mitosis during which chromosomes are at the equator of the mitotic spindle.   387

metastasis (meh-TAS-tuh-sus) Spread of cancer from the place of origin throughout the body, caused by the ability of cancer cells to migrate and invade tissues.   437

MHC (major histocompatibility complex) protein Membrane protein that serves to identify the cells of a particular individual.   151

microtubule Cytoskeletal element composed of 13 rows of globular proteins called tubulin; also found in multiple units within other structures, such as the centriole, cilia, flagella, as well as spindle fibers.   50

microvillus Cytoplasmic projection from epithelial cells, usually containing actin filaments; microvilli greatly increase the surface area of a tissue. 60

midbrain Most superior part of the brain stem that contains traits and reflex centers. 253

middle ear Portion of the ear consisting of the tympanic membrane and the ossicles; where sound is amplified. 277

mineral Homogeneous inorganic substance; certain minerals are required for normal metabolic functioning of cells and must be in the diet. 100

mitochondrion Organelle where ATP is produced during aerobic cellular respiration; the powerhouse of the cell. 50

mitosis Type of cell division in which daughter cells receive the exact chromosome and genetic makeup of the parent cell; occurs during growth and repair. 385

molecule Like or different atoms joined by a bond; the unit of a compound. 16

monoclonal antibody Antibody of one type produced by a single plasma cell. 155

monocyte Type of agranular leukocyte that functions as a phagocyte. 113

monosaccharide Simple sugar; a carbohydrate that cannot be decomposed by hydrolysis. 25

motor neuron Neuron that takes nerve impulses from the central nervous sytem to an effector; also known as an efferent neuron. 242

motor unit Motor neuron and the muscle fibers associated with it. 231

multiple allele Pattern of inheritance in which there are more than two alleles for a particular trait, although each individual has only two of these alleles. 406

muscular tissue Type of body tissue that is specialized for contraction. 65

mutagen (MYOOT-uh-jen) Environmental agent that induces mutations. 438

mutation Change in the genetic material. 418

myelin sheath (MY-uh-lun) Schwann plasma membranes that cover long neuron fibers, giving them a white, glistening appearance. 243

myocardium Cardiac muscle in the wall of the heart. 126

myofibril Contractile portion of a muscle fiber. 227

myoglobin Pigmented compound in muscle tissue that stores oxygen. 233

myosin (MY-uh-sun) One of two major proteins of muscle; makes up thick filaments in myofibrils and is capable of breaking down ATP and pulling actin filaments. See actin. 227

myxedema Condition resulting from a deficiency of thyroid hormone in an adult. 295

N

natural selection Process by which populations become adapted to their environment. 459

Neanderthal Common name for an extinct subspecies of humans whose remains are found in Europe and Asia. 465

negative feedback Mechanism activated by an imbalance that returns conditions to their original value. 73

nephron (NEF-rahn) Anatomical and functional unit of the kidney; kidney tubule. 187

nerve Bundle of fibers outside the central nervous system. 66, 248

nerve impulse Action potential (electrochemical change) traveling along a neuron. 244

neuroglial cell (nuh-ROH-glee-uhl) One of several types of cells found in nervous tissue that supports, protects, and nourishes neurons. 66, 243

neuromuscular junction Point of communication between a neuron and a muscle fiber. 228

neuron Nerve cell that characteristically has three parts: dendrites, cell body, and axon; occurs as sensory neuron, motor neuron, and interneuron. 66, 242

neurotransmitter Chemical stored at the ends of axons that is responsible for signal transmission across a synapse. 246

neutron Subatomic particle that has a weight of one atomic mass unit, carries no charge, and is found in the nucleus of an atom. 15

neutrophil Granular leukocyte that is the most abundant of the white blood cells; first to respond to infection. 113

niche (nich) Functional role of an organism including how it interacts with both the biotic and the abiotic components of an ecosystem. 472

nitrogen fixation Process whereby nitrogen is reduced to ammonia, which is then converted to organic compounds. 479

node of Ranvier Gap in the myelin sheath around a nerve fiber. 243

nondisjunction Failure of homologous chromosomes to separate during meiosis I or daughter chromosomes to separate during meiosis II when gametogenesis is occuring. 392

norepinephrine (NE) (nor-eh-puh-NEH-frun) Neurotransmitter active in the peripheral and central nervous systems; also, a hormone secreted by the adrenal medulla in times of stress. 246, 297

nuclear envelope Double membrane that surrounds the nucleus and is continuous with the endoplasmic reticulum. 47

nuclear pore Opening in the nuclear envelope which permits the passage of proteins into the nucleus and ribosomal subunits out of the nucleus. 47

nucleolus A special region found inside the nucleus where rRNA is produced for ribosome formation. 47

nucleotide Monomer of a nucleic acid that forms when a nitrogen base, a pentose sugar, and a phosphate join. 34

nucleus (NOO-klee-us) Region of a eukaryotic cell, containing chromosomes, that controls the structure and function of the cell. 47

O

obesity Excess adipose tissue; exceeding desirable weight by more than 20%. 103

olfactory cell (ahl-FAK-tuh-ree) A neuron modified as a receptor for the sense of smell. 269

omnivore Animal that feeds on both plants and animals. 472

oncogene (ONG-koh-jeen) Gene that contributes to the transformation of a normal cell into a cancer cell. 444

oogenesis (oh-oh-JEN-eh-sis) Production of an egg in females by the processes of meiosis and maturation. 393

opportunistic infection Infection that has an opportunity to occur because the immune system has been weakened. 336

optic nerve Either of two cranial nerves that carry nerve impulses from the retina of the eye to the brain, thereby contributing to the sense of sight. 271

organelle Specialized membranous structure within cells (e.g., nucleus, mitochondria, and endoplasmic reticulum) that performs specific functions. 42

organ of Corti Hair cells that function as hearing receptors within the inner ear. 280

orgasm Physical and emotional climax during sexual intercourse; results in ejaculation in the male. 313

origin End of a muscle that is attached to a relatively immovable bone. 222

osmosis Movement of water from an area of higher concentration of water to an area of lower concentration of water across a selectively permeable membrane. 45

ossicle One of the small bones of the middle ear—malleus, incus, stapes. 277

osteoblast Bone-forming cell. 206

osteoclast Cell that breaks down and causes reabsorption of bone. 206

osteocyte Mature bone cell. 206

otolith (OH-tul-ith) Calcium carbonate granule associated with stereociliated cells in the utricle and the saccule. 280

outer ear Portion of ear consisting of the pinna and auditory canal. 277

ovarian cycle Monthly changes occurring in the ovary that determine the level of sex hormones in the blood. 319

ovary Female gonad, the organ that produces eggs, estrogen, and progesterone. 317

oviduct Tube that transports eggs to the uterus; also called uterine tube. 317

ovulation Discharge of a mature egg from the follicle within the ovary. 317

oxidation Loss of one or more electrons from an atom or molecule with a concurrent release of energy; in biological systems, generally the loss of hydrogen atoms. 19

oxygen debt Oxygen that is needed to metabolize lactate, a compound that accumulates during vigorous exercise. 233

oxyhemoglobin (ahk-sih-HEE-muh-gloh-bun) Compound formed when oxygen combines with hemoglobin. 172

oxytocin (ahk-sih-TO-sin) Hormone released by the posterior pituitary that causes contraction of the uterus and milk letdown. 292

ozone shield Formed from oxygen in the upper atmosphere, ozone (O_3) protects the earth from ultraviolet radiation. 496

P

pacemaker See SA (sinoatrial) node. 128

pain receptor Sensory receptor that is sensitive to chemicals released by damaged tissues or excess stimuli of heat or pressure. 266

pancreas Elongated, flattened organ in the abdominal cavity that secretes digestive enzymes into the small intestine (exocrine function) and hormones into the blood (endocrine function). 88, 300

pancreatic islets (of Langerhans) Distinctive group of cells within the pancreas that secretes insulin and glucagon. 300

Pap smear Analysis done on cervical cells for detection of cancer. 317

parasympathetic system Part of the autonomic system that usually promotes activities associated with a restful state. 251

parathyroid gland (par-uh-THY-royd) One of four glands embedded in the posterior surface of the thyroid gland; produces parathyroid hormone. 296

parathyroid hormone (PTH) Hormone secreted by parathyroid glands that increases the blood calcium level and decreases the blood phosphate level. 296

partial pressure Pressure produced by one gas in a mixture of gases. 172

parturition (par-too-RISH-un) Birth of a human and the expulsion of the extraembryonic membranes through the terminal portion of the female reproductive tract. 370

pathogen Disease-causing agent. 144, 335

pectoral girdle Portion of the skeleton to which the arms are attached. 214

pelvic girdle Portion of the skeleton to which the legs are attached. 215

pelvic inflammatory disease (PID) Disease state of the reproductive organs usually caused by a chlamydial infection or gonorrhea. 342

penis External organ in males through which the urethra passes and that serves as the organ of sexual intercourse. 313

pentose Five-carbon sugar; deoxyribose is the pentose sugar found in DNA; ribose is the pentose sugar found in RNA. 34

pepsin Protein-digesting enzyme secreted by gastric glands. 90

peptidase Intestinal enzyme that breaks down short chains of amino acids to individual amino acids that are absorbed across the intestinal wall. 90

peptide bond Covalent bond that joins two amino acids. 30

periodontitis Inflammation of the gums. 81

peripheral nervous system (PNS) Nerves and ganglia that lie outside the central nervous system. 242

peristalsis (per-uh-STAWL-sus) Rhythmic contraction that serves to move the contents along in tubular organs, such as the digestive tract. 82

peritubular capillary Capillary that surrounds a nephron and functions in reabsorption during urine formation. 188

pH scale Measure of the hydrogen ion concentration [H^+]; any pH below 7 is acidic and any pH above 7 is basic. 22

phagocytosis (fag-oh-suh-TOH-sis) Taking in of bacteria and/or debris by engulfing; cell eating. 113

pharynx (FAR-ingks) Portion of the digestive tract between the mouth and the esophagus that serves as a passageway for food and also air on its way to the trachea. 82

phenotype (FEE-noh-typ) Outward appearance of an organism caused by the genotype and environmental influences. 398

pheromone Chemical substance secreted by one organism that influences the behavior of another. 305

phospholipid Molecule making up the bilayer of cellular membranes; has a hydrophilic head and hydrophobic tails. 28

photochemical smog Air pollution that contains nitrogen oxides and hydrocarbons, which react to produce ozone and PAN (peroxyacetyl nitrate). 497

photoreceptor Light-sensitive receptor. 266

pineal gland (PY-nee-ul) Gland located near the surface of the body (fish, amphibians) or in the third ventricle of the brain (mammals) that produces melatonin. 303

pituitary gland Small gland that lies just inferior to the hypothalamus; the anterior pituitary secretes several hormones, some of which control other endocrine glands; the posterior pituitary stores and secretes oxytocin and antidiuretic hormone. 292

placenta Structure formed from the chorion and uterine tissue through which nutrient and waste exchange occur for the embryo and later the fetus. 362

plasma Liquid portion of blood. 116

plasma cell Cell derived from a B lymphocyte that is specialized to mass-produce antibodies. 147

plasma membrane Membrane that surrounds the cytoplasm of cells and regulates the passage of molecules and ions into and out of the cell. 42

plasmid Circular DNA segment that is present in bacterial cells that replicates independently from the bacterial chromosome. 426

platelet Fragment of a megakaryocyte; formed element that is necessary to blood clotting. 64, 114

pleural membrane (PLUR-al) Serous membrane that encloses the lungs. 170

polar body Nonfunctioning daughter cell that has little cytoplasm and is formed during oogenesis. 393

pollutant Substance that is added to the environment and leads to undesirable effects for living organisms. 494

pollution Detrimental alteration of the normal constituents of air, land, and water due to human activities. 481

polygenic inheritance Pattern of inheritance in which more than one allelic pair controls a trait; each dominant allele has a quantitative effect on the phenotype. 405

polymerase chain reaction (PCR) Technique that uses the enzyme DNA polymerase to produce millions of copies of a particular piece of DNA. 428

polypeptide Polymer of many amino acids linked by peptide bonds. 30

polyribosome Cluster of ribosomes attached to the same mRNA molecule; each ribosome is producing a copy of the same polypeptide. 47, 422

population All the organisms of the same species in one place. 470

posterior pituitary Portion of the pituitary gland that stores and secretes oxytocin and antidiuretic hormone, which are produced by the hypothalamus. 292

primate Animal that belongs to the order Primates; the order of mammals that includes monkeys, apes, and humans. 463

producer Organism at the start of a food chain that makes its own food (e.g., green plants on land and algae in water). 472

progesterone (proh-JES-tuh-rohn) Female sex hormone secreted by the corpus luteum of the ovary and by the placenta. 318

prolactin (PRL) Hormone secreted by the anterior pituitary that stimulates the production of milk from the mammary glands. 294

prophase Early stage in mitosis during which chromatin condenses so that chromosomes appear. 386

proprioceptor (proh-pree-oh-SEP-tur) Sensory receptor in the muscles that assists the brain in knowing the position of the limbs. 266

prostaglandins (PG) Hormones that have various and powerful local effects. 303

prostate gland Gland located around the male urethra below the urinary bladder; adds secretions to semen. 313

protein Organic compound that is composed of either one or several polypeptides. 29

prothrombin Plasma protein that is converted to thrombin during the steps of blood clotting. 114

proton Subatomic particle found in the nucleus of an atom that has a weight of one atomic mass unit and carries a positive charge; a hydrogen ion. 15

proto-oncogene (PROH-toh-ONG-koh-jeen) Normal gene involved in cell growth and differentiation that becomes an oncogene through mutation. 444

proximal convoluted tubule Highly coiled region of a nephron near the glomerular capsule, where tubular reabsorption takes place. 189

pulmonary artery Blood vessel that takes blood away from the heart to the lungs. 127

pulmonary circuit That part of the circulatory system that takes deoxygenated blood to and oxygenated blood away from the gas-exchanging surfaces in the lungs. 132

pulmonary vein Blood vessel that takes blood to the heart from the lungs. 127

pulse Vibration felt in arterial walls due to expansion of the aorta following ventricular contraction. 128

Punnett square Gridlike device used to calculate the expected results of simple genetic crosses. 400

pupil Opening in the center of the iris of the eye. 270

R

receptor Sensory structure specialized to receive information from the environment and to generate nerve impulses; also, a protein located in the plasma membrane or within the cell that binds to a substance that alters some aspect of the cell. 249, 266

recessive allele Hereditary factor that expresses itself in the phenotype only when the genotype is homozygous for this allele. 398

recombinant DNA DNA that contains genes from more than one source. 426

red blood cell (erythrocyte) Formed element that contains hemoglobin and carries oxygen from the lungs to the tissues. 64, 109

red bone marrow Blood-cell-forming tissue located in the spaces within spongy bone. 144, 204

reduced hemoglobin Hemoglobin that is carrying hydrogen ions. 172

reduction Gain of electrons by an atom or molecule with a concurrent storage of energy; in biological systems, generally the gain of hydrogen atoms. 19

reflex Automatic, involuntary response of an organism to a stimulus. 249

replacement reproduction Population in which each person is replaced by only one child. 493

replication Making an exact copy, as in the duplication of DNA. 418

reproduce To make a copy similar to oneself, as when unicellular organisms divide or humans have children. 2

residual volume Amount of air remaining in the lungs after a forceful expiration. 168

respiratory center Group of nerve cells in the medulla oblongata that sends out nerve impulses on a rhythmic basis, resulting in inspiration.   170

resting potential Polarity across the plasma membrane of a resting fiber due to an unequal distribution of ions.   244

restriction enzyme Enzyme that stops viral reproduction by cutting viral DNA; used in genetic engineering to cut DNA at specific points.   427

retina Innermost layer of the eyeball, which contains the rods and the cones.   271

retrovirus Virus that contains only RNA and carries out RNA to DNA transcription, called reverse transcription.   335

Rh factor One type of antigen on red blood cells.   119

rhodopsin (roh-DAHP-sun) Visual pigment found in the rods, whose activation by light energy leads to vision.   276

rib cage Top and sides of the thoracic cavity; contains ribs and intercostal muscles.   170

ribosomal RNA (rRNA) RNA occurring in ribosomes, structures involved in protein synthesis.   419

ribosome Minute particle that is attached to endoplasmic reticulum or occurs loose in the cytoplasm and is the site of protein synthesis.   47

ribozyme Enzyme that carries out mRNA processing.   421

RNA (ribonucleic acid) Nucleic acid found in cells that assists DNA in controlling protein synthesis.   34, 419

rod Dim-light receptor in the retina of the eye that detects motion but no color.   271

rough ER Endoplasmic reticulum having attached ribosomes.   48

S

SA (sinoatrial) node Small region of neuromuscular tissue that initiates the heartbeat; also called the pacemaker.   128

saccule Saclike cavity in the vestibule of the inner ear; contains receptors for static equilibrium.   280

salivary gland Gland associated with the oral cavity that secretes saliva.   81

sarcolemma (sahr-kuh-LEM-uh) Plasma membrane of a muscle fiber.   227

sarcoma (sar-KOH-mah) Cancer that arises in connective tissue, such as muscle, bone, and fibrous connective tissue.   443

sarcomere (SAHR-kuh-mir) Structural and functional unit of a myofibril; contains actin and myosin filaments.   227

scientific method Step-by-step process for discovery and generation of knowledge, ranging from observation and hypothesis to theory and principle.   8

sclera (SKLER-uh) White, fibrous, outer layer of the eyeball.   270

scrotum Pouch of skin that encloses the testes.   312

selectively permeable Ability of plasma membranes to regulate the passage of substances into and out of the cell, allowing some to pass through and preventing the passage of others.   45

semen Thick, whitish fluid consisting of sperm and secretions from several glands of the male reproductive tract.   313

semicircular canal One of three tubular structures within the inner ear that contains receptors responsible for the sense of dynamic equilibrium.   280

semilunar valve Valve with flap resembling a half moon located between the ventricles and their attached blood vessels.   126

seminal vesicle Convoluted, saclike structure attached to the vas deferens near the base of the urinary bladder in males; adds secretions to semen.   313

sensory neuron Neuron that takes the nerve impulse to the central nervous system; also known as the afferent neuron.   242

septum Partition or wall that divides two areas; the septum in the heart separates the right half from the left half.   126

serum Light yellow liquid left after clotting of blood.   114

sex chromosome Chromosome responsible for the development of characteristics associated with gender; an X or Y chromosome.   380, 408

sex-influenced trait Autosomal trait that is expressed differently in the two sexes. Usually attributed to hormonal influence.   411

sex-linked Allele located on a sex chromosomes, usually the X chromosome.   408

simple goiter Condition in which an enlarged thyroid produces low levels of thyroxin.   295

skeletal muscle Striated, voluntary muscle tissue that comprises skeletal muscles; also called striated muscle.   65

sliding filament theory Movement of actin in relation to myosin; accounts for muscle contraction.   227

small intestine Long, tubelike chamber of the digestive tract between the stomach and large intestine.   85

smooth ER Endoplasmic reticulum that does not have attached ribosomes.   48

smooth muscle Nonstriated, involuntary muscles found in the walls of internal organs.   65

sodium-potassium pump Plasma membrane transport protein that moves sodium ions out of and potassium ions into cells; important in nerve and muscle cells.   244

soft palate Entirely muscular posterior portion of the roof of the mouth.   80

somatic cell In animals, a body cell excluding those that undergo meiosis and become a sperm or egg.   384

somatic system Portion of the peripheral nervous system containing motor neurons that control skeletal muscles.   248

sperm Male sex cell with three distinct parts at maturity: head, middle piece, and tail.   315

spermatogenesis (SPUR-mah-toh-JEN-eh-sis) Production of sperm in males by the process of meiosis and maturation.   393

sphincter Muscle that surrounds a tube and closes or opens the tube by contracting and relaxing.   83

spinal nerve Nerve that arises from the spinal cord.   248

spindle Structure consisting of fibers, poles, and asters (if animal cell) that brings about the movement of chromosomes during cell division.   386

spindle fiber Microtubule bundle in eukaryotic cells that is involved in the movement of chromosomes during mitosis and meiosis.   386

spleen Large, glandular organ located in the upper left region of the abdomen that stores and purifies blood.   143

spongy bone Porous bone found at the ends of long bones where blood cells are formed. 63, 204

squamous epithelium Flat cells found lining the lungs and blood vessels. 60

stapes (STAY-peez) Last of three ossicles of the ear that serves to conduct vibrations from the tympanic membrane to the oval window of the inner ear. 277

starch Storage polysaccharide found in plants that is composed of glucose molecules joined in a linear fashion. 25

stomach Muscular sac that mixes food with gastric juices to form chyme, which enters the small intestine. 84

succession Series of ecological stages by which the community in a particular area gradually changes until there is a climax community that can maintain itself. 470

suppressor T cell T lymphocyte that suppresses certain other T and B lymphocytes from continuing to divide and perform their respective functions. 150

sustainable world Global way of life that can continue indefinitely because the economic needs of all peoples are met while still protecting the environment. 504

sympathetic system Part of the autonomic system that usually promotes activities associated with emergency (fight or flight) situations. 251

synapse (SIN-aps) Region between two neurons where information is transmitted from one to the other, usually from the axon to the dendrite or cell body of the next neuron. 246

synapsis Attracting and pairing of homologous chromosomes during prophase I of meiosis. 390

synaptic cleft Small gap between presynaptic and postsynaptic membranes of a synapse. 246

synovial joint Freely movable joint. 216

synthesis To build up, such as the combining of two smaller molecules to form a larger molecule. 2

syphilis (SIF-ih-lus) Chronic, contagious sexually transmitted disease caused by a bacterium that is a spirochete. 344

systemic circuit That part of the circulatory system that serves body parts and does not include gas-exchanging surfaces in the lungs. 132

systole Contraction of a heart chamber. 128

T

T lymphocyte (LIM-fuh-syt) One of four types of T cells; cytotoxic T cells mature in the thymus and kill antigen-bearing cells outright. 146

taste bud Sense organ containing the receptors associated with the sense of taste. 268

taxonomy Science of naming, classifying, and showing relationships of organisms. 462

telophase Stage of mitosis during which diploid number of daughter chromosomes are located at each pole. 387

template Pattern that serves as a mold for the production of an oppositely shaped structure; one strand of DNA is a template for a complementary strand. 418

tendon Fibrous connective tissue that joins muscle to bone. 62, 222

testis Male gonad, the organ that produces sperm and testosterone. 312

tetanus Sustained maximal muscle contraction. 230

thalamus Part of the brain located in the sides and roof of the third ventricle that serves as the integrating center for sensory input; it plays a role in arousing the cerebral cortex. 254

theory Concept consistent with conclusions based on a large number of experiments and observations, using the scientific method. 8

thermal inversion Temperature inversion that traps cold air and its pollutants near the earth, with the warm air above it. 497

thermoreceptor Sensory receptor that is sensitive to changes in temperature. 266

thrombin Enzyme that converts fibrinogen to fibrin threads during blood clotting. 114

thymine (T) One of four organic bases in the nucleotides composing the structure of DNA and RNA. 417

thymus gland Organ that lies in the neck and chest area and is absolutely necessary to the development of immunity. 144

thyroid gland Organ that is in the neck and secretes several important hormones, including thyroxin and calcitonin. 295

thyroid-stimulating hormone (TSH) Hormone produced by the anterior pituitary that causes the thyroid to secrete thyroxin. 294

thyroxin (thy-RAHK-sin) Hormone secreted from the thyroid gland that promotes growth and development; in general, it increases the metabolic rate in cells. 295

tidal volume Amount of air normally moved in the human body during an inspiration or expiration. 168

tissue Group of similar cells that performs a specialized function. 60

tissue fluid Solution that bathes and services every cell in the body. Also called interstitial fluid. 116

tone Tension that is maintained even when a muscle is at rest. 230

tonicity Degree to which a solution's concentration of solute versus water causes water to move into or out of cells. In isotonic solutions, cells neither gain nor lose water; in hypotonic solutions, cells gain water; in hypertonic solutions, cells lose water. 45

trachea (TRAY-kee-uh) Tube that is supported by C-shaped cartilaginous rings; lies between the larynx and the bronchi; also called the windpipe. 166

trait Phenotypic feature studied in heredity. 397

transcription Process resulting in the production of a strand of RNA that is complementary to a segment of DNA. 420

transfer RNA (tRNA) Molecule of RNA that carries an amino acid to a ribosome engaged in the process of protein synthesis. 419

transgenic organism Free-living organism in the environment that has had a foreign gene inserted into it. 429

translation Process by which the sequence of codons in mRNA directs the sequence of amino acids in a protein. 422

triglyceride Neutral fat composed of glycerol and three fatty acids. 27

triplet code Genetic code in which sets of three bases stand for a specific amino acid. 421

trophic level (TRO-fik) Feeding level of one or more populations in a food web. 474

trypsin Protein-digesting enzyme secreted by the pancreas. 90

tubular reabsorption Movement of molecules from the contents of the nephron into blood at the proximal convoluted tubule. 191

tubular secretion Movement of certain molecules from blood into the distal convoluted tubule so that they are added to urine. 191

tumor Growth that contains cells derived from a single mutated cell that has repeatedly undergone cell division; benign tumors remain at the site of origin, and malignant tumors metastasize. 442

tumor-suppressor gene Genes that, when expressed, prevent abnormal cell division and cancer. 445

tympanic membrane (tim-PAN-ik) Eardrum that receives sound waves and is located at the start of the middle ear. 277

U

umbilical cord Cord through which blood vessels pass, connecting the fetus to the placenta. 362

umbilicus Navel, where the umbilical blood vessels enter the umbilical cord; remains after the umbilical cord has been cut off following birth. 371

urea Primary nitrogenous waste of humans derived from amino acid breakdown. 185

ureter (YUUR-ut-ur) One of two tubes that take urine from the kidneys to the urinary bladder. 184

urethra (yuu-REE-thruh) Tube that takes urine from the urinary bladder to outside. 184

uric acid Waste product of nucleotide metabolism. 185

urinary bladder Organ where urine is stored before being discharged by way of the urethra. 184

uterine cycle Monthly occurring changes in the characteristics of the uterine lining (endometrium). 321

uterus Organ located in the female pelvis where the fetus develops; the womb. 317

utricle (YOO-trih-kul) Saclike cavity in the vestibule of the inner ear that contains receptors for static equilibrium. 280

V

vaccine Antigens prepared in such a way that they can promote active immunity without causing disease. 152

vacuole Membranous cavity, usually filled with fluid. 48

vagina Organ that leads from the uterus to the vestibule and serves as the birth canal and organ of sexual intercourse in females. 317

valve Membranous extension of a vessel or the heart wall that opens and closes, ensuring one-way flow. 125

vas deferens Tube that leads from the epididymis to the urethra in males. 313

vector Carrier, such as a plasmid or a virus, for recombinant DNA that introduces a foreign gene into a host cell. 426

vein Vessel that takes blood to the heart from venules; characteristically having nonelastic walls. 124

vena cava Large systemic vein that returns blood to the right atrium of the heart; either the superior or inferior vena cava. 127

ventilation Breathing; the process of moving air into and out of the lungs. 170

ventricle Cavity in an organ, such as a lower chamber of the heart; or the ventricles of the brain. 126, 253

venule Vessel that takes blood from capillaries to a vein. 125

vernix caseosa (VER-niks kah-see-OH-sah) Cheeselike substance covering the skin of the fetus. 369

vertebral column Backbone or spine through which the spinal cord passes. 212

vertebrate Animal possessing a backbone composed of vertebrae. 2

vestigial structure Remains of a structure that was functional in some ancestor but is no longer functional in the organism in question. 457

villus Fingerlike projection that lines the small intestine and functions in absorption. 85

virus Nonliving, obligate, intracellular parasite consisting of an outer capsid and an inner core of nucleic acid. 334

vital capacity Maximum amount of air moved in or out of the human body with each breathing cycle. 168

vitamin Essential requirement in the diet, needed in small amounts; often a part of a coenzyme. 98

vitreous humor (VIH-tree-us) Clear, gelatinous material between the lens of the eye and the retina. 270

vocal cord Fold of tissue within the larynx; creates vocal sounds when it vibrates. 166

vulva External genitals of the female that surround the opening of the vagina. 317

W

white blood cell (leukocyte) Formed element of which there are several types, each having a specific function in protecting the body from invasion by foreign substances and organisms. 64, 113

X

X-linked Allele located on the X chromosome. 408

Y

yeast Small unicellular fungus. 345

yolk sac Extraembryonic membrane that serves as the first site for blood cell formation. 362

Z

zygote Diploid cell formed by the union of two gametes; the product of fertilization. 359

Credits

TEXT AND LINE ART CREDITS

Chapter opening text prepared by Kathryn Sergeant Brown.

Chapter 4
Figure 4.15 Ancient versus Modern Diet of Native Hawaiians
Source: Data from T. T. Shintani, *Eat More, Weigh Less™ Diet*, 1983.

Chapter 5
Figure 5.6 from John W. Hole, Jr. *Human Anatomy and Physiology*, 6th edition, © 1993 The McGraw-Hill Companies, Inc. All Rights Reserved. Reprinted by permission.

Chapter 6
Figure 6.3 from Kent M. Van De Graaf and Stuart Ira Fox, *Concepts of Human Anatomy and Physiology*, 3rd edition, © 1992 The McGraw-Hill Companies, Inc. All Rights Reserved. Reprinted by permission.

Chapter 7
Figure 7.11 from David Shier et al., *Hole's Human Anatomy and Physiology*, 7th edition. Copyright © 1996 The McGraw-Hill Companies, Inc. All Rights Reserved. Reprinted by permission.
Art on page 157 from Kent Van De Graaff and Stuart Ira Fox, *Concepts of Human Anatomy and Physiology*, 4th edition. Copyright © 1995 The McGraw-Hill Companies, Inc. All Rights Reserved. Reprinted by permission.

Chapter 8
Health Focus, Page 176–77: The Most Often Asked Questions about Tobacco and Health—and The Answers Excerpted from "The Most Often Asked Questions About Smoking Tobacco and Health and…The Answers," revised July 1993. © American Cancer Society, Inc., Atlanta, GA. Used with permission.

Chapter 9
Figure 9.2 from Kent M. Van De Graaff and Stuart Ira Fox, *Concepts of Human Anatomy and Physiology*, 3rd edition. Copyright © 1992 The McGraw-Hill Companies, Inc. All Right Reserved. Reprinted by permission.

Chapter 10
Figure 10.2 from David Shier et al., *Hole's Human Anatomy and Physiology*, 7th edition. Copyright © 1996 The McGraw-Hill Companies, Inc. All Rights Reserved. Reprinted by permission.
Figure 10.5 from David Shier et al., *Hole's Human Anatomy and Physiology*, 7th edition. Copyright © 1996 The McGraw-Hill Companies, Inc. All Rights Reserved. Reprinted by permission.
Figure 10.6 from David Shier et al., *Hole's Human Anatomy and Physiology*, 7th edition. Copyright © 1996 The McGraw-Hill Companies, Inc. All Rights Reserved. Reprinted by permission.

Chapter 11
Figure 11.4 from Kent Van De Graaff and Stuart Ira Fox, *Concepts of Human Anatomy and Physiology*, 4th edition. Copyright © 1995 The McGraw-Hill Companies, Inc. All Rights Reserved. Reprinted by permission.
Figure 11.7 from John W. Hole, Jr. *Human Anatomy and Physiology*, 6th edition, © 1993 The McGraw-Hill Companies, Inc. All Rights Reserved. Reprinted by permission.

Chapter 12
Figure 12.3 from Kent Van De Graaff and Stuart Ira Fox, *Concepts of Human Anatomy and Physiology*, 4th edition. Copyright © 1995 The McGraw-Hill Companies, Inc. All Rights Reserved. Reprinted by permission.

Chapter 13
Figure 13.9 from Joan Creager, *Human Anatomy and Physiology*, 2nd edition. Copyright © 1992 The McGraw-Hill Companies, Inc. All Rights Reserved. Reprinted by permission.

Chapter 14
Figure 14.14 from John W. Hole, Jr. *Human Anatomy and Physiology*, 6th edition, © 1993 The McGraw-Hill Companies, Inc. All Rights Reserved. Reprinted by permission.

Chapter 15
Figure 15.1 from John W. Hole, Jr. *Human Anatomy and Physiology*, 6th edition, © 1993 The McGraw-Hill Companies, Inc. All Rights Reserved. Reprinted by permission.
Figure 15.2 from Kent Van De Graaff and Stuart Ira Fox, *Concepts of Human Anatomy and Physiology*, 4th edition. Copyright © 1995 The McGraw-Hill Companies, Inc. All Rights Reserved. Reprinted by permission.
Figure 15.5 from John W. Hole, Jr. *Human Anatomy and Physiology*, 6th edition, © 1993 The McGraw-Hill Companies, Inc. All Rights Reserved. Reprinted by permission.

Chapter 17
Figure 17.5a from Kent Van De Graaff and Stuart Ira Fox, *Concepts of Human Anatomy and Physiology*, 4th edition. Copyright © 1995 The McGraw-Hill Companies, Inc. All Rights Reserved. Reprinted by permission.
Figure 17.5 from John W. Hole, Jr. *Human Anatomy and Physiology*, 6th edition, © 1993 The McGraw-Hill Companies, Inc. All Rights Reserved. Reprinted by permission.
Figure 17.13 from Kent Van De Graaff and Stuart Ira Fox, *Concepts of Human Anatomy and Physiology*, 4th edition. Copyright © 1995 The McGraw-Hill Companies, Inc. All Rights Reserved. Reprinted by permission.
Figure 17.14 from Kent Van De Graaff and Stuart Ira Fox, *Concepts of Human Anatomy and Physiology*, 4th edition. Copyright © 1995 The McGraw-Hill Companies, Inc. All Rights Reserved. Reprinted by permission.

Chapter 21

Art on page 450 from Lance A. Liotta, "Cancer Cell Invasion and Metastasis," *Scientific American*, February 1992, Illustration by Dana Burns-Pizer, p. 62; modified figure depicting tissue inhibitors of metalloproteinase.

PHOTOGRAPHS

Introduction

I.1: © Jerry Mason/SPL/Photo Researchers, Inc.; **I.A**(rain forest): © Barbara von Hoffmann/Tom Stack & Associates; **I.A**(toucan): © Ed Reschke/Peter Arnold, Inc.; **I.A**(frog): © Kevin Schafer & Martha Hill/Tom Stack & Associates; **I.A**(butterfly): © Kjell Sandved/Butterfly Alphabet; **I.A**(jaguar): © BIOS (Seitre)/Peter Arnold, Inc.; **I.A**(orchid): © Max & Bea Hunn/Visuals Unlimited; **I.B:** © Jacques Jangoux/Peter Arnold, Inc.

Chapter 1

1.1: © The McGraw-Hill Companies, Inc./Jim Shaffer, photographer; **1.3a:** © Biomed Commun/Custom Medical Stock Photos; **1.8a:** © Martin Dohrn/SPL/Photo Researchers, Inc.; **1.8b:** © Comstock, Inc.; **1.8c:** © Marty Cooper/Peter Arnold, Inc.; **1.13:** © The McGraw-Hill Companies, Inc./Bob Coyle, photographer; **1.14:** © The McGraw-Hill Companies, Inc./Carlyn Iverson, photographer; **1.15:** © The McGraw-Hill Companies, Inc./Bob Coyle, photographer; **1.18b:** © Dr. Jeremy Burgess/Photo Researchers, Inc.; **1.19b:** © Don Fawcett/Photo Researchers, Inc.; **1.20b:** © Ulrike Welsch/Photo Researchers, Inc.; **1.27:** © Lawrence Migdale/Photo Researchers, Inc.

Chapter 2

2.1a: © Prof. P. Motta, Dept. of Anatomy, Univ. La Sapienza, Rome/SPL/Photo Researchers, Inc.; **2.1b:** © Pascal Rondeau/Tony Stone Images; **2.2a:** © Biophoto Assoc./Photo Researchers, Inc.; **2.2b:** Courtesy of Stephen L. Wolfe; **2.2c:** © CNRI/SPL/Photo Researchers, Inc.; **2.3a:** © Richard Rodewald/Biological Photo Service; **2.7b:** © Don Fawcett/Photo Researchers Inc.; **2.8a:** ©

W. Rosenberg/Biological Photo Service; **2.9a:** © David M. Phillips/Visuals Unlimited; **2.9c:** © K.G. Murti/Visuals Unlimited; **2.10a:** Courtesy of Dr. Keith Porter; **2.12b:** Courtesy of Kent McDonald, University of Colorado at Boulder; **2.13:** © David M. Phillips/Photo Researchers, Inc.

Chapter 3

3.1: © Julie Houck/Westlight; **3.2a-d:** © Ed Reschke/Peter Arnold, Inc.; **3.4a-d, 3.6a-c, 3.7:** © Ed Reschke

Chapter 4

4.4b: © R.G. Kessel & R.H. Kardon, *Tissues and Organs: A Text Atlas of Scanning Electron Microscopy,* 1979 W.H. Freeman Company; **4.5b:** © Ed Reschke/Peter Arnold, Inc.; **4.5c:** © St. Bartholomew's Hospital/SPL/Photo Researchers, Inc.; **4.6d:** © Manfred Kage/Peter Arnold, Inc.; **4.14:** © The McGraw-Hill Companies, Inc./Bob Coyle, photographer; **4.16a:** © Biophoto Associates/Science Source/Photo Researchers, Inc.; **4.16b:** © Biophoto Associates/Science Source/Photo Researchers, Inc.; **4.16c:** © Ken Greer/Visuals Unlimited; **4.B:** © Larry Brock/Tom Stack & Associates; **4.17:** © The McGraw-Hill Companies, Inc./Bob Coyle, photographer

Chapter 5

5.1: © The McGraw-Hill Companies, Inc./Jim Shaffer, photographer; **5.3a:** © Lennart Nilsson, *Behold Man*, Little Brown and Company, Boston; **5.7b:** © SPL/Photo Researchers, Inc.; **5.10a**(both): Courtesy of Stuart I. Fox

Chapter 6

6.1: © Chris Cole/Tony Stone Images

Chapter 7

7.1: © Damien Lovegrove/SPL/Photo Researchers, Inc.; **7.4a-e:** © Ed Reschke/Peter Arnold, Inc.; **7.8b:** © R. Feldmann/D. McCoy/Rainbow; **7.A:** © John Nuhn; **7.9a:** © Boehringer Ingelheim International/photo Lennart Nilsson; **7.12:** © Matt Meadows/Peter Arnold, Inc.; **7.13:** © Chris Harvey/Tony Stone Images; **7.14b:** Courtesy of Schering-Plough. Photo by Phillip Harrington; **7.B:** © Stuart Franklin/Sygma

Chapter 8

8.1: © Mehau Kulyk/SPL/Photo Researchers, Inc.; **8.5:** © John Watney Photo Library; **8.Ac:** © Bill Aron/Photo Edit; **8.B**(both): © Martin Rotker/Martin Rotker Photography

Chapter 9

9.1: © Jeff Greenberg/Unicorn Stock Photos; **9.6a:** © J. Gennaro, Jr./Photo Researchers, Inc.; **9.B:** Courtesy of the Childrens Hospital of Pittsburgh. Photo by Ed Eckman

Chapter 10

10.1b,c: © Ed Reschke; **10.3b:** © Junebug Clark/Photo Researchers, Inc.; **10.A:** © Royce Bair/Unicorn Stock Photos; **10.6a:** © The McGraw-Hill Companies, Inc./Joe DeGrandis, photographer; **10.11a-d:** © Ed Reschke; **10.11e:** © Ed Reschke/Peter Arnold, Inc.

Chapter 11

11.1: © Jim McHugh/Outline; **11.5b:** © Ed Reschke/Peter Arnold, Inc.; **11.6:** © Ed Reschke; **11.7a:** © Victor Eichler; **11.9a:** International Biomedical, Inc.; **11.11b:** © Tim Davis/Photo Researchers, Inc.; **11.12b:** © G.W. Willis/Biological Photo Service

Chapter 12

12.1a: © Gerhard G. Scheidle/Peter Arnold, Inc.; **12.4b:** © Linda Bartlett; **12.15a:** © W. Frerck/Odyssey Productions; **12.15b:** © Ogden Gigli/Photo Researchers, Inc.

Chapter 13

13.1: © Karen Holsinger Mullen/Unicorn Stock Photos; **13.5:** © Kathy Husemann; **13.12a:** © Lennart Nilsson, *The Incredible Machine* ; **13.A**(both): Courtesy Robert S. Preston and Joseph E. Hawkins, Kresge Hearing Research Institute, University of Michigan

Chapter 14

14.1: © Peter Miller/Photo Researchers, Inc.; **14.7, 14.9:** © Lester Bergman and Associates; **14.10:** From Arthur Grollman, *Clinical Endocrinology and its Physiological Basis,* © 1964 J.B Lippincott Company; **14.11:** © Ken Greer/Visuals Unlimited; **14.16a:** © Custom Medical

Index